# Can Law Make Life (too) Simple?

## From gene patents to the patenting of environmentally sound technologies

*For Brage*

Tine Sommer

# Can Law Make Life (too) Simple?

## From gene patents to the patenting of environmentally sound technologies

DJØF Publishing
Copenhagen 2013

*Tine Sommer*
Can Law Make Life (too) Simple?
From gene patents to the patenting of environmentally sound technologies

1. edition



Photo: Tine Sommer (La Jolla, San Diego)
Cover: Bo Helsted
Print: AKAPRINT, Tilst

Printed in Denmark 2013
ISBN 978-87-574-3119-3

Published with support from Dreyers fond

dreyersfond

This thesis was accepted by the School of Business and Social Sciences at Aarhus University for defence of the doctoral degree in Law (Dr. jur.).

Aarhus School of Business and Social Sciences, Aarhus April 24 2013

Svend Hylleberg
Dean

The public defence will take place at the Department of Law, Aarhus University, on Friday 30 August 2013 at 1 pm
at Juridisk Auditorium, building 1343/auditorium 275, Aarhus University.

*Sold and distributed in Scandinavia by:*
DJØF Publishing
Copenhagen, Denmark
Email: forlag@djoef.dk
www.djoef-forlag.dk

*Sold and distributed in North America by:*
International Specialized Book Services (ISBS)
Portland, USA
Email: orders@isbs.com
www.isbs.com

*Sold in all other countries by:*
The Oxford Publicity Partnership Ltd
Towcester, UK
Email: djof@oppuk.co.uk
www.oppuk.co.uk

*Distributed in all other countries by:*
Marston Book Services
Abingdon, Oxon, UK
Email: trade.orders@marston.co.uk
www.marston.co.uk

# Contents

## PART II. The Human Genome

**Appendices**

# Preface

I started the work which forms the basis for this thesis back in 1998 when I was working at the Aarhus School of Business, first as an assistant professor, then as associate professor and now as professor (temporary). The idea of carrying out an empirical case study arose in 2004 during a four-month sabbatical at Vanderbilt Law School, Vanderbilt University, Tennessee. The idea has evolved over the years as my skills have developed, and the case study has been designed and redesigned accordingly. My theoretical knowledge of US patent law was acquired at Vanderbilt University Law School and particularly at Columbia Law School, Columbia University, New York, where I spent a seven-month sabbatical in 2008 and a three-month sabbatical in 2011. During the periods I spent in New York I was warmly welcomed by Professor Harold Edgar, and my special thanks are due to him for his inspiring teaching and his willingness to discuss the many difficult areas of patent law; he has the ability to ask challenging questions which prompt thoughtful reflection. In 2009 I was also privileged to spend two weeks at the New York University School of Law at the invitation of Professor Rochelle Dreyfuss.

In the course of the work on my thesis I have received valuable help from colleagues. My particular thanks are due to Professor Ellen Margrethe Basse, of Aarhus University, for always being willing to put aside whatever she has been doing in order to help. I would also like to thank Cita Kristensen, who has been ever-ready to help with proof reading and with practical issues, and Steven Harris who has reviewed the use of English. I also owe a considerable debt to Henrik Tværmose, Librarian at Aarhus Universitet; who has always been willing to help with the many demands made or in course to my work on this thesis, ever since its start in 1998. I have also received help from law students Anders Yde, who helped with the initial patent searches, and Anne Hummersgaard Hansen HA(jur), who has compiled the bibliography, table of cases and table of legislation.

Above all, my family has been an invaluable support. Work on this thesis has been a constant part of our lives for many years and it has inevitably influenced our family life. My special thanks are due to Lili, for always believing in me, pointing me in the right direction, and teaching me to persist.

I must also give special thanks to my children, Silje, Emil and Olav. In 2004, 2008 and again in 2011, they have adapted by fitting into the US school system and sporting life. I have had the pleasure of having both my daughter Silje and my son Emil help structure the materials from the search results, not least the many patent documents. I am grateful for the way they have accepted the many years during which I may have been distracted by this work.

This thesis would never have seen the light of day without my supporter and soul-mate, Brage. Thank you for your inspiration, boundless patience and invaluable help in translating the language of genetics into legal concepts. The many evenings with data bases, patent applications and medical textbooks have been essential to the approach to the thesis. Not least the many discussions about the reach of patent law and its justification in biomedical research have been both interesting and instructive.

Needless to say, while I am grateful for all the help I have received, any errors are entirely my own.

Finally, it is my pleasure to thank the many external funds which have supported my work. I have received support from the Danish Council for Independent Research | Social Sciences (2004, 2006 and 2008); from the Genome Research Project: Comparative Genomics focusing on man and pig (2004); Frode and Norma Jacobsens' Foundation (2004); Axel H's Travel Foundation (2004 and 2011); and Julie von Müllen's Foundation (2011).

The references to the literature in the thesis have been updated frequently. However, since April 2012 only references that have come up on alerts (blogs) have been added. The most recent development is that an agreement on a unitary patent has been postponed indefinitely, after the Member States deleted three paragraphs from the draft agreement, following pressure from the United Kingdom, without consulting the European Parliament. The thesis has been updated to take account of the Council Decision of 29 June 2012, but this latest development has not been taken into account.

Aarhus 4 July 2012

Tine Sommer

# Summary in Danish

Afhandlingens kerne er den moderne patentret, således som den er blevet udfordret af især den bioteknologiske udvikling. Patentretten betragtes imidlertid ikke isoleret i afhandlingen, men miljøretlige, handelsretlige og EU-retlige aspekter inddrages med det formål at placere patentretten i en global og dynamisk virkelighed. Det for afhandlingen overordnede spørgsmål er udtrykt i afhandlingens titel. Afhandlingen bærer hovedtitlen '*Can law make life (too) simple?*' I afhandlingen adresseres humane genpatenter, patenter i dyreriget og klimaforandringers pres på patentregimet. Formålet har været at undersøge den ekspansion af patentrettens genstand, som områderne har foranlediget. Afhandlingen er styret af flere underspørgsmål til overskriften såsom patentering af komplekse biologiske systemer hos mennesker og dyr samt teknologiens pres på patentinstitutionen. Skal der reageres i lovteknisk henseende på teknologiske udviklinger, eller skal udviklingen i takt med behovet herfor forenes med den eksisterende lovgivningsmæssige ramme af patentmyndigheder og domstole? Ved eksempelvis komplekse biologiske systemer synes en detailregulering at komme til kort med den følgevirkning, at den forsimplede (detaljerede) lovtekst ikke favner de komplekse biologiske systemer, men får en statisk indvirkning på samspillet mellem teknologi og patentretlig beskyttelse.

Afhandlingen belyser patentretten i en 'multilevel' og fragmenteret kontekst herunder også kritisk i relation til den tiltagende EU-retlige indblanding i den formelle og materielle patentret. Tilgangen er baseret på 'multilevel' regulering, hvorefter patentretten i det globale samspil henter sine kilder fra det internationale, regionale samt nationale niveau. Her har den EU-retlige regulering formået at skabe kaos på det regionale niveau med tab af patentrettens legitimitet til følge. Det anbefales, at patentretten hjemvises til det større forum i Den Europæiske Patent Organisation og videreudvikles gennem globale normer som rettesnor for det internationale samfund. Her har især den amerikanske Supreme Court været inspirerende, og sammenligningen mellem den europæiske detailregulering og den amerikanske mere doktrinbaserede tilgang har bevirket, at afhandlingens anbefaling hælder til brede penselstrøg. Patentrettens forankring i doktriner, som regulerer den patentretlige genstand, har en større fleksibilitet og dynamisk fremtidssikring af patentinstitutionen. Dette illustreres konkret med direktiv 98/44/EC om retlig beskyttelse af bioteknologiske opfindelser, herunder direktivets tilgang og ordlyd i forhold til patentering af DNA sekvenser, embryonale stamceller, og beskyttelsesomfang for DNA sekvenser.

Afhandlingen er inddelt i 5 dele. *Del 1* benævnt 'MULTILEVEL REGULERING OG NYE TEKNOLOGIER' indeholder kapitlerne 1-4. KAPITEL 1 introducerer afhandlingens genstand og udviklingen inden for afhandlingens tre områder: patent på humane gener, patentering af opfindelser vedrørende dyr og teknologier inden for klimaforandring. Kapitlet fastsætter afhandlingens overordnede ramme såvel som teori og metode, herunder afhandlingens kildemateriale. Spørgsmålet i kapitel 1 er hvilke organer, der er bedst egnet til at håndtere kollisioner mellem patentretlig beskyttelse og nye teknologier. Særligt tabel 1.1 i kapitel 1 forsøger at illustrere de mange hensyn, som gør sig gældende i en global sammenhæng baseret på en polycentrisk tilgang til retten. Disse hensyn foreslås forenet i en global patentretlig norm, som udspringer af og baseres på det globale (tidligere miljøretlige) princip *'common concern of mankind'*, som henvender sig til de globale 'commons'. KAPITEL 2 behandler den teknologiske udvikling i kontekst af detailregulering. Kapitlet indfører læseren i den teknologiske verden som baggrund for især kapitel 7's empiriske casestudy. Det forklares, hvorledes lovgivningens ramme skal forstås ud fra en genetisk indførelse i eksempelvis 'hvad er et gen? Hvilken genetisk information indeholder gener? Hvilke funktioner har et gen? Desuden introduceres nye teknologier i biomedicinsk forskning. Særligt forholdet mellem bioteknologi, genteknologi, nanoteknologi, syntetisk biologi og kemikalier forsøges illustreret i figur 2.1. KAPITEL 3 belyser afhandlingens fokus på multilevel regulering ud fra en gennemgang af de for afhandlingen relevante aktører (herunder EU, EPO, WIPO, USPTO). Her er det især EU's optræden på den internationale og regionale scene, som har interesse. Med patentretten som eksempel pointeres det, at EU udvider sin kompetence internationalt såvel som regionalt gennem etableringen af, hvad forfatteren benævner *'distinct European system of public international law and regional law'* forstået således, at EU-retten synes at fremtræde som ophøjet ret på såvel nationalt, regionalt og internationalt niveau inden for patentrettens område. KAPITEL 4 afslutter Del 1 med ud fra patentrettens moralbegreb at illustrere, hvorledes patentretten på det regionale niveau er påvirket af EU hensyn, som kolliderer med den Europæiske Patentkonvention. Hvornår har den ene, og hvornår har den anden myndighed kompetence til at definere indholdet af direktiv 98/44/EC om retlig beskyttelse af bioteknologiske opfindelser? Her er særlig spørgsmålet om embryonale stamceller og direktivets artikel 6, stk. 2 genstand for behandling. Spørgsmålet er, om der ikke *netop her* er behov for en forsimpling (can law make life too simple?). Udviklingen går dog i retning af yderligere 'kompetence uorden' med EU patentet, idet der foruden EPO, herefter bliver tale om et EU i flere tempi, nemlig EU(2) og EU(25) som følge af den valgte fremgangsmåde for forstærket indbyrdes samarbejde, hvorefter hjemmelen for vedtagelsen af EU patentet (TFEU artikel 329, stk. 1) muliggør, at ikke alle medlemsstater tilslutter sig samarbejdet.

Afhandlingens *Del II* er benævnt 'DET HUMANE GENOM'. Kapitlerne 5-7 fokuserer alle på humane genpatenter. KAPITEL 5 diskuterer TRIPS-aftalens *ikke-diskriminationsprincip*, som er kodificeret i TRIPS-aftalens artikel 27, stk. 1. Det antages, at TRIPS-aftalen, som ikke definerer begrebet 'opfindelse', har overladt til medlemmerne at bestemme hvad, der er genstand for patentering. Dog begrænses dette råderum af *ikke-diskriminationsprincippet*, således at hele produktgrupper ikke kan undtages (for eksempel lægemidler). Hermed hindrer TRIPS-aftalen ikke en afgrænsning af hvilke genstande, der udelukkes fra patentbeskyttelse, for eksempel fordi substansen betragtes som værende et naturligt fænomen. Kapitel 5 belyser endvidere humane DNA sekvenser ud fra henholdsvis den europæiske retstilstand og den amerikanske retstilstand. Forskelligt fra den europæiske detailregulering indeholder amerikansk ret ikke detailregler for patenteringen af humane DNA sekvenser, men baseres på domstolspraksis. Den nye udvikling foranlediget af den amerikanske Supreme Court – hvorefter patenteringen af gener revurderes – stilles over for den europæiske praksis, herunder især CJEU, hvorefter afgrænsninger finder sted gennem en præcisering af beskyttelsesomfanget. Afhandlingens empiriske casestudy indledes med baggrundsinformation i KAPITEL 6. Det empiriske studium er baseret på en behandling af patentaktiviteten inden for seks sygdomme, hvoraf de fire første sygdomme Huntington, Spinal Muskel Atrofi, Canavan, og Cystisk Fibrose, som er forårsaget af en mutation i et enkelt gen, er kendetegnet som enkeltgen sygdomme, medens to sygdomme Alzheimers og Parkinsson er karakteriseret ved at være multifaktorielle sygdomme, hvilket indebærer, at flere faktorer, eksempelvis en kombination af små variationer i generne i samspil med miljøfaktorer, kan have indflydelse på, hvorvidt sygdommen kommer til udtryk. Kapitel 6 indeholder en gennemgang af sygdommene, samt angiver patentaktivitetens state-of-the-art inden for hver enkelt sygdom baseret på en gennemgang af den juridiske litteratur. KAPITEL 7 bidrager til den juridiske litteratur med et empirisk studium af i alt 1590 patentansøgninger fordelt på de udvalgte sygdomme. 39 % af de undersøgte patentansøgninger førte til et patent, 38,4 % blev afvist og 22,6 % er stadig verserende. Patentdokumenterne er fremkommet på baggrund af søgninger foretaget af den danske patent- og varemærkestyrelse (hhv. i 2004, 2005, 2007, 2009, 2011, og 2012). Analysen er inden for hver sygdom (eksempelvis Huntington, Canavan osv.) inddelt i fire undergrupper. De fire undergrupper er benævnt sygdomsgenet, sygdomsprocessen, stamceller og en sidste gruppe benævnt highlights, som kan være forskellig fra sygdom til sygdom. De væsentligste konklusioner fra det empiriske studie er tilknyttet de nye RNAi-teknikker (miRNA, siRNA), som udfordrer det traditionelle genpatent og som i afhandlingen tages til indtægt for, at ordlyden i direktiv 98/44/EC er for statisk og uegnet til at imødekomme nye udviklinger i biomedicinsk forskning. Desuden stilles der spørgsmål ved den omfattende brug af 'laundry lists', hvorefter patentansøger ikke indskrænker sin ansøgning til en konkret sygdom, men derimod i langt hovedparten af de undersøgte patenter inddrager en lang række af sygdomme, som kan have mere eller mindre tilfælles med hovedsygdommen.

Afhandlingens *Del III* er benævnt 'DYRERIGET'. Den består af et kapitel – KAPITEL 8, som gennemgår retstilstanden for patentering i relation til dyr, og ligesom kapitel 7 indeholder kapitlet en empirisk analyse, hvor aktivitet i relation til grisen er behandlet. Særligt paralleliteten om forbuddet mod patent på overvejende biologiske fremgangsmåder til forædling af planter eller dyr trækkes videre til kapitel 10. En fremgangsmåde er overvejende biologisk, når den i sin helhed beror på naturlige fænomener som krydsning eller udvælgelse. Derimod er naturlige biologiske processer i relation til mennesker ikke som sådan undtaget fra patentering.

Afhandlingens *Del IV* belyser 'PATENTINSTITUTIONEN SOM 'BELLWETHER' FOR GLOBALE PROBLEMSTILLINGER'. Denne del indeholder to kapitler – kapitel 9 og kapitel 10. KAPITEL 9 ser på retsudviklingen som en bølgemetafor i form af små retsbølger. Lovgivningen gentager sig selv i små bølger, og det konkluderes, at lovgiver må have fortiden præsent før der gribes ind med nye evt. forhastede lovgivningsmæssige tiltag. Konkret behandles klimaforandringernes indvirkning på patentretten. Er det nødvendigt at vedtage klimavenlige incitamenter gennem en ændring af patentlovgivningen – eller er lovgivningen i stand til at absorbere nye udfordringer? Endelig afsluttes afhandlingen i KAPITEL 10 med en opsummering af resultaterne og med et forslag til at bevare patentretten dynamisk og fleksibel. Her er det især multilevel reguleringen og den pluralistiske retskildetilgang, som søges forenet gennem etableringen af en *global offentlig orden* inden for patentrettens område hvilende på det i kapitel 1 introducerede internationale miljøretlige princip *'Common Concern of Mankind'*.

# Table of Abbreviations

| | |
|---|---|
| AD | Alzheimer's disease |
| CAFC | Court of Appeals for the Federal Circuit (US) |
| CBD | Convention on Biological Diversity |
| CD | Canavan's disease |
| CF | Cystic fibrosis |
| CJEU | Court of Justice of the European Union/ Court of Justice |
| DNA | Deoxyribonucleic acid |
| EBA | Enlarged Board of Appeal of the European Patent Office |
| EPC | European patent Convention |
| EPO | European Patent Office/Organisation |
| EU | European Union |
| HD | Huntington's disease |
| hESC | Human embryonic stem cell |
| HUGO | Human Genome Organisation |
| ICTSD | International Centre for Trade and Sustainable Development) |
| IRRs | Intellectual Property Rights |
| MEAs | Multilateral Environmental Agreements |
| miRNA | microRNA |
| NIH | National Institutes of Health |
| PCT | Patent Cooperation Treaty |
| PD | Parkinson's disease |
| PLT | Patent Law Treaty |
| RNA | Ribonucleic acid |
| RNAi | RNA interference |
| siRNA | Short interfering ribonucleic acid |
| SMA | Spinal muscular atrophy |
| SNP | Single nucleotide polymorphism |
| SPLT | Substantive Patent Law Treaty |
| TFEU | Treaty on the Functioning of the European Union |
| TRIPS | Trade Related Aspects of Intellectual Property Rights |
| UN | United Nations |
| UNEP | United Nations Environment Programme |
| UPOV | Convention for the Protection of New Varieties of Plants |
| USPTO | US Patent and Trademark Office |
| WHO | World Health Organisation |
| WIPO | World Intellectual Property Organisation |
| WTO | World Trade Organisation |

PART I

# Multi Level Regulation and New Technologies

'The discovery of genetics will not be imposed on us. Rather, they will be sold to us by the market as something we cannot live without',
Mark S. Frankel. (*Hastings Center Report* 33, No 2 (2003))

CHAPTER 1

# Subject, Method and Theory

Chapter 1 introduces the subject matter of the thesis and developments in the three main areas covered by the thesis: patents for human genes; the patenting of discoveries relating to animals; and technologies in the area of climate change. The chapter sets the overall framework for the thesis as well as the theoretical approach and methodology used, including the source materials. In Chapter 1 the question is raised as to which bodies are best suited to deal with conflicts between patent protection and new technologies. In particular, Table 1.1 in Chapter 1 seeks to illustrate the many considerations that are relevant in a global context, based on a polycentric approach to the law. It is suggested that these considerations should be brought together in a global norm for patent law that takes its inspiration from and is based on the global principle (initially applicable in the field of environmental law) of 'the common concern of mankind' and addresses the issue of 'global commons'.

## 1. Living in a changing society

The focus of this thesis is on the interplay between the legal protection of new technology and multi-level governance. Patent law is an excellent example of a fragmented and complex regulatory regime that is being pushed by developments in global society and is subject to legal requirements and standards from different lawmakers at the international, regional and national levels. Legislators at different levels and with different priorities are empowered to adopt legislation with transnational effects. As a result, when a country is a member of a regional organisation, such as the European Union and the European Patent Organisation, there can be lack of transparency in patent law about the relationship between national legislation and the legislation of the regional organisation and the interfaces between the regional organisations. Sometimes the bodies cooperate, as seen in the efforts to create a Unified Patent Court and Unified Patent Protection,[1] but sometimes they have differing conditions, priorities and interpretations of the law. The role of the courts and their scope for adapting to rapidly changing technology are also important aspects of this thesis. The expertise of specialised patent courts (the US Federal Circuit and the Unified Patent Court (EPO/EU)) may be overruled by the higher courts which do not have patent specialisation (the US Supreme Court, the Court of Justice of the European Union (CJEU)) or courts without jurisdiction over a complete regional organisation (CJEU). In particular, the CJEU adds to the current framework of 'multi-level jurisdiction' in Europe when it rules on what can be considered patentable subject matter. This can be illustrated by the recent 'need' for the CJEU to consider the definition of an embryo.[2] In this fragmented legal framework, the current state of European patent law can be described as an area that is blurred and is without clear distinctions between national, regional and international regimes. In particular, the issues of moral standards and gene patent protection illustrate the lack of coherence in Europe.

This study comprises three major areas that have been narrowed down to a single theme, reflected in the title of this thesis: *'Can law make life too simple?'*. The study started as a case study on morality, human gene patents and patenting within the animal kingdom, but as society has developed an interest in the governance of climate change matters the thesis expanded accordingly. The analysis will primarily comprise patent law, but it will also include the conflicts between patent law, environmental law and trade law at international and regional levels. The environmental law perspective is relevant to patent law as it reflects the *precautionary principle* and '*ordre public*' concerns, as well as *the principle of common concern of mankind* that may be relevant to new global issues such as climate change.[3] The 'new international ecological

1. See working document 15539/11 PI 133 COUR 59, Draft agreement on a Unified Patent Court and draft Statute. Revised presidency text of 26 October 2011. See the press release of the 3133$^{rd}$ Council meeting, 5 December 2011. The Polish Presidency was committed to reaching an agreement before 2011. Adopted 29 June 2012, EUCO 76/12.
2. Case C-34/10 *Oliver Brüstle*, 18 October 2011, cf. Chapter 2, section 3 and Chapter 4, section 4.
3. Jean-Pierre Beurier, Alexandre Kiss and Said Mahmoudi, '*New Technologies and Law of the Marine Environment*', 2000, p. 215 ff.; Kemal Baslar, '*The Concept of the Common Heritage of Mankind in International Law*', 1998; A.C. Tridade and David J. Attard, The Implication of the 'Common Concern of Mankind' concept, in Toru

order' established by the 1992 'paradigm shift' from international environmental law to the international law of sustainable development covers areas of global (common) concern; the UN Framework Convention on Climate Change (UNFCCC) and the 1992 Convention on Biological Diversity (the 2010 Nagoya Protocol on Access and Benefit-Sharing).[4]At a global level, the trade law perspective is unavoidable since the TRIPS Agreement sets the agenda for patent and trade law. Intellectual property rights (IPRs), trade and environmental interests are all actors in the international arena under the auspices of the UN, the WTO and the EU. Thus international law has an impact on patent authorities (e.g. the European Patent Organisation (EPO) and the United States Patent and Trademark Office (USPTO)), and international laws influence the framing of patent law regimes worldwide. The 'too simple law' concept expressed in the title of this thesis embraces the expansion of patentable subject matter that is taking place within this mix of interests. In one way it refers specifically to Article 5 of the EU Directive 98/44/EC on the legal protection of biotechnological inventions, whose wording refers to the patenting of human DNA sequences. Because of the Directive's rather naive design the thesis was initially headed: '*When* law makes life too simple'. Later, in the process of writing, it was changed to: '*Can* law make life (too) simple?' This heading also reflects the fact that any legislation is made within a framework of multi-level governance. Designing substantive patent law today must take account of this context and the key requirements for this process are flexibility and a dynamic framework, without tailor-made specifications for particular technologies – 'the simpler the better' is also what this thesis shows. It is a desirable quality to have simple standards. In fact the heading of the thesis expresses two lessons: first, a tailor-made patent law for new technologies (such as biotechnology) may be too simple to be future-proof; and second, in a fragmented, multi-level regime the standard economic approach is too simple for international patent standards and needs to be supplemented by, for example, environmental standards in order to be future-proof. However, in the latter case simplicity is a desirable quality.

Iwama, '*Policies and Laws on Global Warming: International and Comparative Analysis*', 1991, p. 7 ff. See Chapter 10, section 3.

4. Sandrine Maljean-Dubois and Lavanya Rajamani, *Implementation of International Environmental Law*, 2011, p. 114-116. See also the UN RIO +20 Conference on Sustainable Development, reaffirming the three Rio Conventions; UNFCCC, CBD and the UN Convention to Combat Desertification in Those Countries Experiencing Serious Drought and/or Desertification, Particularly in Africa (CCD), paragraph II, A, A/CONF.216/L.1.

From the regional European Patent Convention in 1973 to the global TRIPS Agreement in 1994,[5] technology developed rapidly. TRIPS was intended to be a global framework convention, setting minimum standards for the protection of intellectual property rights in order to safeguard international free trade in all kinds of tradable technological products.[6] In the first paragraph of the preamble to the TRIPS Agreement it is stated that the overall aim is to reduce distortions and impediments to international trade and to promote effective and adequate protection of intellectual property rights. Thus, the lack of intellectual property rights may constitute a barrier to trade.[7] On the one hand, TRIPS sets a minimum standard for the level of protection that parties are obliged to ensure in their states. On the other hand, as an economic standard, TRIPS balances monopoly and international trade interests and there must be a limit to the level of IPR protection where it becomes an unjustified obstacle to international trade, either by conferring overprotection,[8] or by the abuse of enforcement measures.[9]

5. WTO agreement on Trade-Related Aspects of Intellectual Property Rights, 1994.
6. The main purpose of the TRIPS Agreement was to prohibit free riding on inventions made in developed countries, see Nuno Pires de Carvalho, *The TRIPS Regime of Patent Rights*, 3rd Ed., 2010, p. xix.
7. See Nuno Pires de Carvalho, *The TRIPS Regime of Patent Rights*, 3rd Ed., 2010, p. 79, on a discussion on GATT 1947, Article XX(d), arguing that it is the existence of intellectual property standards and not their absence that creates barriers to trade. Pires states that the drafters of GATT Article XX (d) failed to acknowledge this aspect and that there may be a historical reason for this apparent oversight: 'In the 1940s, after the Second World War ended, international trade was primarily composed of off-intellectual property goods. The rule could be, therefore, that exporters would not assert intellectual property rights. Enforcing those rights in the context of international trade would be the exception, rather than the rule'. See Pires p. 80, referring to Carlos M. Correa, *Trade Related Aspects of Intellectual Property Rights: A Commentary on the TRIPS Agreement*, at p. 3, stating that the paragraph's last sentence only refers to measures and procedures to enforce intellectual property rights as barriers to trade, not to 'IPRs as such'.
8. See e.g. Lorelei Perez Westin, 'Genetic Patents: Gatekeeper to the promised cures', *Thomas Jefferson Law Review*, Fall 2002, 25 TJLR 271, 273-275: 'It is undisputed that all patents have several antitrust-like features that create the potential for monopoly market formation. Patents could provide effective market entry barriers to competing products due to unique product properties as many inventions in the pharmaceutical art show ... However, the balance can be upset by granting patents on broad-reaching inventions. Such patents require others to seek permission in order to manufacture, produce or practice dependent inventions, making licensing or other arrangements essential. Because these 'essential' parts (essential in the sense that they are necessary for the practice of all other downstream inventions) can potentially de-

The scope for Members of TRIPS to adopt substantive patent law for emerging technologies lies between the minimum requirements and the maximum allowable levels in TRIPS.[10] Sometimes Members may differentiate between the legal protection given to different technologies, and the room for manoeuvre or the best way to react to such new concerns is also a topic of this thesis. There is evidence of a pragmatic approach to the economic standard and balanced IPR protection can be seen, for example, in the Doha Waiver.[11] Thus, one can ask what will be the right approach for dealing with global concerns as they appear at international level.

For example, Article 27(1) of TRIPS on patentable subject matter states that patents shall be available, whether for products or processes, in all fields of technology, provided that they are new, involve an inventive step and are capable of industrial application.[12] Members can neither restrict patentable subject matter by introducing further substantive requirements nor grant a patent to an inventor whose invention is not novel. The provision is claimed to be an absolute standard.[13] However, as will be discussed in Chapter 5, it is

termine which inventions are practiced and thereby 'guard' entry into a field, they act as gatekeeper patents ... Although the vast majority of patents are improvements over existing patents in the arts, this dominant, monopolistic scenario is possible with genetic patents because of their potential role as gatekeeper patents'.

9. See Pires de Carvalho, at pp. 80 and 108; and Jonas Ledendal, *Copyright protection of Software under the TRIPS Agreement, Software reengineering and reverse engineering in the context of international trade law*, p. 324, discussing copyright and Article 9(2) of TRIPS as imposing a maximum level.
10. There has been a discussion about whether patents, trademarks and copyright enjoy a stronger position in relation to a WTO member's GATT obligations than other kinds of rights under the TRIPS Agreement, as GATT XX (d) explicitly mentions these three kinds of rights whereas TRIPS lists eight kinds of IPRs. See Jonas Ledendal, *Copyright protection of Software under the TRIPS Agreement, Software reengineering and reverse engineering in the context of international trade law*, p. 162.
11. See Chapter 3, section 3.2.2.
12. Though Nuno Pires de Carvalho, *The TRIPS Regime of Patent Rights*, 3rd Ed., p. 277, point 27.53 writes that: 'As far as substantive requirements are concerned, it seems indeed that Article 27.1 contains a numerous clauses, that is, no other substantive requirements may be used as grounds to reject a patent application (or to invalidate it) ... As it is worded Article 27.1 looks like an exhaustive provision'.
13. See Pires p. 75, and on p. 277 Pires writes that: 'the laws of WTO Members generally make reference to an additional substantive requirement, the so-called unity of invention, according to which patents must cover a single general inventive concept. That is a substantive requirement indeed, because, like novelty, inventiveness, or utility, it regards the nature of the claimed invention. However, no WTO Member may provide for the invalidation of the patent when the patented invention does not comply with the requirement of unity of invention'. However, this statement was not reflected in

questionable whether the TRIPS Agreement is an obstacle to the adoption of a differentiated approach to the scope of protection of DNA sequences.

In Article 27(2) it is stated that 'Members may exclude from patentability inventions, the prevention within their territory of the commercial exploitation of which is necessary to protect *ordre public* or morality, including to protect human, animal or plant life or health or to avoid serious prejudice to the environment'. This exempts Members of the WTO from the obligation to grant a patent on the listed grounds. Is this list exhaustive or it is only a minimum standard? And how can further concerns be recognised within this system?

The TRIPS system has been described as a dynamic construction build on constructive ambiguity.[14] Thus, interpretations and clarifications of most of the TRIPS provisions cannot be supported by examining the preparatory documents, but by examining three other sources as follows:

*One*, the dispute settlement mechanism, which is a central element in providing security and predictability to the multilateral trading system; *two*, the statements of trade ministers at sessions of the Ministerial Conference; and, *three*, the practices of the WTO members in implementing the TRIPS provisions (state practice).[15] *Pires de Carvalho* put it this way:

'This means that the TRIPS Agreement is not a frozen document, and that its provisions are to be constructed against a background of evolving understandings and changing trade environments. It is in this sense that one can affirm that the Agreement, more than the WIPO treaties upon which it has been build (such as the Paris and the Berne Conventions), is of a dynamic nature, which, in practice means that the construction that today is made of a specific provision may change tomorrow.'[16]

The regional parallel is the European Economic Community (EEC) free trade system, which was established by the Rome Treaty of 1957, now replaced by the Treaty on the Functioning of the European Union (TFEU), Articles 30-36. This was initiated as an economic arrangement and economic values have been the overall perspective from the start. The establishment of legitimate room for manoeuvre for the EU Member States with regard to free movement of goods illustrates that the protection of intellectual property rights was originally left to the Member States, and was respected in the Treaty from the start. Derogations from the right of free movement of goods have always been allowed if justified on grounds of the protection of industrial and com-

the case study (see Chapter 7) where patent examiners at the EPO and the USPTO did indeed require unity.

14. Nuno Pires de Carvalho, *The TRIPS Regime of Patent Rights*, 3rd Ed. p. 75.
15. Nuno Pires de Carvalho, *The TRIPS Regime of Patent Rights*, 3rd Ed. p. 75.
16. Nuno Pires de Carvalho, *The TRIPS Regime of Patent Rights*, 3rd Ed. p. 76.

mercial property. Even though the EEC Treaty was intended to be a framework treaty, it was never developed by legislative initiatives as much as by the Court of Justice's creativity. And the jurisprudence on IPRs and free movement of goods is also builds on the Court of Justice's creativity.[17]

As the case law from the Court of Justice[18] developed it became clear that the list of legitimate derogations from free trade now in Article 36 TFEU was complete, and the exceptions were to be narrowly interpreted. In order to compensate for this static inadequacy the Court of Justice developed a doctrine named after a case on a dispute concerning the French Liqueur *Cassis de Dijon.*[19] According to this doctrine, Member States were allowed to restrict imports or exports by using non-discriminatory measures that were justified on the grounds of public health and consumer protection.[20] Later, when environmental protection became a concern, the Court of Justice added this as a ground justifying a Member State's restrictions on imports.[21] The Court has been criticised for its dynamic lawmaking activity.[22] However, derogations justified by the interests of consumer protection, public health, socio cultural and environmental protection and so on were only gradually established on a case-by-case approach, reflecting the concerns of modern society. As the

17. See case law of the Court of Justice of the European Union, balancing free trade and intellectual property rights in e.g. Josephine Steiner & Lorna Woods, *EU Law*, pp. 713-720. See Hjalte Rasmussen, *On Law and Policy in the European Court of Justice, A Comparative Study in Judicial Policymaking*, 1986, pp. 172 and 174, discussing the Court as the prime mover in an integration process and referring to Peter Blok and Claus Gulmann.
18. Following the TFEU, the Court is now named the Court of Justice of the European Union.
19. Case 120/78 *Rewe-Zentral AG v Bundesmonopolverwaltung für Branntwein* (Cassis de Dijon).
20. See Case 120/78 *Cassis de Dijon*, paras 10 and 12.
21. See Case 240/83 *Procecureur de la Republique v ADBHU*, initiating EU environmental protection in general and later Case 302/86 *Commission v Denmark*, all commented on in Tine Sommer, *Affaldsreguleringen i EU-retlig belysning*, 1997, pp. 45 and 302; and Tine Sommer, 'Udførselsrestriktioner og miljøhensyn', *Juristen* No 10, 1998, p. 379 ff.
22. See Hjalte Rasmussen, *On Law and Policy in the European Court of Justice, A Comparative Study in Judicial Policymaking*, 1986; Paul Craig & Gráinne de Búrca (eds), *The Evolution of EU Law*, 2nd Ed. 2011, p. 131 ff.: Alec Stone Sweet, *The European Court of Justice*, p. 144 f. See further Hans-W. Micklitz, 'Judicial Activism of the European Court of Justice and the Development of the European Social Model in Anti-discrimination and Consumer Law', in Ulla Neergaard et al., *The role of Courts in Developing a European Social Model, Theoretical and Methodological Perspectives*, pp. 31-32.

derogation for the protection of intellectual property rights was already listed it was not created but only further shaped by the Court's case law. The regional EU approach can be seen as striking a balance between over-protection and free trade, whereas the existence of intellectual property rights was a matter for the national legal orders of the Member States.[23] However, the overall economic standard concerned safeguarding the free flow of products.[24]

As society changes so does the demand for what should be protected. The point of departure for this thesis is the clash between the legal protection (the patent system) and the need to address global concerns as they emerge.[25] Using free trade as a perspective on patent law, free trade concerns products that are already on the market whereas patent law concerns exclusive rights for products that are not necessarily ready to be placed on the market. Often, patented products in life sciences are required to obtain market authorisation before they can be marketed. In this sense patent protection for emerging technologies is where the public confronts new technologies at an early stage (both in terms of technological development and market access) generating an early opinion for or against a technology long before the legitimate balance between reward and utility has been reached in the patent system.[26] If

23. See e.g. Josephine Steiner & Lorna Woods, *EU Law*, 10th Ed., pp. 398 and 412; and Karsten Engsig Sørensen & Poul Runge Nielsen, *EU-retten*, 5th Ed., 2010, p. 414 ff.
24. There are parallels in the free movement of persons and freedom of establishment. As for the freedom to provide services, in Case C-36/02 *Omega Spielhallen- und Automatenaufstellungs-GmbH* the Court ruled that Community law does not preclude the national prohibition of the commercial exploitation of games simulating acts of homicide on public policy grounds by reason of the fact that the activity is an affront to human dignity.
25. See Tine Sommer, 'Om avlsmarkører, Miljøhensyn og social ansvarligheds betydning for enerettens legitimitet in Miljøretlige emner', in *Festskrift til Ellen Margrethe Basse*, 2008.
26. See e.g. COM(2011) 809 final, on a proposal for a regulation of the European Parliament and of the Council establishing Horizon 2020 – The framework Programme for Research and Innovation (2014-2020), Article 16(3), on ethical principles excluding some technologies from being financed. It is worth noting that the wording of this provision is close to the wording on disputed technologies in Directive 98/44/EC that turned the order upside down in the way it required patent authorities to evaluate the production phase; see Chapter 4. See also COM(2012) 60 final, Communication from the Commission to the European Parliament, the Council, the European Economic and Social Committee and the Committee of the Regions: 'Innovating for Sustainable Growth: A Bioeconomy for Europe', p. 6: Investment in knowledge, innovation and skills on key enabling technologies; 'The Challenges of the 21st Century and perspectives of a Bioeconomy', in National Research Strategy Bioeconomy 2030; 'Our Route towards a biobased economy', Federal Ministry of Education and Research,

legislators react to public concern by adopting substantive patent legislation or providing new ways for dealing with emerging technologies, the legislation may quickly become outdated and only serve to intensify protection at the cost of a loss of patent law legitimacy.

Patent law has always had a social or moral filter where the protection of the commercial exploitation of some forms of creativity can be blocked, but is not the appropriate forum for discrediting science or research into the making of a product or a process.[27] The argument of this thesis is that patent law at the regional level has lost legitimacy in its efforts to establish an EU moral filter respecting human dignity and that unfortunately specific provisions on gene patents have not satisfactorily resolved the clash between legal protection and the need to address global concerns. In the EU, neither legislators nor the Court of Justice have been able to safeguard emerging biotechnologies or address the concerns raised by new technology. As pointed out above, the EU and WTO systems address the essential question: is the clash best safeguarded by legislation or by the courts?

## 2. Patent law challenged by technology?

### 2.1. The dynamic law approach safeguarded by doctrines?

The core aim of patent law is to encourage and promote technology.[28] However, technology has its way of pushing the boundaries of law and emerging technologies in particular tend to leave the law behind. Some of the most significant technologies (biotechnology, software, information and communication technology and so on) do not fit neatly within our current patent systems, and this is often referred to as *the one size fit all dilemma.*[29] TRIPS intro-

Berlin 2011, p. 4; Bio-economy Council Report 2010, 'Bio-Economy Innovation', Berlin 2010; Ronnie Hall, 'Bio-economy versus Biodiversity', Global Forest Coalition, April 2012.

27. See Angus J. Wells, 'Patenting new life forms: an ecological perspective', 1994, European Intellectual Property Review, EIPR 1994, 16 (3), 113. See Chapter 4 and Tine Sommer, 'Interpreting *ordre public* and morality in a patent law context: which is the correct approach?', 2006/2007, 2 BSLR, p. 1.
28. See further Chapter 2 on the technology that is relevant for this thesis.
29. See National research Council (US), Mitchel B. Wallerstein, Mary Ellen Mogee & Roberta A. Schoen (Eds), *Global Dimensions of Intellectual Property Rights in Science and Technology*, 1993, p. 7; and A. Kur & V. Mizaras (Eds.): *The Structure of Intellectual Property Law – Can One Size Fit All?* 2011. See further Paul Sollie & Marcus Düwell, 'Evaluating New Technologies: An Introduction', in Paul Sollie &

duced the doctrine of non-discrimination or technology neutrality in Article 27(1) on patentable subject matter. This provision states that patents shall be available and patent rights enjoyable without discrimination as to the field of technology.[30] The second sentence of the provision (field of technology) repeats what is already provided for in the first sentence of the paragraph, namely that: 'patents shall be available for any inventions, whether products or processes, in all fields of technology'.[31]

The TRIPS minimum standard on non-discriminating patent lawmaking is not absolute, and one may say that the standard has to be further elaborated. In principle the TRIPS discrimination can occur in several ways:

> 'such discrimination occurs in the situation where the substantive requirements that an invention must meet in order to be patentable are less restrictive in respect of one field of technology than other fields of technologies. In other words, one field of technology has advantages which are not given to other fields. Discrimination also occurs when disadvantages relating to an entire technology or part of a technology are present, for example, if patents in the area in question are subject to stricter patentability requirements, or if the monopoly granted is restricted to a purpose-bound protection only. The above interpretation of the doctrine of nondiscrimination expresses the doctrine in an absolute form. However, as a broad minimum standard, this doctrine has to be further elaborated before it can be applied to new areas of technology. The future patent policy for new technologies cannot be based on a doctrine which has not yet been fully defined.'[32]

However, as a guideline the doctrine in Article 27(1) is more flexible than detailed regulation.[33] A *doctrine of technology neutrality* including the concepts

Marcus Düwell (Eds), *Evaluating New technologies, Methodological Problems for the Ethical Assessment of Technology Developments*, 2009, pp. 2-3, on the two horns of dilemma: the prediction horn and the control horn.

30. See C Reed, 'Taking Sides on Technology Neutrality', (2007) 4:3 *SCRIPTed* 263 <http://www.law.ed.ac.uk/ahrc/script-ed/vol4-3/reed.asp>. As stated by Reed, technology neutrality has long been held up as a guiding principle for the proper regulation of technology, particularly information and communications technologies.
31. See Tine Sommer, 'The scope of Gene Patent Protection and the TRIPS Agreement – An exclusively Nondiscriminatory Approach?', (2007) IIC Vol. 38, p. 30 f.
32. See Tine Sommer, 'The scope of Gene Patent Protection and the TRIPS Agreement – An exclusively Nondiscriminatory Approach?', (2007) *IIC* Vol. 38, p. 33 and Chapter 5, section 1.1.1.
33. This viewpoint has been contradicted in Wolrad Prinz zu Waldeck und Pyrmont, 'Special legislation for Genetic Inventions – A violation of Article 27(1)?', in Wolrad Prinz zu Waldeck und Pyrmont, Martin J. Adelmann, Robert Brauneis, Josef Drexl & Ralph Nack (eds), *Patents and Technological Progress in a Globalized World, Liber Amicorum Joseph Straus*, 2009, p. 301, footnote 63.

of invention, novelty, inventive step and industrial application cannot prevent Members from applying a precautionary approach to immature techniques; such as for example has been the case with RNAi techniques.[34] The effectiveness of the TRIPS neutrality doctrine requires the context of the assessment to be open, which is the core issue of this thesis where new technologies, such as nanotechnology, whole genome sequencing, personalised medicine and green technologies, challenge the one-size-fits-all approach in two ways: (1) can patent law be field restricted; and (2) is global stewardship a barrier to precaution?[35] It is worth questioning the appropriateness of legislative provisions whose wording is rapidly outdated, such as EU Directive 98/44/EC regulating DNA sequences and the commercial exploitation of embryos (Directive 98/44/EC, Articles 5 and 6) or addressing climate change as a particular patent law issue.

This thesis puts forward the hypothesis that doctrines and general principles are the most appropriate instruments in patent law for supplementing the international minimum standard of patentable subject matter such as the TRIPS Article 27(1). The core purpose of patent law is to promote new technologies, but emerging technologies arising in different technological contexts result in rapid changes to the legal environment and often cause legal problems. Section 3 below aims at setting out an approach to patent law that builds upon the traditional patentability standard and emphasises the need to draw reasonable boundaries for patentable subject matter. And the case study in Chapters 7 and 8 are intended to support this hypothesis from a practical viewpoint.

## 2.2. Patent as technology indifferent: what is that?

The criticisms of this thesis are directed at and address substantive patent law in the pre-grant phase. The premise of the thesis is that the proper delimitation of the patentable subject matter (pre-grant) can prevent disputes on the scope of protection (post-grant). This is a necessity in order to establish technology neutrality effectively and fairly as technology evolves.[36] When Article 5 of Directive 98/44/EC introduced separate rules for the patenting of human

34. See Charlie Smith, 'Negotiating the RNAi patent thicket', *Nature Biotechnology*, Vol. 25, No 3, March 2007, p. 273.
35. The concept of stewardship, see Roger Brownsword, 'So what does the world need now, reflections on regulating technology', in Roger Brownsword & Kate Yeung, *Regulation technology*, p. 45.
36. Lydia Bennett Moses, 'Recurring Dilemmas: The law's race to keep up with technological chance', *Journal of Law, Technology & Policy*, 2007, No 2, p. 240.

DNA sequences, and created a list of exemptions in Article 6(2), it failed in one very important aspect: it walked into a technology trap. Wording such as 'the sequence or partial sequence of a gene' has now become outdated; it has became obsolete due to new techniques where claims now concern a formula, exon/intron, RNAi and so on, resulting in the same or a broader scope of protection without referring to sequences. The reference to a 'human embryo' in Article 6 blurs the boundaries of what is alive and, given how an invention has been produced, creates more confusion in patent law (as having a Union face, cf. Case C-34/10 *Oliver Brüstle*) than it helps in promoting the progress of technologies. This makes it more difficult for courts, boards of appeals and patent authorities to fine-tune the law as new technologies develop.[37]

In the USA, critics on the gene patenting approach include *Eileene Kane*, *Rebecca S. Eisenberg*, *Rochelle Dreyfuss* and *Arti Rai* (and others), all of whom argue that gene patenting has lost touch with current practice. There has been much debate about human DNA patents, patent pools, patent thickets and compulsory licensing in the US as well as in Europe, where *Gertrui van Owervalle* among others is one of the leading critics of the current practice. Not every critical voice requires a ban on DNA patenting, but all seek a balanced approach either via the pre-grant or the post-grant route or a combination of both.[38]

It is a misunderstanding to interpret the global technology-neutral approach or non-discrimination technology doctrine in the TRIPS Article 27(1) as an absolute standard. Patent law can differ from one technological field to another, but that is not the issue. The issue is to adopt legal frameworks and provisions that are independent of technological changes and thus avoid the need to adopt new regulations when new gene technologies, nanotechnologies, environmentally sound technologies and other new technologies are developed.

In his analysis of the purpose of technology-neutral regulation, *Ledendal* writes in his PhD dissertation that legislators often refer to technology neutrality without distinguishing between different types of neutrality. There are two principal approaches to the application of identical rules to different technologies, he writes, the provision can be identically worded or *the provision can be differently worded with the aim of achieving functionally equivalent treatment.*[39] According to Ledendal, one must identify the further aims of the legisla-

37. See Chapters 3, 4 and 7. See Rebecca S Eisenberg, 'Biotech patents: looking backward while moving forward', *Nature Biotechnology*, Vol. 24, No 3, March 2006, p. 317.
38. See Chapter 5.
39. Emphasis added. The Directive 98/44/EC operates with a mix of the two approaches.

tion in order to properly characterise the respect in which it is or is not technology neutral.[40]

## 2.3. The role of courts and patent offices

The analysis of this thesis will take as its starting point an approach to the rapid changes in technology that takes account not only of the law falling behind technology but also of the roles played by different legal institutions and the methodologies they adopt in easing the law's transition to the future. This approach has been inspired by the legal writing of Lydia Bennett Moses. [41]

In her article on the law's race to keep up with technology, *Bennett Moses* points out that most authors are generally concerned about the law's failure to cope with technological changes, especially rapid or accelerating change.[42] She emphasises that statutory drafting techniques cannot ensure that laws will *continue* to operate fairly and effectively in new technological contexts without reducing the operational impact of at least some types of laws. And she pinpoints four reasons for making legal changes due to technological progresses:

*Special laws:* There may be a need to regulate certain new forms of conduct using new, specially tailored laws. In some cases, it may even be appropriate to ban a particular technology or particular applications of that technology. Alternatively, there may be proposals for mandating or encouraging a new activity. *Uncertainty:* The law may be uncertain when it applies to new forms of conduct. In other words, it may not be clear whether such conduct is prescribed, prohibited, or authorised. Existing legal rules may need to be clarified. *Over-inclusiveness and under-inclusiveness:* Where existing legal rules have not been formulated with new technologies in mind, those rules may inappropriately include or exclude new forms of conduct.[43] *Obsolescence:* Some existing legal rules may be justified, explicitly or implicitly, on the basis of a premise that no longer applies.[44]

40. See Jonas Ledendal, *Copyright protection of software under the TRIPS Agreement*, PhD thesis 2010, pp. 171-172. Laws that are, on the face of it, neutral might be technology-specific in application; see C Reed, 'Taking Sides on Technology Neutrality', (2007) 4:3 *SCRIPTed* 263 <http://www.law.ed.ac.uk/ahrc/script-ed/vol4-3/reed.asp>, referring to Bert-Jaap Koops, 'Should ICT Regulation be Technology-Neutral', in Bert-Jaap Koops, Miriam Lips, Corien Prins & Maurice Schellekens, *Starting Points for ICT Regulation: deconstructing prevalent policy one-liners* (The Hague: TMC Asser Press 2006), pp. 83-90.
41. Lydia Bennett Moses, 'Recurring Dilemmas: The law's race to keep up with technological chance', *Journal of Law, Technology & Policy*, 2007, No 2, p. 285.
42. Lydia Bennett Moses, 'Recurring dilemmas: the law's race to keep up with technological changes', *Journal of Law, Technology & Policy*, 2007, No 2, p. 240.
43. Moses pp. 280-81. Treating DNA sequences as chemical compounds is an excellent example of treating new entities as part of an old category. *Bennett Moses's* reflection

It is not possible to future-proof patent law in a technologically neutral way solely through legislation. It is the premise of the thesis that one must take a broader institutional approach. First, international standards such as Article 27(1) of the TRIPS Agreement need not further specify non-discrimination. The provision is based on traditional patentability standards which may not be subject to changes in order to keep ahead of new technologies. However, non-discrimination, understood as technology neutrality, can be viewed as a future-proofing mechanism being neither over-inclusive nor under-inclusive in new contexts and avoiding most problems associated with technological changes.[45]

For multi-level lawmakers, TRIPS sets the international minimum standard for non-discrimination, whereas USPTO and EPO, to take just two of the major players, complete the standard with their doctrines of 'discovery' and 'products of nature' and more. These concepts can hardly have been given a new content by the TRIPS harmonisation. The higher-level goal is the TRIPS non-discrimination doctrine and this is complemented by the lower-level goal of balancing patentable creation (in current perspective genes and similar creations) and the handiwork of nature, novelty, industrial applicability and so on. And this is where the judicial system plays its part.[46] The number of players in patent law has increased significantly, which is the topic of Chapter 3. However, the balance of the priorities of all the players in patent law can be better promoted and structured by judicial doctrines, than by detailed regulations if the courts are more aware of their responsibility. The courts as well as

on common law rules highlights the problem of treating new entities as part of an old category: 'the judge deciding the scope of a common law rule need not be greatly concerned with the precise form of words in which the rule had previously been articulated. However, while the potential of common law rules in this context may be great, it will not necessarily result in better outcomes. ... Common law rules can become fossilized, in some times and places more than others. ... In particular, there can be a tendency to treat a new entity as part of an old category without thinking through the consequences of that classification. Flexibility in interpretation is thus even more important in the common law context than in the statutory context, where other considerations might apply'.

44. Lydia Bennett Moses, 'Recurring Dilemmas: The law's race to keep up with technological chance', *Journal of Law, Technology & Policy*, 2007, No 2, p. 248.
45. Lydia Bennett Moses, 'Recurring Dilemmas: The law's race to keep up with technological chance', Journal of Law, Technology & Policy, 2007, No 2, p. 274.
46. See also Dan L. Burk & Mark A. Lemley, *The Patent Crisis and how the Courts can solve it*, 2009; and Geertrui Van Owervalle, 'Policy levers tailoring patent law to biotechnology. Comparing US and European approaches', *University of California Irvine Law Review*, 2010.

patent offices are major players in balancing patent protection, new technologies and ensuring patent legitimacy.

In this thesis it is suggested that international principles must evolve in order to future-proof patent law and that dynamic interpretations of patent law standards such as the TRIPS doctrine on non-discrimination may benefit from being interpreted in the light of other international principles.[47] Taking *Bennett Moses's* perspective on future-proofing, it is indeed necessary to supplement the role of the courts. If the courts are unable to advance at an appropriate pace, then it may be necessary to request the patent authorities advance patent prosecution at an appropriate pace.[48] Thus, the international standard must be completed by courts, boards of appeal, Federal Circuit and patent offices. Patent offices and their boards of appeal may also advance the pace of the absorption of new technologies before the courts are invited to confirm or reject the decisions of patent offices.

*Rebecca S Eisenberg* makes two criticisms of the role of the US courts. The first is timing. It typically takes years before the unresolved legal issues associated with patent claims for new technology are resolved by the courts.[49] She gives a very pertinent description (2006) of the breakdown of the patent system as follows:

> 'Only in the past years, for example, has the US Court of Appeals for the Federal Circuit had occasion to consider basic questions concerning the patentability of expressed sequence tags (ESTs), a technology that is more than a decade old, and in deciding that case, the court relied on previous decisions involving technologies that are older still [*re Fisher*, 421 F 3d 1365 (Fed. Cir. 2005)]. In theory, the focus on precedent in the legal system makes decisions more predictable, promising that future cases will be resolved consistently with past cases. But technological change often leads patent applications to pursue novel patent claiming strategies that bring new questions into view, creating uncertainty as to the meaning of the applicable rules.'[50]

47. See Chapter 9 referring to the 'Thomas Cottier' principle of the 'common concern of mankind' and Chapter 10. See 'Policy options for the improvement of the European patent system', *STOA* European Parliament, 2007, p. 33 ff. recommending a preamble inserted in the European Patent Convention stating the economic mission of the patent system.
48. On the question of patent quality, see James Bessen & Michael J. Meurer, *Patent failure, how Judges, Bureaucrats and lawyers put innovation at risk*, 2008, p. 161.
49. Rebecca S. Eisenberg, 'Biotech patents: looking backward while moving forward', *Nature Biotechnology*, Vol. 24, No 3, March 2006, pp. 317-319.
50. Rebecca S. Eisenberg, 'Biotech patents: looking backward while moving forward', *Nature Biotechnology*, Vol. 24, No 3, March 2006, p. 317.

*Eisenberg* describes the time span from pre-*Chakrabarty* (1980) to the Human Genome Project (2000).[51] In the 1970s and 1980s no media attention was paid to DNA sequences. The 'product of nature' doctrine implied that nature had done the heavy lifting and the value added by human invention was relatively trivial. Thus living organisms had been considered non-patentable as products of nature and for example the use of microbial strains to produce antibiotics typically did not result in claims for the strains themselves only for the method of production.[52] And Eisenberg writes by the time the US Institute of Health (NIH) started filing applications for ESTs (express sequence tags) categorical objections to the patenting of DNA seemed untimely and out of touch.[53] However, one of *Eisenberg's* recommendations is that the courts ought to find authority for limiting patentable subject matter. The goal she recalls is the statement in the 1948 case *Funk v Kalo Inoculant* (333 US 127, 130) that the 'storehouse of knowledge of mankind ... [is] free to all men and reserved to none'.[54] The application of such principles as are mentioned above may also serve as guidance at the international level.[55] In the literature *Dan L. Burk & Mark A. Lemley* also suggest how courts can solve the patent crisis and how the patent system can best adapt to the diversity of the modern world.[56] This will be discussed further in this thesis, as the overall recommendation of the thesis is that courts and patent offices are better able to balance patent law and safeguard patent law legitimacy if scrutiny follows a law design that comprise relatively neutral standards.[57] The detailed

51. T. Caskey, R.S. Eisenberg, E.S. Lander, J. Strauss, Belinda J.F. Rossiter (Ed), *HUGO Statement on Patenting of DNA Sequences*, The HUGO Genome Organisation, 1992.
52. Michael John Gulliford, 'Much ado about gene patents: The Role of Foreseeability', *Seton Hall Law Review*, 2004 (34 SHLR 711), pp. 726-727: 'The right of exclusion is particularly harmful in biotechnology because biotechnology research and development involves the use of fundamental, but often patented biochemicals, such as genes, ESTs ... SNPs ... and proteins to which the EST belongs'.
53. Rebecca S. Eisenberg, 'Biotech patents: looking backward while moving forward', *Nature Biotechnology*, Vol. 24, No 3, March 2006, p. 317.
54. Rebecca S. Eisenberg, 'Biotech patents: looking backward while moving forward', *Nature Biotechnology*, Vol. 24, No 3, March 2006, pp. 318-19.
55. See Chapter 9 and the Thomas Cottier reference to international principles such as the principle of 'common concern of mankind'
56. Dan L. Burk & Mark A. Lemley, *The patent crisis and how courts can solve it*, 2009 p. 6.
57. One perceived challenge for new technologies is that the patent offices themselves may take a while to catch up; see Marks & Clerk, *Life Sciences Report*, 2011, p. 22.

regulation of technologies, such as the European Directive on biotechnological invention, serves as an example of what not to do.[58]

## 3. Living with a flexible patent law approach

As indicated by the subtitle of this thesis '*From gene patents to environmentally sound technologies*', today the range of patentable subject matter is broad. The encouragement of technological development has always been a cornerstone of patent law. Originally, the traditional patent law approach was designed to spur mechanical innovation in industry but gradually the chemical and pharmaceutical industries emerged and their imprint on patent law is visible in the protection of biotech inventions.[59] When the author started researching the patentability of biological material, the focus of the precedents was on the patenting of pharmaceuticals and chemical substances. The early approaches to chemical and pharmaceutical patents were cautious. In the Nordic countries, such as Denmark, absolute product protection was introduced step-by-step in the period from 1967-1983, starting with process, use and purpose-bound protection before allowing absolute product protection.[60] Before the TRIPS Agreement entered into force on 1 January 1995, patents for pharmaceuticals and chemicals were not accepted everywhere, and more than 50 countries did not provide for such a protection.[61] The transitional pe-

58. See further the Genomic and Research Accessibility Act proposed by Congressmen Becerra and Weldon, in Arti Rai, 'An Audience with', in *Nature*, April 2007, Vol. 6, p. 264 writing that 'I tend to be wary of legislation, because Congress doesn't necessarily know what unintended consequences might emerge once it has been passed and it's very difficult to undo. In theory, court decisions should be easier to undo, although whether they are in practice is another question.'See Chapter 5, section 3.1.2; Chapter 6, section 5, and Chapter 10, section 2.1, regarding legitimate expectation.
59. Today, Around 30 % of all patent applications filed with the EPO are in the field of chemistry and it is worth noting that chemical engineers have a pivotal role to play in mitigating climate change, see http://www.epo.org/topics/news/2011/20110201.html.
60. In Denmark, chemical patents were allowed by the Danish Patents Act of 1967, but only as a purpose-bound protection. Absolute product protection was introduced by an amendment to the Patent Act in 1978. For pharmaceuticals, absolute protection was granted in 1983; see further Jens Schovsbo & Morten Rosenmeier, *Immaterialret*, 2nd Ed., p. 259.
61. See Nuno Pires de Carvalho, *The TRIPS Regime of Patent Rights*, p. 248; and Tine Sommer, 'The Scope of Gene Patent Protection and the TRIPS agreement – An Exclusively Nondiscriminatory Approach?', 38 IIC 49 (1/2007) p. 30 with footnote 2.

riod for the least developed Members of WTO will not expire before January 2016.[62]

The more cautious approach to pharmaceuticals has not been transferred to biotechnological inventions, and it is clear that this field of technology has no history of patents to build upon. Nevertheless, patent protection for living biological material ('biotechnological inventions'), chemical inventions and pharmaceuticals has been subject to uniform assessment. So far, protection for inventions from the life-science industry has been approached on the same terms as those that are well established in patent law: novelty, inventive step and industrial application (the US law: new, non-obvious and useful). For biotechnological inventions the EU has adopted specific rules on these three concepts without substantively amending their content, and the USA has adopted utility guidelines shaping the content of 'useful' to gene patenting and updating the examination guidelines on obviousness that may have an influence in this area.[63] Thus, as a starting point patent law must be considered rather flexible.

Another kind of flexibility is the exclusion from patentability of methods for the treatment of humans or animals by surgery or therapy, and diagnostic methods used on human or animals. The original reason for this was the lack of susceptibility to industrial application; see the European Patent Convention (EPC) Article 52(4). That the medical profession cannot be regarded as a trade is a dubious reason.[64] Today the EPC 2000 Article 53(3) excludes this category from patentability.[65] US patent law originally prohibited patents on health care inventions. *Lori B. Andrews* writes that throughout the first 150 years of US history, the USPTO did not issue patents for methods used to diagnose and treat sick patients. Medical or surgical methods did not appear to

62. See Nuno Pires de Carvalho, *The TRIPS Regime of Patent Rights*, pp. 227 ff. and 248. See Christopher Heath & Anselm Kamperman Sanders, *Intellectual Property & Free Trade Agreements*, p. 53, footnote 19.
63. See Chapter 3, section 4.1-4.2.
64. See Singer/Stauder, *The European Patent Convention – A Commentary*, 3rd Ed., Vol. 1, Article 52, point 13. Medical Treatment (para. 4), line 61 ff.
65. See Stacey J. Farmer & Dr. Martin Grund, *An Overview of the New European Convention (EPC 2000) and its potential impact on European Patent Practices*, p. 54; and Cornish & Llewelyn, *Intellectual Property: Patents, Copyrights, Trademarks and allied Rights*, 7th Ed., 2010, p. 239, footnote 310. This 'fiction' was averted in EPC 2000 by making the exclusion independent.

fall within the scope of the statutory requirements.[66] In 1862 a New York court decided that the use of ether to anaesthetise patients undergoing surgery was not patentable.[67] A study of the US patent history in this regard takes the reader from the 1883 *Ex parte Brinkerhoff* case, concerning a method for surgical treatment of haemorrhoids, emphasising that patents could not be granted for methods of treating diseases,[68] to the 2002 inclusion of 35 U.S.C § 287 (c), according to which health care providers are not subject to patent infringement suits when using a patented medical or surgical technique.[69]

The following sections provide illustrations of a flexible approach to patent protection from 1927 to the present, covering traditional mechanical products to living materials based on the three main subjects of the thesis: animals, humans and environmentally sound technologies. The focal area of the analysis in this thesis is substantive patent law concerning animal and human gene patents. More precisely, the issue of patenting animal genes concern the enhancement of animal characteristics, cross-breeding and biomedical research, whereas the human gene patenting focuses on inherited diseases. Finally, experiences from the biotechnological area will be used to recommend how environmentally sound technologies ought to be treated in substantive patent law.

66. See Lori B. Andrews, 'The gene patent dilemma: balancing commercial incentives with health needs', *Houston Journal of Health Law and Policy*, 2002, 2HOUJHLP65, p. 73.
67. *Morton v New York Eye Infirmary*, 17 F. Cas. 879 (S.D.N.Y 1862) (No. 9,865). The invention was that ether administered in incremental doses would kill all pain (at 884). However, the new use of ether was not patentable as 'the natural functions of a person or animal was essential part of the method claimed'; see 884.
68. *Ex parte Brinkerhoff*, 24 Dec. Comm'r Pat. 349 (1883). Moreover, the surgical instruments used to perform the surgery had already been patented and to patent the method would be to grant a second patent on the instruments.
    In 1995, an infringement suit for performing cataract surgery using a patented technique without paying a royalty was litigated between two doctors, one as the infringer and one as the patent holder.
69. *Palin v Singer*, 36 U.S.P.Q.2d 1050 (D.Vt 1995); and see Lori B. Andrews, 'The gene patent dilemma: balancing commercial incentives with health needs', *Houston Journal of Health Law and Policy*, 2002, 2HOUJHLP65, page 74. As a result of this case the American Medical Association amended its Code of Ethics to ban physicians patenting medical procedures. American Medical Association, Council on Ethical and Judicial Affairs, *Code of Medical Ethics, Patenting of Medical Procedures*, § 9.095 (2000-2001).

### 3.1. From skinning fish to purebred pigs

Non-human animals, and especially livestock which are being bred for human purposes, has not traditionally been regarded as having an express right to life. Animals may serve different purposes and be subject to different treatments; they can be pets, livestock or animal subjects for use in biomedical research. However, there is currently a rethinking of the traditional legal concepts; for example, the Australian Law Reform Commission has addressed the issue of animal rights which may well be the next big social justice issue for law reformers.[70] In a patent law perspective, patents for non-human animals have developed from being patents on instruments, for example for slaughter, to patents for breeding characteristics (traits for milk production or meat quality), and to genetic modification and cloning of animals for the same purposes. As genetic modification has altered the pattern of livestock breeding it is possible that the traditional pedigree system no longer gives adequate protection. Farmers and breeders have moved away from traditional breeding methods. Furthermore, animal models for biomedical research and for xenotransplantation are also addressed in the new context of patentable subject matter.[71]

A look at the last century may provide a perspective on how flexible patent protection has evolved. In 1927 a Danish fishmonger, Otto Marinus Jacobsen, was granted a Danish product patent on an instrument constructed to hold a fish while it was skinned and gutted.[72] The prior art comprised a procedure for securing fish by pegs. The patent only contained four claims, with very simple wording which in general terms claimed an instrument for securing a fish under skinning and gutting. It followed from the description that the fish could be an eel, a plaice, a cod or any similar fish, and the bars, tracks, chains and the rotatable lever (bracket) of the apparatus could be adjusted to fit the shape of the fish. As the instrument improved the way of fixing fish for skinning and gutting, the method in general was made safer.[73]

70. See Australian Law Reform Commission, *Reform*, Issue 91, Summer 2007-08 – Animals, http://www.austlii.edu.au/au/other/alrc/publications/reform/reform91/ The EU also regards animals as sensitive beings as stipulated in TFEU Article 13, which is new provision inserted by the Lisbon Treaty.
71. See 'The Animal in Us, the latest advances in Xenotransplantation', Editors of *The New Atlantis*, Spring 2003, 143.
72. See Danish Patent No 38142, granted 3 November 1927, protected from 10 November 1926.
73. On a more modern scale this is still an issue, see T 0136/09 of 14 January 2010 *Salmon Brands A/S*, concerning pre-rigor pinbone removal which results in improved odour, texture or gaping. Removing the pinbones at an earlier stage (than in-rigor,

74 years later, in 2001 and 2002 a Swedish researcher, Leif Anderson, and researchers from the UK were granted US and European patents for methods for screening pig nucleic acid to determine pig genotypes with respect to coat colour.[74] It was shown that the *KIT* gene in pigs is involved in coat colour determination. The patents claimed a screening method to determine the presence or absence of duplication of all or part of a *KIT* gene to distinguish the coat colour genotype of a (genomic) DNA sample.[75]

Patenting that testing tool shows how unpredictable developments in society can be. The former patent was for an instrument to provide safer and probably more efficient processing of a dead animal; the latter patent was for a test method for determining the coat colour which has an impact on the breeding processes of living non-human animals. Cross-breeding is based on selection for desired traits, and in the latter patent it is based on genetic information that can be spotted by the use of a genetic marker.

In the first invention – the fishmonger' skinning tool – patent law is applied in its traditional environment, for protection for a physical device. The advance in the method for gutting a fish can be described and claimed in simple terms. The genetic marker is also a substitute to an older method, but a traditional test mating programme was time-consuming and costly. However, the step from claiming the tool as such to claiming for the enhanced animal comprising the improved genetic characteristics subsequently became a development of patentable subject matter, but this has proved to be less successful.

In 2005, the US based multinational corporation Monsanto filed a patent application for a method for the genetic improvement of terminal boars essentially relating to a breeding method for pigs by crossing, selecting desirable traits and using the semen from male pigs to impregnate females, optionally in combination with embryo transfer.[76] The invention particularly concerned a method for improving and producing terminal sires so that these boars have improved genetic merit compared with the average herd animal. The background to the invention states:

post-rigor) shortens processing time, so that fillets can be supplied to the consumer sooner and in a fresher state, (point 2). The Technical Board of Appeal revoked the patent, see http://www.epo.org/law-practice/case-law-appeals/recent/t090136eu1.html (last visited December, 2011).

74. US 6183955 and EP 842296.
75. See Chapter 8, III, section 3.
76. EP1651030 and later US2008028478.

'Owing to the rapidly growing and improving field of genomics, there is a need for a means of using newly available genotypic information to improve the development of commercial swine products. Such a means must allow for the rapid genetic improvement in desirable traits in a swine population (e.g. target herd) without jeopardizing the potential for long-term genetic improvement. Such a method would need to provide a means for quickly and efficiently maximizing the usefulness of new understanding regarding the function of various genes and/or combination of genes; while at the same time optimizing the use of both phenotypic and genotypic information. The instantly disclosed invention solves previously existing problems by providing a method for the rapid genetic change of a target herd, based on phenotypic and genotypic information about both the target herd and GN herd, while maintaining a more balanced approach to selection in the GN herd.'[77]

The communication from the European examination division proceeded with an outline on essential biological processes (EPC Article 53(b)).[78] The subject matter of the invention related to essentially biological processes for the production of animals and the resulting animal varieties, excluded from patentability. The Examiner stated that 'the essence of the claimed 'method of producing swines having improved germplasm' and 'methods of making rapid change in the frequency of an allele' is of non-technical nature. The fact that genetic considerations are used prior to breeding does not change the essentially biological nature of the breeding program into a 'microbial process', as the resulting livestock is not genetically modified, but still the result of selective breeding'.[79] In the US examiner' office communication there was a requirement for election. The application contained groups of inventions e.g. drawn to a target herd, a swine production herd, a genetic nucleus herd and so on, which did not relate to a single general inventive concept and the examination division concluded:

77. See WO 2005/015989 [0003-0008].
78. The concept of essential biological process will be further discussed in Chapter 8. The International search report addressed the application as relating to subject matter: 'as for which no search is required by the International Search Authority (Article 17.2 (a)(i) PCT). Since 'breeding' in animal husbandry and in horticulture has been defined as 'the selection of stock for propagation and the act of insemination by natural or artificial means' (Websters dictionary), claim 1-65 are considered to refer to animal varieties and essentially biological processes for the production of said animals'. See International search report, p. 2. PCT Rule 39: 'No International Searching Authority shall be required to search an international application if, and to the extent to which, its subject matter is any of the following: ... ii) plant or animal varieties or essentially biological processes for the production of plants and animals, other than microbiological processes and the products of such processes.'
79. Annex to communication July, 2008.

'In the instant case, the technical feature is considered to be providing at least one genetic nucleus or target herd for which improvement is desired, selecting a trait(s) for which improvement is desired, proving semen aliquots from an elite sire, using the semen aliquots to impregnating breeding females in a target herd; producing half-sib offspring with improved germplasm, and providing at least one of the half-sib offspring as the terminal swine parent in a swine production herd, or as a replacement animal for the genetic nucleus herd. This technical feature does not make a contribution over the prior art'.[80]

The illustration of patent law development within the animal kingdom narrows the focus of this thesis. As long as an invention in European patent law does not address essential biological processes or the technical feasibility of the invention is not confined to a particular plant or animal variety, then a patent can be issued if the patentability requirements are fulfilled. The borderline between artificial processes and essentially biological processes is of interest. The thesis analysis highlights two issues: (1) the distinction between essential biological processes and the non-essential biological process and (2) the difference between essentially biological processes for the *production* of animals and essentially biological processes for *enhancing* the animal. Both issues comprise a non-patentable area and a patentable area. Considering the challenges from the 1927 'fishmonger skinning tool' to the 2005 pig breeding, one may well say that patent law keeps up with the times.[81]

### 3.2. The human genome – an exclusively chemical compound?

Turning to the human body, Article 4(1) of Directive 98/44/EC prohibits the patenting of animal varieties and essentially biological processes for the production of animals, and a matching prohibition *may be found* Article 5(1) of the Directive relating to humans.[82] Accordingly, the human body, at the various stages of its formation and development, and the simple discovery of one of its elements, including the sequence or partial sequence of a gene, cannot constitute patentable inventions. The use of the wording '*may be found*' has a reason. The Article 5(2) modifies this statement. An element isolated from the human body or otherwise produced by means of a technical process, including the sequence or partial sequence of a gene, may constitute a patenta-

80. US Office communication, requirement for restriction/election June, 2008 and PCT Rule 13, they lack the same or corresponding special technical feature.
81. See Chapter 8.
82. See Chapter 2, section 3, and Chapter 4, and Case C 34/10 *Oliver Brüstle*, paras 32-34. It appears that human dignity is safeguarded in Article 5(1) of the Directive and additional security is offered by Article 6(2). See also Chapter 5, section 2.1. Thus this provision has an ethical reason, verifying the EU stand on human dignity.

ble invention, even if the structure of that element is identical to that of a natural element. Thus, animal varieties and essentially biological processes for the *production* of animals are not patentable unless the technical feasibility of the invention is not confined to a particular animal variety. However, the human body can be patentable if an element is isolated or otherwise produced by means of a technical process. The differences in approach are subject to closer analysis later in this thesis.[83] However, it is noticeable that the provision relating to the human body has no explicit prohibition of the patenting of essentially biological processes copying what is taking place within the human body, as long as the process can be isolated.[84] Thus, we have seen patents claiming protection for fundamental biological phenomena.[85] Next to the isolation doctrine, which is discussed below in Chapter 5, there is only a requirement for the disclosure of the industrial application of a sequence or a partial sequence of a gene in Article 5(3).

The complexity of the functioning of human genes need to be taken into account when assessing genes as a chemical compound. We still need to learn more about the complex environment of genes and proteins if we are ever to be capable to understand the interplay of biological processes in the human body. So far, the solution has been to treat DNA sequences as chemical compounds by inventing the doctrine of isolation, simplifying human biological processes as chemical processes.[86]

83. Chapter 10, section 2.1.
84. The author is aware that it is not the production of a human as such and it is not confined to a human race. Thus, the question is narrowed down to copying biological phenomena which find a legal basis in Directive 98/44/EC Article 5(3).
85. See Christopher M. Holman, 'Patent border wars: defining the boundary between scientific discoveries and patentable inventions, Opinion', *Trends in Biotechnology*, Vol. 25, No 12, 2007, p. 541.
86. See also O. Carter Snead, 'Neuroimaging and Capital Punishment', *The New Atlantis*, Winter 2008, on reducing biological processes to chemistry, p. 40: 'Put simply, the fundamental premise of cognitive neuroscience is 'reductive materialism': it is 'reductive' in that it seeks to explain the 'macrophenomena' of thought and action solely in terms of the 'microphenomena' of the physical brain, and it is 'materialist' in that it postulates that human thought and behavior are caused solely by physical processes taking place inside the brain – a three – pound bodily organ of staggering complexity, but a bodily organ nonetheless. In this way, cognitive neuroscience follows the dominant approach of modern science, which seeks to understand and explain all observable phenomena as functions of their component parts. Under this methodology, questions of biology are thought to be reducible to matters of chemistry, which are, by extension, reducible to problems of physics.' See further Leon R. Kass, 'Ageless bodies and happy souls', *The New Atlantis, a Journal of Technology and Society*, No 1,

Gene patenting in Europe has provoked much debate both before and since the adoption of EU Directive 98/44/EC on the legal protection of biotechnological inventions. Gene patents have played a significant role in the public policy debate and the public perception has particularly focused on the ethical dimension on human patenting. This thesis is mainly concerned with substantive patent law and ethical dimensions are only addressed in two chapters: in Chapter 4 to illustrate the outcome of competing legal authorities at the regional level, and in Chapter 8 on animal patenting.

So far, the debate on gene patents has focused merely on the assumption that genes should be treated differently on moral grounds, with less attention being paid to the need for dynamic, precise and clear substantive patent law norms governing the pre-grant conditions.[87] As a legislative pioneer in biotech patenting the European Community made a first move by adopting Directive 98/44/EC on the legal protection of biotechnological inventions. Compared with the US case law approach it will be argued in this thesis that the EU's substantive biotech legislation is less flexible than dynamic judicial development. It is more complex to amend a controversial directive than to follow technological developments on a case-by-case assessment of neutral standards.

The flexibility of the US system can be illustrated by the way the case law wanders in circles: initially there was an emphasis on the product-of-nature doctrine, which was then modified by the *Diamond v Chakrabarty* decision after which patents on isolated DNA became commonplace,[88] and then there was the focus on utility introduced by the 2001 US Utility Guidelines,[89] and

Spring 2003, p. 9: 'The development in society in general turn itself to understand how even human sensory systems, as for example attention, memory, language, higher cognitive functions, consciousness arise from the structure and function of the brain as it is also the foundational premise of cognitive neuroscience that all aspects of the mind are ultimately reducible to the structure and function of the brain.'

87. See Tine Sommer, 'The Scope of Gene Patent Protection and the TRIPS Agreement – An Exclusively Nondiscriminatory Approach?', 38 IIC 49 (1/2007)
88. Aaron S. Kesselheim & Michelle M. Mello, 'Gene patenting – Is the Pendulum swinging back?', *The New England Journal of Medicine*, 362; 20 May 2010, p. 1856.
89. *May Mowzoon* writes that in fact, research efforts in biotechnology were accelerated after the ruling in *Diamond v Chakrabarty* (447 U.S. 303 (1980)); however, the patentability issue in question became utility, rather than subject matter. The requirement of utility 'is not usually an obstacle for mechanical and electrical applications because engineers do not typically construct devices with no known use. However, the utility requirement is a greater hurdle for chemical and biological inventions ... The 1995 Guidelines provided the following instruction to patent examiners: 'if the applicant has asserted that the claimed invention is useful for any particular purpose

finally there was a return to the focus on patentable subject matter with the ruling of United States District Court, Southern District of New York, in March 2010, where Judge Robert Sweet cited the Supreme Court decision whereby patentable products must have 'markedly different characteristics' from what is found in nature.[90] The claimed isolated DNA was not markedly different from DNA existing in nature, and was therefore not patentable under 35 U.S.C § 101. However, with the Federal Circuit decision in July 2011 Judge Sweet's ruling was set aside, and the academic world left behind with an open door to the reopened doctrine on 'products of nature'.[91]

Viewed in this way, the U.S. courts seems better able to adapt smoothly to the shifting views about the patentability of for example gene technology (35 US §§ 101, 102 or 103). Moreover, micro adjustments are more difficult to achieve if national standards are established by legislation.

> 'using legislative means to rebalance needs is effective in that it sets a national standard; its drawback, however, is that it is a time-consuming process as compared to the fast-paced field of biotechnology'.[92]

Fast-paced fields of technology also develop with a speed that cannot be absorbed by detailed legislation without the risk of it being static or outdated. European patent law (for national courts and patent offices') cannot be as

(i.e., a specific utility) and that assertion would be considered credible by a person of ordinary skill in the art, do not impose a rejection based on lack of utility.' This relaxed 'credible utility' requirement opened the floodgates for gene-related patent applications. Scientists could claim ESTs because they were useful as 'probes' to find the full length of genes involved in various diseases ... The 1995 Guidelines resulted in a 'gold rush' in the biotechnology industry; the PTO responded by issuing the 1999 Revised Interim Utility Guidelines which were later promulgated in January 2001. The credible utility of the 1995 Guidelines was replaced with the requirement of 'specific, substantial, and credible utility' (Examination guidelines for the Utility Requirement, 66 Fed. Reg. 1092 (Jan. 5, 2011)). In 2000, the director of (1089) USPTO, Todd Dickinson, stated 'we have raised the bar to ensure that patent applications demonstrate a 'real world utility''; see May Mowzoon, 'Access versus Incentives: Balancing policies in genetic patents', *Arizona State Law Journal*, Fall, 2003 (35 AZSLJ 1077), page 1088.

90. See Case 1:09-cv-04515-RWS, document 255, filed 3/29/2010: American Civil Liberties Union Foundation and Public Patent Foundation.
91. See Chapter 5, section 3.1 (On April 30th 2012 the Federal Circuit vacated its original decision).
92. May Mowzoon, 'Access versus Incentives: Balancing policies in genetic patents', *Arizona State Law Journal*, Fall, 2003 (35 AZSLJ 1077), p. 1100. See this Chapter, section 2.3.

flexible as the US judicial flexibility towards micro adjustments because Article 5 of Directive 98/44/EC establishes that 'the sequence or partial sequence of a gene may constitute a patentable invention'.[93] The EU standard is clear – DNA sequences are patentable subject matter.

### 3.3. Mitigation and adaptation: Technologies for a cleaner world

Law often repeats itself. New global concerns, such as climate change, are discussed without paying attention to prior concerns. It is the starting point of this thesis that policymakers continue to address new problems while ignoring former failures. The patent law regime has been identified as a major contributor to continuing health problems in the least developed countries and the efforts to transfer developed world patent laws to less and least developed countries have diminished the legitimacy of the patent regime. What will happen when we put food, energy scarcity and further climate change issues in a patent law context and add them to the present experience of safeguarding humans? So far, the United Nations Framework Convention on Climate Change (UNFCCC) COP-meetings (COP-15 in Copenhagen, COP-16 in Cancun and COP-17 in Durban) have failed to address IPR issues.

The TRIPS Agreement's explicit reference to 'serious prejudice to the environment' has been less controversial than its relation *to life-saving medicine.*[94] However, as climate change issues have emerged on the international agenda, patents became a new global concern from both positive and negative perspectives. The positive perspective concerns the efforts to invent new technologies to combat climate change, using the patent system to give an incentive to green industry to make further developments. The negative perspective is seen in the concerns of the developing countries as to the role of

93. Patent offices have responded to criticism by raising thresholds and making it much less attractive for applicants to file speculative, broad claims in the hope of obtaining what many would view as undue rewards. The shift in policy and the observed downward trend in the filing of patents claiming human DNA is, however, unlikely to have much impact on the legacy of specific concerns about the patenting of disease-associated, and other genes for diagnostic purposes; see Michael M. Hopkins, Surya Mahdi, Pari Patel & Sandy Thomas, 'DNA patenting: the end of an era?', *Nature Biotechnology*, Vol. 25, No 2, February 2007, p. 187.
94. See Frederick M. Abbott & Graham Dukes: *Global Pharmaceutical Policy, Ensuring Medicines for Tomorrow's World*, 2009, EE; Thomas Cottier, 'The Doha Waiver and Its Effects on the Nature of the TRIPS System and on Competition Law, The Impact on Human Rights', in Inge Govaere & Hans Ullrich (eds.): *Intellectual Property, Public Policy and International Trade*, at p. 175: 'a permanent regime of *lex specialis* on access to essential drugs in the TRIPS system of intellectual property protection'.

patents in the dissemination of such technologies, which can be compared to their concerns on pharmaceuticals. It is striking how principles adopted for dealing with climate change, e.g. common but differentiated responsibilities, can work as a guideline for combating diseases in the third world as well as natural disasters due to climate change.[95]

It is necessary to focus on the nature of the environmentally sound technologies which ask for legal protection. When referring to environmentally sound technologies, environmental lawyers think of adaptation and mitigation technologies. 'Mitigation' is defined as action taken to permanently eliminate or reduce climate change effects and risk to human life whereas 'adaptation' is efforts to adjust to climate change in order to cope with its consequences. The former includes energy supply based on renewable energy technologies,[96] carbon dioxide capture and storage, and technologies for the reduction of green house gas emissions.[97] The latter, mitigation technologies, include

95. Tuula Honkonen: *The Common but Differentiated Responsibility Principle in Multilateral Environmental Agreements, Regulatory and Policy Aspects*, 2009; Tuula Kolari, The Principle of Common but Differentiated Responsibilities as Contributing to Sustainable Development through Multilateral Environmental Agreements in Hans Christian Bugge & Christina Voigt, *Sustainable Development in International and National Law*, 2008, p. 251; Philippe Sands and Jacqueline Peel, with Adriana Fabra and Ruth MacKenzie, '*Principles of International Environmental Law*', third Ed., 2012, p. 233 ff.
96. Ellen Margrethe Basse, 'The legal design of sustainability criteria on Biofuels used by the European Union?', *Environmental Practice* to be published March 2013: 'all non-fossil sources in the form of wind, solar, aero thermal energy, geothermal energy, hydrothermal energy and ocean energy, hydropower, biomass, landfill gas, sewage plants, biofuels/bio-liquids and biogas'.
97. See Background paper: Climate Change: Technology Development and Technology Transfer, Beijing High-level Conference on Climate Change: Technology Development and Technology Transfer, Beijing, China, 7-8 November 2008, UN Department of Economic and Social Affairs, pp. vii, viii, and 2. See also COM(2012) 60 final, Communication from the Commission to the European Parliament, the Council, the European Economic and Social Committee and the Committee of the Regions: 'Innovating for Sustainable Growth: A Bioeconomy for Europe', p. 6: Investment in knowledge, innovation and skills on key enabling technologies; 'The Challenges of the 21st Century and perspectives of a Bioeconomy', in National Research Strategy Bioeconomy 2030; 'Our Route towards a biobased economy', Federal Ministry of Education and Research, Berlin 2011, p. 4; 'Bio-economy', Council Report 2010, Bio-Economy Innovation, Berlin 2010; Ronnie Hall, 'Bio-economy versus Biodiversity', Global Forest Coalition, April 2012; The Climate Technology Mechanism: Issues and challenges, ICTSD, Information note number 18. March 2011; Technologies for Climate Change and Intellectual Property: Issues for Small Developing Countries, ICTSD, Information note number 12. October 2009.

regional and local climate modelling, early warning, coastal zone management, water resources, agriculture (food security),[98] public health and poverty.

Not only will the patented technology be too costly for most developing countries, but if a green-waiver like the Doha Waiver on pharmaceuticals were to come into force, less developed countries would then be left with the option of producing the technology under licence but without the necessary infrastructure to do so. And if, one country could export the technology under a licence it is hardly likely that *a country-specific-technology for adaptation* will be the first on the developed countries' agenda. As we have seen it with neglected and rare diseases[99] it is probable that the priority in the industrial world will be on mitigation and not as much on adaptation technologies. Outside the IPR system the Clean Development Mechanism (CDM) derived from the 1997 Kyoto Protocol on climate change, Article 12, recognising emissions of greenhouse gases as a global environmental problem. Accordingly, emission reduction can be traded between Parties which have binding limitations and reduction commitments and Parties without (Annex I Parties/non-Annex I-Parties) since the place where the reduction takes place is not important.[100]

98. Abeba Tadesse Gebreselassie, '*The Sustainability of Plants and Plant Intellectual Property Rights*', 2012; David Greer and Brian Harley, '*Blue Genes: Sharing and Conserving the World's Aquatic Biodiversity*', 2004.
99. Regulation (EC) No 141/2000 on orphan medicinal products. See Henry Grabowski, Increasing R&D Incentives for Neglected Diseases: Lessons from the Orphan Drug Act, in Keith E. Maskus and Jerome H. Reichman: *International Public Goods and Transfer of Technology, Under a Globalized Intellectual Property Regime*, 2005, p. 457. As for essential medicines, see: S.K. Verma: The Doha Declaration and access to medicines by countries without manufacturing capacity, in Carlos M. Correa (Ed) *Research Handbook on the Protection of Intellectual Property under WTO Rules*, 2010; Thomas Pogge, Matthew Rimmer & Kim Rubenstein: *Incentives for Global Public Health, Patent Law and Access to Essential Medicines*, 2010.
100. Massimiliano Montini, 'Sustainable Development within the Climate Change Regime', in Hans Christian Bugge & Christina Voigt, *Sustainable Development in International and National Law*, 2008, p. 530; Jan H. Jans & Hans H.B. Vedder, '*European Environmental law, After Lisbon*', fourth ed. 2012, p. 431 f. See also the 2012 United Nations Secretary-General's High-level Panel on Global Sustainability. Resilient People, resilient Planet: A future worth choosing, pp. 52-53, 63, 85: Recommendations for a sustainable economy (Recommendation 37, CDM).

## 4. Method, sources and theory

### 4.1. Introduction

The primary objective of the thesis is to review patent law as a technology-neutral standard, in particular as exemplified by gene technology. There are several reasons for analyzing the current design of patent protection for gene technology. Gene technology has been subject to different legal instruments in the EU and the USA. Both the USA and the EU participate in the international frameworks on patent protection and have ratified several international conventions and agreements; see Table 1.1 below. Gene technology is an example of an emerging technology which can be evaluated from a legal perspective at a stage where it is not too early to review its effects but it is too early to predict its future implications. This inquiry links emerging technologies with the rise of new global concerns that may provoke constant challenges to patent law. The purpose is to inspire the creation of a robust patent law standard, i.e. the future-proofing patent law legitimacy.

The aim is to supplement patent law standards (economic standards) with environmental law standards in order to ensure patent law flexibility. As both patent law and environmental law have an international dimension, it is also appropriate to discuss the prospect of establishing a basic international patent standard (framework law) inspired by environmental standards, thus to some extent displacing the prevailing economic context. Whereas international patent law has existed for well over a century (see the Paris Convention, 1883), environmental law has only relatively recently been established at the international level (starting with the Stockholm Declaration of the United Nations Conference on the Human Environment, 1972) and it is making its way in both written law and soft law, most recently in the context of climate change. Both disciplines exist in a multiple and complex global legal order. The number of players and the legal sources that may be relevant for an inquiry into environmental law influences on patent law standards is illustrated in Table 1.1 below. The aim is to allow ideas of environmental protection to be reflected in patent law already at the *pre-grant* level.

Table 1.1 reflects the idea of legal polycentricity as a counterpart to a single value approach. In his legal writings *Surya Prakash Sinha* wrote that:

'[Legal polycentricity] conceptualizes a complex relationship of interaction between the plural forms of ordering that participate in the same social field. It analyzes this interaction and seeks legal forms of coexistence of these relationships. It rejects the ideology of legal centralism which insists that the state is the only form of ordering. It investigates the relationship between the state and non-state forms of normative ordering with a view to inventing legal forms. Therefore, it is interested in such questions how state law, in turn, is shaped by other normative orders, what are the forms of resistance to this interaction between the two, how viewing law as constitutive of social life leads to an inquiry as to how state law constitutes the normative orders within the state and, in turn, is itself constituted by them. It is interested in the question of how plural normative orders of government, persist, change, or are transformed by historical processes. It is interested in investigating the ideological nature of the various normative orders. It is interested in discovering modes of allocation of decision-making authority that would maximize the coexistence of these normative orders. This allocation of authority occurs among public authorities and private actors, on the one hand, and among different levels of government, on the other.'[101]

*Sinha* also pointed out some of the advantages of adopting legal polycentricity's pluralist approach i.e. maximising the legitimacy of legal order since, as he phrased it, it depends upon how well that order accommodates the demands of liberty, community, and social justice. Pluralism opens up a search for ways to maximise this legitimacy.[102]

This complex interplay between norms and players challenges the traditional understanding of legal sources and value, and it underlines the importance of a dynamic approach to law.[103] The following is a quote from the legal writings of *Inger-Johanne Sand* who gives an accurate overall perspective:

'Coherence and consistency may still be seen as general guidelines or ideal goals for law, but in a more global, dynamic and interdependent world they will have to be combined with a greater degree of acceptance of heterogeneity, change, combinations of competences and reflexive processes within and between institutions.' [104]

The chronological order shown in Table 1.1 has a way of complicating the theoretical foundation. International patent standards, such as the TRIPS standard, refers to concepts of patent law that were in use in the industrialised world (though with differing interpretations) long before the basic norm appeared at a global level. The international harmonisation that occurred in

101. Surya Prakash Sinha, 'Legal Polycentricity', in Hanne Petersen & Henrik Zahle, *Legal Polycentricity: Consequences of Pluralism in Law*, pp. 47-48.
102. Surya Prakash Sinha, 'Legal Polycentricity', in Hanne Petersen & Henrik Zahle, *Legal Polycentricity: Consequences of Pluralism in Law*, p. 48.
103. Martti Koskenniemi: *Fragmentation of International Law: Difficulties arising from the diversification and expansion of international law,* UN Report of the Study Group of the International Law Commission, Fifty-eighth session, 2006.
104. Inger-Johanne Sand, 'From hierarchical national constitutions to combinations of constitutional texts and institutions', in Kaarlo Tuori & Suvi Sankari, *The many constitutions of Europe; The Heroic Saga of Juridical Constitutionalization*, p. 53.

**Table 1.1. International norms and playmakers.**[1] (Global level in grey highlight).

| **1883** | **1963** | **1967-** | **1973 (2000) 1975** | **1982** | **1994** | **1998** | **2000** | **2005** | **2009** |
|---|---|---|---|---|---|---|---|---|---|
| Paris-convention (1883) | The Council of Europe:<br><br>Strasbourg-Convention (1963) | UN/WIPO (established in 1967):<br><br>Patent Cooperation Treaty (PCT, 1970)<br><br>The Budapest Treaty (1977) | European patent organisation:<br><br>European Patent Convention 1973, amended 2000<br><br>London-protocol, 2000<br><br>The EU European Economic Community: (1975-) Failed attempt on Community Patent Convention | Federal Circuit | WTO: TRIPS-Agreement (1994) | EU-Directive 98/44/EC | WIPO: Patent Law Treaty (2000)<br><br>Substantive patent Law Treaty (first drafts, 2000) | New attempts: European patent litigation system (2005) | Lisbon Treaty<br><br>Draft agreement on the European and Community Patent Court and draft statute 2009<br><br>Enhanced co-operation (2012)<br>Unitary patent Package |
| Patent Norms (delimited to the USA and Europe in an international context) ▲ | | | | | | | | | |
| ▲<br>**International Instrument/Norm:** Global exchange of values[2]<br>(The Vienna Convention (1969))[3]<br>▼ | | | | | | | | | |

| Environmental Norms (European/International) ▼ | | | | | | | | |
|---|---|---|---|---|---|---|---|---|
| | The Stockholm Declaration on the human environment (UN) | Brundtland-Commission Report on Our Common Future (UNWCED) | European Single Act (1987) Maastrict Treaty (1993) Amsterdam (1999) Nice (2003) | UNEP: CBD: Convention on Biological Diversity ▲ (1992) Rio-declaration ▼ (1992) UNFCC Framework convention on climate change | Kyoto-Protocol (1997) ▲ (UNFCCC) ▼ Cancun (2011) COP meetings | The Millennium Declaration on Development Goals (2000) ▲ UN[4] | Lisbon Treaty Charter (Art. 6), (Energy) | Cartegena protocol (2003) ▲ CBD ▼ Nagoya Protocol (2010) |
| | **1972** | **1987** | **1987** | **1992** | **1997** | **2000** | **2009** | **2010** |

1. See Chapter 3. Laurence R. Helfner, 'Regime shifting: The TRIPS Agreement and New Dynamics of International Intellectual Property lawmaking', 2004, 29 *Yale Journal of International Law*. In the article Helfner gives an overview of intellectual property lawmaking in four areas: biodiversity, plant genetic resources, public health and human rights. For an overview, see also Michael Blakeney, 'International Intellectual Property jurisprudence after TRIPS', in D. Vaver and L. Bently (eds), *Intellectual Property in the new millennium, Essays in Honour of William R. Cornish*, 2004, p. 8. Table 1.1 is not exhaustive, other forums have relevance such as WHO (see Sixty-First World Health Assembly, 2008 'Global Strategy on public health, innovation and intellectual property, http://www.who.int/mediacentre/events/2008/wha61/en/index.html (last visited May 2012), FAO, CGRFA, and ICTSD.
2. Harold D. Lasswell & Myres S. McDougal, *Jurisprudence for a Free Society, Studies in Law, Sciences and Policy*, Vol. I, pp. 30-32; James Crawford, International Law as an Open System, 2002, pp. 168-173; Konrad Zweigert & Hein Kötz, *Introduction to Comparative Law*, 1987, p. 25.
3. The law of Treaties was established with the Vienna Convention 1969, entering into force in 1980; see Akerhurst Chapter 9.
4. Keeping the promise: a forward-looking review to promote an agreed action to achieve the Millennium Development Goals by 2015, Report from the Secretary-General, Sixty-fourth session, UN general Assembly, 12 february 2010.

1994 with the adoption of the TRIPS Agreement sets the minimum standard for patentable subject matter in Article 27. However, these patentability concepts had been applied by TRIPS Members for decades and the international harmonisation in 1994 cannot have detached the standard from the context of the Members (see below, *Kaarlo Tuori's* deep-structure of law).

The endeavour to create a simple patent law standard within this complexity of legal power and legal order has been inspired by the 'concept of the legal order' developed by *Hans Kelsen.*[105] According to *Kelsen's* theory, one can define a legal order as an aggregate or a plurality of general and individual norms that govern human behaviour, in terms of how one ought to behave. Further, a plurality of norms becomes an order if the norms constitute a unity, and they constitute a unity if they have the same basis of validity.[106] As to the validity of norms, Kelsen claims that the higher norm serves as the basis for the lower norm.[107] However, it is explained: 'The appeal to the basis of the validity of a legal norm cannot go on interminably. It leads in the end to an act whose subject meaning, a norm, cannot be interpreted on the basis of positive law as its objective meaning. That is, this last act cannot be said to be prescribed as an 'ought' in another norm, a positive legal norm, which is in

105. Hans Kelsen, 'The concept of the legal order' (translated by Stanley L. Paulson), *The American Journal of Jurisprudence*, 27 Am. J. Juris. 64 (1982).
106. Hans Kelsen, 'The concept of the legal order' (translated by Stanley L. Paulson), *The American Journal of Jurisprudence*, 27 Am. J. Juris. 64 (1982), p. 64. Acts through which the norms of a legal order are posited or created comprise legislative acts, acts constituting legally binding customs, judicial acts, administrative acts, and private law transactions, in particular contracts; see pp. 64-65.
107. 'Validity' and 'binding' mean that: 'the subjective meaning of the act through which the norm is posited is also interpreted as its objective meaning' (p. 65). And 'the basis of the validity of a norm, an 'ought', can only be another norm, another 'ought'; it cannot be a fact, an 'is', and therefore cannot be the act positing this or another norm either. An 'ought' follows only from an 'ought', not from an 'is', just as an 'is' follows only from an 'is', not from an 'ought' (whereas the conclusion expressing an 'ought' is contained only in the major premise, which also expresses an 'ought'. The conclusion is not contained in the minor premise, which expressed an 'is' ... The validity of a legal norm, a legal 'ought' can indeed be derived only from a norm and not from a fact, an 'is''; Hans Kelsen, 'The concept of the legal order' (translated by Stanley L. Paulson), *The American Journal of Jurisprudence*, 27 Am. J. Juris. 64 (1982) p. 66.

turn the subjective meaning of a legal official's act'.[108] The basic norm in the hierarchy, the basic legal source, is assessed by Kelsen as follows:

'those who interpret the acts in question as legal acts can only *presuppose* that the norm that is the subjective meaning of this last act is a valid legal norm, binding on its addressee (this act is the last act in the intellectual process leading to the basis of the validity of a legal norm, but the first act in the law-making process)'.[109]

What is characterised as 'the basic norm' by Kelsen is thus the 'constitution' of the legal order as the basis for the validity of all other legal norms belonging to a legal order.

*Kelsen* defines international law as a decentralised legal order. Because there is a plurality of states, that is, domestic legal orders, the question arises as to how, on the one hand, domestic legal orders are related to one another and how, on the other hand, they are related to the international legal order.[110]

According to Kelsen one can differentiate international obligations according to two sources of law. The norms created through state treaties (binding only on the parties) belong to particular international law and represent a lower level of the international legal order than that represented by the norms created through the customary practice of states. Further, 'a still lower level comprises the norms posited by those international bodies that are established by state treaty'.[111] Finally, the *international basic norm* is defined as follows:

'Since the highest level of the international legal order comprises the norms of general international law, which are norms created through customary practice of the states, the basic of the validity of all norms of international law is the basic norm: States ought to conduct their affairs as they are accustomed to conducting them. This basic norm establishes customary practice of the state as a law-creating fact. It is the constitution of international law.'[112]

108. Hans Kelsen, 'The concept of the legal order' (translated by Stanley L. Paulson), *The American Journal of Jurisprudence*, 27 Am. J. Juris.(1982), p. 66.
109. Hans Kelsen, 'The concept of the legal order' (translated by Stanley L. Paulson), *The American Journal of Jurisprudence*, 27 Am. J. Juris. (1982), p. 66.
110. Hans Kelsen, 'The concept of the legal order' (translated by Stanley L. Paulson), *The American Journal of Jurisprudence*, 27 Am. J. Juris. (1982), p. 73.
111. Hans Kelsen, 'The concept of the legal order' (translated by Stanley L. Paulson), *The American Journal of Jurisprudence*, 27 Am. J. Juris. (1982), p. 74.
112. Hans Kelsen, 'The concept of the legal order' (translated by Stanley L. Paulson), *The American Journal of Jurisprudence*, 27 Am. J. Juris. (1982), p. 74.

Validity and the relation between two normative legal orders are also considered in Kelsen's work. Two normative orders that do not have the same basis of validity cannot be regarded as simultaneously valid.

Kelsen's norm approach has been further elaborated by himself[113] and by *Kaarlo Tuori* in his critical legal positivism which will be the preferred theory in this thesis. However, Kelsen's basic norm approach is used in this inquiry in searching for international norms comprising expressions of collective values that shape states' rights and responsibilities.[114] It may be possible to recommend doctrines reflecting basic norms that can be legitimised at a global level and which can guide courts and patent authorities.

In the legal order of patent (and environmental) protection there are not only a large number of players (UN, WTO, EU, EPO, WIPO, USPTO, Council of Europe etc.), but there is also a large number of legal sources (treaties, conventions, protocols, agreements, and directives) that cannot easily be placed in a hierarchical legal structure. Traditional tools of interpretation, such as the *lex posterior* or *lex superior*, are less effective when there are fragmented legal sources adopted on the basis of divided or shared competence or by different playmakers. The effect of multi-level and multi-disciplinary regulation of patent law results in a landscape that operates with several basic norms which can be approach separately (patent law, environmental law (e.g. the concepts of access and benefit sharing in the Convention of Biological Diversity)) or which can be integrated. The latter option is the approach of this thesis. Some norms of environmental law, such as sustainability and 'common concern of mankind', may imply a future-proof superior international norm constituting a unity in the plurality of legal norms created in conformity with it.[115] These norms may also prove to be a global social filter, representing dynamic principles in a polycentric approach and aimed at maximising patent law legitimacy.

The current study is based on the fact that there is currently no legal acceptance of the interplay between patent law and environmental law princi-

113. See Hans Kelsen, *General Theory of Norms* (translated by Michael Hartney), 1991; and the criticism in Raymond Wacks, *Understanding Jurisprudence, An introduction to Legal Theory*, 2nd Ed., pp. 123-127: 'Kelsen's approach [centralization of sanctions], like so much that was written by experts on international law and organization in that period, was the product of wishful thinking' (p. 126).
114. See Antje Weiner, 'Enacting meaning-in-use: qualitative research on norms and international relations', Review of International Studies, Vol. 35, p. 176.
115. See Hans Kelsen, 'The concept of the legal order' (translated by Stanley L. Paulson), *The American Journal of Jurisprudence*, 27 Am. J. Juris. (1982), p. 69.

ples, and that the beneficial effect of such interplay on emerging technologies is not in fact measurable or verified.[116] However, due to the mutual dependence of patent law legitimacy and 'common concerns' the interface stated between the two disciplines in this thesis is anticipated to be valuable in a global context. When the common concern of climate change is at stake states are obliged to enjoy sovereignty over their own natural resources within the confines of the global responsibilities set out in the Rio-declaration, the CBD and the UNFCCC and other relevant instruments.[117] The huge number of players and the multi-level regulation is not unusual for modern global governance, but it has a further implication in patent law. The intervention of the EU at the regional level (the European patent system) has caused much confusion; 'the basic (patent) norm', in the terminology of *Kelsen*, has been subject to a '*Europeanisation of international law*',[118] as will be illustrated below and in Chapters 3 and 4, by the adoption of Directive 98/44/EC.

### 4.2. Multi-layered legal order

This thesis is concerned with the complex concept of a global legal order. The global legal issues concerned are not only defined by patent law, but embody environmental, trade, EU and international law in the context of emerging technologies, *primarily* gene technology and *secondarily* environmentally sound technologies. Thus, the overall framework of the thesis is located within global governance and multi-level regulation. Interactions between national and international legal spheres create difficulties in distinguishing between international, regional (EU) and national legal orders.[119] In the work of

116. Inger-Johanne Sand, 'Teoretiske utfordringer for rettsociologien i globaliseringens tid', in Bettina Lemann Kristiansen & Anette Storgaard (eds), *Nordisk retssociologi, status – aktuelle udfordringer – visioner*, 2010, p. 136-137).

117. Patricia Birnie, Alan Boyle, Catherine Redgwell, *International Law & the Environment*, Third ed., 2009, p. 128 ff.

118. Jan Wouters, André Nollkaemper & Erika de Wet (eds), *The Europeanisation of International law, The status of International law in the EU and its Member States*, 2008. See also Inger-Johanne Sand, 'Teoretiske utfordringer for rettsociologien i globaliseringens tid', in Bettina Lemann Kristiansen & Anette Storgaard (eds), *Nordisk retssociologi, status – aktuelle udfordringer – visioner*, 2010, pp. 134 & 135.

119. Håkan Hydén, 'Rättslig pluralism ur ett normvetenskapligt perspektiv', in Bettina Lemann Kristiansen & Anette Storgaard (eds), *Nordisk retssociologi, status – aktuelle udfordringer – visioner*, 2010, p. 59. Hydén writes that among the three levels (supranational, national, subnational/överstatslig, statslig, substatslig niveau) the supranational level has increased the internationalisation of law due to globalisation and that e.g. the WTO also plays a political role as seen with the Doha Declaration. How-

*Follesdal et al.* this has been described as follows: 'international law is increasingly coming to play a role in national (and EU) legal orders, whereas national (and EU) legal developments are exerting a bottom-up influence on the evolution of the international legal order'.[120] And this phenomenon has been introduced by *Wouters et al.* as the 'Europeanisation of international law'. [121] The EU can be placed in between the systems, exerting a top-down influence on Member States at the same time as a bottom-up influence on international law. The latter is especially so in patent law and will be the further elaborated in Chapters 3 and 4.

The definitions of the concepts given by *Follesdal et al.* are used in current thesis. Multi-level governance has its origin in political science and public administration and is the phenomenon of interacting and partly overlapping policy spheres. The concept implies two dimensions: '*governance without government*' and '*governance beyond the State*'. The first dimension refers to the fact that a number of public tasks are carried out by new actors, so that action is no longer solely a matter for the classic government institutions of a national state; and the latter dimension refers to the complexity of governance at distinct but increasingly intertwined levels.[122] Finally, there is the following definition of the concept of *multi-level*:

'a variety of forms of decision making, authority, policy making, regulation, organisation, ruling, steering, etcetera, characterized by complex interweaving of actors operating from the local level, via the national level, to the macro-level and global level ... from a legal

ever, this level is not as powerful as the national level, see p. 60. However, the European Union does have such legitimacy; see p. 62.

120. Andreas Follesdal, Ramses A. Wessel & Jan Wouters, *Multilevel Regulation and the EU, The Interplay between Global, European and national Normative Processes*, 2008, pp. 10-11. See further Frank Hoffmeister, 'Outsider or frontrunner? Recent developments under international and European law on the status of European Union in international organizations and treaty bodies', *CML Rev* 2007 44: 41-68.
121. J. Wouters, A. Nollkaemper & E. de Wet, *The Europeanisation of International Law: The status of international law in the EU its Member States*, 2008.
122. Andreas Follesdal, Ramses A. Wessel & Jan Wouters, *Multilevel Regulation and the EU, The Interplay between Global, European and national Normative Processes*, 2008, p. 11. For the terms used within the EU, see Paul Craig and Gráinne de Búrca, *The Evolution of EU law*, 2$^{nd}$ Ed., 2011, pp. 23-24, on multi-level governance sustained by a two-stage argument. The first considers the circumstances under which national executives might lose their grip on power. The second stage of the argument considers whether policy making accords with that postulated by multi-level government within the EU, focusing on power and dominance and the complex balance between Council, Parliament and Commission.

perspective, the interactions between global, European and national regulatory spheres lead to what one could refer to as the phenomenon of 'multilevel regulation'. We understand 'regulation' in a broad sense here, referring to the setting of rules, standards or principles that govern conduct by public and/or private actors. Whereas 'rules' are the most constraining and rigid, 'standards' leave a greater range of choice or discretion, while 'principles' are still more flexible, leaving scope to balance a number of (policy) considerations.'[123]

The methodological approach of this thesis uses the legal dogmatic view of patent law as a multi-layered legal order. In a multi-layered context, where guidance may be provided by principles that leave scope for balancing political considerations, legal certainty may be complex unless the standards are kept very simple. The legal theories that may provide some guidance in this context have been established by many legal scholars.[124] Thus, there is no prevailing legal theory behind this thesis but a plurality of theories that each contribute to an improved understanding of how to design a global approach to patent law.

The starting point will be critical legal positivism as developed by *Kaarlo Tuori*.[125] According to this, law has two faces: a legal order comprising a set of norms *and* legal practice consisting of a set of specific social practices – and the two aspects of law constantly interact.[126] With a theoretical foundation in critical legal positivism it follows that law comprises three levels: the

123. Andreas Follesdal, Ramses A. Wessel & Jan Wouters, *Multilevel Regulation and the EU, The Interplay between Global, European and national Normative Processes*, 2008, p. 12; Frank Hoffmeister, 'Outsider or frontrunner? Recent developments under international and European law on the status of European Union in international organizations and treaty bodies', *CML Rev* 2007 44: 41-68, writes (before the adoption of the Lisbon Treaty) at p. 68: 'in a nutshell, the European Union is neither an outsider anymore nor has it become a frontrunner in the multilateral arena. Rather it turns into a respected actor in international organizations and treaty bodies with the same speed as the law develops. Under international law that needs years of skilful multilateral diplomacy, under the European law the European Court of Justice may accelerate the process.'
124. See e.g. a Danish research group currently analysing European legal methods in the light of current prevailing legal theories: Towards a European Legal Method: Synthesis or fragmentation? See Ulla Neergaard, Ruth Nielsen & Lynn Roseberry (eds.), *European Legal Method, Paradoxes and Revitalisation*, 2011. And see Raymond Wacks, *Understanding Jurisprudence, An Introduction to Legal Theory*, 2nd Ed., p. 185 on the Scandinavian realists, Alf Ross and valid legal norms as socially binding.
125. Kaarlo Tuori, *Critical legal positivism*, 2002.
126. Kaarlo Tuori, *Critical legal positivism*, 2002, p. 121.

surface level of law, the legal culture and the deep structure of law.[127] *Tuori* has called the surface level the 'turbulent level' of the law.[128] At this level, he writes, the law is in constant flux, as a result of new regulations and new court decisions, as well as new articles and books by legal scholars. At its surface, the law can be conceived of as the constantly changing outcome of an evolving debate to which legislators, judges and legal scholars all contribute. At the surface of law, changes are made almost daily, whereas changes in the legal culture – the intermediate level of law – are slower, and finally the deep structure is the most stable layer of law.[129]

127. See Kaarlo Tuori, *Critical legal positivism*, 2002, p. 147: 'If confronted with the question what constitutes the law, not only laymen but probably even the majority of professional lawyers would refer to express regulation and ... perhaps court decisions ... it also includes deeper layers, which both create preconditions for and impose limitations on the material at the surface level. I call these sub-surface levels the *legal culture* and *the deep structure of the law*. The multi-layered nature of 'mature' modern law is the key to the solution that this type of law can offer to the problem of its limits and criteria of legitimacy'. See further, Ulla Neergaard & Ruth Nielsen, 'Where did the spirit and its friends go? On the European Legal Method(s) and the Interpretation Style of the Court of Justice of the European Union', in Ulla Neergaard, Ruth Nielsen & Lynn Roseberry (eds), *European Legal Method, Paradoxes and Revitalisation*, 2011, p. 101: 'The surface level of the law comprises various pieces of legislation as well as the text of judgements. The legal culture constitutes the middle or mediating level of law'.
128. See further Kaarlo Tuori, *Towards a multi-layered view of modern law, in Justice, Morality and Society* (A Tribute to Aleksander Peczenik on the occasion of his 60th birthday 16 November 1997), p. 439, writes: 'The legal culture and the deep structure of law play a constitutive role in relation to legal regulation, legal decision-making and legal thinking in general. The reverse side of these relations of constitution consists in relation of *limitation*. The deep structure and the legal culture not only create the necessary preconditions for legal regulation, legal decision-making and legal dogmatic research; they even impose restrictions on these practices producing the constant movement at the surface level of law.'
129. Kaarlo Tuori, *Critical legal positivism*, 2002, pp. 191-192 and 316-317, where the author distinguishes between two types of internal contradictions in the law: horizontal and vertical ones. Horizontal contradictions occur at the same level of the law, and may be resolved by classic standards for resolving norm conflicts: *lex superior, lex posteriori* or *lex specialis*. Further he writes: 'However, the contradictions manifested at the surface level of the law can have their source in conflicts between principles justifying, say, contradictory statutory provisions. Then the contradiction is transferred to the level of the legal culture, to the relation between principles and counter-principles. Theoretical legal dogmatics proposes solutions to these contradictions through its general doctrines, through the prima facie order of preference it establishes between legal principles. But the internal contradictions of the legal culture are not exhausted by the relation between principles and counter-principles; we have seen

Combining *Touri's* critical legal positivism with the starting point evolved by *Follesdal et al.* on multi-level regulation and *Wouters et al.* on the 'Europeanisation' of international law one question presses for further consideration. At what level will the Europeanisation have an impact on international law? The multi-level regulations comprise *legal order and legal practices* at international, regional and national levels, each embodying the surface level of law, the legal culture and the deep structure of law. Since the EU absorbs international law and in patent law it also influences regional patent law outside its jurisdiction (which can be called the *Europeanisation of regional law*) one can argue that the EU creates a very turbulent surface level of law but as the turbulence is not limited to the EU level but extends to the EPO level, it seems that contributions to the coherence of law may be found outside the normal frames of critical legal positivism, or any other known theory. Theoretically, EU patent law establishes a *sui generis* system that cannot be supported by established legal theories.[130] In patent law, the EU has a bottom-up influence on international law (TRIPS, the Doha Waiver as implemented in a regulation before TRIPS Members had fully ratified the obligation),[131] a bottom-up influence on regional law (the EPO, the implementation of Directive 98/44/EC in the EPO Rules) and finally, a top-down influence on the EU Member States (Directive 98/44/EC and the enhanced procedure for an EU patent).

that there can be diverging proposals for systematizing the legal order, i.e. alternative general doctrines, which differ not only in their conceptual structure but even in the order of preference they assign to legal principles'. Vertical contradictions are contradictions between elements located at different levels of the law. Individual regulations on the surface level may conflict with the principles of the legal culture and the law's deep structure. See further Tuomas Mylly, *Intellectual Property and European Economic Constitutional law, the trouble with Private Informational power*, 2009, p. 102 ff.

130. A Danish research group currently analysing European legal methods in the light of the currently prevailing legal theories: Towards a European Legal Method: Synthesis or fragmentation? See Ulla Neergaard, Ruth Nielsen & Lynn Roseberry (eds), *European Legal Method, Paradoxes and Revitalisation*, 2011. See also Ruth Nielsen, 'Scandinavian Legal Realism and the EU Law', in Ulla Neergaard, Ruth Nielsen & Lynn Roseberry, *The Role of Courts in Developing a European Social Model, Theoretical and Methodological Perspectives*, 2010, p. 243.

131. Regulation (EC) No 816/2006 of the European Parliament and of the Council of 17 May 2006 on compulsory licensing of patents relating to the manufacture of pharmaceutical products for export to countries with public health problem.

*Table 1.2. Kaarlo Tuori, Follesdal and Wouters et al.: Europeanisation of regional and international law*

| | |
|---|---|
| International Law (TRIPS)<br>▲▼ | |
| Regional Law (e.g. EPC)<br>▲<br>EU Law<br>▼ | ◄ Surface Level<br>Legal Culture<br>Deep Structure |
| National Law | |

The overall question is whether the *Europeanisation* of both regional law and international law (EPC and TRIPS) reaches the deep structure or only the surface level. At the currently stage it seems that the regional level (EPC) will be influenced at the level of legal culture (98/44/EC – especially on moral issues) and enhanced procedures (a change in a uniform patent), whereas international law (TRIPS) only concerns regional law at the surface level unless it is considered as covering the traditions of the Member States. Evaluated this way, concepts such as inventive step, novelty and industrial application are part of the deep structure in the industrialised world whereas the use of these concepts in respect of biological material (98/44/EC) only reaches the surface level. It also follows that, transferred to the developing countries, the patent system only reaches the surface level. In this way, a treaty such as the TRIPS Agreement in fact harmonises as well as regulates and can be assessed at Touri's three levels at the same time. *Ruth Nielsen* describes the structure as follows: 'Legal development is seen as a sedimentation process; i.e. it starts at the surface level and then over the years seeps down to the medium level and the deep structure of law. The surface level of the law comprises various pieces of legislation as well as court and administrative decisions in individual cases. The legal culture constitutes the middle or mediating level of law. The doctrine of the sources of law is part of the expert legal culture at the middle level.'[132] Accordingly, if new sources arise (e.g. environmental law and principles) it will take time before they are accepted and integrated into

132. Ruth Nielsen, 'Scandinavian Legal Realism and the EU Law', *in* Ulla Neergaard, Ruth Nielsen & Lynn Roseberry, *The Role of Courts in Developing a European Social Model, Theoretical and Methodological Perspectives*, 2010, p. 242.

other disciplines (e.g. patent law). Their validity has to reach down into the deep structure of law. The pluralistic view of the sources of law, the fragmentation of law and multi-level regulation has an international surface which can take years to be absorbed and accepted in the regional[133] and national deep structures (such as the broad guideline of the common concern of mankind). Paul Craig writes that multi-level governance is not in itself a theory of integration.[134] It is however interesting to evaluate the overall frame on the basis of multi-level regulation and governance in a global perspective as a kind of promotion of integration or harmonisation. Craig's analysis concerns integration at the EU level. The inquiry in this thesis is searching for a global basic standard to guide patent law and multi-level regulation, and governance can be viewed as a way of integrating and safeguarding all aspects of relevant concerns. Thus, the outcome of the multi-level approach may be one big system evolved by adding to the traditional economic standard the concerns derived from younger disciplines (e.g. environmental law) that have had time to reach down into the deep structure.[135]

*Thomas Mylly* writes that it has become relatively usual to perceive the legal order as constituting a multitude of overlapping legal systems, or at least of legal norms produced in multiple centres.[136] As he phrases it, in this sense law is pluralistic and polycentric and the law's polycentricity is due to the existence of overlapping and parallel legal systems:

133. See Ruth Nielsen, 'Scandinavian Legal Realism and the EU Law', in Ulla Neergaard, Ruth Nielsen & Lynn Roseberry, *The Role of Courts in Developing a European Social Model, Theoretical and Methodological Perspectives*, 2010, p. 259, for a description of a new European legal culture that is significantly less formal, dogmatic and positivistic than the national legal cultures in Europe have traditionally been. In a less formal approach substantive arguments have easier access to legal argument. To the extent that the borderline between the internal and the external perspective is blurred, because legal methods are becoming less formal and admit more substantive arguments, legal analysis can also be said to be moving gradually away from the humanities, which focus on interpretation and meaning, and towards the social sciences, which are more interested in the consequences of human behaviour.
134. See Paul Craig, 'Integration, Democracy and Legitimacy', in Paul Craig & Gráinne de Búrca, *The Evolution of EU law*, 2011, p. 24 ff.
135. See in principle Ruth Nielsen & Christina D. Tvarnø, *Retskilder & Retsteorier*, 3rd Ed., 2011, pp. 474 and 478, who approach EU law as a youthful system (Lisbon Treaty, Maastricht Treaty, a system evolved by putting layer upon layer, and of relatively recent existence compared to most national systems).
136. Tuomas Mylly, *Intellectual Property and European Economic Constitutional law, the trouble with Private Informational power*, 2009, p. 103.

'Their parallel existence relates to legal sphere's polyvalence, which, in turn, is caused by societal value-pluralism. From this perspective, the legal system in any scale or on any level is neither coherent nor consistent, but pluralistic into its deepest levels.'[137]

Discussing the constitutional pluralism of European law, *Thomas Mylly* refers to *Kaarlo Tuori* who maintains that the EU's legal order mostly consists of the laws and individual decisions at the surface level, but a legal culture and deep structure have not yet been developed. Thus he questions the autonomy of the European Union legal order as advocated by the Community Courts. *Tuori* seems to accept that fundamental rights, democracy and the rule of law could ultimately constitute a Union-wide deep structure of law.[138] Lately, Joined Cases C-402/05 and C-415/05 *Yassin Addullah Kadi and others v Council* seemingly confirm that.[139]

'seen from a holistic perspective of European Law, comprising not only the EU treaties and the law emanating directly from them, but also the law based on domestic level and the legal culture based on both, *Tuori's* layered model of law is not incompatible with pluralist perceptions of European law. Yet, contrary to what *Tuori* maintains, the admittedly thin European legal culture of intersystemic interaction is there, and has an important function. It is thus possible to construct this totality as a system of polycentric rules, partly overlapping but capable of compatibility due to deep interlegality, shared understanding of political and legal community, as well as fundamental rights, democracy and the rule of law functioning not only as the law's self-limitation in each sub-system separately, but also as a limitation of other sub-system's legal claims. Evidently, there are diverse interpretations of what fundamental rights, democracy and rule of law mean in practice. This, however, does not prevent them from constituting a common horizon or reference point for the total legal community thus construed: possibility of disagreement about their more detailed con-

137. Tuomas Mylly, *Intellectual Property and European Economic Constitutional law, the trouble with Private Informational power*, 2009, p. 104. He writes on page 105: 'Legal pluralism and polycentricity of law do not have to be seen as new phenomena caused by recent developments related to globalization or accelerating functional differentiation of social fields. They can also be seen to constitute inherent characteristics of law. Yet, their role is accentuated by these developments, making polycentric view of law unavoidable'.
138. Tuomas Mylly, *Intellectual Property and European Economic Constitutional law, the trouble with Private Informational power*, 2009, p. 133.
139. See further Case C-465/07 *Elgafaji* v *Staatssecretaris van Justitie*; and Gráinne de Búrca, 'The Evolution of EU Human Rights Law', in Paul Craig and Gráinne de Búrca (eds), *The Evolution of EU Law*, 2011, pp. 488-489.

tent is part of the common horizon constituting the community in question, also expressed in the comprehensive institutional structure enabling multiple constitutional controls.'[140]

This approach can be acknowledged. However, once there are overlapping systems, such as the European Patent Convention and the EU's mix of authorities, the approach to international law and European constitutional pluralism no longer suffices. The European Patent Office is one internal system (with sovereign rights) and the EU is another.[141] The EPC can be considered the deep structure whereas Directive 98/44/EC is more in the nature of surface law. There is no semi-autonomous field involved, but two autonomous legal systems; one is an autonomous legal system for the granting of European patents and the other is an autonomous legal system with no granting authority but with power to decide on substantive patent law concerning biotechnological patents pursuant to Directive 98/44/EC.[142] Thus, the basic principle of EU protection is independence, not only from national law, but also from other regional supranational organisations of which its Member States are members, such as the European Patent Organisation.

### *4.2.1. More on method and legal theory*

The legal dogmatic method and the theoretical basis in critical legal positivism as described above will be supplemented by a comparative method and the theory will be further elaborated in a legal-sociological perspective. Scandinavian legal realism will also be fundamental to the analysis.

As mentioned in section 1, both the TRIPS Agreement and the rights of free movement of goods within the European Union have established economic standards aimed at allowing the free flow of good as an overall goal, and in the EU this is furthered by a judicial doctrine of effectiveness.[143] Exceptions to freedom of movement are to be interpreted narrowly with case-

140. Tuomas Mylly, *Intellectual Property and European Economic Constitutional law, the trouble with Private Informational power*, 2009, p. 133.
141. Within the EU there has been a tendency from the CJEU to recognise environmental principles, especilly the principle of sustainability (as an ideology) may be regarded as legal culture or deep structure of EU law.
142. See Hanns Ullrich, 'European Intellectual Property protection', in D. Vaver and L. Bently (eds.), *Intellectual Property in the new Millennium*, 2004, p. 27-28.
143. See Paul Craig and Gráinne de Búrca, *EU law, Text, Cases, and materials*, 5th Ed., 2011, p. 581 ff.; and Inger-Johanne Sand, '(Re)Constructing the Boundaries of the market: EU Law and Institutions Analysed through the Lens of Discontinuity', in Hanne Petersen, Anne Lise Kjær, Helle Krunke & Michael Rask Madsen (eds), *Paradoxes of European Legal Integration*, p. 88.

by-case scrutiny of the fulfilment of certain requirements for non-discrimination and proportionality. There is no main rule or general set of rules signalling that exceptions are standards with a status equal to the main rule.[144] Regard for environmental protection, health and other important concerns are accepted as exceptions but they are not given priority over economic standards. The lack of an overall broad and balanced assessment can be due to the inherited legal order in which economic values are given more intense protection than other values in other regulations. In the EU this it obviously connected with the foundation of the EEC in 1958, since when other values have gradually come to be accepted, for example environmental protection in the European Single Act in 1987.

In this context *Inger-Johanne Sand* considers the role of the legal sociologist when it comes to new technologies.[145] She argues that the role of the legal sociologist is to shed light on the imbalance that appears when new technologies are to be regulated in the inherited framework with an inherited mechanism protecting against intervention from external regulations (health, risk and the environment). The regulation of new technologies with an economic potential has an inherited immunity from other considerations, and the starting point is that it outweighs external concerns. However with new technologies come new concerns and social dilemmas which legislators are far from equipped to deal with in terms of detailed law design. This thesis will be based on a mix of patent standards and environmental standards in order to redress this imbalance due the inborn emphasis on economic values in patent laws.[146] Recitals 3 and 5-7 of Directive 98/44/EC on the protection of biotechnological invention contain clear references to economic standards as reason for its enactment (removing obstacles to trade, the smooth functioning of the internal market, encouragement of scientific research and develop-

144. TFEU Article 36 and the doctrine derived from the *Cassis de Dijon* case; see Inger-Johanne Sand, 'Teoretiske utfordringer for rettsociologien i globaliseringens tid', in Bettina Lemann Kristiansen & Anette Storgaard (eds), *Nordisk retssociologi, status – aktuelle udfordringer – visioner*, 2010, p. 138.

145. Inger-Johanne Sand, 'Teoretiske utfordringer for rettsociologien i globaliseringens tid', in Bettina Lemann Kristiansen & Anette Storgaard (eds), *Nordisk retssociologi, status – aktuelle udfordringer – visioner*, 2010, p. 142; Jagdeep S. Bhandari & Alan O. Sykes, *Economic Dimensions in International Law, Comparative and Empirical Perspectives*, 1997, pp. 23-24.

146. Raymond Wacks, *Understanding Jurisprudence, An Introduction to Legal Theory*, 2nd Ed., p. 194: who writes that sociologists 'generally reject the idea (most closely associated with legal positivism) that there can be a value-free explanation of law'.

ments in the field of genetic engineering).[147] Patent law is a very old legal discipline, whereas the standards that have to balance patent law are more recent. One methodological approach to this imbalance is to amend patent law in order to give equal importance to such non-economic standards. This can be burdensome since principles such as sustainability, and the precautionary principle are often too broad and vague to be given a legal quality outside their original application.[148] They are more to be regarded as political statements.[149] However concerns such as human dignity tend to be emphasised but is not always applied.[150] Directive 98/44/EC is a perfect example of legis-

147. See also Case C-34/10 *Oliver Brüstle* v *Greenpeace*, para. 26; and Case C-456/03 *Commission v Italy*, paras 16 and 27. In para. 32 of the *Brüstle* case the balancing of all interest is described as follows: 'In that regard, the preamble to the Directive states that although it seeks to promote investment in the field of biotechnology, use of biological material originating from humans must be consistent with regard for fundamental rights and, in particular, the dignity of the person.' Recital 16 of Directive 98/44/EC emphasises that 'patent law must be applied so as to respect the fundamental principles safeguarding the dignity and integrity of the person'.
148. The precautionary principle has gained legal quality on a case-by-case approach (e.g. Case 94/83 *Heijn*, Case 54/85 *Mirepoix*, Case C-180/96 *UK v Commission*, Case C-473/98 *Toolex Alpha*, C-405/92 *Mondiet*, T-74/00 *Artegodan*, Case C-192/01 *Commission v Denmark*, Case C-443/02 *Screiber*, Case 24/00 *Commission v France*, C-132/03 *Codacons*, Case C-127/02 *Waddenzee*). See Joakim Zander, '*The application of the Precautionary Principle in Practice'*, 2010, p. 103 ff. The principle of sustainability has gained international momentum, see e.g. Philippe Sands and Jacqueline Peel, with Adriana Fabra and Ruth MacKenzie, '*Principles of International Environmental Law'*, third ED., 2012, p. 206 ff. (coined by the 1987 Brundtland Report: *Our Common Future*, the World Commission on Environment and Development, 1987.); The UN Department of Economic and Social Affairs, Division for Sustainable Development, The Johannesburg Plan of Implementation, 2002;' RIO +20', United Nations Conference on Sustainable Development, 20-22 June, 'The future we want', A/CONF.216/L.1.; 'Resilient People, Resilient Planet: A future worth choosing', The report of the United Nations Secretary-General's High-level Panel on Global Sustainability, 2012; Christina Voigt, '*Sustainable Development as a Principle of International Law. Resolving Conflicts between Climate Measures and WTO Law'*, 2009.
149. Nicolas de Sadeleer, *Environmental Principles, from political slogans to legal rules*, 2002; James Harrison, 'to 'CBDR' or not to 'CBDR': The role of common but differentiated responsibilities in international environmental law decision making', The Rule of Law for Nature, Oslo Conference Paper, 2012.
150. See the emphasis on human dignity in Case C-34/10 *Oliver Brüstle v Greenpeace*, paras 30 and 32 -34; and Chapter 4, section 1. In the new US patent legislation, the Leahy Smith America Invents Act, HR 1249 section 33, it is now stated that no patent may be issued on a claim directed to or encompassing a human organism. However, the balance here is also based on pros and cons within an economic standard.

lators' adoption of rules on emerging technologies from an economic standard evaluation. The approach taken in this thesis is that detailed and technology-tailored laws are not good examples and it is suggested that standards based on common values are more future-proof and acceptable, despite their political values.

The comparative method takes a functional approach. The method is aimed at analysing which substantive solutions different legal systems provide for the same legal problem. The subjects of this comparison are the USA and Europe. The comparison will be of two very different systems. The USA has a centralised patent process, one authority, one statute and unification by the Federal Circuit and the Supreme Court system. Europe has several legislators, no uniform patent (not yet effective), one trans-national granting authority (EPO) and the EU/Court of Justice has a say in these matters. Where the USA has no detailed statute on gene technology this is not the case with the EU/EPO. The empirical study in Chapters 7 and 8, together with the analysis in the overall thesis, will shed light on the European approach with this perspective from the USA.[151]

### 4.3. Sources of law and more empirical data

The patent law of today has expanded rapidly in several areas. As for patentable subject matter, the statutory framework has been stretched in order to tailor patent-specific regulation to meet the needs of such as new life sciences technologies, and environmentally sound technologies. The mix of disciplines for inspiring patent law with environmental principles, such as the precaution principle, have been discussed when patents for transgenic animals has been prosecuted for example in Norway.[152]

151. Martijn W. Hesselink, *The New European Private Law, Essays on the future of private Law*, 2002, pp. 47 and 49. See also Alf Ross, *On law and Justice*, 1958, p. 110; and Ruth Nielsen, 'Scandinavian Legal Realism and the EU Law', in Ulla Neergaard, Ruth Nielsen & Lynn Roseberry, *The Role of Courts in Developing a European Social Model, Theoretical and Methodological Perspectives*, 2010, p. 238: where it is said that according to Ross, the legal method varied from legal system to legal system; 'He, for example, underlined the difference in methods used in common law systems where case law is a predominant source of law and civil law systems where legislation is a predominant source of law'. See also Ruth Nielsen & Christina D. Tvarnø, *Retskilder & Retsteorier*, 3rd Ed., 2011, pp. 29-30 and 68-70.

152. See Chapter 8 and the discussion of the principle as a general guideline in 'Recommendations for a Patent System of the Future', Report by a working group under the Danish Board of Technology, 2007, recommendation p. 22, point 2 in section 3.3.2: 'Establish a 'precautionary principle,' to ensure that future reforms and expansions of

The concept of 'sources of law' used in this thesis refers to valid laws in a system of multi-level regulation of patent law in a complex of environmental law principles, as illustrated in Table 1.1 above.[153] Based on Scandinavian Legal Realism, the criterion for a valid law involves an assessment of whether a source of law in a legal system will be applied by the courts because they feel obliged to do so. This approach differs from Kelsen's hierarchical ordering of legal sources, and it accepts a pluralistic view of the sources of law.[154]

In this framework both binding rules and non-binding rules will be addressed. In addressing the validity of laws, the thesis will distinguish between binding law and non legally binding norms (whether social or moral), as well as distinguishing the law *de lege lata* from the law *de lege ferenda.*[155] However, the law as it stands may only provide some principles with political implications and not be legally binding. Some laws have an effect when invoked in an isolated legal context but if invoked outside that context it is uncertain whether they will have a legal impact. That legal impact may also differ when considering the criteria for a law to be valid and accepted in different jurisdictions, whether national, regional or international (the USA, the EU and the EPO, the UN, and WTO). However, mixing the disciplines in their valid forms invokes the law *de lege ferenda.*[156]

Given the lack of a hierarchical structure of international law, Article 38 of the Statute of the International Court of Justice may provide some guidance as to the understanding of or the possible use of international legal principles,

the patent system will be implemented only if they can be shown to benefit technological development' at http://www.tekno.dk/pdf/projekter/p05recommendations _for a_patent_system_of the_future.pdf (last visited November, 2011).

153. Ruth Nielsen, 'Scandinavian Legal Realism and the EU Law', in Ulla Neergaard, Ruth Nielsen & Lynn Roseberry, *The Role of Courts in Developing a European Social Model, Theoretical and Methodological Perspectives*, 2010, p. 237: 'The doctrine of the sources of law provides a basis for a probable prediction of the outcome of hypothetical cases before the courts and other law-applying bodies which in Scandinavian legal realism as defined by Ross is the criterion for valid law ... The doctrine of sources of laws and their interpretation is thus a normative ideology which is shared by judges (and other law-applying bodies)'; see Alf Ross: *On Law and Justice*, 1958, p. 75.
154. Ruth Nielsen, 'Scandinavian Legal Realism and the EU Law', in Ulla Neergaard, Ruth Nielsen & Lynn Roseberry, *The Role of Courts in Developing a European Social Model, Theoretical and Methodological Perspectives*, 2010, pp. 239-240.
155. Akehurst, *Modern introduction to international law*, 1997, p. 34: 'In this sense, the term 'source' has a technical meaning related to the law-making process.'
156. Harold D. Lasswell & Myres S. McDougal, *Jurisprudence for a Free Society, Studies in Law, Sciences and Policy*, Vol. I, pp. 37-38.

such as e.g. environmental principles.[157] Article 38(1)(c) of the Statute states that the Court shall apply 'the general principles of law recognized by civilized nations'.[158]

'According to the first definition (general principles of international law), general principles of law are not so much a source of law as a method of using existing sources – extending exiting rules by analogy, inferring the existence of broad principles from more specific rules by means of inductive reasoning, and so on. According to the second definition of general principles of law (general principles of national law), gaps in international law may be filled by borrowing principles which are common to all or most national systems of law; specific rules of law usually vary from country to country, but the basic principles are often similar ... Not all general principles applied in international practice stem from domestic legal systems and have been transplanted to the international level by recognition. Some are based on 'natural justice' common to all legal systems (...good faith, proportionality), others simply apply logic familiar to lawyers (such as the rules *lex specialis* derogate *legi generali, lex posterior* derogate *legi priori*) and another category is related to 'the specific nature of international community' as expressed in principles of *ius cogens*. Therefore, a real transplantation of domestic law principles to the international level is limited to a number of procedural rules, such as the right to a fair hearing ... The mechanism by which such transformation takes place in practice goes through the mind of the international judge or arbitrator who has to decide on a particular case. This is known as the 'creative role' of the judge, which is not at all peculiar to the international legal system.'[159]

Finally, sources such as judicial decisions from the international court and other tribunals, and national courts (the latter however need to be used with caution)... learned writers can be considered a part of the list of sources whereas other sources of international law such as acts of international organizations, and soft law may also have an impact.[160]

157. Akehurst, *Modern introduction to international law*, 1997, p. 36.
158. According to the Article 1 of the Statute of the International Court, it was established by the Charter of the United Nations as the principal judicial organ of the United Nations. Akehurst, p. 36 writes that Article 38 usually is accepted as constituting a list of the sources of international law.
159. Akehurst, *Modern introduction to international law*, 1997, p. 48-49.
160. See Akehurst, pp. 53-54. On soft law it is stated that: 'soft law may also be relevant from a sociological perspective of international law with regard to the process of the formation of customary law or treaty law and the related issue of 'legitimacy' in the international legal system ... But the result of this law-making process at any given moment of legal decision is either binding law or not. In essence, under any meaningful concept of law, it remains essential to maintain the distinction between the law *de lege lata* and the law *de lege ferenda* ... between the codification of existing law and the progressive development of law, between legal norms and non-legal norms as regards their binding effect, and ultimately between the legal system and the political system. Otherwise, it would be become rather difficult to distinguish ideologically or

The sources used in this thesis also include patents and patent applications. In assessing how patent law adapts to technological changes and to establish a foundation for recommendations for patent standards, the empirical case study in Chapter 7 is based on patent prosecutions at the EPO and in the USA (USPTO) with regard to six diseases: cystic fibrosis, Huntington's disease, Canavan disease, spinal muscular atrophy, Parkinson's disease and Alzheimer's disease. The review of the patent practice serves as an analysis of how the USA and the EPO/EU patent laws adapt to emerging technologies. In other words, the empirical data is a means for understanding the legal framework, unlike the case law of the Boards of Appeal or the Federal Circuit it is not a source of law as such. One of the key issues in this thesis is that patent offices and courts may smoothly adapt to technological changes (micro adjustments), whereas field-restricted legislation may hinder a dynamic approach. Thus, analysing a part of patent practice can illustrate this key issue since the USA has no tailor-made statute, whereas the EPO/EU has implemented Directive 98/44/EC which is tailor-made for biotechnology.

### *4.3.1. The Vienna Convention*

Part III, section 3 of the Vienna Convention on the Law of Treaties 1969 deals with the interpretation of treaties. Article 31(1) states the general rule according to which: 'A treaty shall be interpreted in good faith in accordance with the ordinary meaning to be given to the terms of the treaty in their context and in the light of its object and purpose.'

The Article contains several norms of interpretation, as follows:
2. The context for the purpose of the interpretation of a treaty shall comprise, in addition to the text, including its preamble and annexes:
(*a*) any agreement relating to the treaty which was made between all the parties in connection with the conclusion of the treaty;
(*b*) any instrument which was made by one or more parties in connection with the conclusion of the treaty and accepted by the other parties as an instrument related to the treaty.
3. There shall be taken into account, together with the context:
(*a*) any subsequent agreement between the parties regarding the interpretation of the treaty or the application of its provisions;
(*b*) any subsequent practice in the application of the treaty which establishes the agreement of the parties regarding its interpretation;
(*c*) any relevant rules of international law applicable in the relations between the parties.
4. A special meaning shall be given to a term if it is established that the parties so intended.

otherwise politically motivated claims from accepted rules and principles of international law'; p. 54. See also Akehurst pp. 55-56: on equity, justice and natural law.

Further, Article 32 deals with supplementary means of interpretation:

'Recourse may be had to supplementary means of interpretation, including the preparatory work of the treaty and the circumstances of its conclusion, in order to confirm the meaning resulting from the application of article 31, or to determine the meaning when the interpretation according to article 31:
(*a*) leaves the meaning ambiguous or obscure; or
(*b*) leads to a result which is manifestly absurd or unreasonable'

One can discuss whether the Vienna Convention can fit into the multi-layered patent regime. For example, the Vienna Convention tools can be used to interpret the European Patent Convention and the TRIPS Agreement, but what about Directive 98/44/EC? It is not obvious that EU secondary legislation can fit into the system. However, one can see references to the Vienna Convention in the case law of the EPO Boards of Appeal, despite the mix of the legislation that is subject to interpretation; see further Chapter 4.5.4.[161]

It is obvious that the reference to 'the preparatory work' has no value in a TRIPS assessment. As has been pointed out earlier in this Chapter by reference to *Nuno Pires de Carvalho*, the history of negotiations has no value for the proper interpretation of the agreement.[162] *Pires* and *Pymont*[163] both refer to the Doha Declaration, reaffirming the objectives and purposes of the TRIPS Agreement in Articles 7 and 8, as powerful means of interpretation. Article 5 (a) of the Doha Declaration 2001 states:

161. G 0005/83 (Second medical indication), O.J. EPO 1985, 64.
162. Nuno Pires de Carvalho, *The TRIPS Regime of Patent Rights*, p. 75. However, in Daniel J. Gervais, 'The changing landscape of international intellectual property', in Christoffer Heath & Anselm Kamperman Sanders (eds), *Intellectual property & Free Trade Agreements*, 2007, p. 53, the author writes: 'by contrast, developing countries were forced to accept a package that a number of countries did not fully understand and which contained a complete set of intellectual property norms they now had to implement into their national law'. The TRIPS Agreement cannot be as binding in every aspect since it was not fully understood at the time of its adoption. Therefore, the TRIPS Agreement cannot support any precise intention to change or purpose in changing patent law concepts apart from the fact that it was a response to the lack of patent protection for pharmaceutical products in the developing countries.
163. Wolrad Prinz zu Waldeck und Pyrmont, 'Special legislation for genetic Inventions – A violation of Article 27(1) TRIPS?', in Wolrad Prinz zu Waldeck und Pyrmont, Martin Adelmann, Robert Brauneis, Josef Drexl & Ralph Nack (eds), *Patents and technological progress in a globalized world, Liber Amicorum Joseph Straus*; and Nuno Pires de Carvalho, *The TRIPS Regime of Patent Rights*.

'In applying the customary rules of interpretation of public international law, each provision of the TRIPS Agreement shall be read in the light of the object and purpose of the Agreement as expressed, in particular, in its objectives and principles.'[164]

The Declaration defines itself by the wording in its paragraph 4:

'We agree that the TRIPS Agreement does not and should not prevent members from taking measures to protect public health. Accordingly, while reiterating our commitment to the TRIPS Agreement, we affirm that the Agreement can and should be interpreted and implemented in a manner supportive of WTO members' right to protect public health and, in particular, to promote access to medicines for all.'

*Pires de Carvalho* concludes that the published preamble reflects the negotiations. To use the preamble to attempt to modify the outcome of the negotiations would constitute a renegotiation of the agreement.[165] According to *Pires de Carvalho*, Article 8 has implemented the preamble restrictively; measures may be adopted to pursue socio-economic and technological goals provided that they are consistent with the provisions of the Agreement. *Pyrmont* writes that TRIPS Articles 7 and 8 cannot justify a different approach towards new technologies.[166] Even though the aim of TRIPS was to obtain a joint understanding and acceptance of pharmaceutical and chemical patents, TRIPS was framed to embrace all technologies.

'By its very nature as an instrument to stimulate innovation, patent law was and will be confronted with new technologies that may or may not create specific problems it has to address, and it would be an implausible suggestion that the members were not aware of the fact during negotiations of the TRIPs agreement ... it is very rare that new technologies raise completely new issues – it is rather that certain generally known issues arise more frequently and more visibly in specific technical fields than in others.'[167]

164. See the Declaration on the TRIPS Agreement and public health http://www.wto.org/english/thewto_e/minist_e/min01_e/mindecl_trips_e.htm
165. Nuno Pires de Carvalho, *The TRIPS Regime of Patent Rights*, p. 77.
166. Wolrad Prinz zu Waldeck und Pyrmont, 'Special legislation for genetic Inventions – A violation of Article 27(1) TRIPS?', in Wolrad Prinz zu Waldeck und Pyrmont, Martin Adelmann, Robert Brauneis, Josef Drexl & Ralph Nack (eds.), *Patents and technological progress in a globalized world, Liber Amicorum Joseph Straus*, p. 297.
167. Wolrad Prinz zu Waldeck und Pyrmont, 'Special legislation for genetic Inventions – A violation of Article 27(1) TRIPS?', in Wolrad Prinz zu Waldeck und Pyrmont, Martin Adelmann, Robert Brauneis, Josef Drexl & Ralph Nack (eds), *Patents and technological progress in a globalized world, Liber Amicorum Joseph Straus*, p. 297; and Nuno Pires de Carvalho, *The TRIPS Regime of Patent Rights*, p. 306.

Pyrmont makes several observations that are not compatible with the observations of this thesis. It is correct that new technologies seldom raise new issues. That is exactly the reasoning behind recommendation of this thesis on environmentally sound technologies in Chapter 9. New technologies ought not (always) to be approached with a tailor made law design that may be outdated too fast. However, this thesis suggests that the TRIPS framework should be interpreted on a more dynamic basis than follows from the reading of both *Pyrmont* and *Pires de Carvalho*. Above all, there will be an analysis of whether TRIPS can encompass more than an economic standard which is relevant for guiding courts and patent offices in patent law matters.[168]

## 5. Delimitation

The lifecycles of patents can broadly be divided into two phases. Phase one, the *pre-grant* phase, covers the period from the filing of the patent application to the grant of the patent. Phase two, the *post-grant* phase, covers the period from the grant until the patent term expires (or lapses due to non-payment of fees). This thesis mainly addresses the *pre-grant* phase, the requirements for patentability and the balancing of public and private interests. However, *post-grant* measures are also important and will be briefly discussed.

The thesis is a legal study based on legal methods and legal theory. It can be argued that patent law is not an isolated discipline that can be evaluated solely on legal grounds. In his introductory note, *Nuno Pires de Carvalho* writes as follows:

> 'indeed, intellectual property is a legal tool with an intrinsic economic dimension, and that is perhaps the factor that generates much confusion: economists write about intellectual property without understanding its essentially legal nature, and lawyers write about intellectual property without mastering a few basic notions of microeconomics. And the fact is

168. See Joseph William Singer, *The reliance interest in property revisited*, http://www.legalleft.org/wp-content/uploads/2011/12/Singer_The-Reliance-Interest-in-Property-Revisited.pdf p. 117, expressing some sort of social responsibility associated with property (an economic standard); and the same in *Stanford Law Review*, Vol. 40, No 3 (Feb., 1988), pp. 611-751, especially at p. 652 on property and social relations, the social responsibility of property ownership, despite its context of plant closings, the article can be considered as giving a broader perspective on property rights as such.

that policy makers in both developed and developing countries have approached intellectual property without a clear vision of either of its two dimensions.'[169]

However, the study has been shaped as a legal review of international patent law as it operates in the EU and the USA. Its purpose is to consider whether the courts and boards of appeal, assisted by patent authorities, can ensure the legitimacy of the patent system. Thus, Chapters 7 and 78 evaluate the prosecution of patent applications by a case study on the granting process on human genes exemplified with six genetic diseases (Chapter 7) and on animal genes exemplified by the pig (Chapter 8).

The economic approach is absent from this thesis. Thus, there is no review of the patent system questioning its existence. The thesis is not a study on pros and cons of the patent system. It has been conducted in the light of Machlup's famous statement that:

'If one does not know whether a system 'as a whole' (in contrast to certain features of it) is good or bad, the safest 'policy conclusion' is to 'muddle through' ... If we did not have a patent system, it would be irresponsible ... to recommend instituting one. But since we have had a patent system for a long time, it would be irresponsible, on the basis of our present knowledge, to recommend abolishing it. This last statement refers to a country such as the U.S. – not to a small country and not a predominantly nonindustrial country, where a different weight of argument might well suggest another conclusion.'[170]

According to the Machlup's economic review of the patent system, patent protection is advocated *either* on ethical grounds, in the name of 'justice' or 'natural rights', *or* on pragmatic grounds in the name of the 'promotion of the public interests'. He gives the chief arguments for patent protection and as the four best-known propositions for patent protection he examines the 'natural-law' thesis, the 'reward-by-monopoly' thesis, the 'monopoly-profit-incentive' thesis, and the 'exchange-for-secrets' thesis. Machlup's review can be summarised as follows:

'The '*natural-law*' thesis assumes that man has a natural property right in his own ideas and that Society is morally obligated to recognize and protect this property right ... The '*reward-by-monopoly*' thesis assumes that justice requires that a man receive reward for

169. Nuno Pires de Carvalho, *The TRIPS Regime of Patent Rights*, p. 2.
170. See Fritz Machlup, 'An Economic Review of the Patent System', Study No 15, Subcommittee Patents, Trademarks and Copyrights of the U.S. Senate Judiciary Comm. (1958), pp. 20-23. http://onlinebooks.library.upenn.edu/webbin/book/ search?author=&amode =&title=An%20Economic%20Review%20of%20the%20Patent%20System&tmode=start&c=x

his services in proportion to their usefulness to society, and that, where needed, society must intervene to secure him such reward ... The '*monopoly-profit-incentive*' thesis assumes that industrial progress is desirable, that inventions and their industrial exploitation are necessary for such progress, but that inventions and/or their exploitation will not be obtained in sufficient measure if inventors and capitalists can hope only for such profits as the competitive exploitation of all technical knowledge will permit. The '*exchange-for-secrets*' thesis presumes a bargain between inventor and society, the former surrendering the possession of secret knowledge in exchange for the protection of a temporary exclusivity in its industrial use.'

All four arguments for patent protection have been severely criticised, both by opponents of any sort of patent protection, and by advocates who support one argument but reject the others.[171]

## 6. Introduction to the thesis

The thesis is divided into four parts. Part I, Multi-level regulation and new technologies, comprises Chapters 1-4. Part II, 'The human genome', comprises Chapters 4-7. Part III, 'The animal kingdom', comprises Chapter 8. Part IV, 'Patent law: A bellwether for global concerns?' comprises Chapters 9-10.

The thesis elaborates previous articles written from 1997, when the author defended her PhD thesis on environmental law and waste regulation in the EU. Thus, in Chapters 1 and 4 elaborate the author's 2007 IIC article: 'Scope of Gene Patent Protection and the TRIPS Agreement – An Exclusively Non-discriminatory Approach?' Chapter 4 elaborates the author's 2008 *Bioscience Law Review* article 'Interpreting Ordre Public and Morality in a Patent Law Context: Which is the Correct Approach?', which in turn was based on the 2007 report to the Norwegian Advisory Board on Ethical Aspects of Patenting, 'Ordre Public and Morality in a Patent Law Context'.[172]

Chapters 4 and 5 also elaborate the author's two articles and one monograph chapter written in Danish, namely: the 2008 chapter 'Genpatentet og patentetik i EU – en harmoniseret retsvildfarelse?'; the 2004 article 'Patente-

171. See Fritz Machlup, 'An Economic Review of the Patent System', Study No 15, Subcommittee Patents, Trademarks and Copyrights of the U.S. Senate Judiciary Comm. (1958), pp. 20-23. http://onlinebooks.library.upenn.edu/webbin/ book/search?author=&amode=&title=An%20Economic%20Review%20of%20the%20Patent%20System&tmode=start&c=x, p. 22.

172. See also Tine Sommer, 'Sædelighedsbegrebet i relation til bioteknologi', *Juristen* (10), December 2001; and 'Patentmyndigheder som bioteknologiens moralske vogter', *Nordisk Immateriellt Rättsskydd*, 2005 (74), Vol. 4, pp. 333-344.

ring af det humane genom'; and the 2000 article 'Retlig beskyttelse af bioteknologi'.[173]

Chapter 8 contains the 2008 IIC article: 'Patenting the Animal Kingdom? From Cross-Breeding to Genetic Make-up and Biomedical Research'; and finally Chapter 9 the 2010 NELJ article: 'Designing substantial patent law: From life sciences to climate change?' The articles are further addressed in Chapter 10.

The thesis is also based on a comprehensive case study on human gene patent for six diseases: Huntington's disease, cystic fibrosis, Canavan disease, spinal muscular atrophy, Alzheimers and Parkinson's disease (Chapter 7) and on the pig gene patents (Chapter 8).

173. See 'Retlig beskyttelse af bioteknologi, den danske lovgivning vedrørende implementeringen af direktiv 98/44/EF af 6. juli 1998 om retlig beskyttelse af bioteknologiske opfindelser', *UfR* 2000 B 457; and 'Patentering af det humane genom', *UfR* 2004B 214.

CHAPTER 2

# Protection of Biotechnology: A Frozen Statute?

Chapter 2 deals with technological development in the context of regulation in detail. The chapter introduces the reader to the technological background, in particular to the empirical case study in Chapter 7. There is an explanation of how the legislative framework should be understood from a genetics perspective, for example: What is a gene? What genetic information do genes contain? What is the function of a gene? New technologies used in biomedical research are also introduced. In particular, Figure 2.1 illustrates the relationships between gene technology, nanotechnology, synthetic biology and chemicals.

## 1. Biotechnology as a concept

The purpose of this chapter is to introduce the reader to the emerging technologies that are the subject of this thesis and to illustrate that the relationships between the major technologies cannot be disregarded when regulating their patentability. The aim of this interdisciplinary approach is threefold. First, it provides the background knowledge of the technology that legislators or courts have framed in a patent law context in order to highlight the complexity that has either been narrowed down in regulation in detail (the EU/EPO approach) or absorbed in judicial doctrines (the US approach). Second, it introduces the reader to new technologies that challenge the current legislation. And third, it highlights the problems that such tailor-made regulation can face due to scientific developments. As such it can be considered as an introduction to the empirical case study in Chapter 7.

The breeding of plants and animals has been known for thousands of years and its techniques may be regarded as *traditional* biotechnology.[1] Modern biotechnology goes further using new techniques to alter the genetic material of plants, animals and humans or to reveal genetic causes and biological pathways that go beyond the traditional methods of cross-breeding or the improvement of crops. New possibilities in healthcare and life sciences due to scientific advances and technological developments has challenged (such as cloning of genes) and new techniques and methods (such as full genome sequencing and RNAi techniques), already challenge and will undoubtedly continue to challenge legislators at a global level.[2] The question is whether patent laws should be tailor-made for each new technology, or whether it would be better to adapt to new developments and global concerns other than by using specifically tailored solutions, as discussed in Chapter 1, section 2.[3] One can categorise technologies under different headings such as biotechnology, gene technology, nanotechnology and synthetic biology, but regulating for each technology would embrace a broad range of disciplines, e.g. healthcare, the motor industry, the energy sector, waste management and the food industry.[4] 'Biotechnology' is a broader concept than 'gene technology'. Bio-

1. David A. Micklos & Greg A. Freyer with David A. Crotty, *DNA Sciences, A first Course*, 2nd Ed., 2003, p. 3: 'literally translated biotechnology means 'life technology' applying knowledge about living things for the practically use of mankind. The ancient uses of yeast in making bread and alcoholic beverages and of bacteria in making cheese are, in the broadest sense, biotechnology. However, the modern biotechnology revolution is based on a deep understanding of the technology of life, the mechanics of living machines'. See Robert Patrick Mergers & John Fitzgerald Duffy, *Patent Law and Policy: Cases and materials*, 5th Ed., 2011, p. 159, on Louis Pasteur's 1873 US patent for an improvement in the manufacture of beer and yeast. See also Michael John Gulliford, 'Much ado about gene patents: the role of foreseeability', *Seton Hall Law Review*, 2004 (34 (SHLR 711), pp. 714-715.
2. R.S. Crespi, *Patents: A basic guide to patenting in Biotechnology*, 1988, p. 97 ff., on the new biotechnology. European Parliament, *STOA*: 'Making perfect life', Bioengineering (in) the 21st Century, September 2012: 'concepts 'biology becoming technology' and 'technology becoming biology''.
3. Fred Warshofsky, *The Patent wars, the battle to own the world's technology*, 1994. As patent protection has expanded patent law has been challenged by two important developments: the rapid pace in technological change, and the globalization of the marketplace; see Jeffrey B. Hawkins, *Harvard Journal of Law & Technology*, Vol. 9, No 1 1996.
4. Synthetic biology often combines approaches from different disciplines, such as biotechnology, chemistry, nanotechnology, computing and engineering; see Berthold Rutz, 'Synthetic biology and patents, A European perspective', *EMBO Reports*, Vol. 10, special issue, 2009, p. S15. Nanotechnology involves e.g. Neurophysiology, bota-

technology covers both gene technology and nanotechnology and it is relevant for almost every aspect of the modern global society. Gene technology is concerned with the science of the building blocks of life itself.[5] Medical genetics seeks to apply a large-scale analysis of the human genome, as well as interactions between genes and the environment, to improve health care.[6]

*Figure 2.1. Technology*

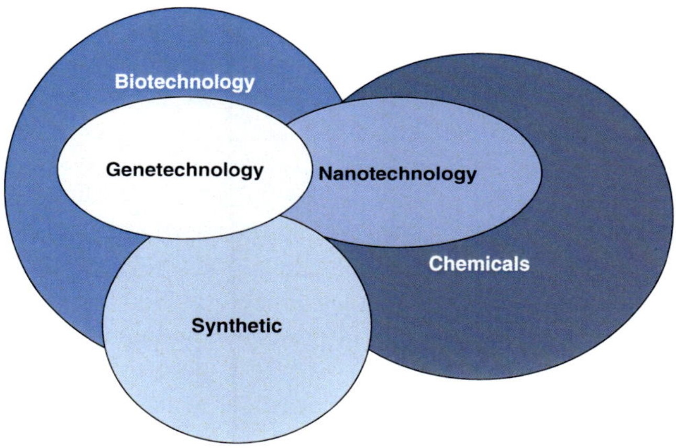


In 1984, biotechnology was defined as including 'all techniques using animals, plants, micro-organisms and any type of biological material which can be assimilated to micro-organisms, or which can create organic changes the-

ny, immunology or forensic science; see David A. Micklos & Greg A. Freyer with David A. Crotty, *DNA Sciences, A first Course*, 2nd Ed., 2003, p. 4; molecular biology is a hybrid discipline. See also Timothy Caulfield, 'Underwhelmed: Hyperbole, Regulatory Policy, and the Genetic Revolution', 45 *McGill Law Journal* 2000, p. 437, section headed 'What is the genetic revolution?: Science, Commerce and Culture' at p. 441; and Charles Tahan, 'The Nanotechnology (R)evolution', in Fritz Allhoff, Patrick Lin, James Moor & John Weckert (eds), *Nanoethics, The Ethical and Social Implications of Nanotechnology*, 2007, p. 91.

5. Michael John Gulliford, 'Much ado about gene patents: the role of foreseeability', *Seton Hall law Review*, 2004 (34 (SHLR 711), pp. 714-715.
6. Robert L. Nussbaum, Roderick R. McInnes & Huntington F. Willard, *Thompson & Thompson Genetics in Medicine*, 7th Ed., 2007, p. 1.

rein'.[7] In 1992, Article 2 of the Convention on Biological Diversity, adopted at the United Nations Conference on Environment and Development (the Rio 'Earth Summit'), included definitions of biotechnology and genetic material. Accordingly, 'biotechnology' means any technological application that uses biological systems, living organisms, or derivatives thereof, to make or modify products or processes for specific use. 'Genetic material' means any material of plant, animal, microbial or other origin containing functional units of heredity. In a way both definitions seem to embrace both gene and nanotechnology. However, nanotechnology essentially refers to biotechnology at a minor scale, and is more than biotechnology (e.g. chemicals).

In patent law there are definitions of biological material, microbiological processes, essentially biological processes (traditional biotechnology) and of biotechnological invention. In Article 2(1) of European Union Directive 98/44/EC '*biological material*' is defined as 'any material containing genetic information and capable of reproducing itself or being reproduced in a biological system' and '*microbiological process*' is defined as 'any process involving or performed upon or resulting in microbiological material'. According to the European Patent Office (EPO) Guidelines for Examination, June 2012, Part G, Chapter II-13, point 5.1, ''biotechnological inventions' are inventions which concern a product consisting of or containing biological material or a process by means of which biological material is produced, processed or used and further that 'biological material' means any material containing genetic information and capable of reproducing itself or being reproduced in a biological system'.[8] The definition of '*essentially biological processes*' will be further analysed in Chapters 8 and 10. It follows from both Article 4(1)(b) of Directive 98/44/EC and Article 53(b) of the European Patent Convention (EPC) that essentially biological processes for the production of plants and animals are not patentable. Furthermore it follows from Article 2(2) of Directive 98/44/EC, Rule 26(5) of the EPC, and Part G, Chapter II-16, point 5.4.2 of the EPO Guidelines for Examination that 'a process for the production of

7. World Intellectual Property Organization Committee of experts & Angus J. Wells, 'Patenting new life forms: an ecological perspective', *European Intellectual Property Review*, EIPR 1994, 16 (3), p. 111.
8. The Guidelines for Examination June 2012 (prior amendment 30 May 2011 regarding processes for the production of plants and animals); Part G, Guidelines for Substantive Examination, point 5.1, General Remarks and Definitions; see http://www.epo.org/law-practice/legal-texts/guidelines.html. The Guidelines for Examination refer to Rules 26-29 EPC, which in turn refers to Directive 98/44/EC. See further Part G, Chapter II, Patentability, point 5.2, on patentable biotechnological inventions.

plants or animals which is based on the sexual crossing of whole genomes and on the subsequent selection of plants or animals is excluded from patentability as being essentially biological, even if other technical steps relating to the preparation of the plant or animal or its further treatment are present in the claim before or after the crossing and selection steps'.[9] Will that change when/if the natural phenomena are replaced by synthetic biology in order to create artificial life? If for instance a synthetic new living organism (capable of self-reproducing with new genes that do not exist in nature) has artificial biological processes (new biological pathways); can it be patented as a whole? Can the individual genes or the whole genome of such an organism be patented? Which specific criteria will distinguish between existing or new essentially biological processes? How can we draw the borderline?

The general patent law definitions on technology in the EU and the EPC are unfortunately further elaborated in respect of gene technology. Articles 5 and 6 of Directive 98/44/EC refer to 'the sequence or partial sequence of a gene' and 'human embryos', and the use of such terms may turn out to ossify legislation as technological developments race ahead of legislators.

The definitions contained in both the EPC and the EU systems are tailor-made and thus field restricted, but at the same time they were enacted to be used in the broad area of biotechnology. The more narrowly the wording describes a technology (e.g. gene sequences or embryos), the greater is the possibility of it becoming out of date and of there being discussions of whether further field-restricted areas (e.g. within nanotechnology) require a specific patent legislation, see below section 4.

## 2. Gene technology

### 2.1.a. What is a gene?

Genes or the chemical deoxyribonucleic acid (DNA) from which genes are made have been studied in life sciences for a long time, from both purely scientific and industrial points of view. But what is a gene and where are genes expressed? In order to understand the challenges that lawmakers face when regulating property rights in genomics it is appropriate to explain briefly the

9. For scope of protection of biological material containing genetic information, see Tine Sommer, 'Patenting the Animal Kingdom? From Cross-Breeding to Genetic Make-Up and Biomedical Research', 2/2008 *IIC* Vol. 39, p. 161 ff.

core concepts of genetics.[10] It must be emphasised that all cells in the human body work together, and this complex nature cannot be simplified.

By increased structural knowledge of DNA sequences we learn more about the composition of human beings and the parameters of human life. Scientific ideas about DNA and its role in heredity were confirmed in 1962 by James D. Watson, Maurice Wilkins and Francis Crick's deduction of the molecular structure of DNA.[11] DNA is the substance that embodies the genetic code. Their finding of the double helix in 1953 was the starting point for future scientific research into the genetic code. It had already been propounded in 1946 that genes are the key components of living cells and that, to understand what life is, we must know how genes act.[12]

The Human Genome Project was initiated in 1990 with the goal of analyzing the structure of human DNA and determining the location of what were then estimated to be 100,000 human genes.[13] When the Human Genome Project was completed in 2003, the complete set of genes required to build and run a human being was found to be approximately 20,000-25,000 genes.[14]

10. This presentation is based on Robert L. Nussbaum, Roderick R. McInnes & Huntington F. Willard, *Thompson & Thompson Genetics in Medicine*, 7th Ed., 2007; Bruce Alberts et al., *Essential Cell Biology*, 3rd Ed., 2010; David A. Micklos, Greg A. Freyer with David A. Crotty, *DNA Science, A first course*, 2nd Ed., 2003; Anna Claybourne, *The Usborne Introduction to Genes & DNA*; and The Danish Council of Ethics, '*Patenting Human Genes and Stem cells, A report*', 2004.
11. Bruce Alberts et al., *Essential Cell Biology*, 3rd Ed., 2010, p. 173 ff.; and David A. Micklos, Greg A. Freyer with David A. Crotty, *DNA Science, A first course*, 2nd Ed., 2003, p. 3. See also Jennifer L. Schuster, 'Combining the Components of life: The application of Patent Extraterritoriality Doctrine to Biotechnology', *Indiana Law Journal*, Winter 2008, 83 Ind. L. J. 363.
12. James D. Watson, *The Double Helix, A personal Account of the Discovery of The Structure of DNA*, 1968, p. 13.
13. Understanding our genetic inheritance the U.S. Human Genome Project, The first five years: Fiscal years 1991-1995, published April 1990, NIH Publication No 90-1590: *http://www.ornl.gov/sci/techresources/Human_Genome/project/5yrplan/summary.shtml.* (visited August 2011).
14. HUGO Statement on Patenting of DNA Sequences, Belinda J.F. Rossiter (Ed.) 1997; Bruce Alberts et al., *Essential Cell Biology*, 3rd Ed, 2010, p. 181. See further http://www.ornl.gov/sci/techresources/Human_Genome/project/about.shtml. The race to sequence the human genome between competing private and public interests began when Craig Venter/Celera/Applied Biosystems Inc. (ABI) announced a joint venture to sequence the human genome. On 26 June 2000, the public and private groups announced jointly at a White House-sponsored event that each had succeeded in producing an initial draft of the human sequence. The international Human Genome Project published a full and significantly more accurate human genome sequence in

The reduction by 75,000 genes nicely illustrates the complexity of the human genome. Even with the 25,000 genes it is still difficult to understand how these genes are regulated and interact. Moreover, new types of genes with distinctly different functions have been discovered over the years. An example of this is the discovery of microRNA genes in 1993.[15]

Knowing that (almost) every cell in a human body contains approximately 25,000 genes, we begin to understand the complexity of genetics. DNA provides the genetic blueprint for all proteins in the body.[16] Genes ultimately influence all aspects of the body's structure and functions, and specifying the many proteins requires a great deal of genetic information.[17]

2004; see The National Research Council of the National Academics, *Reaping the Benefits of Genomic and Proteomic Research, Intellectual Property Rights, Innovation and Public Health*, 2006, p. 36. See also Anna Claybourne, *The Usborne Introduction to Genes & DNA*, p. 4.

15. RC Lee, RL Feinbaum & V Ambros (December 1993). 'The C. elegans heterochronic gene lin-4 encodes small RNAs with antisense complementarity to lin-14'. *Cell* 75 (5): 843-54. doi:10.1016/0092-8674(93)90529-Y. PMID8252621 (last visited February 2012).
16. In David A. Micklos, Greg A. Freyer with David A. Crotty, *DNA Science, A first course*, 2nd Ed. 2003, pp. 186 and 183, it is stated: 'Imagine a genomic book of life, the entire sequence of the human genome would comprise a stack of 1406 volumes. At 3.5 cm per volume, the stack would measure 49.21 meters – the height of a 15-story building'. See also Lynn B. Jorde, John C. Carey, Michael J. Bamshad, & Raymond L. White, *Medical Genetics*, 3rd Ed., 2006, p. 6.
17. Harvey Lodish et al., *Molecular Cell Biology*, 6th Ed., p. 623: 'No cell lives in isolation; cellular communication is a fundamental property of all cells and shapes the function and abilities of every living organism. Even single-celled organisms have the ability to communicate with each other or other organisms ... More important in plants and animals are extracellular signaling molecules that function *within* an organism to control metabolism of sugar, fats, and amino acids, the growth and differentiation of tissues, the synthesis and secretion of proteins, and the composition of intracellular and extracellular fluids.' See The Nuffield Council on Bioethics, '*The ethics of patenting DNA*', 2002, a discussion paper, p. 3: 'Many of these genes play a role in human diseases and disorders. Their identification may be the first step in the development of new diagnostic tests and treatments. Research in the rapidly expanding field of genomics aims to discover the biological function of particular genes, and how sets of genes and proteins work together in health and disease. Research is also focusing on identifying and understanding the proteins produced by the genes'. See further Minna Allarakhia & Anthony Wensley, 'Innovation and Intellectual property rights in systems biology, A framework for characterizing system-based knowledge is needed to understand the new complexities arising from the assignment of IP rights to biological information', *Nature Biotechnology*, Vol. 23, No 12, December 2005, p. 1485.

Each person has their own personal set of genes inside their cells, which is why we all have a different genetic make-up (genotype) resulting in different phenotypes.[18] Each human cell has a control unit called the nucleus, and genes are stored inside a cell's nucleus. [19] DNA is stored in long strands called chromosomes, and these chromosomes carry the coded hereditary information. Humans have 46 chromosomes in 23 pairs. A gene is part (or a section) of a chromosome.[20] Thus each cell's nucleus contains 46 chromosomes. Together, the 46 chromosomes in each cell nucleus contain a full set of human genes, with approximately 25,000 human genes in every cell. The size of genes also shows great variability.[21]

DNA consists of four different nucleotides linked together in various combinations in two long anti-parallel strands. Each nucleotide has an identical sugar group (deoxyribose), but differs from the other nucleotides by the nature of its base constituent. The adenosine (A) nucleotide has an adenosine base, cytidine (C) has a cytosine base, guanosine (G) has a guanine base and thymidine (T) has a thymine base group. In each of the two strands of DNA the individual nucleotides are linked together by phosphodiester bonds between the sugar moieties of the nucleotides. The two strands in the DNA molecule are held together by hydrogen bonds formed between the bases of the nucleotides. The nucleotides can only form hydrogen bonds in one way: adenosine can form two hydrogen bonds with thymidine in the other strand and guanosine can form three hydrogen bonds with cytidine in the other strand. These rules are referred to as the base pairing rules and explain why the two strands in a DNA molecule are referred to as complementary. This

18. The observed characteristics of an individual, produced by the interaction of genes and environment.
19. Anna Claybourne, *The Usborne Introduction to Genes & DNA*, p. 8.
20. Bruce Alberts, et al., *Essential Cell Biology*, 3rd Ed., 2010, pp. 172-173. Sections in between genes are called 'junk DNA'. These are long strings of base pairs in seemingly random patterns. A chromosome containing several genes has a lot of junk DNA as only 5% of our DNA is actually taken up with genes. Whether junk DNA has any significance is still unknown; see Anna Claybourne, *The Usborne Introduction to Genes & DNA*, p. 13. See also Rochelle C. Dreyfuss & James P. Evans, 'From Bilsky back to Benson: Preemption, Inventing Around, and the Case of Genetic Diagnostics', June 2011 63 *Stanford Law Rev.* 1362: 'Much of the non-coding DNA appears to simply be 'junk' left over from evolution; unnecessary to the function of the genome. But scientists continue to sort out the meaning (and lack thereof) of non-coding DNA.'
21. Peter Turnpenny & Sian Ellards, *Emery's Elements of Medical Genetics*, 13th Ed., p. 15; and Robert L. Nussbaum, Roderick R. McInnes & Huntington F. Willard, *Thompson & Thompson Genetics in Medicine*, 7th Ed., 2007, p. 6.

means that if the nucleotide sequence of one strand is known, the nucleotide of the other (complementary) strand can be inferred on the basis of the base pairing rules. The principles of specific base pairing and complementarity of DNA strands forms the underlying principle for how DNA is replicated from generation to generation in living cells, but it is also the principle underlying the majority of genetic tests, assays for gene expression and recombinant gene technology.

The bases used to make a gene are arranged in groups of three. Each group of three bases works as a code. The four letters A, C, G and T can be arranged in 64 different groups of three, so there are 64 possibilities, for example.

ATG GAT TTA TCT GCT CTT CGC GTT ← **Nucleotides (codons)**

Each three-letter group (also called a triplet codon) encodes an amino acid. Amino acids are the chemicals used to make proteins (body substances, e.g. a hormone or enzyme). There are 20 amino acids that are essential for human life, such as alanine, lysine and proline. They come from food or are synthesised in the body, and are carried to the cells by blood.[22] When a cell reads a gene, the code tells it which amino acids to use and the order in which to join them together. Using this system, cells can make thousands of body proteins, each made up of different arrangements of amino acids. Since there are 64 possible three-letter codes, but only 20 amino acids, some codes stand for the same thing; for example TTT and TTC both encode the amino acid lysine.[23]

22. Anna Claybourne, *The Usborne Introduction to Genes & DNA*, p. 12; and David A. Micklos, Greg A. Freyer with David A. Crotty, *DNA Science, A first course*, 2nd Ed. 2003, pp. 53-54.
23. David A. Micklos, Greg A. Freyer with David A. Crotty, *DNA Science, A first course*, 2nd Ed., 2003, p. 55; and Robert L. Nussbaum, Roderick R. McInnes & Huntington F. Willard, *Thompson & Thompson Genetics in medicine*, 7th Ed, 2007, pp. 7-8. See also Lorelei Perez Westin, 'Genetic Patents: Gatekeeper to the promised cures', *Thomas Jefferson Law Review*, Fall 2002, 25 TJLR 271, p. 275, who writes: 'Within every living being's cell is a system capable of directing all physiological processes necessary for the development and subsequent sustenance of all life forms. This master system, also known as the genome, contains the blueprint for the human body. Chemical constituents known as deoxynucleotides store this blueprint. Deoxynucleotides, in turn, carry the code for amino acids, the building blocks for proteins. The cell, according to the code that the genome carries, literally strings together chains of amino acids into proteins, which carry out many functions and tasks the body needs. Although capable of high complexity, the genome code comprises only four letters: A, G, C and T. The unique combination, or sequence, of these letters de-

| | | | | | | | | |
|---|---|---|---|---|---|---|---|---|
| Met | Asp | Leu | Ser | Ala | Leu | Arg | Val | ← **Amino-acids in a protein** |
| ATG | GAT | TTA | TCT | GCT | CTT | CGC | GTT | ← **Nucleotides (codons)** |

In the 2004 report from the Danish Council of Ethics there is a short definition of a gene as a functional unit of the genome, meaning that a DNA sequence codes for RNA or a protein (a hormone, enzyme and much more).

'In terms of gene patenting, a distinction needs to be made between the tangible gene concept and the information content of the genes. The gene is a biological substance; this substance however, is less interesting in its own right. The interesting feature in a patent context is the function of the substance – i.e. to code for RNA/protein.

According to what is called the central dogma of biology, the cell's information flow from DNA to RNA and on to protein, never the other way around. Broadly speaking, proteins are the molecules that control which chemical processes are to take place and how the cells are to appear and function. Thus, in reality, they determine the entire [architecture and] behavior of the organism. Taking the central dogma as a point of departure, the notion of the genes as the molecules that determine our fate seems straight forward. But the central dogma is subject to intense modification in current years. In turns out, for instance, that many genes code for RNA, which is never translated into proteins. The RNA molecules in themselves make up a large and complicated network of great importance to the function of the body. The RNA molecules are crucially important, for example, to the way in which the genes are activated [gene expression is regulated]. The determinism from gene to function thus becomes less clear than stated by the central dogma.'[24]

The scientific view is that DNA functions as life's information molecule, embodying the organising plan of molecular biology; understanding the structure of a molecule gives a clue to its biological function.[25]

'the information for making a protein is provided by a gene and the information is transcribed into a messenger RNA, which then associates with the ribosome where protein synthesis occurs. But what happens then? How is the information in the DNA actually stored, and how is it translated into a protein? The answer lies in the fact that DNA is a linear information molecule, which is transcribed into a linear mRNA molecule, which then

termines the amino acids strung together, and thus the type of proteins made. See also Tine Sommer, 'Retlig beskyttelse af bioteknologi', *UfR* 2000B 458.

24. The Danish Council of Ethics, '*Patenting Human Genes and Stem cells, A report*', 2004, pp. 35-36.
25. David A. Micklos, Greg A. Freyer with David A. Crotty, *DNA Science, A first course*, 2$^{nd}$ Ed., 2003, p. 37.

is translated into a linear protein sequence. Thus, it makes sense that the linear order of nucleotides in DNA is a code for the genetic information.'[26]

In gene banks, where different information on genes is stored, there are different types of DNA sequences, for instance: full-length DNA sequences, expressed sequence tags (ESTs) and single nucleotide polymorphisms (SNPs). ESTs were generated when DNA sequencing technology was less advanced and they were meant to be a quick way to obtain bits of a sequence from expressed genes by sequencing part of all cDNAs from a cDNA library. EST sequences could then be used as a starting point for researchers to identify the whole sequence of the gene where the EST sequence originated. ESTs could also be used to monitor gene expression, since identifying an EST in a cDNA library prepared from a certain cell type shows that the underlying gene is expressed in that cell type. Moreover the number of ESTs from a gene in a certain cell type provides information about the expression level of the gene in that cell type.

Single nucleotide polymorphism (SNP) is a single nucleotide variation at a certain location in the genome which occurs with a frequency of more than 1% of the chromosomes in a given population. For example, 2 out of every 100 analysed individuals (200 chromosomes) from Northern Europe show a G nucleotide at a certain position, whereas all others have an A nucleotide at the same position. SNPs are typically neutral, and are simply used as genetic markers, but some SNPs do have an effect on gene function, and may even be directly causative for disease development. The possible effect of the individual SNP's in the human genome is only known for a very minor fraction. SNPs are deposited in the SNP database.[27]

'Two human genomes chosen at random from the world's population will differ by approximately 2.5 x $10^6$ SNPs that are scattered throughout the genome. Because SNPs are present at such a high density, they provide useful markers in genetic analyses in which one attempts to link a specific trait (such as disease susceptibility) with a particular pattern of SNPs ... This type of analysis should lead to improvements in health care by allowing doctors to determine whether an individual is susceptible to a disease, such as heart dis-

26. David A. Micklos, Greg A. Freyer with David A. Crotty, *DNA Science, A first course*, 2nd Ed., 2003, p. 53.
27. http://www.ncbi.nlm.nih.gov/snp/. See also Nuffield Council on Bioethics, '*The ethics of patenting DNA*', 2002, p. 32. Bruce Alberts et al., *Essential Cell Biology*, 3rd Ed., 2010, pp. 319-320.

ease, long before he or she shows any symptoms. The person can then change his or her behavior to help prevent the disease before it arises.'[28]

### 2.1.b. What kind of genetic information?

In the case studies of gene patents in Chapter 7, the genetic information of a single-gene defect and of multifactorial inheritance has been subject to a review of patentability. A single-gene defect relates to diseases that are caused by individual mutant genes. The cause is a critical error in the genetic information carried by a single gene.[29]

Gene mutation in the protein coding part of a gene can be illustrated as follows:

| | |
|---|---|
| Pro Leu **Cys** Lys | (partial sequence for the normal protein) |
| CCT TTA **T**GT AAG | (partial sequence for a normal gene) |
| CCT TTA **G**GT AAG | (partial sequence for a disease gene) |
| Pro Leu **Gly** Lys | (partial sequence for the mutant protein) |

In addition, mutations in regulatory sequences can lead to the faulty expression of a gene. The identification of a mutation makes it possible to predict its effect on the gene product (the protein that is encoded by the protein).

28. Bruce Alberts et al., *Essential Cell Biology*, 3rd Ed., 2010, pp. 320-321. See Nuffield Council on Bioethics, *'The ethics of patenting DNA'*, 2002, pp. 32-33. At p. 25 in the Nuffield report, Table 3.1 (Patents and DNA) one can find examples of the form in which genes or DNA sequences appear in patent claims (2002). However what can be patented has changed since the report was written. It lists: DNA sequences (complete or partial gene); promoters; enhancers; individual exons; ESTs or cDNAs; whole transcribed genes as cDNAs; individual mutations known to cause disease; variations between people not associated with disease (polymorphisms); cloning vectors, formed from bacterial DNA, which are used to replicate DNA sequences, expression vectors, also formed from bacterial DNA, which are used to express proteins in replicated DNA sequences; isolated host cells transformed with expression vectors, which are cells that have been created to express particular proteins; amino acid sequences (proteins); the use of proteins as medicines; antibodies which are used as markers; nucleic acid probes, which are fragments of DNA that are used to locate particular parts of DNA sequences; methods of identifying the existence of a DNA sequence or a mutation or deletion in an individual; testing kits for detecting genetic mutations; whole genomes. See Tine Sommer, 'Patentering af det humane genom', *UfR*. 2004 B. 218. See The Danish Council of Ethics, *'Patenting Human Genes and Stem cells, A report'*, 2004, p. 32; Jorge A. Goldstein, and Elina Golod, 'Human Gene Patents', *Academic Medicine*, vol. 77, no. 12, December 2002, Part 2, p. 1318-1319.
29. Robert L. Nussbaum, Roderick R. McInnes & Huntington F. Willard, *Thompson & Thompson Genetics in Medicine*, 7th Ed., 2007, p. 2.

Whether or not the mutant protein causes the disease (phenotype) depends on its effect at the protein level.[30] One of the diseases in the Chapter 7 case study is cystic fibrosis, which is a single-gene disorder. The cystic fibrosis gene (CFTR, cystic fibrosis transmembrane conductance regulator) is required for the normal secretion of ions and mucus in the lung. A mutation in this gene causes cystic fibrosis. Thus, patients born with the disease can be diagnosed by finding the mutation in the gene.[31]

All genetic tests employ DNA or RNA probes and primers, which are short (typically approximately 20 nucleotides long) synthetic oligonucleotides that are complementary to a sequence in their target gene.

```
DNA strand of target gene:             -----TAAAGTTCATTGGAACAGTTAGTCTCCATGCCTATCG----
Basepairs:                                           ::::::::::::::::::::
Hybridized oligonucleotide probe:                    AACCTTGTCAATCAGAGGTA
```

Typically the DNA in the test sample (for instance genomic DNA purified from a persons blood cells) is denatured. This means that all the hydrogen bonds, which hold the two nucleotide strands of the DNA molecule together are broken by heating (or treatment with alkali), so that the DNA becomes single stranded. An added single stranded probe or primers, can then due to the base pairing rules bind to its complementary sequence in the target DNA. The binding of a DNA or RNA strand to its complementary sequence is referred to as hybridization. The identification of a gene or a specific DNA sequence is thus based on hybridization of a probe or DNA primers to their complementary sequence according to the base pairing rules. Just like two complementary DNA strands can hybridize, a single stranded RNA molecule can also hybridize to its complementary DNA strand. Furthermore, two complementary RNA strands can also hybridize. The binding of microRNAs and siRNAs (see below) to their target gene mRNAs is an example of hybridization of one RNA strand (for instance a siRNA) to a complementary strand (part of the target gene mRNA) that follows the base pairing rules.

In contrast, multifactorial inheritance has no simple explanation. The case study in Chapter 7 includes the examples of Alzheimer's disease and Parkinson's disease. These diseases are not simply caused by a mutation of a single

30. Gert Matthijs, 'DNA diagnostics in practice', in Geertrui Van Overwalle (ed), *Gene Patents and Public Health*, 2007, p. 31.
31. Gert Matthijs, 'DNA diagnostics in practice', in Geertrui Van Overwalle (ed), *Gene Patents and Public Health*, 2007, p. 30.

gene. They can be caused by one, two or more different mutant genes and can be triggered by environmental factors.[32]

Progress in molecular genetics has improved the understanding of the pathways and biological processes in individuals, thus enabling the development of individually designed drugs and therapies. Advances in personalised medicine have made it one of the 'hot' achievements of the last decade.[33] Personalised medicine uses the latest genomic knowledge and technologies to tailor treatments to individuals, e.g. drugs that have been designed to target a particular molecular pathway that has gone wrong. However there are two major obstacles, namely: complex biology and complex economics. Not only are the pathways involved not well understood, but most targeted drugs are too expensive.[34]

Systematic investigation of human phenotypes requires full knowledge of DNA sequence variation in their genomes. The 1000 Genomes Project is described in an article 'A map of human genome variation from population-scale sequencing', where it is pointed out that understanding the relationship between a genotype and a phenotype is one of the central goals in biology and medicine.[35] The aim of the 1000 Genomes Project is thus to discover genotypes and provide accurate haplotype information on all forms of human DNA polymorphism in multiple human populations.[36] This information on all the possible genetic variants is essential if personalised medicine is ever to be possible. In the article in *Nature* it is stated that:

32. Robert L. Nussbaum, Roderick R. McInnes & Huntington F. Willard, *Thompson & Thompson Genetics in Medicine*, 7th Ed., 2007, p. 2.
33. Marks & Clerk, '*Life Sciences Report*', 2011, p. 20: 'a lot of effort going into the identification of small genetic variations (SNPs) between individuals at key sites being associated with particular disease outcomes or susceptibility to certain drugs. Whilst chips allow for identification of individual sites, admittedly often en masse, rapid sequencing of an individual's genome has the capacity to cast a great deal more light on a host of interrelated genes, and their variations, helping to build a fuller picture of how a particular disease may be treated. It may also allow more predictive studies to be carried out, so that those predisposed to developing disease can be identified and carefully watched.'
34. Editorials, *Nature*, 19 May 2011, Vol. 473, p. 253.
35. '1000 Genomes Project. The third phase of human genomics – sequencing the whole genome of 179 individuals and characterizing all the protein-coding sequences of many others', *Nature*, Vol. 467, 1061-1073, October 2010, including Rasmus Nielsen, 'Genomics: in search of rare human variants', at pp. 1050-51.
36. Haplotype is a specific combination of SNP variants in a genomic region.

'the aspect of whole-genome patenting that lends itself most readily to investigation is conceptual. All the patents we have identified raise important questions about how genomes are conceptualized, especially in regard to how the utility of a genome can be specified. Is the genome just a concept that is used to unite several nucleotide sequences into a single invention, or is it a causally efficacious phenomenon that does something more than an aggregation of genes can do? What is the relationship between the utility of a part (a gene) and any utility associated with the whole (the genome)? The answer to these questions will be different depending on whether the genomes are thought of in terms of biochemistry or bioinformatics.'[37]

## 2.1.c. More than one function?

Immediately following the Human Genome Project there were many concerns about the future uses of genetics.[38] Despite this first step, there is still a long way to fulfil the promises of biological and biomedical research. Technology has developed from the sequencing of the human genome to focusing on proteomics seeking to understand all proteins, their biological functions and the mechanisms by which they interact.

However, in recent years a dramatically increasing number of proteins have been shown to have more than one function and they may thus function in very different and unexpected ways. This makes it difficult to classify proteins. For instance the *ACAD-9* gene is classified as a classic enzyme with a role in fatty acid oxidation (as a long chain acyl-CoA-dehydrogenase). Recently, it was discovered, that *ACAD-9* also plays a very important role as a structural protein, which is fundamental for the integrity of the respiratory chain complex.[39] Another example that is directly relevant to the case study of patents in this thesis is the *TARBDP* gene.[40] This gene was thought to have its primary function as an RNA binding protein, regulating alternative splicing of e.g. the *CFTR* gene (See Chapter 6, 2.a and Chapter 7.2). Recently, it was discovered that *TARDBP* gene is associated with neurodegenerative dis-

37. Maureen A. O'Malley, Adam Bostanci & Jane Calvert, 'Whole-genome patenting', *Nature Reviews Genetics*, June 2005, Vol. 6, p. 506.
38. Patricia Baird, 'Will genetics be used wisely', *ISUMA: Canadian Journal of Policy Research*, Vol. 2n 1, Spring 2001, ISSN 1492-0611; Francis S. Collins, Eric D. Green, Alan E. Guttmacher & Mark S. Guyer, 'A vision for future genomics research', *Nature*, Vol. 422, 835-847 (24 April 2003).
39. http://www.ncbi.nlm.nih.gov/pubmed/20929961. M Gerards et al., 'Riboflavin-responsive oxidative phosphorylation complex I deficiency caused by defective ACAD9: new function for an old gene', *Brain*, January 2011, 134(Pt 1):210-9.
40. TAR DNA-binding 43 (*TARDBP*) cellular protein which is encoded in humans by the *TARDBP* gene; see http://en.wikipedia.org/wiki/TARDBP (last visited January 2011).

orders, including frontotemporal lobar degeneration. *TARDBP* is the major disease protein in ubiquitin-positive, tau-, and alpha-synuclein-negative (FTLD-U).[41] Its clinical significance is that *TARDBP* has been shown to bind both DNA and RNA and have multiple functions in transcriptional repression, pre-mRNA splicing and translational regulation.[42]

In the case-study it was found that most applications either claimed or referred to a 'laundry list' in the description of the invention. A laundry list is a list of diseases that may also benefit from the invention. The case study is based on a search for patent activity on the disease-gene (e.g. *CFTR*) as well as on the disease name (cystic fibrosis). A surprisingly large number of applications used the laundry list technique, despite their relevance for e.g. a single-gene disorder. In other words a distinction between a single-gene disorder and multifactorial disorders did not reveal an increase in the use of laundry lists. However, if the invention claims a defect gene as such, e.g. a mutation, it is less obvious that a laundry list should also be claimed. Thus, claiming a single-copy gene coding for polypeptides that are involved in or carry out a variety of cellular functions: e.g. enzymes, hormones, receptors and structural and regulatory proteins,[43] opens up for multiple use for a variety of diseases.

The case study also found examples of applications for multigene families which are genes with similar functions. Multigene families can be split into two types: classic gene families that show a high degree of sequence homology and gene super-families that have limited sequence homology but are functionally related, having similar structural domains.[44]

Pleiotropy speaks against the grant of broad patents and may be related to the 'laundry list' (see Chapters 6 and 7). Pleiotropy has been known for many years and its strict definition is that a gene may have different effects in different tissues of the body. Traditionally pleiotropy has been explained by the fact that a gene may affect different processes depending on the cell/tissue type. Other examples are when a gene product, protein, may have fundamen-

41. M Neumann et al., 'Ubiquitinated TDP-43 in frontotemporal lobar degeneration and amyotrophic lateral sclerosis', *Science*, 314, October 2006 (5796): 130-3. doi:10.1126/science.1134108 and http://en.wikipedia.org/wiki/TARDBP.
42. P Kuo, LG Doudeva, Y Wang, CJ Shen, HS Yuan, 'Structural insights into (*TARDBP*) in nucleic-acid binding and domain interactions', *Nucleic Acids Research*, (2009) 37 (6): 1799-1808; and http://en.wikipedia.org/wiki/TARDBP. The protein encoded by this gene is a transcriptional repressor that binds to chromosomally integrated TAR DNA and represses HIV-1 transcription. In addition, this protein regulates alternate splicing of the *CFTR* gene.
43. Peter Turnpenny & Sian Ellards, *Emery's Elements of Medical Genetics*, 13th Ed., pp. 15-16.
44. Peter Turnpenny & Sian Ellards, *Emery's Elements of Medical Genetics*, 13th Ed., pp. 15-16.

tally different functions. Typically, proteins are classified according to their function as for instance an enzyme, hormone, receptor or a structural protein.

**2.1.d. Not all genes are proteins**

Proteins as building blocks embrace a myriad of functions, from special-purpose proteins to more general proteins such as motor proteins, receptor proteins, transport proteins and signal proteins. Understanding how genes work implies understanding proteins.[45] The term 'gene' has been subject to several evolutions and it is clear that there is no longer a hypothesis of 'one gene-one protein' whereby each gene specifies the synthesis of one particular protein.

Not all genes encode proteins. Many different cellular organelles and functions are dependent on RNAs. The ribosomes, which are the cellular organelles responsible for translating the mRNA into proteins, are made up of ribosomal RNAs (rRNA). Ribosomal RNAs are encoded by several copies of the rRNA genes present on five different chromosomes. Likewise, the transfer RNAs (tRNA), which are responsible for transporting the amino acids used in protein synthesis, are encoded by numerous transfer RNA genes. There are also numerous other non-coding RNA genes in our genome, and it is likely that non-coding genes, with functions that are not yet understood, are also present in our genome. That the genome is not fully understood and most probably still contains unknown types of genes is illustrated by the fact that the mere existence of microRNAs and consequently their genes was not realised until 1993.[46] MicroRNAs are short double-stranded RNAs, which are encoded in the human genome and which, after transcription and processing, can modulate the expression of other genes by binding to their mRNAs, inhibiting their translation to protein.

SiRNAs are short RNA molecules which are complementary to a proportion of an mRNA from a gene. siRNAs can be delivered to cells via different systems as short double-stranded RNA oligonucleotides. Once inside the cells the cellular systems that normally handle the human microRNAs encoded by the human genome will process the siRNAs and mediate binding to and degradation of their target mRNA.

45. Bruce Alberts et al., *Essential Cell Biology*, 3rd Ed., 2010, pp. 119-120.
46. Lee RC, Feinbaum RL, Ambros V (December 1993): 'The C. elegans heterochronic gene lin-4 encodes small RNAs with antisense complementarity to lin-14'. *Cell* 75 (5): 843-54.

### 2.1.d.1. RNAi

Patents including RNAi (siRNA/microRNA) are included in the case study (Chapter 7). RNA interference (RNAi), which means small interference RNA is one of the most promising scientific breakthroughs and it has been described in the literature as a powerful method for the targeted silencing of gene expression including the inhibition of disease-causing genes. The technique has surpassed other gene silencing approaches, such as antisense oligonucleotides and ribozymes, and not only is it relevant for gene function analysis it is also predicted that it will have a significant impact on genetic therapy.[47]

In an article in 2008, Daniel Grinberg reviewed recently issued patents on the use of siRNAs as therapeutic tools for human diseases, with a main focus on patents of siRNA strategies for monogenic inherited diseases and multifactorial disorders.[48] Grinberg emphasises that one of the challenges facing therapeutic strategies concerns the side effects of siRNAs. Of particular importance, is how to deliver the siRNA to the specific tissue (for instance the central nervous system). And of particular interest for the present study, Grinberg makes some references to Alzheimer's, Parkinson's and Huntington's diseases. For example the use of siRNAs has been mentioned as a therapeutic strategy for Alzheimer's,[49] a patent has been claimed for siRNAs for the treatment of Parkinson's[50] and finally there is mention of the use of siRNAs and other molecules targeted at the Huntington's disease gene.[51] In his article Grinberg mentioned several other siRNA applications.[52] From the

47. Derek M. Dykxhoorn, 'RNA interference as an anticancer therapy: a patent perspective', *Informa, Expert opinion*, Expert opin.Ther. Patents (2009) 19(4): 475-491, pp. 485-486.
48. Daniel Grinberg, 'Recent Patents on DNA & Gene Sequences', *Bentham Science Publisher*, 2008, 2, p. 40.
49. US2006178328.
50. US2005137155. The patent claimed compounds and methods useful for modulating Parkinson genes (e.g. *PARK1/SCNA*, *PARK2*, *PARK7*, and/or *PARK5* gene expression) in a variety of applications, including use in therapeutic, diagnostic, target validation, and genomic discovery applications by RNA interference (RNAi) using small nucleic acid molecules.
51. Daniel Grinberg, 'Recent Patents on DNA & Gene Sequences', *Bentham Science Publisher*, 2008, 2, p. 42; and US Patents US05255086 and US2005096284.
52. Erika Check, 'Patent flurry casts cloud over gene silencing', *Nature*, Vol. 417, 779, June 2002. One of the first scientists working in the RNAi area is a leading scientist named Thomas Tuschl. See WO04007718 and WO04065600. See further Adam Bostanci, Jane Calvert & Pierre-Benoit Joly, 'Regulating gene regulation: patenting small RNAs', in Matthew Rimmer & Alison Mclennan (eds), *Intellectual Property and*

above it is clear that RNAi technology is promising and since it is crucial to many biological processes it can be expected to have a major impact on patent practice. The case study in Chapter 7 confirms developments in this direction and adds to the concerns about the considerable uncertainty over the precise nature of the inventions for which protection is being sought. If we allow the patenting on every molecular entity, we disregard the complexity of the biological processes and the possibility of biological molecules participating in different processes.[53]

*2.1.d.2. What are synthetic genes and BioBricks?*

Genes can be manufactured synthetically to be identical to naturally occurring genes and synthetically produced genes can be modified and changed from the natural form. The function may or may not change. However, from a patent law perspective one can ask whether a claim for a formula which, on the surface, looks like a chemical formula but in fact mirrors a naturally occurring gene will fulfil the patentability requirements. This is questioned in Chapter 7, however, the formulas analysed in the empirical study differ from synthetic genes, which is the core issue in this section. The European Commission has defined synthetic biology as 'the engineering of biology: the synthesis of complex, biologically based (or inspired) systems which display functions that do not exists in nature. This engineering perspective may be applied at all levels of the hierarchy of biological structures – from individual molecules to whole cells, tissues and organisms. In essence, synthetic biology will enable the design of 'biological systems' in a rational and systematic way'.[54] One can therefore say that synthetic biology is not a new concept but

*Emerging Technologies, The new biology*, 2012, pp. 215-218, on the Carnegie patent initially claiming the very principle of RNAi, and the Tuschl I and II patents, also initially claiming fundamental aspects of RNAi, however restricted to a method for examining the function of the gene.

53. See further Adam Bostanci, Jane Calvert & Pierre-Benoit Joly, 'Regulating gene regulation: patenting small RNAs', in Matthew Rimmer & Alison Mclennan (eds), *Intellectual Property and Emerging Technologies, The new biology*, 2012, p. 227.
54. Report of a NEST High-level Expert Group, 'Synthetic Biology: Applying Engineering to Biology', 2005, p. 5, available at ftp://ftp.cordis.europa.eu/pub/nest/docs/syntheticbiology_b5_eur21796_en.pdf (last visited January 2012); European Group on Ethics in Science and New Technologies to the European Commission, 'Ethics of synthetic Biology', Brussels, 17 November 2009, Paragraph 3.2.4: 'Synthetic biology provides a new set of tools for using biology, either for the purpose of pure research with an intention to understand the manner in which living systems have developed, including their interactions, or for producing new processes or products. Three areas are pointed out: Environmental applications, energy and sustainable chemical indus-

is a combination of known technologies, and as it is within the realm of biotechnology it is covered by Directive 98/44/EC on the legal protection of biotechnological inventions and the EPC 2000 regime.[55] The BioBricks Foundation distances itself from the established system by aiming to ensure that the fundamental building blocks of synthetic biology are freely available for open innovation.[56]

A question arises in the wake of a recent report from the J. Craig Venter Institute in San Diego, USA, on the bringing to life of a synthetic cell.[57] The engineering of a self-replicating bacterium with a chemically synthesised genome is a step towards artificial life and raises potential ethical concerns. Since the announcement, the team has been criticised for calling the resulting cell 'synthetic' when the genome was essentially a replica of a natural genome and required an existing recipient cell.[58] A critical editorial in the British Medical Journal stated that it is 'unclear how the difficult operation of synthesizing an entire genome offers any advantages over the increasingly powerful and directed techniques of genetic manipulation that are already in daily use for generating new and useful life forms, or for gene therapy'.[59] However, does the emergence of new technologies regularly raise the question of whether the patent system is suitable for protecting inventions in fields such as new synthetic biology?[60] Also it seems difficult to distinguish synthetic biology from established fields of biotechnology such as metabolic engineering or recombinant DNA technology. This is a difference that exists more in the scale of modification of naturally occurring organisms 'involving tens to hundreds of genes in synthetic biology – and in the underlying concepts – an engineering approach or the creation of artificial life, than in the

try and biomedicine and biopharmaceuticals production.' See also 'Synthetic biology, enabling sustainable solutions for food, feed, bio-fuel and health: new potentials for European bio-economy', *STOA*, 2012 Copenhagen workshop.

55. Berthold Rutz, 'Patentability of synthetic biology inventions in Europe', *Biotechnology Journal*, 2010, 5, p. 11.
56. http://biobricks.org/ (last visited January 2012). See further Alison McLennan, 'Building on BioBricks: constructing a commons for synthetic biology research', in Matthew Rimmer & Alison Mclennan (eds), *Intellectual Property and Emerging technologies, The new biology*, 2012, p. 180.
57. Roberta Kwok, 'DNA's master craftsmen', 22, *Nature*, Vol. 468, 4 November 2010.
58. Roberta Kwok, 'DNA's master craftsmen', 22, *Nature*, Vol. 468, 4 November 2010.
59. Sarah Chan, 'Patents in synthetic biology, May hinder future research and restrict access to innovation', *British Medical Journal* (BMJ), 2010, Vol. 340; c1315 editorials.
60. Berthold Rutz, 'Synthetic biology and patents, A European Perspective, European Molecular Biology Organization', *EMBO Reports*, Vol. 10, special issue, 2009, p. S14.

actual technology used'.[61] The similarity or even identity of building parts used for synthetic biology and those used in other areas of biotechnology make synthetic biology merely a sub-unit of biotechnology.

In the Commission's 2005 report on synthetic biology mentioned above, there are some examples of non-natural phenomena. In particular, the idea of producing organisms with an expanded genetic alphabet may result in molecules that show much increased therapeutic potency (e.g. for anti-sense or RNAi based therapies):

> 'One very promising way to expand the chemistry of life is by extending the genetic alphabet so that it contains more than four characters (A, T, G, C). This could allow new types of information to be genetically encoded, with wide-ranging applications in nucleic acid and protein chemistry'[62]

Uses of synthetic biology for environmental protection are also exemplified by the Commission. The report has references to possible future uses in this regard. Environmentally friendly production of chemicals is one example, as the world's fossil fuel reserves are exhausted. Chemistry needs new raw materials and synthetic biology may provide these new materials, embodying microorganisms that produce efficiently the bulk chemicals that supply today's raw materials, and progressing to microorganisms that make even more complex chemicals in ever more complicated combinations of synthetic pathways. The Commission's report also refers to bioremediation (the rational modification of bacteria and other microorganisms such as fungi to eliminate toxic waste from soil) and the production of energy where the challenge is to design a set of converging chemical pathways that allow a quantitative

61. Berthold Rutz, 'Synthetic biology and patents, A European Perspective, European Molecular Biology Organization', *EMBO Reports*, Vol. 10, special issue, 2009, p. S15.
62. Synthetic Biology: Applying Engineering to Biology, Report of a NEST High-level Expert Group, 2005, p. 15, available at ftp://ftp.cordis.europa.eu/pub/nest/docs/syntheticbiology_b5_eur21796_en.pdf (last visited January 2012). Another example on p. 16 is organisms with an expanded genetic code (i.e. in which more than the canonical 20 amino acids can be incorporated in proteins) which will allow the manufacture of protein drugs with novel or enhanced properties – for example, enhanced serum half-life – which might revolutionise monoclonal antibody applications in therapy.

conversion of readily available solar energy and natural or waste materials into biofuels, for example.[63]

## 3. Stem cells

One of the questions raised in this chapter is whether field-restricted statutes become too rapidly outdated or are too static to be able to give proper protection to new technologies as they mature. This is a question of legitimacy, as discussed in Chapter 1. The *WARF* decision on embryonic stem cells in Case T-1374/04 (the Board of Appeal, EPO) and G 2/06 (the Enlarged Board of Appeal) make it possible to follow one rapid development of technology that is relevant to the Parkinson case study in Chapter 7.7 and emerging technologies in general.[64] At the time of the enactment of the Directive (1998), stem cell technology was in its infancy. During the time from filing of the patent application in question in 1996[65] until the end of the appeal on the decision in 2008, new embryonic stem cells techniques had been developed. At the time of filing, compositions containing pluripotent hES (human embryonic stem cells) could only be made by a process involving the destruction of human embryos, whereas at the time of the appeal, stem cells could also be obtained from human embryonic stem cell lines. Scientific progress in this technology seems to have just started. This development potentially has an impact on Directive 98/44/EC, Article 6(2)(c) which states that use of human embryos for industrial or commercial purposes is excluded from patentability. As interpreted by the EPO Enlarged Board of Appeal and by Advocate General Bot in Case C-34/10 *Oliver Brüstle*, the issue was whether the application of the technical process for which the patent was filed necessitated the prior destruction of human embryos or their use as base material, even though the description of the process did not contain any reference to the use of human embryos.[66] The case was decided by the Court of Justice of the European Union, 18 October 2011, which decided that inventions derived from human embryonic stem cells are largely non-patentable, see further Chapter 4.5.4.

63. Synthetic Biology: Applying Engineering to Biology, Report of a NEST High-level Expert Group, 2005, p. 16, available at ftp://ftp.cordis.europa.eu/pub/nest/docs/syntheticbiology_b5_eur21796_en.pdf (last visited January 2012).
64. T-1374/04 O.J. EPO 2007. 313, and G 2/06 OJ EPO 2009, 306.
65. European Patent Application No 96903521.1 of Wisconsin Alumni Research Foundation, in 1996.
66. Case C-34/10, point 8.

The background to the case was described in Case C-34/10, paragraphs 15-18, as follows: Mr Brüstle is the holder of a German patent, filed on 19 December 1997, which concerns isolated and purified neural precursor cells, processes for their production from embryonic stem cells and the use of neural precursor cells for the treatment of neural defects. In the patent specification it is claimed that the transplantation of brain cells into the nervous system is a promising method of treatment of numerous neurological diseases. The first clinical applications have already been developed, in particular for patients suffering from Parkinson's disease. In order to remedy such neural defects, it is necessary to transplant immature precursor cells, still capable of developing. In essence, that type of cell exists only during the brain's development phase. The use of cerebral tissue from human embryos raises significant ethical questions and means that it is not possible to meet the need for the precursor cells which are required to provide publicly available cell treatment. As human embryonic stem cells for the production of cells for transplantation are pluripotent, they can develop into all types of cells and tissues and can be conserved during many passages in the state of pluripotentiality and can multiply. The patent at issue seeks to make it possible to resolve the technical problem of producing an almost unlimited quantity of isolated and purified precursor cells having neural or glial properties, obtained from embryonic stem cells.

The case illustrates that legal discussions and definitions in new biological technologies may not be relevant. Even, before the Court of Justice gave its decision, stem cell research techniques had moved on. The Parkinson case study in Chapter 7 points to new ways for generating embryonic stem cells. The application initially described current state of the art as follows:

[0007] Human embryonic stem cells have the potential to give significant therapeutic benefit to patients, provided that the problem of immune rejection can be solved. Embryonic stem cells that are genetically related to the recipient may overcome such rejection problems. Currently, human embryonic stem cells (hES) are derived from *three sources*: *blastocysts* remaining after infertility treatments and donated for research, *blastocysts* generated from donated gametes (oocytes and sperm), and the *products* of nuclear transfer (NT). Cadaveric fetal tissue is the only source of human embryonic germ cells (hEG). hES and hEG cells offer remarkable scientific and therapeutic possibilities, involving potential for generating more specialized cells or tissues. Ethical concerns about the sources of hES and hEG cells, however, and fears that use of NT for research could lead to use of NT to produce a human being, have fostered a great deal of public discussion and debate.[67] (Emphasis added)

The new technique is then described as follows:

[0008] Parthenogenetic activation of mammalian oocytes may be used as an alternative to fertilization by sperm/NT to prepare oocytes for embryonic stem cell generation. Parthe-

67. US2008299091 (EP2188861).

nogenetic activation is the production of embryonic cells, with or without eventual development into an adult, from a female gamete in the absence of any contribution from a male gamete.[68]

Thus, given the debate on human dignity and the definition of a human embryo, the interpretation of Directive 98/44/EC may have to be altered accordingly.[69] With the pace of technological development and scientific progress, it may well be impossible to express the requirement in a detailed patent law

68. US2008299091 (EP2188861). As for the new techniques regarding inherited mitochondrial disorders recent years have seen increasing attention paid to two techniques both currently at the research stage: pronuclear transfer (PNT) and maternal spindle transfer (MST). According to a new Nuffield Council report these two IVF-based techniques seem to have the potential to prevent transmission of maternally inherited mitochondrial disorders caused by mutations in the genes of mitochondria, Nuffield Council on Bioethics, 'Novel techniques for the prevention of mitochondrial DNA disorders: An ethical review', 2012, Introduction and paragraph 4.53-4.60.
69. Chapter 4.5.4. and Report from the Commission to the European Parliament and the Council – Development and implications of patent law in the field of biotechnology and genetic engineering (COM(2002) 545 final); and Report from the Commission to the Council and the European Parliament – Development and implications of patent law in the field of biotechnology and genetic engineering (COM(2005) 312 final). In paragraph 2.2 of the latter report, on the patentability of human embryonic stem cells and cell lines obtained from them, the Commission refrained from defining embryonic stem cells as it was too early in the technological development to establish a definition: 'the Commission considers that it is premature to give further definition or provide for further harmonisation in this area'. See further Opinion of the Group of Advisors on the Ethical Implications of Biotechnology in Science and New Technologies to the European Commission, Opinion No. 8, 'Ethical aspects of patenting inventions involving elements of human origin', 25 September 1996; and Opinion of the European Group on Ethics in Science and New Technologies to the European Commission, Opinion No 15, 'Ethical aspects of human stem cell research and use', 14 November 2000; Opinion No 16, 'Ethical aspects of patenting inventions involving human stem cells', 7 May 2002; and European Group on Ethics in Science and New Technologies to the European Commission, 'Study on the patenting of inventions related to human stem cell research', 2002; Aurora Plomer, 'Stem Cell Patents: European Patent Law and Ethics Report', 2006. In the Edinburgh case described in Chapter 8, the patent was narrowed down from encompassing animal stem cells to non-human (embryonic) stem cells, although based on a stem cell culture. See Cornish, Llewellyn & Aplin, *Intellectual Property: Patents, Copyrights, Trademarks and Allied Rights*, 7$^{th}$ Ed., 2010, p. 933. See also Glenn McGee & Elisabeth Banger, 'Ethical Issues in the patenting and control of stem cell research', in David Magnus, Arthur Caplan Glenn McGee (eds), *Who owns life?*, 2002, p. 255 ff.; Dr. Tade Matthias Spanger, 'Legal Status and Patentability of Stem Cells in Europe', 21 *Biotechnology Law Report* 105, Number 2, 2002.

provision.[70] Adopting patent provisions before a technology has become stabilised cannot be recommended, and once it has become stabilised the need for its regulation is less demanding.[71] From a patent law perspective, administrative and judicial control is much more flexible and suitable for safeguarding the legitimate interest of patent law in promoting technologies while balancing the interests of the surrounding environment. This can again be illustrated by Case C-34/10 *Oliver Brüstle.* The technique described above ([0008] Parthenogenetic activation) was evaluated in terms of the directive in paragraphs 35-37:

Accordingly, any human ovum must, as soon as fertilised, be regarded as a 'human embryo' within the meaning and for the purposes of the application of Article 6(2)(c) of the Directive, since that fertilisation is such as to commence the process of development of a human being. In paragraph 36 the Court went on to state that the classification must also

70. COM(1995) 312 final, paragraphs. The question of whether the Directive specifically addressed the patentability of stem cells was raised in the first 16c Report, and referred to the Group of Experts in May 2003. European Group on Ethics in Science and New Technologies (Opinion No 16, paragraphs 13-16) considered that there was no ethical reason for a complete ban on patenting of inventions relating to stem cells or stem cell lines, although the normal requirements for patentability would have to be met. The provisions of Directive 98/44/EC are clear in relation to totipotent stem cells, since each cell could develop into a human being on its own and under Article 5(1) the human body at the various stages of its formation and development cannot constitute a patentable invention. This principle was reiterated in the practice notice issued by the UK Patent Office in April 2003. For pluripotent embryonic stem cells the situation is more complex. The Group of Experts on Ethics in Science and New Technologies considered that the question of patenting was closely linked to the definition of what constitutes an embryo and the scope of research allowed, which is determined by national legislation. As regards research funding under the European framework programmes, such funding may be available for projects involving human embryonic stem cell research on a case-by-case basis. In the light of the clear differences existing between the Member States as regards the acceptability of research using embryonic stem cells, the continuing and rapid developments in this field, and the fact that the Directive itself provides for Member States to refuse patents on grounds of *ordre public* or morality under Article 6(1), the Commission considers that it is premature to give a further definition or provide for further harmonisation in this area. At the same time the Commission will monitor developments taking into account both the ethical aspects and the potential impact on competitiveness. It has launched a study looking at the ethical and legal aspects of stem cell patenting.
71. Paul Sollie & Marcus Düwell, 'Evaluating New Technologies: An Introduction', in Paul Sollie & Marcus Düwell (eds), *Evaluating New technologies, Methodological Problems for the Ethical Assessment of Technology Developments*, 2009, pp. 2-3, on the two horns of dilemma – the prediction horn and the control horn.

apply to a non-fertilised human ovum into which the cell nucleus from a mature human cell has been transplanted and a non-fertilised human ovum whose division and further development have been stimulated by parthenogenesis. Although those organisms have not, strictly speaking, been the object of fertilisation, due to the effect of the technique used to obtain them they are, as is apparent from the written observations presented to the Court, capable of commencing the process of development of a human being just as an embryo created by fertilisation of an ovum can do so.

In paragraph 37 the Court gave the following guidelines to the national court: 'As regards stem cells obtained from a human embryo at the blastocyst stage, it is for the referring court to ascertain, in the light of scientific developments, whether they are capable of commencing the process of development of a human being and, therefore, are included within the concept of 'human embryo' within the meaning and for the purposes of the application of Article 6(2)(c) of the Directive.'

This part of the ruling clearly illustrates that the courts are in a position to interpret the law in the light of technological developments. However, the guidance in paragraph 37 (in the conclusion paragraph 38) does not actually point to the research purposes. The Court emphasised, in paragraph 34, that the wide interpretation of a human embryo, with regard to respect for human dignity, is far from precise. As it stated in paragraph 37, it is for the referring court to ascertain, in the light of scientific developments, whether stem cells obtained from a human embryo at the blastocyst stage are capable of commencing the process of developing into a human being, and therefore being covered by Article 6(2)(c). In paragraph 30, the Court referred to the fact that the definition of human embryo is a very sensitive issue and that the Court is not called upon to broach questions of a medical or ethical nature, but must restrict itself strictly to legal interpretation. However, biomedical research is not always the most extreme scenario and in particular research into neurological diseases is not aimed at 'commencing the process of the development of a human being'. Thus, despite an evaluation concerning the commencement of the process of developing a human being and the protection of human dignity in order to obtain protection under Article 6(2)(c) that is not at all the purpose of the biomedical research.[72] Science would have been better served for a legitimate assessment of how to address stem cell research at the

72. Chapter 4, section 1, referring to the Court's statement in paragraphs 43-46 that one cannot separate or distinguish the use of human embryos for the purposes of the scientific research which is the subject matter of a patent application from industrial and commercial use, and thus avoid exclusion from patentability.

regional level if the field-restricted provision in Article 6(2)(c) had not been enacted.[73]

Referring again to the empirical study in Chapter 7.3, the case of spinal muscular atrophy (SMA) is a good example of a disorder where the patho-physiological processes that underlie the main symptoms are still not fully understood, and where stem cells may help to target disease-specific mechanism at cellular and physiological levels in the early stage of the disease.[74] SMA is considered a model case for other neurodegenerative diseases, and it is also a very good example of different strategies for increasing protein amounts (in motor neurons) to develop a therapy, e.g. antisense oligonucleotide technologies (modulate mRNA splicing), and viral gene transfer of SMN used for RNAi.[75] Evidently, therapy based on human induced pluripotent stem cells (iPS), which has been shown to be a promising resource for the study of disease mechanisms, screening new drug components and developing new therapies, is also a major topic for scientific advances for treatment of the disease.

'Induced pluripotent stem cells (iPSCs) are adult cells that have been genetically reprogrammed to an embryonic stem cell–like state by being forced to express genes and factors important for maintaining the defining properties of embryonic stem cells. Although these cells meet the defining criteria for pluripotent stem cells, it is not known if iPSCs and embryonic stem cells differ in clinically significant ways. Mouse iPSCs were first reported in 2006, and human iPSCs were first reported in late 2007. Mouse iPSCs demonstrate important characteristics of pluripotent stem cells, including expressing stem cell markers, forming tumors containing cells from all three germ layers, and being able to contribute to many different tissues when injected into mouse embryos at a very early stage in develop-

73. Nuffield Council on Bioethics, 'Novel techniques for the prevention of mitochondrial DNA disorders: An ethical review', 2012 emphasises in Paragraph 3.29 the 2005 preamble to UNESCO's Universal Declaration on Bioethics and Human Rights 'that 'based on the freedom of science and research, scientific and technological developments have been, and can be, of great benefit to humankind in increasing, inter alia, life expectancy and improving the quality of life.' It continues at Art.16: 'The impact of life sciences on future generations, including on their genetic constitution, should be given due regard'.''
74. Michael Sendtner, 'Therapy development in spinal muscular atrophy', *Nature Neuroscience*, Vol. 13, No 7, July 2010, p. 795.
75. Chapters 6 and 7 for further information on the genetics of the disease. And see Michael Sendtner, 'Therapy development in spinal muscular atrophy', *Nature Neuroscience*, Vol. 13, No 7, July 2010, pp. 795 and 797.

ment. Human iPSCs also express stem cell markers and are capable of generating cells characteristic of all three germ layers.'[76]

The first study to show that human iPS can be used to model the specific pathology of genetically inherited diseases (in this case SMA) used iPS from skin fibroblast samples taken from a child with spinal muscular atrophy. These iPS cells expanded robustly in a culture, maintained the disease genotype and generated motor neurons that showed selective defects compared to those derived from the child's unaffected mother.[77] In other words, taking cells from an affected child and from its unaffected mother, and reprogramming each set of cells to become iPS cells to derive motor neurons of the same genetic make-up as each donor was shown to be an invaluable tool for studying the disease.[78] Since human iPS cells can be obtained from somatic cells that have undergone a process of 'reprogramming' by genetic manipulation so as to develop pluripotency, this does not involve the destruction of embryonic stem cells which the patent case law opposes. However, the patent landscape is not fully defined either in the USA or in Europe.[79]

The relevance of stem cell technology for scientific advances and for biomedical progress is not disputed. However, the discussion of a ban on the patenting of embryonic stem cells in European patent law, initiated by Directive 98/44/EC, has made such research a hot topic. In the USA some restrictions were imposed under President Obama by an executive order in 2009, as well as by restrictions on federal funding for the use human embryonic stem cells obtained by destroying human embryos.[80] There are different practices in the

76. http://stemcells.nih.gov/info/basics/basics10.asp (last visited January 2012).
77. Allisson D. Ebert et al., 'Induced pluripotent stem cells from a spinal muscular atrophy patient', *Nature*, Vol. 457, January 2009, p. 277.
78. Michael Sendtner, 'Tailor-made diseased neurons', *Nature*, Vol. 457/15, January 2009, p. 269.
79. For the US, see Bilyana P Georgieva & Jane M Love, 'Human Induced pluripotent stem cell: a review of the US patent landscape, special report', *Regen. Med.* (2010) 5(4), pp. 581-591, where it is stated at p. 582: 'However, therapeutic methods that use iPS cells are still in development. Thus, unsurprisingly, patent applications not only reflects the development of iPS cell technology but also possible future commercial aspects of iPS cells'; Antoinette F. Konski & Doris J.F. Spielthenner, 'Stem cell patents: a landscape analysis. An evaluation of the development of innovation in stem cell technologies by network analysis of stem cell patent filings', *Nature biotechnology*, Vol. 27, August 2009, p. 722.
80. Meredith Wadman, 'Stem-cell arguments heard in US appeal courts' *Nature*, news blog, 23 April 2012. Executive Order of 9 March 2009, removing barriers to respon-

different EU Member States, all related to ethical considerations.[81] However, this discussion ought not to take place within the realm of patent law as much as in the realm of the decisions of scientific communities. Struggling to determine the precise interpretation of the wording of the Directive is not the right way to debate this issue, and it leaves open the question of what happens when new techniques circumvent the static wording of the Directive. Are we to start a new debate when scientific results are based on new techniques or will the scientific communities address new questions in forums that are separate from patent law?

## 4. Nanotechnology

There can be a discussion about how the concept nanotechnology is best described. Metaphorically it has been said that 'one nanometre is to a human head what a human head is to the planet Earth'.[82] The focusing of biotechnology on substances measured in 'nano' as a matter of size is one approach. Thus, a 'nanometre' refers to the base unit in quantum metrology: the science of measurement at the nanoscale. One nanometre is one billionth of a metre.[83] According to the Oxford English Dictionary, 'nano' is used as a prefix to the names of units of measurement to denote a factor of $10^{-9}$.[84] *Hailemichael*

sible scientific research involving human stem cells, 74 Federal Register 10667 11 March (2009) and NIH guidelines for human stem cell research http://stemcells.nih.gov/policy/2009guidelines.htm (last visited January, 2012).

81. See also the European Commission report on human embryonic stem cell research, SEC (2003) 441, April 2003; and Chapter 4, section 1.
82. A quote of Nobel laureate Sir Harold Kroto. See Hailemichael Teshome Demissie, 'Taming matter for the welfare of humanity: Regulating Nanotechnology', in Roger Brownsword & Karen Yeung (eds), *Regulating Technologies, Legal Futures, Regulatory Frames and Technology Fixes*, 2008, p. 329. Its etymology is given as classical Latin *nānus* (also *nannus*) or its etymon ancient Greek *νᾶνος* (also *νάννος*) meaning dwarf, http://www.oed.com.ez.statsbiblioteket.dk:2048/view/Entry/124974?rskey=mbM73k&result=2&isAdvanced=false#eid, visited June 2011.
83. Hailemichael Teshome Demissie, 'Taming matter for the welfare of humanity: Regulating Nanotechnology', in Roger Brownsword & Karen Yeung (eds), *Regulating Technologies, Legal Futures, Regulatory Frames and Technology Fixes*, 2008, p. 329.
84. 1947 *Compt. Rend. de la 14me Conf.* (Union Internat. de Chimie) 115, http://www.oed.com.ez.statsbiblioteket.dk:2048/view/Entry/124974?rskey=mbM73k&result=2&isAdvanced=false#eid and http://en.wikipedia.org/wiki/Nano-, visited June 2011.

*Teshome Demissie* discusses the definition and writes that nanotechnology is often defined as science and technology operating at the nanoscale and that scale is defined in the range of 1-100 nm. However, this is a somewhat arbitrary approach:

'as materials within this scale range do not necessarily behave in the strange ways that differ them from their macro or micro-scale state ... therefore, the novel property rather than the scale range of 1-100 nm becomes the central defining concept of nanotechnology ... it is indeed the strange properties of matter at the nanoscale and the possible applications thereof that made nanotechnology a subject of enormous interest.'[85]

The European Patent Organisation (EPO) has issued the following definition on nanotechnology in order to monitor the impact of nanotechnology on the development of new products and identify patents in this area:

'The term nanotechnology covers entities with a controlled geometrical size of at least one functional component below 100 nanometres (nm) in one or more dimensions susceptible of making physical, chemical or biological effects available which are intrinsic to that size.'[86]

In 2011, patent offices around the world started classifying nanotechnology uniformly under the International Patent Classification (IPC) system (introducing a new symbol B82Y in January 2011). From this one can quickly get an impression of the diversity of nanomaterial such: as nanobiotechnology or nanomedicine; nanotechnology for information processing, storage and transmission; nanotechnology for interacting, sensing and actuating; nanotechnology for optics; nanomagnetism; nanotechnology for materials and surface science, methods or apparatus for measurement or analysis of nanostructures, and the manufacture or treatment of nanostructures.[87]

85. Hailemichael Teshome Demissie, 'Taming matter for the welfare of humanity: Regulating Nanotechnology', in Roger Brownsword & Karen Yeung (eds), *Regulating Technologies, Legal Futures, Regulatory Frames and Technology Fixes*, 2008, pp. 329-330: 'It is like you shrink a cat, and keep shrinking it, and then at some point, all at once, it turns into a dog. It is, thus, understandable why the exploitation, actual or potential, real or purported, of these strange properties has become the crucial concept in the defining of nanotechnology.'
86. http://www.epo.org/news-issues/issues.html (last visited January 2012), *Nanotechnology and Patents*, 2011 (brochure).
87. http://www.epo.org/news-issues/issues/classification.html. See further Alison McLennan & Matthew Rimmer, 'Cosmo, Cosmolino: Patent law and nanotechnolo-

According to *Hailemichael Teshome Demissie*, nanotechnology can be an umbrella term for scientific and technological activities where the minimum conditions of length, scale and novel properties are met, but he also refers to another approach suggesting a departure from the strict definition of the phenomenon: 'strict definitions may be irrelevant as perspectives on how it is best pursued and what it can achieve become more important'.[88]

Nanotechnology is a new and still emerging technology. While at an early stage, it nevertheless poses a legislative challenge as a cross-disciplinary field of technology. Single products can be applied in many industries.[89]

No doubt bio-nanotechnology can simply be a matter of biotechnology on a smaller scale, and from a patent law perspective the particle size has to be defined.[90] A recent case from the EPO Board of Appeal that is relevant to all nanotechnologies, including bio-nanotechnology, stated that a claim characterising a copper catalyst precursor by indicating a range for the average particle size from about 0.1 to about 600 nanometres did not meet the requirement of Article 84 EPC that the claims must be clear.[91] The Board found that the application disclosed several different methods for determining the average particle size and refused the application.

From a biotechnological perspective, nanotechnology is concerned with understanding and manipulating matters at the molecular level.[92] This can best be illustrated by a couple of examples from the empirical study, once again using spinal muscular atrophy (SMA) in Chapter 7.3 as an example.

gy', in Matthew Rimmer & Alison Mclennan (eds), *Intellectual Property and Emerging technologies, The new biology*, 2012, p. 260.

88. Hailemichael Teshome Demissie, 'Taming matter for the welfare of humanity: Regulating Nanotechnology', in Roger Brownsword & Karen Yeung (eds), *Regulating Technologies, Legal Futures, Regulatory Frames and Technology Fixes*, 2008, p. 331.
89. European Parliament, 'Nanotechnology in the food sector', *STOA*, 2009; European Parliament, 'Nano*Safety* – Risk Governance of Manufactured Nanoparticles', 2012; Wim Helwegen, 'The research exemption from a nanotechnology perspective', *EIPR*, 2010, Issue 7, 2010, p. 342.
90. See as an example of uses, Paul Burgess et al., 'On firm ground: IP protection of therapeutic nanoparticles', *Nature Biotechnology*, Vol. 28, No 12, December 2010, p. 1267.
91. T 1819/07 – Decision of the Technical Board of Appeal 3.3.01 of March 2011. See further Patrick M. Boucher, *Nanotechnology, legal aspects*, p. 16 on obviousness and scale factors using a nanocar as an example (it is not simply a matter of taking the blueprints for a Ferrari and scaling them down).
92. Adam Keiper, 'Nanethics as a Discipline?', *The New Atlantis*, Spring 2007, p. 56.

Some of the applications comprise claims on replacement therapy where lipid coated *nanoparticles* comprising the therapeutic agent (in suit a protein) and *microparticles* comprising the therapeutic agent (in suit a protein) encapsulated in a matrix, are carriers for delivery.[93] Others using the carrier model claim relates to drug delivery and specifically to the preparation and use of functionalised carriers such as nanopolymers and nanovesicles for improved delivery of nucleic acid agents (NAA) to tissues and cells.[94] Thus from this perspective the nanoparticles are designed to contain and carry the therapeutic agent and deliver it to the destination. In a way one can say that the technology is a mix of fundamental gene technology where the therapeutic agents, e.g. proteins and oligonucleotides, that modulate RNA splicing are found, and then nanotechnology is added in order to invent a small scale container to carry the agents and penetrate barriers on the delivery routes and get to the end destination where the agent is designed to do its work.

Nanotechnology has presented policymakers and legislators with new challenges, but this time there has been no directive on patentability to further complicate things.[95]

Inventions based on nanotechnology are to be found in every aspect of society. They can embrace materials science, including consumer products such as cosmetics, stain-resistant clothing, and antibacterial food containers. *Keiper* writes that nanotechnology in:

'the years ahead it is expected to yield powerful medical diagnostic tools, ultra-efficient water-filtration systems, strong and lightweight materials for military armor, and numerous breakthroughs in energy, computing, and medicine. The expectations for those nanomaterials, however, pale in comparison to the hope and hype surrounding another kind of nanotechnology: theoretical nanomachines. This more radical vision, first described in detail by Eric Drexler in the 1980s, involves molecular manufacturing – building things 'from the bottom up' by precisely placing atoms. Personal nanofactories the size of a mi-

93. EP 2244697, US 2011059172, US 2011052673, EP 2244694.
94. WO 2007133812, US 2009011004.
95. Regulation takes place outside the area of patent law. Patrick M. Boucher gives the example of nanosized particles in many products due to their antibacterial effects, making them useful in e.g. shoe liners to destroy odours. However, many environmentalists have voiced concerns that the resulting introduction of nanoscale silver particles into the environment at large may kill beneficial bacteria as well as other organisms. In the USA there is regulation of the use of nanoparticles for antimicrobial applications; see Patrick M. Boucher, *Nanotechnology, legal aspects*, p. 111; The European Group on Ethics in Science and New Technologies to the European Commission, Opinion N° 21 on 'the ethical aspects of Nanomedicine, 17 January, 2007.

crowave oven could replace or repair damaged cells in the body, helping to stave of aging. Still others could make terrible new world-destroying weapons.'[96]

Global concerns have different headings, and *Keiper* also address the tendency to discuss social justice concerns for all new technologies:

'Similar concerns have been raised about a broad array of new technologies, and they are now being dressed up with a 'nano' prefix as in the past they have worn 'bio,' 'digital,' or 'eco.' So, for instance, the 'digital divide,' the 1990s-era fear that the computer age might leave millions of poor people behind, has a nano-analogue: the worry of a potential 'nano-divide' between nano-haves and nano-have-nots. Another byword for activists and academics for the last two decades, 'sustainable development,' is now being applied to nanotech. And countless other social and economic concerns, including the effects of globalization and the old Luddite worry about job displacement caused by technology, are now getting linked to nanotechnology.'[97]

*Hailemichael Teshome Demissie* highlights the promise of nanotechnology in relation to sustainability and the remediation of the environment. Nano-enabled economic growth, he writes, will not be the kind of growth that is achievable only at the expense of the environment.[98] Clean energy, waste-free

96. Adam Keiper, 'Nanoethics as a Discipline?', *The New Atlantis*, Spring 2007, p. 56. On pp. 64-65, Nanoethics and the knowable future, Keiper highlights four broad categories of inquiry as follows: 'The *first* is the question of safety, the most fundamental concern of the modern state. Although, as described above, researchers still understand very little about the health and environmental effects of nanoparticles, it seems likely that, as they learn more, some kind of regulatory regime will be developed. The *second* category of concern to nanoethicists could, for the sake of convenience, be given the heading 'social justice.' This category includes questions about equity, access, and socioeconomics, as well as questions about how much control governments, militaries, or corporations should have over nanotechnology. The *third* area of nanoethical inquiry relates to vast and genuinely novel social changes that the development of nanotechnology might wreak. Could nanotechnology play a part in the development of artificial intelligence, and if so, would that development be as socially transformative as many of its advocates hope? The *final* category is also revolutionary; it involves the use of nanotechnology not just to transform society but to redefine our very humanity. Some of the scientists and policymakers involved in the federal government's nanotechnology initiative have made extreme claims about the medical benefits nanotechnology will bring relatively soon. It is, for instance, slated to play a key role in the National Cancer Institute's plan to 'eliminate suffering and death from cancer' by 2015'.
97. Keiper, p. 65.
98. Hailemichael Teshome Demissie, 'Taming matter for the welfare of humanity: Regulating Nanotechnology', in Roger Brownsword and Karen Yeung (eds), *Regulating*

production and raw material independency are among the nanotech revolutions. However, the uncertainty associated with new technologies calls for a precautionary approach and risk assessments. The precautionary principle was included in the 1992 Rio Declaration on the Environment and is included in Article 191(2) of the Treaty on the Functioning of the European Union.[99] Nanotechnology poses manifold challenges, but the challenges cannot be resolved by patent law. The risk to human health and to the environment from exposure to nanoparticles will exist with or without patent protection.

*Technologies, Legal Futures, Regulatory Frames and Technology Fixes*, 2008, p. 341.

99. Hailemichael Teshome Demissie, 'Taming matter for the welfare of humanity: Regulating Nanotechnology', in Roger Brownsword & Karen Yeung (eds), *Regulating Technologies, Legal Futures, Regulatory Frames and Technology Fixes*, 2008, p. 341.

CHAPTER 3

# The Power of the Playmakers

Chapter 3 concerns the focus of the thesis on multilevel regulation based on a review of the most relevant actors, including: the European Union (EU), the European Patent Organisation (EPO), the World Trade Organization (WTO/TRIPS), the World Intellectual Property Organization (WIPO) and the US Patent and Trademark Office (USPTO). In this context it is particularly the EU's activities, both internationally and regionally, that are of most interest. Taking patent law as an example, it is pointed out that the EU has extended its competence at both international and regional levels by establishing what the author calls a 'distinct European system of public international law and regional law' in the sense that, in the area of patent law, Union law appears to be some form of superior law at national, regional and international levels.

## 1. Multilateral legal framework and interfaces

The task of granting patents has changed from the standard administrative apparatus of late fifteenth century Venice to the modern administrative patent authorities.[1] Not only have technological changes created new aspects of property, but the growing trend towards multilateral legal frameworks has also influenced the patent system and has increased the numbers of playmakers involved in patent law developments. The perspective of the current thesis is that on the one hand multilevel governance has challenged patent legitimacy, and on the other hand the subject matter to be governed at a multilateral level has increased due to new interfaces, thus further challenging legitimacy. In-

1. Robert Patrick Merges & John Fitzgerald Duffy, *Patent Law and Policy: Cases and Materials*, 5th Ed., 2011, p. 3; and Sa Yu, 'Political Privilege, Legal Rights, or Public Policy Tool? A History of the Patent System', p. 8, ATRIP Essay Competition 2009: http://www.atrip.org/26.htm.

terfaces such as those between intellectual property rights and health, climate change and poverty are priorities on the agenda of the World Intellectual Property Organization (WIPO).[2]

The current playmakers in patent law are to be found in multilateral fora at different levels.[3] As the topics of this thesis embrace biomedical research, animal breeding and climate change from a patent law perspective, the playmakers and their legal powers may differ from area to area. For example, the recent focus on patents and clean technologies has introduced a new stakeholder on the intellectual property rights (IPR) agenda – the climate lawyer.[4] There may also be grassroots movements, such as Greenpeace, and NGOs that can influence and be important at various levels. And the modern biomedical climate is shaped by the significant advocacy of patients' organisations with disease-specific constituencies; these play a role in funding disease-specific genetic research and contributing tissue samples.[5] There is also industry lobbying when legislation is drafted.[6] In the USA, the National Institutes of Health (NIH) can also be viewed as an active player which has issued Principles and Guidelines on Sharing Biomedical Research Resources (e.g.

2. See 'Intellectual Property watch: Echoes of global patent war at WIPO Annual General Assembly', where interfaces are mentioned as the new buzzword, see the Annual Meeting http://www.wipo.int/pressroom/en/articles/2010/article_0038.html.
3. See Follesdal et al. (eds.), *Multilevel Regulation and the EU, The interplay between Global, European and National Normative Processes*, p. 1; and Chapter 1, section 4.1, table 1.1.
4. EPO, UNEP patent law study, 'Patents and clean technology bridging the gap between evidence and policy', Report from September 2010, http://www.epo.org/news-issues/issues/clean-energy/study.html (last visited, December 2011).
5. Report of the NIH Secretary's Advisory Committee on Genetics, Health, and Society (SACGHS), April 2010, pp. 25-26: 'In case of spinal muscular atrophy (SMA), the patent holder did not even bear the financial burden of the discovery, rather an advocacy group and patients and family suffering from the disease donated funds and tissue samples to a researcher who then patented her discovery and sold it'. See Eileen M. Kane, 'Molecules and conflict: Cancer, patents, and women's health', *American University Journal of Gender, Social Policy & the Law*, 2007 (15 AM. U.J. Gender Soc. Pol'y & L. 305, p. 307; and Jon F. Merz, David Magnus, Mildred K. Cho & Arthur L. Caplan, 'Protecting Subjects' Interests in Genetics Research', *Am. J. Hum. Genet*. 70:965-971, 2002. See also Chapter 6.
6. Susan K. Sell, 'Industry strategies for intellectual property and trade: the quest for TRIPS, and post-TRIPS strategies', 10 *Cardozo J. Int'l & Comp. L.* 74 2002, p. 94.

NIH-funded research tools) and Best Practices for the Licensing of Genomic Inventions.[7]

At the *international level* there are intergovernmental organisations with different kinds of supranational powers, such as the United Nations, the World Health Assembly of the WHO, the UN hosting WIPO as the instigator of several international patent conventions. Furthermore, at the international level there are intergovernmental bodies with specific mandates; the Framework Convention on Climate Change (UNFCCC, UNEP) and the Kyoto Protocol, the Convention on Biological Diversity (CBD, UNEP) and the Nagoya Protocol on Access and Benefit-sharing (UNEP), and the WTO establishing the TRIPS Agreement.

At *the regional level* the most important player is the European Patent Organisation (EPO) which has supranational power over its 38 members on the pre-grant phase and prosecution for European patents. Furthermore, since the adoption of Directive 98/44/EC, the European Union has had a say in patent matters, and in the near future the EU will extend its powers as the unified EU patent, which has been on the agenda since 1975, has been finalised.[8]

The organisations mentioned above cooperate on matters that are relevant for this thesis. In the area of climate change there has been collaboration, for example, between the United Nations Environment Programme (UNEP), the EPO and the International Centre for Trade and Sustainable Development (ICTSD).[9]

It is at the *national level* that governments, patent authorities and courts influence international policy and address the impact of the international policy and law on national law. In Europe patent law is logical in this regard; the post-grant phase is still broadly left to the national level, given the relation-

7. See NIH, Principles and Guidelines for Recipients of NIH Research Grants and Contracts on Obtaining and Disseminating Biomedical Research Resources: Final Notice, Federal Register 64 (246), 23 December 1999. Notices. P. 72090, http://ott.od.nih.gov/pdfs/64FR72090.pdf and http://ott.od.nih.gov/policy/genomic_invention.html; and Report of the Secretary's Advisory Committee on Genetics, Health, and Society (SACGHS), April 2010, p. 75.
8. See Regulation (EU) No 1257/2012 of the European Parliament and the Council of 17 December 2012 implementing enhanced cooperation in the area of the creation of unitary patent protection; Regulation (EU) No 1260/2012 of 17 December 2012 implementing enhanced cooperation in the area of the creation of unitary patent protection with regard to the applicable translation arrangements; and Agreement on a Unified Patent Court (16351/12).
9. See 'Patents and Clean Energy: bridging the gap between evidence and policy', http://www.epo.org/topics/issues/clean-energy/study.html.

ship between EPO and its Members. However, it is not so simple, as the powers of the EU affect this relationship. In practice, Directive 98/44/EC both interferes with the powers and autonomy of the EPO, and muddies the distinction between pre-grant and post-grant measures. This 'multilevel' structure has a missing element in so far as it comprises a mixture of jurisdictions, thus blurring the boundaries of the competences and sovereign rights of two regional legal orders.

Furthermore at the national level this thesis addresses the US patent law system as well as the US patent regime's participation in a multilateral legal framework. However, the particular perspective of Chapters 3 and 4 is that of the EU. The centre of analysis of multilevel governance is that of the EU legal order expanding its power into a sort of *EU international legal order*. Issues such as biomedical research, animal breeding and climate change have to be addressed at a global level, so the legal powers of the European Union are not only supranational powers over its Member States, but they also concern its legitimacy in international law and as an actor in multilateral cooperation.

The creation of intellectual property rights is greatly affected by scientific and technological developments. As patentable subject matter expands, patent authorities in Europe and in the USA find themselves in a new situation. Instead of merely applying the law, they create law. Both the European patent law regime and the US patent law regime are faced with the same issue of legitimacy; who should have the predominant role in fashioning patent law?[10]

The issue of legitimacy is viewed broadly. First, it embraces the question of the institutional legitimacy: who is the gatekeeper, who decides what patent law is meant to protect? Is it the EPO or its Boards of Appeal? Is it the European Union Commission, the European Council or the Court of Justice of the European Union? And finally, is it the legislators of the Member States and their patent authorities? As this thesis compares the current state of patent law in European with the current state of patent law in the USA, the equivalent questions have to be asked: is it Congress, the Supreme Court, the Federal Circuit or the US Patent and Trademark Office (USPTO) which decides what patent law is meant to protect?

10. See e.g. Robert Patrick Merges & John Fitzgerald Duffy, *Patent Law and Policy: Cases and Materials*, 5th Ed., 2011, p. 68.

## 2. Expanding the power of the European Union

The development of the European legal order was indeed the handiwork of the European Court of Justice. All EU textbooks start with a description of the first step towards the supremacy of EC law which was made with the ruling of the Court of Justice in Case 26/62 *Van Gend en Loos*.[11] The oft quoted sentence from the case is that 'the Community constitutes a new legal order in international law, for whose benefit the States have limited their sovereign rights, albeit within limited fields'.[12] *Steiner & Woods* write that the reference to a new legal order was a first step from treaty organisation to independent legal order, and that it had a greater impact on the national legal systems than that of a 'normal' organisation under international law.[13] In Case 6/64 *Costa v ENEL*,[14] the Court elaborated on the established *Van Gend* 'new legal order' doctrine and referred to Article 189 of the EC Treaty (Article 288 of the Treaty on the Functioning of the European Union (TFEU)), stating that there had been a transfer of powers to the Community institutions. The Court clearly affirmed the principle of the supremacy of Community law,[15] thus substantiating the principle of loyalty in Article 5 (Article 4 TFEU) and emphasising the Member States' commitment to respect Community law.[16]

In its activism the Court established its own style of interpretation, taking account of the *nature* and the *purposes* of the Community and legitimising the new legal order with its permanent restrictions on the Member States' le-

11. In addition to textbooks, e.g. Paul Craig & Gráinne de Búrca, *EU Law, Text, cases and materials*, 5th Ed., 2011, p. 256 ff., see also Kaarlo Tuori, 'The many Constitutions of Europe', in Kaarlo Tuori & Suvi Sankari, *The many constitutions of Europe; The Heroic Saga of Juridical Constitutionalization*, p. 17 ff.
12. See para. II.B. The case concerned a higher rate of duty on a product. A higher rate of duty was charged on the same product, which was held illegal irrespective of whether the increase arose from a higher rate of customs duty or from a rearrangement of the tariff resulting in the classification of the product under a more highly taxed heading.
13. See Josephine Steiner & Lorna Woods, *EU Law*, 10th Ed., p. 87.
14. The case concerned a treaty provision and an Italian statute, which was adopted after the treaty provision, and which nationalised the electricity company of which the defendant, Costa, was a shareholder.
15. On grounds of the contrast with ordinary international treaties, the EEC treaty created its own legal system which became an integral part of the legal systems of the Member States on the entry into force of the Treaty and which their courts are bound to apply; see Case 6/64 *Flamingo Costa v Enel*, [1964] ECR 585 and 593; and Paul Craig & Gráinne de Búrca, *EU Law, Text, cases and materials*, 5th Ed., 2011, p. 257.
16. See *Josephine Steiner & Lorna Woods*, *EU Law*, 10th Ed., p. 87.

gal powers.[17] The Court developed a teleological style of interpretation, emphasising the need to make Community law effective by focusing on the objectives of integration and the measures necessary for its development and implementation.[18]

In establishing the EU legal order the Court has assessed almost every kind of conflicting legal act. In Case 11/70 *International Handelsgesellscaft*,[19] the Court stated that the supremacy of Community law also embraced provisions enshrined in a Member State's constitution.[20] *Steiner & Woods* summarise the Court's line of reason as a 'need to ensure the effectiveness of Community law whatever the cost to the national legal order'.[21] *Lenaerts and Gutiérrez-Fons* refer to the 'gap-filling' function of general principles thus ensures the autonomy and coherence of the EU legal system and to 'mutual cross-fertilization' between the Court and the national courts, which has both a top-down and a bottom-up dimension.[22] Elaborating on this, general princi-

17. See Alan Dashwood et al., *Wyatt and Dashwood's European Union Law*, 6th Ed., 2011, p. 270; and Hjalte Rasmussen, *On Law and Policy in the European Court of Justice*, 1986, p. 10 ff.
18. See Josephine Steiner & Lorna Woods, *EU Law*, 10th Ed., p. 88; Hjalte Rasmussen, *On Law and Policy in the European Court of Justice*, 1986, pp. 98 and 400; and Allan Rosas, 'The European Court of Justice: Sources of Law and Methods of interpretation', in Giorgio Sacerdoti, Alan Yanovich & Jan Bohanes, *The WTO at ten: the contribution of the dispute settlement system*, 2006, p. 487. Thus there was a departure from the subjective intention of the Treaty's founding fathers however, when it comes to secondary legislation the EU courts may refer to the *travaux préparatoires*.
19. The claimant argued that the principle of proportionality enshrined in the German constitution had been breached, and sought to have the regulation annulled on those grounds. EC law had been incorporated into German law by the Act of Ratification.
20. See para. 3: 'The protection of fundamental rights in the community legal system recourse to the legal rules or concepts of national law in order to judge the validity of measures adopted by the institutions of the community would have an adverse effect on the uniformity and efficacy of community law. The validity of such measures can only be judged in the light of community law'. And further in para. 4: 'however, an examination should be made as to whether or not any analogous guarantee inherent in community law has been disregarded. In fact, respect for fundamental rights forms an integral part of the general principles of law protected by the court of justice.'
21. See Josephine Steiner & Lorna Woods, *EU Law*, 10th Ed., pp. 88-89.
22. See Lenaerts & Gutiérrez-Fons, 'The Constitutional Allocation of Powers and General Principles of EU law', (2010) 47 *Common Market Law Review*, Issue 6, pp. 1629-1669, at p. 1629-1631 and Jan Wouters, Dominic Coppens & Dylan Geraets, 'The Influence of General principles of Law', in Sanford E. Gaines, Birgitte Egelund Olsen & Karsten Engsig Sørensen (eds), *Liberalising Trade in the EU and the WTO, A legal Comparison*, p. 65.

ples are defined as 'internal principles of EU law' and are not merely a common denominator deduced from national legal systems but are actively constructed on the basis of, and with the aim of strengthening, the Union's *sui generis* character.

The European legal order can be illustrated as follows:

*Figure 3.1. European legal order*

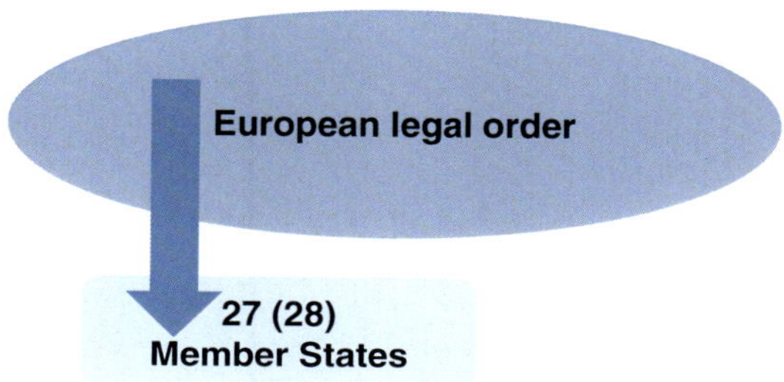


The Community legal order differs from public international law due to its body of autonomous law-making institutions and especially due to the role of the Court of Justice of the European Union which is not an international court in the traditional sense and has not limited itself to applying the law in a traditional legal dogmatic sense.[23]

However, the next thing to do is to determine how this legal order operates from the broader perspective of international law. In Case 22/70 *Commission v Council* (*ERTA*), the Court stated that once the Community lays down common rules for implementing a common policy, the Member States no longer have the right, individually or collectively, to enter into obligations with third countries affecting these common rules. This means that where a Member State attempts to exercise concurrent authority, its laws will be overridden to the extent that they conflict with Union law.[24]

23. Paul Craig & Gráinne de Búrca, *EU Law, Text, Cases and Materials*, 5th Ed. 2011, p. 338; Marisa Cremona, 'External relations and external competence of the European Union', in Paul Craig & Gráinne de Búrca, *The Evolution of EU Law*, 2nd Ed., 2011, p. 260; Paul Craig in same, p. 51, quoting Wallace that the Community had moved beyond other international regimes; and Andrea Ott, 'The ECJ and International law, Thirty years of case-law by the European Court of Justice on International law: A pragmatic approach towards its integration', in Vincent Kronenberger (ed.), *The European Union and the International Legal Order, Discord or Harmony?*, 2001, p. 98.
24. See Case 22/70 *European Agreement on Road Transport (ERTA)*, para. 17; Josephine Steiner & Lorna Woods: *EU Law*, 10th Ed., p. 89; and Paul Craig & Gráinne de Búr-

The EU legal order has been elaborated further by the Court and the latest statement in this regard came with the Joined Cases C-402 and 415/05 *Kadi v Commission and Council,*[25] *de facto* stating supremacy of Union law over international law.[26] The Union is based on the rule of law and it cannot avoid judicial review of the conformity of its acts with the TFEU and with the principles of human rights.[27] However, in paragraphs 287 and 288 of its judgment the Court of Justice clearly accepted the primacy of the UN Charter, stating that the Court's review would not entail any challenge to the primacy of an UN resolution in international law. Agreements binding on the Union take precedence over secondary Union acts but not over primary Union Law.[28]

The background to the dispute was the fight against terrorism. Condemning the fact that Afghan territory continued to be used to shelter and train terrorists and to plan terrorist acts, the UN Security Council adopted a number of resolutions for the freezing of the assets of people thought to be involved in terrorism.[29] The cases were joined under appeal

ca, *EU Law, Text, Cases and Materials*, 5th Ed., 2011, p. 307. See further Jan Klabbers, 'Safeguarding the Organizational *Acquis*: The EU's External Practice', *International Organizations Law Review* (2007), p. 58: 'external powers are not so much implied in the existence of international powers but rather serve to shield the *acquis communitaire*: 'The purpose of the exclusive competence of the Community is primarily to preserve the effectiveness of Community law and the proper functioning of the systems established by its rules', in the words of Opinion 01/03, Lugano Convention, Opinion of 7 February 2006, para. 131'.

25. Case T-315/01 *Kadi* v *Council and Commission* [2005] ECR II-3649; Case T-306/01 *Yusef and Al Barakaat International Foundation* v *Council and Commission* [2005] ECR II-3533; and Joined Cases C-402/05 P and C-415/05 P *Yassin Abdullah Kadi and Al Barakaat International Foundation* v *Council and Commission* [2008] ECR I-6351; see Kaarlo Tuori, 'The many constitutions of Europe', in Kaarlo Tuori & Suvi Sankari, *The many constitutions of Europe*, 2010, p. 21.
26. Joined Cases C-402/05 P and C-415/05 P, *Kadi and Al Barakaat*, para. 303 et seq
27. See Niels Blokker, 'Reviewing the review: Did the European Court of Justice in Kadi indirectly review Security Council resolutions? On the downside of a courageous judgment', in Mielle Bulterman, Leigh Hancher, Alison McDonnell & Hanna Sevenster (eds), *Views of European Law from the Mountain, Liber Amicorum Piet Jan Slot*, 2009, p. 317.
28. See Jan Wouters, Dominic Coppens & Dylan Geraets, 'The Influence of general principles of law', in Sanford E. Gaines, Birgitte Egelund Olsen & Karsten Engsig Sørensen (eds), *Liberalising Trade in the EU and the WTO, A legal Comparison*, p. 54.
29. Joined Cases C-402/05 P and C-415/05 P, *Kadi and Al Barakaat*, para. 11 et seq.

from the Court of First Instance (now the General Court)[30] and the Court of Justice overruled the decision of the Court of First Instance rejecting the claims brought by Kadi and Al Barakaat against Council Regulations which imposed certain restrictive measures on them (prohibiting the export of certain goods and services to Afghanistan, freezing assets/funds and so on). Both Kadi and Al Barakaat (Barakaat International Foundation) challenged the legality of the Union measures, among other things on grounds of the breach of fundamental rights.

In paragraphs 285 and 286 of its judgment the Court of Justice stated that the obligations imposed by an international agreement 'cannot have the effect of prejudicing the constitutional principles of the EC Treaty, which include the principle that all Community acts must respect fundamental rights, that respect constituting a condition of their lawfulness which it is for the Court to review in the framework of the complete system of legal remedies established by the Treaty'. The Court further emphasised that, in circumstances such as those in this case, 'the review of lawfulness thus to be ensured by the Community judicature applies to the Community act intended to give effect to the international agreement at issue, and not to the latter as such'.

In their analysis of the case *Steiner* and *Woods* write that the impact of this judgment has been to reassert the coherence and autonomy of the *Community* legal order, and to re-emphasise its constitutional basis. As regards the relationship of the EU and other international organizations, it suggests that the EC, if not the EU constitutes an impermeable layer protecting its own inner organizational structure; it also reminds us that to the extent that the EU is still an international law body; it sees itself in a different light from other such bodies.[31] Commenting on the decision of the Court of First Instance, *Eeckhout* writes that the UN Charter is a cornerstone of the international law edifice, and the Security Council performs a particularly important function. Union law must respect international law, and in particular UN Resolutions. However, Eeckhout emphasises that the principle of respect for fundamental

30. See Piet Eeckhout, 'EC law and UN Security Council Resolution – In search for the right fit', in Alan Dashwood & Marc Maresceau (eds), *Law and Practice of EU External Relations, Salient Features of a Changing Landscape*, 2008, pp. 104-110.
31. See Josephine Steiner & Lorna Woods, *EU law*, $10^{th}$ Ed., 2009, p. 89. See also Gráinne de Búrca, 'The Evolution of EU Human Rights Law', in Paul Craig & Gráinne de Búrca, *The Evolution of EU Law*, $2^{nd}$ Ed., 2011, pp. 488-489; and Case C-465/07 *Elgafaji* v *Staatssecretaris*, on the formal autonomy of the ECJ to interpret provisions of EU law.

rights is, on balance, the stronger imperative, and if it clashes with international law it should override it.[32]

It cannot be ignored that the Court distinguished the European legal order, including the UN EU Member States, from the other UN Members.[33] One can discuss whether, in the context of globalisation and harmonisation, this indirect 'export of EU norms' has an implicit impact on the overall assessment of the UN solution or the legitimacy of the EU as a global actor.[34] The *sui generis* character of the EU legal order (EU's autonomous legal order) in terms of 'cross-fertilisation' can be characterised as one-sided, a kind of bottom-up relationship to international law as if international law has a tailor-made EU context (thus the *sui generis* approach). This is also what this thesis emphasises in relation to the EU and the regional order (EPO). The Court's decision in *Kadi* in terms of 'the bottom-up approach' may also reflect that this is an unending process.[35]

32. See Piet Eeckhout, 'EC law and UN Security Council Resolution – In search for the right fit', in Alan Dashwood & Marc Maresceau (eds), *Law and Practice of EU External Relations, Salient Features of a Changing Landscape*, 2008, p. 126.
33. See also Niels Blokker, 'Reviewing the review: Did the European Court of Justice in Kadi indirectly review Security Council resolutions? On the downside of a courageous judgment', in Mielle Bulterman, Leigh Hancher, Alison McDonnell & Hanna Sevenster (eds), *Views of European Law from the Mountain, Liber Amicorum Piet Jan Slot*, 2009, p. 326.
34. For the terms 'import of norms' and 'export of norms', see Wessel W. Geursen, 'Harmonisation in a Globalizing Market Place', in Mielle Bulterman, Leigh Hancher, Alison McDonnell & Hanna Sevenster (eds), *Views of European Law from the Mountain, Liber Amicorum Piet Jan Slot*, 2009, p. 21. See also Piet Eeckhout, EU External Relations Law, 2nd Ed, 2011, p. 326 ('whether Article 3(5) TEU will affect the findings, remains to be seen'). See further Christopher Harding, 'The Identity of European Law: Mapping out the European Legal Space', *ELJ*, Vol. 6, No 2, June 2000, pp. 128-147, especially p. 135 f. on exportation.
35. See Jan Wouters, Dominic Coppens & Dylan Geraets, 'The Influence of General principles of Law', in Sanford E. Gaines, Birgitte Egelund Olsen & Karsten Engsig Sørensen (eds), *Liberalising Trade in the EU and the WTO, A legal Comparison*, p. 65, referring to Lenaerts & Gutiérrez-Fons, 'The Constitutional Allocation of Powers and General Principles of EU law', (2010) 47 *Common Market Law Review*, Issue 6, pp. 1629-1669, at p. 1630. On p. 57 Wouters et al. write that 'the more legal systems such as the WTO become rule-oriented and borrow notions from EU law, the more need there will be to emphasise the Union's sui generis nature. See further Gráinne de Búrca, The European Court of Justice and the International legal order after *Kadi*, 51 *Harv. Int'l L.J. 1 2010*, p. 5.

That the Union sees itself as a global player is further mirrored in Article 3(5) of the Treaty on European Union (TEU), which states as one of the core objectives:

'In its relations with the wider world, the Union shall uphold and promote its values and interests and contribute to the protection of its citizens. It shall contribute to *peace, security, the sustainable development of the Earth*, solidarity and mutual respect among peoples, free and fair trade, *eradication of poverty and the protection of human rights*, in particular the rights of the child, as well as to *the strict observance and the development of international law*, including respect for the principles of the United Nations Charter.'[36] [Emphasis added]

Thus the traditional discussion of whether states approaches international law on a monistic or dualistic basis is not relevant to the EU legal order which at best can be classified as a *sui generis* system in both its internal and external relations.[37] The question of the direct effect of international agreements and treaties has also been raised, but none of these cases has changed the status of the Union. International treaties entered into by the Union become part of the EU legal order and no further implementation is required. If the requirements

36. TFEU Article 1(2) TFEU states that: 'This Treaty and the Treaty on European Union constitute the Treaties on which the Union is founded. These two Treaties, which have the same legal value, shall be referred to as 'the Treaties'', and there is a similar statement in Article 1, third paragraph, TEU. See José Maria Beneyto, 'From Nice to the Constitutional Treaty: Eight theses on the (Future) Constitutionalisation of Europe', in Stefan Griller & Jacques Ziller (eds.), *The Lisbon Treaty, EU Constitutionalism without a Constitutional Treaty*, 2008, p. 11, emphasising the word 'contribute'. Further, Jan Wouters, Dominic Coppens & Bart de Meester write that the provision requires, for example, 'that the Common Commercial Policy ('CCP') not only pursues trade-related objectives as stated in the current Article 133 TEC, but taken into account and even contributes to other dimensions, such as human rights and sustainable development. After the Lisbon Treaty, there is thus no doubt anymore as to whether there is a legal basis for including human rights or other 'essential elements' clauses in association agreements or international trade agreements'; see 'The European Union's external relations after the Lisbon Treaty', in Stefan Griller & Jacques Ziller (eds.), *The Lisbon Treaty, EU Constitutionalism without a Constitutional Treaty*, 2008, p. 149.
37. See also Allan Rosas, 'The European Court of Justice: Sources of Law and Methods of interpretation', in Giorgio Sacerdoti, Alan Yanovich & Jan Bohanes (eds.), *The WTO at Ten, The Contribution of the Dispute Settlement System*, 2006, p. 482; and Andrea Ott, 'The ECJ and International law, Thirty years of case-law by the European Court of Justice on International law: A pragmatic approach towards its integration', in Vincent Kronenberger (ed.): *The European Union and the International Legal Order, Discord or Harmony?*, 2001, p. 98.

for direct effect are fulfilled, then individuals can invoke the provisions of such treaties before the national courts of the Member States. On the question of the direct effect of the WTO rules (and thus also the TRIPS agreement[38]) it has been established by the Court of Justice that WTO regime (and GATT) cannot have direct effect due to the structure and nature of the agreement as a whole.[39] Furthermore, as well as UN Convention on the law of the Sea (UN-

38. See Case 104/81 *Hauptzollamt Mainz* v *CA Kupferberg*, which is sufficiently precise and unconditional to apply to all EU legal sources, regardless of their origin. See also Jan Wouters, Dominic Coppens & Bart de Meester, 'The European Union's external relations after the Lisbon Treaty', in Stefan Griller & Jacques Ziller (eds.), *The Lisbon Treaty, EU Constitutionalism without a Constitutional Treaty*, 2008, p. 187; and Paul Craig & Gráinne de Búrca, *EU Law, Text, Cases and Materials*, 5th Ed., 2011, p. 344 ff. With regard to Article 50 of the TRIPS Agreement and direct effect, see Case C-53/96 *Hermes*, para. 35, where the European Court of Justice avoided making an express ruling; and Joined Cases C-300/98 and 392/98, *Dior*, paras. 44 and 49: 'in a field to which TRIPs applies and in respect of which the Community has already legislated, the judicial authorities of the Member States are required by virtue of Community law, when called upon to apply national rules with a view to ordering provisional measures for the protection of rights falling within such a field, to do so as far as possible in the light of the wording and purpose of Article 50 of TRIPS, but in a field in respect of which the Community has not yet legislated and which consequently falls within the competence of the Member States, the protection of intellectual property rights, and measures adopted for that purpose by the judicial authorities, do not fall within the scope of Community law. Accordingly, Community law neither requires nor forbids that the legal order of a Member State should accord to individuals the right to rely directly on the rule laid down by Article 50(6) of TRIPS or that it should oblige the courts to apply that rule of their own motion.' Thus, while the Court denied the direct effect of the TRIPS Agreement in Community law, the agreement does form part of Community law without individuals being able to invoke it directly. See Piet Eeckhout, *EU External Relations Law*, 2nd Ed, 2011, pp. 349-350; and Trevor Cook, *EU Intellectual Property Law*, 2010, p. 3. See also Case C-431/05 *Merck Genéricos*.
39. For literature on direct effect see Piet Eeckhout, *EU External Relations Law*, 2nd Ed., 2011, p. 343 (especially footnote 68); Panos Koutrakos, *EU International Relations Law*, 2006, p. 217 ff. and 251 ff.; and J. Klappers, 'International Law in Community Law: The Law and Politics of Direct Effect', 2002 *YEL* 263; Francis G. Jacobs, 'Direct effect and interpretation of international agreements in the recent case law of the European Court of Justice', in Alan Dashwood & Marc Maresceau, *Law and Practice of the EU External relations, salient features of a changing landscape*, p. 15 ff.; Birgitte Egelund Olsen & Michael Stenicke, 'The WTO and the EU', in Birgitte Egelund Olsen, Michael Stenicke & Karsten Engsig Sørensen (eds.), *WTO Law – from a European Perspective*, 2005, pp. 115-122; Matthew Elsmore, 'Comparing regulatory treatment of intellectual property at WTO and EU level', in Sanford E. Gaines, Bir-

CLOS) did not establish rules intended to apply directly and immediately to individuals or confer rights upon them.[40]

The European Union Court of Justice has confirmed that international environmental agreements (multilateral agreements) can be invoked by a broader range of parties. In Case C-377/98 *Netherlands v Council*, paragraph 54, the Court ruled that even if the Biodiversity Convention 'contains provisions which do not have direct effect, in the sense that they do not create rights which individuals can rely on directly before the courts, that fact does not preclude review by the courts of compliance with the obligations incumbent on the Community as a party to that agreement'.[41]

Where the lack of direct effect gives room for manoeuvre, the duty of loyal interpretation is an important limitation. Article 4(3) TEU imposes a general duty of loyal cooperation, and based on this the EU Member States have a duty of consistent interpretation (first identified in Case 14/83 *von Colson*). If a legislative provision does not provide for direct effect, the Court of Justice may find that it has indirect effect by interpreting national law/EU law in the light of the wording and purpose of international law.[42] In order to prevent conflicts, the primacy of the EU's international agreements over secondary EU legislation must be interpreted as far as possible, in a manner that is consistent with those agreements.[43]

gitte Egelund Olsen & Karsten Engsig Sørensen (eds), *Liberalising Trade in the EU and the WTO, A legal Comparison*, p. 431 ff.

40. See Case C-308/06 *Intertanko*, paras. 59 and 64.
41. See also para. 53: 'However, such an exclusion cannot be applied to the CBD, which, unlike the WTO agreement, is not strictly based on reciprocal and mutually advantageous arrangements'; and Paul Craig & Gráinne de Búrca, *EU Law, Text, Cases and Materials*, 5th Ed., 2011, p. 349.
42. See Wyatt & Dashwood, *European Union Law*, 2011, p. 239. Lenaerts & Gutiérrez-Fons, 'The Constitutional Allocation of Powers and General Principles of EU law', (2010) 47 *Common Market Law Review*, Issue 6, pp. 1636 ff.
43. Case C-61/94 *Commission v Germany*, para. 10. See Piet Eeckhout, *EU External Relations Law*, 2nd Ed., 2011, p. 356, on the reservation of the EU Courts if the case involves WTO dispute decisions adopted against the EU. See Panos Koutrakos, *EU International Relations Law*, 2006, p. 263 ff.; Francis G. Jacobs, 'Direct effect and interpretation of international agreements in the recent case law of the European Court of Justice', in Alan Dashwood & Marc Maresceau, *Law and Practice of the EU External relations, salient features of a changing landscape*, p. 23 ff.

The relationship between EU and international law can be illustrated as follows:

*Figure 3.2. European Legal Order and International Law*

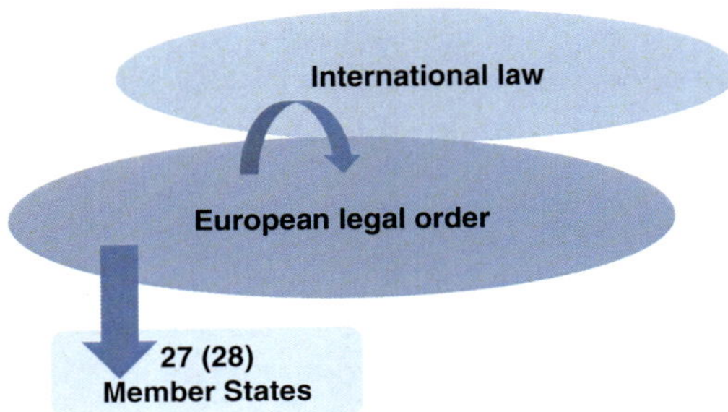


This approach has developed over time (and is still developing) both as a judicial creation of the Court and through treaty amendments. After the failure in 2001 to establish a Constitution for Europe, with enhanced political power, an anthem and so on, the Lisbon Treaty of 2007 was expected to transform the Union into an efficient global actor.[44]

The fundamental principles of Union competences are based on Articles 4 and 5 TEU. Competences that are not conferred upon the Union in the Treaty remain with the Member States (Articles 4(1) and 5 (1)(2)). In the same vein, the principles of subsidiarity and proportionality frame the Union's competences. It follows from Article 5(3) that in areas which do not fall within its exclusive competence, the Union may act only if and in so far as the objectives of the proposed action cannot be sufficiently achieved by the Member States (and be better achieved at Union level) (subsidiarity). Article 5(4) adds the principle

44. Jan Wouters, Dominic Coppens & Bart de Meester, 'The European Union's external relations after the Lisbon Treaty', in Stefan Griller & Jacques Ziller (eds), *The Lisbon Treaty, EU Constitutionalism without a Constitutional Treaty*, 2008, p. 144, state at p. 146: 'One of the principal aims behind the constitutional process was to provide the EU with the tools to enable it to become a global actor not only in the economic sphere but also in the political sphere.' This aim was not fulfilled by the Lisbon Treaty, following the abandonment of the idea of a constitution. See further Gráinne de Búrca, 'The EU on the Road from the Constitutional Treaty to the Lisbon Treaty', *Jean Monnet Working Paper* 03/08; Paul Craig, 'The treaty of Lisbon, process, architecture and substance', E.L. Rev. 2008, 33 (2), 137-166; and Ingolf Pernice, 'Salvaging the Constitution for Europe, A Reform Treaty for the EU', 2007, *WHI*, Paper 4/07.

of proportionality, according to which form and content of Union action may not exceed what is necessary to achieve the objectives of the Treaties.[45]

The question of competence in internal as well as external matters has concerned everything in the treaty system from exclusive Member State competence, shared or complementary competence, obligations to coordinate policies, to exclusive Union competence, thus aiming at the progress of exclusive and shared competences as the treaty system has evolved; see Article 2(1)(2) TFEU.[46]

Exclusive competence conferred on the Union in a specific area means that regulatory authority belongs to the Union alone. This is the exception under the Treaty but it is relevant to this thesis that the common commercial policy falls under this heading (see below). Shared competence means that the Member States may only exercise their competence to the extent that the Union has not exercised its competence. Competence to support, coordinate and supplement Member State action follows from Article 2(5) TFEU. Examples in this category are the issue of guidelines, incentive measures and exchange of information about best practice.[47]

The terms 'express' and 'implied' external competences of the EU relate to the principle of conferral Article 5(1) TEU in the context of external competence. Authorisation must be conferred by the Treaty (substantive legal basis). However, this may not be exclusive authorisation (Member States may exercise external power themselves). Article 3(1) TFEU is a codification of the status quo (except for the inclusion of competition law) and Article 3(2) TFEU provides that the Union shall also have exclusive competence to enter into international agreement when provided for in a Union legislative act or it is necessary to enable the Union to exercise its internal competence or in so far its conclusion may affect common rules or alter their scope. With the adoption of the Lisbon Treaty (Articles 3(2) and 216(1) TFEU) further development of this issue can only be expected.[48]

Despite the fact that the EU has become party to a huge number of international agreements, there is still a lack of explicit power to conclude international agreements, and the Lisbon Treaty did not remedy this defect.[49] There

45. See TFEU Protocol (No 2) on the application of the principles of Subsidiarity and Proportionality.
46. The latest amendment to the EU Treaty system came with the Lisbon Treaty and the subsequent codification with the consolidated version of the Treaty on the Functioning of the European Union entering into force in December 2009.
47. Wyatt & Dashwood, *European Union Law*, 2011, pp. 100-104.
48. See Wyatt & Dashwood, *European Union Law*, 2011, pp. 909.
49. See Paul Craig & Gráinne de Búrca, *EU Law, Text, Cases and Materials*, 5th Ed., 2011, p. 307 on international legal personality and the principle of conferred powers; the EU may only act when the Treaties provide a legal basis for action, footnote 18 referring to Declaration No 24 annexed to the EU Treaties by the Lisbon Treaty

is also no explicit Treaty provision that allows the Union to become a member of international organisations.

*Wouters et al.* describe the current position on the competence to conclude international agreements, emphasising the principle of conferral in Article 5(2) TEU, which has been further elaborated by the Lisbon Treaty. They write that the external competences of the Union have mainly been clarified by the Court: 'the general thrust of the case-law boils down to the principle of 'parallelism (*in foro interno, in foro externo*): if there is an internal competence of the Union, there is also external competence''.[50]

Article 216 TFEU now mirrors this position, as follows:

> 1. The Union may conclude an agreement with one or more third countries or international organisations where the Treaties so provide or where the conclusion of an agreement is necessary in order to achieve, within the framework of the Union's policies, one of the objectives referred to in the Treaties, or is provided for in a legally binding Union act or is likely to affect common rules or alter their scope.
>
> 2. Agreements concluded by the Union are binding upon the institutions of the Union and on its Member States.[51]

Prior to the Lisbon Treaty, the EC had competence in the area of energy on the basis of Articles 95, 100, 155-156 and 175 of the Treaty Establishing the European Community (TEC), but with the Lisbon Treaty this competence was transferred to the EU, enabling it to act in the area of energy; see Article 194 TFEU.[52] On health matters, Article 168(7) of the Lisbon Treaty restricted EC competence over the management of health services and medical care, which reflects the fact that the arrangement of competences can be flexible.

which confirms 'the fact that the European Union has a legal personality will not in any way authorise the Union to legislate or to act beyond the competences conferred upon it by Member States in the Treaty.'

50. Jan Wouters, Dominic Coppens & Bart de Meester, 'The European Union's external relations after the Lisbon Treaty', in Stefan Griller & Jacques Ziller (eds), *The Lisbon Treaty, EU Constitutionalism without a Constitutional Treaty*, 2008, p. 144 at p. 168.
51. See Paul Craig & Gráinne de Búrca, *EU Law, Text, Cases and Materials*, 5th Ed., 2011, p. 319; the case law has now been loosely codified so that every EU policy has a potential external dimension.
52. See further Article 122(1) TFEU, concerning the supply of energy; and Jean-Claude Pires, *The Lisbon treaty, A legal and political analysis*, 2010, p. 319, who argues that the competence of the EU is made clearer but not wider, reflecting the growing importance attached to this area by the EU.

Before turning to the elaboration of international patent norms, the next section will elaborate on the regional patent norms as they currently exist in Europe. The core question in this regard concerns the role of the European Union. It is the view of the author that the Union has led patent law onto an unsuccessful road in establishing an 'EU international patent law', thus establishing a regional level that has blurred powers both outside and inside the Union. It remains to be seen what effect the current 'EU patent package' will have on the regional system.[53]

## 3. The patent law perspective

### 3.1. EU Treaties: the regional level

From a patent law perspective it is important to realise that the EU 'patent legal order' has gained authority on *a step-by-step* approach, thus creating a new legal system with a kind of scattered legal power. The original Treaty was silent on the IPR issues, stating in Article 295 TEC that the Treaty in no way prejudiced the rules of Member States governing the system of property ownership.

However, as also mentioned in Chapter 1, section 1, the protection of IPRs vis-à-vis free trade, was considered in the original Treaty structure. One of four major freedoms established by the 1958 Rome Treaty was the free movement of goods. According to the Rome Treaty Articles 30 and 34 (now Articles 34 and 35 TFEU), no obstacles were to be allowed to the import, transit or export of goods. However, Article 36 of the Treaty (now Article 36 TFEU) did give a legal basis for one exception. According to this, Member States were allowed to restrict the free movement of goods if this was justified by an exhaustive list of grounds:

'public morality, public policy or public security, the protection of health and life of humans, animals or plants, the protection of national treasures possessing artistic, historic or

53. The debate on the unitary patent package in the European Parliament has been rescheduled for the afternoon/evening of 3 July 2012 and the scheduled vote is thought to be taking place the following day on the auspicious Fourth of July, but as of yet there is no vote listed on the agenda (information from the IPKat readers 25 June: 'will there be fireworks in Brussels on the Fourth of July?)). Jochen Pagenberg, 'The EU Patent Package, Politics vs. Quality and the New Practice of Secret Legislation in Brussels, June 2012, to be published. Pagenberg writes that the entire project in the last six months has been subject to a completely unusual secrecy.

archaeological value, or the protection of industrial and commercial property' (now Article 36, first sentence, TFEU).

Thus, in 1958 the Member States' laws on intellectual property rights could in principle legitimately restrict the free movement of goods in the European Community. A balance of rights in this regard was based on economic standards between too much protection and the right of free movement rights. The Court soon interpreted the provision as a question of non-discrimination in the grant and protection of IPRs in the Community.[54] It developed the doctrine of the regional exhaustion of rights by the placing of a product on the market in the Community.[55] Without direct Treaty authority, the Court managed to develop a set of rules on patent law on the way to a non-discriminatory trade law. Thus, while the European Community allows restrictions on trade to be justified in the name of the protection of industrial and commercial property, such restrictions may not constitute a means of arbitrary discrimination or a disguised restriction on trade between Member States. Members States may hinder trade as long as it can be justified under Article 36 TFEU.[56]

The furthest point so far, at treaty level,[57] is the new Article 118 TFEU, which states:

54. For copyright, see e.g. Joined Cases C-92 and 326/92 *Phil Collins*, Case C-360/00 *Land Hessen*; compulsory licence Case C-30/90 *Commission* v *United Kingdom*. See Wyatt and Dashwood's, European Union Law, Alan Dashwood, Michael Dougan, Barry Rodger, Eleanor Spaventa and Derrick Wyatt, sixth edition, 2011, 499 ff.
55. See Case 15/74 *Centrafarm* v *Sterling Drug*. See Trevor Cook, EU Intellectual Property Law, 2010, p. 35; and Catherine Seville, *EU Intellectual Property Law and Policy*, 2009, p. 322 ff.
56. See further Raimund Raith, 'The European community, The WTO – TRIPS agreement and the WIPO Conventions', in Vincent Kronenberger (ed.), *The European Union and the International Legal Order: Discord or Harmony*, 2001, p. 240. See Jens Hartig Danielsen, *Parallelhandel og varernes frie bevægelighed*, 2005. See Chapter 1, section 1.
57. Thus, before the Lisbon Treaty there were no corresponding articles, though articles had been proposed. In the negotiations for the Amsterdam Treaty the Commission tried to extend Article 133 TEC to IPR but without success. However, at the Nice Intergovernmental Conference in December 2000, a new paragraph 5 to Article 133 TEC extended the Community's exclusive competence to trade in services and to trade-related aspects of intellectual property. The draft establishing the European Constitution (The European Convention Article III-68, OJ 18.7.2003 C 169/39) proposed a legal basis for IPR protection.

> 'In the context of the establishment and functioning of the internal market, the European Parliament and the Council, acting in accordance with the ordinary legislative procedure, shall establish measures for the creation of European intellectual property rights to provide uniform protection of intellectual property rights throughout the Union and for the setting up of centralised Union-wide authorisation, coordination and supervision arrangements.
>
> The Council, acting in accordance with a special legislative procedure, shall by means of regulations establish language arrangements for the European intellectual property rights. The Council shall act unanimously after consulting the European Parliament.'[58]

This apparently shift is about to be given a practical application and no doubt the future development of EU patent law will depend on this Treaty authority.[59] Prior to the Lisbon Treaty, the EU adopted acts in the fields of IPR on the basis of Article 308 TEC.

The Treaty of Lisbon has changed the future perspectives in another way too. It has introduced several important amendments to the former Article 133 TEC.

The new Article 207(1) TFEU (ex Article 133(1) TEC) states:

> 'The common commercial policy shall be based on uniform principles, particularly with regard to changes in tariff rates, the conclusion of tariff and trade agreements relating to trade in goods and services, and the commercial aspects of intellectual property, foreign direct investment, the achievement of uniformity in measures of liberalisation, export policy and measures to protect trade such as those to be taken in the event of dumping or subsidies. The common commercial policy shall be conducted in the context of the principles and objectives of the Union's external action.'

58. Furthermore, the new Article 262 TFEU (ex Article 229 A TEC) provides legal authority for the EU with regard to a patent court system: 'Without prejudice to the other provisions of the Treaties, the Council, acting unanimously in accordance with a special legislative procedure and after consulting the European Parliament, may adopt provisions to confer jurisdiction, to the extent that it shall determine, on the Court of Justice of the European Union in disputes relating to the application of acts adopted on the basis of the Treaties which create European intellectual property rights. These provisions shall enter into force after their approval by the Member States in accordance with their respective constitutional requirements'.
59. The Article has received little attention despite the fact that it gives the power to Brussels. William Kingston, European Intellectual Property Review, E.I.P.R. 2008, 30 (11), p. 440, writes that: 'the Lisbon treaty changes 67 areas from unanimity to qualified majority voting. Article 118 is one of these for its substantive provision, but the part of it which deals with languages still requires unanimous approval. This clearly reflects the three decades of disagreement which have so far prevented the Community patent from becoming a reality.'

The Union is explicitly allowed to enter into international agreements in the field of common commercial policy, which Article 3(1)(e) TFEU lists as an exclusive competence. The Lisbon statement that common commercial policy in its entirety is exclusive is a major simplification of Article 133 TEC, and the Member States are thus excluded from entering into international agreements on the commercial aspects of intellectual property.[60] It appears that all matters dealt with by the WTO/TRIPS fall within the exclusive competence of the Union.[61] However, as pointed out by *Wyatt and Dashwood*, it remains to be seen whether States remain individually competent in respect of TRIPS since it was concluded in 1994 as a mixed agreement (referring to the wording of Article 207 TFEU on the *conclusion* of agreements). The question is whether the Member States' competence remains unimpaired in the case of agreements that were already in existence when the Lisbon Treaty entered into force.[62]

### 3.2. International conventions and the EU: the international level

There are several international conventions on patent law and bearing in mind Table 1.1 in Chapter 1, one can imagine the scenario of the EU integrating international conventions into EU law by way of cross-fertilising (bottom-up-relationship)/indirectly exporting EU norms.

60. Jan Wouters, Dominic Coppens & Bart de Meester, 'The European Union's external relations after the Lisbon Treaty', in Stefan Griller & Jacques Ziller (eds.), *The Lisbon Treaty, EU Constitutionalism without a Constitutional Treaty*, 2008, p. 144 at p. 170. Referring to Opinion 1/94 in footnote 126: 'similarly, with regard to intellectual property, according to the ECJ only those commercial aspects of intellectual property that concern the prohibition of the release info free circulation of counterfeit goods fell within the scope of common commercial policy, ECJ, Opinion 1/94, Competence of the Community to conclude international agreements concerning services and the protection of intellectual property (WTO), [1994] *ECR* I-5267, para. 56. The Nice Treaty extended the scope of the CCP, with respect to the negotiation and conclusion of international agreements, to trade in service and trade related aspects of intellectual property (Article 133.5 TEC). However, the *nature* of the Union's competence was disputed.'
61. Jan Wouters, Dominic Coppens & Bart de Meester, 'The European Union's external relations after the Lisbon Treaty', in Stefan Griller & Jacques Ziller (eds.), *The Lisbon Treaty, EU Constitutionalism without a Constitutional Treaty*, 2008, p. 180; and Panos Koutrakos (ed.), 'The European Union's external relations a year after Lisbon', CLEER Working Papers 2011/3, p. 17.
62. See Wyatt & Dashwood, *European Union Law*, 2011, p. 947.

There are three important international systems of patent regulation: *WIPO (UN),*[63] *TRIPS (WTO), and the European Patent Organisation (EPO).* Even though the Strasbourg Convention has had an impact on international patent law development, it will not be discussed further in this chapter.[64] The European Union, the Council of Europe and the EPO all have observer status in WIPO,[65] and the European Union became a Member of the WTO on 1 January 1995. All 27 Member States of the EU are also Members of the WTO of their own right. The United Nations is a key player both in patent law and in environmental law. In respect of patent law, Article 1 of the Universal Declaration on the Human Genome and Human Rights (UNESCO) states that: 'The human genome underlies the fundamental unity of all members of the human family, as well as the recognition of their inherent dignity and diversity. In a symbolic sense, it is the heritage of humanity.'[66]

### *3.2.1. WIPO*

Practically all international conventions concerning IPR protection are administered or co-administered by WIPO.[67] Important treaties such as the Paris Convention (1883), The Patent Cooperation Treaty (1970), The Patent Law

63. The World Intellectual Property Organization created by the Convention Establishing the World Intellectual Property Organization (Stockholm, 14 July 1967, as amended on 28 September 1979) is the successor organisation to the Bureaux Internationaux Réunis pour la Protection de la Propriété Intellectuelle dating back to 1883.
64. The 1963 Strasbourg Convention entered into force on 1 August 1980; see e.g. Christopher Wadlow, 'Strasbourg, the Forgotten Patent Convention, and the origins of the European Patent Jurisdiction', *IIC* Vol. 41, 2/2010, p. 123. The article describes the Strasbourg Convention as a neglected convention. 'All the substantive provisions of the Strasbourg Convention were reproduced a decade later when they were incorporated into the European Patent Convention, and it is in the works on the latter that detailed commentaries on individual provisions are to be found, though typically divorced from their original context.'
65. According to Article 5 of the WIPO Convention, only States can obtain member status, however there is an unpublished working agreement between WIPO and the EU; see Raimund Raith, 'The European community, The WTO – TRIPS agreement and the WIPO Conventions', in Vincent Kronenberger (ed.), *The European Union and the International Legal Order: Discord or Harmony*, 2001, p. 248, footnote 50.
66. Thus comprising basic guidelines on research in genetics and biology, as well as the application of the results.
67. See http://www.wipo.int/treaties/en/ (last visited December 2011); and Raimund Raith, 'The European community, The WTO – TRIPS agreement and the WIPO Conventions', in Vincent Kronenberger (ed.), *The European Union and the International Legal Order: Discord or Harmony*, 2001, p. 240.

Treaty (2000), and the Budapest Treaty (1977) are administered by WIPO.[68] WIPO's core area is mainly procedural (and policy oriented), such as administering the procedures of the Patent Cooperation Treaty for international patent applications and filing dates,[69] the Patent Law Treaty's[70] harmonisation of formal procedures, and the depositing of microorganisms according to the Budapest Treaty.[71] However, there have also been attempts to establish a global patent system, such as the Substantive Patent Law Treaty.[72] While the EU's involvement in most IPR areas within the WIPO regime has been ac-

68. See http://www.wipo.int/treaties/en/ (last visited December 2011). See also WIPO Standing Committee on the Law of Patents, twelfth session, Geneva, 2008: Report on the international patent system, point 135: 'a number of regional agreements have contributed to the development of the international patent system through the harmonization and simplification of regional patent laws. Examples of such regional agreements are: (i) the European Patent Convention (34 member States); (ii) the Eurasian Patent Convention (9 member States); (iii) the Harare Protocol (African Regional Intellectual Property Organization (ARIPO)) (16 member States); (iv) the Bangui Agreement (African Intellectual Property Organization (OAPI)) (16 member States); (v) the Patent Regulation of the Cooperation Council for the Arab States of the Gulf (Gulf Cooperation Council (GCC)) (6 member States); (vi) Decision 486, a common intellectual property regime of the Andean Community'.
69. See Article 11 of the Patent Cooperation Treaty (PCT) 1970, which entered into force on 1 April 2002.
70. See Patent Law Treaty, Geneva, 1 June 2000.
71. See Budapest Treaty on the International Recognition of the Deposit of Microorganisms for the Purposes of Patent Procedure, 28 April 1977, and amended on 26 September 1980.
72. Jerome H Reichman & Rochelle Cooper Dreyfuss, 'Harmonisation without consensus: Critical reflections on Drafting a Substantive patent law treaty', in Daniel Gervais (ed.) *Intellectual Property, Trade and Development, Strategies to Optimize Economic Development in the TRIPS-Plus Era*, 2007, pp. 443-473; Graham Dutfield, 'Is the world ready for substantive patent law harmonization?', in Peter Draho (ed.), *Death of Patents*, 2005, pp. 228-249; Morten Walløe Tvedt, 'One Worldwide Patent System: What's in it for developing countries?', *Third World Quarterly*, Vol. 31, No 2, 2010, pp. 277-293; and Morten Walløe Tvedt, 'The Path to One Universal Patent', *Environmental Policy and Law*, Vol. 37, No 4, 2007, pp. 297-305. As an alternative, the IP5 initiative (cooperation between EPO, and the patent offices of Japan, Korea, China and the USA) aims to achieve greater operational integration and engage in 'the creation of a sound, viable, sustainable global patent system, focusing on work-sharing'; see Alison Brimelow, 'Plea for a courageous move towards a truly global patent system', interview with the President of the European Patent Office (EPO), March 2010 (Archive 2010, 20100302a), see further: http://www.epo.org/news-issues/news/2010/20100419.html

cepted, this is not the case in area of patents, where the EU has failed to adopt a significant Community law.[73]

The TRIPS Agreement (1994) requires the TRIPS Council to work with WIPO and in 1995 an agreement on a mutually supportive relationship was established.[74]

### *3.2.2. TRIPS*

At the GATT/WTO level, the EU has succeeded in being an active global player. *Raimund Raith* writes that, in representing the Community, the Commission 'has played a very active role from early on' and in the Uruguay Round in the mid-1980s the Commission 'took the lead in regard to new issues (IPRs, service, investment ...) ... it followed somewhat 'naturally' from the distribution of tasks between the Community and its Members in the GATT, that the Community, represented by the Commission, took the lead also in relation to these new topics, despite legal doubts as to the question of Community Competence in these areas'.[75] *Marise Cremona* describes the role of the Community as follows:

'The EU's role as a rule generator is not just as an exporter of its own regulatory norms to specific third countries as part of a market-opening or integration package. The EU also plays a role as a key participant in multilateral rule-setting within the WTO and in other fora, on issues such as competition, intellectual property and investment ... Here too, as well as being a player, the EU may, with its experience, provide a model for regulating across national boundaries, tackling the need to balance open markets with national regulatory preferences ... The EU's own rhetoric presents this role as a matter of responsibility

73. E.g. the proposed software directive (COM(2002) 92 final), which was rejected in 2005; and the EC proposed patent law (COM(2010) 350 final); see Raimund Raith, 'The European community, The WTO – TRIPS agreement and the WIPO Conventions', in Vincent Kronenberger (ed.), *The European Union and the International Legal Order: Discord or Harmony*, 2001, p. 248.
74. See TRIPS Articles 63 and 68, Agreement between the World Intellectual Property Organisation and the World Trade organization, WIPO Doc WO030 (22 December 1995 available at www.wipo.int/treaties/en/agreement/pdf/trtdocs_wo030.pdf ; see Rochelle Dreyfuss, 'TRIPS and Essential Medicine', in Thomas Pogge, Matthew Rimmer & Kim Rubenstein (eds.), *Incentives for Global Public Health, Patent Law and access to Essential Medicines*, 2010, p. 48.
75. Raimund Raith, 'The European Community, The WTO – TRIPS agreement and the WIPO Conventions', in Vincent Kronenberger (ed.), *The European Union and the International Legal Order: Discord or Harmony*, 2001, p. 243.

and not just advantage; the Laeken Declaration, for example, speaks of the need for Europe 'to shoulder its responsibilities in the governance of globalization.'[76]

However, in the ratification of the TRIPS Agreement a clear distinction had to be established between Member State competence and Community competence. The Court of Justice gave its opinion on the matter pursuant to Article 228(6) TEC (now Article 218(11) TFEU). The Court stated in the Opinion 1/94, point 103, that the harmonisation achieved within the Community in certain areas covered by TRIPS was only partial and that in other areas no harmonisation had been envisaged. Trademarks had only been subject to partial harmonisation, other areas such as industrial designs and patents had not yet been regulated.[77] It followed that:

'With regard to patents, the only acts referred to by the Commission are conventions which are intergovernmental in origin, and not Community acts: the Munich Convention of 5 October 1973 on the Grant of European Patents (JORF, Decree No 77-1151, of 27 September 1977, p. 5002) and the Luxembourg Agreement of 15 December 1989 relating to Community Patents (OJ 1989 L 401, p. 1), which has not yet, however, entered into force.'[78]

In response to observations submitted by some governments, that most of the TRIPS areas fall within the competence of the Member States, the Court pointed out in point 104 that 'if that argument is to be understood as meaning that all those matters are within some sort of domain reserved to the Member States, it cannot be accepted'. The Community is certainly competent to harmonise national rules on those matters, in so far as, in the words of Article 100 of the Treaty, they 'directly affect the establishment or functioning of the common market'. Thus, in point 105 of the opinion the Court stated that the

76. Marise Cremona, 'The Union as a global actor: Roles, Models and Identity', *Common Market Law Review* 41, 2004, p. 557.
77. See further Case C-431/2005 *Merck Genéricos – Produtos Farmacêuticos* v *Merck & Co. Inc. and Merck Sharp & Dohme*, paras. 39-40, concerning TRIPS Article 33: 'it is now appropriate to examine whether, in the particular sphere into which Article 33 of the TRIPS agreement falls, that is to say, that of patents, there is any Community legislation. As Community law now stands, there is none'.
78. Opinion 1/94 of the Court of 15 November 1994, point 103, on the competence of the Community to conclude international agreements concerning services and the protection of intellectual property – Article 228 (6) of the EC Treaty. See Meinhard Hilf, 'The ECJ's Opinion 1/94 on the WTO – No Surprise, but Wise?', 6 *EJIL* (1995) 245-259, especially page 247 ff., on the continuing 'total disagreement' between the Commission and the Council, the European Parliament and the Member States which had submitted observations.

Community and its Member States were jointly competent to conclude TRIPS.[79] In a new reference for a preliminary ruling, Case C-414/11 *Daiichi Sankyo Company Ltd., Sanofi-Aventis Deutsland GmbH v DEMO Anonimos Viomikhaniki kai Emporiki Etairia Farmakon*, three questions have been asked about patentable subject matter under TRIPS Article 27, about the scope of protection for chemical and pharmaceutical products, and about the relation between the European Patent Convention and TRIPS for the protection of process patents. In the national case, the claimant, Daiichi Sankyo, a pharmaceutical company, holds a Greek patent granted in 1986 for the chemical compound Levofloxacin hemihydrates, which is an active ingredient of the original new medicine Tavanic. The chemical compound constitutes a new invention and is protected by a supplementary protection certificate for pharmaceutical patents. The defendant, Demo Aveef, is a Greek pharmaceutical company holding a marketing authorisation from the Greek National Organisation of Pharmaceuticals for a medicine named Talerin. Talerin contains quantitatively and qualitatively the same active ingredients as Tavanic, i.e. the patented Levofloxacin hemihydrate. The question concerns the scope of protection, as the patent for Tavanic (Greek patent No 801028) is subject to the Greek reservation provided for in Article 167(2)(a) of the European Patent Convention (EPC), according to which protection of the chemical as such is ineffective or revocable (effective until 7 October 1992).[80] This transition period established by EPC thus conflicts with obligations under Article 27 of the TRIPS Agreement.

The questions referred by the Greek Court to the Court of Justice of the European Union for preliminary rulings are as follows:

1. Does Article 27 of the TRIPS Agreement setting out the framework for patent protection fall within a field for which the Member States continue to have primary competence and, if so, can the Member States themselves accord direct effect to that provision, and can the national court apply it directly subject to the requirements laid down by national law?

79. The Court assessed Case 22/70 *ERTA* and held for the first time that the competence of the Community to conclude international agreements 'arises not only from an express conferment by the Treaty – but may equally flow from other provisions of the Treaty and from measures adopted, within the framework of those provisions, by the Community institutions'.
80. See OJ EPO 1992, 301. For information on the IP cat, see http://ipkitten.blogspot.com/2011/10/daiichi-now-at-least-we-know-what.html.

2. Under Article 27 of the TRIPS Agreement are chemical and pharmaceutical products patentable subject matter provided that they satisfy the requirements for the grant of patents and, if so, what is the scope of their protection?

3. Under Articles 27 and 70 of the TRIPS Agreement, do patents covered by the reservation in Article 167(2) of the 1973 Munich Convention which were granted before 7 February 1992, that is to say, before the above agreement entered into force, and concerned the invention of pharmaceutical products, but which, because of the aforementioned reservation, were granted solely to protect their production process, fall within the protection for all patents pursuant to the provisions of the TRIPS Agreement and, if so, what is the extent and content of that protection, that is to say, have the pharmaceutical products themselves also been protected since the above agreement entered into force or does protection continue to apply to their production process only or must a distinction be made based on the content of the application for grant of a patent, that is to say, as to whether, by describing the invention and the relevant claims, protection was sought at the outset for the product or the production process or both?

This referral is of the utmost importance in several ways. First, Question 1 asks the Court of Justice whether decisions on patentable subject matter are within or outside the Member States' competence. As there is not yet an effective EU patent, this question has to be answered in the positive. Furthermore, the question is isolated from the EPO, which is a very good example of the mixed competences and highlights the overall approach of this thesis in asking for mandatory global standards and thus eliminating the multi level scenario, strictly speaking, especially this regional mix asking the CJEU about EPO/TRIPS matters. Question 1 involves several issues: (a) is TRIPS an absolute standard?; (b) as a Member of TRIPS, does the EU have a mandate to decided on patentable subject matter; and (c) Greece, as a member of the EPO, cannot neglect its obligations with regard to what may be patentable subject matter. It will be interesting to see how far the Court of Justice will stretch the EU mandate in its ruling. In his opinion of 31 January 2013 Advocate General Villalón emphasised that, as Union law now stands, matters governed by Article 27 of the TRIPS Agreement (patentable subject matter) do not fall within the scope of the 'commercial aspects of intellectual property' within the meaning of Article 207(1) TFEU, with consequences for jurisdiction to interpret that provision. Alternatively, Villalón recommends it should be found that Article 27 does not have direct effect. [81]

The TRIPS Agreement has been subject to immense criticism and proposals for its amendment have been put forward from the start (from TRIPS

81. Opinion of Advocate General Cruz Villalón in case C-414/11 *Daiichi Sankyo and Sanofi-Aventis Deutschland.*

to TRIPS (PLUS) and TRIPS (PLUSPLUS)).[82] These discussions will continue but in the current context of EU competence it is worth noting the WTO Doha Ministerial Declaration in 2001,[83] which is the first (and only) measure of change since 1994.[84] In point 17 of the Declaration it is stressed that public health and access to existing medicine as well as research and development into new medicines is of TRIPS importance. Furthermore, a review of TRIPS Article 27(3)(b) includes an examination of the relationship between the TRIPS Agreement and the Convention of Biological Diversity, with reference to TRIPS Articles 7 and 8. Paragraphs 17 and 19 of the Doha Ministerial Declaration state:

> 17. We stress the importance we attach to implementation and interpretation of the Agreement on Trade-Related Aspects of Intellectual Property Rights (TRIPS Agreement) in a manner supportive of public health, by promoting both access to existing medicines and research and development into new medicines and, in this connection, are adopting a separate declaration

82. See Daniel Gervais, 'TRIPS and Development', in Daniel Gervais (ed.) *Intellectual Property, Trade and Development, Strategies to Optimize Economic Development in the TRIPS-Plus Era*, 2007, pp. 23 and 26. See further, Annette Kur & Marianne Levin (eds.), *Intellectual property Rights in a fair world trade system, proposals for reform of TRIPS*, 2011; and also Annette Kur & Vytautas Mizaras (eds.), *The Structure of Intellectual Property Law, Can One Size Fit All?*, 2011. See also Abeba Tadesse Gebreselassie, *The Sustainability of Plants and Plant Intellectual Property Rights*, 2012, p. 117, quoting Michael Handler, *Intellectual Property, in Bilateral and Regional trade Agreements: Commentary and analysis*, 2009: 'In general, TRIPS-PLUS provisions require the inclusion of additional protectable subject matter, broader and more extensive standards of protection, strong enforcement mechanisms, and a weakening of 'flexibilities' and 'special and differential treatment' granted to developing and least developed countries in the TRIPS Agreement'. See further, Steve Charnovitz, *Mapping the law of WTO accession*, p. 11 ff., available at: http://ssrn.com/abstract=957651.
83. Known as the Doha Developing Agenda, the Ministerial Declaration WT/MIN (01)/DEC/1 20 November 2001 is available at http://www.wto.org/english/thewto_e/minist_e/min01_e/mindecl_e.htm. It followed an earlier UN declaration, the Millennium Declaration, 2000, comprising millennium goals to be reached by 2015 including combating diseases and developing a global partnership for development; see Graeme B. Dinwoodie, William O. Hennessey, Shira Perlmutter & Graeme W. Austin, *International Intellectual Property Law and Policy*, 2nd Ed., 2008, p. 446.
84. See Daniel J. Gervais, 'The Changing Landscape of International Intellectual Property', in Christopher Heath & Anselm Kamperman Sanders (eds.), *Intellectual Property and Free Trade Agreements*, p. 55.

> 19. We instruct the Council for TRIPS, in pursuing its work programme including under the review of Article 27.3(b), the review of the implementation of the TRIPS Agreement under Article 71.1 and the work foreseen pursuant to paragraph 12 of this declaration, to examine, inter alia, the relationship between the TRIPS Agreement and the Convention on Biological Diversity, the protection of traditional knowledge and folklore, and other relevant new developments raised by members pursuant to Article 71.1. In undertaking this work, the TRIPS Council shall be guided by the objectives and principles set out in Articles 7 and 8 of the TRIPS Agreement and shall take fully into account the development dimension.[85]

Paragraph 31 of the Declaration states the importance of enhancing the mutual supportiveness of trade and environment, for example as in the relationship between existing WTO rules and specific trade obligations set out in multilateral environmental agreements (MEAs).

The separate Doha Declaration on the TRIPS Agreement and Public Health repeats the message in paragraph 17 of the Doha Ministerial Declaration. The problems of infrastructure were highlighted in paragraphs 6 and 7, as follows:[86]

85. See now the Nagoya Protocol on Access to Genetic Resources and the Fair and Equitable Sharing of Benefits arising from their Utilization to the Convention on Biological Diversity of the United Nations Convention on Biological Diversity (CBD), adopted 29 October 2010 and open for signature 2 February 2011 to 1 February 2012.
86. Document WT/MIN (01)/DEC/2 available at http://www.wto.org/english/thewto_e/minist_e/min01_e/mindecl_trips_e.htm. See Daniel Gervais, 'TRIPS and Development', in Daniel J. Gervais (ed.), *Intellectual Property, Trade and Development, Strategies to Optimize Economic Development in the TRIPS-Plus Era*, 2007, p. 8 ff.; Frederick M. Abbott & Jerome H. Reichman, 'The DOHA round's public health legacy: strategies for the production and diffusion of patented medicines under the amended TRIPS provisions', *Journal of International Economic Law* 10(4), 921-987, 2007; Thomas Cottier, 'The Doha Waiver and its Effects on the Nature of the TRIPS System and on Competition Law: The Impact of Human Rights', *NCCR Trade Regulation Working Paper* No 2006/21; and Sisule F. Musungu, 'The TRIPS Agreement and Public Health', in Carlos M. Correa & Abdulqawi A. Yusuf (eds.), *Intellectual Property and International Trade, The TRIPS Agreement*, 2nd Ed., 2008, p. 453 ff. According to Thomas Cottier 'the Doha Waiver and the amendment introduce a new dimension to intellectual property and competition law: They reflect concerns for human rights, in particular the right to health and the right to life. I regret that these rights are not explicitly mentioned, not even in preambular terms. But they implicitly render these concerns operational in terms of compulsory licensing. They mark the beginning of a new triangular relationship. It will need further research and development. The right to health and life entail components of distributional justice which intellectual property and competition laws per

Point 6. We recognize that WTO members with insufficient or no manufacturing capacities in the pharmaceutical sector could face difficulties in making effective use of compulsory licensing under the TRIPS Agreement. We instruct the Council for TRIPS to find an expeditious solution to this problem and to report to the General Council before the end of 2002.

7. We reaffirm the commitment of developed-country members to provide incentives to their enterprises and institutions to promote and encourage technology transfer to least-developed country members pursuant to Article 66.2. We also agree that the least-developed country members will not be obliged, with respect to pharmaceutical products, to implement or apply Sections 5 and 7 of Part II of the TRIPS Agreement or to enforce rights provided for under these Sections until 1 January 2016, without prejudice to the right of least-developed country members to seek other extensions of the transition periods as provided for in Article 66.1 of the TRIPS Agreement. We instruct the Council for TRIPS to take the necessary action to give effect to this pursuant to Article 66.1 of the TRIPS Agreement.

The Ministerial Declaration was signed in November 2001, and in Cancun in August 2003 the WTO General Council adopted a temporary Decision on Implementation of paragraph 6 of the Doha Declaration on the TRIPS Agreement and Public Health, the 'Paragraph 6 waiver'.[87] According to its paragraph 2, the Paragraph 6 waiver waives the obligations of an exporting Member under Article 31(f) of the TRIPS Agreement 'with respect to the grant by it of a compulsory license to the extent necessary for the purposes of production of a pharmaceutical product(s) and its export to an eligible importing Member(s) in accordance with the terms set out' in the Decision.[88] The

se should support and assist in realization. A closer examination may reveal that the component is not only of relevance to developing countries, but to industrialized countries alike ... Beyond the Doha Waiver and exception, other rights may become relevant in coming years and should be taken into account as well'; see Thomas Cottier, 'The Doha Waiver and its Effects on the Nature of the TRIPS System and on Competition Law, The impact of human rights', Chapter 7, in Inge Govaere & Hans Ullrich (eds.), *Intellectual Property, Public Policy and International Trade*, 2007, p. 198.

87. Decision of the General Council of 30 August 2003, Document number WT/1/540 available at http://www.wto.org/english/tratop_e/trips_e/implem_para6_e.htm.
88. Paragraph 1 (b) allows any eligible importing WTO Member (any least-developed country Member, and any other Member) 'that has made a notification to the Council of TRIPS of its intention to use the system as an importer, it being understood that a Member may notify at any time that it will use the system in whole or in limited way, for example only in the case of a national emergency or other circumstances of extreme urgency or in cases of public non-commercial use'.

waiver was adopted as a permanent TRIPS amendment (Article 31 *bis*) (Cancun Declaration) in December 2005.

The waiver was implemented in the EU legal order by Regulation (EC) No 816/2006 of the European Parliament and of the Council of 17 May 2006 on compulsory licensing of patents relating to the manufacture of pharmaceutical products for export to countries with public health problems.[89] However, it has not yet been incorporated in the TRIPS Agreement (as a new Article 31 bis)).[90] The decision on this amendment of 5 December 2005 requires the support of two-thirds of WTO Members, and so far only few Members have supported it.[91] Thus, the EU has accepted as a valid norm in the regional EU legal order an international obligation that has not yet entered into force at the international level and is recognised to be at stand still, a moratorium.[92]

89. See Duncan Matthews, 'Lessons from negotiating an Amendment to the TRIPS Agreement: Compulsory licensing and access to medicine', in Guido Weatkamp (ed.), *Emerging Issues in Intellectual Property, Trade, Technology and Market Freedom: Essays in Honour of Herchel Smith*, page 231.
90. The TRIPS Council met in March 2011 (second meeting after the annual review according to Paragraph 8 of the 2003 declaration on the temporary waiver) to discuss the waiver; see Catherine Saez, 'Patentable Subject Matter, IP Waiver for health Discussed at WTO', *Intellectual Property Watch*, March 2011, http://www.ip-watch.org/
91. Two-thirds of the WTO membership (102 Members out of 153 Members) must ratify the change for it to go into effect. So far only 39 appear to have done so; see William New, *Intellectual Property Watch*, 1 December 2011. On 30 November 2011 the WTO General Council again agreed to extend the deadline for countries to adopt the amendment at national level. The deadline is now 31 December 2011; see http://www.wto.org/english/tratop_e/trips_e/amendment_e.htm (last visited January 2012). See Rochelle Dreyfuss, 'TRIPS and Essential Medicine', in Thomas Pogge, Matthew Rimmer & Kim Rubenstein (eds.), *Incentives for Global Public Health, Patent Law and access to Essential Medicines*, p. 48: 'Change in the Agreement would require a vote of the TRIPS Council, followed by a vote of the General Council. Since, under current procedures, both Councils operate by consensus, states interested in maximum protection could veto proposals to relax the standards.' See Jens Schovsbo & Morten Rosenmeier, *Immaterialret*, 2nd Ed., 2011, p. 331.
92. Frederick M. Abbott, 'Worst Fears Realised: The Dutch Confiscation of Medicines Bound from India to Brazil', *Comment – Bridges*, International Centre for Trade and Sustainable Development, No 1, February – March 2009, www.ictsd.net p. 13, on Council Regulation (EC) No 1383/2003 concerning customs actions against good suspected of infringing certain intellectual property rights, and the interplay with responsible global trade and public health policy: 'In light of the critical report recently issued by the European Competition Directorate regarding originator pharmaceutical

### 3.2.3. EPO

The European Patent Organisation (EPO) was established in 1977 based on the European Patent Convention (EPC) of 1973.[93] The Convention was amended in 2000 (EPC 2000), and supplemented by the London Protocol on the application of Article 65 EPC, comprising an optional agreement aimed at reducing the costs for translating European patents.[94] In a way, the EPC is a compromise as the political landscape at the time of the adoption of the European Patent Convention did not allow for a unitary European Community patent. Thus it was decided to follow the European patent route.[95] A Europe-

industry patenting practices, it is paradoxical that the EU is taking extraordinary steps to extend the effect of those patents so as to block legitimate trade among developing countries. But this is a long-standing paradox of the European Union. EU authorities take a hard line against pharmaceutical industry abuse within the Union, but apply entirely different standards to external trade. The European Parliament, apparently to little avail, has been trying to alleviate this internal/external dichotomy-in-standards. The recent actions directly contradict sentiments expressed by the Parliament.' See e.g. European Parliament resolution of 12 July 2007 on the TRIPS Agreement and access to medicines, PA_TA(2007)0353, at paras. 8 and 16; and recommendation of the Committee on International Trade to the European Parliament on the proposal for a Council decision on the acceptance of the Protocol amending the TRIPS Agreement, done at Geneva on 6 December 2005, PE 390.562v02-00, 22 Oct. 2007. See further, Frederick M. Abbott, 'Seizure of Generic Pharmaceuticals in Transit Based on Allegations of Patent Infringement: A Threat to International Trade, Development and Public Welfare', *W.I.P.O.J.* No 1, 2009, p. 49.

93. The European Patent Organisation is an intergovernmental organisation established on 7 October 1977 on the basis of the European Patent Convention (EPC), signed in Munich in 1973; see Singer & Stauder, *Europäisches Patentübereinkommen, kommentar*, 5th Ed., 2010, p. 20 ff.
94. See http://www.epo.org/law-practice/legal-texts/london-agreement.html. The EPC system also comprises the EPC Implementing Regulations, latest amended 1 January 2011, and protocols which are integral parts of the EPC (Protocol on the Interpretation of Article 69 EPC, Protocol on Centralisation, Protocol on Recognition, Protocol on Privileges and Immunities, Protocol on the Staff Complement). See Catherine Seville, *EU Intellectual Property Law and Policy*, 2009, p. 143; and Rob J. Aerts, 'The legitimacy of patent law-making in Europe and the U.S. – A tentative Comparison', 2/2007 *IIC* Vol. 38, p. 168 (EPC 2000).
95. Convention for the Grant of European patents (Munich, 5 October 1973) and later, the Convention for the European Patent for the Common Market (Luxembourg, 15 December 1975). However, the latter never entered into force and it was followed by the 1989 Luxembourg agreement relating to Community patents, including a protocol on the settlement of litigation concerning infringement and validity of Community Patents (89/695/EEC: Agreement relating to Community patents). This agreement never entered into force either. The purpose of the different systems has been described as follows: 'The Community patent (introduced by the Community Patent

an patent is a patent that is granted by the European Patent Office and it is important to distinguish between the European Patent Organisation and the European Union. EPC Article 1 establishes a substantive and procedural European law for the grant of patents as a 'system of law, common to the Contracting States, for the grant of patents for invention'. In their commentary on the European Patent Convention, *Singer/Stauder* write, that:

> 'EPC law is uniform international law applicable both to the activity of the EPO and in the Contracting States. Together with the Implementing Regulations (Article 164 (1)) the Convention almost exclusively embodies for the grant phase (pre-grant and grant) including opposition, 'a system of law that is both independent of national legal systems and autonomous'. The fact of incorporation of the text of the EPC into a national law ... does not deprive the unified law of its international character.' [96]

The Convention established a unique transnational patent process creating the European patent. When granted a European patent is in fact a bundle of European patents if more than one Contracting State is designated, and it remains so.[97] The European Patent Convention provides a process for the prosecution of patents (the grant phase) as well as opposition to grant,[98] whereas the post-grant phase is mostly an exclusively national matter.[99]

Convention) is intended to bring together the bundle of protection rights resulting from the grant of a European patent and merge them into a single, unitary and autonomous, protection right valid throughout the Community of Twelve and governed only by the provisions of the 1989 Agreement relating to Community patents. However, this Convention has yet to take effect owing to delays in ratification by the twelve signatory Member States. The aims of the European Patent Convention and the Agreement relating to Community patents differ but complement each other. The European Patent Convention is intended to rationalize the grant of patents by establishing a centralized procedure managed by the European Patent Office in Munich. It is thus open to accession by any European State, in some cases at the invitation of the Administrative Council of the European Patent Organization. The Agreement relating to Community patents is designed to contribute to the achievement of the aims of the single market, with special reference to fair competition and the free movement of goods' (COM(1997) 314 final, para. 2.1).

96. See Singer/Stauder, *The European Patent Convention, A Commentary*, 3rd Ed., Vol. 1, Article 1, points 5 & 6.
97. See Singer/Stauder, *The European Patent Convention, A Commentary*, 3rd Ed., Vol. 1, Article 2, point 6.
98. EPC 2000, Part V, Article 99 ff.
99. E.g., when possible the English Courts have sought to align their interpretations of patent laws with the approach taken by the European Patent Office, but there have always been some inconsistency; see Patrick Reimer & Jonathan Radcliffe, *Patent World*, Issue # 211, April 2009, p. 26.

It is important to repeat that the European Patent Organisation and the European Union are two separate bodies. The European Patent Organisation grown over time and today has 38 Members.[100] Some of the Members are also Member States of the European Union (in fact all EU Member States are EPO Members). The others include the non-EU Members Switzerland, Albania, Liechtenstein, Monaco, Serbia, Norway, and San Marino as well as the candidate EU countries Croatia, Iceland, Turkey, and Macedonia.

Thus, the EPO is a centralised system administered by a new regional institution – the European Patent Office[101] – operating on behalf of 38 Contracting States. Since, the EPO establishes the only European regional granting authority it will be discussed in more detailed below. Thus it differs from both WIPO and TRIPS which have no granting authority.[102]

100. See http://www.epo.org/about-us/organisation/member-states.html; these are Albania (May 2010), Austria (May 1979), Belgium (October 1977), Bulgaria (July 2002), Croatia (January 2008), Cyprus (April 1998), Czech Republic (July 2002), Denmark (January 1990), Estonia (July 2002), Finland (March 1996), France (October 1977), Germany (October 1977), Greece (October 1986), Hungary (January 2003), Iceland (November 2004), Ireland (August 1992), Italy (December 1978), Latvia (July 2005), Liechtenstein (April 1980), Lithuania (December 2004), Luxembourg (October 1977), Former Yugoslav Republic of Macedonia (January 2009), Malta (March 2007), Monaco (December 1991), Netherlands (October 1977), Norway (January 2008), Poland (March 2004), Portugal (January 1992), Romania (March 2003), San Marino (July 2009), Serbia (October 2010), Slovakia (July 2002), Slovenia (December 2002), Spain (October 1986), Sweden (May 1978), Switzerland (October 1977), Turkey (November 2000), and United Kingdom (October 1977).

101. The European Patent Office is supervised by the Administrative Council; the Office is the executive arm of the European Patent Organisation; see http://www.epo.org/about-us/office.html. See Singer/Stauder, *The European Patent Convention, A Commentary*, 3rd. Ed., Vol. 1, Articles 3 and 4, point 9 ff.; and Singer & Stauder, *Europäisches Patentübereinkommen, kommentar*, 5th Ed., 2010, pp. 21-22.

102. In the EPO Enlarged Board of Appeals decision in G 2/02 and G3/02, 26 April 2004, *India/AstraZeneca* it was stated that the obligations under the TRIPS Agreement do not directly bind the European Patent Organisation or the European Patent Office, they only bind Members of the WTO.

## 4. The institutional framework of patent authority – reverse developments

### 4.1. The US framework

The US Constitutional Convention of 1789 resolved to establish a US patent system rooted in the Constitution itself granting authority to Congress to: 'promote the Progress of Science and useful Arts, by securing for limited Times to Authors and Inventors the exclusive Rights to their respective Writings and Discoveries'.[103]

The first patent statutes were drafted in 1790 and 1793 and one of the key players was Thomas Jefferson. As Secretary of State his views on the implementation of the statutes were expressed in letters which the Supreme Court subsequently used as means of interpretation.[104] An amendment in 1836 introduced a formal system of examination, replacing the former system of registration.[105] However, since 1836 the development of US patent legislation has been relatively static. The first major revision occurred with the 1952 Patent Act which in fact restated the basic 1790 principles.[106] 56 years later, under the Bush administration, the 2008 Patent Law Reform was expected to make major changes to US patent law, though it was unlikely that such changes would be adopted within the lifetime of the Bush administration. On 23 June 2011, under the Obama administration, the Leahy-Smith America Invents Act was passed by the House of Representatives (with 407 votes to 117). From an international perspective the major change is the transition

103. US Constitution, Article 1, sec. 8, clause 8. See Robert Patrick Merges & John Fitzgerald Duffy, *Patent Law and Policy: Cases and Materials*, 5th Ed., 2011, p. 7; and Edward C. Walterscheid, 'Patents and the Jeffersonian Mythology' 29, *J. Marshall L. Rev.* 269 (1995), p. 271.
104. See Robert Patrick Merges & John Fitzgerald Duffy, *Patent Law and Policy: Cases and Materials*, 5th Ed., 2011, p. 8 and footnote 24, referring to Walterscheid's work on Jefferson questioning this 'godfather myth' as Jefferson would have been more likely to reject the notion of government-created monopolies; see Edward C. Walterscheid, 'Patents and the Jeffersonian Mythology', 29, *J. Marshall L. Rev.* 269 (1995), pp. 269-270, the latter quoting the Supreme Court: 'because of his active interest and influence in the early development of the patent system, Jefferson's views on the general nature of the limited patent monopoly under the Constitution as well as his conclusions as to conditions for patentability under the statutory scheme, are worthy of note', *Graham,* 383 U.S. at 8.
105. See Robert Patrick Merges & John Fitzgerald Duffy, *Patent Law and Policy: Cases and Materials*, 5th Ed., 2011, pp. 8-9.
106. Robert Patrick Merges & John Fitzgerald Duffy, *Patent Law and Policy: Cases and Materials*, 5th Ed., 2011, p. 10.

from a first-to-invent system to a first-to-file system.[107] However, the amendment of the grace period from a full grace period to a great shrinkage in combination with the first-to-file amendment has been criticised.[108] The US system does not operate with an opposition procedure corresponding to that of the EPO, but the Leahy-Smith Act (House bill HR 1249) introduced a Chapter 32 on post-grant review.[109]

In HR 1249 Section 33, headed 'Limitation on issuance of patents', in paragraph (a) it is explicitly stated that: 'Notwithstanding any other provision of law, no patent may issue on a claim directed to or encompassing a human organism'. At the same time as the Leahy-Smith Act was enacted, the administration announced additional initiatives to move ideas from the laboratory to

107. The Leahy-Smith America Invents Act (House bill (HR. 1249) and as passed by the Senate (S.23)). Two versions of the patent reform act amending 35 U.S.C. The America Invents Act (AIA, HR 1249) were enacted in law on 16 September 2011, according to Section 35 of the Act, it is to take effect 'upon expiration of the 1-year period beginning on the date of the enactment of this Act and shall apply to any patent issued on or after that effective date', except as otherwise provided in the Act. The effective date for the human organism prohibition is 16 September 2011 (AIA section 33); the effective date for the genetic testing report is 16 June 2012 (AIA Section 27); the effective date for the priority examination for important technologies is 16 September 2012 (AIA section 25); and the effective date for the first-inventor-to-file procedure is 16 March 2013 (AIA Section 3).

108. See Interview With Chief Judge Paul R. Michel, On US Patent Reform, 14 July 2011 in *Intellectual Property Watch*. For an earlier comment on the bill with regard to biotech innovation, see 'An Audience with Arti Ray', Nature, April 2007, Vol. 6, p. 264. On the Act in general, see Toshiko Takenaka, 'Harmony with the rest of the world? The America invents Act', *Journal of Intellectual Property Law and Practice Advance*, 11 November 2011, available at http://ssrn.com/abstract=1940008 .

109. The Patent Reform Act of 2011 makes changes to improve patent quality. It has a first window post-grant opposition proceeding, open for nine months after the grant of a patent, and it improves the current system for making administrative challenges to the validity of a patent. However, it may be that the major legal impediment to filing a post-grant review is the tough new threshold that requires it to be shown 'that there is a reasonable likelihood that the petitioner will prevail with respect to at least one claim challenged.' If the threshold is not met, the petition for review will be rejected. This threshold is substantially higher than the USPTO's historic standard requiring a 'substantial new question of patentability.' The EPO does not have a similar threshold; see Dennis Crouch, 'Reviewing the New USPTO Post Grant Review System with Reference to EPO Oppositions', 6 January 2012, http://www.patentlyo.com/patent/2012/01/uspto-post-rant.html?utm_source=feedburner&utm_medium=email&utm_campaign=Feed%3A+PatentlyO+%28Dennis+Crouch%27s+Patently-O%29 (last visited January 8). For more on patent offices see, e.g. Peter Drahos, *The Global Governance of Knowledge, Patent Offices and their Clients*, 2010, p. 148.

the market, e.g. the launch of new National Institute of Health (NIH) centre to assist biotech entrepreneurs, and the development of a National Bioeconomy Blueprint by January 2012 to detail administration-wide steps to harness biological research innovations to address national challenges in the areas of health, food, energy and the environment.[110]

There have been major structural changes since the 1952 Patent Act. In 1982 Congress passed legislation establishing the Federal Circuit, highlighting the fact the US patent law is federal law. Thus all appeals involving patents from the lower federal district courts and final rejections of patent applications by PTO concerning patentable subject matter are decided by the Federal Circuit.[111] The position of the Federal Circuit ensures that there is a powerful voice on the interpretation and administration of US patent law.[112] In the literature, *Mergers & Duffy* state that the Federal Circuit has enhanced the stature of the patent system and that it is to be considered a watershed in the history of the US patent system.[113] Most remarkably, initially (from 1982-2002) the Supreme Court only very reluctantly intervened in Federal Circuit decisions. Today the Supreme Court has resumed its traditional role in polic-

110. William New, 'US Patent Reform signed into law', *Intellectual Property Watch*, 16 September 2011, http://www.ip-watch.org/. Joe Matal, 'A guide to the legislative history of the American Invents Act (Part I & II), 21, *Fed.Cir. B.J.* 435, 539 (2012).
111. The decisions of the PTO are reviewed by an internal review board, the Federal Circuit and finally the Supreme Court; see 35 USC § 141. Thus the Federal Circuit has power to determine issues of prosecution as well as litigation. See, e.g. Janice M. Mueller, *Patent Law*, 3rd Ed., p. 35 ff.; Chapter 1, Adjudicatory Forums for Patent Matters.
112. See Kenneth J. Burchfiel, *Biotechnology and the Federal Circuit*, 2nd Ed., 2010, p. 4 ff.: 'The Federal Circuit has diverse jurisdiction over administrative appeals and specialized courts', p. 4, footnotes 4 & 5; and p. 5: 'this has made patent law the most important area of specialized federal appellate jurisdiction in U.S. history'. See further Robert Patrick Merges & John Fitzgerald Duffy, *Patent Law and Policy: Cases and Materials*, 5th Ed., 2011, p. 10: 'while the Federal Circuit was ostensibly formed strictly to unify patent doctrine, it was no doubt hoped by some (and expected by others) that the new court would make subtle alterations in the doctrinal fabric, with an eye towards enhancing the stature of the patent system'. The authors conclude that this was exactly what happened.
113. Robert Patrick Merges & John Fitzgerald Duffy, *Patent Law and Policy: Cases and Materials*, 5th Ed., 2011, p. 11 and footnote 36, referring to the granting of certiorari to review seven cases, each of which presented important issues of patent law and policy. See further Christopher M. Holman, 'Bilski: assessing the Impact of a Newly Invigorated Patent-Eligibility Doctrine on the Pharmaceutical Industry and the Future of Personalized Medicine', available at http://ssrn.com/abstract=1424493 (last visited January 2012), p. 14.

ing patent law and policy; it has recently decided in three cases (*Bilsky v Kappos* and *Prometheus Labs. v Mayo Collab. Servs.* and the *Myriad* case (*Association for Molecular Pathology (AMP) v USPTO*).[114] Patent law development in the USA clearly has the advantage of institutional coherence, which is a most vulnerable point of European patent law.

The USPTO is active in the framing of the application of patent legislation by issuing guidelines such as the updated 2001 guidelines for the examination of applications for compliance with the 'utility' requirement (35 U.S.C. 101) and the 'written description' requirement (35 U.S.C. 112).[115] The USPTO has also updated the examination guidelines on the core patentability issue of obviousness.[116] The new guidelines primarily focus on the Federal Circuit opinion that interprets and implement the reasoning in *KSR v Teleflex*.[117]

*Conley & Makowski* write that the USPTO saw the utility requirement as a way to control the flood of gene and gene-related patents and tightened its interim utility guidelines in 2001 after having invited comments in December 1999.[118] Facing the combination of early upstream patenting and a broad util-

114. Chapters 5 and 10. *Prometheus Labs. v Mayo Collab. Servs.*, 628 F.3d 1347 (Fed. Cir. 2010), *cert. granted sub nom. Mayo Collab. Servs. v Prometheus Labs*, 131 S.Ct. 3027 (2011); *Ass'n for Molecular Pathology v USPTO*, 653 F.3d 1329 (Fed. Cir. 2011); and *re Bilski,* 545 F.3d 943 (2008). Cert. granted sub nom., *Bilsky v Doll*, 129 S. Ct 2735 (2009). *Bilsky v Kappos*, 561 U.S., 130 S.Ct. 3218, 2010. In its 2010 *Bilski* decision, the Supreme Court stated that the machine-or-transformation test only offers an important clue. Finally, *Association for Molecular Pathology v USPTO*, 702 F. Supp. 181 (S.D.N.Y. 2010), *AMP v USPTO*, Fed. Cir. 2011, 2010-1406, decided July 2011. See Robert Patrick Merges & John Fitzgerald Duffy, *Patent Law and Policy: Cases and Materials*, 5th Ed., 2011, p. 12 and footnote 37, and p. 170 ff.

115. See Lisa A. Karczewski, 'Biotechnological gene patent applications: the implications of the USPTO written description requirement guidelines on the biotechnology industry', 31 *McGeorge Law Review*, 2000, 1043; and Daniel P. Chisholm, 'Note: The effect of the USPTO's written Description Guideline on Gene Patent applications', 35 *Suffolk University Law Review*, 543.

116. See 53644, Federal Register, Vol. 75, No 169, Wednesday, 1 September 2010, Notices, see further Manual of Patent Examining Procedure (MPEP), MPEP http://www.uspto.gov/web/offices/pac/mpep/mpep.htm.

117. See *KSR v Teleflex* 550 U.S. 398 (2007). See Kenneth J. Burchfiel, *Biotechnology and the Federal Circuit*, 2nd Ed., 2010, p. 159 ff.; and Robert Patrick Merges & John Fitzgerald Duffy, *Patent Law and Policy: Cases and Materials*, 5th Ed., 2011, p. 692 ff.

118. See the revised utility examination guidelines, request for comments, 64 Fed. Reg. 71440 (Dec. 21, 1999) and discussion of Public comments (1092, Federal Register,

ity requirement, traditional economic theories supporting patent protection no longer adequately legitimised the protection of basic research tools.[119]

The guidelines entered into force by 5 January 2001. The legal status of the guidelines is described as purely governing internal practices and according to Section II A, Introduction, of the guidelines, they do not alter the substantive requirements of 35 U.S.C. 101 and 112, nor are they designed to obviate the examiner's review of applications for compliance with all other statutory requirements for patentability. It follows from the wording *guidelines* they do not constitute substantive rulemaking and hence do not have the force and effect of law.[120]

Section B contains guidelines on utility. An invention has a well-established utility: (1) if a person of ordinary skill in the art would immediately appreciate why the invention is useful based on the characteristics of the invention (e.g. properties or applications of a product or process); and (2) the utility is *specific, substantial, and credible*. The requirement for utility to be *specific, substantial, and credible* is further elaborated in the guidelines, point B.2 and the courts have used the terms 'practical utility' and 'real world utility' interchangeably in determining whether an invention offers 'substantial' utility.[121] The utility

vol. 66, no. 4, January 5, 2001, Notices and the Manual of Patent Examining Procedure (MPEP), Eighth Edition, August 2001, Revision 8, July 2010. http://www.uspto.gov/web/offices/pac/mpep/documents/2100_2107_01.htm. See further, John M. Conley & Roberte Makowski, 'Back to the future: rethinking the Product of Nature Doctrine as a barrier to Biotechnology Patents', 85 *Journal of the Patent and Trademark Office Society* 301 (2003), http://www.whoownsyourbody.org/conley_article.pdf p. 66.

119. Teresa M. Summers, 'The Scope of Utility in the Twenty-first Century: New Guidance for Gene-related Patents', 91 *Georgetown Law Journal*, January 2003, p. 485; Mary Breen Smith, 'An end to Gene Patents? The Human Genome Project versus the United States Patent and Trademark Office's 1999 Utility Guidelines', 73 *University of Colorado Law Review*, 2002, p. 747 ff.; 'The fate of gene patents under the new utility guidelines', 2001, *Duke Law & Technology Review*, 28 February 2001, 8; and Tine Sommer, 'Patentering af det Humane Genom', *UfR* 2004 B 219.
120. Furthermore the guidelines state that any perceived failure by Office personnel to follow the guidelines is neither appealable nor petitionable, 1098 Federal Register, Vol. 66, No 4, Friday, 5 January 2001, Notices. See Robert Patrick Merges & John Fitzgerald Duffy, *Patent Law and Policy: Cases and Materials*, 5th Ed., 2011, p. 243; and Kenneth J. Burchfiel, *Biotechnology and the Federal Circuit*, 2010, 2nd Ed., p. 88. See Cherylyn A. P. Esoy, 'The PTO's 2001 revised utility examination guidelines for gene patent applications: has the PTO exceeded the scope of authority delineated by the Court's interpretation of a 'useful' invention?', 33 *Seton Hall Law Review*, 2002, 127.
121. Robert Patrick Merges & John Fitzgerald Duffy, *Patent Law and Policy: Cases and Materials*, 5th Ed., 2011, p. 248. It follows from statement in *Nelson v Bowler*, 626

guidelines can be seen as a response to the NIH's attempt to patent expressed sequence tags (ESTs), which are short stretches of sequences from randomly selected complementary DNA (cDNAs) that identify bits of the coding region of genes, without identifying the function of the gene segment.[122] One of the public comments addressed the question, see comment 6 (page 1094), other comments (7) stated that patents should not be issued for genes, or (8) that DNA should be considered unpatentable because a DNA sequence by itself has little utility. Comment (10) stated that the scope of patent claims directed to DNA should be limited to applications or methods of using DNA and should not encompass the DNA itself. According to comment 11: 'A patent on a composition gives *exclusive* rights to the composition for a limited time, even if the inventor disclosed only a single use for the composition. Thus, a patent granted on an isolated and purified DNA composition confers the right to exclude others from *any* method of using that DNA composition, for up to 20 years from the filing date. This result flows from the language of the statute itself. When the utility requirement and other requirements are satisfied by the application, a patent granted provides a patentee with the right to exclude others from, *inter alia,* 'using' the patented composition of matter. *See* 35 U.S.C. 154. Where a new use is discovered for a patented DNA composition, that new use may qualify for its own process patent, notwithstanding that the DNA composition itself is patented'. Finally, comment 13 stated that: 'A DNA sequence is not patentable because a sequence is merely descriptive information about a molecule. An isolated and purified DNA molecule may be patentable because a molecule is a 'composition of matter', one of the four classes of invention authorized by 35 U.S.C. 101. A DNA molecule must be *nonobvious* in order to be patentable. Obviousness does not depend on the amount of work required to characterize the DNA molecule.[123] *See* 35 U.S.C. 103(a) ('Patentability shall not be negatived by the manner in which the invention was made').' In *re Fisher* the Federal Circuit affirmed that ESTs are unpatentable under 35 U.S.C. Section 101.[124] The patent concerned 5 purified nucleic acid sequences encoding proteins and protein fragments in maize plants, but where the precise structure or function of the genes and proteins they encoded was unknown. The Federal Circuit held that the application must disclose an invention that is useful to the public currently and not merely at some future date after further research, and that the utility must be significant

F.2d, 853, 856 (CCPA 1980) that 'practical utility is a shorthand way of attributing 'real-world' value to claimed subject matter. In other words, one skilled in the art can use a claimed discovery in a manner which provides some immediate benefit to the public'.

122. Kenneth J. Burchfiel, *Biotechnology and the Federal Circuit*, 2nd Ed., 2010, p. 78.

123. As the non-obviousness requirement has been interpreted by the US Court of Appeals for the Federal Circuit, whether a claimed DNA molecule would have been obvious depends on whether a molecule having the particular *structure* of the DNA would have been obvious to a person of ordinary skill in the art at the time the invention was made. See, e.g., *In re Deuel,* 51 F.3d 1552, 1559, 34 USPQ2d 1210, 1215 (Fed. Cir. 1995) ('[T]he existence of a general method of isolating cDNA or DNA molecules is essentially irrelevant to the question whether the specific molecules themselves would have been obvious.'); see also, *In re Bell,* 991 F.2d 781, 26 USPQ2d 1529 (Fed. Cir. 1993).

124. 421 F.3d 1365, 76 USPQ2d 1225 (Fed.Cir.2005).

and presently available.[125] Thus the analogy to *Brenner v Manson* (no disclosure of how to use the end product)[126] was relied on. There must be an immediate well-defined, real world benefit to the public which merits protection. Thus, the Federal Circuit and the Supreme Court, together with the USPTO's guidelines, have been at the continuous adaption of US Patent law to technological changes.

## 4.2. The European framework

Without focusing on the global harmonisation of patent law, such as the negotiations on the TRIPS Agreement and the draft Substantive Patent Law Treaty, which concerns Europe as well as the USA,[127] Europe has its own battlefield which is crowded with far too many playmakers as the regional institutional framework has several levels.

Patent law has emerged differently in the different territories of the European countries. Italy protected the arts of Venice and Florence as early as in the fourteenth century, and other countries started to create protection; in England in the early seventeenth century the crown granted privileges (by letters patent); Denmark had its first patent statute in 1894 but before that, privileges were granted as early as 1671.[128] Other European countries, for example Switzerland and the Netherlands were more reluctant and gave no patent

125. See also Robert Patrick Merges & John Fitzgerald Duffy, *Patent Law and Policy: Cases and Materials*, 5th Ed., 2011, p. 248; and Robert Fitt & Edward Nodder, 'Specific, Substantial and Credible? A New Test for Gene Patents', *Bio-Science Law Review*, Vol. 9, issue 5, p. 187. The utility requirement is further analysed in Chapter 6, section 5.
126. *Brenner v Manson*, 383 U.S. 519, 86 S.Ct. 1033, 16 L.Ed.2d 69 (1966).
127. See also the Cooperative Patent Classification project (CPC), launched in 2011 with the aim of harmonising the patent classification systems of the EPO and the USPTO. The US and European patent offices are not the only stakeholders at a global level, there are also regional stakeholders in the African and Asia-Pacific areas, and there are further questions such as the establishment of a World IP litigation court or a truly global patent system initiated by IP5 (a cooperation between the EPO with the patent offices of Japan, Korea, China and the USA). See further the WTO Public Forum e.g. September 2011, 'Seeking answers to global trade challenges': http://www.wto.org/english/forums_e/public_forum11_e/public_forum11_e.htm (last visited January 2012).
128. See EPO, 'Scenarios for the future, how might IP regimes evolve by 2025? What global legitimacy might such regimes have?', 2007, p. 15 ff. See Jeremy Phillips, 'The English Patent as a Reward for Invention, The Importation of an Idea', 1983, 2 *EIPR* 41 ff. As for the Danish development see Leo A. Damm, *Opfinderrettens genstand*, 1937; Jens Schovsbo & Morten Rosenmeier, *Immaterialret*, 2nd Ed., 2011, p. 214; Jens Schovsbo, 'Om pandaens tommeltot, gamle dage, Kina men – især – patentretten', in Per Andersen, Pia Letto-Vanamo, Kjell Åke Modéer & Helle Vogt (eds.), *Liber Amicorum Ditlev Tamm, Law, History and Culture*, p. 629 ff.

protection. Later, by 1963 the Nordic Countries (Denmark, Norway, and Sweden) harmonised their procedural and substantive patent laws.[129]

From 1958 to 2008 the European Union developed by a step-by-step approach with several treaty amendments not only increasing the numbers of Member States from 7 to 27, but also, as mentioned above, expanding the areas over which the EU has gained legislative powers.[130] As mentioned above in section 2, from the first decade of the establishment of the European Economic Community, the Court of Justice became a surprisingly dynamic force for expanding the Community's powers. By interpreting the Treaty provisions in unforeseeable ways, the Court created a more far-reaching Community than could have been expected by its founders. This situation even characterises the Union today. Somehow patent protection seems to have slipped through the net, and judicial review and the establishment of a patent court has been a problem child in the political endeavours to establish an EU patent system. Ever since the Rome Treaty in 1958 there have been discussions about a project to establish an EU patent as a unitary right. The EU Commission has published communications, reports, and proposed several drafts for legislation for an EU patent. Today the solution to the EU patent problem has almost been narrowed down to a question of the location of the patent court,[131] and unfortunately the EU patent seems to have been rushed through.[132] The road has been full of obstacles, none of which seems to have been taken seriously.[133]

129. Fellesnordisk betænkning NU 1963:6.

130. As for the enlargements, see e.g. Josephine Steiner & Lorna Woods, *EU Law*, 10th Ed., p. 4 ff.; and for the development of European integration, see Paul Craig & Gráinne de Búrca, *EU law, Text, Cases and Materials*, 5th Ed., 2011, p. 1 ff. The treaty amendments are as follows: the Single European Act (1987), the Maastricht Treaty (1993), the Amsterdam Treaty (1999), the Nice Treaty (2003) and the Lisbon Treaty (2009).

131. The court location of Paris appears to be backed by the President of the EU Council, Herman Van Rompuy, with specialised clusters in Munich and London; http://blog.ksnh.eu/en/category/eu-law/eu-unified-patent-court/ (last visited June 2012). The Unified Patent Court system will consist of a Court of First Instance with a central division (with its seat in Paris and two sections in London and Munich) and several local and regional divisions. The Court of Appeal will be located in Luxembourg.

132. A date for signature by the end of June 2012 and still pressing reservations; industry, user associations and the academic world are asking for attention.

133. First Community Patent Convention (OJ 1976 L 17/1); Second Community Patent Convention (Agreement relating to the Community patent, including a protocol on

The 2008 Communication from the Commission to the European Parliament, the Council and the European Economic and Social Committee: *An Industrial Property Rights Strategy for Europe*, introduces a 'fifth freedom' to the Union system of freedoms – that of the free movement of *knowledge*. The Commission states that a clear regime for intellectual property rights is an essential condition for the single market which may also contribute as part

the settlement of litigation) (OJ 1989 L 401/1); Proposal for a Council regulation on the Community Patent (COM(2000) 412 final), (OJ 2000 C 377E/278), revised 6 times in 2003 (Doc. Nos 15086/03, 14233/03, 12219/03, 10728/03, 10404/03 (PI 53), 8535/03); Council document 6874/04, 11 March 2004. Interinstitutional file 2000/0177(CNS)(7199/04); proposal for a Council Regulation on the Community Patent, revised text of October 2009, Interinstitutional file 2000/0177 (CNS)(15149/09), and December Council document 16113/09, Add 1. See also proposals for a Council decision conferring jurisdiction on the Court of Justice in disputes relating to the Community patent and concerning appeals before the Court of First Instance (COM(2003) 827 & 828 final); a unified patent litigation system in small business act (COM(2008) 394 final); first suggested in a Commission Communication of April 2007 (IP/07/463). As for enhanced cooperation, see COM(2010) 790 final and COM(2011) 215 final, and Council Decision 2011/167/EU authorising enhanced cooperation in the area of the creation of unitary patent protection (OJ 2011 L 76/53); Draft report on the proposal for a Council regulation implementing enhanced cooperation in the area of the creation of unitary patent protection with regard to applicable translation arrangements (COM(2011) 216 final), (2011/0094 (CNS); and Proposal for a Council Regulation (EU) on the translation arrangements for the European Union patent (COM(2010) 350 final). As for strategy papers, see Commission working document on the planned Community patent jurisdiction (COM(2002) 480 final); Communication from the Commission: Enhancing the Patent System in Europe (COM(2007) 167 final); the Green Paper on innovation (COM(1995) 688 final); the Green Paper on the Community patent and the patent system in Europe (COM(1997) 314 final); and the First Action Plan for Innovation in Europe – Innovation for growth and employment (COM(1996) 589 final). See more generally Commission Communication from the Council, Europe 2012, A strategy for smart, sustainable and inclusive growth (COM(2010) 2020 final). For the literature, see e.g. Sir Robin Jacob, 'Creating the community patent and its courts', in D. Vaver & L. Bently, *Intellectual Property in the new millennium, Essays in Honour of William R. Cornish*, 2004, p. 79 ff.; Jürgen Schade, 'Synergies created by international cooperation in the patent area?', in Wolrad Prinz zu Waldeck und Pyrmont et al. (eds.), *Patents and technological Progress in a Globalized World, Liber Amicorum Joseph Straus*, p. 619 f.; Michael Schneider, 'Patents in Europe and their Courts – is there light at the end of the tunnel?', in Wolrad Prinz zu Waldeck und Pyrmont et al. (eds.), *Patents and technological Progress in a Globalized World, Liber Amicorum Joseph Straus*, p. 640 at note 30 with further references to literature on an EU patent court. See also Jürgen Schade, 'Is the Community (EU) patent behind the times? – globalization urges multilateral cooperation', *IIC* 7/2010, p. 806 ff.; and Thomas Jaeger, 'The EU Patent: Cui bono et quo vadit?', *Common Market Law Review* 47: 63-115, 2010. See further Stefan Luginbuehl, *European Patent Law, Towards a Uniform Interpretation*, 2011.

of a wider policy to finding solutions that could address global issues of increasing significance such as climate change.[134] The 2010 Communication from the Commission: EUROPE 2020, a strategy for smart, sustainable and inclusive growth launched a flagship initiative: 'Innovation Union' with the aim of creating a single EU Patent and a specialised Patent Court.[135] However, in 2012 the major hurdle still remains the continued lack of a unitary patent system despite the major progress in 2011.[136] Already in 1975 the first initiative was taken for an EC patent convention, and in 1989 a second convention text was attempted. In 1997, the Commission launched a Green Paper on promoting innovation containing a proposal for a Community Patent.[137] In 2000, an EC patent regulation was proposed and in 2004 a second regulation was proposed. In January 2006 the Commission launched a public consultation on an EU-wide system of protection taking account of stakeholders' needs.[138] In 2007 another strategy paper was published for an EU patent and a patent court (jurisdiction for European and Community patents), but without success.[139] Finally, in October and December 2009 the EU Council reached agreement on further revised proposal for a Council Regulation on the Community patent, now called the EU patent under the Lisbon Treaty.[140] With the Lisbon Treaty came the legal basis in Article 118 TFEU distinguishing between authority for establishing a unitary patent (first paragraph) and authority for establishing translation arrangements (second paragraph). Thus, the question of translation was excluded from the considerations until 2010 when the Commission proposed a system for translation.[141] However, massive disagreements on the question of translation stopped further negotiations, and by the end of 2010 12 Member States asked the Commission to propose a system for enhanced cooperation, later followed by the 15 other Member States.[142]

134. COM(2008) 465 final of 16 July 2008, p. 3, where there is also reference to the ageing world population and possible energy crises.
135. COM(2010) 2020 final.
136. See Jürgen Schade, 'Is the Community (EU) Patent behind the times? Globalisation Urges Multilateral Cooperation', *IIC* Vol. 41 7/2010, p. 818, who concludes that the creation of the Community (EU) patent no longer makes sense and suggests that WIPO should act as a regulatory authority of the international community of nations. The requirements for European and international patent systems in the first quarter of the 21st century are fundamentally different from the requirements after the Second World War, above all in Europe. The present polycentric patent system is characterised by the existence of a large number of transnational forms of cooperation.
137. COM(1997) 314 final, Promoting Innovation through Patents – Green Paper on the Community patent and the patent system in Europe.
138. Public hearing on Future Patent Policy in Europe, http://ec.europa.eu/internal_market/indprop/patent/hearing_en.htm (last visited January 2012).
139. COM(2007) 165 final, Communication Enhancing the Patent System in Europe.
140. Council document 16113/09, Add 1. revising the text from 2000 (COM(2000) 412 final).
141. COM(2010) 350 final.
142. See COM(2010) 790 final, COM(2011) 215 final, COM(2011) 216 final, and the Draft report on the proposal for a Council regulation implementing enhanced cooper-

The Unified Patent Litigation System has also slipped through the net. The basic concept for the proposed patent system in Europe was that the EPO should grant patents pursuant to the European Patent Convention (EPC), and EU should accede to the EPC. According to the main conclusions for an enhanced patent system in Europe, the new EEUPC (The European and EU Patents Court) should have exclusive jurisdiction over civil litigation relating to the infringement and validity of EU and European patents. On the other hand, the European Court of Justice should ensure the principle of primacy of EU law and its uniform interpretation.[143] If questions of interpretation of Union law were raised before the EEUPC this court (but not the national courts) could refer a question to the Court of Justice.

The Council submitted a request to the Court of Justice in July 2009 for an opinion on the establishment of a new EEUPC (Opinion 1/09, concerning the 2000 proposal with amendments) and the Advocates General delivered their opinion on 2 July 2010, stating in paragraph 60 that judicial competence in matters concerning the European patent had always been exercised by the national courts and that the Member States were therefore free to assign their competences to an international body. However, Advocate General Kokott (for the Advocates General) found that the question was more delicate if a future 'Community patent' were to be adopted. Advocate General Kokott found that the agreement in its current state was incompatible with the Treaty system since it did not satisfy the requirement for effective judicial control and the correct and uniform application of Union law in administrative proceedings concerning the granting of Community patents, see points 71-72, where it is pointed out that the EPO has its own appeals system with no external judicial review. There is no possibility of the Court of Justice to ensure the correct and uniform application of Union law. The Court came to the same conclusion in its opinion of March 2011.[144] In point 78, it stated that the international court 'is to be called upon to interpret and apply not only the provisions of that agreement but also the future regulation on the Community patent and other instruments of European Union law, in particular regulations and directives in conjunction with which that regulation would, when neces-

ation in the area of the creation of unitary patent protection with regard to the applicable translation arrangements (COM(2011) 216-C7-0145/2011-2011/0094(CNS)), and documents from 26 October 2011 (revised Presidency text, 15539/11 PI 133 COUR 59, and 15472/11 PI 131 CODEC 1676) *and* 19 October and 26 October 2011 (draft agreement on a Unified Patent Court and draft statute). For the background description, see e.g. Council Decision 2011/167/EU authorising enhanced cooperation in the area of the creation of unitary patent protection, recitals 3 to 5 (OJ 2011 L 76/53).

143. See Council of the European Union, Conclusions on an enhanced patent system in Europe, 2982$^{nd}$ Competitiveness, Brussels, 4 December 2009. The language regime and the patent litigation system are still outstanding questions.
144. Opinion of the Court (Full Court) of 8 March 2011. Opinion delivered pursuant to Article 218(11) TFEU – Draft agreement – Creation of a unified patent litigation system – European and Community Patents Court – Compatibility of the draft agreement with the Treaties. Opinion 1/09 available at http://eur-lex.europa.eu/LexUriServ/LexUriServ.do?uri=CELEX:62009V0001:EN:NOT (last visited January 2012).

sary, have to be read, namely provisions relating to other bodies of rules on intellectual property, and rules of the TFEU concerning the internal market and competition law. Likewise, the Patent Court may be called upon to determine a dispute pending before it in the light of the fundamental rights and general principles of European Union law, or even to examine the validity of an act of the European Union'. This would set aside all Community law remedies, the supremacy of Union law, its direct effect and state liability, and it would deprive the national courts of the right to refer questions to the Court of Justice, and the Court ruled the following in point 89:

'Consequently, the envisaged agreement, by conferring on an international court which is outside the institutional and judicial framework of the European Union an exclusive jurisdiction to hear a significant number of actions brought by individuals in the field of the Community patent and to interpret and apply European Union law in that field, would deprive *courts of Member States* of their powers in relation to the interpretation and application of European Union law and the Court of its powers to reply, by preliminary ruling, to questions referred by those courts and, consequently, would alter the essential character of the powers which the Treaties confer on the institutions of the European Union and on the Member States and which are indispensable to the preservation of the very nature of European Union law.'[Emphasis added][145]

Immediately after the Court gave its opinion, the Council adapted a Decision of 10 March 2011 authorising enhanced cooperation in the area of the creation of unitary patent protection (25 Member States) with only two articles. Article 1 states:

'The Kingdom of Belgium, the Republic of Bulgaria, the Czech Republic, the Kingdom of Denmark, the Federal Republic of Germany, the Republic of Estonia, Ireland, the Hellenic Republic, the French Republic, the Republic of Cyprus, the Republic of Latvia, the Republic of Lithuania, the Grand Duchy of Luxembourg, the Republic of Hungary, Malta, the Kingdom of the Netherlands, the Republic of Austria, the Republic of Poland, the Portuguese Republic, Romania, the Republic of Slovenia, the Slovak Republic, the Republic of Finland, the Kingdom of Sweden and the United Kingdom of Great Britain and Northern Ireland are hereby authorised to establish enhanced cooperation between themselves in the area of the creation of unitary patent protection, by applying the relevant provisions of the Treaties.'[146]

145. Gavin Lawson writes that the opinion appears to kill the 2009 draft regulation. Unless the Court of justice is a court of reference on all matters, including substantive patent law, there will be no union patent; see Gavin Lawson, 'The ECJ says 'no' to the Community patent Court, Opinion 1/09 Of the ECJ (Full Court)', *Journal of Intellectual Property Law & Practice*, 2011, Vol. 6, No 8, p. 512.
146. Decision 2011/167/EU (OJ 2011 L 76/53). Article 2 states: 'This Decision shall enter into force on the day of its adoption'. In the Commission's proposal (IP/11/469) of 13 April 2011, it was reasoned that the current cost of a European patent is far too expensive and that the proposal would reduce the costs in Europe by up to 80 %. Vali-

On 26 June 2011 the Council agreed on a general approach to the draft Regulation implementing the enhanced cooperation (see Article 329(1) TFEU) in the area of the creation of unitary patent protection and the draft Council Regulation implementing enhanced cooperation in the area of the creation of unitary patent protection with regard to the applicable translation arrangements.[147] The draft for the enhanced procedure for unitary patent protection follows the co-decision procedure, whereas the translation arrangements require a unanimous decision. Accordingly, the unitary patent is expected to be ready in 2013. It is based on Article 118(1) TFEU and aims to achieve such protection by giving unitary effect to European patents granted by the EPO in the post-grant phase. According to recital 7 of the draft regulation of 23 June 2011, the purpose is: 'to ensure the uniform substantive scope of protection conferred by unitary patent protection, only European patents that have been granted for all the participating Member States with the same set of claims should benefit from unitary effect. Finally, the unitary effect attributed to a European patent should have an accessory nature and should cease to exist or

dation in each EPO member country, involving translation and administrative costs, is approximately EUR 32,000 (EU 27) with EUR 23,000 for translation costs alone (which is not machine translation). A US patent cost on average the equivalent of EUR 1850.

147. The document (9224/11 PI 31 CODEC 671 9226/11 PI 32) is available on http://www.epo.org/law-practice/legislative-initiatives/eu-patent.html (last visited January 2012). As for the language issues after the grant of a European patent, only high-quality machine translation would be available; the internet search engine Google has made an agreement with the EPO on translation (without legal force). See the latest Draft report on the proposal for a Council regulation implementing enhanced cooperation in the area of the creation of unitary patent protection with regard to the applicable translation arrangements (COM(2011) 216-C7-0145/2011-2011/0094(CNS)), and documents from 26 October 2011 (revised Presidency text, 15539/11 PI 133 COUR 59, and 15472/11 PI 131 CODEC 1676) *and* 19 October and 26 October 2011 (draft agreement on a Unified Patent Court and draft statute). See: st13751/11 of 2. September 2011, st15289/11 of 7. October 2011, st16741 of 16. November 2011, st17539/ of 24. November 2011, st17578/11 of 1.12. 2011, st18239/11 of 6. December 2011. *Pagenberg* 2012, writes that of each of these drafts, there are a number of versions, the relevance of which cannot be assessed from outside because the entire project has in the last six months been subject to a completely unusual secrecy, cf. 'The EU Package, Politics vs. Quality and new Practice of Secret Legislation in Brussels'.

be limited to the extent that the basic European patent has been revoked or limited'. [148]

If agreement cannot be reached by all 27 EU Member States, the Lisbon Treaty allows legislation to be adopted by a minority of Member States. The Lisbon Treaty reformed the procedure providing for 'flexible integration' which is now based on Articles 326 and 327 TFEU and Article 20 TEU.[149] The Lisbon Treaty modified the procedure for enhanced cooperation in order to make it feasible for the EU 27. The cooperation must be aimed at furthering the objectives of the EU, protecting its interests, and reinforcing the integration process. There must be a minimum of nine Member States in the cooperation.[150] Enhanced cooperation is a last resort (Article 20(2) TEU) and it does not apply if the area concerned falls within the Union's exclusive competence. As stated above in section 3.1, pursuant to Article 3(1)(e) and Article 207(1) TFEU, the commercial aspects of intellectual property seem to fall within the exclusive competence of the Union.

Italy and Spain are not taking part in the enhanced cooperation. On the contrary, on 3 June 2011, Spain reacted by bringing proceedings for the annulment of the Council Decision in Case C-274/11 *Spain v Council*, claiming misuse of powers since there was recourse to enhanced cooperation even though the purpose was not to achieve integration of all the Member States – the mechanism having been used instead to avoid negotiating with a Member State, imposing upon it an opt-out solution – and even though the objectives pursued in this instance could have been achieved by means of a special agreement as provided for in Article 142 of the European Patent Convention.[151] Further, Spain claimed that the decision failed to respect the judicial system of the EU in that no dispute resolution system was provided for in relation to certain legal rights subject to EU law.[152] Italy followed immediately

148. Recital 7 of 1 December 2011: 'The creation of unitary patent protection should be achieved by giving unitary effect to European patents in the post-grant phase by virtue of this Regulation and in respect of *all* the participating Member States. The main feature of European patents with unitary effect should be their unitary character, i.e. providing uniform protection and having equal effect in all participating Member States. Consequently, a European patent with unitary effect should only be limited, transferred, revoked or lapse in respect of all the participating Member States'.
149. Alan Dashwood et al., *Wyatt and Dashwood's European Union Law*, 6th Ed., 2011, pp. 27 and 127 (secondary flexibility).
150. Paul Craig & Gráinne de Búrca, *EU Law, Text, Cases and Materials*, 5th Ed. 2011, p. 142.
151. According to Intellectual property magazine, January 2011, p. 6, with English, French and German as the default languages for the EU patent, countries such as Spain remain bitterly opposed to exclusion from the language regime.
152. Further claims relate to Article 20 TEU and Articles 326 to 327 TFEU.

after on 10 June 2011 with proceedings for the annulment of the Council Decision in Case C-295/11 *Italy v Council*. Italy too argued for the annulment of Council Decision 2011/167/EU authorising enhanced cooperation in the area of the creation of unitary patent protection. Italy's four main arguments are in line with those of Spain, as follows:

1. The enhanced cooperation procedure (Article 20(1) TEU) is to be allowed only within the framework of the European Union's non-exclusive competences. The European Union has an exclusive competence to create 'European rules' which have Article 118 TFEU as their legal basis.
2. The authorisation of enhanced cooperation in the present case is contrary to – or, in any event, not compatible with – the objectives in view of which such cooperation is provided for by the Treaties. In so far as that authorisation is contrary to, if not the letter, at least the spirit of Article 118 TFEU, it infringes Article 326(1) TFEU, in that the latter requires enhanced cooperation to comply with the Treaties and with EU law.
3. The authorisation decision was adopted without an appropriate inquiry with regard to the *last resort* requirement and without an adequate statement of reasons on that point.
4. The authorisation decision infringes Article 326 TFEU in that it adversely affects the internal market, introducing a barrier to trade between Member States and discrimination between undertakings, causing distortion of competition. Furthermore, it does not help to reinforce the EU's integration process, and is thus contrary to the second subparagraph of Article 20(1) TEU.[153]

Advocate General Bot delivered his Opinion on 11 December 2012. In points 56 ff. and 66-67 of the Opinion he emphasised that since the Treaty of Lisbon Article 118 TFEU has provided an appropriate legal basis for the creation of intellectual property rights and that the provision refers expressly to the establishment and functioning of the internal market, which is an area within the

153. http://curia.europa.eu/jurisp/cgi-bin/form.pl?lang=en&alljur=alljur&jurcdj=jurcdj&jurtpi=jurtpi&jurtfp=jurtfp&numaff=C-295/11&nomusuel=&docnodecision=docnodecision&allcommjo=allcommjo&affint=affint&affclose=affclose&alldocrec=alldocrec&docdecision=docdecision&docor=docor&docav=docav&docsom=docsom&docinf=docinf&alldocnorec=alldocnorec&docnoor=docnoor&docppoag=docppoag&radtypeord=on&newform=newform&docj=docj&docop=docop&docnoj=docnoj&typeord=ALL&domaine=&mots=&resmax=100&Submit=Rechercher (last visited August, 2011).

shared competences of the Union and the Member States. He also considered that the allegation of misuse of powers was unfounded (point 80 ff.), and finally in point 114 he stated that the Council has wide discretion to determine whether enhanced cooperation is indeed adopted as a last resort and whether the objectives of such cooperation cannot otherwise be attained within a reasonable period by the Union as a whole. Elaborating on that argument, in point 124 he referred to the years of discussion which had always ended in failure. Furthermore, since the authorisation of the enhanced cooperation was not based on a manifest error of assessment, he concluded that the Court should reject the argument of lack of competence to set up enhanced cooperation with a view to creating the unitary patent. The Advocate General proposed that the Court should dismiss the claims, which it did. The Court dismissed the actions and Court of Justice Press Release No 47/13 of 16 April 2013 gives a summary of its reasons. It was stated that the creation of EU intellectual property rights was authorised in the context of the internal market and was within the ambit of competences shared between the EU and the Member States. The non-exclusiveness of those competences meant that the Council was competent to authorise such enhanced cooperation. Once again, integration appeared to be the main motivation of the Court of Justice. Having found that the single European patent and its language arrangements could not be established by the EU as a whole within a reasonable period, the Council's decision to authorise enhanced cooperation was by no means a circumvention of the requirement for unanimity or an exclusion of those Member States that did not join in requesting enhanced cooperation. The Court found that, in the light of the impossibility of reaching common arrangements for the whole EU within a reasonable period, the contested Decision contributed to the *process of integration*. Furthermore, the Court found that the Council had carefully and impartially ascertained whether the condition that the measure should be adopted as a 'last resort' had been met (the legislative process had begun in 2000 and carried out in several stages). As for whether the uniform patent would be advantageous in terms of uniformity and so of integration (compared to the operation of the rules laid down by the EPC), the Court stated that the EU patent would confer uniform protection in the territory of all the Member States taking part in the enhanced cooperation. The Council decision did not damage the internal market or the economic, social and territorial cohesion of the EU, nor did it prejudice any competence, right or obligation of those Member States not participating in the enhanced cooperation. The Member States participating in the cooperation can prescribe rules with which the non-participating States would not agree if they did take part in it but such rules do not negate the opportunity for non-participating Member States to participate in the future.

In 2011 the Polish presidency of the EU Council did not succeed in concluding the EU patent package (unitary patent, language regime and unified patent court) based on enhanced cooperation and it was up to the Danish presidency to negotiate the remaining hurdles and have the proposal enacted. Thus, despite the green light from the European Parliament legal affairs Committee in December 2011 it was yet to be endorsed by the whole Parliament and the Council.[154] Several critical voices have been raised in the aftermath of the drafting of the EU patent package. The critic addresses competences in multiple ways; the referral of preliminary rulings to the CJEU on substantive patent law (Articles 6 to 8-9 demanded to be removed from the draft Unitary Patent Regulation), the referral of power from the EU to the EPO, and the drafted Unified Patent Court Agreement; an international agreement outside the EU might never be ratified by EU 27.[155]

One critic, *Dr. Jochen Pagenberg*, had the following comment on the draft regulation:

> 'If one wants a really unattractive, inefficient, unpredictable and probably extremely expensive patent court system, then we will get it; one must only give the ECJ a chance to receive as many referrals in patent law as possible. If one wants to see substantive patent law in Europe to be decided by judges without any solid knowledge and experience in this field, then one must involve the ECJ whenever possible. And if somebody intended to lay a solid ground for failure of this – at some time very promising – project, then he will probably succeed.'[156]

154. http://www.europarl.europa.eu/news/en/pressroom/content/20111219IPR34540/html/EU-patent-gets-Legal-Affairs-Committee-green-light.
155. Draft Agreement on the creation of a Unified Patent Court – Suggested way forward for a political agreement, doc. 10059/12, Brussels, 24 May 2012.
156. https://webmail.asb.dk/OWA/?ae=Item&a=Open&t=IPM.Note&id=RgAAAAAbTAYXq6BHR5ovsEIU4tqfBwAweB0N4pg1SILLm%2bI2vkutAAAAH6UCAAA7obmcw8psTY6l8FrLoHyrAAAdQzhbAAAA&pspid=_1326060162545_493532843 (last visited January, 2012). See also http://patlit.blogspot.com/2011/09/draft-agreement-on-unified-court-unwise.html and https://www.unitary-patent.eu/content/presentation-about-unitary-patent. More critics: The Confederation of Swedish Enterprises, Stockholm 30 August 2011; and 'A disaster in the making', Three New Square, December 2011; and 'a disastrous agreement', Chartered Institute of patent Attorneys, UK, 16 December 2011, see e.g. https://docs.google.com /viewer?a=v&pid=explorer&chrome=true&srcid=0B_U9nV8-MjxrODViMzc1ZTYtODZjOS00 MjI4L WJhNjgtMWM3YmVjODhhNjZh&hl=en_US and https://docs.google.com/viewer?a=v&pid=explorer&chrome=true&srcid=0B_U9nV8-MjxrNDE3ZTI

On 29 June 2012 the European Council agreed on the solution for 'the last outstanding issue' of the patent package, namely the seat of the Central Division of the Court of First Instance of the Unified Patent Court (UPC) to be located in Paris with thematic clusters in London ad Munich. The contested Articles 6 to 8 of the Regulation implementing enhanced cooperation are to be deleted.[157] The whole package was adopted into the EU legal order in 2012. The new term *European patents with a unitary system* completes the mix of competences at the regional level.[158] With the new enhanced cooperation, we are now faced with European patents, European patents with a unitary system and European patents with EU2 (EU remaining group). Not only will the regulation make substantive patent law part of the EU legal order, but preliminary rulings concerning this *new* European patent with a unitary system will be subject to the jurisdiction of the Court of Justice of the European Union. Recital 26 and Article 1 of Regulation (EU) No 1257/2012 of the European Parliament and the Council of 17 December 2012 implementing enhanced cooperation in the area of the creation of unitary patent protection[159] read as follows:

Recital 26. This Regulation should be without prejudice to the right of Member States to grant national patents and should not replace Member States' laws on patents. Patent applicants should remain free to obtain either *a national patent*, a *European patent with unitary effect*, *a European patent* taking effect in one or more of the Contracting States to the EPC or a European patent with unitary effect validated in addition in one or more other Contracting States to the EPC which are not among the participating Member States [emphasis added]

Article 1 This Regulation implements the enhanced cooperation in the area of the creation of unitary patent protection authorised by Council Decision 2011/167/EU. This Regulation constitutes a special agreement within the meaning of Article 142 of the Convention on the

0NWEtZTc2Ni00MjZmL WI0ZjYtNmM2MGE2ODM1NDAw&hl=en_US, and http://www.publications.parliament.uk/pa/cm201012/cmselect/cmeuleg/428-xlv/42804.htm. Jochen Pagenberg, 'The EU Patent Package, Politics vs. Quality and the New Practice of Secret Legislation in Brussels', 2012.

157. EUCO 76/12, Conclusions, 29 June 2012, point 3.
158. First reading, 4 October 2011, (COM(2011)0215 – C7-0099/2011 – 2011/0093 (COD)).
159. OJ L 361, 31.12.2012, p. 1. And Regulation (EU) No 1260/2012 of December 2012 implementing enhanced cooperation in the area of the creation of unitary patent protection with regard to the applicable translation arrangements.

Grant of European Patents (European Patent Convention), as amended (hereinafter 'the EPC').[160]

And the mix of judicial competences and jurisdiction at the regional level follows from the revised version of the draft agreement on unified patent court, now adopted as the Agreement on a Unified Patent Court:[161]

The Agreement includes the following recitals and articles:

CONSIDERING that the Unified Patent Court should be a court common to the Contracting Member States and thus part of their judicial system, with exclusive competence in respect of European patents with unitary effect and European patents granted under the provisions of the EPC;
CONSIDERING that, as any national court, the Unified Patent Court must respect and apply Union law and, in collaboration with the Court of Justice of the European Union as guardian of Union law, ensure its correct application and uniform interpretation; the Unified Patent Court must in particular cooperate with the Court of Justice of the European Union in properly interpreting Union law by relying on the latter's case law and by requesting preliminary rulings in accordance with Article 267 TFEU;
CONSIDERING that infringements of Union law by the Unified Patent Court, including the failure to request preliminary rulings from the Court of Justice of the European Union, are directly attributable to the Contracting Member States and infringement proceedings can therefore be brought under Article 258, 259 and 260 TFEU against any Contracting Member State to ensure the respect of the primacy and proper application of Union law;
CONSIDERING that this Agreement should be open to accession by any Member State of the European Union; Member States which have decided not to participate in the enhanced cooperation in the area of the creation of unitary patent protection may participate in this Agreement in respect of European patents granted for their respective territory;
Article 20: Primacy of and respect for Union law
The Court shall apply Union law in its entirety and shall respect its primacy.
Article 24: Sources of law
(1) In full compliance with Article 20, when hearing a case brought before it under this Agreement, the Court shall base its decisions on:

(a) Union law, including Regulation (EU) No 1257/2012 and Regulation (EU) No 1260/2012;
(b) this Agreement;
(c) the EPC;
(d) other international agreements applicable to patents and binding on all the Contracting Member States; and

160. Proposal for a Regulation of the European Parliament and the Council implementing enhanced cooperation in the area of the creation of unitary patent protection – Analysis of the final compromise text (Doc. 1758/11, 1 dec. 2011, Working doc. for Coreper (Part 1 on 2 Dec. 2011)
161. Brussels, 11 January 2013 no 16351/12.

(e) national law.

(2) To the extent that the Court shall base its decisions on national law, including where relevant the law of non-contracting States, the applicable law shall be determined:
(a) by directly applicable provisions of Union law containing private international law rules, or
(b) in the absence of directly applicable provisions of Union law or where the latter do not apply, by international instruments containing private international law rules; or
(c) in the absence of provisions referred to in points (a) and (b), by national provisions on private international law as determined by the Court.

(3) The law of non-contracting States shall apply when designated by application of the rules referred to in paragraph 2, in particular in relation to Articles 25 to 28, 54, 55, 64, 68 and 72.

Article 34: Territorial scope of decisions
Decisions of the Court shall cover, in the case of a European patent, the territory of those Contracting Member States for which the European patent has effect.

Article 66: Powers of the Court concerning decisions of the European Patent Office
(1) In actions brought under Article 32(1)(i), the Court may exercise any power entrusted on the European Patent Office in accordance with Article 9 of Regulation (EU) No 1257/2012, including the rectification of the Register for unitary patent protection.

(2) In actions brought under Article 32(1)(i) the parties shall, by way of derogation from Article 69, bear their own costs.

As written at the beginning of this chapter, Europe has too many players and the scenario of a unified patent comprising EU 25, the EPO members and the further EU 2 is far from ideal. Using a Union law procedure without having the whole EU on board clearly creates a political split, and then mixing this unitary approach with a completely different system (inter-governmental)[162]

162. Proposal for a Regulation of the European Parliament and the Council implementing enhanced cooperation in the area of the creation of unitary patent protection – Analysis of the final compromise text (Doc. 1758/11, 1 December 2011, Working doc. for Coreper (Part 1 on 2 December 2011), recital 6. Part IX of the EPC provides that a group of Member States of the European Patent Organisation may provide that European patents granted for those States may have a unitary character. This Regulation constitutes a special agreement within the meaning of Article 142 of the EPC, a regional patent treaty within the meaning of Article 45(1) of the Patent Cooperation Treaty of 19 June 1970 and a special agreement within the meaning of Article 19 of the Convention for the Protection of Industrial Property, signed in Paris on 20 March 1883 and last revised on 14 July 1967.

is indeed to go in the wrong direction. One can only regard this bottom-up relationship to regional law or 'export of norms' with concern, especially since the newly established Patent Court, based on an international agreement and intended to be a special Court, will be required to refer questions of interpretation to the General Court (CJEU). Judicial law making e.g. gap-filling at the regional level is now accepted in an international agreement.

Article 21: Requests for preliminary rulings
As a court common to the Contracting Member States and as part of their judicial system, the Court shall cooperate with the Court of Justice of the European Union to ensure the correct application and uniform interpretation of Union law, as any national court, in accordance with Article 267 TFEU in particular. Decisions of the Court of Justice of the European Union shall be binding on the Court.

The past 11 years with Directive 98/44/EC has shown that patent legitimacy has suffered, and it is to be expected this new set-up will worsen this situation, especially in regard to future global negotiations (with the USA, Japan and China) where it would have been fruitful to stand as a united European region with additional purposes to those of trade policy and the functioning of the internal market.[163] In an article in 2010 *Jürgen Schade* wrote 'the EU is precisely not a centralized state, nor a federal state. The response to globality in Europe is not to be found in a large supranational State, but rather in the closing of the ranks of the national State ... the protection of inventions is not to be found in a world patent and not necessarily in major supranational rights such as the Community patent but rather in international, multinational and bilateral collaboration and implementation, of which the EPO based on the European Patent Convention is a good example'.[164] Both the EPO system and the EU system are relatively young and the experiences of the EPO regime can be more fruitfully elaborated in the EPO 38 system than in a new fragmented solution with EPO 13, EPO 11, EU 25 and EU 2. If we have not learnt that we are not better off with an experienced regime, comprising a granting authority and specialised boards of appeal than with rushed through Union interference, then we are blind to the benefit of a complete solution where patent legitimacy detached from economic standards can be much valued. However, this did not turn out to be the case.

163. Recitals 1 and 2 in presidency compromise text (2011/0093 (COD)) of 23 June 2011.
164. Jürgen Schade, 'Is the Community (EU) Patent behind the times? Globalisation Urges Multilateral Cooperation', *IIC* Vol. 41 7/2010, p. 811.

The multilevel governance and global structure of patent law is evidently challenged by EU law and the Court of Justice of the European Union which has given the regional level of patent law a variety of sources of law which is now indirectly subject to the supremacy of Union law. The promotion of integration, which legitimates enhanced cooperation (cf. Joined Cases C-274/11 and C-295/11 *Spain and Italy v Council*), and the current possibility of a shift of competence from shared to exclusive competence by future WTO/TRIPS amendments (TRIPS (PLUS) and TRIPS (PLUSPLUS) must be view in the light of the *Kadi* cases. Despite Advocate General Villalón's Opinion in Case C-414/11 *Daiichi Sankyo Company, Sanofi-Aventis Deutsland* v *DEMO Anonimos Viomikhaniki kai Emporiki Etairia Farmakon*, the *sui genesis* character of Union law can only point in one direction. Export of EU norms is to be expected from future patent law developments. Already, before this future starts, we have seen that the Court's regard for human dignity is one such standard (see Chapter 4 on morality in Europe).

With the above elaboration of the EU legal 'international/regional' order, and with the perspective of Court of Justice's Opinion 1/09 of March 2011, the competences of the Union may be illustrated in figure 3.3 as follows:

The current unitary patent package illustrated in figure 3.3.a will increase the number of players, challenge the protection of general public interests and pushing back patent legitimacy:[165]

While the European Union has failed to design of a proper EU patent solution, it has nevertheless adopted three pieces of legislation that interfere with the Member States' legislative powers solely with regard to patent law. Two of the pieces have been quite uncontroversial and the third has created more disharmony than harmony in the Member States and beyond. The first piece of legislation is the 1994 Regulation on the Community plant variety right.[166] The plant variety Regulation seems quite uncontroversial as it only implements international obligations derived from the International Convention for

165. Chapter 4. See also Schovsbo, Jens, 'Constructing an Efficient and Balanced European Patent System: 'Muddling Through'' (September 23, 2011). Available at SSRN: http://ssrn.com/abstract=1932690 or http://dx.doi.org/10.2139/ssrn.1932690.

166. See Regulation (EC) No 2100/94 as amended by Council Regulation (EC) No 15/2008 of 20 December 2007 as regards the entitlement to file an application for a Community plant variety right. An application for a Community plant variety right may be filed by any natural or legal person, or anybody ranking as a legal person under the law applicable to that body, jointly by two or more such persons and also earlier applications that were filed in another State; Article 12.

*Figure 3.3. EU New legal order*

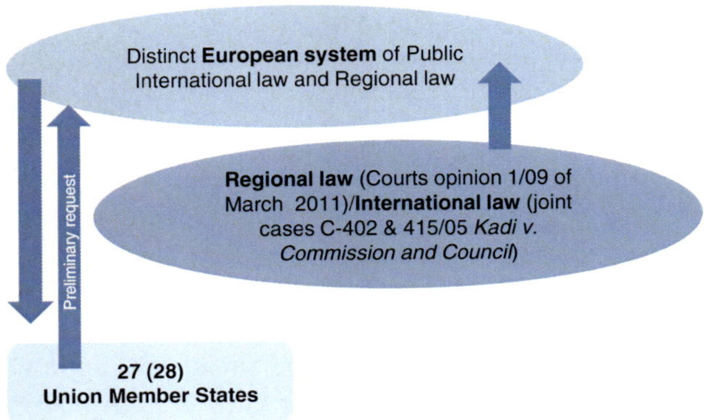


*Figure 3.3.a. New EU New legal order*

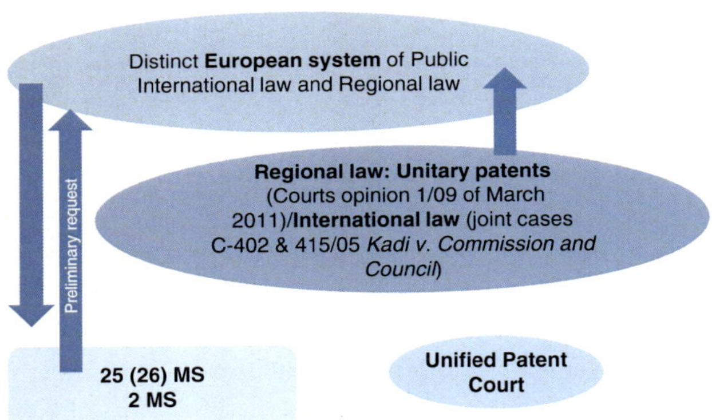


the Protection of New Varieties of Plants (UPOV Convention) from 1961 and 1991.[167] The latter revision of the UPOV Convention now explicitly mandates double protection for plant varieties. Thus, in principle there can be parallel plant variety rights and patents rights. However, in Europe, plant varieties can only obtain a variety right.[168]

167. Questionnaire on the patent system in Europe, closing date was 31 March 2006.
168. See Abeba Tadesse Gebresalassie, *The Sustainability of Plants and Plant Intellectual Property Rights*, 2012, p. 121 ff.

The latest move is the adoption of Directive 2004/48/EC on the enforcement of intellectual property rights.[169] The aim of this Directive is to ensure that intellectual property rights enjoy an equivalent level of protection throughout the Community, see recital 8. The Directive is intended to harmonise provisional measures in particular to preserve evidence, calculate damage or the arrangements for applying injunctions, but it does not harmonise the rules for judicial cooperation; see recital 11 and Article 3. The Directive codifies the judicial doctrine created decades ago by the Court of Justice, that procedural remedies in a Member State may not discriminate between national rights and rights which derived from Union law, for example due to a Member State's failure to implement a directive. With regard to patent law, this means that the legal protection of a patented gene sequence must be enforceable in a German Court, for example, (with the same procedural guarantee, with equal procedural protection) without consideration as to the nationality of the holder of the right. In other words, the Directive does not establish substantive rights; its only purpose is to secure procedural guarantees. This is not a new invention in the European Union.

In many ways Directive 98/44/EC on the legal protection of biotechnological invention, which comes in-between the plant variety Regulation and Directive 2004/48/EC, is the most controversial. This is primarily due to the subject matter it regulates. However, the controversy described in this chapter relates to the complete frustration and disorder which this EU interference has created in the European patent law regime.[170]

The first party to implement Directive 98/44/EC on the legal protection of biotechnological invention was the European Patent Organisation. The Di-

169. Directive 2004/48/EC reflects e.g. the Berne, Brussels and Paris Conventions and the TRIPS Agreement. It harmonises rules on standing, evidence, interlocutory measures, seizure, injunction damage, costs and judicial publication. In a statement concerning Article 2 of Directive 2004/48/EC on the enforcement of intellectual property rights, the Commission has said that it considers that at least the following intellectual property rights are covered by the Directive: copyright, rights related to copyright, *sui generis* rights of a database maker, rights of the creator of the topographies of a semiconductor product, trademark rights, design rights, patent rights, including rights derived from supplementary protection certificates, geographical indications, utility model rights, plant variety rights, trade names, in so far as these are protected as exclusive property rights in the national law concerned (OJ 2005 L 94/37).

170. Tine Sommer, 'Interpreting Ordre Public and Morality in a Patent Law Context: Which is the correct Approach?', *Bio-Science Law Review*, Vol. 9, issue 2, 2006/2007 p. 62 ff.; and Tine Sommer, 'Genpatentet og patentetik I EU – en harmoniseret retsvildfarelse?', Chapter 26 in Birgitte Egelund Olsen & Karsten Engsig Sørensen (eds.), *Europæisering af dansk ret*, 2008. See further Chapter 4 below.

rective was simply inserted as new Implementing Regulations, Rules 23b to 23e.[171] This is surprising as the EPO is not a member of the EU. On the other hand, the Member States of the EU have an obligation to implement the Directive. The deadline of implementation expired in July 2000. As there was strong opposition to the Directive, some Member States, e.g. Germany, France and Italy, have implemented in ways that differs substantially from other Member States.[172]

The European patent regime is young compared with the US patent system, but the systems of the separate European states have as much length of service behind them as the US system. While Europe has scattered authority, the USA has concentrated it. The resulting total confusion in Europe is in remarkable contrast to the USA, as the European Union does not have jurisdiction to rule on patentable subject matter or on the scope of protection under the European Patent Convention. The interpretation of Directive 98/44/EC may in principle be diverted from the Court of Justice to the Boards of Appeal of the European Patent Organisation. The EU Member States which have the power to enforce these rights have no coherent guidance as to which right to enforce: should it be according to the interpretations of the Court of Justice of the European Union or the Boards of Appeal of the European Patent Organisation?[173]

171. See Chapter 5 and Tine Sommer, 'Interpreting Ordre Public and Morality in a Patent Law Context: Which is the correct Approach?', *Bio-Science Law Review*, Vol. 9, issue 2, 2006/2007, p. 64.
172. See Chapter 5 and Tine Sommer, 'The Scope of Gene patent Protection and the TRIPS Agreement – An Exclusively Nondiscriminatory Approach?', *IIC* Vol. 38 1/2007, p. 43 f., describing the long-standing process of transposing Directive 98/44/EC into national laws, including the huge number of cases on the implementation of the Directive; see Case C-377/98 *The Netherlands v European Parliament and Council*; Case C-395/03 *Commission v The Netherlands*; Case C-448/03 *Commission v France*; Case C-450/03 *Commission v Luxembourg*; Case C-454/03 *Commission v Belgium*; Case C-456/03 *Commission v Italy*; Case C-518/03 *Commission v Sweden*; Case C-4/04 *Commission v Austria*; and Case C-5/04 *Commission v Germany*. The case against Lithuania was closed on 12 October 2005 (Commission Case No C-973/04), and the case against Latvia was closed on 4 April 2006 (Commission Case No C-1061/04).
173. See Tine Sommer, 'Interpreting Ordre Public and Morality in a Patent Law Context: Which is the correct Approach?', *Bio-Science Law Review*, Vol. 9, issue 2, 2006/2007, p. 62 ff.

*Figure 3.4. Directive 98/44/EC on the legal protection of biotechnological inventions*

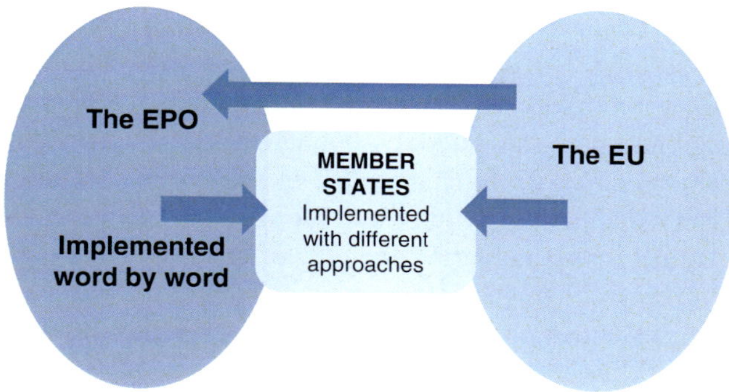


This confusion is only added to by the enhanced cooperation; a backdoor competence following the unitary patent at the expense of a more structured and transparent patent legal order.

Unlike the US framework, the European framework has developed into disorder. It is obvious that the frameworks have developed in opposite directions[174] and the proper solution to this problem is to make European patent law a unitary system centred on the experienced body which has a geographical jurisdiction that is wider than that of the European Union.

## 5. Brief introduction to environmental players and more

As the issue of climate change has emerged so has discussion on patents as an instrument for combating climate change. The president of the European Patent Organisation, *Mr. Benoît Battistelli*, has said that 'although patents and other IP rights are not mentioned in the Copenhagen Accord, experts attending the UN conference maintain that patents have become the most polarised and controversial item on the agenda.'[175]

The major players on the climate change scene are mainly to the same as the players mentioned above. However, in respect of the EU is worth noting

174. See also Rob J. Aerts, 'The legitimacy of patent law-making in Europe and the U.S. – A tentative Comparison', 2/2007 *IIC* Vol. 38, p. 165 ff.

175. See full text of speech 22 July 2010 at http://www.epo.org/news-issues/news/2010/20100727.html (last visited July 2011).

that the Union has had a strong influence on the development of environmental law since the beginning of the 1970s and the adoption of the European Single Act in 1987 established a formal legal basis for this. Today the legal authority of the Union is given in Title XX (Environment) TFEU, Articles 191-193. According to Article 191(1), fourth indent, Union policy on the environment shall contribute to 'promoting measures at international level to deal with regional or worldwide environmental problems, and in particular combating climate change'. In paragraph 2 it is stated that: 'Union policy on the environment shall aim at a high level of protection taking into account the diversity of situations in the various regions of the Union and that it shall be based on four environmental principles namely, the precautionary principle and on the principles that preventive action should be taken, that environmental damage should as a priority be rectified at source and that the polluter should pay.' Furthermore, Article 11 TFEU states that: 'Environmental protection requirements must be integrated into the definition and implementation of the Union's policies and activities, in particular with a view to promoting sustainable development.' The new Title XXI (Energy) comprises Article 194. Article 194(1)(c) says that the Union aims to promote energy efficiency and energy saving and the development of new and renewable forms of energy.

In environmental matters the Union takes external action and thus Article 191(4) provides that: 'Within their respective spheres of competence, the Union and the Member States shall cooperate with third countries and with the competent international organisations', and that such arrangements may be the subject of agreements between the Union and the third parties concerned.

According to Article 4(2)(e) and (i) TFEU, on the environment and on energy matters there is shared competence is shared between the Union and the Member States.

Article 13 TFEU is a new provision on animal welfare; it states:

> 'In formulating and implementing the Union's agriculture, fisheries, transport, internal market, research and technological development and space policies, the Union and the Member States shall, since animals are sentient beings, pay full regard to the welfare requirements of animals, while respecting the legislative or administrative provisions and customs of the Member States relating in particular to religious rites, cultural traditions and regional heritage.' [176]

176. See Tine Sommer, 'Patenting the Animal Kingdom? From Cross-Breeding to Genetic Make-Up and Biomedical Research', *IIC* 2/2008 Vol. 39, p. 159, footnote 77.

Pursuant to the Lisbon Treaty, Article 6(1) of the Treaty on European Union states that the Union recognises the rights, freedoms and principles set out in the Charter of Fundamental Rights of the European Union of 7 December 2000, as adapted at Strasbourg, on 12 December 2007, which shall have the same legal value as the Treaties. The Lisbon text adapts the wording of the Charter as proclaimed on 7 December 2000, and replaces it as from the date of entry into force of the Treaty of Lisbon. According to the Charter, human dignity (Article 1) is inviolable and it must be respected and protected. In the fields of medicine and biology (Article 3) the prohibition of making the human body and its parts as such a source of financial gain and the prohibition of the reproductive cloning of human beings must be respected in particular. Further, in Article 37 of the Charter it is stated that: 'A high level of environmental protection and the improvement of the quality of the environment must be integrated into the policies of the Union and ensured in accordance with the principle of sustainable development', thus repeating the purposes already stated in the Treaty chapter on the environment. While these fundamental rights are reasonable and do not differ from the consensus as to what constitutes protectable fundamental rights, the Union has its own fundamental right when it comes to intellectual property.

Article 17 of the Charter of Fundamental Rights, on the right to property, states in paragraph 2 that intellectual property shall be protected. As argued by *Christopher Geiger*, the IP world would have been better off without this badly drafted provision.[177] Thus, the growing importance of IP justifies the fundamental quality on economic arguments as well as from the increased Community activity in the field. One can argue that the traditional understanding of the shaping of what can become fundamental rights is not binding upon the Union legal order.[178]

The idea of IPR protection as a human right has been indirectly addressed by the European Court of Human Rights.[179] The ECtHR stated that it was

177. Christophe Geiger, 'Intellectual Property shall be protected? Article 17(2) of the Charter of Fundamental Rights of the European Union: A mysterious provision with an Unclear Scope', *European Intellectual Property Review* (EIPR) 2009, 31 (3), p. 319.
178. Christophe Geiger, 'Intellectual Property shall be protected? Article 17(2) of the Charter of Fundamental Rights of the European Union: A mysterious provision with an Unclear Scope', *European Intellectual Property Review* (EIPR) 2009, 31 (3), p. 313. On p. 315 he writes that 'the drafters took care to specify that the guarantees laid down in paragraph 1 (of Article 17) shall apply as appropriate to intellectual property', so that IPR protection can be limited by the need to safeguard public interests.
179. Case No 40382/04 *Rambus Inc.* v *Germany*, 16 June 2009.

doubtful whether it had jurisdiction over the alleged procedural shortcomings of international tribunals such as the Boards of Appeal of the European Patent Office. The applicant company complained that it had been denied a fair trial with regard to its patent rights. However, the ECtHR stated that the European Patent Convention (and the German Constitution) provides for protection equivalent to the European Convention on Human Rights.[180]

At the international level climate change has been discussed and new approaches initiated by the USPTO, EPO, UN (/UNEP) (UNFCCC, CBD), and International Centre for Trade and Sustainable Development (ICTSD).[181] WIPO has taken a first step by establishing a 'WIPO Green', a technology exchange platform launched in the second part of 2011. This initiative supports the United Nations Framework Convention Climate Change (UNFCCC).[182]

One of the contested provisions in the TRIPS Agreement is that of Article 27(3)(b), which states:

> (b) plants and animals other than micro-organisms, and essentially biological processes for the production of plants or animals other than non-biological and microbiological processes. However, Members shall provide for the protection of plant varieties either by patents or by an effective *sui generis* system or by any combination thereof. The provisions of this subparagraph shall be reviewed four years after the date of entry into force of the WTO Agreement.

The disputes of biodiversity protection, access to genetic resources and benefit sharing can be evaluated in the interface of environmental protection and intellectual property. Thus, the part of this thesis that analyses the current patent regime from the climate change perspective will again involve EU as a global actor, as well as the legitimacy of patent law as described in the former section. The Europeanisation of international law is also reflected in the EU's actions in environmental protection matters.[183]

180. *Rambus Inc. v Germany* in 41 IIC Vol. 1/2010 99, concerning the European Convention on Human Rights, Article 6.
181. Ricardo Meléndez-Ortiz (ICTFS Chief Executive), 'Patents and Clean Energy: Bridging the gap between evidence and policy. A first glance at the Methodology and findings, presenting an analysis of brand new data on patenting and licensing trends in climate chance technologies', EPO press release, November 2009.
182. Catherine Saez, 'WIPO, The 'Sleeping Beauty' of Climate Change Policy, Urged to Awaken', *Intellectual Property Watch*, 14 July 2011.
183. Ulrich Brand, Christoph Görg, Joachim Hirsch & Markus Wissen, *Conflicts in Environmental Regulation and the Internationalisation of the state, contested terrains*,

*Figure 3.5. EU, Public International Law & MS*

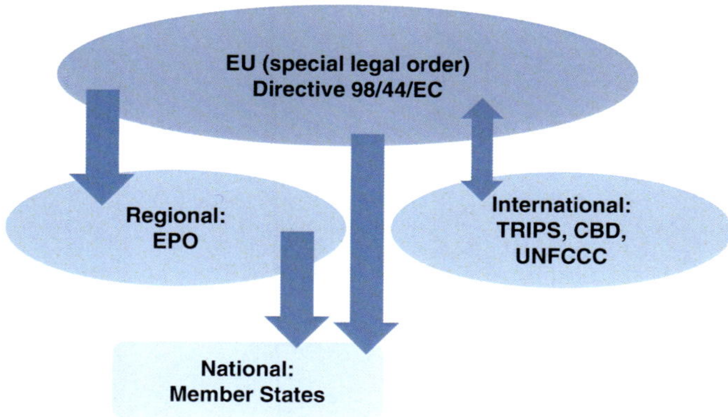


To conclude, the EU legal order seems to be above both regional and international law, with special licence to enter into areas where there are already established inter-governmental arrangements.

2008, p. 116 ff.; Mario Teló, *The European Union and global governance*, 2009, p. 205; and Wayne Sandholtz & Alec Stone Sweet, *European Integration and Supranational Governance*, 2004, p. 298 ff.

CHAPTER 4

# Morality in Europe

Chapter 4 concludes Part I of the thesis by illustrating, on the basis of the patent law concept of morality, how at the regional level patent law is affected by the interests of the EU which may conflict with the European Patent Convention. There is a question of when one authority or the other has the competence to define the content of Directive 98/44/EC on the legal protection of biotechnological inventions. Here it is particularly embryonic stem cells and the interpretation of Article 6(2) of the Directive that are examined. The question is whether in this area there is a need for simplification (can law make life (too) simple?). However there is a development towards greater confusion of competences in connection with the European patent, as in addition to the EPO there are different EU regimes moving at different speeds (the EU2 and the EU25), due to the process chosen for authorising enhanced cooperation, according to which the authority for adopting the EU patent (pursuant to Article 329(1) TFEU) means that not all Member States are required to take part in the cooperation.

## 1. Overview of the legal statutes

Despite the persistent lack of express legal authority on IPR matters in the EC and EU Treaties prior to the Lisbon Treaty, this did not stop the EU's legislators from adopting rules on almost every aspect of intellectual property rights.[1] However, the Union has been less successful within the patent law

1. On the basis of Article 95 of the Treaty establishing the European Community (TEC) (on qualified majority) or Article 308 TEC (on unanimity), the EU has adopted rules on plant varieties, designs, trademarks, geographical indications, copyright, supplementary protection certificates for plants and pharmaceuticals, enforcement and more; see Hanns Ullrich, 'Harmony and unity of European Intellectual property protection', in D. Vaver & L. Bently (eds.), *Intellectual Property in the new millennium,*

area in which so far only one directive has been adopted, on the protection of biotechnological inventions, as mentioned in Chapter 3.[2] The present chapter essentially focuses on the interpretation and scope of the ethical standard in that Directive which, in its classic form states: 'Inventions shall be considered unpatentable where their commercial exploitation would be contrary to *ordre public* or morality'.[3]

As an introduction, the chapter outlines the patent law framework for moral issues. The legal moral standards of patent law have been shaped by different international institutions, for example the Council of Europe (the Strasburg Convention 1963), the European Patent Organisation (the European Patent Convention 1973), the WTO (the TRIPS Agreement 1994) and the European Community (Directive 98/44/EC), and the content and interpretation of the legal standard may differ from one system to another.[4] Furthermore,

*Essays in Honour of William R. Cornish*, 2004, p. 38 ff.; and Trevor Cook, *EU Intellectual Property Law*, 2010, p. 6 ff., Table 1.1, on new unitary Community wide intellectual property rights. See further Catherine Seville, *EU Intellectual Property Law and Policy*, 2009.

2. Directive 98/44/EC of the European Parliament and of the Council of 6 July 1998 on the legal protection of biotechnological inventions (OJ 1998 L 213/13).
3. Chapter 4 is mainly based on an article published in 2008; Tine Sommer, 'Interpreting ordre public and morality in a patent law context: Which is the correct approach?', *Bio-Science Law Review* [2006/2007], 2 BSLR, pp. 62-74. The background paper for that article was a report presented to the Norwegian Advisory Board on Ethical Aspects of Patenting in October 2007 in Brussels; see the report 'Patentnemnd uten portefølje? En analyse av etiske utfordringer ved patentering', Oslo 2008, available at http://etikkom.episerverhotell.net/Documents/Publikasjoner-som-PDF/Patentnemnd%20uten%20portef%C3%B8lje%20%282008%29.pdf (accessed July 2011). Some aspects are also covered in the author's articles (in Danish): 'Sædelighedsbegrebet i relation til bioteknologi (Ordre public and morality in biotechnology)', *Juristen*, 2001, pp. 394-409; 'Patentmyndigheder som bioteknologiens moralske vogter (Patent authorities as moral guardians of biotechnology) in *Nordisk Immateriellt Rättsskydd* (NIR), 2005, pp. 333-343; 'Genpatentet og patentetik i EU – en harmoniseret retsvildfarelse?', in Birgitte Egelund Olsen & Karsten Engsig Sørensen (eds.), *Europæiseringen af dansk ret*, 2008; and 'Om avlsmarkører, miljøhensyn og social ansvarligheds betydning for enerettens legitimitet', in Helle Tegner Anker & Birgitte Egelund Olsen (eds.), 'Festskrift til Ellen Margrethe Basse', 2008.
4. The twin concepts *ordre public* and morality are used to define the outer limits of what present society tolerates and to delimit the absolutely unacceptable: see Geertrui Van Overwalle, 'Legal and Ethical Aspects of Bio-Patenting', in Peter Drahos, *Death of Patents* (Queen Mary Intellectual Property Institute and Lawtext Publishing Limited, 2005), p. 221. The concepts are treated more as two distinct qualities rather than as a collective name for a single quality. See Amanda Warren-Jones, 'Finding a 'Common Morality Codex' for Biotech – A question of Substance', 6/2008, *IIC* Vol.

international society weighs ethical aspects differently in different contexts, and it is relevant to consider whether patent authorities and national courts are equipped to deal with every such aspect during the examination procedure and the post-grant phase.

The classic standard, embracing *ordre public* or morality, has traditionally been supplemented by excluding from patentability methods for the treatment of humans or animals by surgery or therapy, or diagnostic methods practised on humans or animals.[5] Such methods have been defined as lacking industrial application and, more importantly from an ethical perspective, the medical profession is not regarded as a trade.[6] Another ethical concern addressed by the legislators with regard to biotechnology is that the human body as such cannot constitute a patentable invention.[7] On the other hand, plants and animal varieties are also excluded from patentability, but this exclusion is not based solely on ethical reasons.[8]

Today, the area of ethical concern has expanded. Accordingly, patent authorities are required to consider whether a patent application offends against new ethical concerns, even though there is no clear guidance in patent law on the subject. For example, new concerns were raised in the 1992 UN Convention on Biological Diversity (CBD) (the 2020 Bonn guidelines, and the 2010

39, discusses the terms as truly separate concepts; morality is equated with 'ethical values' whereas public order is equated with 'legal values', p. 646.

5. Article 53(c) of the European Patent Convention (EPC) 2000 (formerly EPC Article 52 (4)).
6. See Chapter 1, section 3. Singer & Stauder, *The European Patent Convention, A Commentary*, 3rd Ed., 2003, Vol. 1, Article 52, notes 61-62 and Article 53 note 5.3. The EPC has abandoned the fiction that methods of diagnosis and treatment lack commercial application; now EPC Article 53(c). See further Chapter 1, section 3.2.
7. Directive 98/44/EC, Article 5(1), and Rule 29 of the EPC Implementing Regulations. The concern is not new, and it is further established in the Universal Declaration on Human Genome and Human Rights, adopted on 11 November 1997 at UNESCO's 29th General Conference; pursuant to Article 4 of this Convention the human genome in its natural state shall not give rise to financial gains. See also Chapter 1, section 3.2 and Chapter 5.2. See further Case C 34/10 *Oliver Brüstle*, paras 32-34, where the Court ruled that human dignity is safeguarded in the Directive's Article 5(1), with additional security provided by Article 6(2).
8. Plant varieties can be protected by an institutional plant variety right (for example, the International Convention for the Protection of New Varieties of Plants (UPOV Convention)) but there is no corresponding system for animals. The ban on patent protection for plant varieties was initially accepted in order to prevent double protection.

Nagoya Protocol),[9] such as the fair and equitable sharing of the benefits arising out of the use of genetic resources. Key aspects of the CBD, such as the disclosure of the origin of genetic resources, prior informed consent and technology transfer to developing countries, are also discussed in modern pre-grant patent law.[10] Although these issues relate to the exploitation of a patent, they have been highlighted as pre-grant conditions so as to ensure enforceability. The patent law system has also been affected by ethical issues concerning the post-grant phase of a patent, such as access to medicines via compulsory licensing.[11] Finally, climate change concerns about food scarcity, GMO techniques and environmentally sound technologies have been put in a patent law context, adding to the concerns about pharmaceuticals.

Ethical concerns related to science and scientific progress have also become hot topics in patent law. *Mary Warnock* has written that occasionally science makes procedures possible that are so radical that those at the interface between science and politics are called on to define moral standards for society.[12] She gives the example of the year of 1978 when the first baby was born by *in vitro* fertilisation (IVF). Generalising the case of IVF to other cases, such as embryonic stem-cell research,[13] therapeutic cloning and the con-

9. Bonn Guidelines on Access to Genetic Resources and Fair and Equitable Sharing of the Benefits Arising out of their Utilization (2002) http:// www.cbd.int/programmes/socio-eco/benefit/bonn.aspx#; and the Nagoya Protocol on Access to Genetic Resources and the Fair and Equitable Sharing of Benefits Arising from their Utilization to the Convention on Biological Diversity, 29 October 2010, http://www.cbd.int/abs.
10. On plant genetic rights and such mechanisms, see Abeba Tadesse Gebreselassie, *The sustainability of Plants and Plant intellectual property rights*, 2012.
11. See COM(2004) 258 final, now Regulation (EC) No 816/2006 on compulsory licensing of patents relating to the manufacture of pharmaceutical products for export to countries with public health problems, and Chapter 3.3.2.2.
12. See Mary Warnock, 'The Ethical Regulation of Science', *Nature*, Vol. 450/29 November 2007, p. 615.
13. So far there have been several European reports on the issue, for example, Aurora Plomer (project co-ordinator), 'Stem Cell Patents', *European Patent Law and Ethics Report*, Nottingham, July 2006; Geertrui Van Overwalle, 'Study on the patenting of inventions related to human stem cell research', European Commission, 2002; and The Second Report of the Commission (2005) given in accordance with Article 16(c) of Directive 98/44/EC (COM(2005) 312 final). See also Singer & Stauder, *The European Patent Convention, A Commentary,* Article 53, note 20, where it is argued that this exception raises a number of questions. The EPO has also addressed the stem cell issue in the *Edinburgh* Case, where the dispute concerned claims in which the wording used the term 'animal' in general terms, thus including humans, where it should have included the qualification 'non-human'; see Edinburgh Patent No EP 695351,

struction of mixed species embryos for research purposes, she states that it is important that ignorance and prejudice should not be allowed to dictate the outcome.[14] *Warnock* concludes that without a broad understanding of science and an appreciation of its potential for good, we cannot responsibly erect barriers to scientific advance. From a patent lawyer's viewpoint, there continue to be questions about whether patent law is an appropriate place to discredit science and research. Patent law grants exclusive rights without assessing their commercial exploitation. By having a broad interpretation of ethical standards we allow patent authorities to become the moral guardians of science, sometimes even before the question of commercial exploitation has been resolved by the relevant authorities. The period from the first use and discussions of research techniques, to approval of the technique in scientific communities or the stabilisation of a new technology, and to the grant of marketing authorisation somehow embodies the patent examiner's presumed capacity to make ethical assessments (sometimes with different outcome in different countries). In the view of the Court of Justice of the European Union in Case C-34/10 *Oliver Brüstle*, paragraphs 43-46, one cannot separate or distinguish the use of human embryos for purposes of scientific research, which is the subject-matter of a patent application, from industrial and commercial use, and thus avoid exclusion from patentability.[15]

now final since the appeal was withdrawn 14 years after the patent was filed, see Case T 1079/03, opposition division (not reported). The case study in Chapter 7 includes several patents and patent applications awaiting the *WARF* decision. The Enlarged Board of Appeal of the European Patent Office handed down its decision on 25 November 2008 on European Patent Application No 96903521.1 of Wisconsin Alumni Research Foundation (WARF), Decision G-2/06 (2009) OJEPO 306. The *WARF* decision was published after the 2008 article on Morality, which current chapter refers to, was published in 2008.

14. See Mary Warnock, 'The Ethical Regulation of Science', *Nature*, Vol. 450/29 November 2007, p. 615.
15. Case C-34/10 *Oliver Brüstle* v *Greenpeace*. In para. 45, the Court also referred to the interpretation adopted in Decision G 2/06 of the Enlarged Board of Appeal, 25 November 2008 (306 OJEPO 5/2009). As discussed in Chapter 2.3 the purpose of making a strictly legal interpretation is hard to see when the Court in this case expressly stated that it is not to separate the issues. See also *R* v *Leland Stanford*, Modified Animal Opposition Division 16 August 2001 [2002]; and Amanda Warren-Jones, 'Finding a 'Common Morality Codex' for Biotech – A question of Substance', 6/2008, *IIC* Vol. 39; morality is equated with the 'ethical values' whereas public order is equated with 'legal values', p. 643. It is to be noted that the Opposition Division, in a statement of wider principle, indicated that: 'it cannot be the role of the EPO to act as a moral censor and invoke the provisions of Article 53(a) of the EPC to refuse on ethi-

In order for patent law to remain transparent and for it to stay within its core function, especially when the European moral standard is split between the EPO and the EU, blurring the distinction between the EPO (and its Contracting States) and the EU (and its Member States), it is relevant to argue for a simpler construction of the moral standard in European patent law. It is also necessary to point out that the result of the moral approach in combination with the future unitary patent protection (EU25)[16] promises further disharmony at a higher level. The multilevel regulation that has been described in Chapter 1, Table 1.1, and the power of the playmakers described in Chapter 3, all concern the different levels in global society (national, regional and international), each with their playmakers. Sadly, at the regional level in Europe there is no need to involve other levels in order to complicate things. The competition between authorities to establish the moral standard, blurring the distinction between supranational and inter-governmental powers, is without precedent.

## 2. Morality as a legal standard: the legal framework

Ethical concerns are often complex and difficult to harmonise, and the attitude has often been that patent law is not the right place to harmonise the moral issues raised by an invention.[17] However, since the Strasbourg Convention of 1963,[18] moral considerations have had a role as a legal standard in European patent law. Article 2 of the Strasbourg Convention states:

> 'The Contracting States shall not be bound to provide for the grant of patents in respect of:
> (a) inventions the publication or exploitation of which would be contrary to 'ordre public' or morality, provided that the exploitation shall not be deemed to be so contrary merely because it is prohibited by a law or regulation.'

The morality standard has been elaborated in Europe since 1963. The first step in this elaboration at the regional level was the adoption of the 1973 Eu-

cal grounds to grant a patent on legal research and directed to an invention indisputably associated with medical benefits'.

16. See Chapter 3, section 4.2.
17. See, e.g., Gerald Kamstra et al., *Patents on Biological inventions: The EC Directive*, 2002, p. 41.
18. Convention on the Unification of Certain Points of Substantive Law on Patents for Invention of 27 November 1963.

ropean Patent Convention (EPC), with wording very close to the Strasbourg version. The wording of the relevant provision, Article 53(a), states:

> 'Exceptions to patentability.
> European patents shall not be granted in respect of:
> (a) inventions the publication or exploitation of which would be contrary to 'ordre public' or morality, provided that the exploitation shall not be deemed to be so contrary merely because it is prohibited by law or regulation in some or all of the Contracting States.'

At a broader international level, the 1994 TRIPS Agreement refers to a moral legal standard which is framed in more modern terms than Strasbourg and the EPC. Article 27(2) of TRIPS reads:

> 'Members may exclude from patentability inventions, the prevention within their territory of the commercial exploitation of which is necessary to protect *ordre public* or morality, including to protect human, animal or plant life or health or to avoid serious prejudice to the environment, provided that such exclusion is not made merely because the exploitation is prohibited by their law.'

Thus TRIPS Article 27(2) has interpreted the twin concepts of *ordre public* and morality as including the lives and health of humans, animals and plants as well as the avoidance of serious prejudice to the environment.

In the EU, the 1998 Directive 98/44/EC on the legal protection of biotechnological inventions established the conditions for granting a patent when the subject of the invention consists of elements isolated from the human body and from plants and animals. The Directive clearly states the requirement for respect for human dignity, which is referred to in several of its provisions and recitals.[19] The morality standard in Article 6 is a further development of the classic standard as it is placed within, and thus elaborated in an exclusively biotechnological context.

Besides the usual references to *ordre public* or morality, there is a non-exhaustive list referring to different kinds of biotechnological processes which are considered non-patentable per se. The first proposal only referred to a list containing:

19. See recitals 20, 21 and 38, and Article 5(1) and (3) and Article 6 of Directive 98/44/EC; see Antonina Bakardjieva Engelbrekt, 'Institutional and jurisdictional aspects of stem cell patenting in Europe (EC and EPO): Tensions and Prospects', in Aurora Plummer & Paul Torremans (eds.), *Embryonic Stem Cell Patents, European Law and Ethics*, 2009, p. 245.

(a) the human body or parts of the human body per se;[20]
(b) processes for modifying the genetic identity of the human body for a non-therapeutic purpose which is contrary to the dignity of man;
(c) processes for modifying the genetic identity of animals which are likely to inflict suffering or physical handicaps upon them without any benefit to man or animal.[21]

This proposal was rejected in March 1995. The proposal that followed in December 1995 referred to methods of human treatment involving germ-line gene therapy, and processes for modifying the genetic identity of animals only.[22] The 1997 draft changed the wording of subsection (a) to procedures for the reproductive cloning of a human, and of subsection (b) to procedures for modifying the genetic identity of human reproductive tissue. A new subsection (c) was inserted to cover procedures involving the use of human embryos, and, finally, the previous subsection (b) on procedures for modifying the genetic identity of animals was retained as a new subsection (d). The wording which applies today amended the 1997 draft, since there is no express reference to reproductive cloning in the enacted provisions, but there is a reference to it in recital 41. The prohibition of the patenting of methods involving the use of human embryos has been softened to exclude commercial use. The Directive, as adopted, has clearly given priority to the exclusion of inventions related to the human body, but it only excludes the patentability of inventions related to animals if the invention causes suffering to animals without any benefit to humans or animals.[23] In its final version Article 6 states:

1. Inventions shall be considered unpatentable where their commercial exploitation would be contrary to *ordre public* or morality; however, exploitation shall not be deemed to be so contrary merely because it is prohibited by law or regulation.

2. On the basis of paragraph 1, the following, in particular, shall be considered unpatentable:
(a) processes for cloning human beings;
(b) processes for modifying the germ line genetic identity of human beings;

20. Now Directive 98/44/EC, Article 5(1).
21. COM(1992) 589 final.
22. See COM(1995) 661 final, Article 9(2); and Tine Sommer, 'The Scope of Gene Patent Protection and the TRIPS Agreement – An Exclusively Nondiscriminatory Approach', 38 *IIC* 43 (2007).
23. See Tine Sommer, 'The Scope of Gene Patent Protection and the TRIPS Agreement – An Exclusively Nondiscriminatory Approach?', 38 *IIC* 43 (2007).

(c) uses of human embryos for industrial or commercial purposes;
(d) processes for modifying the genetic identity of animals which are likely to cause them suffering without any substantial medical benefit to man or animal, and also animals resulting from such processes.[24]

As mentioned in Chapter 3, the Directive became effective in the Member States in July 2000, but before that in 1999 it was implemented in EPC 1973 by being inserted as new Implementing Regulations 23b to 23e (now Rules 26-29).[25] Today, the amended Article 53 EPC 2000, states as follows:

European patents shall not be granted in respect of:
(a) inventions the commercial exploitation of which would be contrary to '*ordre public*' or morality; such exploitation shall not be deemed to be so contrary merely because it is prohibited by law or regulation in some or all of the Contracting States;
(b) plant or animal varieties or essentially biological processes for the production of plants or animals; this provision shall not apply to microbiological processes or the products thereof.
(c) methods for treatment of the human or animal body by surgery or therapy and diagnostic methods practised on the human or animal body; this provision shall not apply to products, in particular substances or compositions, for use in any of these methods.

Thus Article 53 EPC 2000 contains an exhaustive list of exceptions to the general rights to patents in Article 52 EPC 2000.[26] However, Rule 26(1) states that Directive 98/44/EC of 6 July 1998 on the legal protection of biotechnological inventions shall be used as a supplementary means of interpretation. The Implementation Regulations (EPC 2000), Chapter V, Biotechnological inventions, Rule 28 reads:

24. Graeme Laurie writes in *A Commentary on Ethical Aspects of Patenting to the Norwegian Advisory Board on Ethical Aspects of Patenting,* p. 8: 'difficulties have arisen because of a consistent failure to distinguish between objections to an invention itself and objections about its commercial exploitation. We have arrived at this situation because we have never been clear about what the morality provision in patent law actually mean nor about what purposes they should serve. This has been compounded by the adoption of the ill-conceived terms of Article 6 which in reality represent little more than a snapshot of the hottest ethical controversies at the time when the Directive was adopted'.
25. Implementing Regulations to the Convention on the Grant of European Patents of 5 October 1973 as adopted by decision of the Administrative Council of the European Patent Organisation of 7 December 2006 and as last amended by decision of the Administrative Council of the European Patent Organisation of 26 October 2010.
26. See Singer & Stauder, *The European Patent Convention, A Commentary*, Article 53, note 5.3.

Exceptions to patentability
Under Article 53(a), European patents shall not be granted in respect of biotechnological inventions which, in particular, concern the following:
(a) processes for cloning human beings;
(b) processes for modifying the germ line genetic identity of human beings;
(c) uses of human embryos for industrial or commercial purposes;
(d) processes for modifying the genetic identity of animals which are likely to cause them suffering without any substantial medical benefit to man or animal, and also animals resulting from such processes.

To conclude, it must be emphasised that the scope of the legal statutes in relation to the moral standard has been expanded. From Strasbourg in 1963 to the 1998 Directive, the legal provisions have expanded to express concerns about the environment, protection of human, animal and plant life (TRIPS) and four scientific techniques: the cloning of human beings, modifying human germ lines, the use of human embryos for commercial purposes, and animal suffering without substantial medical benefit (Directive 98/44/EC, Article 6(2)). While the concepts of *ordre public* and morality are often used in other legal contexts, as in Article 36 TFEU, there is no agreement about its correct interpretation.[27]

Although the wording of the standards are formulated similarly at different regulatory levels (Strasbourg, EPC, EC), their interpretations at the different levels are legally independent of each other. The wording of the Strasbourg Convention was mirrored in the EPC. The TRIPS Agreement comprises a minimum standard, safeguarding the content in all systems. However, the vulnerable point in European patent law is the increasingly mixed EPO and EU competences. In Opinion (1/09) of March 2011 of Court of Justice, the Union law system by which the Court gives preliminary rulings was stated to be indispensable to the preservation of the very nature of European Union law.[28] Thus, a system which mixes together the competences of the EU and the EPO, and with a patent court which deprives the Court of Justice and the courts of the Member States of their jurisdiction, is not an option. Correspondingly, in its Decision G 2/06 of 25 November 2008, the Enlarged Board

27. It is notable that the concepts are also used in legal contexts other than the patent system, such as in the Article 36 TFEU. Despite several Treaty amendments (Treaties of Rome, Maastricht, Amsterdam, Nice and Lisbon) the wording of Article 36 TFEU has never changed. Thus the protection of the environment and animal welfare are not relevant concerns for the purposes of this provision, but are subject to a judicial doctrine (*Cassis de Dijon*); see Tine Sommer, 'Sædelighedsbegrebet i relation til bioteknologi' (Ordre public and morality in biotechnology), *Juristen* 2001, pp. 394-409.
28. OJ 2011 C 211/2, see point 89; and Chapter 3, section 4.2.

of Appeal stated that 'the Boards of Appeal of the EPO are not courts or tribunals of a member state of the EU, and there is no power under the EPC for a Board of Appeal to refer questions to the ECJ'.[29] Thus, the Implementing Regulations affect the case law of the Boards of Appeal of the EPO and complement Article 53 EPC in their own way; and at the same time, Directive 98/44/EC affects the EU Member States in the same way that a directive normally works in the Union legal order, without regard for the parallel regional patent system outside the EU.

## 3. The relationships between Article 6(2) of Directive 98/44/EC and Rule 28 of the Implementing Regulations, and between Article 6(1) of Directive 98/44/EC and Article 53(a) of the EPC

The wording of Rule 28 in the Implementing Regulations for the EPC corresponds to the wording of Article 6(2) of the Directive, and the wording of Article 53(a) of the EPC corresponds to the wording of Article 6(1) of the Directive. As the EPC already contains the classic moral standard in Article 53(a), only Article 6(2) is implemented in Rule 28.

In 1999, the Standing Advisory Committee before the EPO (SACEPO) outlined the approach to the implementation of the Directive in European patent law.[30] SACEPO examined those parts of the Directive which required implementation and it is notable that the Committee mentioned a comprehensive list of recitals which have to be taken into account when interpreting and implementing the Directive and applying the provisions adopted in accordance with it. In point 8 it is further stated that:

> ... the principles set forth in the Directive regarding the patentability of biotechnological inventions are based on the relevant provisions of the EPC and essentially reflects current practice as developed by the Office and its boards of appeal in applying the Convention, some extensions and clarifications are required in this area to ensure that the patentability provisions of the EPC also continue to be interpreted in keeping with the Directive.

*Point 11:* The provisions in the Directive, Chapter 1 are essentially based on considerations of the boards and on their interpretation of the relevant provisions of the EPC.

29. Decision G 2/06 of the Enlarged board of Appeal of 25 November 2008 (306 OJEPO 5/2009).
30. See SACEPO 4/99 of 25 February 1999.

*Point 15:* The Implementing Regulations to the EPC are thus the appropriate instrument for implementing the requirements of the Directive in European patent law on schedule and with binding force. Under Article 164(1) EPC the Implementing Regulations are an integral part of the Convention and hence are equally binding on the EPO's boards of appeal (Art. 23(3) EPC) and on national courts. The practice of making general provisions of the Convention specific in the form of implementing regulations is accepted as an essential instrument for the development of legislation.

The relationship between the list (derived from Article 6(2) of the Directive) as implemented in the Implementing Regulations and Article 53(a) was characterised by the Board of Appeal in decision in Case T 315/03, point 6.1, as follows:

'It follows from the preamble, i.e. from the use of the words 'in particular' that the rule is not intended to provide an exhaustive list of inventions excluded from patentability but that, on the contrary, it is limited to four categories – its effect is simply to ensure that inventions which fall within sub-paragraphs (a), (b), (c) or (d) of Rule 23 d EPC must not be granted patents under Article 53(a) EPC. A case not falling within the Rule does not thereby escape Article 53(a) EPC: there might well be biotechnological inventions not falling within sub-paragraphs (a) to (d) which, nonetheless, must not be granted patents under Article 53(a) EPC. In short, a case falling within one of the four categories must ipso facto be denied a patent under Article 53(a) EPC and there is no need to consider that Article further; but, on the contrary, a case not falling within one of those categories must be considered further under Article 53(a) EPC.'[31]

## 4. The European Patent Convention, the EPO Boards of Appeal, the Implementing Regulations and the Court of Justice

The European Patent Convention and the Implementing Regulation to EPC are to be interpreted by the European Patent Office Boards of Appeal and the Enlarged Board of Appeal. At the regulatory level, the relationship between the European Patent Convention and the Implementing Regulations are regulated in Article 164(1) EPC, which states that the Implementing Regulations shall be integral parts of the Convention. Article 164(2) EPC states as follows:

'in case of conflict between the provisions of this Convention and those of the Implementing Regulations, the provisions of the Convention shall prevail.'

31. *Harvard/Transgenic animals*, Case T 315/03, OJ EPO 1/2006, pp. 15-82.

The relationship between national patent authorities and the European Patent Organisation is transparent. European patents are granted by the EPO, whereas national patents are granted by national authorities. In Article 2(2) EPC it is stated that the 'European patent shall in each of the Contracting States for which it is granted, have the effect of and be subject to the same conditions as a national patent granted by that State, unless otherwise provided in this Convention'. Furthermore, some of the EPC provisions on European patents take precedence over national law, in particular the text of a European patent application is the authentic text (Article 70 EPC) and according to Article 69 and its associated Protocol of Interpretation, the extent of protection of a European patent application and of a European patent is determined on an exclusively European basis.[32]

As mentioned above, until 1998, patent law harmonisation was only an issue at the national, regional and international levels (EPC, TRIPS, and WIPO), excluding the EU. However, the only substantive Union patent legislation, concerning only one (but a major) industry, changed the landscape.[33] As the Directive is addressed to the 27 EU Member States, it intervenes in the legal arrangements between the European Patent Organisation and its Contracting States, or at least those of its Contracting States that are part of the EU. Since the Directive complements Article 53 EPC, as implemented in Rule 28 of the Implementing Regulations (EPO, 2000), as well as determining substantive EU patent law, conflicts can be expected between the EU, the European Patent Organisation and the Member States/Contracting States subject to both regimes. Such risk of conflict can only be increased by the enhanced cooperation on EU unitary patent protection, separating the EU unitary patent protection from that of the EPO (without unitary protection).[34] As emphasised in Chapter 3.2 and 3.4.2 (e.g. the reference to Opinion 1/09 of March 2011[35]), the Court of Justice is the final interpreter of Union law on

32. See Singer & Stauder, *The European Patent Convention, A Commentary*, 3rd Ed., 2003, Vol. 1, Article 2, note 12.
33. See also Directive 2004/48/EC of the European Parliament and of the Council of 29 April 2004 on the enforcement of intellectual property rights.
34. Art. 3: 'European patents granted with *the same set of claims* in respect of all *the* participating Member States shall benefit from unitary effect in the participating Member States provided that their unitary effect has been registered in the Register for unitary patent protection *referred to in Article 2(da)*. European patents that were granted with different sets of claims for different participating Member States shall not benefit from unitary effect', draft proposal 1 December, 2011, see Chapter 3, section 4.2.
35. Opinion of the Court (Full Court) of 8 March 2011. Opinion delivered pursuant to Article 218(11) TFEU – Draft agreement – Creation of a unified patent litigation sys-

EU patent questions brought before it. In the event of a referral for a preliminary ruling, the Court determines how the provisions of the Directive, including the preamble and therefore the recitals, are to be applied. The Member States are obliged both to implement the Directive in their national law and to comply with the rulings of the Court of Justice. Apart from there being an obligation to implement the Directive and thus harmonise national patent laws in accordance with the Directive, the Union has no authority with regard to the Member States' patent laws.[36] However, the Directive exerts a Union influence on the interpretation of every aspect of biotech patenting of human, animal and plant material. That which it was not possible to harmonise under the heading of substantive unitary patent protection became, under the Directive, a uniform obligation to protect biotech patents in the Member States. So, far the effect of this legislation has not been seen in practice as the Commission has been very reluctant to take action in this area and the Court has so far been very cautious in its rulings. However, the EU's new approach with enhanced cooperation may change this.

However, one can doubt whether that will continue to be the case. Given the Report from the Commission to the Council and the European Parliament – Development and implications of patent law in the field of biotechnology and genetic engineering (COM (2005) 312 final); the decisions in Case C-377/98 Commission v The Netherlands and Case C-456/03 Commission v Italy, where it was first noted that the Commission was reluctant to define the content in biomedical patenting and the Court emphasised that Article 6(2) of Directive 98/44/EC is an EU invention with an EU monopoly on the content; Case C-34/10 *Oliver Brüstle v Greenpeace*, where in paragraph 26 the Court of Justice stated that the same provision (Article 6(2)) designated an autonomous concept of Union law which must be interpreted in a uniform manner throughout the territory of the Union; it is hard to speak of a soft approach. With the enhanced procedure (EU25) and with the Court's Opinion 1/09 highlighting the mandate of preliminary rulings, it is even harder to see how Article 69 EPC and its associated protocol of interpretation, together with the Enlarged Board of Appeal's Opinion in G 2/06, can come close to achieving a new regional patent. If the EU 25 unified patent takes one direction and the other EPO Contracting States take another, then the EPO 11 can follow that

tem – European and Community Patents Court – Compatibility of the draft agreement with the Treaties. Opinion 1/09 available at http://eur-lex.europa.eu/LexUriServ/LexUriServ.do?uri=CELEX:62009V0001:EN:NOT (last visited January 2012).

36. Except for the doctrine of the exhaustion of rights according to TFEU Article 36 (as developed by the ECJ).

of the EPO Board of Appeal's (the new patent court's) interpretations, the EU 25 can follow the direction of the Court of Justice, and finally the EU 2 will eventually have to follow the direction of the Court of Justice, not by virtue of being bound by the unified patent, but as a EU Member States bound by Directive 98/44/EC. A new discussion follows: The expertise of specialised patent courts (the EPO Boards of Appeal and the new Patent Court) may be overruled by the 'higher court' which do not have patent specialisation (CJEU, without jurisdiction in some of the EPO Members) but seeks to protect specific EU policy interests.

In 2011, in Case C-34/10 *Oliver Brüstle*, the German Bundesgerichtshof referred questions to the Court of Justice for preliminary rulings on the moral standard in Europe. In his opinion, Advocate General Bot argued that concept from the directive moral clause list in Article 6(2) of Directive 98/44/EC (a human embryo[37]) must be subject of a common understanding in all Member States of the European Union. The Governments of the Member States claimed that the concept of a human embryo must be left solely to their discretion. Advocate General Bot argued that the concept must be defined *autonomously specifically* for Union law (point 53, emphasis added).

> 56. If it were left to the Member States to define the concept of a human embryo, in view of the differences which exist in this regard, this would mean, for example, that an invention like that of Mr Brüstle could be granted a patent in some Member States, while the patentability of such an invention would be excluded in others. This would run counter to the main objective of the directive, which is to establish effective and harmonised legal protection of biotechnological inventions.

> 60. Thirdly, as regards the scope accorded to the Member States by Article 6(2) of the directive, the Court ruled that that provision allows the Member States no discretion with regard to the unpatentability of the processes and uses which it sets out. This binding aspect of one of the key provisions of the directive would also seem to call for a uniform interpretation of the concept of a human embryo within the Union. I cannot see how such a categorical prohibition, applying to all the Member States, could exist on the basis of concepts which were not common.

Thus, despite the lack of general substantive EU patent law, the Directive has effectively harmonised the Union biotech-specific moral standard in Article 6(2) of Directive.[38]

37. Point 61: 'I therefore take the view that the concept of a human embryo must have a Community understanding.'
38. See Hanns Ullrich, 'Harmony and unity of European Intellectual property protection', in D. Vaver and L. Bently (eds.), *Intellectual Property in the new millennium, Essays*

The Court of Justice ruled on Article 6 of the Directive in two cases prior to the *Brüstle* case. These were Case C-377/98 *Commission v The Netherlands* and Case C-456/03 *Commission v Italy*; these will be discussed below. The decisions in these cases need not be followed by the EPO when granting patents, nor will the European Patent Office Boards of Appeal be affected by such rulings, even though the Implementing Regulations implement the Directive, as the Enlarged Board of Appeal made clear in the above mentioned Decision G 2/06 *WARF*. As mentioned by SACEPO, the comprehensive list of recitals is also to be taken into account when interpreting the Directive, but so far as the Implementation Regulations Rule 26(1) is concerned, the EPC only states that the Directive shall be used as a supplementary means of interpretation. The Enlarged Board of Appeal also elaborated on the competences in its Decision G 2/06, points 6-11, where it stated:

6. The mere identity of the wording of Rule 28(c) (formerly 23d(c)) EPC and of Article 6(2)(c) of the Directive cannot lead to the conclusion that the ECJ now has jurisdiction to decide matters for the EPO under the EPC. The Boards of Appeal apply the provision because it is law under a specific Rule of the Implementing Regulations to the EPC, and not because the Directive is a source of law to be applied directly. This is corroborated by the fact that Rule 26(1) (formerly 23b(1)) EPC only states that the Directive shall be used as a supplementary means of interpretation of Rules 26 to 29 (formerly 23b to 23e) EPC.

7. Article 23(3) EPC provides that in their decisions the members of the Boards shall not be bound by any instructions and shall comply only with the provisions of this Convention. While Article 23(3) EPC is in its present form, the Enlarged Board concludes that neither it, nor any Board of Appeal of the EPO, has the power to bind itself to follow a ruling of the ECJ on the interpretation of Article 6(2)(c) of the Directive and apply this to Rule 28(c) (formerly 23d(c)) EPC.

8. The Enlarged Board has not been made aware of any precedent for asking the ECJ for a consultative opinion and it must be questionable whether the ECJ would entertain such a request in a situation where it would be unclear as to who would be entitled to make submissions to the ECJ on any questions submitted.

9. The Appellant seeks to rely on ECJ Case C-337/95 ('*Dior v Evora*') on the basis that the situation of the Benelux Court being allowed to make references to the ECJ in relation to matters referred to it by the highest courts of Belgium, the Netherlands and Luxembourg

*in Honour of William R. Cornish*, 2004, p. 25: 'It was mainly the combined effect of the rapidity of technological progress and slowness of progress in patent law unification which caused the Community astonishingly to intervene by harmonization of national law in an area of patent law where protection will hardly ever be sought only on a national basis: biotechnology'.

suggested a precedent for a referral by the Enlarged Board to the ECJ. Closer consideration destroys this as a suitable precedent. If the Enlarged Board can make a referral under Article 234 of the EC Treaty to the ECJ, then this possibility would also have to apply to each of the EPO Boards of Appeal, because against their decisions too there is no appeal, and a referral to the Enlarged Board by them is optional. The position of a Board of Appeal cannot be compared to that of one of the highest courts of Belgium, the Netherlands and Luxembourg, each of which is clearly a national court of an EU member state entitled to make a referral to the ECJ. The referral in *Dior v Evora* was by the Netherlands Hoge Raad, which took it as a premise that either it or the Benelux Court could make a reference to the ECJ but wished to know whether it was obliged itself to make a reference or could leave it to the Benelux Court as the highest court for matters governed by the Benelux treaty. Further the latter is a special treaty set up with permission of the European Community authorities as a regional treaty. The Benelux Court is composed of three judges from each of the highest courts of Belgium, the Netherlands and Luxembourg, and this position as a national judge is a requirement for being a member of the Benelux Court. The Benelux Court can thus be considered in relation to matters for which it has jurisdiction under the Benelux treaty as the highest national court of each of these three EU states. In contrast to this, some or possibly even all the members of a Board of Appeal might not even be nationals of an EU state. The Enlarged Board of Appeal is unable to deduce from *Dior v Evora* anything suggesting that the ECJ would regard a reference by an EPO Board of Appeal as permissible under Article 234 of the EC Treaty.

10. That the seat of the EPO Boards of Appeal is in a member state of the EU, Germany, cannot alter their status as part of an international organisation with jurisdiction conferred under the EPC.

The EPO Boards of Appeal are not and have never been treated as courts or tribunals of their host country.

11. For the above reasons the Enlarged Board concludes that it has no power to ask the ECJ for a preliminary ruling under the existing provisions of the EPC, so that the request for referral of the questions for a preliminary ruling by the European Court of Justice must be refused as inadmissible.[39]

In Case T 83/05 (point 57),[40] which was an interlocutory decision of May 2007 of the Technical Boards of Appeal of the European Patent Office, it was emphasised that pursuant to Article 164(2) EPC, the true interpretation of an Article of the EPC as established by a ruling of the Enlarged Board of Appeal

39. Decision G 2/06 of the Enlarged Board of Appeal of 25 November 2008 (306 OJEPO 5/2009). At point 31 the Enlarged Board of Appeal concluded that that the legislators (*both the legislator of the Implementing Regulations to the EPC and of the Directive)* wanted to exclude from patentability inventions such as that in the appeal, and that in doing so they have remained within the scope of Article 53(a) EPC and of the TRIPS Agreement (emphasis added).
40. T 83/05 (OJ EPO 2007, 644).

cannot be overturned by a new rule of the Implementing Regulations which conflicts with the interpretation. In Case T 315/03, the Board also stated its view on the balancing test in Case T 19/90.[41] The Board showed clearly that the balancing test as developed by the Board in Case T 19/90 embraced a wider range of benefits and was thus broader than the Rule 23(d) test. The relationship between Rule 23(d) EPC (now Rule 28(d)), Directive Article 6(2)) and the EPO case law is summarised as follows: 'the effect of this interpretation [of Rule 23(d) EPC] is to insert a test which, depending on the facts and thus on the outcome of the test, may be either additional or alternative to that previously established by case law'.[42] And the decision of the Enlarged Board of Appeal (G 2/06), point 31, suggests the balancing test may be further elaborated in the future:

'In view of this result, it is not necessary nor indeed appropriate to discuss further arguments and points of view put forward in these proceedings such as whether the standard of ordre public or morality should be a European one or not, whether it matters if research in certain European countries involving the destruction of human embryos to obtain stem cells is permitted, *whether the benefits of the invention for humanity should be balanced against the prejudice to the embryo,* or what the point in time is to assess ordre public or morality under Article 53a EPC. The legislators have decided, remaining within the ambit of Article 53(a) EPC, and there is no room for manoeuvre' (emphasis added).[43]

The decision in G 2/06 reflected a shift in the EPO's approach to morality. Certain 'norms' cannot be balanced even if they comprise potential benefits. One such norm is respect for human dignity. Balancing human dignity is thus inappropriate.[44] In principle, this may result in diverse morality standards, depending on which norms are classified as superior norms by different interpreters. The first standard may cover the European Union, the second may cover the Contracting States of the European Patent Conventions, and the third may cover countries that find themselves in between the EU and the EPC. It can be questioned whether the last of these will survive the new ap-

41. T 19/90 (OJ EPO 1990, 476).
42. Case T 315/03, (OJ EPO 2006, 15), point 7.7.
43. Decision G 2/06 of the Enlarged Board of Appeal of 25 November 2008 (306 OJEPO 5/2009). In T-1374/04 *Stem cells/WARF*, Technical Board of Appeal (*OJ* EPO 2007, 313), point 55, the Board of Appeal stated: 'The Board has doubts whether, when it comes to human life, it would be ethically acceptable to make a decision by weighing the interests of human beings which could potentially benefit from the exploitation of the technology against a right if any of human embryos'.
44. See also Sigrid Sterckx & Julian Cockbain, *Exclusions from patentability, How far has the European Patent Office eroded Boundaries?*, 2012, pp. 243 and 286.

proach with the enhanced cooperation for a unitary patent protection for the EU25. Only Spain and Italy fall within the third group, and only partly due to their EU membership, but this make the distinction between the EPO Contracting States and the EU Member States (minus Spain and Italy) even more artificial. Whether the prevailing standards in the EPC, the EU and the national states differ is discussed below.

## 5. Interpretation of the legal morality standard: *Ordre Public* and morality – General guidelines from the EPO Boards of Appeal

The most important provision on the morality standard is Article 53(a) EPC, as interpreted by the EPO over several decades.[45] As mentioned above, Article 53(a) EPC contains the traditional exception to patentability, embracing modern social ethical concerns. The provision has had renewed relevance in relation to biotechnological developments, especially with the *Oncomouse* decision, the *Relaxin* case and the *Plant Genetic System* case.[46] Thus, in the words of the former president of the EPO, Alain Pompidou, Article 53(a) EPC constitutes a blanket clause, which ensures the incorporation and enforcement of fundamental ethical and legal principles in patent law, and is framed so as to accommodate future developments.[47] Furthermore, he argues:

45. In his legal writing, Oliver Mills finds that the EPO is unclear in its guidance, and that Article 53(3) is not an appropriate mechanism for determining exceptions to patentability. He argues that what is required is a definition of *ordre public* and morality that can be controlled by specialised agencies. See Oliver Mills, *Biotechnological Inventions, Moral Restraints and Patent Law*, 2005, pp. and 54. An example of a specialised agency is the Norwegian Advisory Board on Ethic Aspects, established in 2004. According to Section 15a of the Norwegian Patent Act, the Patent Office may obtain an advisory statement from the Board before making a decision. The Board cannot question the law as it stands, but according to section 24(2) the effect of the Norwegian system is that the 'closing date for submitting an opposition in this regard has been extended from the usual nine months to three years from the date the patent was granted. If however the opposition is filed later than nine months after this date the opponent shall pay a prescribed fee.'
46. Case T 19/90 (OJ EPO 1990, 476), V 8/94 (OJ EPO *1995*, 388) and Case T 356/93 (OJ EPO 2006, 15).
47. See G 2/06, Comments by the President of the European Patent Office, September 2006, p. 14. See also Tine Sommer (in Danish): 'Sædelighedsbegrebet i relation til bioteknologi' (*Ordre public* and morality in biotechnology); *Juristen* 2001, pp. 394-

The generality of Article 53(a) EPC ensures that it can be applied to inventions in the fast evolving field of biotechnology that cannot at present be foreseen. It also ensures that possible changes in fundamental legal and ethical principles can be taken into consideration and immediately incorporated into patent law.[48]

It is the view of this author that, as a fundamental and high-ranking norm, the core of Article 53(a) EPC comprises a provision that is future-proof in the sense of being technology neutral. The addition of a detailed set of technologies (Rule 28) to its scope is an example of the 'export of EU norms' that has been discussed in Chapter 3. The EPO's reluctance to balance new concerns and classify impermissible areas, such as human dignity, need not be derived from Union law.

### 5.1 A uniform moral standard in Europe?

In the EPO case law and in the legal literature[49] it is assumed that Article 53(a) EPC refers to the moral norms of the individual Contracting States. *Singer and Stauder* describe the situation as follows:

'European Patent law contains no reason for abandoning the concepts of *ordre public* and morality that apply to national patent law (however T 356/93 [the culture in question is the culture inherent in European society and civilisation]). The relevant standards largely correspond in all Contracting States, but not without exceptions.

European patent law provides two means of dealing with the differences in the legal and moral systems of the Contracting States:

If the invention is contrary to *ordre public* or morality prevailing in only one or a part of the designated States, the applicant may, on his own initiative or following a corresponding ruling of the examination division, withdraw the designation of the States in question while maintaining it for the others. In this way he acquires protection in the other

409; and 'Patentmyndigheder som bioteknologiens moralske vogter' (Patent authorities as moral guardians of biotechnology) in *Nordisk Immateriellt Rättsskydd* (NIR), 2005, pp. 333-343.

48. See G 2/06, Comments by the President of the European Patent Office, September 2006, also citing case G 1/98, stating that laws are not restricted in their application to situations known to the legislator. See Amina Agovic, 'Stem cell patents on a knife edge', *Journal of Intellectual Property Law and Practice*, 2008, Vol. 3, No 11, pp. 724-725.
49. See Ulrich Schatz, 'Patentability of Genetic Engineering in European Patent Office Practice', 1 *IIC* 1998.

States. If, however, the applicant does not make use of that opportunity, the application must be refused as a whole.'[50]

Article 53(a) EPC was subject to the interpretation of the Boards of Appeal of the EPO in Case T 356/93 *Plant Genetic Systems.*[51] The Board referred to the historical documentation, where the EPC Working Party recognised that there was no European definition of morality (or *ordre public)* and that the interpretation of the concepts should be a matter for European institutions (point 4). In points 5 and 6, the Board elaborated on the concepts as follows:

'5. It is generally accepted that the concepts of 'ordre public' covers the protection of public security and the physical integrity of individuals as part of society. This concept encompasses also the protection of the environment. Accordingly, under Article 53(3) EPC, inventions the exploitation of which is likely to breach public peace or social order (for example, through acts of terrorism) or to seriously prejudice the environment are to be excluded from patentability as being contrary to 'ordre public'.

6. The concept of morality is related to the belief that some behaviour is right and acceptable whereas other behaviour is wrong, this belief being founded on the totality of the accepted norms which are deeply rooted in a particular culture. For the purpose of the EPC, the culture in question is *the culture inherent in European society and civilisation.* Accordingly, under Article 53(a) EPC, inventions the exploitation of which is *not* in conformity with the conventionally-accepted standard of conduct are to be excluded from patentability as being contrary to morality.' (Emphasis added)

Thus, in an EPC context the moral standard refers to a harmonised inherent culture in European society and civilisation, which indicates that the Contracting States' room for manoeuvre in this regard has been narrowed. The philosophy of the EPC is based on the idea that the morality provision should be based on a unitary European standard.[52] However, it is currently an open question whether the European standard comprises a two or three class system as it is obvious from the Decision G 2/06 *WARF* and Case C 34/10 *Oliver Brüstle* that European morality is not a class of its own. If one counts Spain and Italy as a separate class, it can easily be a three class system. However, in the case of embryonic stem cells 'the harmonisation' follows from

50. See Singer & Stauder, *The European Patent Convention, A Commentary*, 3rd Ed., 2003) Vol. 1, Article 53, notes 22 and 23.
51. Decision of 21 February 1995 (OJ EPO 2006, 15).
52. Case G 2/06, Comments by the President of the European Patent Office, 28 September 2006, p. 4. http://www.cipa.org.uk/download/EPO_WARF.pdf (last visited January 2012).

the newly amended Examination Guidelines, 20 June 2012, Part G-Ch. II-15, 5.3 (iii):

> 'A claim directed to a product, which at the filing date of the application could be *exclusively* obtained by a method which necessarily involved the destruction of human embryos from which the said product is derived is excluded from patentability under Rule 28(c), even if said method is not part of the claim (see G 2/06). The point in time at which such destruction takes place is irrelevant.'[53]

### 5.2. Proof of accepted norms

As guidance, in Case T 356/93, point 13, the Board emphasised that the right approach is to look at the particular facts of each case and to examine whether, in the light of those facts, the case ought to stand. In this regard one cannot rely on a survey or opinion poll as proof of public opinion to the effect that patents should not be granted for certain kinds of inventions. It was the view of the Board that surveys and so on do not necessarily reflect *ordre public* concerns or moral norms that are deeply rooted in European culture, but tend to reflect special interests or biased beliefs. If they were to be relied upon, they would have to be made ad hoc on the basis of specific questions and in relation to the particular subject-matter claimed, which would scarcely be feasible.

Furthermore, in point 15 the Board reasoned as follows:

> 'Like national law(s) and regulation(s) approving or disapproving the exploitation of an invention ... a surveyor an opinion poll showing that a particular group of people or the majority of the population of some or all of the Contracting States opposes the granting of a patent for a specified subject matter, cannot serve as a sufficient criterion for establishing that the said subject matter is contrary to 'ordre public' or morality.'[54]

As to the validity of opinion polls as barometers of the public perception of morality, this was also discussed in Case T 356/93, point 10.4, (with reference to Case T 315/03), and the Board confirmed its earlier view as follows:

53. Taking the entire teaching of the application and the relevant disclosure in the description into account: 'in order to establish whether products such as stem cell cultures are obtained exclusively by the use, involving the destruction, of a human embryo or not. For this purpose, the disclosure of the description has to be considered in view of the state of the art at the date of filing.' Thus an invention must not employ stem cell lines.
54. Case T 356/93 (OJ EPO 2006, 15).

'10.4. [T 356/93] ... a decision which identified many drawbacks ranging from the type and the number of questions posed within one poll, through the size and representative nature of the cross-section of the population polled, to the manner of interpretation of the result obtained. The Board is in full agreement with the views on opinion poll evidence expressed in that earlier decision.'

In the *Howard Florey/Relaxin* case, concerning a genetically engineered human hormone, the Opposition Division formulated a test whereby 'a fair test to apply to consider whether it is probable that the public in general would regard the invention as so abhorrent that the grant of patent rights would be inconceivable'.[55] As pointed out by Oliver Mills, the 'overwhelming consensus' test limits the scope of the EPO to refuse patentability. The decision supports the proposition that the consensus among Contracting States is the controlling factor. Neither the EPO nor any other European institution has inherent power to decide what constitutes an abhorrent invention.[56]

The Board's moral considerations concerning the patent in Case T 356/93 focused on plant genetic engineering techniques which, of course, would be capable of being used for constructive as well as destructive purposes.[57] Since none of the claims of the patent referred to subject matter which related to a misuse or destructive use of plant biotechnological techniques, in the light of the conventionally accepted standards of conduct of European culture the technique claimed could not be considered to be wrong as such.

### 5.3. The environment

According to TRIPS Article 27(2), Members may exclude from patentability inventions so as to 'avoid serious prejudice to the environment'. The wording of TRIPS Article 27(2) suggests a broader protection than Article 53(a) EPC. Despite the differences in wording, through its interpretation of *ordre public*

55. See V-8/94 Opposition Division (OJ EPO *1995*, 388); Hans Henrik Lidgard (ed.), *Protecting and Transferring Biotech Inventions,* 2004; and Oliver Mills, *Biotechnological Inventions, Moral Restraints and Patent Law,* 2005, p. 59.
56. Oliver Mills, *Biotechnological Inventions, Moral Restraints and Patent Law,* 2005, p. 60, Note 21. See the EPO Examination Guidelines (April 2010, Chapter IV-8, Part C, point 4.1), which repeat the abhorrent criteria and state that the provision is likely to be invoked only in rare and extreme cases. See Amanda Warren-Jones, 'Finding a 'Common Morality Codex' for Biotech – A question of Substance', 6/2008, *IIC* Vol. 39: morality is equated with 'ethical values' whereas public order equated with 'legal values', p. 653.
57. Case T 356/93 (OJ EPO 2006, 15).

EPO case law does make references to the environment.[58] Thus Case T 356/93 establishes that Article 53(a) EPC constitutes a bar to patentability under the *ordre public* standard for inventions whose exploitation is likely to seriously prejudice the environment.

A relevant question in this regard is the extent to which environmental protection is taken into account by the patent authorities. Are they to apply the precautionary principle and risk assessment? The question is receiving more attention from the climate change perspective, which will be discussed in Chapter 9.

In the analysis of the environmental perspective in Case T 356/93, the Board held, in point 18.5, that the *ordre public* standard presupposes that the threat to the environment is sufficiently substantiated at the time when the decision to refuse the grant of a patent is taken by the EPO. According to the Board, this view is consistent with the requirement that exceptions to patentability under Article 53(a) EPC must be narrowly construed. The Board further emphasised that most of the applicant's arguments were based on the *possible* occurrence of harmful events (for example, the transformation of crops into weeds, the spread of herbicide-resistance genes to other plants, damage to the ecosystem). However, the documentary evidence must sufficiently substantiate the existence of a threat to the environment so as to be a bar to patentability under Article 53(a).[59]

Thus it seems that the EPO Boards of Appeal do not apply the precautionary principle so as to refuse the grant of a patent before it can be proved that a product or process would in fact cause harm.[60]

58. For another opinion, see Overwalle, 'Legal and Ethical Aspects of Bio-patenting', in Peter Drahos (ed.), *Death of Patents*, 2005, p. 221. She points out that the specifications in TRIPS have not been taken up in the EPC, and that the grant of a European patent has never been refused on the basis that it might cause damage to human, animal or plant life, health or the environment. See Ulrich Schatz, 'Patents and Morality', in Sigrid Sterckx (ed.), *Biotechnology, Patents and Morality*, 2nd Ed., 2000, pp. 219, 221-222. See Amanda Warren-Jones, 'Finding a 'Common Morality Codex' for Biotech – A question of Substance', 6/2008, *IIC* Vol. 39: morality is equated with the 'ethical values' whereas public order is equated with 'legal values'; p. 645.
59. As concluded by Singer & Stauder, there is only an infringement of *ordre public* if the commercial exploitation of the invention actually endangers *ordre public*, Article 53, note 13. Warren-Jones: 'Finding a 'Common Morality Codex' for Biotech – A question of substance', *IIC* 2008 Vol. 6, p. 645: however, according to the rationale of the Technical Board of Appeal in the PGS case, dumping radioactive waste would be an issue of 'ordre public', but 'litter-bugging' is a question of morality.
60. In the report 'Recommendations for the patent System of the future', by a Danish working group (including the author) under the Danish Board of Technology, it is

In its decision in Case T 356/93, the EPO Boards of Appeal also addressed the question of scientific risk arising from the application of genetic engineering techniques to plants and in particular the production of herbicide-resistant plants. In points 18.2 to 18.4 of the Decision, the Board emphasised that the patented invention may only be exploited within the framework defined by national laws and regulations for the use of the invention:

> 'patent offices are placed at the crossroads between science and public policy. However, at this crossroads patent offices are not alone, but find themselves side-by-side with increasing number of other authorities and bodies, in particular regulatory authorities and bodies, whose function is inter alia to ensure that the exploitation of a given technology, regardless of whether it is protected by a patent or not, takes place within the regulatory framework provided by laws, international treaties, administrative provisions.'

The Board points out that the assessment of the hazards stemming from the exploitation of a given technology is one of the important duties of such regulatory authorities and bodies. In point 18.4, it further emphasised that the risks of exploiting a patented invention cannot be anticipated merely on the basis of the disclosure of the invention in the patent specification. The actual approval by the authorities is often obtained after the grant of the patent. Only specialised authorities and bodies are in a position to carry out a realistic assessment of risks or hazards on the basis of the regulations in force, objective criteria and scientifically valid parameters:

> 'This is because a realistic assessment of therapeutic operability requires a comprehensive and time consuming programme of testing and evaluation of the products. The results of such tests are usually *not available* to patent offices during the prosecution of a case. During this time, the exploitation of the claimed products is most likely to be in the initial phase when risk and safety assessment by the competent authorities or bodies has either not yet taken place or not yet been completed' (Emphasis added).[61]

recommended that a form of precautionary principle should be established in the patent system, so that future modifications and extensions should only be made at a more precautionary rate. Apart from that, one might have thought that the principle which has been recognised in the EU treaties since 1987, as well as in the UN Cartagena Protocol, could serve as a legislative and administrative guideline. The precautionary principle is not generally accepted in global society. The EU Commission has issued a communication on the use of the principle (COM(2000) 1 final).

61. Case T 356/93 (OJ 1995, 545), and Ulrich Schatz, 'Patents and Morality', in Sigrid Sterckx (ed.), *Biotechnology, Patents and Morality*, 2nd Ed., 2000, p. 221.

In Norway, the scenario changes from agriculture to aquaculture, in the form of a genetically modified salmon.[62] The patent in question claimed an 'all fish' chimeric growth hormone gene construct.[63] The issue of risk assessment and precaution (as well as animal welfare/suffering) was raised by the Norwegian Advisory Board on Ethical Aspects of Patenting, as the growth hormone had problematic side effects on the fitness (behavioural, physiological and genetic) of such transgenic fish. The Norwegian Board addressed the questions of the precautionary principle and risks to the environment from the potential negative impact of transgenic fish on wild populations (for example, farm fish escaping).[64] The patent was granted in Norway in 2006.

As pointed out in the *travaux préparatoires* for Directive 98/44/EC, it is important to make it quite clear to the general public that a system of legal protection merely confers negative rights and does not grant any positive rights regarding an invention. Thus the grant of a patent does not impede the procedures for examining compatibility with existing rules outside the patent regime, and in particular such procedures as apply to upstream research activities and to the downstream marketing of products.[65]

### 5.4. Different approaches to the legal standard

In Case T 315/03, which followed Case T 356/93, the Board emphasised that 'in effect there are two quite different Article 53(a) objections – on the one hand, a 'Rule 23d-type' Article 53(a) objection which requires only that the invention is assessed as to whether or not it falls in one of the four limited categories set out in the Rule and, on the other hand, a 'real' Article 53(a) objection which requires an assessment as to whether or not exploitation of the invention in question would be contrary to morality or 'ordre public'.'[66]

62. Proposals for ethical boards that are separate from patent offices (like the Norwegian one) have often been made, see e.g. a proposal for tribunals made up of specialists in ethics, research and economics in E. Richard Gold & Timothy A. Caulfield, 'The moral tollbooth: a method that makes use of patent system to address ethical concerns in biotechnology', *The Lancet*, Vol. 359, 29 June 2002, p. 2269 ff.
63. Patent Nos EP 321650, EP 578653, and WO9216618.
64. One question raised by the Norwegian Committee following the grant of the patent was the documentation or burden of proof with regard to the risk assessment or suffering of the transgenic animal. If the applicant has to provide documentation of the lack of risk/uncertain effects, then the documentation has to been reviewed in the light of the best existing information and thus have regard to scientific standards and findings.
65. See OJ 1996 C 295/13, point 3.3.3.
66. See point 10.1 in Case T 315/03 2006 OJ EPO 15.

When making the latter assessment, the EPO case law offers some guidance. In Case T 315/03, point 10.2, the Board referred to Case T 356/93, points 5 and 6 (quoted above). It held that those definitions confirm the view, which appears from the wording of Article 53(a) EPC itself, that *ordre public* and morality form the basis of two separate objections, either or both of which can be raised in a particular case.

The interpretation of the twin concepts within the EPC regime are thus to be assessed in two main categories:

the 'real' Article 53(a) objection, and
the 'Rule 23d-type' Article 53(a) objection.

In Case T 356/93, the Board referred to the historical documentation when justifying the narrow construction of the exceptions to patentability under Article 53(a) EPC. As a starting point, exceptions to patentability under Article 53(a) (as all other exceptions) must be narrowly interpreted, since the concept of patentability in the EPC must be as wide as possible.[67] This approach has been questioned by the former President of the EPO, Alain Pompidou,[68] when invited to comment under Article 11a of the Enlarged Board's rules of procedure on Case G 2/06 *WARF*, concerning human embryonic stem cells. It was argued that the principles of the Vienna Convention clearly indicate that the interpretation of an international treaty does not have to rely solely on retracing the original intentions of the contracting parties. Such interpretation is a complex exercise and must be supported by an evolving or dynamic approach.[69] From the conclusions, point 1, it follows that:

'Article 52(1) EPC does not enshrine a general principle of narrow interpretation of exclusions. A presumption in favour of a narrow interpretation of exceptions to patentability would unduly limit the significance of the moral jurisdiction under Article 53(a) EPC and Rule 23d(c), the purpose of which is the incorporation of higher ranking legal and moral principles into European patent law, and would thus be in conflict with the general objective of those norms. When a decision is taken on how Rule 23d(C) EPC has to be interpreted, the normal rules of treaty interpretation are to be applied, as found in the Vienna

67. http://www.cipa.org.uk/download_files/epo_warf.pdf
68. See http://www.cipa.org.uk/download_files/epo_warf.pdf
69. See Case G 2/06, Comments by the President of the European Patent Office, 28 September 2006, p. 36 http://www.cipa.org.uk/download/EPO_WARF.pdf (last visited January 2012).

Convention. Accordingly, the principle of narrow interpretation of exceptions should not be generally invoked without stating any reasons as to its applicability.'[70]

The comments of the former President of the European Patent Office disregard the question of the status of the moral clause after the implementation of the Directive. Despite the fact that the Implementation Rules are part of the Convention (EPC Article 164), the Rule on biotech originates from a Directive that only 27 Contracting States of the EPO have an obligation to follow. The Vienna Convention does not address a situation like the mixed-up relations of the EPO and the EU. However, while the Enlarged Board of Appeal and the Court of Justice dissociate and stress the distinction between the two systems, in its decision in Case G 2/06 the Enlarged Board of Appeal refers to the *travaux préparatoires* for the Directive as well as the Vienna Convention, as follows:

'16. When looking at the travaux préparatoires relating to the introduction of Rules 26 to 29 (formerly 23b to e) EPC, it becomes apparent that the aim *was to align the EPC to the Directive*. This follows from the Notice dated 1 July 1999 concerning the amendment of the Implementing Regulations to the European Patent Convention (OJ 1999, 575) and is also evidenced by the fact that, according to Rule 26(1) (formerly 23b(1)) EPC, the Directive shall be used as a supplementary means of interpretation. Therefore, the Enlarged Board of Appeal *turns to the interpretation of Article 6(2) of the Directive*, which corresponds to Rule 28 (formerly 23d) EPC. Although the *Directive is not a treaty*, the Enlarged Board of Appeal will, in view of the reference in Rule 26(1) EPC just mentioned and in line with the established case law (see e.g. G 5/83, OJ EPO 1985, 064, G 1/84, OJ 1985, 299, J 16/96, OJ 1998, 347) apply mutatis mutandis the general rules laid down in the Vienna Convention on the Law of Treaties. It will thus look at the ordinary meaning to be given to the terms of a provision in its context and in the light of its object and purpose, including the preparatory documents.'

The parallel Case C-34/10 *Brüstle*,[71] where the German Federal Supreme Court referred a question of interpretation to the Court of Justice for a preliminary ruling, provides more input on the interpretation of Article 6(2).[72] Is the Enlarged Board of Appeal ready to address that too or will the similarities stop before that? The Enlarged Board of Appeal gave a narrow answer in

70. http://www.cipa.org.uk/download_files/epo_warf.pdf at p. 2.
71. See Case DE 19756864 C1.
72. See Nick Bassil, 'Developments in patentability of inventions relating to Human Embryonic Stem cells', *Bioscience Law Review*, Vol. 12 issue 1, p. 5 ff., with an analysis of Case T 1374/04, Case G 2/06, and the Opinion of Advocate General Bot in Case C-34/10.

Case G 2/06, point 35, though limited by the formulation of the questions, when stating:

'In view of the questions referred, this decision is not concerned with the patentability in general of inventions relating to human stem cells or human stem cell cultures. It holds unpatentable inventions concerning products (here: human stem cell cultures) which can only be obtained by the use involving their destruction of human embryos.'[73]

However, in his Opinion in Case C-34/10, Advocate General Bot gave much more detail. Without going into details at this point, the opinion concluded as follows in point 119:

'119. In view of all the foregoing considerations, I propose that the Court give the following answers to the questions asked by the Bundesgerichtshof: Article 6(2)(c) of Directive 98/44/EC of the European Parliament and of the Council of 6 July 1998 on the legal protection of biotechnological inventions must be interpreted as follows:

- The concept of a human embryo applies from the fertilisation stage to the initial totipotent cells and to the entire ensuing process of the development and formation of the human body. That includes the blastocyst.
- Unfertilised ova into which a cell nucleus from a mature human cell has been transplanted or whose division and further development have been stimulated by parthenogenesis are also included in the concept of a human embryo in so far as the use of such techniques would result in totipotent cells being obtained.
- Taken individually, pluripotent embryonic stem cells are not included in that concept because they do not in themselves have the capacity to develop into a human being.
- An invention must be excluded from patentability where the application of the technical process for which the patent is filed necessitates the prior destruction of human embryos or their use as base material, even if the description of that process does not contain any reference to the use of human embryos.

73. (OJ EPO 2009, 306). See e.g. Aurora Plummer, 'Towards systemic legal conflict: Article 6(2)(c) of The EU Directive on Biotechnological Inventions', in Aurora Plummer & Paul Torremans (eds.), *Embryonic Stem Cell Patents, European Law and Ethics*, 2009, p. 193: 'When given its natural meaning, the wording of Article 6(2)(c) would, for instance, justify the exclusion of patents on inventions in which the claims are directed to processes applied to a human embryo, other than those which benefit the embryo itself. Such processes would, a fortiori, include activities causing destruction of the embryo.'

- The exception to the non-patentability of uses of human embryos for industrial or commercial purposes concerns only inventions for therapeutic or diagnostic purposes which are applied to the human embryo and are useful to it.'[74]

And in contrast to the EPO Boards of Appeal, the Court of Justice stated that an autonomous definition is necessary for the EU legal order. Further, in paragraph 30 of its judgment the Court stated that on a sensitive social issue such as the definition of a human embryo, the Court must restrict itself to a legal interpretation, and that this definition must be understood in a wide sense in order to safeguard human dignity. The Court stated in paragraph 38:

'any human ovum after fertilisation, any non-fertilised human ovum into which the cell nucleus from a mature human cell has been transplanted and any non-fertilised human ovum whose division and further development have been stimulated by parthenogenesis constitute a 'human embryo' within the meaning of Article 6(2)(c) of the Directive'

As stated by Alain Pompidou, although they are not binding as regards the EPO, the judgments of the Court of Justice regarding the Directive appear to be relevant to the nature and purpose of the illustrative list in Article 6(2) of the Directive and Rule 28(c) EPC. But, given the first indication of how this may turn out, it seems that the future will be interesting for patent lawyers. The Enlarged Board of Appeal refrained from defining an embryo in point 20, which reads:

'20. Neither the EU legislator nor the EPC legislator have chosen to define the term 'embryo', as used in the Directive or now in Rule 28 (formerly 23d) EPC. This contrasts with the German law (Gesetz zum Schutz von Embryonen of 13 December 1990, § 8) where embryo is defined as including a fertilized egg, or the UK law (Human Fertilisation and Embryology Act 1990, Section 1(1)) where embryo includes the two cell zygote and an egg in the process of fertilisation.

The EU and the EPC legislators must presumably have been aware of the definitions used in national laws on regulating embryos, and yet chose to leave the term undefined. Given the purpose to protect human dignity and prevent the commercialization of embryos, the Enlarged Board can only presume that 'embryo' was not to be given any restrictive meaning in Rule 28 (formerly 23d) EPC, as to do so would undermine the intention of the

74. There was the same conclusion in Case G 2/06, para. 27: 'Thus, if anything, these reasons point in the direction of the opinion of this Board that in the present case human embryos are used for industrial or commercial purposes, since patentability was only considered if the invention was to the benefit of the embryo itself (compare also decision of the BPatG of 5 December 2006, loc. cit., point IV 3). That this is not the case here is evident, since the embryos used to perform the invention are destroyed.'

legislator, and that what is an embryo is a question of fact in the context of any particular patent application.'

The ruling of the Enlarged Board of Appeal on the definition question has been 'overturned' by the Court of Justice, since an EU concept was considered necessary. But, given the EU legal order, this is also a way of avoiding a conflict. Had the Board defined an embryo, it is likely that the EPO definition would not have been upheld in a review by the Court of Justice. However, the question still remains as to which approach is the best solution for the purposes of patent law and patent legitimacy. The approach of the Court of Justice could soon be seen as being be more static than the Directive provision itself. That approach of the EPO Board is more flexible and leaves room for science-based progress to safeguard human lives in ways which it would be difficult to categorise as being an offence to human dignity.

Turning to the views of scientists, an evaluation of the judicial review of technology has already been touched on in Chapter 2.3. One life-science lawyer says 'even a restrictive interpretation should allow companies to patent the technologies needed to turn human ES cells into treatment, rather than patenting procedures involving the cells themselves'.[75]

On behalf of six co-authors, Jeremy B.A. Green of King College London writes that the Court of Justice ruling contains crucial errors concerning the underlying science. 'The new ruling misleading classifies pseudo-fertilized eggs, or parthenotes (lumping together those made with and without nuclear transfer), as requiring the protections of personhood. This classification was made on the grounds that these eggs are 'capable of commencing the process of development of a human being just as an embryo created by fertilization of an ovum can'. However, mammalian parthenogenetic embryos do not develop in the same way as normal embryos; nor are they developmentally viable if made without nuclear transplant. They are far down the sliding scale of developmental potential shared by all cells (including cells in the adult human body). It is therefore misguided for the ruling to put all parthenote-derived cells, which have technological potential at the same end of the scale as cells that can fully differentiate (totipotent cells). The ruling sensibly prevents a second confusion: inventions derived from human embryonic stem cells that require prior destruction of embryos are, it says, unpatentable. Because the technology already exists to make human embryonic stem-cell lines that preserve the viability of the donor embryo, embryo destruction is unnecessary. The scope of the ruling may therefore be narrower than some might conclude'. In conclusion the author writes that it will be unfortunate if the

75. Julian Hitchcock, quoted in Ewen Callaway, 'European ban on stem-cell patents has silver lining', *Nature*, 478, 441, October 2011.

wording of the Court's ruling should inadvertently inhibit a potentially useful, ethical technology using parthenotes just because of a lapse in scientific understanding.[76]

The main question concerned the ban on the patentability of techniques involving the destruction of embryos, based on the wording of Directive 98/44/EC, Article 6(2)(c), which prohibits 'uses of human embryos for industrial or commercial purposes'. If such used implicated a destruction the provision activates and the CJEU stated in point 52, that 'the Directive excludes an invention from patentability where the technical teaching which is the subject matter of the patent application requires the prior destruction of human embryos or their use as base material, whatever the stage at which that takes place and even if the description of the technical teaching claimed does not refer to the use of human embryos.'

The obvious question is therefore whether a mere reference to the use of a human embryo will lead to the refusal of a patent. Apparently that is the case in European patent law.[77]

76. Jeremy B.A. Green, 'Patenting: European stem-cell ruling is misleading', Nature, 479, 41, November 2011.
77. See also Sven J.R. Bostyn, 'A decade after the birth of the biotech directive: was it worth the trouble?', in Emanuela Arezzo & Gustavo Ghidini (eds.), *Biotechnology and Software Patent Law, A Comparative Review of New Developments*, 2011, p. 239 stating that the EPO Boards of Appeal in 2/06 has left us in the dark as to which reasoning should be followed for cases where the human embryo is not destroyed; see same on page 242: 'the use of human embryos and human embryonic stem cells derived from human embryos can from an ethical point of view be valued differently.' See Ulrich Schatz, 'Patents and Morality', in Sigrid Sterckx (ed.), *Biotechnology, Patents and Morality*, 2nd Ed., 2000, p. 218: 'the moral permissibility of the act of patenting as such is not the issue at all in Art. 53 (a), EPC. It is not patenting the invention but exploiting the invention which has to be contrary to morality or 'ordre public' in order to justify the rejection of the patent application'. See also Sigrid Sterckx & Julian Cockbain, *Exclusions from patentability, How far has the European Patent Office eroded Boundaries?*, 2012, p. 288, footnote 14. Subsequent to G2/06 WARF it has been the practice of the EPO to allow patents for stem cell inventions in circumstances where the inventions can be put into practice at the filing date without requiring the destruction of human embryos (e.g. stem cells which have been grown in the laboratory).

## 6. General guidelines from the Court of Justice

Directive 98/44/EC expressly states that the traditional moral standards that are recognised in a Member State shall apply, and the Directive addresses the moral standards in two ways, namely by the traditional clause in Article 6(1) and the list in Article 6(2). As pointed out below, the Court of Justice has essentially referred to the EPC regime as interpreting traditional standard codified in Article 6(1).[78]

In the Directive's recitals the moral standard is recognised as follows:

'(37) Whereas the principle whereby inventions must be excluded from patentability where their commercial exploitation offends against ordre public or morality must also be stressed in this Directive;

(38) Whereas the operative part of this Directive should also include an illustrative list of inventions excluded from patentability so as to provide national courts and patent offices with a general guide to interpreting the reference to ordre public and morality; whereas this list obviously cannot presume to be exhaustive; whereas processes, the use of which offend against human dignity, such as processes to produce chimeras from germ cells or totipotent cells of humans and animals, are obviously also excluded from patentability;

(39) Whereas ordre public and morality correspond in particular to ethical or moral principles recognised in a Member State, respect for which is particularly important in the field of biotechnology in view of the potential scope of inventions in this field and their inherent relationship to living matter; whereas such ethical or moral principles supplement the standard legal examinations under patent law regardless of the technical field of the invention.'

The value of the recitals is as an aid to interpretation. If, for example, chimeras are claimed on the basis of a mix of human and animal embryonic stem cells or germ cells, the application can be refused with reference to the Directive Article 6(1), as interpreted pursuant to recital 38.

In Case C-456/03 *Commission v Italy*, the Court of Justice was called upon to interpret Directive 98/44/EC Article 6(2). Since the Italian patent legislation did not lay down that certain specific processes, such as the cloning of

78. See also Opinion of the Economic and Social Committee on the 'Proposal for a European Parliament and Council Directive on the legal protection of biotechnological inventions', OJ 1996 C 295/11, at p. 13, point 3.3.2: 'It is essential here for the legislative authority to be quite clear that legal patent protection system cannot in any way condone practice or processes that are deemed by law to be contrary to public policy and morality, notably as laid down by the Munich Patent Convention.'

human beings and the use of human embryos for industrial and commercial purposes, are not patentable, the question was whether Italian law adequately implemented the principles laid down in the Directive Article 6(2). On the relationship between Article 6(1) and 6(2) the Court of Justice stated as follows:

'Unlike Article 6(1) of the Directive, which allows the administrative authorities and courts of the Member States a wide discretion in applying the exclusion from patentability of inventions whose commercial exploitation would be contrary to ordre public (public policy) and morality, Article 6(2) allows the Member States no discretion with regard to the unpatentability of the processes and uses which it sets out, since the very purpose of this provision is to give definition to the exclusion laid down in Article 6(1) (see, to this effect, *Netherlands v Parliament and Council*, paragraphs 37 to 39).[79] It is apparent from the 40th recital in the preamble to the Directive that processes for cloning human beings must be excluded 'unequivocally' from patentability, since there is a consensus on this question within the Community. *It follows that, by expressly excluding from patentability the processes and uses to which it refers, Article 6(2) of the Directive seeks to grant specific rights in this regard.*'[80]

The Court of Justice referred to the Opinion of Advocate General Colomer, point 55, which is as follows:

'Article 6(1) of the Directive requires that inventions be considered unpatentable where their exploitation would be contrary to ordre public (public policy) and provides that 'exploitation shall not be deemed to be so contrary merely because it is prohibited by law or regulation'.

That statement may be interpreted as requiring the transposition of the principle that commercial processes involving the use of human embryos are not patentable. In any event, it is clear from a careful reading of the Directive that the principle must be transposed.'

It can be concluded from the statements of the Court of Justice quoted above, that in principle Article 6(1) allows the Member States 'wide' discretion, but that the scope in Article 6(1) has been narrowed by the list in Article 6(2) (paragraph 39 of the judgment in Case C-377/98 *The Netherlands v Parliament and Commission).* The Member States are obliged to implement the list in Article 6(2) in their national patent legislation (paragraph 79 in Case C-456/03).

79. Emphasis added.
80. See Case C-456/03 *Commission v Italy,* paras 78 and 79.

In Case 377/98, *The Netherlands v Parliament and Commission,* paragraphs 37 to 39, the Court of Justice outlined the room manoeuvre left to Member States on the question of the moral standard:

'37. As regards, first, Article 6 of the Directive, which rules out the patentability of inventions whose commercial exploitation would be contrary to ordre public or morality, it is common ground that this provision allows the administrative authorities and courts of the Member States a wide scope for manoeuvre in applying this exclusion.

38. However, that scope for manoeuvre is necessary to take account of the particular difficulties to which the use of certain patents may give rise in the social and cultural context of each Member State, a context which the national legislative, administrative and court authorities are better placed to understand than are the Community authorities. That sort of provision, which allows patents to be refused where there, is a threat to ordre public or morality is, moreover, a *well known one in patent law* and appears inter alia in the relevant international legal instruments, such as the EPC.[81]

39. Furthermore, the scope for manoeuvre left to Member States is not discretionary, since the Directive limits the concepts in question, both by stating that commercial exploitation is not to be deemed to be contrary to ordre public or morality merely because it is prohibited by law or regulation, and by giving four examples of processes or uses which are not patentable. Thus, the Community legislature gives guidelines for applying the concepts at issue *which do not otherwise exist* in the general law on patents.'[82]

In Case 377/98 (paragraph 38, quoted above) the Court of Justice refers to the EPC regime where the moral standard is well known. It is thus necessary to interpret the standard in the light of EPC case law. Interpretations of the 'real' Article 53(a) objection are a matter for the EPO. Article 6(1) of the Directive is not implemented in the EPC regime and, if one takes paragraph 38 literally, the Court of Justice is reluctant to interpret the traditional standard. Thus the proof of accepted norms, the broad balancing test and the protection of the environment are more dictated by the EPC than by the EU. Since only Article 6(2) has been implemented in the EPC Implementing Regulations, and since the Court of Justice in paragraph 39 refers to the Community legislature giving guidelines not previously seen, the list may be subject to a conflict, as mentioned above, where the interpretation of the Implementing Regulations by the Enlarged Board of Appeal may conflict with the Court of Justice's interpretation. In that case, the acceptable norm will differ at different levels (EU and EPC). Furthermore, the recitals of the Directive may be implied au-

81. Emphasis added.
82. Emphasis added.

thorisation for national legislators to introduce stricter patent law requirements.[83] Thus, the room for manoeuvre of Member States has to be viewed from the dual perspectives of the EPC regime (with the 'real' Article 53(a) objection and the 'Rule 23d-type' Article 53(a) objection) and the EU regime (Article 6(2) of the Directive. Finally, the new EU unitary patent approach may leave EPO Members (except from) Spain and Italy behind and establish a new group in which post-grant assessment is made by the EU alone (for the EU 25) or by the group of Members signing the agreement on a unified patent court.

Besides the conflict described above concerning the WARF Case, there are the concerns reflected in Directive 98/44/EC Article 6(2)(d) and in Rule 28(d) of the Implementing Regulations EPC 2000 which addresses 'processes for modifying the genetic identity of animals which are likely to cause them suffering without any substantial medical benefit to man or animal, and also animals resulting from such processes'. Article 6(2)(d) and recital 45 of Directive 98/44/EC clearly state that processes for modifying the genetic identity of animals which are likely to cause them suffering without any substantial benefit in terms of research, prevention, diagnosis or therapy to man or animals resulting from such processes, must be excluded from patentability.[84]

However, guidance on interpretation of the standard is also to be found in the decisions of the EPC Boards of Appeal. Thus the decision of the EPO Boards of Appeal in Case T 315/03 in July 2004 is the authoritative decision on the *Oncomouse* case.[85] The *Oncomouse* patent comprised a method for producing transgenic animals. In this decision the Board referred to the test suggested in Case T 19/90 as being only starting point for a 'real' Article 53(a) EPC assessment of the genetic manipulation of animals:

> 'the decision as to whether or not Article 53(a) EPC is a bar to patenting the present invention would seem to depend mainly on a careful weighing up of the suffering of animals and possible risks to the environment on the one hand, and the invention's usefulness to mankind on the other'.[86]

83. See, for example, Geertrui Van Overwalle, 'Legal and Ethical Aspects of Biopatenting', in Peter Drahos (ed.), *Death of Patents*, 2005, p. 221.
84. See Tine Sommer, 'Patenting the Animal Kingdom, From Cross-Breeding to Genetic Make-up and Biomedical Research', 2/2008 *IIC* Vol. 38, pp. 139-177.
85. Case T 315/03 [2006] OJ EPO 15.
86. OJ EPO 1990, 476, reasons, para. 5.

This test is further qualified in paragraph 10.5 in Case T 315/03, which states as follows:

> 'This test may be employed in different cases using evidence relating to particular times and conditions and thus appears sufficiently flexible to allow for the ... (current at the effective date ... ) views as to social order, environmental risk and accepted standards of behaviour in European culture. In contrast to the test in Rule 23(d) EPC, it balances suffering of animals not against likely substantial benefit to man or animal, but against usefulness to mankind. It also suggests balancing possible risks to the environment against usefulness to mankind. Thus, it offers a test for use in 'ordre public' cases, morality cases and both.'

The use of the word 'mainly' in Case T 19/90 allows for other considerations to be taken into account, such as a threat to evolution, possible increased trade in genetically manipulated animals, and the moral acceptability of such manipulation. There is a newly opened window, in the decision in Case G 2/06 mentioned above, in the balancing test concerning humanity benefitting from human embryo stem cell inventions. The assessment of a 'real' Article 53(a) EPC objection is summarised in paragraph 10.10 as follows:

> 'The many bases (economic, religious, etc) for definitions of morality suggested by the appellants are of no assistance since no single such basis represents an accepted standard in European culture. Opinion poll evidence is of very limited value for the reasons given in T 356/93. In animal manipulation cases, the test in T 19/90 is appropriate. This differs in several respects from the test in Rule 23(d) EPC, most importantly by allowing for matters other than animal suffering and medical benefit to be taken into account. Since the T 19/90 test is 'mainly' the basis of assessment, other arguments as to the appropriate standard of morality or 'ordre public' can additionally be considered but, like any other argument, must be supported by evidence.'[87]

One can ask why, if that is/or will be the case, regional patent law lacks coherence. The answer to this can be found in the multilevel approach that mixes competences, operates with detailed technology-specific wording and thus, put simply, 'exports EU norms'. As for the mix of competences, it is also questionable whether Article 6 can be separated between EPC matters (Article 53(a) EPC/Article 6(1), and EU matters (Article 6(2)).

The EPO has implemented Directive 98/44/EC in Rules. Using *travaux préparatoires* from Union law to interpret EPC/Rules highlights the substance of the export of EU norms. As suggested by *Sterckx* and *Cockbain*, the moral extension of Article 53(a) EPC found in Rule 28 should only serve as a

87. See the decision of the Enlarged Board of Appeal of 25 November 2008 in Case G 2/06 (306 OJEPO 5/2009).

recommendation and its proper place would have been in the Guidelines of Examination.[88] The section below adds a further aspect – the moral concern in patent law has been moved (exported) from exploitation to research processes.

### 6.1. Views on Article 6(2)

The terms *ordre public* and morality are not defined either in the Directive or in the EPC. They are standards that have traditionally been disconnected from prohibitions in national law.[89] These standards have been developed through EPO case law from an exclusively non-living/mechanical sphere into a biotechnological sphere involving living matter. Thus the standards have been shown to be relatively flexible, being shaped by a step-by-step approach in line with the rapid technological developments and demands from society for balanced legal protection. However, due to intervention of the European Union legislator, these flexible and yet strict moral standards have been diluted. The list in Article 6(2) has been nicknamed 'the catalogue of horrors' and seems to have caused more confusion as a guideline than the traditional Article 6(1) standard.[90] *Geertrui Van Overwalle* has discussed the moral exclusion in Article 6(2) and, based on the view that one should decide whether it is science or the invention that is objectionable, part of her conclusion is as follows:

> 'On the other hand, the more fundamental question arises of whether the list of exclusions should be taken up in patent law. There is a strong feeling that Article 6(2) does not aim at limiting the patent implications of certain biotech inventions, but wishes to exclude certain fields of research as such. Patent law should take into account ethical concerns. It can and should act as a 'moral toolbooth', but it should only do so to the extent that it concerns matters which are directly and inextricably linked with patents and the exercise of patents: patent law should not interfere when research is ethically undesirable. Since a direct link is

88. See also Sigrid Sterckx & Julian Cockbain, *Exclusions from patentability, How far has the European Patent Office eroded Boundaries?*, 2012, pp.292-293.
89. See e.g., para. 7 in Case T 356/93: 'Thus, approval or disapproval of the exploitation by national law(s) or regulation (s) does not constitute per se a sufficient criterion for the purposes of examination under Article 53 (a) EPC.'
90. The term is used by Sven J.R. Bostyn, 'A decade after the birth of the biotech directive: was it worth the trouble?', in Emanuela Arezzo & Gustavo Ghidini (eds.), *Biotechnology and Software Patent Law, A Comparative Review of New Developments*, 2011, p. 235.

missing between ethics and patents in Article 6(2), we take the view that this provision should be abolished and the exclusion should be treated in research regulations.'[91]

In the legal literature one can see that the decision of the Enlarged Board of Appeal in Case G2/06 *WARF* represents an important modification of the principle that the morality clause is confined to future commercial exploitation. It is said that the argument in favour of including the nature or the making of the invention in the object of the moral clause examination is to avoid situations in which the reference to commercial exploitation merely functions as a convenient legal artifice.[92] *Malene Rowlandson* writes that Case G 2/06 is an extremely important decision due to the acknowledgement that the making of the invention is taken into consideration, and she suggests that the nature of the patent owner's post-grant monopoly position also should play an important role in the morality assessment.[93]

In a comment on the (Case G 2/06) *WARF* decision, *Sigrid Sterckx and Julian Cockbain* criticise the reasoning of the Enlarged Board. First, they argue that the moral clause providing in EPC Article 53(a) which stipulates the exclusion from patenting is not motivated by the need to prevent the commercialisation, i.e. a commercial monopoly of a particular technical method.[94] They quote *Dolder* as follows:

'It is not the actual process of the objectified use of the human body in carrying out a technical teaching that infringes the legal interests protected in Art. 53(a) in the field in question, but the legal process of exercising an exclusive right to this teaching.

91. See Geertrui van Owervalle, 'Legal and Ethical Aspects of Bio-patenting', in Peter Drahos (ed.), *Death of Patents*, 2005, p. 222, footnote 18. See also Gerald Kamstra et al., *Patents on Biological inventions: The EC Directive*, 2002, p. 46: one might also wonder how appropriate it is to 'micro-manage patent law in this way'.
92. Malene Rowlandson, 'WARF/Stem cells (G2/06): the ordre public and morality exception and its impact on the patentability of human embryonic stem cells', *European Intellectual Property Review*, 2010 (E.I.P.R) 32(2), p. 75 'the present case has illustrated that it is not impossible to imagine a scenario whereby the actual commercial exploitation of an invention – for instance the use of a medical product, is not immoral if viewed in isolation, but the making of which on the contrary could be viewed to be'.
93. Malene Rowlandson, 'WARF/Stem cells (G2/06): the ordre public and morality exception and its impact on the patentability of human embryonic stem cells', *European Intellectual Property Review*, 2010 (E.I.P.R) 32(2), p. 75.
94. See S. Sterckx & J. Cockbain, 'Assessing the Morality of the Commercial Exploitation of Inventions Concerning Uses of Human Embryos and the Relevance of Moral Complicity: Comments on the EPO's WARF Decision', (2010) 7:1 *SCRIPTed* 83, http://www.law.ed.ac.uk/ahrc/script-ed/vol7-1/sterckx.asp

The goal of an ethically motivated ban on patenting ... is to prevent the commercialisation, i.e. the commercial monopolisation, of certain technical teachings'[95]

They find this view too narrow, arguing that the Article excludes patenting where *commercial exploitation* of the invention, *whether monopolistic or not*, would be contrary to morality:

'To *limit* the operation of the exclusion to circumstances where commercial monopolisation would be wrong is to say that what is at issue is whether the *patenting* of the invention is wrong. This is not the assessment the Article prescribes. By reference to Moufang[96] they find that, the act of applying for a patent places exploitation of the invention within a commercial context, but the question to be addressed remains whether *commercial exploitation* rather than monopolisation is contrary to morality. And they continue that the Enlarged Board of Appeal accordingly mentions the 'instrumentalization of the human body (as opposed to parts of it)' as being considered a 'barrier to patentability'.

'While there is probably general agreement that instrumentalisation of a conscious human being is wrong, there is no such agreement in relation to 'the human body' as such. The widespread acceptability of *post mortem* transfer of human organs, i.e. the use of cadavers as organ sources and hence the instrumentalisation of cadavers, clearly supports this point.'[97]

The authors conclude that that commercial exploitation of the claimed subject-matter would involve the commercialisation and commodification of human embryos, which would violate human dignity. Indeed, given that the principles of non-commercialisation and non-commodification are key dimensions of human dignity, and that human dignity applies to embryos, the subject-matter of the *WARF* patent application clearly contravenes the Article, since obtaining a patent would have allowed *WARF* to reap financial reward from the use of human embryos. Thus, the objection relates to human dignity whether or not its violation is caused by processes or products.[98]

95. F. Dolder, 'Schranken der Patentierbarkeit biotechnologischer Erfindungen nach dem Europäischen Patentübereinkommen', 1984 *Mitteilungen der deutschen Patentanwält* 1-7. The title translates as 'Barriers to patentability of biotechnological inventions under the EPC'.
96. R. Moufang, 'Patenting of Human Genes, Cells and Parts of the Body: the Ethical Dimensions of Patent Law', 1994 25 *IIC* pp. 487-515, at p. 504.
97. See S. Sterckx & J. Cockbain, 'Assessing the Morality of the Commercial Exploitation of Inventions Concerning Uses of Human Embryos and the Relevance of Moral Complicity: Comments on the EPO's WARF Decision', 2010, 7:1 *SCRIPTed* 83, http://www.law.ed.ac.uk/ahrc/script-ed/vol7-1/sterckx.asp
98. The concept of human dignity is acknowledged by the overwhelming majority of the Contracting States of the EPC and, unlike the concept of human rights, it can be ar-

*Paul L.C. Torremans* discusses the reference to the raw material on which the invention is based and argues that this implies that patent offices go beyond their appointed role and police overall moral standards in society. Patent offices, he claims, should restrict themselves to assessing the invention for which an application is made.[99]

*Aurora Plummer's* view on the Union's contribution to the patent law moral clause is in line with that of *Van Owervalle.*[100] Plummer refers to the purpose of the Directive, and that it was not intended to alter existing patent law and render unpatentable inventions that are based on activities which were considered morally permissible and were lawful in Member States at the time of the adoption of the Directive (recital 8). She continues by pointing to recital 17 of the Directive which reads:

> 'Whereas significant progress in the treatment of diseases has already been made thanks to the existence of medicinal products derived from elements isolated from the human body and/or otherwise produced, such medicinal products resulting from technical processes aimed at obtaining elements similar in structure to those existing naturally in the human body and whereas, consequently, research aimed at obtaining and isolating such elements valuable to medicinal production should be encouraged by means of the patent system'

Plummer emphasises that the Directive is expressly aimed at encouraging the development of regenerative therapies. A broad construction of the exclusion in Article 6(2)(c) of the Directive would have an effect contrary to the Directive's aim of encouraging biomedical research. Plummer writes as follows:

> 'If Article 6(2)(c) is constructed as precluding patents on inventions the practice of which necessitates destruction of human embryos and reaches through to downstream patents on hESC-based inventions, the effect is to potentially render unpatentable cell-based therapies including product development and more generally inventions which lawfully practised today, not only in the UK but in a substantial number of states in Europe and which are regarded as morally worthwhile by a large numbers of citizens in the EU. This would be

gued to apply to human beings at all stages of development and whether dead or alive; see S. Sterckx & J. Cockbain, 'Assessing the Morality of the Commercial Exploitation of Inventions Concerning Uses of Human Embryos and the Relevance of Moral Complicity: Comments on the EPO's WARF Decision', 2010, 7:1 *SCRIPTed* 83, http://www.law.ed.ac.uk/ahrc/script-ed/vol7-1/sterckx.asp.

99. See also Paul L.C. Torremans, 'Patentability of human stem cell or synthetic biology based inventions', in Emanuela Arezzo & Gustavo Ghidini (eds.), *Biotechnology and Software Patent Law*, pp. 306-307.
100. Aurora Plummer, 'Towards systemic legal conflict: Article 6(2)(c) of The EU Directive on Biotechnological Inventions', in Aurora Plummer & Paul Torremans (eds.), *Embryonic Stem Cell Patents, European Law and Ethics*, 2009, p. 191.

contrary to the aim of the Directive which was to prevent damage to the unity of the internal market and not to change patent law and turn patent offices into moral censors.'[101]

As emphasised in Chapter 1, section 3.2, and in Chapter 2, section 3, this focus on human dignity by the Court of Justice seems to be contrary to the promotion of biomedical research aimed at saving human life. The overall conclusion in Case C 34/10 *Oliver Brüstle,* paragraph 28, aiming at a coherent EU system (unitary without/before the existence of an EU patent) even conflicts with the EPO's overall distinction between pre-grant and post-grant. And the reasoning in paragraph 32, that the use of biological material originating from humans must be consistent with regard for fundamental rights, in particular the dignity of humans (recital 16 and Article 5(1), and recital 38 making the list in Article 6(2) non-exhaustive for all processes which offend against human dignity) can be elaborated and the same argument can be applied to gene patenting; see Article 5(2). Why would the Court of Justice mix patent law with research questions that belong to the scientific community? In other areas such as same-sex marriage or abortion, a Member State can make its own moral evaluation, and under the EPO an inventor can refrain from designating patent cover in an EPO Contracting State which has a particular moral policy. It can only be recommended that Directive 98/44/EC Article 6(2) should be deleted from the patent law regime, or that the whole regional patent regime should be sorted out before the Court of Justice starts upon its assessment of what patents the EU should ban.

## 7. Europe has too many players

The aim of this chapter is to analyse the legal standards of *ordre public* and morality in patent law. European patent law allows exclusions by the application of these legal standards at different levels. For instance, the EPC has interpreted the standards as having essentially two approaches: the 'real' Article 53(a) objection and the 'Rule 23(d)-type' Article 53(a) objection.[102] With regard to the 'real' Article 53(a) objection, issues such as uniform moral standards, proof of accepted moral norms, the balancing test, and environ-

101. Aurora Plummer, 'Towards systemic legal conflict: Article 6(2)(c) of The EU Directive on Biotechnological Inventions', in Aurora Plummer & Paul Torremans (eds.), *Embryonic Stem Cell Patents, European Law and Ethics*, 2009, p. 192, referring to recitals 5 and 6.
102. Now 'Rule 28-type' Article 53(a) objection.

mental risks are being discussed and shaped in a patent law context. The 'real' Article 53(a) objection is solely determined by the European Patent Organisation, but this also reflects the harmonisation in TRIPS Agreement, Article 27(2).

However, regarding the 'Rule 23(d)-type' Article 53(a) objection, it seems that the Court of Justice may intervene in the interpretation of the standard. In Case 377/98 *The Netherlands v The Parliament and Commission* the Court emphasised that, in contrast to Article 6(1) Directive 98/44/EC, the guideline in Article 6(2) was a masterpiece of the Community legislator. When interpreting the list in Article 6(2), the Court will, if necessary, take the recitals into account. Thus there is a third approach: the 'Article 6(2) EC-type' objection. If this non-exhaustive list expands in the future, because of new concerns of the Court of Justice or of the EPO Boards of Appeal, perhaps adopted in the Implementing Regulation (EPC) or in the Directive (EU), who will then be the final arbiter? The language of the Court of Justice in Opinion 1/09 on the EU patent, as well as in Case C-34/10 *Oliver Brüstle v Greenpeace*, clearly supports the EU legal order at any price; this is misplaced in a European patent regime where the EU's intervention is premature – both in terms of a field restricted Directive narrowed down to wordings such as gene sequences and embryos, as well as in terms of enhanced cooperation before a proper foundation has been established.[103] The current scenario is not transparent and the arguments of the Board and the Court indicate that there is a mixture of levels which is difficult to straighten out. Law can make life too simple but it can also make life unnecessarily complicated. The EPO system is built upon a distinction between pre-grant and post-grant and its foundation and integrity in patent law matters are undisputed. The EU has never established a patent law and yet it has succeeded in making a mess out of biotech patenting. Before sorting out this mess, the EU has struggled to establish yet another regional level, with the setting up of a unitary patent protection for the EU 25. This leaves Europe with different scenarios, all with the starting point that the EPO (and not the EU) has the granting authority at a supranational level:

103. In the 'new' legal patent order based on enhanced cooperation it is also stipulated that the Court of Justice's rulings would be binding on the unified patent court, and that Member States that are part of the cooperation would be jointly and severally liable for damages arising from infringement of EU law by the unified patent court. The European legal order would not be altered and would remain intact; see David Paul Sant & Christoph de Coster, 'The Unitary patent and the European patent court: A reality check', Vol. 12 Issue 1 *BSLR,* p. 30.

EPO: pre-grant as usual to the EPO 38.
EPO: pre-grant as usual but granted as an EU patent to the EU25; unitary patent law (pre- as well as post-grant) as EU competence only; the Court of Justice can interpret all matters affecting the EU 25. The EU 25 cannot rely on the EPC if it differs from Union law.
EPO: pre-grant to the EPO 13: EPO interpretation only.
EPO: pre-grant to the EU 2 (Spain and Italy); EPO interpretation but also Directive 98/44/EC apply. The EU 2 may in principle be subject to conflicting interpretations from the EPO and the Court of Justice.

Europe has several authorities for moral standards. There is the EPC, the EU, and the Member States themselves in between the EPC and the EU (the 'state-type' objection). The balancing test, for example, whether concerning salmon or human embryos, has been codified in Article 6(2)(d) of Directive 98/44/EC, but is not suitable in this connection. In principle objections can be based on concerns for the environment or animal suffering without medical benefit (nutrition only), and the balancing test is not applicable under to Article 6(2)(c), but under Article 6(1). Listed in this way, it is doubtful whether these concerns apply under all types of objections:

(a) the 'real' Article 53(a) objection
(b) the 'Rule 23(d)-type' Article 53(a) objection
(c) the 'Article 6(2) EC-type' objection
(d) the 'state-type' objection[104]

How will an application or a granted patent for fish, for example, be examined if the applicant argues that: the fish are quite healthy; that there is no suffering whatsoever and that the fish merely grow faster; that ultimate size of the fish is not relevant since they are harvested well before maturity; that there is no issue of suffering as may be the case when animals are used for laboratory research;[105] and finally that there is no risk involved in commercially culturing these fish in ocean pens? Questions could also be raised about the balancing of human stem cell inventions. A more suitable approach to

104. Warren-Jones, 'Finding a 'Common Morality Codex' for Biotech – A question of substance', *IIC*, 2008, Vol. 6, p. 651, also referring to Straus, 'Patenting Human Genes in Europe – Past developments and prospects for the future', 26 (6) *IIC* 920 1995. He asserts that the moral provisions should be administered principally by individual Member States according to standards specific to their territories, and that the EPO provision could represent a choice for Member States in determining the scope of issues to be regulated under their respective patent systems.
105. See the reply to the examination report of 3 April 1996 (EP578653).

balancing human dignity and biomedical progress might be made under EPC Article 53(a).[106]

If subject to the EPC, objection (a) will be sufficient, as the balancing test in Case T 19/90 will be effective. Since risk has to be present during the examination, this question will have to be addressed, with proper documentation provided by the applicant. If the Court of Justice were asked to give a preliminary ruling, then in principle only objection (c) or new concerns added to the list by case law (substantial medical benefit) would apply as particular Union arguments. If the question of risk assessment and suffering were examined by the Court of Justice, it is most likely that the Court would rule by reference to the EPO Board of Appeals (EPC Article 53a), but since the list is non-exhaustive, risk assessments could in principle be added by the Court. So far only objections (b) and (c) are clearly disconnected and the state objection (d) is bound by the interpretations of objections (a), (b) and (c).

However, the mix of competences is far from acceptable and the future scenario only add to the confusion as the EU 25 will conform more to Union law than to EPO law. Furthermore, as the list concerns objections against scientific method more than against a final product with an industrial application, the outcome of the European Union legislator's intervention in the creation of a field restricted patent law moral standard (EU27) is less obvious. As the systems compete to guide the patent systems on moral considerations and legitimacy, it is time to look again at the words of *Mary Warnock* in a patent law context. We cannot responsibly erect barriers to scientific advance through the patent system, and the patent system has only been weakened by there being too many players at the regional level. There is a clash between emerging technologies and ethical values, and in Europe we are faced with a solution that has only succeeded in aggravating this clash. Chapter 7.2.b, on Parkinson's disease, shows how the current position on embryonic stem cells might have moved research in another direction, towards induced pluripotent stem cells (iPSCs), and asks whether in this respect patent law has promoted or harmed biomedical research. In any event, the norm of human dignity has turned out to be a norm invoking the export of EU norms due to the interplay between legislative made law (Article 6(2)) and Judge-made law.

106. Warren-Jones, 'Finding a 'Common Morality Codex' for Biotech – A question of substance', *IIC*, 2008, Vol. 6, p. 654; and Case G 2/06, comments by the President of the EPO, 28 September 2006, pp. 54-55: adoption of a balancing exercise at the initial examination stage involves an ethical assessment beyond the ability and mandate of patent offices. In addition, he argued that it was both ethically and evidentially inappropriate to balance issues relating to humans.

PART II

# The Human Genome

*'A patent simply cannot reach what it does not teach'*
*(Foster Wheeler Corp. V. Babcock & Wilcox Co.*
512 F.Supp. at 809, 210 U.S.P.Q.)

CHAPTER 5

# Field Restrictions

Part II of the thesis is headed 'The human Genome'. Chapters 5 to 7 focus on all human gene patents. Chapter 5 discusses the doctrine of non-discrimination which is codified in Article 27(1) of the TRIPS Agreement. It is shown that the TRIPS Agreement, which does not define the concept of an 'invention', has left it to each Member to determine what is patentable. However, this room for discretion is limited by the doctrine of non-discrimination so that it is not possible to exclude whole product groups (such as medicines). Thus the TRIPS Agreement does not restrict what can be excluded from being granted patent protection, for example because some material is considered to be naturally occurring. Chapter 5 also looks at the patentability of human DNA sequences from both the perspectives of European law and US law. In contrast to European law's detailed regulation, US law does not contain detailed regulation of the patenting of human DNA sequences and its law is based on the decisions of the courts. The recent developments prompted by the US Supreme Court, according to which the patenting of genes is re-evaluated, is contrasted with the practice of the European courts, including the Court of Justice of the European Union, according to which limits are placed by the definition of the scope of protection.

## 1. Keeping hands off?

The rationale behind the patent system is to promote the progress of science and useful arts. These are the words of the 1789 US Constitution, Article I, Section 8, and we can see a similar rationale behind the patent system in Europe. The purpose of the European Patent Convention (EPC) is to strengthen cooperation between the States of Europe and to promote technical progress by ensuring appropriate protection of inventions.[1] Both the US patent system and the European Patent Organisation (EPO) patent system also encourage the disclosure of valuable technical information to the public, and require specification of the invention in the form of a description and claims.[2]

In the same vein the TRIPS Agreement states that its objective is the promotion of technological innovation.[3] The TRIPS Agreement's substantive minimum standards for patent laws, as discussed in Chapter 1, require WTO Members to comply with a broadly defined concept of patentable subject matter.[4] The patentability standard in TRIPS Article 27(1) was known in established patent regimes long before the TRIPS Agreement entered into force. Regardless of the TRIPS harmonisation, the idea that a patent must be available for any invention, provided it is new, involves an inventive step and is capable of industrial application, is not an isolated TRIPS norm. The TRIPS provision refers to standards elaborated in most (developed) Members' patent systems and, as stated by the WIPO Standing Committee on the Law of Patents, TRIPS allows Members to provide more extensive protection of intellectual property if they so wish and allows some flexibility for the Members to design their patent system since the Agreement does not address all issues.[5] One example of this lack of detail is the definition of an 'inven-

*This Chapter is partly based on the article Tine Sommer, 'The Scope of Gene Patent Protection and the TRIPS Agreement – An Exclusively Nondiscriminatory Approach?', 2007, *IIC* Vol. 38, p. 30 ff.

1. See the preamble to the EPC; and Singer/Stauder, *The European Patent Convention, A Commentary*, 3rd Ed., preamble, points 10-11.
2. EPC Articles 83 and 84; see W. Cornish, D. Llewelyn & T. Aplin, *Intellectual Property: Patents, Copyright, Trademarks and Allied Rights*, 7th Ed., p. 162. The abstract is a separate text, serving search purposes and giving the public immediate notice of the fact and nature of the application. See also Lionel Bently & Brad Sherman, *Intellectual Property Law*, 3rd Ed., 2008 p. 4. For the US, 35 U.S.C § 112, requiring adequate disclosure which directs enablement, written description, definiteness of claims and best mode, see Robert Patrick Merges & John Fitzgerald Duffy, *Patent Law and policy: cases and materials*, 5th Ed., 2011, p. 264; Marc Tritel, 'Obtaining useful patent protection for nucleotide and amino acid sequences in the USA', *Journal of Intellectual Property Law & Practice*, July, 2009, 4(9): 623-628.
3. See TRIPS Article 7.
4. See Holyoak & Torremans, *Intellectual Property Law*, 6th Ed., 2010, p.47.
5. See WIPO Standing Committee on the Law of Patents, Twelfth session, Geneva, 23 to 27 June 2008, Report on the International Patent System, point 146. The developed countries already had established patent systems whereas the undeveloped countries had to establish a patent system (the North/South perspective).

tion'.[6] The global TRIPS norm guiding international patent law can be narrowed down to a requirement for technology-neutral patent protection, based on the well known concepts of invention, novelty, inventive step and industrial application without further guidelines as to their interpretation or a requirement to conform to a TRIPS interpretation. As discussed in Chapter 1, it is not possible to interpret TRIPS Article 27(1) in the light of the *travaux préparatoires* and it is clear that there was no agreed understanding of the complete patent package that came with TRIPS ratification.[7]

'Patentable subject matter' has been a *one-size-fit* all heading and much argument has been directed to the requirement in Article 27(1) of the TRIPS Agreement for there to be patent protection in all fields of technology. However, the technology-neutral requirement has to be balanced against known doctrines in the Members' patent laws such as the US 'laws of nature' principle distinguishing the patentable subject matter from the unpatentable, and in Europe the distinction between patentable inventions and non-patentable discoveries.[8]

The argument frequently made in favour of unrestricted patentable subject matter is the justification based on economic incentives, without which a technological field would not be developed. In itself and by the judicial doctrines developed over time, patent law has been field restricted but not technology specific, however lately there has been a ban on patenting techniques involving the destruction of an embryo.[9] The US judicial doctrines on laws of

6. Joseph Strauss, 'Genomics and the food industry', in D. Vaver & L. Bently, (eds.), *Intellectual Property in the New Millennium, Essays in Honour of William R. Cornish*, p. 132: 'this does not in principle, throw any doubt upon the patentability of naturally occurring biological material, such as DNA, viruses, plasmids, cell lines, etc.'; Carlos M. Correa, 'Patenting Human DNA', *The Journal of World Intellectual Property*, 2007, Vol. 10, No 6, p. 424: 'the extension of the concept of invention to genes and other matters found in nature cannot be deemed universally accepted or incorporated into the TRIPS agreement'. See more generally Justine Pila, *The Requirement for an Invention in Patent Law*, 2010.
7. See Chapter 1, section 1; and Daniel J. Gervais, 'The Changing Landscape of International Intellectual Property', in Christopher Heath & Anselm Kamperman Sanders (eds.), *Intellectual Property & Free Trade Agreements*, 2007, p. 53.
8. On EPC Article 52(2), and the US judicial doctrine on 'laws of nature, physical phenomena, and abstract ideas', see Robert Patrick Merges & John Fitzgerald Duffy, *Patent Law and Policy: Cases and Materials*, 5th Ed., 2011, p. 107.
9. Europe has had restrictions on chemical patents and pharmaceutical patents and now on techniques involving the destruction of embryos; see Case C-34/10 *Oliver Brüstle v Greenpeace* and Chapters 1 and 3. See further Grubb & Thomsen, *Patents for Chemicals, Pharmaceuticals, and Biotechnology, Fundamentals of Global law, Prac-*

nature, physical phenomena and abstract ideas exclude such fields from patent eligibility. Field restrictions are not unknown in European statutes. Many exclusions of discoveries of mathematical methods, business methods, programs for computers, animal and plant varieties, essential biological processes and the human body have been good law in European statutes.[10]

There tends to be greater awareness of technology-specific restrictions, such as for biotechnology or nanotechnology. If it is argued that the scope of protection should be narrowed by purpose-bound protection (rather than product protection in US terms; use-patents instead of compound patents) or that DNA sequences as such are not to be patentable, this will run up against the TRIPS requirement for non-discrimination.[11] This part of the thesis will address the patenting of genes in the pre-grant phase and it will be argued that narrowing the scope of protection of substantive patentable matter in the pre-grant phase does not conflict with the overall principle of technological non-discrimination.[12]

### 1.1. The TRIPS agreement and the doctrine of non-discrimination

The doctrine of non-discrimination has been implemented in new areas of patent legislation, including amendments to international conventions, without

*tice, and Strategy*, 2010, pp. 77-78; Sommer, 'The Scope of Gene Patent Protection and the TRIPS Agreement: An Exclusively Non-discriminatory Approach', 2007, *IIC* Vol. 38, p. 30 ff; Dr. Kelvin W. Willoughby, 'How much does technology really matter in patent law? A comparative analysis of Doctrines of appropriate patentable subject matter in American and European Patent law', 18, Fed. Cir. B.J. 63 2009, 63 ff.; Dan L. Burk & Mark A. Lemley, 'Is patent law Technology Specific?', 17 *Berkeley Tech. L. J.* 1155 (2002), emphasising that 'the fundamental rules of patent law were set in a world in which inventions were mechanical. Because inventions in the past were far more homogenous than they are today it made sense to have a unified set of rules for dealing with those inventions' (p. 7 in the draft), concluding that patent law is becoming technology specific. See also Jeffrey M. Kuhn, 'Patentable subject matter matters: new uses for an old doctrine', 22 *Berkeley Tech. L. J.* 2007, p. 111, addressing the fact that changes in patentable subject matter cannot be resolved by tweaking the bright-line rules by which patent law usually operates. Rather, Congress, the courts, and scholars should mould these rules in new ways to take account of new developments.

10. EPC Article 52(2) and (4), and Article 53; and Directive 98/44/EC Articles 4 and 5.
11. See section 2.1 below and Tine Sommer, 'The Scope of Gene Patent Protection and the TRIPS Agreement – An Exclusively Nondiscriminatory Approach?', 2007, *IIC* Vol. 38, p. 30 ff.
12. See the discussion of the non-discrimination principle in Tine Sommer, 'The Scope of Gene Patent protection and the TRIPS Agreement – An Exclusively Nondiscriminatory Approach', 2007, *IIC* Vol. 38, p. 30 ff.

an in-depth analysis of the impact of such a general principle. The doctrine has been inserted in the national patent laws of the TRIPS Members, including in implementations of Directive 98/44/EC, as well as in the revised text of the European Patent Convention (EPC 2000). EPC 2000, Article 52(1) as amended, brings the provision into line with TRIPS Article 27(1).[13]

As mentioned in Chapter 1, the application of the doctrine of non-discrimination to patentable subject matter in Article 27(1) of the TRIPS Agreement is often referred to as one of the major achievements of the Agreement. The wording of the provision is as follows:

> Subject to the provision of subparagraphs 2 and 3, patents shall be available for any inventions, whether products or processes, in all fields of technology, provided that they are new, involve an inventive step and are capable of industrial application. Subject to paragraph 4 of Article 65, paragraph 8 of Article 70 and paragraph 3 of this Article, patents shall be available and patent rights enjoyable without discrimination as to the place of invention, the field of technology and whether products are imported or locally produced.

The wording of the doctrine of non-discrimination encompasses three areas of discrimination. Patent rights must be available without discrimination as to:

1. the place of invention,
2. the field of technology, and
3. whether products are imported or locally produced[14]

13. EPC 2000 Article 52; and Singer & Stauder, *The European Patent Convention, A Commentary*, Vol. 1, 67, 3rd Ed., 2003); Gerald Paterson, 'The European Patent System, The law and practice of the European Patent Convention 2. Ed., 2001, p. 22 states 'that the TRIPS agreement has created and strengthened the international obligations of Member States to provide a minimum level of intellectual property protection and enforcement. Some of the main provisions relating to patents are clearly derived from the corresponding provisions of the EPC: for example, the provisions in Article 27 concerning patentable subject matter'. On the other hand the non-discrimination doctrine is not inherited from the EPC.
14. TRIPS Article 27(1) also provides a basis for limiting the power of Members to differentiate the treatment of locally produced and imported products. Though not explicitly mentioned in this provision, the main aim of the proponents of such a non-discrimination clause was to restrict the use of compulsory licences where there is lack of local exploitation; see 'UNCTAD-ICTSD Project on IPRs and Sustainable development, Resource Book on TRIPS and Development' 353 (2005).

The second area (the field of technology) repeats what is already provided for in the first sentence of the paragraph, namely that: 'patents shall be available for any inventions, whether products or processes, in all fields of technology'. Combining these two statements, it follows that patent protection must be available in all fields of technology without discrimination regarding substantive patent law and the scope of protection.

This thesis focuses solely on discrimination related to the field of technology. In principle, such discrimination occurs where the substantive requirements that an invention must fulfil in order to be patentable are less restrictive in one field of technology than in other fields of technology. In other words, one field of technology has advantages which are not given to others. Discrimination also occurs when there are disadvantages relating to all or part of a field of technology, for example, if patents in the area in question are subject to stricter patentability requirements, or if the monopoly granted is restricted to a purpose-bound protection only. This interpretation of the doctrine of non-discrimination expresses the doctrine in an absolute form. However, as a broad minimum standard, this doctrine has to be further elaborated before it can be applied to new areas of technology. The future patent policy for new technologies cannot be based on a doctrine which has not yet been fully defined.[15]

The legal literature (for example, the writings of *Nuno Pires de Carvalho*) assumes that patentability cannot be excluded and patent rights cannot be diminished in areas which have traditionally been subject to discrimination, such as pharmaceuticals and chemicals.[16] Given the lack of *travaux préparatoires* for the TRIPS Agreement, and the fact that the discussions were related to chemicals and pharmaceuticals, which had been subject to in-depth discussions for years, it is relevant to ask why gene technology has immediately been subject to the same non-discrimination doctrine? The doctrine of non-

15. See Keith E. Maskus & Jerome H. Reichman, 'International Public Goods and Transfer of Technology, Under a Globalized Intellectual Property Regime', p. 36, which states, in relation to a moratorium on stricter international intellectual property standards: 'From this perspective, further harmonization is not an improper goal, but rather a premature exercise under the new and uncertain conditions that attend the development of cutting-edge technologies generally and information-based technologies in particular'. See also Jerome Reichman, 'Patent Law Harmonization and the Draft SPLT', Paper presented to the World Intellectual Property Organization's (WIPO) Open Forum on the Draft Substantive Patent Law Treaty (SPLT), International Conference Center (ICC), Geneva, Switzerland, 1-3 March 2006.
16. See Nuno Pires de Carvalho, 'The TRIPS Regime of Patent Rights', 3rd Ed., 2010, p. 278.

discrimination goes beyond the areas where there has traditionally been discrimination. Nevertheless, it is doubtful whether this doctrine can prevent Members from taking a precautionary approach, in line with the introduction of patents for pharmaceuticals and chemicals.[17]

The provision in Article 27(1), whereby patents must be available for any invention, whether products or processes, in all fields of technology, implies that at least two general types of patent claims (for products and processes) must be provided for in the Members' patent laws.[18] Since the extent of protection is determined by the patent claims (EPC 2000 Article 69),[19] the non-discrimination requirement in Article 27(1) is connected to the scope of protection. A product patent claim provides the maximum protection because it prevents any application of the invention. As formulated by *Bently et al.*: 'products patents or, as they are also known, claims to a product per se confer protection over all uses of that product, no matter how the product was derived'.[20] A process patent protects the particular process (e.g. activities and actions) by which an invention is produced.[21]

### *1.1.1. The non-discrimination doctrine in context*

Article 27 is divided into three paragraphs. Paragraph 1 comprises all three substantive patentability requirements for there to be an invention (novelty,

17. E.g., Nuno Pires de Carvalho also argues that nothing prevents WTO Members from dealing with different fields of technology differently, such as using procedures like those of the Budapest Treaty. See the Budapest Treaty on the International Recognition of the Deposit of Microorganisms for the Purposes of Patent Procedure, 1977; and see Nuno Pires de Carvalho, 'The TRIPS Regime of Patent Rights', 3rd Ed., 2010, p. 278.
18. Also TRIPS Article 28. Concerning use patents, see UNCTAD-ICTSD Project on IPRs and Sustainable development, Resource Book on TRIPS and Development, 2005, p. 356: 'An important interpretation question is whether this Article [27( 1) 1 obliges Members to protect uses as such, for instance, new uses of known products, in addition to products and processes', and at p. 357: 'Under TRIPS, WTO Members are free to decide whether to allow the patentability of the uses of known products'.
19. See Pagenberg & Cornish, *Interpretation of Patents in Europe, Application of Article 69 EPC*, 2006, p. 9.
20. See Bently & Sherman, *Intellectual Property Law*, 3rd Ed., 2009, p. 365.
21. With the combination of the explicit inclusion of both product and process inventions and the part of the last sentence which prohibits any distinction between fields of technology, it might be said that a general principle of patent eligibility is established. Any exclusion from patentability would therefore be an exception to that rule. See Daniel Gervais, *The TRIPS Agreement, Drafting History and Analysis*, 3rd Ed., 2008, p. 337; Bently & Sherman, *Intellectual Property Law*, 3rd Ed., 2009, p. 365.

an inventive step, and an industrial application), as well as the doctrine of non-discrimination. Paragraph 1 focuses on the nature of a patentable invention without considering its exploitation. It follows from paragraph 1 of TRIPS Article 27 that the non-discrimination doctrine is subject to the provisions of paragraphs 2 and 3. While discussions relating to paragraph 1 (and presumably 3) can be limited to strictly legal matters, discussions relating to paragraph 2 will be of a more ethical nature. According to Article 27(2), parties to the TRIPS Agreement can refuse to grant a patent by reference to *ordre public* or morality. The wording of Article 27(2) uses the concepts of *ordre public* and morality collectively, expressly including the protection of the life or health of humans, animals and plants, and the avoidance of serious harm to the environment. These considerations are closely related to the exploitation of a patent, so exclusions from patentability should not be based merely on the ground that exploitation of such a patent would be prohibited by the country's own laws.[22]

There is another way in which a WTO Member can exclude inventions from patent protection in Article 27(3)(a). According to this provision, Members may also exclude from patentability diagnostic, therapeutic and surgical methods for the treatment of humans or animals. In the TRIPS Agreement the general exclusion from patentability of therapeutic methods is not connected to the three substantive requirements.[23] Therapeutic methods are generally considered to be non-patentable on the grounds of professional ethics.[24] The exception for methods of medical treatment has been regarded by *Cornish et al.* as a last redoubt against the sweep of the patent system into the territory of health care.[25] Finally, Article 27(3)(b) allows Members to exclude from patentability plants and animals other than microorganisms, and essentially biological processes for the production of plants or animals other than non-

22. See Chapter 4.
23. See Nuno Pires de Carvalho, *The TRIPS Regime of Patent Rights*, 3rd Ed., 2010, p. 304.
24. See Chapter 1, section 3.2, Chapter 4, section 1, and Nuno Pires de Carvalho observes: 'Both gene therapy and therapeutic transplantation tend to become standardized and repetitive procedures, and, therefore, they tend to be no longer performed individually by doctors and surgeons in their private offices, but rather become mass-used methods ... If society is interested in promoting the inventive activities of private companies in the field of biotechnological therapeutic methods, *the* time may come soon for a review of the scope of patent protection.' See *The TRIPS Regime of patents rights*, p. 307.
25. W. Cornish, D. Llewelyn & T. Aplin, *Intellectual Property: Patents, Copyright, Trademarks and Allied Rights*, 7th Ed., p. 218.

biological and microbiological processes. The list in paragraph 3(b) implies that WTO Members may exclude the following biotechnological inventions from patentability: whole animals, humans and plants; parts of animals or humans; animal and plant varieties and essentially biological processes for the production of plants and animals. Article 27(3) (b) has been subject to review since December 1998.

This provision reflects the considerable uncertainty about the field of biotechnology at the time of the Uruguay Round negotiations. Given that most contracting parties to GATT had not acquired experience in this area, they were naturally afraid to commit themselves to obligations when they did not know whether it would be advisable or even feasible to implement. Canada, for example, was reluctant to accept the mandatory patentability of higher life forms in Article 27(3)(b) because this was a matter of current national debate.[26]

Article 27(2) and (3) essentially protects the patent system from immoral inventions. It is obvious that the patenting of human beings as such would offend against the opinion of society. The doctrine of non-discrimination cannot be applied in support of a technology which would effectively nullify the exceptions provided for in subparagraphs 2 and 3. The aim of TRIPS is to promote the effective and appropriate protection of intellectual property rights. In addition, there is a desire to limit distortions of trade and barriers to international trade. The objectives and principles of TRIPS are set out in Articles 7 and 8. The status of these articles was referred to in the Doha Ministerial Declaration which stated that 'the TRIPS Council shall be guided by the objectives and principles set out in Articles 7 and 8 of the TRIPS Agreement and shall take fully into account the development dimension'.[27]

*Daniel Gervais* has classified the special importance (i.e. higher legal status) of the provisions in relation to three situations:

- negotiations,
- interpreting the Agreement in the context of e.g. dispute-settlement procedures, and
- as a basis for the interpretation of certain provisions of the Agreement.[28]

26. See Nuno Pires de Carvalho, *The TRIPS Regime of Patent Rights*, 3rd Ed., 2010, pp. 306-307.
27. See Daniel Gervais, *The TRIPS Agreement, Drafting History and Analysis*, 3rd Ed., 2008, p. 208.
28. See Daniel Gervais, *The TRIPS Agreement, Drafting History and Analysis*, 3rd Ed., 2008, p. 50 (used to stretch the exceptions).

The provisions can support a dynamic interpretation of the purpose of the TRIPS Agreement in order to ensure its current and reasonable interpretation. The use of preambles, objectives and principles to support the dynamic interpretation of laws is often seen in the European Union and is known as the *effet utile* doctrine, as developed by the European Court of Justice.

Article 7 of the TRIPS Agreement is as follows:

> The protection and enforcement of intellectual property rights should contribute to the promotion of technological innovation and to the transfer and dissemination of technology, to the mutual advantage of producers and users of technological knowledge and in a manner conducive to social and economic welfare, and to a balance of rights and obligations.

The provision is very vague, but it can still be useful in defining and shaping the doctrine of non-discrimination. It is not a mandatory provision, as shown by the statement that intellectual property rights 'should' contribute to the promotion and transfer, etc. Article 8 encourages the Members to take specific action, and it expressly narrows the scope for national measures to those which are consistent with other TRIPS provisions.[29] Article 7 refers to the balance of rights and obligations (promotion of technological innovation, transfer and dissemination of technology, mutual advantages of producers and users) as a prerequisite for TRIPS. Measures taken by the international community to avoid upsetting the balance might be supported by Article 7 at a more general level, for example a requirement that the industrial application of a gene must be disclosed in a patent application. Article 8 could be considered a safeguard clause,[30] whereas Article 7 is more likely to serve as a basis

29. Daniel Gervais defines Article 8 as essentially representing a policy statement in relation to Articles 30, 31 and 40. Article 8 is as follows: '1. Members may, in formulating or amending their laws and regulations, adopt measures necessary to protect public health and nutrition, and to promote the public interest in sectors of vital importance to their socioeconomic and technological development, provided that such measures are consistent with the provisions of this Agreement. 2. Appropriate measures, provided that they are consistent with the provisions of this Agreement, may be needed to prevent the abuse of intellectual property rights by right holders or the resort to practices which unreasonably restrain trade or adversely affect the international transfer of technology.'
30. See UNCTAD-ICTSD Project on IPRs and Sustainable development, Resource Book on TRIPS and Development' (2005), pp. 126-27: 'it would appear that Article 8.1 is to be read as a statement of TRIPS interpretative principle: it advises that Members were expected to have the discretion to adopt international measures they consider necessary to protect public health and nutrition, and to promote the public interest in sectors of vital importance to their socio-economic and technological development.

for interpretation (i.e. a dynamic, guiding principle used as a tool of interpretation) within the areas of substantive patent requirements and the scope of protection (see EPC Article 69, and the grant of purpose-bound protection only).[31]

If the doctrine of non-discrimination in Article 27(1) were to be applied to its full extent, no Member would be allowed to restrict the grant of patents for new technologies, since patents must be available for any invention in all fields of technology. However, if a new technology were to be offered full traditional patent protection and this resulted in an imbalance of rights and obligations only benefiting producers to the disadvantage of users, then Article 7 could apply. If only user interests were favoured, Article 7 would also apply. In other words, the doctrine of non-discrimination is subject to an assessment of the balance of rights and obligations, representing public interests as well as the rights of patent holders.[32]

### 1.2. Patentable subject matter, scope of protection and technology-neutral patent law

The reward and promotion of technological developments via patent law has been an important instrument and a major force for ensuring improvements in life sciences. By the grant of property rights the holder has been entitled to exclude others from the commercial use of the protected invention. However, this encompasses an inherent dilemma. Can patents be issued for inventions that are so basic and fundamental to future advances that patents for them would unduly burden subsequent inventors?[33] This question is especially rel-

The constraint is that the measures they adopt should not violate the terms of the agreement. This suggests that measures adopted by Members to address public health, nutrition and matters of vital socio-economic importance should be presumed to be consistent with TRIPS, and that any Member seeking to challenge the exercise of discretion should bear the burden of proving inconsistency.'

31. See UNCTAD-ICTSD Project on IPRs and Sustainable development, Resource Book on TRIPS and Development' 353 (2005), p. 126: 'Article 7 provides guidance for the interpreter of the Agreement, emphasizing that it is designed to strike a balance among desirable objectives ... Article 7 makes clear that TRIPS negotiators did not mean to abandon a balanced respective on the role of intellectual property in society. TRIPS is not intended only to protect the interests of right holders. It is intended to strike a balance that more widely promotes social and economic welfare.'
32. Peter K. Yu, 'The objectives and principles of the TRIPS Agreement', in Carlos M. Correa (ed.), *Research Handbook on the Protection of Intellectual Property under WTO Rules, Intellectual Property in the WTO*, Vol. I, 2010, p. 161 ff.
33. See Robert Patrick Merges & John Fitzgerald Duffy, *Patent Law and Policy: Cases and Materials*, 5th Ed., 2011, pp. 141 and 118. On p. 141 the authors write that: 'Fun-

evant to human gene patenting and it can be questioned whether patents ought to be issued for fundamental biological phenomena.

A most quoted study is that of *Michael Heller* and *Rebecca Eisenberg* from 1998,[34] using the term '*the tragedy of the anticommons*', adapting the term '*the tragedy of the commons*' first introduced by *Garrett Hardin* in 1968[35] to help explain overpopulation, air pollution, and species extinction. Later in 2008, a further elaborated form of this argument was given by Michael Heller in his book 'The Gridlock Economy'.[36] The first two terms refer to the overuse of commons as opposed to underuse in anticommons.[37] The gridlock thesis applies when privatisation is taken too far, to the point where it destroys rather than creates wealth. *Heller* argues that privatising a commons may cure the ill of wasteful overuse, but it may unintentionally have the opposite effect. He argues that well-functioning systems of private property are fragile balances, poised between the extremes of overuse and underuse.[38] One example is the gridlock in organ donation, due to fragmented ownership

damental inventions with a very wide range of applications might perhaps be the most valuable types of inventions; providing incentives to make such discoveries is a logical goal of the patent system. On the other hand, patenting the fundamental tools of science does decrease the incentives for improvers and also creates difficult problems for negotiations where future improvements are all covered under a single early pioneer patent'. See also John S. Leibovitz, 'Inventing a nonexclusive Patent System', 111 *Yale Law Journal*, 2002, p. 2254, discussing the controversy over granting exclusive rights to new technologies and balancing interests.

34. See Michael A. Heller, 'The Tragedy of the Anticommons; Property in transition from Marx to markets', 111 Harvard Law Review, 111, 621 (1998); Michael A. Heller & Rebecca S. Eisenberg, 'Can Patents Deter Innovation? The Anticommons in Biomedical Research', *Science,* 1 May 1998, Vol. 280, No 5364, pp. 698-701; Merges, Mendell & Lemley, *Intellectual property in the new technology age*, 4th Ed., 2007, p. 153 (recent empirical evidence finds little support for the anticommons idea so far); and Dan L. Burk & Mark A. Lemley, 'Policy Levers in Patent law', 2003, University of Minnesota Law School, Public Law & Legal Theory Research Paper series, Research paper No 03-11, p. 90 ff.
35. *Science* 162, 1243 (1968).
36. Michael Heller, *The Gridlock Economy. How too much ownership wrecks markets, stops innovation and costs lives*, 1998.
37. Cf. Michael Heller, Harvard Law Rev. 111, 621 (1998), F. Michelman, Nomos 24, 3 (1982), W. N. Hohfeld, *Fundamental Legal Conceptions as Applied in Judicial Reasoning and Other Legal Essays* (Yale Univ. Press, New Haven, CT, 1923). This work introduced the now standard vocabulary of 'privileges of use' and 'rights of exclusion', Cf. Can Patents Deter Innovation? The Anticommons in Biomedical Research, *Science,* 1 May 1998, Vol. 280. No 5364, pp. 698-701.
38. See Michael Heller, *The Gridlock Economy*, pp. 19 and 20.

rights. Organs go to waste and potential recipients get sicker or die while doctors make sure they have all the necessary permissions.[39]

In the literature there has been and still are extensive discussions of the patenting of genes, focusing on genes patented as product-coding information, methods using such information or the ethical issues of gene patenting.[40] In

39. See Michael Heller, *The Gridlock Economy*, pp. 19-20. In writing about gridlock, Heller makes a joke illustrating how this fundamental law of economic and social life has escaped our notice until now: 'there is a joke about two economists who are walking down the street when they see some hundred-dollar bills scattered on the sidewalk. As one of them bends down to pick up the money, the other says, 'leave them alone! If those bills were real, someone would have picked them up already'.' According to Heller, simple truths are real but we overlook them because they don't fit with our existing theories, and gridlock is such a truth, a feature of our everyday world that nobody stops to notice.
40. See Chapter 1, section 3.2; and e.g. Josephine Johnston & Angela A. Wasunna, 'Patents, Biomedical Research, and treatments: Examining Concerns, Canvassing Solutions', *Hasting Center Report*, January-February, 2007, p. 11. On p. 8 there is a reference to a 2005 survey by Kyle Jensen & Fiona Murray in *Science* 310, No 5746, 2005, pp. 239-40, indicating that some important genes are heavily patented, including genes associated with increases in breast cancer, diabetes, and obesity – three diseases that exert a heavy burden in the developed world (and therefore represent potentially profitable markets for new treatments.) Their statement that 20 % of human genes have been patented has been countered in Christopher M. Holman, 'Will gene patents impede whole genome sequencing?: Deconstructing the myth that 20% of the human genome is patented', *IP Theory*: Vol. 2: Issue 1; and in Christopher M. Holman, 'Debunking the myth that whole-genome sequencing infringes thousands of gene patents', *Nature Biotechnology*, 2012, 30, pp. 240-244; *Lorelei Perez Westin,* 'Genetic Patents: Gatekeeper to the promised cures', *Thomas Jefferson Law Review*, Fall 2002, 25 TJLR, p. 279; J. Paradise, L. Andrews, & T. Holbrook, 'Patents on Human Genes: An analysis of Scope and Claims', *Science*, 307, No 5715, 2005, pp. 1566-67. James Bradshaw writes that information-allocation models based on dedication of the discovery to the public domain (by unselfish scientists) has now been followed by funding concerns and thus the patenting and licensing of such results; see 'Gene Patent Policy: Does issuing gene patents accord with the purposes of the U.S. Patent System', (37 WMTLR 637), *Willamette Law Review*, pp. 644-645; Eileen M. Kane, 'Patent-Mediated Standards in Genetic testing', 2008, *Utah L. Rev.* 837, No 3, p. 846; Darryl Macer, 'Whose Genome Project?', in Sheila A. M. Mclean (ed.), *Genetics and Gene Therapy*, p. 661; Mark S. Frankel, 'Inheritable Genetic Modification and a brave new world, Did Huxley have it wrong?', *Hasting Center Report*, in Sheila A. M. Mclean (ed.), *Genetics and Gene Therapy*, p/. 397; David Magnus, Arthur Caplan & Glenn Mcgee (eds), Who *owns life?*, 2002; 'The Ethics of patenting human genes and stem cells', 2004, *The Danish Council of Ethics*; Donna M. Gitter, 'International conflicts over patenting human DNA sequences in the United states and the European Union: An argument for compulsory licensing and a fair use exemption',

the USA a *Hastings Center Report* in 2007 emphasised the criticisms of the patenting of life forms: 'some critics assert that human biological substances should not be patentable because they already belong to 'humanity' – they are 'our common heritage' – and patenting them inappropriate commodifies the human body'.[41] In Europe in 2002 a paper from the Nuffield Council on Bioethics and a 2004 report from the Danish Council of Ethics also raised con-

76 *New York University Law Review*, 2001, p. 1636, the fundamentally different US and EU perspectives on human DNA patents creates international tension; Jordan Paradise, Lori Andrews & Timothy Holbrook, 'Patents on Human genes, an analysis of scope and claims', 2005, *Science*, Vol. 307, p. 1566 ff; John M. Conley, 'Gene patents and the product of nature', 2010, 84 *Chicago-Kent Law Review*, p. 132: 'But I fear that genes may be different ...'. Michael John Gulliford, 'Much ado about gene patents: The role of Foreseeability', 2004, 34 *Seton Hall Law Review*, p. 711 ff.; James Bradshaw, 'Gene patent policy: Does issuing gene patents accord with the purposes of the U.S. patent system?', *Willamette Law Review*, 2001, 37, 637 ff.; Michael M. Hopkins, Surya Mahdi, Pari Patel, & Sandy M. Thomas, 'DNA patenting: The end of an era?', *Nature Biotechnology*, 2007, Vol. 25, No 2, p. 185 ff.; M. Scott Mcbride, 'Patentability of Human genes: our patent system can address the issue without modification', 85 *Marquette Law Review*, 2001, p. 511 ff.; Lori B. Andrews, 'The gene patent dilemma: balancing commercial incentives with health needs', 2002, *Houston Journal of Health Law and Policy*, p. 65 ff.; May Mowzoon, 'Access versus Incentives: Balancing policies in genetic patents', 2003, 35 *Arizona State Law Journal*, p. 1077 ff.; Lorelei Perez Westin, 'Genetic patents: Gatekeepers to the promised cures', 2002, *Thomas Jefferson Law Review*, 25, p. 271 ff.; Lori B. Andrews, 'Genes and patent policy: rethinking intellectual property rights', *Nature Review, Genetics*, 2002, Vol. 3, p. 803 ff.; Stephen G. Whiteside, 'Patents claiming genetically engineered inventions: A few thoughts on obtaining broad property rights', 1996, *New England Law Review*, p. 1019 ff.; Aaron Xavier Fellmeth, 'Reinventing the double helix: A novel and nonobvious reconceptualization of Biotechnology patents', 55, *Stanford Law Review*, p. 303 ff, 2002; Antony L. Ryan & Roger G. Brooks, 'Innovation vs. Evasion: Clarifying Patent rights in second-generation genes and proteins', 17 *Berkeley Technology Law Journal*, 2002, p. 1265 ff; Rebecca S. Eisenberg, 'Re-examining the role of patents in appropriating the value of DNA sequences', 2000, *Emory University School of Law*, p. 783 ff; John Murray, 'Owning genes: disputes involving DNA sequence patents', 1999, *Chicago-Kent Law Review*, p. 231 ff; Andrew W. Torrance, 'Metaphysics and patenting life', *UMKC Law Review*, 2007, Vol. 76: 2, p. 364 ff.; Andrew W. Torrance, 'Patenting human evolution', *Kansas Law Review*; 2008, Vol. 56, p. 1075 ff.; and F. Scott Kieff, *Perspectives on properties of the human genome project*, 2003.

41. See Josephine Johnston & Angela A. Wasunna, 'Patents, Biomedical Research, and treatments: Examining Concerns, Canvassing Solutions', *Hasting Center Report*, January-February, 2007.

cerns about the patenting of genes.[42] Patenting human genes in fact gave rise to the gene manifesto '*the tragedy of the anticommons*'.

As pointed out in the 2007 Hastings Center report, inventions and discoveries resulting from biomedical research have always been patented, some for profit others not. According to the report there has been a dramatic increase in the number of biomedical patents world-wide in the past 25 years. This increase includes patents for biological materials and for inventions or discoveries that are still at an early stage of development.[43] As seen in literature and in the case law for example from the UK House of Lords, new technology may achieve a broad scope of protection at the early stage of the development but it may be subject to a more limited scope once the technology becomes better known to a skilled person. In the case *Biogen Inc v Medeva PLC*, on the patenting of a DNA sequence, the 'Lord Hoffman principle' was given as a guideline in this regard: 'the first person to find a way of achieving an obviously desirable goal is not permitted to monopolise every other way of doing so'.[44] In the US Supreme Court case of *Brenner v Manson* the claim comprised a process for the production of an already known steroid hormone, whose utility had not yet been shown.[45] The Court stated that 'such a patent may confer power to block off whole areas of scientific development, without compensating benefit to the public'.[46] A frequently quoted passage from the

42. 'Patenting human genes and stem cells, A report', *The Danish Council of Ethics*, 2004; 'The Ethics of patenting DNA, a discussion paper', 2002.
43. See the *Hastings Center Report*, p. 8.
44. See [1997] RPC 1, 41-41, concerning a recombinant DNA molecule (DNA sequence encoding a protein displaying HVB antigen specificity). In *Chiron* [1996] RPC 535, concerning isolated polypeptides encoding for hepatitis C virus (HVC) on speculative claims (claiming all ways of achieving a result, but in fact only describing one way). *Kirin-Amgen Inc v Hoechst Marion Roussel Ltd*, [2001] EWHC 433, Pat. (DNA sequence encoding for human protein erythroprotein), appealed to the Courts of Appeal/later appealed to the House of Lords [2004] UKHL 46, [77] giving a monopoly for the use of the disclosed information as such. Finally, in *Generics v Lundbeck* (isolated chemical compound enantiomer) the House of Lords considered the 'Lord Hoffmann' principle derived from *Biogen* as distinguishable on its facts as involving a different type of claim. The judgment in *Biogen* only applied only to 'product-by-process' type claims; it did not apply to simple product claims. See also Justine Pila, *The Requirement for an Invention in Patent Law*, p. 298 ff., 2010; and chemical products and proportionate patents, before and after *Generics v Lundbech*, available at http://ssrn.com/abstract=1355524.
45. *Brenner v Manson*, 383 U.S. 519, 148 U.S.P.Q. (BNA) 689 (1966)
46. *Brenner v Manson*, 383 U.S. 519, 528-36, 148 USPQ 689, 693-96 (1966). See also 2106 Patent Subject Matter Eligibility [R-6] – 2100 Patentability II.A on the require-

case by *Justice Fortas* stated: 'This is not to say that we mean to disparage the importance of contributions to the fund of scientific information short of the invention of something 'useful', or that we are blind to the prospect that what now seems without 'use' may tomorrow command the grateful attention of the public. But a patent is not a hunting license. It is a not reward for the search, but compensation for its successful conclusion. 'A patent system must be related to the world of commerce rather than to the realm of philosophy''.[47]

The concept of gridlock can apply to new technologies if patents are claimed in the early phase of a developing technology, and it can also come into question with the patenting complex biological processes. Furthermore, in high-technology cases, as expressed by *Rochelle Dreyfuss*, there can be a 'wow factor' and a tendency for juries to be overawed by technological complexity.[48] If a patent is granted too broad a scope of protection it may hinder subsequent access to materials and processes needed for further research and treatment.[49] Gridlock can be prevented by balancing overuse and underuse with a strict approach to patentable subject matter and the scope of protection.

ment that the invention as a whole must be useful and accomplish a practical application. The purpose of this requirement is to limit patent protection to inventions that possess a certain level of 'real world' value, as opposed to subject matter that represents nothing more than an idea or concept, or is simply a starting point for future investigation or research http://www.uspto.gov/web/offices/pac/mpep/documents/2100_2106.htm#sect2106.

47. *Brenner v Manson*; 383 U.S. 519, 148 U.S.P.Q (BNA) (1966).
48. Rochelle Cooper Dreyfuss, 'Nonobviousness: a comment on three learned papers', 12 *Lewis & Clark Law Review* 435 (2008). The author focuses on nonobviousness with type II errors, taking material out of the public domain, increasing patent tickets and transaction costs, acting as barriers to entry and to cumulative research, and encouraging trolling (pa. 435) and on p. 437 on overprotection.
49. Robert Cook-Deegan & Christopher Heaney, 'Patents in genomics and human genetics', *Annu. Rev. Genom. Human Genet.*, 2010, 11, pp. 383-425; Ester van Zimmeren, Birgit Verbeure, Gert Matthijs & Gertrui Van Overwalle, 'A clearing house for diagnostic testing: the solution to ensure access to and use of patented genetic inventions', *Bulletin of the World Health Organization*, May 2006, 84 (5), p. 352; Zimmeren, Verbeure, Matthijs & Van Overwalle, 'Patent pools and diagnostic testing', *Trends in Biotechnology*, Vol. 24, No 3, March 2006, p. 115 ff.; Zimmeren, Verbeure, Matthijs & Van Overwalle, 'Model for facilitating access to patents on genetic inventions', *Nature Review*, 2006, Vol. 7, p. 143 ff.; Isabelle Huys, Nele Berthels, Gert Matthijs & Gertrui Van Overwalle, 'Legal uncertainty in the area of genetic diagnostic testing', *Nature Biotechnology*, Vol. 27, No 10, October 2009, p. 903 ff.; B. Verbeure, G. Matthijs & G. Van Overwalle, 'Analysing DNA patents in relation with diagnostic genetic testing', *European Journal of Human genetics*, 2006, 14, pp. 26-33; Julia

The US legal principle that excludes 'laws of nature, physical phenomena and abstract ideas' from patentability is a tool by which the courts can limit patentable subject matter.[50] As pointed out by *Merges and Duffy*, the precise application of this doctrine has not been uniform.[51] It is also questionable whether the doctrine has three separate categories or whether it is all in one?

In *Laboratory Corporation of America Holdings v Metabolite Laboratories, Inc.*, on writ of certiorari to the United States Court of Appeals for the

Carbone et al, 'DNA patents and diagnostics: not a pretty picture', *Nature Biotechnology*, Vol. 28, No 8, 2010, p. 784 ff.; Gertrui Van Overwalle, 'Turning patent swords into shares', *Science*, 2010, Vol. 330, p. 1630; S Aymé, G Matthijs & S Soini, 'Patenting and licensing in genetic testing, Recommendations of the European Society of Human genetics', *European Journal of Human Genetics*, 2008, 16, p. S3-S9; Lori Andrews, Jordan Paradise, Timothy Holbrook & Danielle Bochneak, 'When patents threaten science', *Science*, Vol. 314, 2006, p. 1395 ff; Cara Koss, 'Oysters & Oligonucleotides, Concerns and proposals for patenting research tools', 25 *Cardozo Arts & Ent LJ*, 747, 2007; Dan L. Burk, 'Copyrightability of recombinant DNA sequences', 1989, *Jurimetrics J, Section of Science & Technology, American Bar Association*, pp. 490-491; Natalie M. Derzko, 'In search of a compromised solution to the problem arising from patenting biomedical research tools', 2004, 20 *Santa Clara Computer and High Technology Law Journal*, p. 347 ff.; Peter Yun-hyoung Lee, 'Inverting the logic of scientific discovery: applying common law patentable subject matter doctrine to constrain patents on biotechnology research tools', 19 *Harvard Journal of Law and Technology*, 2005, p. 79 ff. (stem cells); Beth E. Arnold & Eva Ogielska-Zei, 'Patenting genes and genetic research tools: Good or bad for innovation?', *Annu. Rev. Genomics Hum. Genet*, 2002, 3:415-32; David C. Hoffman, 'A modest proposal: towards improved access to biotechnology research tools by implementing a broad experimental use exception', 89 *Cornell L. Rev*. 993 ff.

50. *Rubber-Tip Pencil Co. v Howard*, 87 U.S. (20 Wall) 498, 507 (1874) (an idea of itself is not patentable); *O'Reilly v Morse*, 56 U.S. 62 (1854) (the mere discovery of a new scientific principle is not patentable, however Justice Grier dissenting stated that limits arose from the utility requirement); *The telephone cases* 126 U.S. 1 (1888) (Justice Waite delivered the opinion of the court: 'In this art – or what is the same thing under patent law, this process, this way of transmitting speech – electricity, one of the forces of nature, is employed; but electricity, left to itself, will not do what is wanted'); *Funk Brothers Seed Co v Kalo Inoculant* Co., 333 U.S. 127, 130 (1948); *Parker v Flook*, 437 U.S. 584, 585 (1978); and *Diamond v Chakrabarty*, 447 U.S. 303, 309 (1980). See Robert Patrick Merges & John Fitzgerald Duffy, *Patent Law and Policy: Cases and Materials*, 5th Ed., 2011, pp. 107 and 115. On page 127, Merges and Duffy write that the nineteenth century law on patentability and natural principles was mainly concerned with ensuring specified utility and that the claims bore some relation to the inventors' contribution to the useful arts.
51. See Robert Patrick Merges & John Fitzgerald Duffy, *Patent law and Policy: Cases and Materials*, 5th Ed., 2011, p. 107.

Federal Circuit,[52] the justification of the 'doctrine of laws of nature' was elaborated by *Justice Breyer* (with whom Justices Stevens and Souter joined in dissenting). *Justice Breyer* stated that the reason for the exclusion is that sometimes *too much patent protection can impede rather than 'promote the Progress of Science and useful Arts'.* Patents can both encourage research by giving an economic incentive and *discourage research by impeding the free exchange of information.*

> 'Patent law seeks to avoid the dangers of overprotection just as surely as it seeks to avoid the diminished incentive to invent that underprotection can threaten. One way in which patent law seeks to sail between these opposing and risky shoals is through rules that bring certain types of invention and discovery within the scope of patentability while excluding others.'[53]

The above given explanation by Justice Breyer points at several pathways when discussing gridlock, its interface with new technologies (early stage) and technology neutrality.[54] First the Breyer quote refers to *a basic scientific*

52. 545 US 126 S.Ct. 2921 (2006), of 22 June 2006: 'This case involves a patent that claims a process for helping to diagnose deficiencies of two vitamins, folate and cobalamin. The process consists of using any test (whether patented or unpatented) to measure the level in a body fluid of an amino acid called homocysteine and then noticing whether its level is elevated above the norm; if so, a vitamin deficiency is likely. We granted certiorari in this case to determine whether the patent claim is invalid on the ground that it improperly seeks to '*claim a monopoly over a basic scientific relationship*,' Pet. for Cert. namely, the relationship between homocysteine and vitamin deficiency'. See further Robert Patrick Merges & John Fitzgerald Duffy, *Patent Law and Policy: Cases and Materials*, 5th Ed., 2011, p. 129.
53. The case is analysed in Christopher M. Holman, '*Bilski*: Assessing the impact of a newly invigorated patent-eligibility doctrine on the pharmaceutical industry and the future of personalized medicine', *Current Topics in Medicinal Chemistry*, 10:1945. 2010; available at SSRN: http://ssrn.com/abstract=1424493, 'the vocal minority comprising three of the Courts nine Justices dissented from the majority's decision not to decide the case. Moreover, it was clear from the tenor of Justice Breyer's opinion that these Justices were generally concerned that the lower courts and PTO were applying an overly permissive interpretation of the patent-eligibility doctrine, which had resulted in the proliferation of too many patents broadly claiming fundamental principles and that threatened to impede science and innovation.' See further Philip McGarrigle & Vern Norviel, 'Laws of nature and the business of biotechnology', 24, *Santa Clara Computer & High Tech. L.J.* 275; Daniel T. Marvin, 'Comment', 26. *Temp. J. Sci. Tech. & Envtl. L.* 113.
54. See Martin J. Adelman, *Patent Law Perspectives*, 2nd Ed., Chapter 1, Statutory Subject Matter, § 1.1, 1-8.14: 'The above makes little sense. Scientific principles are discoveries which may lead to patentable inventions if someone uses a scientific princi-

*relationship* and not an early stage technology as such. However, it is emphasised that the justification for the principle does not lie in any claim that 'laws of nature' are obvious, that their discovery is easy, or that they are not useful. The exclusion is based on a concern of granting too much or too little patent protection thus seeking to maintain a delicate balance between over and under protection. The rule *brings certain types of invention and discovery within the scope of patentability while excluding others* and it must be evident that this room for manoeuvre is allowed to TRIPS Members when framing what is considered a patentable invention without conflicting with the overall TRIPS norm of non-discrimination between technologies.

The judicial discussion continues to be directed at questions of over protection and under protection, and striking a balance seems to be the key issue in the recent US Supreme Court cases which are reviewed below in section 3.[55]

In Europe a distinction is made between whether a creation is considered a discovery or an invention. The EPC framework requires there to be an invention; see EPC Article 52(1). The EPC system is discussed in the section below. There is an analysis of whether gridlock can be avoided by addressing two issues: patentable subject matter and the scope of protection.

It has been argued that in the early-stage of biotechnology, many patents granted were too broad due to a lack of identifiable prior art.[56] In *Marks & Clerk's 2011 Life Sciences Report* there is a reflection on the scope of protection of inventions in the area of nanotechnology:

'the ship is expected to right itself in due course as the law and the patent office's catch up. Perhaps with this in mind, the offices have assigned particular class codes to nanotechnol-

ple in a nonobvious way. However, if the principle was never disclosed until it was used in an obvious way to obtain a useful product, then the useful product is unobvious over the prior art and should be patentable. The same goes for useful processes. By way of contrast, how could one give a patent on a discovery itself? Would we say that anybody who uses the discovery *to make a useful product or process is an infringer?* But how will we know when we examine the product or process that the inventor used the discovery? We will have to interrogate him or her. This would make hash of patent law. Hence, no major patent system permits the patenting of discoveries no matter how significant. It has nothing to do with incentive effects or a fear of stopping the spread of useful knowledge. In any event' [emphasis added].

55. And the literature follows this question, see Kenneth G. Chahine, 'Anchoring gene patent eligibility to its constitutional mooring, Commentary', *Nature Biotechnology*, Vol. 28, No 12, December 2012, p. 1253 ff., proposing five criteria useful for assessing patentable subject matter.
56. Marks & Clerk, *Life Sciences Report 2011*, p. 22.

ogy and are drawing in Examiners with a wealth of backgrounds into the new Examining Divisions to cope with the increased filings. It is hoped that this will lead to faster prosecution of cases, but it is also expected that the examination will be much more rigorous and extremely broad patents of the past are less likely to be followed'.[57]

This approach to early stage technologies has evidently also been adopted by the European Patent Office. It follows from Article 84 EPC that 'the claims shall define the matter for which protection is sought. They shall be clear and concise and be supported by the description'. This is addressed in the EPO Guidelines for Examination, June 2012, Part F, Chapter IV-23, point 6.1, which states that 'there must be a basis in the description for the subject-matter of every claim and that the scope of the claims must not be broader than is justified by the extent of the description and drawings and also the contribution to the art'. As for the extent of generalisation according to point 6.2:

Most claims are generalisations from one or more particular examples. The extent of generalisation permissible is a matter which the examiner must judge in each particular case in the light of the relevant prior art. *Thus an invention which opens up a whole new field is entitled to more generality in the claims* than one which is concerned with advances in a known technology. A fair statement of claim is one which is *not so broad that it goes beyond the invention nor yet so narrow as to deprive the applicant of a just reward* for the disclosure of his invention. The applicant should be allowed to cover all obvious modifications of, equivalents to and uses of that which he has described. In particular, if it is reasonable to predict that all the variants covered by the claims have the properties or uses the applicant ascribes to them in the description, he should be allowed to draw his claims accordingly. After the date of filing, however, he should be allowed to do so only if this does not contravene Art. 123(2).'[58]

The EPO approach is thus to taking for a differentiated approach to new technologies (inventions which open up a whole new field). Thus, on paper European patents provide wider protection for pioneering inventions, but the ship is supposed to right itself once the technology matures.

However this thesis addresses the point that a ship cannot right itself if the legal basic conditions are wrong. It may adjust the scope of protection, but if to start with the patentability of an invention (e.g. human DNA sequences) is

57. Marks & Clerk, *Life Sciences Report 2011*, p. 22.
58. Guideline for substantive examination, June 2012 Part F, IV-Claims (Article 84 and formal requirements). See further Geertrui Van Owerwalle, 'Policy Levers Tailoring Patent Law to Biotechnology. Comparing US and European Approaches', *UC Irvine Law Review*, Vol. 1, No 2, p 41.

within the scope of what must be excluded under a 'Justice Breyer assessment' due to an overprotection, then the ship cannot right itself and gridlock will ensue. To illustrate: In the early days of gene technology the discovery of new genetic information in itself was complicated.[59] Today this is a routine matter and there has been remarkable progress in genetics since the Human Genome Project announced the blueprint of Human genome in 2003.

> 'the scientific breakthrough – the development in 1972 of recombinant DNA technology spawned further advancements in genetic research, including the discovery in 1983 of a generally applicable method for cloning genes for polypeptides where the amino acid, DNA and mRNA sequences were not completely known, the availability beginning in 1986 of computer controlled sequencing machines for the DNA base pairs that forms genes and the development of polymerase chain reaction technology the same year. These advancements have powerfully boosted the ability of scientists to locate and sequence genes. As the president of one major biotechnology company noted, a few decades ago it might have taken ten years to find a particular gene, but, with modern gene maps, a gene can now often be found with a fifteen second computer search. Sequencing has also become far less laborious. The ability of scientists to rapidly sequence DNA has resulted in an explosion of discoveries of DNA sequences – both meaningful and meaningless scientifically – that, in turn, has caused a deluge of patent applications claiming DNA sequences and the proteins and other biomedicals for which these sequences code'.[60]

Another example is nanotechnology. As described in Chapter 2, section 4, nanotechnology is first and foremost a matter of size. One can question whether bio-nanotechnology should be assessed differently from molecular technology (bio-/gene technology) if the underlying processes are the same.

59. See some of the early articles on the DNA patents: Nada Jain, 'To patent or not to patent: Gene Therapy in the European Union and the United States', *Cardozo Journal of International and Comparative Law*, Winter 1996, 4 CDZJICL 103; Byron V. Olsen, 'The biotechnology balancing act: Patents for gene fragments, and licensing the 'useful arts'', *Albany Law Journal of Science and Technology*, 7 ALBLJST 295, 1997; John K. Flanagan, 'Gene Therapy and patents, Journal of the patent and trademark office society', 80 *JPTOS* 739, 1998; Stuart A. Newman, 'Averting the clone age. Prospects and perils of human developmental manipulation', *Journal of Contemporary Health Law and Policy*, 19 JCHLP 431, 2003; Amanda S. Pitcher, 'Contrary to first impression, genes are patentable: should there be limitations?', 6 JHTHCLP 284, *Journal of Health Care Law and Policy*, 2003; Donna M. Gitter, 'International conflicts over patenting human DNA sequences in the United States and the European Union: An argument for Compulsory licensing and a fair-use exemption', 76 N.Y.U.L. Rev 1623, *New York University Law Review*, 2001.
60. Aaron Xavier Fellmeth & Linda J. Demaine, 'Reinventing the Double helix: A novel and nonobvious reconceptualization of the Biotechnology patent', *Stanford Law Review*, November 2002 55 STNLR 306.

The *Marks & Clerk's Life Sciences Report 2011* highlights two EPO cases. The first was Case T-547/99 *BASF v Orica Australia*[61] from 2002, concerning claims for 'very small water-insoluble particles' of up to 100 nm in diameter with certain properties, as well as preparation processes therefore. The Board found that while the claims comprised a combination of separately known features, this combination was non-obvious. The second case highlighted in the *Life Sciences Report 2011* was Case T-552/00 *Smithkline Beecham Biologicals v Wyeth Holdings*[62] concerning a vaccine comprising antigen, a carrier and certain lipid 'whose particle size does not exceed 120 nm'. The technical feature was to find improved adjuvants for the lipid, and the Board held that particles of the size in question were inventive, as none of the prior art showed that a reduction in size would be an advantage.[63] However, the interesting part is that the process described does not differ from what could be implicated by molecular biology – as mentioned in Chapter 2, section 4, nanotechnology is biotechnology at a smaller scale. This view can be supported by the comments in the *Life Sciences Report 2011* where it is said that the type of inventive step reasoning (in Case T-552/00 *Smithkline Beecham Biologicals v Wyeth Holdings*) is becoming less and less common in any jurisdiction as the volume of prior art on nanosising increases to the point where just making a known feature (such as a particle) into a smaller (nano) size is now deemed to be an obvious step. Additional and unexpected advantages, not existing in the prior art, will have to be shown.[64] This raises the question of whether from a patent law perspective bio-nanotechnology can be better dealt with than biotechnology in terms of the scope of protection as long as the basic question of patentability of human DNA sequences has not been sorted out. It is clear from both the illustrations above from biotechnology and bio-nanotechnology that technology will enable further progress to be made, but the balanced scope of protection depends on a 'Justice Breyer assessment', and one way in which patent law seeks to sail between these opposing and tricky shoals is through rules that bring certain types of invention and discovery within the scope of patentability while excluding others. It is obvious that this is not the only way, but it is important to draw a line between the patentable subject matter and the non-patentable subject matter.

The case study on which Chapter 7 is based found that the patent examiner often referred to the early stages of a claimed technology. An example group

61. Case T-547/99 *BASF v Orica Australia*.
62. Case T-552/00 *Smithkline Beecham Biologicals v Wyeth Holdings*.
63. Marks & Clerk, *Life Sciences Report 2011*, p. 22.
64. Marks & Clerk, *Life Sciences Report 2011*, p. 21.

(7) in the HD study involved siRNA techniques. The US application was filed in 2004 (US2004147475) and the US patent examiner stated that:

'a claim rejection (35 USC § 112) because the specification, while being enabling for a method of attenuating the expression of target genes *in Zebrafish cells in culture or zebrafish embryos*, in avian neural crest tissue explant culture and in rat cell culture, does not reasonably provide enablement for attenuating gene expression *in vivo* in any non-embryonic animal. Moreover, the specification does not reasonably provide enablement for *a method for treating a disease or infection in an organism*. The specification does not enable any person skilled in the art to which it pertains, or with which it is most nearly connected, to use the invention commensurate in scope with these claims. ... Applicant traverses the scope of enablement rejection by arguing that *the references relied upon by the examiner to describe the unpredictability of the RNA interference art at the time of filing* are not supported by later findings, which support the opposite conclusion. Applicant describes the teachings of post-filing art that gene silencing with double stranded RNA is specific. While the examiner agrees that much progress has been made in the field of RNA interference, these apparently contradictory teachings, as well as the references describing issues such as degree of attenuation, toxicity and possible immune responses, actually supports the argument that at the time *of filing the field of RNA interference was in its infancy and that several questions regarding mechanism, specificity and side effects relating to predictability remained to be determined.*'[65]

During examination (Copy of the international preliminary report on patentability, 2008), *group (20)* in the HD study (EP1941059) also met the argument that although the application disclosed a method for inhibiting Huntington's disease in a cell in vitro, by administering a double-stranded ribonucleic acid (dsRNA) for inhibiting the expression of a human HD gene, *such a disclosure would not be considered enabling since the state of antisense-mediated gene inhibition in vivo is highly unpredictable*. In general the unpredictability of a technique seems to make the requirement of enablement stricter, instead

65. Non-final rejection from January 2006, page 12: 'Thus, while the specification is enabling for the examples set forth in the specification, *the specification is not enabling for the broad claims of introducing dsRNA in any animal, as the art of attenuating gene expression by introducing dsRNA into a cell or organism is neither routine nor predictable.* Thus, one of skill in the art could not practice the invention commensurable in scope with the claims without undue, trial and error experimentation'. See also Charles Schmidt, 'Negotiating the RNAi patent thicket', *Nature Biotechnology*, Vol. 25, No 3, March 2007, p. 274, on the Carnegie patent from 2003: 'Scientists had nearly given up on RNAi for drug development until four researchers – Thomas Tuschl at Max Planck Institute (currently at Rockefeller University in New York), Philip Sharp at MIT, Philip Zamore at UMass Medical School and David Bartel at the Whitehead Institute – found that by making the molecules smaller, they could avoid immune reactions while disabling mRNA.'

of allowing for generalisation as mentioned above in the EPO Guidelines for Examination; see further the PD case study *group (2)*, US 2006239966, in the 2008 non-final rejection stating that the applicant's argument failed to overcome the unpredictability in the art with regard to the administration route for gene therapy for treating PD.

The following sections give illustrations of two patent issues namely patentable subject matter and the scope of protection. The review of the scope of protection for gene patents will be made from a European perspective, whereas the patentability of genes will be discussed from a US perspective. While the US courts have initiated a revival of the patentable subject matter issue, there is a discussion of whether the Court of Justice of the European Union has introduced a limited scope of protection for gene patents. These approaches on different sides of the Atlantic are questionable, as the lack of distinction between what can be patented (pre-grant) and to what extent (post-grant) cannot be viewed as two separate questions. There is a need for a fundamental global understanding on gene patenting, making the patchwork of solutions redundant.

## 2. Pushing back: DNA sequences as patentable subject matter and the narrow scope of protection. Europe and the USA: static legislation or dynamic judicial review?

### 2.1. The EU/EPO framework

In 1998, after 10 years of dispute and in a form much modified from the original proposals, the European Community adopted Directive 98/44/EC on the legal protection of biotechnological inventions. The Directive, which became effective in the Member States in July 2000, was implemented in the EPC regime in 1999.[66] Despite the opposition to its adoption, based more on ethical considerations than on substantive patent law, the current Directive was adopted without any reservations regarding the use of absolute product protection for genetics, as was the case for pharmaceutical and chemical inventions for some Countries; see Chapter 1. Had the Commission taken a more precautionary approach instead, it could have proposed a directive in which claims for product patents for gene-technological inventions were *either* prohibited, *or* limited to an expressed industrial application (*de facto* that was not the outcome of Article 5(3)), perhaps even during a transitional period, and it

66. See Implementation Rules 23(b-e) (now Rule 26-34).

could have avoided embodying DNA sequences as patentable subject matter in a legal statute long before gene techniques had matured. As the Directive entered into force, it was clear that the legal requirements only differ from traditional patent law by the introduction of specific requirements for human DNA sequences, whereas genes from plants or animals are only subject to traditional patentability requirements (the definition of patentable subject matter is found in Articles 3 and 4, and is subject to the limitations in Article 6 of the Directive). Despite the wording of Article 5(3), the strict requirement in relation to human DNA has been more or less ignored by the Commission in its subsequent assessment of the Directive.

The Directive thus establishes the conditions for granting a patent when the subject of the invention consists of elements isolated from the human body (gene technology, gene sequences, stem cells, cell lines, etc.) or from plants or animals. Given the potential scope of such inventions, and given their inherent relationship to living matter, a Member State can exclude inventions from patentability by reference to ethical or moral principles recognised in that Member State. The morality exemption has always been part of traditional patent law; see Chapter 4.

In order for a patent to be granted, there must be an invention. The examination of a patent application consist of three elements, to establish whether there is an invention which is new, which involves an inventive step, and which is susceptible of industrial application.[67] It has been argued by *Luigi Palombi* that Directive 98/44/EC departs from TRIPS Article 27(1) in one very important respect:

> It mandates that *'a product* consisting of or containing biological material or *a process* by means of which biological material is produced, processed or used' be deemed an 'invention' within the terms of the EPC and the national patent laws of EC members. This departure, which 'shall be' undertaken by all EC members is, however, questionable because a product consisting of biological material which has been 'isolated' or 'produced by means of a technical process' that is *identical* to a natural biological material arguably is not an 'invention' within the meaning of the word in Art. 27.1 TRIPS. It is questionable because such biological material, although it is per se artificial or is in an artificial stasis, is indistinguishable from the natural counterpart.[68]

67. See Directive 98/44/EC Article 3(1), EPC Article 52(1), and TRIPS Article 27(1).
68. Luigi Palombi, 'The Patenting of Biological Materials In the Context of the Agreement on Trade-Related Aspects of Intellectual Property Rights', PhD Thesis, University of New South Wales, Sydney, 2004, pp. 4-5; and Luigi Palombi, 'The Impact of TRIPS on the Validity of the European Biotechnology Directive', I *JIBL* Vol. 2,

According to the UNTAD-ICTSD Project on IPRs and Sustainable Development, TRIPS Article 27(1) makes it clear that patents are to be granted for inventions but, as mentioned above, TRIPS does not define an 'invention', it only specifies the requirement an invention should fulfil in order to be patentable (Article 27(1)). This leaves Members considerable freedom to determine what should be deemed an invention and to exclude from patentability any substance which exists in nature as being a mere discovery and not an invention.[69]

It is the view of this author that Article 3(1) of Directive 98/44/EC only determines that inventions, which fulfil the three requirements of being new, involving an inventive step and being susceptible to industrial application are patentable, even if they concern a product consisting of or containing biological material or a process by means of which biological material is produced, processed or used. This only implies that naturally occurring material is not excluded from patentability per se. On the contrary, it should be treated in the same way as traditional mechanical inventions. The first proposal for the Directive included *a draft Article 8*, which declared that: 'The subject of an invention concerning a biological material shall not be considered a discovery or lacking in novelty merely on the grounds that it already formed part of the natural world.'[70]

Luigi Palombi argues against the generally held view that artificially isolated biological materials produced by human intervention constitute inventions. The Directive operates with a distinction between biological material occurring in nature and biological material that has been isolated artificially.[71] In Article 3(2) it is stated that biological material which is isolated from its natural environment or produced by means of a technical process may be the subject of an invention even if it previously occurred in nature. This distinc-

2005, p. 63 f.. See EPC 2010 guidelines, Part C – Chapter IV, point 2.2: 'It must also be borne in mind that the basic test of whether there is an invention within the meaning of Article 52(1) is separate and distinct from the questions whether the subject-matter is susceptible of industrial application, is new and involves an inventive step.'

69. UNTAD-ICTSD, Project on IPRs and Sustainable Development, Resource book on TRIPS and Development, see p. 357.
70. Products of nature have also been dealt with in the past as a problem of novelty. The application here of the normal novelty rules was thus already established in the first draft of the Directive; see COM(1995) 661 final.
71. *See* Luigi Palombi, 'The Impact of TRIPS on the Validity of the European Biotechnology Directive', I *JIBL*, Vol. 2, 2005.

tion is referred to as the *doctrine of isolation.*[72] The traditional distinction between a discovery and an invention has been diluted by the application of the traditional framework of patent law to gene-technology, because of gene-technology's relationship to nature. So far, this has mainly been an ethical discussion about whether parts of the human body could ever be the subject of an 'invention'.[73] However, this should not divert attention from the need to consider these issues within the context of substantive patent law.

Since the adoption of the Directive, ongoing legal discussions about the doctrine of isolation have lifted isolated biological material per se to the level of an invention. It has become legitimate to operate with natural parts and artificial parts as though they represented two different substances which could be objectively distinct from each other.[74]

With regard to gene technology in particular, one can argue that the traditional distinction between discoveries and inventions (e.g. in EPC Article 52(1) and (2)(a)) contains a distinction that is very important to the scope of protection. If the discovery consists of a gene sequence, then an invention will be found to exist if the inventor can prove that the gene sequence in question is new, involves an inventive step, and can specify its industrial application (a technical effect). In this case, Article 3(1) ensures that the inventor can claim product protection for the gene sequence as a discovery, and not

72. The first sketch of this doctrine of isolation appears in Common Position (EC) No 4/1994, recitals 10 and 11, referring to isolated elements of the human body and DNA sequences. It was further developed in COM (1995) 661 final, Article 3(2): 'Notwithstanding paragraph 1, the subject of an invention capable of industrial application which relates to an element isolated from the human body or otherwise produced by means of a technical process shall be patentable, even if the structure of that element is identical to that of a natural element'. This amendment was based on ethical considerations.
73. See e.g. 'Patenting Human Genes and Stem Cells', a report by The Danish Council of Ethics (2004).
74. As discussed, e.g., in the Danish Council of Ethics report, 'Patenting Human Genes and Stem Cells' (2004), p. 44. See also Luigi Palombi, 'The Impact of TRIPS on the Validity of the European Biotechnology Directive', I *JIBL*, Vol. 2, 2005, p. 63 f.: 'TRIPS requires that the patentability criteria in Art. 27(1) be applied across 'all fields of technology ... *without discrimination* as to ... the field of technology'. Given this requirement, the Directive violates TRIPS because by requiring ... Member countries to enact amendments to their patent laws so that technologies that come within Arts. 3 and 5 [of the] Directive are presumed to be 'inventions', it positively discriminates against all other technologies which do not benefit from such a presumption.'

merely for the product (the core invention) using the gene sequence.[75] If the product claim is allowed to be absolute, then the use of the gene sequence in unrelated inventions will require the approval of the first patentee. To this extent, absolute product patent protection expands traditional patent protection beyond the protection conferred to chemical compounds.[76]

Article 3(1) of the Directive, providing for all types of patent claims, could have been dealt with by the European Commission in a more sophisticated way. With regard to the introduction of chemical and pharmaceutical patents, it would not have been necessary to think along entirely new lines in order to be able to adopt a patent system in which the expansion of patent rights was thought through. The use of *a step-by-step approach*, without absolute product protection initially, rather than the rapid, one-size-fits-all approach has the value of sending a signal to patent authorities, patents applicants and society in general. In the following, this precautionary approach will be considered as taking account of the need to safeguard the balance between rights and obligations. Securing a patent system which is based on such a balance will to ensure society's acceptance of the legitimacy of the system.

What has happened instead is that the European legislation has ranked genes and chemical substances side by side. The patent system for chemicals was adapted gradually from 1958 to 1994 (TRIPS), a long time after the establishment of the elements in the periodic table. Nobody would apply for a patent for nitrate as a chemical product patent, but they might apply for a patent for a preservative substance or a fertiliser containing nitrate. Gene patents have previously been granted long before even the simplest function of the gene was known. This is not the case today, but it illustrates that the early

75. As formulated by Bently & Sherman, *Intellectual Property Rights*, 2009, p. 365; 'Product patents confer protection over all uses of that product' which also cover future uses of a *DNA* sequence.
76. Cornish & Llewelyn, *Intellectual Property: Patents, Copyrights, Trademarks and Allied Rights*, 2011, p. 926, state that patent laws everywhere protect 'inventions' as distinct from 'discoveries' and at p. 831 f. they write that 'behind [Art. 5(2)] lies the view that stripping back the DNA helix to RNA and its translation into cDNA involves an artificial chemical process. It gives a chemical substance that is not the same as that which occurs in nature. Accordingly, so it is argued, there is patentable subject matter distinct from living matter in its natural state for two reasons; because a laboratory process has intervened, and because the result is chemically different. In any case, the Directive does not even require the second factor of chemical change. It is enough to make an element isolated from the body by a technical process. From this it may be assumed that there does not have to be anything inventive about carrying out the process.'

years of gene patenting were characterised by carelessness or even recklessness. In the EU, the reason for introducing a legal framework for gene patenting was the eagerness to eliminate barriers to trade and safeguard the proper functioning of the internal market.

A gene is generally more complex than the average chemical compound. Genes are based on evolution and are more dynamic and complex in nature. While the number of chemical compounds is unlimited, the tally of human genes is limited to around 25,000. A claim for a product patent on a gene, which remains a gene whether naturally occurring or isolated, will be a claim for natural substance regardless of its technical effect.

According to Article 4(1) of the Directive, plant and animal varieties, and essentially biological processes for the production of plants or animals, are not patentable. This provision is identical to both TRIPS Article 27(3)(b), first sentence, and EPC 53(b). However, Article 4(2) of the Directive goes on to state that inventions which concern plants or animals are patentable if the technical feasibility of the invention is not confined to a particular plant or animal variety.

One of the most controversial provisions of the Directive is to be found in Article 5. Article 5 consists of a very wide-ranging application of the doctrine of isolation. Article 5 (1) provides that the human body at the various stages of its formation and development, and the simple discovery of one of its elements, including the sequence or partial sequence of a gene, cannot constitute patentable inventions. However, pursuant to Article 5(2), *an element isolated from the human body or otherwise produced by means of a technical process,* including the sequence or partial sequence of a gene, may constitute a patentable invention, even if the structure of that element is identical to that of a natural element. The function of the isolation requirement/doctrine is obviously to narrow the scope of the doctrine of nature (the handiwork of nature is not considered patentable). If isolated and if fulfilling the three steps required for patentability, the scope of protection will cover naturally occurring substances as well. Of course, nobody is prevented from using the product in nature (a gene performing its function in the human body cannot be patentable), but when used commercially and in other connections, a gene will be subject to the patent holders' rights and a patent will thereby prevent others from using a naturally occurring gene commercially. By extending the scope of protection to cover discovery as well as invention, the legislature has acted without caution. If only the use of the isolated gene in a specific function were patentable, then the discovery (the gene sequence as such) would remain a discovery and therefore unpatentable.

Article 5(3) introduced a requirement specific for human gene sequences.[77] The industrial application of a sequence or a partial sequence of a gene must be disclosed in the patent application. This stricter requirement is close to the wording used for the introduction of chemical and pharmaceutical patents referred to in Chapter 1, section 3. For many years inventors were only granted purpose-bound protection in these areas. However, the requirement only applies to human genes/DNA; genes of plants or animals are not subject to an explicit requirement for gene-specific industrial application. One could argue that human gene technology is therefore discriminated against compared with other fields of gene technology and other technologies in general. It is more than likely that the ethical debate about gene patenting has focused more on human ethics than on animal ethics, and even less on plants which have traditionally been subject to plant breeders' rights. This is reflected in the wording of the Directive, Articles 5(3) and 6.

The morality clause in Article 6 (see Chapter 4) is a further development of the traditional standard of European patent law in the biotechnology context. The non-exhaustive list referring to different kinds of biotechnological inventions/technologies which are considered non-patentable per se has clearly given priority to inventions related to the human body or inventions whereby animals are used for the benefit of humans. Nowhere in the literature is this list referred to as being inconsistent with the doctrine of non-discrimination. Since TRIPS Article 27(2) allows Members to exclude the patentability of inventions in order to protect human, animal, or plant life or health, or to avoid serious prejudice to the environment, the list cannot as such be questioned from the perspective of the doctrine of non-discrimination.

### *2.1.1. The long-standing process of transposing Directive 98/44/EC into national laws*

In principle, Directive 98/44/EC entered into force in July 2000. However, transposing the Directive has been one of the most controversial tasks for EU

77. Winfriend Tilmann, 'Scope of protection for gene sequence patents', *IIC*, Vol. 36, 8/2005, p. 909: 'one could also speak of a 'Janus-headed' form of substance protection in one direction, according to which the patent enjoys absolute substance protection, whereas it is restricted to the disclosed function in the other direction, albeit encompassing equivalence'; Andreas Oser, 'Patenting (partial) gene sequences taking partial account of the EST issue', *IIC*, Vol. 30, No 1/1999, p. 1 ff.; S.J.R. Bostyn, 'The prodigal son: The relationship between patent law and health care', *Medical law Review*, 11, 2003, p. 78 ff.

institutions in modern EC law, and it became subject to several proceedings before the Union Court of Justice immediately after its adoption. The first proceeding was for the annulment of the Directive under Article 230 of the EC Treaty (now Article 263 of the Treaty on the Functioning of the European Union (TFEU)) brought before the Court of Justice by the Dutch Government.[78] The Court of Justice dismissed the application and maintained the Directive. By the summer of 2003, only seven Member States had implemented the Directive and the EU Commission started proceedings against the other Member States for failing to fulfil their obligations under Article 226 of the EC Treaty (now Article 258 TFEU).[79] In 2005, cases were still pending against Latvia and Luxembourg, which had not yet implemented the Directive.[80] Luxembourg implemented the Directive on 23 April 2006 and Latvia on 29 December 2005. According to the second report of the Commission (2005) given pursuant to Article 16(c) of Directive 98/44/EC, as of June 2005, 21 Member States had notified the Commission of their instruments implementing the Directive,[81] 19 Member States had implemented provisions practically identical to the provisions of the Directive, and only Germany and France had transposed the Directive into national law with limited patent protection for gene patents. Furthermore, the implementation of the Directive in

78. See Case C-377/98, *The Netherlands v European Parliament and Council.*
79. See Case C-395/03 *Commission v The Netherlands*; Case C-448/03 *Commission v France*; Case C-450/03 *Commission v Luxembourg*; Case C-454/03 *Commission v Belgium*; Case C-456/03 *Commission v Italy*; Case C-518/03 *Commission v Sweden*; Case C-4/04 *Commission v Austria*; and Case C-5/04 *Commission* v *Germany*.
80. The Commission decided to send a further reasoned opinion to Luxembourg, pursuant to Article 228 of the EC Treaty, requesting it to comply immediately with a previous ruling of the Court of Justice requiring it to implement Directive 98/44/EC on the legal protection of biotechnological inventions in national law (Case C-450/03) *see* http://www.eu.int/rapid/pressReleasesAction.do?reference=IP/06/14&format=HTML&aged=O&language=EN&guiLanguage=da. Proceedings were initiated against Lithuania and Latvia in December 2004. By that time they had not completed the transposition of the Directive. On 18 October 2005, the Commission decided to send a reasoned opinion asking Latvia for a full explanation of the measures it has taken to implement Directive 98/44/EC on the legal protection of biotechnological inventions in its national law. The case against Lithuania was closed on 12 October 2005 (Commission Case C-973/04), and the case against Latvia was closed on 4 April 2006 (Commission Case C-1061/04).
81. Report from the Commission to the Council and the European Parliament, development and implications of patent law in the field of biotechnology and genetic engineering, p. 2 (COM(2005) 312 final).

Italy (March 2006) seems to follow the German and the French approach concerning the function of the sequence to be claimed in a product patent, but goes even further with regard to human embryos, where any use is unpatentable.[82] So far, the implementation had caused more confusion than harmony.

At this stage Europe had five different approaches to human gene patenting, the approaches in France (1), Italy (2), Germany (3), and the other Member States (4), and the approach of the EPO (5). Prior to the Directive's entry into force, gene patenting in Europe had primarily been a matter for the EPO.

*2.1.1.a. Implementation: Different countries – different views*

Limited patent protection for gene and partial gene sequences has clearly been expressed in the laws implementing the Directive in France[83] and in Germany.[84] Pursuant to Article L. 611-18 of the French Intellectual Property Code, only an invention relating to the technical application of a function of an element of the human body may be protected by a patent. Such protection covers the elements of the human body only to the extent necessary to carry out and exploit that specific application. The technical application must be disclosed in concrete and detailed terms in the patent application. The French code also explicitly states that whole or partial gene sequences as such are not patentable.[85] The code states that, as a discovery a gene can never be patentable, and it also limits the scope of protection, since the scope of a claim covering a gene sequence is limited to the part of the sequence that is directly linked to the specific function disclosed in concrete terms.[86] The rights created by the grant of a patent which includes a gene sequence cannot be relied upon against a later claim relating to the same sequence, if that later claim

82. See the Italian implementation of the Directive *(Legge n. 78 del 22 febbraio 2006)* Article 3, published on 10 March 2006.
83. Intellectual Property Code, Articles L. 611-18 and L. 613-2-1, both inserted by Article 17 (a)( III ) of Law No 2004-800 of 6 August 2004.
84. See BCBI No *6/2005,* at 146.
85. See Article 5(1) of the Directive. This approach could be covered by that provision if there were political support for it.
86. Intellectual property code, Article L. 611-18, (2) and (2)(d) is as follows: 'Only an invention relating to the technical application of a function of an element of the human body may be protected by a patent. Such protection covers the element of the human body only to the extent necessary to carry out and exploit that specific application. The application must be disclosed in concrete and detailed terms in the patent application. The following, in particular, shall not be patentable: ... (d) entire or partial gene sequences as such)'; see Translation: Implementation of the EC BioTechDirective, unpublished report.

satisfies the requirements of Article L 611-18 and discloses a different specific application of the sequence.[87]

The amended German Patent Act (PatG) now provides that the industrial application of a sequence or partial sequence of a gene must be disclosed in the patent application, together with an indication of the function performed by the sequence or partial sequence (PatG. Sec. 1(a)(3)). Subsection 4 of the same provision further provides a mandatory requirement that if the subject matter of the invention is a sequence or partial sequence of a gene whose structure is identical to the structure of a naturally occurring sequence or partial sequence of a human gene, the industrial application which is disclosed according to paragraph 3 must be incorporated in the patent *claim*.[88]

Furthermore, countries outside the European Community, such as the USA, Australia[89] and Switzerland,[90] have made or have suggested measures to enhance the patent system in relation to biotechnology. The United States Patent and Trademark Office (USPTO) has published a revised version of the guidelines to be used by office personnel for reviewing patent applications for

87. Article L. 613-2-1, 'The scope of a claim covering a gene sequence shall be limited to the part of that sequence which is directly linked to the specie function disclosed in concrete terms in the description. The rights created by the grant of a patent which includes a gene sequence cannot be relied upon against a later claim relating to the same sequence, if that later claim itself satisfies the requirements of Article L. 611-18 and discloses a different specific application of the sequence'; see Translation: Implementation of the EC BioTechDirective, unpublished report.
88. See Martin Grund & Martin R. Burda, 'The New German Patent Law Fails to Transpose EC Directive 98/44/EC Correctly', 3 *BSLR* 129 (2004/2005). They argue that, by requiring disclosure in the patent claim, the German Patent Act is stricter than Article 5(3) of the Directive, which only requires disclosure in the patent application. And they argue that it is an accepted practice in German patent law that indications of purpose, effect and function do not limit the scope of protection of product claims. The owner of a product patent can prevent any use of the product of the invention, irrespective of whether he has identified possible individual uses.
89. Australian Law Reform Commission, http://www.austlii.edu.au/
90. The proposed amendment to the Swiss Patent Act implies a requirement for a concrete description of the function of the nucleotide sequence and operates with an expression where only gene sequences which are not synthetically developed are subject to a limitation to the function of the sequence specifically described in the patent application. According to Stefan Kohler: 'Getting the legal system ready for Biotechnology, Country Report', I JIBL Vol. 2 (2005), 'patent protection for these natural gene sequences may cover only those specific uses that are explicitly indicated in the patent application. However, such purpose-restricted protection is not suggested for gene-products (i.e. proteins) derived from the genetic code of natural gene sequences'.

compliance with the utility requirements of 35 USC 101.[91] On the other hand, broad and absolute product patent protection is given effect by the Danish Patent Law. The more rigorous requirement of Article 5(3) of the Directive, concerning disclosure of the industrial application, has not been implemented in substantive Danish Patent Law, but it has been implemented in separate legislation on formal requirements for patent applications.[92]

*2.1.1.b. Different institutions – different views?*

*2.1.1.b.1. The Second Report of the Commission (2005) pursuant to Article 16(c) of Directive 98/44/EC*

The second report of the Commission (2005) pursuant Article 16(c) of Directive 98/44/EC[93] focuses on two questions raised in the first report (2002).[94] The first question concerns the scope of human gene patents and the second question concerns the patentability of human embryonic (pluripotent) stem cells and stem cell lines obtained from them. The difficulties in connection with human stem cells have been discussed in Chapter 4. The only point made here is that a stem cell is not a gene and is therefore eligible for the widest form of patent protection (classical patent protection). Purpose-bound protection (if Article 5(3) can be so understood), as it appears today, is directed only at human genes.

The first question raised in the second report concerns *whether* patents for gene sequences should be allowed according to the classical model of patent claims (absolute product protection covering future uses of that sequence) or *whether* they should be restricted so that only the specific use disclosed in the patent application can be claimed (purpose bound protection).[95] This question

91. US Utility Guidelines, effective from 5 January 2001. See Chapter 3, section 4.1. and e.g. Donald L. Zuhn. Jr, 'DNA patentability: shutting the door to the utility requirement', 34 *John Marshall Law Review*, 2000-2001, p. 995: 'The revised utility guidelines seemingly restore the requirement from *Brenner* that an asserted utility be 'substantial' (heightened utility standard, and restoring the lock to the door of section 101, page 998 and 999).
92. See Executive Order No 6 of 6 January 2003, *om patenter* og *supplerende beskyttelsescertifikater,* Sec. 17(5) (and Act No 1430 of 21 December 2005, Sec. 8).
93. See Development and implications of patent law in the field of biotechnology and genetic engineering (COM(2005) 312 final).
94. See the first report: Development and implications of patent law in the field of biotechnology and genetic engineering (COM(2002) 545 final).
95. For a discussion of the breadth of claim, too narrow or too broad, see e.g. Philip W. Grubb & Peter R. Thomsen, *Patents for Chemicals, Pharmaceuticals and Biotechnology*, 2010, p. 286.

could be considered under the non-discrimination doctrine of TRIPS Article 27(1). For patent protection to be granted to a naturally occurring gene, it must be isolated. If *absolute product protection* were given to a gene that has been isolated and given a technical effect, the scope of its protection would be far-reaching, including covering the discovery and the future use of the naturally occurring gene, whereas *purpose-bound protection* would create the balance of rights and obligations referred to in TRIPS Article 7.

In the report it is concluded that none of Articles 8-11 in Chapter II of the Directive, designating the scope of protection, addresses the concept of protection restricted to the scope of the specific use identified for the gene sequence concerned. Article 9 provides that:

> 'The protection conferred by a patent on a product containing or consistent of genetic information shall extend to all material, save as provided in Article 5(1), in which the product is incorporated and in which the genetic information is contained and performs its function.'

There are no divergent views about the substantive patent law of the Directive covering inventions based on plants, animals and humans and the Directive's assessment of the scope of protection. Whereas Article 5(3) only relates to human DNA sequences, requiring their industrial application to be disclosed in the patent application, Article 9 can be seen as a supplement, indicating that the protection must be extended to all materials in which the product is incorporated and in which the genetic information is contained and *performs its function.* Because of the functional requirement, the purpose-bound protection limits the scope of protection for human DNA sequences.

On the other hand, the standard requirement in Article 3(1) provides for absolute product protection for other isolated parts of the human body. Article 5(3) implies a stricter requirement, and does not merely repeat the standard requirement of traditional patent law.[96] In its second report, the Commission concludes that it does not presently intend to take a position between the classical protection and the limited scope of protection for gene sequences.

96. See the second report of the Commission (2005) pursuant to Article 16(c) of the Directive; recitals 23-25 of Directive 98/44/EC; and Case C-377/98 *The Netherlands* v *European Parliament and Council*, para. 75.

The Commission intends to monitor the economic impact of the outcomes of the different approaches of the Member States.[97]

*2.1.1.b.2. Court of Justice of the European Union*

The view of the Court of Justice in Case *C-377/98 The Netherlands v European Parliament and Council* could be regarded as a stricter interpretation of the assessment in Article 5(3). The Court discussed the scope of protection in paragraphs 74 and 75 of its judgment:

> 74. That distinction [that the human body may not, in its natural environment, be appropriated] applies to work on the sequence or partial sequence of human genes. The result of such work can give rise to the grant of a patent only if the application is accompanied by both a description of the original method of sequencing which led to the invention and an explanation of the industrial application to which the work is to lead, as required by Article 5(3) of the Directive. In the absence of an application in that form, there would be no invention, but rather the discovery of a DNA sequence, which would not be patentable as such.
>
> 75. Thus, the protection envisaged by the Directive covers only the result of inventive, scientific or technical work, and extends to biological data existing in their natural state in human beings only where necessary for the achievement and exploitation of a particular industrial application.

Despite the relatively vague formulation in paragraph 74, 'and an explanation of the industrial application *to which the work is to lead',* paragraph 75 stipulates that 'the protection only extends to discovery [biological data existing in their natural state] *where necessary for the achievement and exploitation of a particular industrial application.*'

The isolation doctrine or the requirement for artificial creation explains how only certain discoveries (biological material as such) can be patentable.

97. Before reaching this conclusion, the Commission put forward the arguments for and against purpose-bound protection, and considered the following: 1. The fact that human gene sequences *have* been isolated from the human body implies that they should be treated differently from chemical substances on ethical grounds (France and Germany); 2. The balance between investments and potential rewards for the first inventor in a field compared to subsequent innovators (broad protection or limited scope); and 3. The market will regulate itself as a specific field of technology matures – it is questionable whether attempts to further refine the scope of gene patents will have any significant effect on actors in the field. These questions are often asked when discussing future amendments to the patent system; see for example, 'Recommendations for a Patent System of the Future', a report by a working group under the Danish Board of Technology, 2005/7, www.tekno.dk.

Only inventions with a technical effect and parts from the natural world which are given a technical effect are no longer treated as 'discoveries' and are thus patentable, but the natural parts as such can never be patented. This might explain the reasoning behind Article 5(3) with regard to gene sequences. Isolation moves an item away from its natural environment. If a biological material that forms part of the natural world were patentable, society's access to nature would be restricted. When an element is artificially created but is identical to nature, nature is still freely accessible. In order to maintain this freedom, there must be a strict requirement for purpose-bound protection, ensuring that access to nature is still allowed for other purposes.[98]

In Case C-428/08 *Monsanto v Cefetra*, a reference for a preliminary ruling under Article 267 TFEU, the Dutch Rechtbank's-Gravenhage (Court of The Hague) referred four questions to the Court of Justice for preliminary rulings. The dispute concerned the European patent (EP 0546090) granted to Monsanto claiming a *plant* gene patent (an isolated DNA sequence) with the capacity for rendering the RR (Roundup Ready) soybean plant herbicide resistant. The RR soybean plant is cultivated on a large scale in Argentina, where there is no patent protection for the plant gene, and then sold in the Netherlands as soya meal by the European companies Cefetra and Toepfer. According to paragraphs 20 to 22 of the decision, three consignments were detained by the customs authorities because of suspected infringement of Monsanto's patent. The test carried out verified that the soya meal contained traces of the RR soybean characteristics, indicating that the meal had been produced using RR soybeans. The Rechtbank's-Gravenhage, asked the Court of Justice whether the presence alone of the patented product (the DNA sequence) in the traded product (the soya meal) was sufficient to constitute an infringement of Monsanto's European patent. The key issue was that the DNA sequence was present in the processed product (the soya meal processed from a soy plant) but it did not perform its function previously in the soy plant. However, the DNA information could be isolated from the soya meal and then continue to perform its function if inserted into a plant cell. The first question, which is the most relevant with regard to this section, concerned the scope of protection pursuant to Article 9 of Directive 98/44/EC.[99]

98. *See* Pires de Carvalho, *The TRIPS Regime of Patents Rights*, 2010, p. 261, point 27.25. See also Michael Blakeney, *Trade related Aspects of Intellectual Property Rights: A Concise Guide to the TRIPS Agreement*, (Intellectual Property in Practice), 1996, p. 81.
99. Question 3 concerned *lex posterior* principles on absolute product protection, and question 4 concerned TRIPS Articles 27 and 30.

The referring court asked whether the Directive precluded national patent legislation from offering absolute product protection, regardless of whether the protected DNA sequence performed its function.

The Court of Justice stated, in paragraph 34, that the protection conferred in Article 9 is subject to the condition that the genetic information contained in the patented product or constituting that product *performs* its function in the material in which that information is contained. This implies that the function is *being performed* in the present and in the actual material in which the DNA sequence containing the genetic information is found. Thus, patent protection is no longer available when the genetic information has ceased to perform the function it performed in the initial material from which the material in question is derived.[100]

As for the protection of the DNA sequence as such (absolute product protection) the Court of Justice did not accept Monsanto's argument. Article 9 of Directive 98/44/EC, interpreted in the light of recitals 23 and 24 as well as Article 5(3), do not support the protection of DNA sequences without a function, and do not protect a patented DNA sequence which is not able to perform the specific function for which it was patented. Further, the wording of Article 9 requires the DNA sequence to perform its function in the material in which it is incorporated.[101] In conclusion, the Court of Justice refrained from absolute product protection; a sequence as such does not enjoy patent right protection. Protection only applies when the DNA sequence performs its function, and the Court of Justice thus ruled that there was a purpose-bound protection for DNA sequences.[102]

It has been argued that, contrary to what many commentators have suggested, the Court of Justice does not give a clear answer as to the exact scope of the protection for DNA sequences.[103] *Overwalle* asks how far the protec-

100. Paragraphs 37 and 40 address the soybean material specifically. It is not an argument that the information found in a residual state in soy meal as a dead material can perform its function again in another material.
101. Paras 43-47.
102. See further para. 63, addressing absolute product protection with regard to Article 9 as harmonising the scope of protection with in the EU. Article 9 of the Directive effects an exhaustive harmonisation of the protection it confers, with the result that it precludes national patent legislation from offering absolute protection to the patented product as such, regardless of whether it performs its function in the material containing it.
103. Gertrui van Overwalle, 'Editorial, The CJEU's Monsanto soybean decision and patent Scope – as clear as mud?', *IIC* Vol. 42, No 1/2011, p. 2. On p. 3 she writes that if the Court meant to say that the scope of protection of DNA sequences is to be re-

tion of a patented DNA sequence stretches, assuming that the indicated function is performed? A closer look at the decision (paragraph 36) may answer that question:

Paragraph 36. In the case of genetic information such as that at issue in the main proceedings, the function of the invention is performed when the genetic information protects the biological material in which it is incorporated against the effect, or the foreseeable possibility of the effect, of a product which can cause that material to die.

Paragraph 39. It also follows that that protection [genetic information ceased to perform its function] cannot be relied on in relation to the material in question on the sole ground that the DNA sequence containing the genetic information could be extracted from it and perform its function in a cell of a living organism into which it has been transferred. In such a scenario, the function would be performed in a material which is both different and biological. It could therefore give rise to a right to protection only in relation to that material.

It cannot be said that the Court did not give a clear answer to the exact scope of protection for the patent in the case, but one can question whether the Court's decision has a clear impact other than in this particular case, where there was a clear distinction between when the genetic information performed its function as claimed and described in the patent (within a plant – living organism) or in a dead material (soy meal). As will be discussed in Chapter 10, one can question the exact scope of protection for miRNA's and for proteins/genes with polygenic traits.

stricted to the functions mentioned in the patent application, it would be one of the very first manifestations of the European courts calling for a halt to the ongoing expansion of patent law in the area of biotechnology. See Philip W. Grubb & Peter R. Thomsen, *Patents for Chemicals, Pharmaceuticals, and Biotechnology, Fundamentals of Global Law, Practice and Strategy*, 5th Ed., 2010, p. 310; Gareth Morgan & Lisa A Haile, 'A shadow falls over gene patents in the United States and Europe', *Nature Biotechnology*, Vol. 28, No 11, 2012, p. 1172; Robert Fitt & Edward Nodder, 'An uncertain future for gene patents: the view from Europe', 29 *Biotechnology Law Report*, No 6, 2010, p. 619; Joseph Strauss wrote about Article 9 before the Monsanto-case, saying that use of end-products in which the protected substance is not present anymore does not constitute an act of patent infringement (case from Düsseldorf 1985, *Grasherbizid*, 1987 GRUR 896), 'The scope of protection conferred by European Patents on transgenic plants and on methods for their protection', in *Festskrift til Marianne Levin*, 2008, pp. 648-649; Michael A. Kock, 'Purpose-bound protection for DNA sequences: in through the back door?', *Journal of Intellectual Property Law and Practice*, 2010, Vol. 5, No 7, p. 495 ff.; Michael A. Kock, 'Court of Justice of the European Union limits patents on DNA sequences: Much ado about nothing or the beginning of erosion for biotech patents?', Vol. 11, *BSLR*, issue 1, p. 3 ff.

It is notable that the Court of Justice rules on DNA sequences in general. Even though the patent in the case concerned a *plant gene sequence*, the Court of Justice referred to Article 5(3) of Directive 98/44/EC and recitals 22, 23 and 24, which all address *human* genes. The context of the recitals implies that all of recitals 16-27 concern humans. Article 5(3) and recitals 22, 23 and 24 concern patentability requirements, whereas Article 9 regulates the scope of protection conferred on a patented product. This mixture of issues has been criticised in the legal literature, where it has been pointed out that it is an improper importation of patentability rules into the scope of protection.[104] The 'loss' of protection derived from Article 5(3) was transferred to the analysis of infringement, and according to some authors this has resulted in an incomplete and confusing interpretation. The Court of Justice's function-limited scope of protection is in serious conflict with EPC Article 69, because it requires patent offices and courts to read a limitation into the claims based on a function described in the specification.[105]

*2.1.1.b.3. EPC framework: Implementation Rules, guidelines*

This section gives a brief overview of the EPC regime on patenting genes. The requirement for an invention to be susceptible of industrial application is stated in Article 52(1) and elaborated in Article 57 EPC. As for biotechnological inventions, Rule 29(3) of the EPC Implementing Regulations repeats Article 5(3) of Directive 98/44/EC word for word. However Rule 29(3) cannot take precedence over EPC Article 57; see Article 164(2) EPC.[106] When it is not obvious from the description or nature of the invention, Rule 42(1)(f) requires the description to indicate explicitly, how the invention is industrially applicable.

An invention is considered as being susceptible of industrial application if it can be made or used in any kind of industry. It follows from Case T 870/04 *Max Planck*, that inventions involving the elucidation of a structure or function and disclosing a method of making it in commercial quantities, or recov-

104. Jan B. Krauss & Toshiko Takenaka, 'A special rule for compound protection for DNA-sequences – impact of the ECJ 'Monsanto' decision on Patent Practice', 93 *J. Pat. & Trademark Off. Soc'y* 189, 2011.
105. Jan B. Krauss & Toshiko Takenaka, 'A special rule for compound protection for DNA-sequences – impact of the ECJ 'Monsanto' decision on Patent Practice', 93 *J. Pat. & Trademark Off. Soc'y* 189, 2011, at p. 203. See 11-796, *Bowman v Monsanto* et al, 657.F3d.1341, Cert.granted May 5 2012.
106. Case T-1191/01 *ICOS* (a first attempt to clarify the relationship of Article 5(3), Rule 29(3) and Article 57 EPC.

ering it from its natural environment, may be industrially applicable if the manufacture or use of the natural product is enabled.[107] Thus, a practical application with some profitable use should be disclosed and the term 'function' is not the same as 'biological function', as it may also embraces for example the structure of the genetic material having a function as a primer, a diagnostic or forensic marker, a probe, a promoter or an enhancer.[108]

As mentioned earlier in this Chapter and in Chapters 3 and 4, European patent law comprises a mixture of competences. This section focuses on the European Patent Organisation (EPO) competences manifested in the EPC and in secondary legislation derived therefrom. The legal basis for gene patents is to be found in the Implementing Regulations to the Convention on the Grant of European Patents, Rule 29, which mirrors the exact wording of Article 5 of Directive 98/44/EC.[109] The Implementing Regulations are supplemented by the Guidelines for Examination.[110] The Guidelines for Examination are not legal provisions, but general instructions addressed to the EPO examiners, and parties can expect the EPO to act in accordance with the Guidelines; see 2012 Guidelines for Examination 2012 – General Part, paragraph 3.1 and 2.[111]

Part G, Chapter II gives instructions on the concept of an invention. It follows from the general remarks in paragraph 1, that the non-exhaustive list of things listed in EPC Article 52 (2) is all either abstract and/or non-technical.

107. Richard Hacon & Jochen Pagenberg (eds.), *Concise European Patent Law*, 2nd Ed., p. 62.
108. See Rob J. Aerts, 'Biological patents in Europe – Functions of Recombinant DNA and expressed proteins and satisfaction of the Industrial Applicability Requirement', *IIC* 3/2008, Vol. 39, p. 286. In *HGS Inc v Eli Lilly and Company*, 2011, UKSC 51, points 43 f, Lord Neuberger examines the EPO Board's jurisprudence on EPC Article 57 and biological material (Case T 870/04 *BDP1 Phosphate/Max Planck*; Case T 1329/04 *Factor 9/John Hopkins*; Case T 0604/04 *PF4A receptors/Genentech*; Case T 1452/06 *Serine protease/Bayer*; Case T 1165/06 *IL-17 related polypeptide/Schering*.
109. See Chapter 4 on the legal effect of the Rules and the interplay between the Rules and Directive 98/44/EC.
110. From June 2012 the new guidelines enter into force According to the concordance table for the 2010 and 2012 guidelines, there has been no change in Part G. A new Part H has been inserted on amendments and corrections during prosecution. Part G contains Guidelines for Examination for patentable biotechnological inventions.
111. The Guidelines for Examination further state that the sources of law are primarily the European Patent Convention itself, including the Implementing Regulations, the Protocol on the Interpretation of Article 69 EPC (and further protocols) and secondarily the interpretation of the EPC by the Boards of Appeal and the Enlarged Board of Appeal.

Thus, the concept of an invention must be of both a concrete and a technical character. An example of the distinction between a discovery and an invention is given in paragraph 3.1, where it is explained that a previously unrecognised substance occurring in nature is considered a discovery, whereas the same substance shown to produce a technical effect may be patentable. It is further explained that a gene which is discovered to exist in nature may be patentable if a technical effect is revealed, e.g. its use in making a certain polypeptide or in gene therapy.

The examination guideline given to the patenting of the human body (Part G, Chapter II-13, paragraph 5.2) very briefly states that a sequence or partial sequence of a gene structure may be patentable even if the structure of that element is identical to that of a natural element (Article 5(2) and Rule 29 (1)). The advice given in the guideline is:

> Such an element is not a priori excluded from patentability since it is, for example, the result of technical processes used to identify, purify and classify it and to produce it outside the human body, techniques which human beings alone are capable of putting into practice and which nature is incapable of accomplishing itself (EU Dir. 98/44/EC, rec. 21).

This guideline is closely linked to the justification of the distinction between discovery and invention and is less focused on guidance on the outdated wording focusing on only gene sequences. The empirical case study in Chapter 7 questions whether RNAi patents are properly addressed in a gene patent context however the new Guidelines for Examination did not clarified this outdated approach to genes and functions that is still a main feature of the new 2012 Guidelines for Examination.

The guidance on Rule 29(3) can be viewed as a clear departure from the questioning of whether gene patents require tailor-made rules. It is stated that the examination of a patent application or a patent for gene sequences or partial sequences should be subject to the same patentability criteria as all other areas of technology (EU Dir. 98/44/EC, rec. 22). However, the industrial application of a sequence or partial sequence must be disclosed in the patent application as filed. This guideline is supplemented in Part G, Chapter III-1 – Industrial application, addressing Rule 42(1)(f) and Rule 29(3). See paragraph 4:

> 'In general it is required that the description of a European patent application should, where this is not self-evident, indicate the way in which the invention is capable of exploitation in industry. The invention claimed must have such a sound and concrete technical basis that the skilled person can recognise that its contribution to the art could lead to practical exploitation in industry (see T 898/05). In relation to sequences and partial sequences of genes, this general requirement is given specific form in that the industrial application of

a sequence or a partial sequence of a gene must be disclosed in the patent application. A mere nucleic acid sequence without indication of a function is not a patentable invention (EU Dir. 98/44/EC, rec. 23). In cases where a sequence or partial sequence of a gene is used to produce a protein or a part of a protein, it is necessary to specify which protein or part of a protein is produced and what function this protein or part of a protein performs. Alternatively, when a nucleotide sequence is not used to produce a protein or part of a protein, the function to be indicated could e.g. be that the sequence exhibits a certain transcription promoter activity.'

There are several cases on the issue of function under the EPC regime. The following cases on polypeptides and their encoding polynucleotides have contributed to the interpretation on function under the EPC regime:[112]

Case T 338/00 on various members of the steroid/thyroid superfamily of receptors: invention with a direct technical result; Case T 870/04 on the house-keeping role of BDP-1 (brain-derived claimed as a polypeptide and nucleic acid molecule encoding such a polypeptide): there is no evidence for a passive, an active, a positive or a negative role of BDP-1 in cancer;[113] Case T 604/04 on G-protein-coupled superfamily: PF4A superfamily of cytokines related to IL-8 receptor gene: important to the pharmaceutical industry;[114] and Case T 898/05 on cytokine receptors: plausible therapeutic role is disclosed: the receptor has a plausible application in an industrial activity.[115] In Case T 989/05 the Board emphasised

112. Rob J. Aerts, 'Biological patents in Europe – Functions of Recombinant DNA and expressed proteins and satisfaction of the Industrial Applicability Requirement', *IIC* 3/2008, Vol. 39, p. 286.
113. See para. 11. See also para. 2: the fact that a polypeptide can be merely made does not necessarily fulfil the requirements of Article 57 EPC unless there is also some profitable use of the polypeptide. Para. 6 of the decision 'there must be a borderline between what can be accepted, and what can only be categorized as an interesting research result which per se does not yet allow a practical industrial application to be identified', and paras 19 and 21 'it is considered that a vague and speculative indication of possible objectives that might or might not be achievable by carrying out further research with the tool as described is not sufficient for fulfillment of the requirement of industrial applicability'. See Rob J. Aerts, 'Biological patents in Europe – Functions of Recombinant DNA and expressed proteins and satisfaction of the Industrial Applicability Requirement', *IIC* 3/2008, Vol. 39, p. 290.
114. Although there was no absolute certainty that the two receptors were receptors for the PF4A superfamily of cytokines, in the Board's judgement the above mentioned structural features made it plausible that this was indeed the case, para. 6 of the decision; and Rob J. Aerts, 'Biological patents in Europe – Functions of Recombinant DNA and expressed proteins and satisfaction of the Industrial Applicability Requirement', *IIC* 3/2008, Vol. 39, p. 291.
115. See para. 22, discussing that the fact that the invention was based on computer-assisted methods rather than traditional wet-lab techniques did not mean that the invention was automatically disregarded or excluded from a careful examination. The

that the function of a protein could be seen at different levels, such as molecular function, cellular function and biological function in a broad sense. All levels are suitable for industrial application.[116] In Case T 641/05 the claims were for a polynucleotide encoding a G-protein coupled receptor (GPCR)-like polypeptide. The application disclosed the nucleotide and amino-acid sequence of CEGPCR1 from Caenorhabditis elegans, possibly useful for invertebrate pest control. The Board stated that 'although under certain conditions, the board is well prepared – following the case-by-case approach adopted in decision T 898/05 – to acknowledge a possible function based on computer-assisted methods ..., in the present case the probative value of these (sequence homology) methods is completely lacking'.[117] In Case T-1452/06 the Board found no experimental evidence in the application in support of a serine protease activity for a polypeptide comprising the amino acid sequence and thus no 'immediate concrete benefit' within the meaning of the decision in Case T-898/05 could be acknowledged for this use. Accordingly, the Board stated that a basic principle of the patent system is that the full disclosure of the invention (in exchange for the grant of exclusive rights).[118] It follows from Article 57 EPC that there must be an indication of how to exploit the invention. This indication must have 'a sound and concrete technical basis', as a 'speculative indication of possible objectives that might or might not be achievable by carrying out further research with the tool as described was not sufficient to fulfil of the requirement of industrial applicability' (see Case T 898/05 and Case T 870/04).[119]

If not all members of a protein family share the same biological function then it follows from Case T 1452/06 that the attribution of function to an alleged

invention (Zcytor 1 receptor) was clearly identified as a putative member of the hematopoietin receptor family, and according to the prior art members of that family have considerable comparative activities (para. 27).

116. Case T 898/05, paras 29 and 30.
117. Para. 14. See Rob J. Aerts, 'Biological patents in Europe – Functions of Recombinant DNA and expressed proteins and satisfaction of the Industrial Applicability Requirement', *IIC* 3/2008, Vol. 39, p. 294.
118. See also Case T 1191/01 on the disclosure of a predicted function of a protein combined with a method used to verify this function which was not adequate to satisfy the criterion of industrial applicability; and Richard Hacon & Jochen Pagenberg (eds.), *Concise European Patent Law*, 2nd Ed., p. 441.
119. See further http://www.epo.org/law-practice/legal-texts/html/caselaw/2010/e/clr_i_e_1_2.htm (last visited April 2012); Hans-Rainer Jaenichen, 'Reduction to practice, All inventions are created equal: refinement of T 1329/04 in favour of the completeness of DNA inventions without wet biology experiments', *Biotechnology Law Report*, Vol. 26, No 1, at p. 6, hypothetical function cannot be a sufficient basis for industrial applicability (Case T 870/04); and Friederike Stolzenburg, Barbara A. Ruskin & Hans-Rainer Jaenichen, 'Of incomplete inventions', *EPL Information*, 1/2006, p. 15 (Case T-1329/04, plausible incompleteness of inventions).

member of the family amounts to mere speculation.[120] However, Case T-0018/09 *HGS v Eli Lilly* which is discussed in Chapter 6, section 5 has a more relaxed approach to Article 57 EPC. The case concerned pleiotropic cytokines which display a wide range of activities and have distinctive, but also overlapping biological functions.

On the basis of its structural properties, Neutrokine-α was said to have been correctly identified as a new member of the TNF ligand superfamily. In paragraph 18, it was stated that the key question was whether the patent specification disclosed in sufficient terms the nature of Neutrokine-α, its function and how it could be used in industrial practice to solve a given technical problem. Paragraph 26, in fine, state as follow:

> 'In view of the known broad range of possible activities of such a molecule, the skilled person is aware of the fact that the full elucidation of all properties requires further investigations which will gradually reveal them. In this context, the skilled person regards the long listing of possible actions of Neutrokine-α and of medical conditions in which it might take part as the enumeration or generalisation of the properties of the members of the TNF ligand superfamily. This is seen as the frame in which the newly found molecule has to be placed as one could *prima facie* have a reasonable expectation that most of them could in fact be present.'

The Board concluded (paragraph 27) that the 'description of of the patent delivers sufficient technical information, namely the effect of Neutrokine-α on T-cells and the tissue distribution of Neutrokine-α mRNA, to satisfy the requirement of disclosing the nature and purpose of the invention and how it can be used in industrial practice'.

### 2.2. US statutes and judicial doctrine – What is natural?

The US statutes can be distinguished from those of the European Patent Convention and the EU Directive as the US legislation contains only broad formulations from the Constitution on *promoting the progress of science and useful art*, and the basic principles of the Patent Acts of 1790 and 1793 have remained relatively stable since their adoption; see Chapter 3, section 4.1.[121]

120. See para. 17; and Rob J. Aerts, 'Biological patents in Europe – Functions of Recombinant DNA and expressed proteins and satisfaction of the Industrial Applicability Requirement', *IIC* 3/2008, Vol. 39, p. 297.
121. Aaron J. Zakem, 'Rethinking patentable subject matter: Are Statutory categories useful?', 30 *Cardozo L. Rev.* 2009, p. 2990: 'the Patent Act 1790 contained substantially broader reach in the statutory language itself, allowing an inventor to seek a patent for 'any useful art, manufacture, engine, machine, or device, or any improvement therein not before known or used' [Act of April 10, 1790, ch. VII, § 1,1 Stat. 109]. This lan-

The issue of field restriction has surfaced in bills, but without ever being passed into law. The *Becerra-Weldon* bill (HR 110-997) would prohibit patents being obtained for a nucleotide sequence, or its functions or correlations, or the naturally occurring products it specifies.[122] Today, the HR1249 bill would ban on claims encompassing a human organism; see Chapter 3, section 4.1.

In US law, patentable subject matter is defined in 35 U.S.C § 101 as 'any new and useful process, machine, manufacture, or composition of matter'. A § 101 inquiry concerns whether the claimed subject matter falls into one of these categories. The US assessment of patentable subject matter is a dynamic assessment originating in judicial scrutiny of the § 101 categories. The content of patentable subject matter in USC 35 § 101 is flexible in all respects. It can be interpreted broadly, in order to cover anything made by man and it can be narrowly interpreted, even overruling prior cases.[123] It has been empha-

guage was modified a scant three years later to encompass 'any new and useful art, machine, manufacture, or composition of matter' [Act of Feb. 21, 1793, ch. XI, § 1,1 Stat.318]. That language remained until the Patent Act of 1952, which exchanged the term 'art' for 'process', and defined a process as a 'process, art or method'.'

122. The Genomic and Research Accessibility Act proposed by Congressmen Becerra and Weldon. For a discussion, see Arti Rai, 'An Audience with ...', *Nature*, April 2007, Vol. 6, p. 264. Arti Rai writes, that 'a significant portion of the genome – at least 20 % of all human genes – is thought to have already been patented, and it is therefore likely that many of the genes with significance in biomedical research have been patented already. And if one looks at patents on nucleotide sequences, which is what the Becerra-Weldon bill purports to cover, there are also a huge number of such patents – tens of thousands according to some estimates'. Arti Rai highlights synthetic biology as a field in which the Bill could have had an impact. See further Eileen M. Kane, 'Patenting Genes and Genetic Methods: What's at Stake?', *Journal of Business & Technology Law*, 2011, Vol. 6, No 1, p. 17, also referring to the Secretary's Advisory Committee on Genetics, Health, and Society (SACGHS) report recommending certain liability exclusions that would alleviate the impact of gene patents on research and patient care.

123. See *Chakraberty* (447 U.S at 303, 206 U.S.P. Q (BNA 193 (1980); and e.g. *re Comiskey* (499 F.3d 1365 (Fed. Cir. Sep. 20, 2007); *re Nuijten* (500 F. 3 d 1346 (Fed. Cir. Sept. 20, 2007); and *re Bilski* (F.3.d Federal Circuit 2008, Bilski v Kappos, Supreme Court 2010 (08-964)). See Kenneth J. Burchfiel, *Biotechnology and the Federal Circuit*, 2nd Ed., 2010, p. 41; and Daniel T. Marvin, 'The Supreme Court's Missed Opportunity to Settle the Handiwork of Nature Exception to Patentable Subject Matter in Laboratory Corporation of America v Metabolite Laboratories, 126 S. Ct. 2921 (2006)', 26 *Temple Journal of Science, Technology & Environmental Law*, Spring 2007, p. 116.

sised by *Merges and Duffy* that the law on patentable subject matter should be viewed as containing a high degree of uncertainty.[124]

However, broad as it may seem, the Supreme Court has stated that the provision is not unlimited.[125] The judicially created §101 exceptions are twice three-fold, comprising '*laws of nature, physical phenomena, and abstract ideas*'[126] as well as '*phenomena of nature, mental processes and products of nature*'.[127] The justification for these non-statutory exceptions has been formulated by the Supreme Court as follows: 'the concepts covered by these exceptions are part of the storehouse of knowledge of all men ... free to all men and reserved exclusively to none'.[128]

### *2.2.1. Laws of nature, physical phenomena and abstract ideas*

As indicated above, the 35 U.S.C. § 101 prohibition of patenting '*laws of nature, physical phenomena and abstract ideas*' is a primary doctrinal instrument.[129] The Supreme Court has been criticised for its reluctance to clarify the Federal Circuit's wayward and confused patentable subject matter jurisprudence in this regard.[130] US law is not entirely clear on the core values that

124. Robert Patrick Merges & John Fitzgerald Duffy, *Patent Law and Policy, Cases and Materials*, 2008-2009 supplement, p. 1; see 5th Ed. pp. 81, 107 and 189 (clumsy tool for deciding whether permitting patents in a particular area is sound policy). See also Eileen M. Kane, 'Splitting the gene: DNA Patents and the Genetic Code', 71 *Tennessee Law Review* 707, Summer 2004, pp. 745-746: 'while there is general agreement about what is off-limit, there can be disagreement as to whether an individual patent claim has crossed the line. The debate is not structured by a simple distinction between product and process claims.'
125. *Bilski v Kappos*, 130 S.Ct. 3218, 3225 (2010); *Chakrabarty*, 447 U.S. at 308.
126. *Chakrabarty*, 447 U.S. at 309.
127. *Gottschalk v Benson*, 409 U.S. 63, 67 (1972); *Chakrabarty*, 447 U.S. at 313; *Diamond v Diehr*, 450 U.S. 175, 189, n. 12 (1981): 'if carried out to its extreme, make all inventions unpatentable because all inventions can be reduced to underlying principles of nature which, once known, make their implementation obvious'. The question of how much human manipulation is necessary before an invention is sufficient artificial thus recognised by the US Supreme Court; see Peter M. Kohlhepp, 'When the invention is an invention: Revitalizing Patentable Subject Matter to exclude unpredictable processes', 93 *Minnesota Law Review*, 2008, p. 808.
128. *Bilski*, 130 S Ct. at 3225; and *Funk Brothers Seed Co v Kalo Inoculant Co.*, 333 U.S. 127, 130 (1948).
129. See Robert Patrick Merges & John Fitzgerald Duffy, *Patent law and policy: Cases and Materials*, 4th Ed., 2011, p. 79.
130. See Daniel T Marvin, 'The Supreme Court's Missed Opportunity to Settle the Handiwork of Nature Exception to Patentable Subject Matter in Laboratory Corporation of America v Metabolite Laboratories, 126 S. Ct. 2921 (2006)', 26 *Temple Journal of*

lead to these exclusions. As a parallel exclusion to the product of nature doctrine which is analysed below, this category (or categories) is focused on natural processes (law of nature as one category) and a patent can be granted for an invention which uses a natural law, principle or process, but which does not prevent access by others to the same laws or processes.[131]

In the literature it has been stressed that by defining what is unpatentable, the Court necessarily refers to a public domain where these excluded categories of invention reside, and it is argued that an affirmative definition of the public domain in molecular biology would help to define the boundaries of patentable subject matter for this field.[132]

Two early decisions from the US Supreme Court have elaborated on the dilemma concerning classes of inventions (electromagnetism and sound waves). The question of whether whole classes of laws of nature, scientific principles and mathematic formulas etc. can be deprived patent protection has been elaborated since the beginning of the grant of legal protection.[133] As

*Science, Technology & Environmental Law*, Spring 2007, p. 147; see Aaron J. Zakem, 'Rethinking patentable subject matter: Are Statutory categories useful?', 30 *Cardozo L. Rev*. 2009, p. 3000.

131. See Eileen M. Kane, 'Splitting the gene: DNA Patents and the Genetic Code', 71 *Tennessee Law Review* 707, Summer 2004, p. 745.
132. See Eileen M. Kane, 'Splitting the gene: DNA Patents and the Genetic Code', 71 *Tennessee Law Review* 707, Summer 2004, p. 747; Eileen M. Kane, 'Patent ineligibility: maintaining a scientific public domain', 80 *St. John's L. Rev.* 519; Helen M. Berman & Rochelle C. Dreyfuss, 'Reflections on the Science and Law of Structural biology, Genomics, and Drug development', 53 *UCLA. Rev.*, p. 889: raw information about biological endowments without end-product functionality belongs to the public domain; Antony Taubman, 'The public domain and international intellectual property law treaties', in Charlotte Waelde & Hector Macqueen, *Intellectual Property, The many faces of the public Domain*, 2007, p. 63, viewed as a binary tension, the policy-makers make a strict trade-off between public and private domains; on multiple public domains, see Pamela Samuelson, 'Enriching discourse on public domains', 2006, 55 *Duke L. J.*, pp. 819, 829; Lee Anne Fennell, 'Common interest tragedies', 98 *Nw. U.L. Rev.* 907, p. 907 ff. A call for a global/international public domain is addressed in Chapter 10.
133. See e.g. *Bilski v Kappos*, Supreme Court 2010 (08-964). See Robert Patrick Merges & John Fitzgerald Duffy, *Patent Law and Policy: Cases and Materials*, 5th Ed., 2011, p. 79: these exceptions have defined the reach of the statutes as a matter of statutory *stare decisis* going back 150 years ('laws of nature, physical phenomena, and abstract ideas'). Rebecca S. Eisenberg, Bargaining over the transfer of property research tools: Is this market failing or emerging? In Rochelle C. Dreyfuss, Diane Leenheer Zimmerman & Harry First (eds.), *Expanding the boundaries of intellectual property, innovation policy for the knowledge society*, 2001, p. 223 ff.

emphasised in the literature, the phenomena of nature were carefully scrutinised by the Supreme Court which adopted the overall approach that exclusive control of a scientific principle had a unique dimension affecting public interests 'that invited careful application of patent principles'.[134] The first decision *O'Reilly v Morse* is important in many ways. This old case (1854) was decided when the administrative procedure was changed from registration to administrative examination, see Chapter 3, section 4.1.[135] The manifest of prohibition of patenting a natural principle has its foundation in this decision and has developed ever since.[136]

The *Morse* case is still valuable with regard to issues such as the inherent patent dilemma, e.g. the proper scope of protection (do not claim things that you did not invent),[137] incen-

134. See Eileen M. Kane, 'Splitting the gene: DNA Patents and the Genetic Code', 71 *Tennessee Law Review* 707, Summer 2004, p. 749, noting that the two cases demonstrate the tension between the expansiveness pronounced in *Chakrabarty* and the continued insistence by the Court on certain zones of exclusion within patentable subject matter.
135. *O'Reilly v Morse*, 56 U.S. 62 (1854). The patent in question concerned the electric telegraph, which was one of the most significant inventions of the nineteenth century. By a new combination of known powers, of which electro-magnetism was one, Morse, who was a painter and the founder of the Academy of Fine Arts in New York, had been able to discover a method by which intelligible marks or signs could be printed at a distance; see Eileen M. Kane, Splitting the Gene: DNA patents and the Genetic Code, 71 *Tennessee Law Review* 707, at p. 747; and Robert Patrick Merges & John Fitzgerald Duffy, *Patent Law and Policy: Cases and Materials*, 5th Ed., 2011, pp. 112 and 113.
136. The *Morse* case provided the foundation for the prohibition of the patenting natural principles. Both the majority and the dissenting judges agreed that the mere discovery of a new scientific principle is not patentable. For the dissenting view that limit arose from the utility requirement. The justices in *Morse* were unanimous that the mere discovery of a new scientific principle is not patentable; at the very least there had to be utility, (§101) p. 90.
137. Claim 8 was as follows: 'I do not propose to limit myself to the specific machinery or parts of machinery described in the foregoing specification and claims; the essence of my invention being the use of the motive power of the electric or galvanic current, which I call electro-magnetism, *however developed* for marking or printing intelligible characters, signs, letters, at any distance, being a new application of that power of which I claim to be the first inventor or discoverer'. As pointed out by Mr. Chief Justice Taney, Morse claimed the exclusive right to every improvement where the motive power was the electric or galvanic current and the result was the marking or printing intelligible characters, signs, or letters at a distance, *O'Reilly v Morse*, 56 U.S. 62 (1854); and Merges & Duffy, *Patent Law and Policy: Cases and Materials*, 5th Ed., 2011, p. 109. Justice Taney referred to the Neilson case (*Nielson v Harford*,

tives for making improvements, and the utility requirements,[138] which in this thesis are mainly discussed in section 2 above concerning the EPO/EU approach. The practice on undue claim breadth has followed this Morse pattern. While in the Morse case the majority invalidated the disputed claim 8, due to its broad scope, the dissenting Justice Grier challenged the majority. Accordingly, no future application of the natural principle would be outside the legitimate scope of the invention as claimed (the natural principle being the discovery of electromagnetism) to a natural phenomenon.[139] In other words, future technology depended on the original discovery.[140]

The second important case in US patent law is the *Bell* case (the *Telephone* cases).[141] Like Morse, Alexander Graham Bell filed a patent application containing a very broad claim. He claimed 'a method of, and apparatus for, transmitting vocal or other sounds telegraphically. However, the disputed fifth claim did not claim all uses of electricity converted to sound waves.[142]

Webster's Patent Cases 295, 371 (Exch. 1844)), where the defendant claimed that the Nielson patent, for pushing hot air into the furnace instead of cold and thereby increasing the intensity of the heat, was a patent for a principle. This argument was addressed by Baron Parke who delivered the opinion in *O'Reilly v Morse*, 56 U.S. 62 (1854), and *Patent Law and Policy*, 2011, p. 111. 'Undoubtedly, the principle that hot air will promote the ignition of fuel better than cold, was embodied in the machine. But the patent was not supported because this principle was embodied in it. He would have been equally entitled to a patent, if he had invented an improvement in the mechanical arrangements of the blowing apparatus, or in the furnace, while a cold current of air was still used. The patent was supported, because he had invented a mechanical apparatus, by which a current of hot air, instead of cold, could be thrown in'; Mr Justice Grier dissenting on the objection to claim 8.

138. See Robert Patrick Merges & John Fitzgerald Duffy, *Patent Law and Policy: Cases and Materials*, 5th Ed., 2011, p. 115.
139. Dissenting Justice Grier: 'The mere discovery of a new element, or law, or principle of nature, without any valuable application of it to the arts, is not the subject of a patent. But he who takes this new element or power, as yet useless, from the laboratory of the philosopher, and makes it the servant of man; who applies it to the perfecting of a new and useful art, or to the improvement of one already known, is the benefactor to whom the patent law tenders its protection', *O'Reilly v Morse*, 56 U.S. 62 (1854), and *Patent Law and Policy*, 2011, p. 117.
140. Eileen M. Kane, 'Splitting the gene: DNA Patents and the Genetic Code', 71 *Tennessee Law Review* 707, Summer 2004, p. 748.
141. The *Telephone* cases 126 U.S. 1 (1888); Justice Waite delivered the opinion of the Court.
142. Eileen M. Kane, 'Splitting the gene: DNA Patents and the Genetic Code', 71 *Tennessee Law Review* 707, Summer 2004, p. 749; and Merges & Duffy, *Patent Law and Policy: Cases and Materials*, 5th Ed., 2011, p. 124. Like the electric telegraph, the telephone did not suddenly spring into being through the efforts of a single inventor; it

As discussed in the literature by narrowing the claim Bell did not claim the principle of controlling the force of transmitting speech. He was claiming his particular method of doing so and by the claim wording he did not claim all future developments.

The themes of *laws of nature, physical phenomena, abstract ideas and products of nature* directs patent authorities and courts in the USA to assess whether an invention is outside the scope of patentability or is patentable. The first to discover a practical use for substances within these themes, or the inventor who controls the art or discovers the practical use may be granted a patent. *Merges and Duffy* question whether non-obviousness provides a more coherent framework for explaining the cases than distinctions between natural and non-natural?[143] The 'non-obviousness rule' was not firmly established in modern doctrine before the 1952 statute. Inquiries before that tended to mix the two issues of patentability and non-obviousness which are now separated by modern courts.[144]

### 2.2.2. *Product of nature doctrine*

As described above, the product-of-nature doctrine puts a limit on patentable subject matter. The idea that an invention can be off-limits for the patent system can give the impression that the natural world cannot be patented and the judicially created *product of nature doctrine* has been a formidable obstacle to inventions derived from the natural world.[145] The doctrine originates from the *Ex parte Latimer* case (1889),[146] with the rejection of a patent claim to an unadulterated fibre needle recovered from a plant (separated from the sili-

evolved over a period of two decades. But unlike Morse, Bell limited even his broad fifth claim in a way that corresponded to his specific contribution to the technology.

143. Merges & Duffy, *Patent Law and Policy: Cases and Materials*, 5th Ed., 2011, p. 126. In sum, the nineteenth century law on the patentability of natural principles was mainly concerned with ensuring: (1) that the inventor specified a practical utility; and (2) that the patent claims bore some relation to the inventor's contribution to the useful arts. It was these pragmatic inquiries, and not the metaphysical distinctions between the natural and the artificial, that were at the heart of the law governing patentability.
144. Merges & Duffy, *Patent Law and Policy: Cases and Materials*, 5th Ed., 2011, p. 125.
145. Eileen M. Kane, 'Splitting the gene: DNA Patents and the Genetic Code', 71 *Tennessee Law Review* 707, Summer 2004, p. 732.
146. 889 Dec. Comm'r Pat. 123 (1889); see Merges & Duffy, *Patent Law and Policy, Cases and Materials*, 5th Ed., 2011, p. 71.

cious, resinous, and pulpy parts of the needle) leaving the public domain as the proper sphere for natural products.[147]

'The patent applicant may have discovered that these particular pine needles were very valuable because they 'possess more or less strength or fineness' than known fibers. But if they were, it was because 'nature made them so' and the discovery of these properties no more entitled the applicant to a patent on them 'than to find a new gem or jewel in the earth would entitle the discoverer to patent all gems which should be subsequently found.'[148]

The precedence given to the public domain in the *Latimer* case was followed in the 1948 case *Funk Bros Seed co v Kalo Inoculant Co*, concerning a mixture of bacterial species with the surprising property of mutual non-inhibition (mixed by the inventor in a useful combination). The subject matter did not pass the test of patentability.[149] One often quoted sentence from that case reads:

The qualities of these bacteria, like the heat of the sun, electricity, or the qualities of metals 'are part of the storehouse of knowledge of all men. They are manifestations of laws of nature, free to all men and reserved exclusively to none. He who discovers a hitherto unknown phenomenon of nature has no claim to a monopoly of it which the law recognizes. If there is to be invention from such a discovery, it must come from the application of the law of nature to a new and useful end.'[150]

The *Funk Bros* judgment was not based on the product of nature doctrine alone but was a mixture of that doctrine and the laws of nature category (see

147. See Eileen M. Kane, 'Splitting the gene: DNA Patents and the Genetic Code', 71 *Tennessee Law Review* 707, Summer 2004, p. 733.
148. 889 Dec. Comm'r Pat. 125 (1889); and Merges & Duffy, *Patent Law and Policy: Cases and Materials*, 5th Ed., 2011, p. 161. In the *Latimer* case it was suggested by the Commissioner that if the pine needle had been given a new function or quality or altered from its natural state, a patent could have been issued. Eileen M. Kane points to the analytical difficulty that arises when it is not clear whether an inventor has simply manipulated naturally occurring processes (which argues against patentability) or when an inventive step has been made to generate a non-naturally occurring organism. She also stresses that it is necessary to determine whether an inventor is claiming an innate quality of the product as an actual invention; see Eileen M. Kane, 'Splitting the gene: DNA Patents and the Genetic Code', 71 *Tennessee Law Review* 707, Summer 2004, p. 732.
149. 333 U.S. 127, 76 U.S.P.Q (BNA) 280 (1948).
150. *Funk*, 333 U.S, at 130. See Eileen M. Kane, 'Splitting the gene: DNA Patents and the Genetic Code', 71 *Tennessee Law Review* 707, Summer 2004, p. 734; and Merges & Duffy, *Patent Law and Policy: Cases and Materials*, 5th Ed., 2011, p. 161.

the section above). By mixing natural phenomena and the laws of nature (noting that natural phenomena result from the operations of the underlying laws of nature) the court suggested that the criterion for the patentable use of a law of nature is that the invention applies the law of nature to a specific purpose.[151] The bacteria still performed in their natural way.[152]

The first Supreme Court decision on patentable subject matter in the area of biotechnology concerned genetically engineered organisms.[153] The case concerned a bacterium which had been genetically engineered, and by altering the genetic make-up of the organism it was not a naturally occurring.

The case is a landmark decision and much quoted, discussed and elaborated in jurisprudence. The assessment of the invention had differing outcomes on its way to Supreme Court. The USPTO rejected the application (product of nature and as a living organism not included in 35 U.S.C. § 101). The Board of Appeal also rejected it, but for a different reason (as it was genetically engineered it was not a product of nature, but as a living matter it was not included in 35 U.S.C. § 101). The Court of Customs and Patent Appeals (C.C.P.A) held that 35 U.S.C. § 101 did not constitute a bar to the patenting of a living organism. Finally, the Supreme Court affirmed the C.C.P.A's position on 35 U.S.C. § 101. Microorganisms were either manufactures or compositions of matter.[154]

The famous statement from the case that 'Congress intended statutory subject matter to 'include anything under the sun that is made by man''[155] has been quoted in the legal patent literature ever since. The phrases from the *Funk Bros Seed* case that 'they are manifestations of laws of nature, free to all men and reserved exclusively to none' and the 'handiwork of nature' are still good law for distinguishing between patentable and non-patentable subject matter, and two categories have emerged to distinguish between nature's handiwork

151. 333 U.S. 132 (1948). See Eileen M. Kane, 'Splitting the gene: DNA Patents and the Genetic Code', 71 *Tennessee Law Review* 707, Summer 2004, p. 734.
152. The court reasoned that once nature's secret was discovered, the state of the art made the production of mixed inoculants a simple step; see also Merges & Duffy, *Patent Law and Policy: Cases and Materials*, 5th Ed., 2011, p. 163.
153. *Diamond v Chakrabarty*, 447 U.S. 303, 305 (1980), addressed the § 101 terms 'manufacture' and 'compositions of matter'. The decision is further analysed in Chapter 9.
154. *Diamond v Chakrabarty*, 447 U.S. 306, 318, 310 (1980); and Eileen M. Kane, 'Splitting the gene: DNA Patents and the Genetic Code', 71 *Tennessee Law Review* 707, Summer 2004, pp. 735-736.
155. The quote refers to the Committee Report on the 1952 Patent Act; see Eileen Kane, 'Splitting the gene: DNA Patents and the Genetic Code', 71 *Tennessee Law Review* 707, Summer 2004, at p. 736; and Aaron J. Zakem, 'Rethinking patentable subject matter: Are Statutory categories useful?', 30 *Cardozo L. Rev*. 2009, p. 2983.

or the inventor's own handiwork.[156] What is at issue is not whether matter is living, but whether it is *man-made.*[157]

In the recent and much discussed *In re Bilski* case, the Federal Circuit further defined patentable subject matter under 35 U.S.C. § 101 by scrutinising the statutory category of 'process', and a new standard emerged that is a departure from previous cases and is having an impact on biotechnological inventions. The claimed invention consisted of a business method (instruction to buyers/sellers on how to protect against the risk of price fluctuations in the energy marked). Although that the claimed invention qualified as a process, the Federal Circuit stated that the statutory process category (§101) is narrower than its ordinary meaning, and excludes 'laws of nature, natural phenomena, abstract ideas, mental processes and abstract intellectual concepts.[158] The patent in question (a business method for commodity trades) claimed fundamental principles in the form of an abstract idea (not as laws of nature/natural phenomena) or a mental process.[159] The Federal Circuit stated that, according to *the machine-or-transformation test*, a process is only pa-

156. *Diamond v Chakrabarty*, 447 U.S. 309 (1980); *Funk Bros. Seed Co. v Kalo Inoculant Co.*, 333 U.S. 127 (1948); and Eileen Kane, 'Splitting the gene: DNA Patents and the Genetic Code', 71 *Tennessee Law Review* 707, Summer 2004, at p. 737: 'The microbial cases Funk and Chakrabarty illustrate the range of outcomes to living subject matter patentability: unpatentable subject matter when there has been no alteration of natural characteristics (Funk) or patentable subject matter when there has been a clear alteration of natural characteristics' (Chakrabarty).
157. The *Chakrabarty* case impact on biotechnological patenting is analysed by *Eileen M. Kane* who addresses the impact of the case in authorising the grant of patents to recombinant organisms; see Eileen M. Kane, 'Splitting the gene: DNA Patents and the Genetic Code', 71 *Tennessee Law Review* 707, Summer 2004, pp. 736-737: 'the case did not specifically address the patent eligibility of the molecular components of the biotechnology revolution, including genes, a fact that is often overlooked in discussions evaluating the significance of the case. Chakrabarty did emphasise the essentially paradoxical nature of the patentable subject matter inquiry as it has developed historically – while it is inherently expansive and embracing of new technologies, there is a zone of unpatentability that excludes those ideas and processes which belong in the public domain so that the useful arts may advance.'
158. *In re Bilski* (F.3.d Federal Circuit 2008). See Kenneth J. Burchfiel, *Biotechnology and the Federal Circuit*, 2nd Ed., 2010, p. 41.
159. A process is defined under the Patent Act, 35 U.S.C § 100(b). The only test that is relevant for determining the patentability of a process under the Patent Act, 35 U.S.C § 101 is the machine-or-transformation test (M-or-T). See also Merges & Duffy, *Patent Law and Policy: Cases and Materials*, 5th Ed., 2011, p. 79, thus it was not relevant to assess whether it involved a method of conducting business or it was merely an abstract idea.

tentable according to the statutory category in § 101 if: (1) it is tied to a particular machine or apparatus; or (2) if it transforms a particular article into a different state or thing.[160] The Supreme Court (writ of certiorari) modified this narrow approach by stating that it had more than once cautioned that courts should not read into patent laws limitations and conditions which legislators had not expressed. The machine-or-transformation test is a useful and important clue, an investigative tool, but not the sole test for determining whether an invention is a patentable process.[161]

In the *Classen Immunotherapies v Biogen* case[162] the Federal Circuit had to decide whether an invention claiming methods for evaluating and improving the safety of immunisation schedules was patentable. There was a further claim for screening schedules, broadly related to methods for conducting clinical trials and developing vaccination protocols. The Court stated that the patent did not claim a specific technique or technical process for testing vaccine safety, but only described a general inquiry into the existence of the proposed correlation between an immunisation schedule and the incidence of chronic disease. According to the Court, the immunisation patent was an indirect attempt to patent the idea that there is a relationship between vaccine

160. Kenneth J. Burchfiel, *Biotechnology and the Federal Circuit*, 2nd Ed., 2010, p. 43, write that the Federal Circuit effectively restored the nineteenth-century standard for patentable subject matter, under which a process is required to act upon the subject matter to reduce it to a different state or thing; cf. *Cochrane v Deener*, 94 U.S 780, 788 (1876).

161. *Bilski v Kappos*, 130, S.Ct. at 3227 (561 U.S. (2010), with no indication of how that clue should be used. See Rochelle C. Dreyfuss & James P. Evans, 'From Bilski back to Benson: preemption, inventing around, and the case of genetic diagnostics', *Stanford Law Review*, Vol. 63, issue 6, June 2011, p. 1349; Brian P. Murphy & Daniel P. Murphy, 'Bilski's machine-or-transformation test: Uncertain prognosis for diagnostic methods and personalized medicine patents', 20 *Fordham Intell. Prop. Media & Ent. L.J.*, 755, 2009-2010; Christopher M. Holman, 'Bilski: Assessing the impact of a newly invigorated patent-eligibility doctrine on the pharmaceutical industry and the future of personalized medicine', *Current Topics in Medicinal Chemistry*, 10:1937-49. 2010, available at SSRN: http://ssrn.com/abstract=1424493; Timo Minssen & Robert M Schwartz, 'US patent Eligibility in the wake of *Bilski v Kappos*: 'Business as Usual' in the age of New technologies?', 30 *Biotechnology Law Report*, 3, No 1, 2011 (sequel in 30 *Biotechnology Law Report* 57, No 1, 2011); Peter S. Menell, 'Forty years of wondering in the wilderness and no closer to the promised land: Bilski's superficial textualism and the missed opportunity to return patent law to its technology mooring', 2011, 63 *Stanford Law Review*, 1289 ff.

162. *Mayo Collaborative Services DBA Mayo Medical Laboratories, et al., Petitioners v Prometheus Laboratories, Inc. 566 US (2012), No 10-1150.*

schedules and chronic immune-mediated disorders and an attempt to patent an unpatentable natural phenomenon.[163]

The series of cases dealing with classes of natural phenomena peaked with the Supreme Court's decision in *Mayo Collaborative Services v Prometheus Labs. Inc.*[164] This decision, on writ of certiorari, was delivered by Judge Breyer and was unanimous in overturning a medical treatment patent.

The two patents at issue claimed the use of thiopurine drugs to treat autoimmune diseases such as Crohn's disease and ulcerative colitis. When ingested, the body metabolises the drugs, producing metabolites in the bloodstream. Because patients metabolise these drugs differently, doctors have found it difficult to determine whether a particular patient's dose is too high, risking harmful side effects, or too low and so likely to be ineffective. The patent claims set forth processes embodying researchers' findings identifying correlations between metabolite levels and likely harm or ineffectiveness with precision. Each claim comprised an *administering step*, a *determining step* and a *wherein step* – all embodying an assessment of the correlation between the dose, the effect and the side effects.[165]

In *Gottschalk v Benson* (409 U.S. 63, 67 (1972)), Judge Bryon stated: 'phenomena of nature, though just discovered, mental processes, and abstract intellectual concepts are not patentable, as they are the basic tools of scientific and technological work'. Monopolising such tools might impede innovation, but too broad an interpretation of these exclusionary principles may weaken patent protection. Bryon emphasised that the Court has made it clear that in order to transform an unpatentable law of nature into a patentable application

163. *Classen Immunotherapies, Inc v Biogen* IDEC, Fed. Cir. Dec. 19, 2008. See Kenneth J. Burchfiel, *Biotechnology and the Federal Circuit*, 2nd Ed., 2010, p. 48. See further *In re Abele*, 684 F. 2d 902, 908-09 (C.C.P.A. 1982), involving a method claim for the electronic transformation of X-ray data to produce images of bones, organs and other tissues; and *Arrhythmia Research Technology, Inc. v Corazonix Corp.* 958 F 2.d 1053, 1055 (Fed. Cir. 1992), for a diagnostic process to convert the measurements of a person's heartbeat from analogue to digital form, assessing a person's risk of a certain type of arrhythmia. Rebecca S. Eisenberg, 'Wisdom of the ages or dead-hand control? Patentable subject matter for diagnostic methods after In re Bilski', 3 *Case W. Res. J.L. Tech. & Internet* 1(2012), pp. 27-29 (forthcoming) writes that: 'In both *Classen* and *Prometheus*, the Federal Circuit took its machine-or-transformation test to be entirely dispositive of the issue of patentable subject matter for the claimed methods, and did not consider whether claims to analysis of biological markers might call for a different analysis than claims to business methods.'

164. *Mayo Collaborative Services DBA Mayo Medical Laboratories, et al., Petitioners v Prometheus Laboratories, Inc. 566 US (2012), No 10-1150.*

165. *Mayo Collaborative Services DBA Mayo Medical Laboratories, et al., Petitioners v Prometheus Laboratories, Inc.* 566 US (2012), No 10-1150. Syllabus p. 1.

of such law, one must do more than simply state the law of nature and add the words 'apply it' (referring to *Gottschalk v Benson*, 409 U.S. (1972) at page 71-72). And he stated that the patent in question 'lies at the intersection of these basis principles. The claims purport to apply natural laws describing the relationships between the concentration in blood of certain thipurine metabolites and the likelihood that the drug dosage will be ineffective or induce harmful side-effects. We must determine whether the claimed processes have transformed these unpatentable natural laws into patenteligible applications of those laws. We conclude that they have not done so and that therefore the processes are not patentable'.[166]

> First, the *administering step* simply refers to the relevant audience, namely doctors who treat patients with certain diseases with thiopurine drugs. ... Second, the wherein clauses simply tell a doctor about the relevant natural laws, at most adding a suggestion that he should take those laws into account when treating his patient. Third, the determining step tells the doctor to determine the level of the relevant metabolites in the blood, through whatever process the doctor or the laboratory wishes to use (patent in suit known methods) scientists routinely measured metabolites as part of their investigations into the relationships between metabolite levels and efficacy and toxicity of thiopurine compounds.[167]

The upshot, Bryon concluded, was that the three steps simply told doctors to gather data from which they might draw an inference in the light of correlations. The claims informed about certain laws of nature – additional steps were already known and routine – and viewed as a whole nothing significant was added beyond the sum of the parts taken separately.[168] Thus, the steps were not sufficient to transform unpatentable natural correlations into patent-

166. *Mayo Collaborative Services DBA Mayo Medical Laboratories, et al, Petitioners v Prometheus Laboratories, Inc.* 566 US (2012), No 10-1150, p. 3.
167. *Mayo Collaborative Services DBA Mayo Medical Laboratories, et al, Petitioners v Prometheus Laboratories, Inc.* 566 US (2012), No 10-1150, p. 10.
168. 'Unlike, say, a typical patent on a new drug or a new way of using an existing drug, the patent claims do not confine their reach to a particular application of those laws. The presence here of the basic underlying concern that these patents tie up too much future use of laws of nature simply reinforces our conclusion that the processes described in the patents are not patent eligible, while eliminating any temptation to depart from case law precedent'; see *Mayo Collaborative Services DBA Mayo Medical Laboratories, et al., Petitioners v Prometheus Laboratories*, Inc. 566 US (2012), No 10-1150, pp. 18-19.

able subject matter – one cannot apply known technology to natural phenomena and claim to have made an invention.[169]

The decision narrowed down *35 U.S.C § 101 patentable subject matter*, and it has already had an impact on gene patenting, as mentioned below in section 3. It was followed on 26 March 2012 by the US Supreme Court reversing the Federal Circuit decision of July 2011 in *AMP v USPTO and Myriad genetics.*[170]

### 2.2.3. *The purification doctrine*

The European doctrine of isolation as discussed above in regard to the European framework has a parallel within the US framework. However, there is no statutory basis for the doctrine in US law.[171] In the literature, DNA patents are also discussed from the perspective of the 'purification doctrine'. Many claims for DNA will specifically identify their subject matter as isolated or purified, or both. The weight of 'the product of nature' jurisprudence supports the conclusion that such claims do not correspond to naturally occurring ge-

169. Methods of medical treatment are not patentable in Europe; see EPC Article 53(c). See Richard Hacon & Jochen Pagenberg, *Concise European Patent Law*, 2nd Ed., 2008, p. 37. See the Enlarged Board of Appeal in Case G-1/04 on diagnostic methods; and two decisions of the Enlarged Board of Appeal: Treatment by surgery in Case G 1/07 (Treatment by surgery/*Medi-Physics*); and of more indirect importance for this section ('once a day prior to sleep') decision in Case G2/08 (Dosage regime/*Abbott respiratory*). Commented on by Dr. Franz-Josef Zimmer & Dr. Benjamin Quest, 'EPO Enlarged Board of Appeal decide on dosage regimes (G2/08) and Treatment by Surgery (G 1/07)', 29 *Biotechnology Law Report* 485, No 5, 2010, p. 485 ff.; and Julian Cockbain & Sigrid Sterckx, 'Is the Enlarged Board of Appeal of the European Patent Office authorized to extend the bounds of the patentable? The G-3785 second medical indication/EISAI and G-2/08 Dosage Regime/Abbott respiratory cases', 3/2011 *IIC*, p. 257 ff. See a critical comment from Paul Cole, Patent law Blog Patently-O, 3 April 2012, comparing US patents with European patents (EP 1114403 corresponding to US 6355623, and EP 1695092 corresponding to US 7524851, arguing that the key ruling of the US Supreme Court is that the reasoning of the EPO Appeal Boards (in Case G3/08) cannot be followed.
170. See Stephen Pessango, 'Prometheus and Bilski: Pushing the bounds of patentable subject matter in medical diagnostic techniques with the machine-or-transformation test', *American Journal of Law and Medicine*, 36, 2010, p. 650: 'With the Supreme Court's recent disposition in Bilski, Prometheus, and Classen, the Federal Circuit faces the difficult challenge of deciding how these principles properly apply to medical diagnostic techniques'.
171. Apart from the 2001 USPTO Utility Examination Guidelines, 66, Fed. Reg. 1092, 1093 (2001): 'A patent on a gene covers the isolated and purified gene but does not cover the gene as it occurs in nature.'

nomic DNA if the claims are drafted so as to clearly define an isolated and purified DNA molecule that is categorically different from the natural molecule.[172]

Both *Arti Rai* and *Rebecca S. Eisenberg* have been keen to push the approach to DNA away from a merely chemical classification. *Eileen Kane* suggests a dual approach to shaping the scope of property rights – a distinction between the DNA sequence as a static chemical and a dynamic template executed by the genetic code. Kane's suggestion in fact mirrors the discussion above in section 2, concerning absolute product protection and purpose-bound protection. She argues that a DNA as a chemical as well as a template invites the evaluation of both product-based and process-based exclusions, thus dictating a bifurcated legal analysis.[173] Kane's suggestion of a split of DNA into a static chemical and a dynamic template leaves the DNA sequence to rely on underlying laws of nature for its dynamic function as a template for converting DNA to protein, thus evaluating the DNA gene sequence and its reliance on the genetic code.[174] In a recent article it is argued that the dichotomy of the gene being both a carrier of abstract information and a physical compound exposes both sides as being wrong due to over simplification.[175] It is argued that the mere presence of information in a man-

172. See Eileen M. Kane, 'Splitting the gene: DNA Patents and the Genetic Code', 71 *Tennessee Law Review* 707, Summer 2004, pp. 741-742 and 744, and p. 743 referring to Arti K. Rai's argument that DNA is more properly characterised as a carrier of information.
173. See Eileen M. Kane, 'Splitting the gene: DNA Patents and the Genetic Code', Tennessee Law Review 71 *Tennessee Law Review* 707, Summer 2004, p. 744; Eileen M. Kane, 'Patenting Genes and Genetic Methods: What's at Stake?', *Journal of Business & Technology Law*, 2011, Vol. 6, No 1, p. 10 and p. 16 referring to Judge Sweet: 'the weight of the opinion is that any exclusive 'use' of DNA obtained through patent protection is to be avoided, precisely because its critical template functions require that it remain in the public domain'; Rebecca S. Eisenberg, 'Molecules vs. Information: should patents protect both?', 8 *B.U.J.Sci. & Tech. L.* 190, 196 (2002); and Michael Risch, 'Everything is patentable', *Tennessee Law Review*, Vol. 75, 2008, p. 616.
174. See Eileen M. Kane, 'Splitting the gene: DNA Patents and the Genetic Code', 71 *Tennessee Law Review* 707, Summer 2004, p. 744.
175. Eric J. Rogers, 'Can you patent genes? Yes and no', 93 *J. Pat. & Trademark Off. Soc'y*, 2001, p. 39 ff.; at p. 40 the author explains that the differences in the compositions of the human chromosome and the man-made DNA molecule with regard to DNA binding proteins, RNA and contextual DNA sequences are not incidental, but have a direct impact on the function of DNA sequences. The differences are substantial because e.g. the DNA binding proteins, regulatory RNAs, contextual DNA sequences, epigenetic regulation, and spatial arrangements of chromosomes, can all have an impact on gene expression. 'The DNA fragment cloned into a vector has new

made object does not render the object ineligible for patentability. DNA sequences are not unique in their ability to contain abstract and natural information.[176]

The 1912 the *Parke Davis & Co v H. K. Mulford & Co* case[177] elaborated the standard theory of patenting purified natural substances which has had an effect on the patenting of genes.[178] The inventor, J. Takamine, purified the hormone adrenalin in base form (as an extract from the suprarenal glands of living animals).[179] Takamine obtained two patents. One patent covered the purified adrenalin base[180] and one patent included a claim for adrenalin-based salt, i.e. the purified base covered with an acid.[181] Even if purification does not alter the substance it could not be used without purification, and Justice Hand stated as follows: 'But even if it were merely an extracted product without change, there is no rule that such products are not patentable. Takamine was the first to make it available for use by removing it from the other

fundamental properties. ... the physicochemical properties of synthesized DNA molecules and naturally occurring DNA sequences are very different.'

176. Eric J. Rogers, 'Can you patent genes? Yes and no', 93 *J. Pat. & Trademark Off. Soc'y*, 2001, p. 39 ff., at p. 46, where the author compares DNA to a chair made of wood (providing information through tree-rings (age, higher-order information about rainfall levels and other general climatic conditions).
177. 189 F. 95 (S.D.N.Y. 1911)/196 F 496 (2d Cir 1912).
178. See Merges & Duffy, *Patent Law and Policy: Cases and Materials*, 5th Ed., 2011, p. 169. *Parke-Davis* is now a standard citation for the theory permitting patents for DNA sequences. See p. 169 on the requirement for a more stringent substantial transformation test for patenting DNA material. Eileen M. Kane, 'Splitting the gene: DNA Patents and the Genetic Code', 71 *Tennessee Law Review* 707, Summer 2004, p. 113: constructive pre-emption of the genetic code ought to be found contrary to the Supreme Court's ruling that the laws of nature are not patentable. P. 170: under a *Parke-Davis* theory the substance as produced and used in the human body is viewed as different from the substance in its isolated and purified state.
179. Takamine also was able to isolate adrenalin in the form of a chemical base as opposed to a salt (i.e. the base combined with an acid). The purified base adrenalin overcame the negative side effects associated with prior art salts, which were also extracts from animal suprarenal glands.
180. US730176, 176-patent: Claim 1 read as follows: 'A substance possessing the herein-described physiological characteristics and reactions of the suprarenal glands in a stable and concentrate form, and practically free from inert and associated gland tissue'. See also Merges & Duffy, *Patent Law and Policy: Cases and Materials*, 5th Ed., 2011, p. 164.
181. US753177, 177-patent: 'The substance consisting of a salt of the herein-described product of the suprarenal glands; said salt being soluble in water and possessing the physiological and therapeutic characteristics and reaction of said product in substantially stable and concentrated form.'

gland-tissue in which it was found, and, while it is of course possible logically to call this a purification of the principle, it became for every practical purpose *a new thing commercially and therapeutically*. That was a good reason for a patent. That the change here resulted in ample practical differences is fully proved.'[182]

The case law both before and after *Parke Davis* case has elaborated on the purification issue.[183] In the 1874 *American Wood Paper Co v Fibre Disintegrating Co* case, the Court reasoned that many well known and valuable things may be extracted from divers' substances, but the extract (wood pulp in the case in question) would be the same no matter what it had been taken from. The process to obtain it may differ, but the output (extract) when obtained cannot be called a new manufacture.[184]

The 1889 *Ex parte Latimer* case (referred to above in section 2.2.2) concerned a patent claim for the cellular tissue of a particular species of pine needle (product of nature). *Kuehmsted v Farbenfabriken of Elberfeld* considered whether a purified aspirin compound was sufficiently distinguished from the impure compound to make it patentable.[185] In the 1958 *Merck & Co. v Olin Mathiieson Chemical Corp* case, on purified vitamin B12, by limiting the 'product of nature' defence the decision relied on the utility doctrine; the

182. In the most important passage in *Parke Davis*, Judge Learned Hand decided that a purified substance from a living creature can be patented even though 'it is of course possible logically to call this a purification of the principle'. For the purposes of the 176 patent that passage is dicta because, as Hand noted, even the defendant's expert had 'supposed that the natural substance existed in the form of a salt'. But the 177 patent included six claims that covered adrenalin-based salts and adrenalin-based salts in solution. Judge Hand found four of these claims valid and infringed, including the broad claim 5 set out above. Sustaining these claims required Hand to reason that purified and concentrated adrenalin salt is 'a new thing commercially and therapeutically and therefore patentable' notwithstanding the existence of 'natural salt' in living organisms. See also Merges & Duffy, *Patent, Law and Policy: Cases and Materials*, 5th Ed., 2011, p. 168.
183. See Judge Sweet in the *Myriad* case (*AMP v USPTO*, 702 F. Supp. 2d 181 (S.D.N.Y 2010) stating that *Parke Davis* concerned a question on novelty (modern §102) not of patentable subject matter; see Merges & Duffy, *Patent Law and Policy: Cases and Materials*, 5th Ed., 2011, p. 180, for a discussion of who is right or whether the views of Hand and Sweet are incompatible.
184. 90 U.S.566, 593-94 (1874); and see Merges & Duffy, *Patent Law and Policy: Cases and Materials*, 5th Ed., 2011, p. 168.
185. 179 F. 701, 705 (7th Cir. 1910) and further *Union Carbide Co v American Carbine Co*, 181 Fed. 106 (2d Cir. 1910), concerning a patent for crystalline calcium (the amorphous form was in the prior art). See Merges & Duffy, *Patent Law and Policy: Cases and Materials*, 5th Ed., 2011, p. 169. Both cases decided on the novelty of artificial substances compared with the prior art.

claimed composition had been transformed by purification, in its natural state it had no utility, therapeutically or commercially: 'The new products are not the same as the old, but new and useful compositions entitled to the protection'.[186] In *Am. Fruit Growers, Inc. v Brogdex Co.*, the Supreme Court stated that the addition of borax to the rind of an orange to increase its longevity did not confer a patentable distinction, compared with an unadulterated orange ('does not produce from the raw material an article for use which possesses a new or distinctive form, quality, or property).[187] The *Re Bergstrom* case in 1970 concerned purified prostaglandins from human tissue; the Court's analysis of the distinction between pure and impure forms of a compound required there to be a strict compositional difference between the pure and impure compounds on order to confer patentability.[188] The 1928 *General Elec. Co v De Forest Radio Co* case concerned a claim for pure tungsten. However, what the inventor had discovered was the natural qualities of pure tungsten. The inventor manifestly did not create pure tungsten, nor did he create its characteristics. These were created by nature.[189] The *Schering Corp. v Geneva Pharmaceuticals, Inc.* case[190] concerned a claim for a molecule that was inherently produced by the metabolism of a prior art compound. The Court indicated that there was no bar per se to the patenting of a metabolite produced by the normal metabolism of a known compound, as long as the metabolite was clearly claimed in a non-naturally occurring form.[191]

The US purification doctrine concerns the conversion of natural qualities and thus the subject matter concerned does not fall foul of the doctrine of nature. *Parke Davis* and *Chakrabarty* are both cases of major importance, and the purification theory in *Parke Davis* is analogous to the doctrine of isolation in Article 5(2) (human genes) and Article 3(2) (biological material) of the European Directive 98/44/EC.

186. 183 253 F 2d 156 (4th Cir 1958) at 164. See Eileen M. Kane, 'Splitting the gene: DNA Patents and the Genetic Code', 71 *Tennessee Law Review* 707, Summer 2004, p. 740; and Merges & Duffy, *Patent Law and Policy: Cases and Materials*, 5th Ed., 2011, p. 169.
187. 283 U.S. 1 (1931). And see Eileen M. Kane, 'Splitting the gene: DNA Patents and the Genetic Code, 71 *Tennessee Law Review* 707, Summer 2004, footnote 187.
188. 427 F.2d 1394 (C.C.P.A 1970). See also Merges & Duffy, *Patent Law and Policy: Cases and Materials*, 5th Ed., 2011, p. 169; and Eileen M. Kane, 'Splitting the gene: DNA Patents and the Genetic Code', 71 *Tennessee Law Review* 707, Summer 2004, p. 740.
189. 28 F.2d 641, 643 (3d cir. 1928). According to Merges & Duffy, *Patent Law and Policy: Cases and Materials*, 5th Ed., 2011, pp. 169-170, this has little impact on modern law.
190. 339 F.3d 1373 (Fed. Cir.2003). And Eileen M. Kane, 'Splitting the gene: DNA Patents and the Genetic Code', 71 *Tennessee Law Review* 707, Summer 2004, p. 740.
191. See Eileen M. Kane, 'Splitting the gene: DNA Patents and the Genetic Code', 71 *Tennessee Law Review* 707, Summer 2004, p. 741, who writes that the case may have implications for the scope of a patent claim for a DNA molecule and whether metabolites inherently produced from a template may anticipate patent claims to isolated and purified forms of the metabolites, such as proteins.

# 3. Developing gene patenting: The case of Myriad Genetics

'DNA is not 'life', but a chemical substance which carries genetic information ... The patenting of a single human gene has nothing to do with the patenting of human life. Even if every gene in the human genome were cloned (and possibly patented) it would be impossible to reconstitute a human being from the sum of its genes.'[192]

## 3.1. Spotlight on Myriad Genetics

Every scholar of patent law has come across the never-ending story of Myriad Genetics. The *BRCA* patents never seem to escape attention.[193] Breast cancer is a common disease and 9 % of all women in North America and Western Europe will develop breast cancer in their lifetime.[194] There are different kinds of breast cancer. In the case of familial breast and ovarian cancer the genetic component, mutations in the *BRCA1* and *BRCA2* genes, cause an increased susceptibility to early onset breast cancer.

From the facts presented in the case before the US District Court, Southern District of New York, one can also see that mutations in the *BRCA1* and *2* genes correlate with an increased risk of breast and ovarian cancer. Women with *BRCA1* and *BRCA2* mutations face an up to 85% cumulative risk of breast cancer, as well as an up to 50% cumulative risk of ovarian cancer. In addition, among the 10-15% of ovarian cancer cases in which there is a genetic component, 80% of women diagnosed under the age of 50 carry mutations in their *BRCA1* genes and 20% carry mutations in their *BRCA2* genes. Women with inherited *BRCA1* mutations have a 40-52% cumulative risk of ovarian cancer by the time they reach 70. For women with inherited *BRCA2*

192. Quote from *Howard Florey/Relaxin* [1995] EPOR 541 (EPO Opposition Division). See Philip W. Grubb & Peter R. Thomsen, *Patents for Chemicals, Pharmaceuticals and Biotechnology*, 5th Ed., 2010, p. 318.
193. EP 699754, 705902 and EP 705903. Whereas the EP 705902 patent essentially claimed the BRCA1 DNA and polypeptide sequence as product protection, EP 699754 relates to a use claim for the diagnostic methods ('germline mutations in the BRCA1 gene and their use in the diagnosis of predisposition to breast and ovarian cancer'). See further Gert Matthijs & Gert-Jan Van Ommen, 'Gene Patents: from discovery to invention, A geneticist's view', in Geertrui Van Overwalle (ed.), *Gene Patents and Collaborative Licensing Models, Patent pools, Clearinghouses, Open Source Models and Liability Regimes*, p. 35; and Gert Matthijs, 'DNA diagnostics in practice', in Geertrui Van Overwalle (ed.), *Gene Patents and Public Health*, p. 35 f.
194. Nussbaum, McInnes & Willard, *Thompson & Thompson, Genetics in Medicine*, 7th Ed., 2007, p. 472.

mutations, the risk is approximately 15-25%. Male carriers of mutations are also at increased risk of breast and prostate cancer.[195]

A Belgian geneticist, *Gert Matthijs*, who has fought vigorously against the Myriad patents, has written that the breakthrough in the research leading to identification of the breast cancer gene *BRCA1* was accomplished by the scientist Mary Claire King and her team at Berkeley, California. She discovered that in some families the defect was localised on the long arm of chromosome 17.[196] The *BRCA2* was located on the long arm of chromosome 13 by the Institute of Cancer Research in the UK. Then in 1994, the *BRCA1* gene was cloned and sequenced by Myriad, whereas the *BRCA2* gene was identified in 1995 by the Institute of Cancer Research.[197] *Matthijs* also points out the division of work and emphasises the international initiative of the Breast Cancer Linkage Consortium (BCLC), set up in 1989 with the aim of pooling family sample materials and the results of linkage analysis and other studies. The BCLC was supported from 1993-1998 by the European Commission (Biomed1 & Biomed2 programme). Its database contained genetic data from more than 700 breast cancer families from Europe, Canada and the USA, and it involved nearly one hundred research centres. Accordingly, *Matthijs* states that the 99 other organisations which also provided the basis for this scientific work were left behind when the patents were granted solely to one organisation: Myriad Genetics of Utah University.[198]

195. US District Court for the Southern District of New York, Judgement of 29 March 2010, *Association for Molecular Pathology et al. (AMP) v USPTO*, Case 09 Civ. 4515, pp. 58-59.
196. See further Gert Matthijs & Gert-Jan Van Ommen, 'Gene Patents: from discovery to invention, A geneticist's view', in Geertrui Van Overwalle (ed.), *Gene Patents and Collaborative Licensing Models, Patent pools, Clearinghouses, Open Source Models and Liability Regime*, p. 35.
197. See further Gert Matthijs & Gert-Jan Van Ommen, 'Gene Patents: from discovery to invention, A geneticist's view', in Geertrui Van Overwalle (ed.), *Gene Patents and Collaborative Licensing Models, Patent pools, Clearinghouses, Open Source Models and Liability Regime*, p. 35.
198. Gert Matthijs & Gert-Jan Van Ommen, 'Gene Patents: from discovery to invention, A geneticist's view', in Geertrui Van Overwalle (ed.), *Gene Patents and Collaborative Licensing Models, Patent pools, Clearinghouses, Open Source Models and Liability Regime*, p. 36. See further Eileen M. Kane, 'Symposium: The third annual IP/gender: The unmapped connections symposium: Molecules and conflict: Cancer, Patents, and woman's health', 15 *American University Journal of Gender, Social Policy & the Law*, 2007, p. 305, at pp. 328-329; John Murray, 'Owning genes: Dispute involving DNA sequence patents', *Chicago-Kent Law Review*, 1999 (75 CHIKLR 231 ff.).

In 2001, the European Parliament adopted a resolution on the patenting of *BRCA1* and *BRCA2* genes.[199] Recital F of the resolution emphasises that the granting of patents by the EPO, similar to that of the USPTO, could create a monopoly for the patentee within the European Union, which could seriously impede or even completely prevent the further use of existing inexpensive and more effective tests for the breast cancer genes *BRCA1* and *BRCA2*. Furthermore, it was stated that this development could have an unacceptable detrimental effect on the women concerned and constitute a serious drain on the funds of public health services, and that it could seriously impede the development of and research into new methods of diagnosis.

Patents on the *BRCA* genes have provoked tremendous opposition.[200] As the initial patents were drafted and claimed broadly it is interesting to follow their prosecutions and the oppositions. First the patents were prosecuted and granted in two different patent systems, under the USPTO and the EPO regimes.

The EPO patents (three on *BRCA1*, and two on *BRCA2*) were filed in 1995. The first patent on BRCA1 (EP 699754 – a use patent) asserted rights over the diagnostic use of the *BRCA1* gene, whatever the technique chosen by the laboratory. The patent was filed on 11 August 1995 and it was initially granted on 10 January 2001. Due to the oppositions, the oral hearing took place in May 2004 and it was finally appealed in November 2008. The patent was maintained in amended form and finally granted in July 2009. As finally granted it does not include the normal or mutated forms of the *BRCA1* gene itself.

199. See European Parliament resolution of 3 October 2001 on the patenting of BRCA1 and BRCA2 ('breast cancer') genes. The resolution recalled the former resolution of 30 March 2000 on the decision by the European Patent Office with regard to patent No EP 695 351 granted on 8 December 1999, and had regard to the 'Advice on the patentability of the human genome', adopted by consensus by the International Bioethics Committee of Unesco (IBC) at the conclusion of its Eighth Session on 14 September 2001.

200. In Europe: Case T 156/08, 14 January 2011 (BRCA2, EP 1260520); Case T 902/07, 7 September 2010 (interlocutory decision EP0858467); Case T 666/05, 13 November, 2008 (EP0705903); and Case T 80/05, 19 November 2008 (EP 699754). See e.g. Gert Matthijs, *The European opposition against the BRCA gene Patents, Familial Cancer*, 2006, 5:95-102; Jordan Paradise, 'European opposition to exclusive control over predictive breast cancer testing and inherent implications for U.S. patent law and public policy: A case study of the Myriad genetics BRCA patent controversy', *Food and Drug Law Journal*, 2004, p. 133 ff.; Isabelle Huys, Geertrui Van Overwalle & Gert Matthijs, European Journal of Human genetics, 2011, 19, 1104-1107. Aaron S. Kesselheim and Michelle M. Mello, 'Gene patenting – is the pendulum swinging back?', *New England Journal of Medicine*, 2010, 362, 20, p. 1855 ff.; Jeffery Bernhardt & James Fox, 'Gene sequences are still patentable: the federal circuit decides the Myriad genetics case', *BSLR*, Vol. 12, issue 2, p. 45 ff..

Initially, the EP 705903 patent had broad claims, e.g. for 34 specific mutations plus a method for diagnosis; but as granted it only claimed a probe of 15 to 30 nucleotides for the detection of the 185 delAG.[201] The EP 705902 patent, a broad product patent, claimed all uses of the *BRCA1* gene and protein (diagnosis, therapeutic and prevention). The patent as maintained claimed a probe comprising 300 bp of the *BRCA1* coding sequence, or a few other fragments.[202] Patents EP 705903 and EP 705902 were both filed on 11 August 1995, they were granted in May and November 2001 respectively, and the oral hearings took place in January 2001. Both patents were maintained in 2009 July and December respectively. The University of Utah Research Foundation, Myriad and the United States of America were applicants for all three patents.

With respect to the *BRCA2* gene, the applications EP 785216 and EP 858467 were filed in December and November 1996 respectively and granted in January 2003 and February 2004 respectively. Hearings on EP 785216 took place in June 2003 and the patent was maintained in an amended form and finally granted in August 2008. Patent EP 858467, claimed by the Cancer Research Technology Limited (UK), was revoked in October 2010.[203]

As mentioned above, the research prior to the grant of the first patent originated from several organisations. In the USA this concerned the National Institute for Environmental Health Science (NIEHS, a subdivision of the NIH). The NIH had provided approximately USD 2 million to the research as well as the funding for six NIEHS researchers. Subsequently, a dispute arose between Myriad and NIH over the exclusion, as co-inventors, of NIH and NIEHS. NIH maintained that its scientists had conducted some of the most important work leading up to the sequencing of the gene.[204]

201. EP 705903.
202. EP 705902.
203. Decision of the Technical Board of Appeal, T 0902/07 – 3.3.04 of 7 September 2010. The revocation was claimed by Myriad as Appellant II (the appeal concerned the entitlement to priority and was decided in accordance with Opinion of the EBA G 2/98 and decision T 923/92). In point 13 of the decision, the Technical Board of Appeal stated that: ‘the Board arrives at the decision that the subject-matter of claim 2 of Appellants I’s main request [the applicant: Cancer Research Technology limited] is not entitled to claim priority from (P1) or (P2). Further in point 14 on continuation it is stated that documents (1) and (8) belong to the state of the art under Article 54(2) EPC and document (9) belongs to the state of the art under Article 54(3) EPC. According to point 15 prior art discloses the full length BRCA 2 CDNA sequence and the complete BRCA2 coding sequence. Thus, the subject matter of claim 2 of the main request is not novel and does not meet the requirements of Article 54 EPC.’
204. See US District Court for the Southern District of New York, judgment of 29 March 2010, *Association for Molecular Pathology et al. (AMP) v USPTO*, 09 Civ. 4515, p. 55.

In the US literature *Robert Cook-Deegan*, among many others, has been closely involved in the debate on gene patenting and about this Myriad background he has said:

'In the case of Myriad, initial research took place at the University of Utah – with public funding from the US National Institute of Health (NIH ...). The researchers then spun off Myriad, which attracted investment from Eli Lilly (... USA) and succeeded in patenting BRCA1 and a diagnostic test for breast cancer (patents were ultimately jointly assigned to the University of Utah, Myriad and the NIH). Rather than licensing out the test to clinical genetics and laboratories around the world, Myriad required initial testing in each family to be performed at its laboratories in Salt Lake City. In the United States, the company sent out cease-and desist letters to laboratories – both academic and commercial – already performing tests when the patent was issued.

Threatened patent enforcement resulted in a backlash around the world from public laboratories, clinicians, molecular geneticists and some patient groups – against both the patenting of human genes and what they viewed as Myriad's strong-arm tactics. These groups feared that by closing down public laboratories, Myriad would thwart research identifying weaknesses in Myriad's test or distinguishing the effects of different mutations in the genes on disease severity or progression, and prevent the integration of breast and ovarian cancer genetic tests into genetic health services. Although some of these fears were clearly exaggerated, Myriad's aggressive initial patent enforcement affected practice in the clinical genetics community and stirred long-standing resentment. Furthermore, in countries with public health care systems, health administrators objected to Myriad's business models because it removed their ability to deploy genetic tests to their citizens in the manner that they viewed as most efficient [Gold, E.R. & Carbone, J. Genet. Med. 12 Suppl, S39-S70 (2010)].

... However, as a result of Myriad's enforcement actions coupled with broad patent claims, its fairly narrow conception of what constituted acceptable research and its failure to clearly state that it would not pursue those conducting such research, university and private laboratories ceased to offer the test publicly in the United States. Outside the United States, resistance to Myriad's model – particularly from health care administrators and government departments – caused the company to lose most of its markets. Furthermore, Myriad's relationship with scientists and policymakers around the world was seriously damaged.'[205]

Eventually, 12 years after the EC Biotech Directive introduced the first law on gene patenting, the Myriad adventure backfired in the USA with the decision of the Federal District Court for the Southern District of New York in *AMP v USPTO* in March 2010.[206] However, on July 29, 2011, Judge Sweet's

205. See Julia Carnone et al., 'DNA patents and diagnostics: not a pretty picture', Commentary, *Nature Biotechnology*, Vol. 28, No 8, August 2010, p. 786.

206. US District Court for the Southern District of New York, Judgment of March 29, 2010 in *Association for the Molecular Pathology et al. (AMP) v USPTO*, Case 1:09-cv-04515-RWS, document 255, filed 03/29/2010, Addressing seven patents: U.S. pa-

detour was corrected by US Court of Appeal for the Federal Circuit, but the case law *may* have given a push to the patenting landscape in the USA. The US Supreme Court has corrected the Federal Circuit on several occasions since the decision in *Bilski v Kappos*, reshaping US public policy. In the *Mayo v Prometheus* case, decided on 20 March 2012 on writ of certiorari, it pull back patentable subject matter pursuant to 35 U.S.C § 101 which has been echoed at a global level. The case focused on method claims. The US Supreme Court recognised the role of Congress in crafting more finely tailored rules where necessary, and stated that in this case it did not need to determine whether, from a policy perspective, increased protection for discoveries of diagnostic 'laws of nature' is desirable.

As a spin off from the *Mayo v Prometheus* decision, already on 26 March 2012 the Supreme Court overruled the Federal Circuit's decision in *AMP v Myriad Genetics* (on appeal from the Southern District of New York), ordering the Appeal Court to reconsider its decision in light of the *Mayo v Prometheus* decision.[207] The core issue in this case is a product claims, and one can expect the US Supreme Court *either* to outlaw gene patents (which is perhaps unlikely if one reads the oral arguments submitted on 15 April 2013) *or* rule that isolated genes are not patent-eligible – but that gene related inventions are ('is the isolation or extraction of natural products patentable?').[208] Briefly, the EU/EPO is lagging behind and is stuck with its outdated provision in Article 5(2) of Directive 98/44/EC, while the US courts are dynamically reshaping the area of patentable subject matter. As the case study in Chapter 7 indi-

tent 5,747,282 (the 282 patent), U.S. patent 5,837,492 (the 492 patent), U.S. patent 5,693,473 (the 473 patent), U.S. patent 5,709,999 (the 999 patent), U.S. patent 5,710,001 (the 001 patent), U.S. patent 5,753,441 (the 441 patent), and U.S. patent 6,003,857 (the 857 patent).

207. The *Myriad* cases have been subject to intense debate in legal literature, e.g. Ester Van Zimmeren & Geertrui Van Overwalle, 'A Paper Tiger? Compulsory License Regimes for Public Health in Europe', *IIC* 1/2011, Vol. 42, pp. 4-40; Amanda Odell-West, 'The legacy of Myriad for gene-based diagnostics: a new policy and regulatory option', *Journal of Intellectual Property Law and Practice*, 2009, Vol. 4, No 4, pp. 267-277; Eileen M. Kane, 'Patenting Genes and Genetic Methods: What's at Stake?', *Journal of Business & Technology Law*, 2011, Vol. 6, No 1, p. 11; Eileen M. Kane, 'Patent-Mediated Standards in Genetic testing', 2008 *Utah L. Rev*. 837, No 3, p. 848; and 'Wisdom of the ages or dead-hand control? Patentable subject matter for diagnostic methods after In re Bilski', 3 *Case W. RES. J.L. Tech. & Internet* 1(2012), p. 45 (forthcoming).

208. See Steven Seidenberg for *IP Watch*: US Supreme Court may invalidate gene patents, but create little change, 23 April 2013 (last visited 30 April 2013)

cates, one can question whether the design of Article 5(2) will be able to cope with all emerging scientific developments in gene technology.

*3.1.1. The Federal District Court for the Southern District of New York*
*3.1.1.a. Laws of nature and purification*

The paradox of gene patenting peaked with the ruling of Judge Robert W. Sweet of the Federal District Court for the Southern District of New York in the *AMP v Myriad Genetics* case. The plaintiffs, the American Civil Liberties Union Foundation and the Association for Molecular Pathology (AMP) et al., argued that the disputed claim[209] and the disputed patent,[210] relating to the BRCA1 and BRCA2 genes, should be declared invalid because they covered products of nature, laws of nature and/or natural phenomena, and abstract ideas or basic human knowledge or thought. The defendant was the United States Patent and Trademark Office (USPTO) and the patentees Myriad Genetics and the University of Utah Research Foundation. The plaintiffs challenged the validity of the patent claims, and asking a question that previously has been answered in the affirmative: are isolated human genes and comparisons of their sequences patentable?

The disputed claims concerned DNA and isolated DNA. In his judgment Judge Sweet sought to resolve the opposing views of the parties on the claim terms 'DNA' and 'isolated DNA'. He addressed the terms as follows:

'As its name implies, DNA, or deoxyribonucleic acid, is an acid – a tangible, chemical compound. As Myriad correctly notes, the specifications make clear that 'DNA' as used in the patents, refers to the physical manifestation of the acid, one that may be 'substantially separated from other cellular components which naturally accompany a gene [...] Isolated DNA is [therefore] construed to refer to a segment of DNA nucleotides existing separate from other cellular components normally associated with native DNA, including proteins and other DNA sequences comprising the remainder of the genome, and includes both

209. The challenged patent claims comprises two types of claims (composition claims and methods or processes): (1) isolated DNA containing all or portions of the BRCA1 and BRCA2 gene sequence; and (2) methods for 'comparing' or 'analyzing' BRCA1 and BRCA2 gene sequences to identify the presence of mutations correlating with a predisposition to breast or ovarian cancer; see *AMP v USPTO*, pp. 83-84, in the US District Court for the Southern District of New York, Judgment of 29 March 2010.

210. The subjects are 15 claims contained in seven patents: U.S. patent 5,747,282 (the 282 patent), U.S. patent 5,837,492 (the 492 patent), U.S. patent 5,693,473 (the 473 patent), U.S. patent 5,709,999 (the 999 patent), U.S. patent 5,710,001 (the 001 patent), U.S. patent 5,753,441 (the 441 patent), and U.S. patent 6,003,857 (the 857 patent); see *AMP v USPTO*, the US District Court for the Southern District of New York, Judgment of 29 March 2010, p. 84.

DNA originating from a cell as well as DNA synthesized through chemical or heterologous biological means.'[211]

The core of the dispute can be narrowed down to the examination of the judicially created doctrine of nature and purification, as it has been developed in US case law. According to *Funk Bros. Seed Co. v Kalo Inoculant Co.*, manifestations of laws of nature are free to all men and reserved exclusively to none.[212] The bar to patenting laws of nature is flexible, and Judge Sweet initiated his legal analysis by referring to a long line of cases. The Supreme Court has emphasised the need for some transformation requiring a new and different article to emerge with a distinctive name, character or use compared with the naturally-occurring article.[213] The quality of the article cannot be a work of nature.[214] On purification, it was stated in *re Merz* that mere purification of known materials does not result in a patentable product unless the product obtained has properties and characteristics which differ in kind rather than in degree from those of the known product.[215] Judge Sweet also referred to the *American Wood-Paper* case, where it was specifically stated that purification of a natural compound, without more, is insufficient to render a product of nature patentable.[216] The quote from the *American Wood-Paper* decision that clearly establishes the requirement for patentability of purification is mentioned above in section 2.2.3.[217]

Even if the patent applicant was the first to uncover the article and bring it to view, they cannot have a patent because a patent cannot be awarded for a discovery or for a product of nature or for a chemical element.[218] Already in

211. *AMP v USPTO*, US District Court for the Southern District of New York, Judgement of 29 March 2010, p. 95.
212. 333 U.S. 127, 130 (1948).
213. *American Fruit Growers*, 282 U.S. at 12-13, concerning patent claims for fruit whose skin had been treated mould-resistant borax.
214. See *Funk Brothers*, 333 U.S. at 128-31; and *Diamond v Chakrabarty*, 447 U.S. 303.
215. *In re Merz*, 97 F.2d 599, 601 (C.C.P.A. 1935).
216. *The American Wood-Paper Co v The Fibre Disintegrating Co*, 90 U.S. (23 Wall.) 566 (1874), concerning purified pulp derived from wood and vegetable matter.
217. *The American Wood-Paper Co v The Fibre Disintegrating Co*, 90 U.S. (23 Wall.) at 593 (1874); and *AMP v USPTO*, page 115. Also *Cochrane v Badische Anilin & Soda Fabrik*, 111 U.S. 293 (1884) concerning a patent for an artificial version of a natural red dye called alizarine, quoted at page 114, in *AMP v USPTO*, the US District Court for the Southern District of New York, Judgment of March 29, 2010.
218. *General Electric v De Forest Radio*, 28 F.2d 642 (1928) concerning purified tungsten possessing superior characteristics and utility over its brittle, naturally-occurring form.

*ex parte Latimer*, concerning pine needle fibres that were better suited for textile production and subject to a process to remove the needle from its sheath and other resinous materials, they were still products of nature, since 'nature made them so and not the process by which they were taken from the leaf or needle'.[219]

This line of cases clearly supports the argument that purification of human genes, such as the BRCA 1 and 2 genes, is not enough in itself. However, relying on *Parke-Davis* and Judge Hand, Myriad argued that the 'purification of 'naturally occurring' compounds that 'do not exist in nature in pure form' renders such compounds patent-eligible.'[220] Judge Sweet clearly stated that the opinion of Judge Hand failed to support such a conclusion since it concerned a question of novelty and not a question of patentable subject matter. Furthermore, the *Parke-Davis* case is no longer good law in the light of subsequent Supreme Court decisions requiring that a claimed invention should possess 'markedly different characteristics' from products existing in nature in order for it to constitute patentable subject matter.[221]

After examining the Myriad arguments in the light of further cases (*In re Kratz*, 592 F.2d 1169 (C.C.P.A. 1979); *In Bergstrom*, 427 F.2d at 1398; *Bergy*, 596 F.2d at 961; *Merck & Co., Inc. v Olin Mathieson Chem. Corp.*, 253 F.2d 156; and *Am. Fruit Growers*, 283 U.S. at 11), Judge Sweet stated that 'every slight step in purification does not produce a new product. What is gained may be the old product, but with a greater degree of purity'. Judge Sweet stated that the clear line of Supreme Court precedent and accompanying lower court authorities, stretching from *American Wood-Paper* through to *Chakrabarty*, established that purification of a product of nature, without more, cannot transform it into patentable subject matter. Rather, 'the purified product must possess 'markedly different characteristics' in order to satisfy the requirements of § 101'.[222]

219. *Ex Parte Latimer*, Com. Pat. 123 (1889).
220. *AMP v USPTO* pp. 117-118 (quoting *In re Bergstrom*, 427 F.2d 1394, 1401 (C.C.P.A. 1970)). *Parke-Davis & Co. v H.K. Mulford* 189 F.2d 95 (S.D.N.Y. 1911) concerning an adrenalin compound that had been isolated and purified from animal suprarenal glands. The isolated adrenalin, however, possessed the desired therapeutic properties and could be administered to humans.
221. *AMP v USPTO*, the US District Court for the Southern District of New York, Judgment of 29 March 2010, pp. 120-121.
222. *AMP v USPTO*, the US District Court for the Southern District of New York, Judgment of 29 March 2010, p. 125.

*3.1.1.b. Marked differences between isolated and native DNA?*

Only if the isolated BRCA 1 and 2 genes possess 'markedly different characteristics' from the product of nature, can the native BRCA 1 and 2 genes be patentable. In supporting this argument Myriad's central premise was that '[i]solated DNA molecules should be treated no differently than other chemical compounds for patent eligibility' and that the alleged 'difference in the structural and functional properties of isolated DNA' rendered the claimed DNA subject matter patentable.[223] Judge Sweet, however, stated that DNA is more than a chemical compound:

> 'Myriad's focus on the chemical nature of DNA, however, fails to acknowledge the unique characteristics of DNA that differentiate it from other chemical compounds. As Myriad's expert Dr. Joseph Straus observed: 'Genes are of double nature: On the one hand, they are chemical substances or molecules. On the other hand, they are physical carriers of information, i.e., where the actual biological function of this information is coding for proteins. *Thus, inherently genes are multifunctional.'* ... This informational quality is unique among the chemical compounds found in our bodies, and it would be erroneous to view DNA as 'no different' than other chemicals previously the subject of patents.
>
> Myriad's argument that all chemical compounds, such as the adrenaline at issue in Parke-Davis, necessarily convey some information ignores the biological realities of DNA in comparison to other chemical compounds in the body. The information encoded in DNA is not information about its *own* molecular structure incidental to its biological function, as is the case with adrenaline or other chemicals found in the body. *Rather, the information encoded by DNA reflects its primary biological function: directing the synthesis of other molecules in the body – namely, proteins, 'biological molecules of enormous importance' which 'catalyze biochemical reactions' and constitute the 'major structural materials of the animal body.'* ... DNA, and in particular the ordering of its nucleotides, therefore serves as the physical embodiment *of laws of nature* – those that define the construction of the human body. Any 'information' that may be embodied by adrenaline and similar molecules serves no comparable function, and none of the declarations submitted by Myriad support such a conclusion. Consequently, the use of simple analogies comparing DNA with chemical compounds previously the subject of patents cannot replace consideration of the distinctive characteristics of DNA.'[224]

Judge Sweet concluded that, in the light of DNA's unique qualities as a physical embodiment of information, none of the structural and functional differences cited by Myriad between native *BRCA1 and 2* DNA and the isolated *BRCA1 and 2* DNA claimed in the disputed patents rendered the claimed

223. *AMP v USPTO*, the US District Court for the Southern District of New York, Judgment of 29 March 2010, pp. 126-127.
224. *AMP v USPTO*, the US District Court for the Southern District of New York, Judgment of 29 March 2010, pp. 126-128 (emphasis added).

DNA 'markedly different'. The conclusion was based, Sweet stated, on the overriding importance of DNA's nucleotide sequence to both its natural biological function and the utility associated with DNA in its isolated form. Thus the claims were for products of nature and thus unpatentable.[225]

Myriad further argued that the assessment of the patentability of isolated DNA should focus on the differences between native and isolated DNA rather than the similarities:

'[T]he observation that isolated DNA and native DNA share this single property [i.e. the same protein coding sequences] is irrelevant to the critical issue of whether there are *differences* in their properties. It is the *differences* that are legally relevant to the novelty inquiry under Section 101, not the properties held in common.' (Emphasis in original)[226]

Again Judge Sweet emphasised that novelty considerations are irrelevant for § 101 purposes, and that according to the Federal Circuit the claimed invention must be considered as a whole and evaluated for what it is.[227] Quoting *Funk Brothers* and *Chakrabarty* again, Sweet stated that Myriad's approach to the law would make it difficult to fail the test. While the *Funk Brothers* bacterial mixture most certainly differed from any pre-existing bacterial mixture, as a whole it was considered the handiwork of nature. Despite identifiable differences, the fundamental characteristics of the claimed invention as a whole must be sufficiently distinct from natural phenomena to possess the required 'distinctive name, character, and use', as required in *Chakrabarty*.[228]

Myriad argued that that association of chromosomal proteins with native DNA established the existence of '*structural differences*' between native and isolated DNA. Judge Sweet however, stated that this relied on an incorrect comparison between isolated DNA and *chromatin,* which are indeed different insofar as chromatin includes chromosomal proteins normally associated with DNA. The proper comparison is between the claimed isolated DNA and the corresponding native DNA, *and the presence or absence of chromosomal proteins merely constitutes a difference in purity* that cannot serve to establish subject matter patentability ... Myriad also claimed that that native DNA contains intron sequences

225. *AMP v USPTO*, the US District Court for the Southern District of New York, Judgment of 29 March 2010, p. 125.
226. *AMP v USPTO*, the US District Court for the Southern District of New York, Judgment of 29 March 2010, pp. 129-130.
227. See *Diamond v Diehr*, 450 U.S. at 18S; *re Grams*, SSS F.2d 835, 839 (Fed. Cir. 1989); and *AMP v USPTO*, the US District Court for the Southern District of New York, Judgment of 29 March 2010, p. 130.
228. *AMP v USPTO*, the US District Court for the Southern District of New York, Judgment of 29 March 2010, p. 131.

that are absent in the claimed *BRCA1/2* DNA. Judge Sweet stated that for the claimed small DNA fragments, the existence of introns in native *BRCA1* DNA was completely irrelevant to the question of structural differences when comparing these short DNA molecules with native *BRCA1* DNA.[229]

Furthermore, Myriad claimed a difference in functionality. The functional differences between native and isolated DNA demonstrated that they are 'markedly different'. Isolated DNA may be used for diagnostic tests, in biotechnological processes and in medical treatments such as gene therapy. Native DNA would be unsuitable for these uses.[230]

Concerning the purification process Judge Sweet stated:

'While the absence of proteins and other nucleotide sequences is currently required for DNA to be useful for the cited purposes, the purification of native DNA does not alter its essential characteristic – its nucleotide sequence – that is defined by nature and central to both its biological function within the cell and its utility as a research tool in the lab. The requirement that the DNA used be 'isolated' is ultimately a technological limitation to the use of DNA in this fashion, and a time may come when the use of DNA for molecular and diagnostic purposes may not require such purification (emphasis added). The nucleotide sequence, however, is the defining characteristic of the isolated DNA that will always be required to provide the sequence-specific targeting and protein coding ability that allows isolated DNA to be used for the various applications cited by Myriad. For these reasons, the use of isolated DNA for the various purposes cited by Myriad does not establish the existence of differences 'in kind' between native and isolated DNA that would establish

229. *AMP v USPTO*, the US District Court for the Southern District of New York, Judgment of 29 March 2010, pp. 131-132: 'More generally, the fact that the *BRCA1/2* cDNA molecules covered by the composition claims-in-suit contain only the protein coding exons and not the introns found in native DNA does not render these cDNAs and their native counterparts 'markedly different'. The splice variants represented by these cDNAs are the result of the naturally-occurring splicing of pre-mRNA into mature mRNA. Therefore, not only are the coding sequences contained in the claimed DNA identical to those found in native DNA, the particular arrangement of those coding sequences is the result of the natural phenomena of RNA splicing. Finally, at least in the case of *BRCA1,* the claimed cDNA sequences are actually found in the human genome in the form of a naturally occurring pseudogene.' (Quote from p. 129)
230. P. 134 ff.: 'as an Isolated DNA's utility as a *primer or a molecular probe* (for example, for Southern blots) arises from its ability to 'target and interact with other DNA molecules', that is, the ability of a given DNA molecule to bind exclusively to a specific DNA target sequence ... the cited utility of the isolated DNA as a primer or probe is primarily a function of the nucleotide sequence identity between native and isolated *BRCA1/2* DNA. Similarly, the utility of isolated DNA as a *sequencing target* relies on the preservation of native DNA's nucleotide sequence ... The entire premise behind Myriad's genetic testing is that the claimed isolated DNA retains, in all relevant respects, *the identical nucleotide sequence found in native DNA.*'

the subject matter patentability of what is otherwise a product of nature. See Am. Fruit Growers, 283 U.S. at 11.231

Judge Sweet made the surprising decision that the claimed isolated DNA was not markedly different from native DNA as it exists in nature, and constituted unpatentable subject matter under 35 U.S.C. § 101.[232]

*3.1.1.c. Method for comparing gene sequences?*

Judge Sweet also found the claims for 'analysing and comparing' DNA sequences invalid under § 101.

'Claim 1 of the '999 patent is directed to the process of 'analysing' a BRCA sequence (in order to identify whether a naturally-occurring mutation exist). Claim 1 of the '001, '441, and '857 patents as well as claim 2 of the '857 patents are directed to comparing two gene sequences to see if any differences exists. The processes are not further described, specified or limited in the patent application.'[233]

Myriad argued that the process incorporated a transformation step and it therefore satisfied the Bilski 'machine or transformation' test.[234] Myriad based its argument on the Federal Circuit's ruling in the *Prometheus* case[235]

231. P. 137 ff.: Finally, Judge Sweet stated that the isolated *BRCA1 and 2* DNA claimed in Myriad's patents bore comparison with the bacterial mixture in *Funk Brothers*: 'It is no more than the discovery of some of the handiwork of nature and hence is not patentable'; 33 U.S. at 131. 'The identification of the BRCA1 and BRCA2 gene sequences is unquestionably a valuable scientific achievement for which Myriad deserves recognition, but that is not the same as concluding that it is something for which they are entitled to a patent.'
232. *AMP v USPTO*, the US District Court for the Southern District of New York, Judgment of 29 March 2010, p. 139.
233. *AMP v USPTO*, the US District Court for the Southern District of New York, Judgment of 29 March 2010, p. 141.
234. *Bilski*, 545 F.3d at 953 (quoting *Diamond v Diehr*, 450 U.S. at 187). In *Bilski*, the Federal Circuit set out 'the definitive test to determine whether a process claim is tailored narrowly enough to encompass only a particular application of a fundamental principle rather than pre-empt the principle itself', at p. 954. The 'transformation must be central to the purpose of the claimed process', at p. 962; 'adding a data-gathering step to an algorithm is insufficient to convert that algorithm into a patent-eligible process'.
235. *Prometheus Labs. Inc. v Mayo Collaborative Servs.* 581 F.3d 1336, 92 USPQ2d 1075, 1082-83 (Fed. Cir. 2009). The fact that a process relies on natural processes does not disqualify it from being patentable. Nearly every transformation of physical matter can be described as occurring according to natural processes and natural law. The *Prometheus* court reasoned that the process at issue was patentable because the

(discovery of a correlation between metabolite levels in a patient's blood and the therapeutic efficacy of a dose of the drug). The Federal Circuit held that the relevant transformation was the transformation of the human body as well as the chemical and physical changes of the drug's metabolites. The Federal Circuit ruled the methods qualified as patentable subject matter under Section 101.[236] However, this decision was reviewed under the Supreme Court's grant of *certiorari* in *Prometheus* (20 June 2011) and its subsequent decision in March 2012, thus reopening the assessment of patentable subject matter, as mentioned above.[237]

Judge Sweet stated that the disputed claims in the *AMP v USPTO and Myriad* case related to a law of nature[238] and they were therefore invalid pursuant to 35 U.S.C. § 101:

> 'the language of the method claims-in-suit and the plain and ordinary meanings of the terms 'analyzing' or 'comparing' establish that the method claims-in-suit *are directed only to the abstract mental processes of 'comparing' or 'analyzing' gene sequences.* Although Myriad asserts that the challenged method claims are directed to comparing *DNA molecules rather than DNA sequences,* the language of the claims belies such an interpretation ... the *method* actually claimed is 'analyzing a *sequence* of a BRCA1 gene ... *The essence of what is claimed is the identification of a predisposition to breast cancer based on 'analyzing' or 'comparing' BRCA1/2 gene sequences. ... As in Grams, isolation and sequencing of DNA from a human sample, even if incorporated into the method claims-in-*

natural process occurs as the result of a physical act, i.e. administering a drug; *Robert L. Harmon, Cynthia A. Homan & Charles M. McMahon, Patents and the Federal Circuit*, 2010 Supplement, 9th Ed., p. 9, footnote 22a.

236. PP. 1339-40: The correlation between metabolite levels in a patient's blood and the therapeutic efficacy of a drug dose enabled the inventor to optimise therapeutic efficiency while minimising side effects, thus identifying a need to adjust drug dosage upwards or downwards, based on the metabolite levels.
237. The issue raised on appeal was whether the requirements of 35 U.S.C. § 101 are satisfied by a patent claim that covers observed correlations between blood test results and patient health, so that the claim effectively pre-empt all uses of naturally occurring correlations, simply because well-known methods for administering prescription drugs and test blood may involve 'transformations' of body chemistry; see *Mayo Collaborative Services v Prometheus Laboratories, Inc.,* 556 U.S.___(2012), available as slip opinion.
238. Geertrui Van Overwalle, 'Policy levers tailoring Patent Law to Biotechnology. Comparing US and European Approaches', *UC Irvine Law Review*, Vol. 1, No 2, p. 10: in the USA, the debate on patentable subject matter in the realm of biotechnology was recently reopened. Exactly thirty years after the leading *Chakrabarty* decision, the Federal District Court of New York ruled that genes are to be regarded as discoveries and should be excluded from patentability. Is the pendulum swinging back again?

*suit, would represent nothing more than data gathering steps to obtain the DNA sequence information on which to perform the claimed comparison or analysis.'*[239]

### 3.1.2. AMP et al. v USPTO and Myriad Genetics inc. et al. before the Federal Circuit

'In its simplest form, the question in this case is whether an individual can obtain patent rights to a human gene. From a common-sense point of view, most observers would answer, 'of course not'. Patents are for inventions. A human gene is not an invention. The essence of Myriad's argument in this case is to say that it has not patented a human gene, but something quite different – an *isolated* human gene, which differs from a native gene because the process of extracting it results in changes in its molecular structure (although not in its genetic code). We are therefore required to decide whether the process of isolating genetic material from a human DNA molecule makes the isolated genetic material a patentable invention. The court concludes that it does; I conclude that it does not.' (Circuit Judge Bryson, 2010-1406, page 2-3).

The appeal from the United States District Court for the Southern District of New York (Senior Judge Robert W. Sweet) was decided on 29 July 2011.[240] The United State Federal Appeal Court reversed Judge Sweet's ruling. However, the Courts' decision was not unanimous; Circuit Judges Lourie and Moore *found that DNAs are* patentable but Circuit Judge Bryson did not. His opposition is indicated in his quote above on human gene patenting.

With reference to a Supreme Court ruling, Judges Lourie and Moore cautioned that courts should not read into patent laws limitations and conditions which the legislature has not expressed, and that Judge Sweet had in effect created a categorical rule excluding isolated genes from patentability.

'Plaintiffs argue that because the claimed isolated DNAs retain the same nucleotide sequence as native DNAs, they do not have any 'markedly different' characteristics. This approach, however, looks not at whether isolated DNAs are markedly different – have a distinctive characteristic – from naturally occurring DNAs, as the Supreme Court has directed, but at one similarity: the information content contained in isolated and native DNAs' nucleotide sequence. Adopting this approach, the district court disparaged the patent eligibility of isolated DNA molecules because their genetic function is to transmit information. We disagree, as it is the distinctive nature of DNA molecules as isolated compositions of matter that determines their patent eligibility rather than their physiological

239. *AMP v USPTO*, the US District Court for the Southern District of New York, Judgment of 29 March 2010, p. 145 ff.
240. *AMP v USPTO and Myriad Genetics*, Case No 2010-1406 on appeal from the United States District Court for The Southern District of New York, in case No 09-CV-4515, Senior Judge Robert W. Sweet.

use or benefit. Uses of chemical substances may be relevant to the non-obviousness of these substances or to method claims embodying those uses, but the patent eligibility of an isolated DNA is not negated because it has similar informational properties to a different, more complex natural material that embodies it. The claimed isolated DNA molecules are distinct from their natural existence as portions of larger entities, and their informational content is irrelevant to that fact. We recognize that biologists may think of molecules in terms of their uses, but genes are in fact materials having a chemical nature and, as such, are best described in patents by their structures rather than their functions.'[241]

Finally, Judges Lourie and Moore stated that their decision that isolated DNA molecules are patentable comports with the long-standing practice of the USPTO. They pointed out that the Supreme Court has repeatedly stated that changes to long-standing practice should come from Congress, not the courts.

241. *AMP v USPTO and Myriad Genetics*, Case No 2010-1406 on appeal from the United States District Court for The Southern District of New York, in Case No 09-CV-4515, Senior Judge Robert W. Sweet, part II, pp. 44-45. However, *Circuit Judge Moore* concurring-in-part: 'Given the chemical differences highlighted by Judge Lourie's opinion and discussed *supra*, the mere fact that the larger chromosomal polymer includes the same sequence of nucleotides as the smaller isolated DNA is not enough to make it per se a law of nature and remove it from the scope of patentable subject matter. The actual molecules claimed in this case are therefore not squarely analogous to unpatentable minerals, created by nature without the assistance of man. Instead, the claimed isolated DNA molecules, which are truncations (with different ends) of the naturally occurring DNA found as part of the chromosome in nature, are not naturally produced without the intervention of man. *Cf. Chakrabarty*, 447 U.S. at 312-13 ... (page 15) .. As discussed above, I believe many species within this genus – the shorter isolated DNA fragments – are clearly patentable subject matter based on their new structure and corresponding enlarged range of utility. Yet that still leaves species that include most or all of the isolated gene sequence. While I ultimately conclude that these longer isolated sequences, including the isolated gene sequence as a whole, are also patentable subject matter, I do so for a reason different than for the shorter sequences. All of the same structural arguments apply to any length of isolated DNA so, like the shorter strands, an isolated DNA coding for a gene does have a literal chemical difference from the gene as it appears on the chromo-some. Different ends in a 15 nucleotide sequence have greater significance than different ends in a 6000 nucleo-tide sequence. Unlike the shorter strands of isolated DNA, the chemical and structural differences in the isolated gene do not clearly lead to an 'enlargement of the range of ... utility' as compared to nature', p. 18. And Judge Moore after having explained the differences in the length of genes concluded: 'If I were deciding this case on a blank canvas, I might conclude that an isolated DNA sequence that includes most or all of a gene is not patentable subject matter. Despite the literal chemical differences, the isolated full length gene does not clearly have a new utility and appears to simply serve the same ends devised by nature, namely to act as a gene encoding a protein sequence', p. 18.

In this case, they said, the USPTO has issued patents for DNA molecules for almost thirty years, so that if the law is to be changed and DNA inventions are to be excluded from the broad scope of § 101, contrary to the settled expectation of the inventing community, the decision must come from Congress and not from the courts.[242]

Judge Bryson, dissenting, agreed with Judges Moore and Lourie that claims to the *BRCA* cDNA were eligible for patenting as it could not be isolated from nature but had to be created in the laboratory. However, Bryson did not support the argument that the claimed short segments of DNA were patentable (claim 6 of '282 patent covers any sequence of the *BRCA* cDNA that is at least 15 nucleotides long). Bryson stated that the claim encompasses each *BRCA1* exon, even though each exon is naturally defined by transcription. Further, he emphasised that small sequences of DNA are repeated throughout the three billion nucleotides of the human genome, so the claim covered portions of the cDNA of more than 4% of human genes. It also covered portions of the DNA of nearly all human genes. Accordingly, efforts to sequence almost any gene could infringe claim 6, even though Myriad's specification contributed nothing to human understanding of our genes.[243] Judge Bryson did not accept the view that courts should keep their hands off thirty years of administrative practice. On the contrary, he quoted *Chakrabarty* on *Marbury v Madison* where it was explained that 'Congress has performed its constitutional role in defining patentable subject matter in § 101; we perform ours in construing the language Congress has employed'.[244] Judge Bryson did not leave any doubt about the evaluation of the PTO guidelines, and found that the comments that the USPTO issued at the time of its 2001 guidelines were perfunctory and lacking a thorough consideration and study of the issue. To take the opposite view would in effect give the USPTO lawmaking authority that Congress had not accorded it.[245]

242. *AMP v USPTO and Myriad Genetics*, Case No 2010-1406 on appeal from the United States District Court for The Southern District of New York, in Case No 09-CV-4515, Senior Judge Robert W. Sweet, part II, p. 48, and the dissenting Moore, Part II, p. 19.
243. *AMP v USPTO and Myriad Genetics*, Case No 2010-1406 on appeal from The United States District Court for The Southern District of New York, in Case No 09-CV-4515, Senior Judge Bryson, part II, p. 16.
244. *AMP v USPTO and Myriad Genetics*, Case No 2010-1406 on appeal from The United States District Court for The Southern District of New York, in Case No 09-CV-4515, Senior Judge Bryson, part II, p. 18; *Chakrabarty*, 477 U.S. at 315.
245. *AMP v USPTO and Myriad Genetics*, Case No 2010-1406 on appeal from The United States District Court for The Southern District of New York, in Case No 09-CV-4515, Senior Judge Bryson, part II, pp. 17 and 19.

All judges affirmed that Myriad's process claims for a method of 'comparing' or 'analyzing' two gene sequences fell outside the scope of § 101 because they claimed only abstract mental processes, but they reversed the decision on Myriad's claim for screening potential cancer therapeutics via changes in cell growth rates. The claims covering the drug screening method were patentable.[246]

## 4. Conclusion

On gene patents, it may be said that the ship is about to right itself. Two different systems have been used to illustrate that the courts on each side of the Atlantic are engaged in narrowing down the patentability of gene patents (genes as such, full length sequences and so on, in the USA) and the scope of protection in the EU. This illustrates quite well that there is lack of flexibility once rules have been adopted providing a legal basis for genes to be considered patentable subject matter, and while the case law of the Supreme Court may have wandered around uncertainly for thirty years, it has still found it possible to bypass what was thought impossible. The ship may right itself without further notice at a time when no one expected changes to be made by the courts. However, if the legal basis clearly mandates patents, the only room for manoeuvre is to narrow down the scope of protection. The case study in Chapters 6 and 7 illustrates that thirty years of gene patenting debates and disputes is not enough. New technologies are emerging and new dilemmas appear. Technological developments continue to push patentable subject matter and the scope of protection. However, compared with the EU system where there are far too many players, the US system, where the courts are the key players, is well suited for addressing the problems at its root (patentable subject matter). The EU can only address the problem by addressing its symptoms (scope of protection). However, the TRIPS non-discrimination obligation cannot hinder Members from making these kinds of field restrictions.

246. *AMP v USPTO and Myriad Genetics*, Case No 2010-1406 on appeal from The United States District Court for The Southern District of New York, in Case No 09-CV-4515, Senior Judge Robert W. Sweet, part II, p. 54. See further Rebecca S. Eisenberg, 'Wisdom of the ages or dead-hand control? Patentable subject matter for diagnostic methods after *In re Bilski*', 3 *Case W. RES. J.L. Tech. & Internet*, 1, 2012, p. 46 ff. (forthcoming).

CHAPTER 6

# Inherited Diseases

The empirical case study of the thesis is introduced by giving background information in Chapter 6. The empirical study is based on an examination of patent activity relating to the treatment of six diseases, of which four (Huntington's disease, spinal muscular atrophy, Canavan's disease, and cystic fibrosis), which are caused by the mutation of a single gene, are characterised as being single gene diseases, while Alzheimer's disease and Parkinson's disease are characterised by being multifactorial diseases so that several factors, for example a combination of small genetic variations together with environmental factors, can have an influence on whether the disease is expressed. Chapter 6 contains a review of the diseases as well as a review of the state-of-the-art for patent activity for each disease, based on a review of the legal literature.

## 1. Background

This chapter looks at a case study of inherited disorders in a patent law context. The case study focuses on the following diseases: Huntington's disease, spinal muscular atrophy, Canavan's disease, cystic fibrosis, Alzheimer's disease and Parkinson's disease. Before reviewing the patents and patent applications in Chapter 7, this chapter will give a brief introduction to the diseases, their clinical features, the genetics involved and, where possible, the patent history. All of the diseases selected for this case study involve a human disease gene, are thus illustrative for the analysis of the scope of DNA-sequence protection. Some of the diseases can be classified as traditional single-gene-disorders, namely Huntington's disease, spinal muscular atrophy, Canavan's disease, and cystic fibrosis, while Alzheimer's disease and Parkinson's disease are genetically heterogeneous diseases (or multifactorial). While the identification of the mutational basis and the characterisation of the associated protein product are important features in single-gene-disorders, there can be several gene loci (and environmental causes) identified with diseases which exhibit genetic heterogeneity; see Chapter 2.2.1.b.

## 2. Single-gene disorders

### 2.1. Huntington's chorea and juvenile Huntington's – An introduction

Huntington's disease (HD), which was originally called Huntington's chorea, has been described as one of the most feared and unpleasant hereditary diseases in humans. Huntington's disease is inherited as an autosomal dominant disease, meaning that a single defective copy of the gene inherited from one parent is sufficient to cause the disease. It is associated with increases in the length of a CAG triplet repeat (encoding the amino acid glutamine) present in a gene called 'huntingtin'. The HD gene maps to chromosome 4p16.3.[1] The disease is characterised by slow progressive selective cell death in the central nervous system which is associated with a slowly progressive movement disorder (choreic movements),[2] impairment of intellectual function with psychiatric disturbance,[3] and dementia. Huntington's usually appears as an adult-onset disease (average age of onset is around 40 years) or as a juvenile disease (presented before the age of 20).[4] There is no treatment, no cure and the average duration of the disease is 10-15 years.[5] However, a recent breakthrough by NeuroSearch gives hope for a treatment of the motor symptoms of

1. See OMIM, http://www.ncbi.nlm.nih.gov/entrez/dispomim.cgi?id=143100.
2. Chorea is the most common movement abnormality. This takes the form of subtle involuntary movements such as facial grimacing, twitching of the face and limbs, folding the arms and crossing the legs. As the disease progresses the gait becomes very unsteady and speech unclear; see Peter Turnpenny & Sian Ellard, *Emery's Elements of Medical Genetics*, 13th Ed., p. 282; and OMIM, http://www.ncbi.nlm.nih.gov/entrez/dispomim.cgi?id=143100.
3. Intellectual changes in the early stages of HD include memory impairment and poor concentration span. Anxiety and panic attacks, mood changes and depression, aggressive behaviour, paranoia, irrationality, increased libido and alcohol abuse can also occur; see Peter Turnpenny & Sian Ellard, *Emery's Elements of Medical Genetics*, 14th Ed., p. 293.
4. Instead of chorea, juvenile Huntington's disease is characterised by rigidity, with a slowing of voluntary movement and clumsiness and a decline in school performance heralds the onset of severe progressive dementia, often in association with epileptic seizures; see Christian P. Schaaf, Johannes Zschocke & Lorraine Potocki, *Human Genetics: from Molecule to Medicine*, 2012, p. 295.
5. See Peter Turnpenny & Sian Ellard, *Emery's Elements of Medical Genetics*, 14th Ed., p. 293; and OMIM, http://www.ncbi.nlm.nih.gov/entrez/dispomim.cgi?id=143100.

Huntington's disease. Data from the 437 Huntington's patients, who participated in the MermaiHD study (the European phase III study with Huntexil (pridopidine) in Huntington's disease) showed that 26 weeks treatment with Huntexil resulted in improvements in a broad range of voluntary and involuntary motor symptoms associated with the disease.[6] However, in May 2011 it was decided that a new phase III study was needed before submitting a European marketing application for Huntexil.[7]

### *2.1.1. Huntington's disease – a literature case study*

#### *2.1.1.1. State of the art*

The first US patent (828-patent) was issued in 1987. It claimed diagnostic uses of DNA markers to detect the HD gene.[8] Because part of the work leading to the invention was supported with US Government funds, the US Government had certain rights in this invention. The isolation of Huntington's disease gene has been described from an intellectual property right perspective in the 2006 US report *Reaping the benefits of Genomic and Proteomic Research*:

6. Published 4 February 2010: http://www.newstatesman.com/healthcare-and-pharmaceuticals/2010/02/huntexil-study-disease. 'Pharmacology of the dopaminergic stabilizer pridopidine' reports findings from a number of pharmacological studies suggesting that pridopidine stabilises psychomotor activity and may therefore offer clinical relief from psychomotor symptoms arising from dopaminergic dysfunction in conditions such as a HD; see http://www.huntington-assoc.com/neurosearch.htm. Furthermore, the design of Europe's first large scale comprehensive study measuring the socioeconomic burden of Huntington's disease (Euro-HDB) was revealed [iv]. The primary objective of the Euro-HDB study is to assess the cost of HD across six European countries: France, Germany, Italy, Spain, Sweden and the UK. The study aims to recruit 1,000 patients, to collect information on clinical characteristics, health-related quality of life (HR-QoL) and healthcare resource utilisation. The main aim of the study is to demonstrate the true cost of the illness and to identify any relationships between clinical status, patient management and patient outcomes; see http://www.huntington-assoc.com/neurosearch.htm.
7. http://www.reuters.com/article/2011/05/24/neurosearch-huntexil-idUSLDE74N0IN20110524. The decision was made on advice from the US Food and Drug Administration and the European Medicines Agency on the clinical development plans for Huntexil, Neurosearch.
8. US patent # 4.666.828; National Research Council of the National Academies, 2006 (Washington D.C.), 'Reaping the benefits of Genomic and Proteomic Research, Intellectual Property Rights, Innovation and Public Health, p. 66, that treatment of intellectual property related to the HD gene and the genetic test has been largely shaped by patient and family concerns. Massachusetts General Hospital (MGH) has been granted three patents related to the HD gene.

'In 1979, the Hereditary Disease Foundation (HDF) organized a workshop at NIH to discuss using DNA markers to find the HD gene. HDF subsequently funded a grant to David Housman, a molecular biologist from the Massachusetts Institute of Technology, and his graduate student James Gusella[9] to develop and use restriction fragment length polymorphisms (RFLPs) for genetic linkage analyses to locate the HD gene. Gusella continued this research at the Massachusetts General Hospital (MGH), where he began collaborating with Nancy Wexler, then at the National Institute of Neurological Disorders and Stroke. Wexler and colleagues traveled to a region in Venezuela that is home to the largest number of kindreds in the world, more than 18,000 people, in order to collect pedigree information and DNA for these genetic studies. The collaboration was successful; in 1983 Gusella, Housman, and colleagues discovered a RFLP marker [restriction fragment length polymorphisms] tightly linked to the HD gene on the short arm of chromosome 4 (Gusella et al., 1983). This discovery marked the first time a disease gene of unknown chromosomal locale had been localized using anonymous DNA markers'[10]

The '828 patent claimed a method for detecting the presence in a subject of the gene for Huntington's disease which consists of analysing the human chromosome 4 of the subject for a DNA polymorphism linked to Huntington's disease. The field of the invention is described as a detection test, especially a pre-symptomatic test, for Huntington's disease, which uses recombinant DNA techniques in genetic linkage analysis.

'In 1983, scientists who discovered the HD gene marker, as well as investigators who had developed technologies for cloning genes from linked markers, organized the Huntington's Disease Collaborative Research Group (HDCRG), under the auspices of HDF. After 10 years of collaboration by more than 50 scientists, the HD gene was isolated in 1993 (MacDonald et al., 1993). The HD gene encodes a protein called 'huntingtin', the function of which remains unknown. The protein normally contains a stretch of 7 to 34 glutamines in a row. In HD, a DNA expansion causes an increase in repeats of CAG, the nucleotides encoding glutamine. In late middle age, when 40 or more CAGs occur in a row, HD almost invariably appears, usually around 30 to 40 years of age. Juvenile onset results with 60 CAG repeats or more.'[11]

9. Gusella was listed as the sole inventor in patent # 4.666.828. Gusella was one of four inventors from his own laboratory on US Patents #5,686,288 and #5,693,757 issued in 1997 (MacDonald et al., were also the inventors of the related EPO 614977).
10. National Research Council of the National Academies, 2006 (Washington D.C.), 'Reaping the benefits of Genomic and Proteomic Research', Intellectual Property Rights, Innovation and Public Health, pp. 65-66.
11. National Research Council of the National Academies, 2006 (Washington D.C.), 'Reaping the benefits of Genomic and Proteomic Research', Intellectual Property Rights, Innovation and Public Health, pp. 65-66.

Furthermore, two US patents were issued in 1997 and one European Patent in 2002. The first patent (issued in November 1997) mainly claimed an isolated nucleic acid encoding the huntingtin protein;[12] the second patent (issued December 1997) mainly claimed a purified protein comprising an amino acid sequence corresponding to a mammalian huntingtin protein. Both patents describe the field of the invention as the detection and treatment of human diseases.[13] In particular, the invention related to the huntingtin gene (IT15 gene), huntingtin protein encoded by such gene, and the use of this gene and protein in assays:

*for* the detection of a predisposition to develop Huntington's disease,
*for* the diagnosis of Huntington's disease,
*for* the treatment of Huntington's disease, and
*for* monitoring the course of such treatment.

Unfortunately, the US patents image file wrappers were not available. Both patents expired long before their expected expiration dates due to non-payment of the maintenance fees.

The related European patent was granted in December 2002. The patent lapsed in all 17 designated states in 2003 due either to a failure to submit a translation of the description or to pay a fee. The communication from the European examining division during patent prosecution reveals the objections which the applicant had to overcome in order to obtain the patent:

'The application concerns the huntingtin gene (IT15 gene), the protein encoded by this gene and the use of the gene and the protein for detecting predisposition to develop Huntington disease (HD), for the diagnosis of HD, for treating the disease and for monitoring the course of the treatment. In reference to the definition 'huntingtin protein', Art. 84 EPC in combination with Rule 29 EPC establishes that the claims shall define the matter for which protection is sought in terms of technical features. Thus, a peptide, polypeptide, pro-

12. US patent # 5.686.288. Massachusetts General Hospital (MGH) was granted three patents related to the HD gene (#5,693,757 was issued in 1997). The 1997 patents were broadly written and described diagnostic and therapeutic uses of the HD gene; however, MGH did not pursue further patents claiming these methods. When Gusella and MGH filed the first application, their main concern was that the marker should not be used inappropriately. To date, MGH has not exerted its patent rights or licensed the patents to others for financial gain; National Research Council of the National Academies, 2006 (Washington D.C.), 'Reaping the benefits of Genomic and Proteomic Research', Intellectual Property Rights, Innovation and Public Health, p. 67.
13. US Patent #5,686,288.

tein, oligonucleotide, gene, etc. being chemical products, must be clearly and unambiguously characterized by their amino acid and/or nucleotide sequence, i.e. by reference to their specific SEQ.ID No. The characterization of a product only by *the desired function* or by an arbitrary abbreviation *without any real technical meaning* does not fulfil the requirements of Art. 84 EPC, in combination with Rule 29 EPC.' [emphasis added].[14]

The first claim was found unclear in that the huntingtin protein was not defined by technical features. Claim 1 also used the terms *mutant, fragment or variant thereof* without any further limitation. This was considered too vague and the examiner stated that any protein (or nucleic acid) can be considered to be a variant/mutant of another by making a suitable number and type of modifications (by deleting, inserting and/or substituting). As for the definition used in claim 1 *having substantially the same activity as huntingtin protein*, this was stated to be obscure as the term 'substantially' is vague and the activity of huntingtin not defined.

The definition in claim 3 '*essentially purified*' provided no technical teaching to a skilled person about the degree of purity required for the polypeptide for which protection was sought. Claim 4 referred to '*substantially homologous*', which did not sufficiently define the extent of homology required. An acceptable range of homology should be indicated. The examiner found that Claim 5 was unclear in that it referred to a vector according to Claim 4 or 5, but these claims did not define a vector. As for Claim 6 the examiner stated, that it was not clear which features a nucleic acid molecule should possess (*comprising being at least 5, such as at least 15, contiguous nucleotides).*[15]

Examining claims 14 and 15, the examiner stated that parts of claim 14, e.g. on what basis an individual is considered as not having or as not being suspected of having HD, was not clear. The characteristics of a sample from a non-affected individual to be used as reference should have been specified. The applicant was furthermore reminded that the claims comprise the step of

14. See Communication pursuant to Article 96(2) and Rule 51(2) EPC of 14 July 1999, pp. 1-2.
15. Claims 5 and 7 were also objected to. Claim 5 referred to '*a nucleic acid according to Claim 1*' constituting part of the subject matter of the claim. However, Claim 1 referred to a polypeptide, so this formulation made the claim unclear. Moreover, the use of '*and/or*' in the claim did not sufficiently clearly define the matter for which protection was sought. The sentence in Claim 7 '*comprising a transcriptional region ... operably linked to a sequence ... encoding a protein*' was not clear as a transcriptional region could already comprise a sequence encoding a protein, whose nature was not defined in the application.

taking a sample from a subject. This step was considered a diagnostic method and thus not patentable according to Article 52(4) EPC.

In Claims 16 and 21 the term '*functional and (or non-defective)*' was not clear. As the function of the huntingtin protein was not known, the activity of the protein could not be measured and the mere fact that its amino acid sequence might be different from the wild type did not imply that the protein is non-functional. The term '*antagonist*' was vague and unclear. No specific activity having been assigned to the huntingtin protein, the characteristics of an antagonist to this activity could not be defined. Moreover, the term '*a compound that binds to*' was unclear as it was not defined by technical features; it could be a chemical compound, an antibody or another protein etc. In the interests of conciseness, claim 16 would have been better formulated as a dependent claim of claim 21, whose subject matter it limits to the delaying or preventing of a neurodegenerative disorder.

The nucleotide and amino acid sequence of the gene linked to the development of Huntington's disease (HD), antibodies to the protein and uses of the nucleic acid and protein for the diagnosis and/or treatment of HD were not known in the prior art. Before the priority date of the application, only a tentative chromosomal mapping of the gene had been disclosed in the prior art and cosmids had been isolated which map to regions likely to contain the gene.

'Prior art discloses two regions of chromosome 4 which are candidates to contain the gene as well as the most likely localization of the HD gene id restricted to a region of 700 Kb within the internal region. Despite this, no hint can be found in the document as to how the HD gene can be located with certainty within these fragments and/or cosmids and isolated. Claim 20 can also be considered novel as it refers to the host cell of claim 8, however, no meaningful opinion on the inventivity of this claim can be expressed at this time due to its lack of clarity.'[16]

16. See Communication pursuant to Article 96(2), 5 September 2000.

*Box 6.a. EP 614977: Amendments in the claim from application to grant*

| Application (1994) Number of claims 23 | Amended application (1999) | Amended application (2001) | Druckexemplar (2002) Number of claims 19 |
|---|---|---|---|
| *Claim 1.* An isolated, purified or recombinant polypeptide comprising a huntingtin protein *or a mutant, fragment or variant thereof having substantially the same activity as huntingtin protein.* | *Claim 1.* An isolated, purified or recombinant huntingtin polypeptide comprising the amino acid sequence shown in SEQ ID NO:6. | As 1999 | *Claim 1.* An isolated, purified or recombinant huntingtin polypeptide comprising the amino acid sequence shown in SEQ ID NO:6. |

*Box 6.b. US patent family (no image file wrapper available)*

| US 4666828 | US5686288 | US 5693757 |
|---|---|---|
| *Claim 1.* A method for detecting the presence in a subject of the gene for Huntington's disease which comprises: analyzing the human chromosome 4 of said subject for a DNA polymorphism linked to Huntington's disease. | *Claim 1.* An isolated nucleic acid encoding huntingtin protein. *Claim 2.* The nucleic acid of claim 1, wherein said huntingtin protein has the amino acid sequence shown in SEQ ID NO:6. | *Claim 1.* A purified protein comprising an amino acid sequence corresponding to a mammalian huntingtin protein. *Claim 2.* The purified protein according to claim 1 having the amino acid sequence in SEQ NO:6. |

During the prosecution there was a clear demand for a precise claim. However, the patents as granted remained quite broad, comprising polypeptide, a nucleic acid or its complementary strand, vectors and antibodies. It can be questioned whether patents like this will prevent subsequently developed techniques claiming RNAi. This will be discussed in the case study in the next chapter.

In the description of the EP patent it is emphasised that the teaching of the use of huntingtin for diagnostic and treatment purposes is directed at human patients in particular, but that such use is also applicable to any animal that expresses huntingtin and in which an alteration of huntingtin, especially the amplification of the CAG repeat copy number, leads to a defect in the huntingtin gene (structure or function) or huntingtin protein (structure or function or expression).[17] The description also emphasises that the methods referred to in the patent are applicable to any patient suspected of developing or having Huntington's disease, whether manifested at a young age (juvenile Huntington's) or at a more advanced age (Huntington's chorea). Pre-natal diagnosis can also be made.[18]

## 2.2. Spinal muscular atrophy – an introduction

Spinal muscular atrophy (SMA) is a common autosomal recessive neuromuscular disease. The disease is among the most common genetic causes of death in childhood. There are three common forms of spinal muscular atrophy, all of which show autosomal recessive inheritance.[19] In type I SMA (Werdnig-Hoffmann disease) disease manifestation is present at birth or in the first 6 months of life. Children with Type I SMA show normal intellectual activity but their profound muscular weakness affects swallowing and respiratory functions, and leads to death within the first 2 years of life.[20] There is no effective cure or means of delaying its rate of progression.[21] The age of onset for children with Type II SMA is between 6 and 18 months. The main presenting features of Type II SMA are also muscle weakness and hypotonia. Most affected children survive into early adult life.[22] Type III SMA (Kugelberg-Welander disease) is described as a relatively mild form of childhood or juvenile-onset SMA. The age of onset is after 18 months and all patients are able to walk without support. Long-term survival can be jeopardised by re-

17. Druckexemplar, 2002, p. 19.
18. Druckexemplar, 2002, p. 21.
19. Christian P. Schaaf, Johannes Zschocke & Lorraine Potocki, *Human Genetics: from Molecule to Medicine*, 2012, p. 297. According to Emery's Elements of Medical Genetics, 14 th edition, Peter Turnpenny and Sian Ellard, Elsevier, page 306 type I SMA generally show a high degree of intra familial concordance, with affected siblings showing an almost identical clinical course. In type II and III SMA, intrafamilial variations can be quite marked.
20. See Peter Turnpenny & Sian Ellard, *Emery's Elements of Medical Genetics*, 14th Ed., p. 306.
21. See Peter Turnpenny & Sian Ellard, *Emery's Elements of Medical Genetics*, 14th Ed., p. 306.
22. See Peter Turnpenny & Sian Ellard, *Emery's Elements of Medical Genetics*, 14th Ed., p. 306.

current respiratory infections and weakness of the spinal muscles (development of scoliosis).[23]

### 2.2.1. Spinal muscular atrophy – a literature case study

#### 2.2.1.1. State of the art

Spinal muscular atrophy (SMA) involves two genes (*SMN1*, *SMN2*) which only differ in five base pairs (two of them within exons).[24] SMA is caused by homozygous mutations of the disease gene (*SMN1*- survival motor neuron 1, the primary functional gene), while the severity of the disease is primarily determined by the number of *SMN2* copies.[25] The diagnosis and genetics of the disease are as follows:

The gene product, the SMN protein, participates in RNA processing in the neurons. *SMN1* produces full-length mRNA transcripts, while *SMN2* mainly produces mRNA transcripts lacking exon 7 due to a C>T transition in exon 7 of *SMN2*, which disrupts splicing of this exon. The resulting SMN protein lacks the C-terminal 16 amino acid residues, which are replaced by four amino acids encoded by the exon 8 sequence. The corresponding protein is less stable than the full-length SMN protein. Due to the reduced efficiency in exon 7 inclusion, the *SMN2* gene cannot fully compensate for deficiency of the *SMN1* gene. However, some transcripts from the *SMN2* gene undergo correct alternative splicing and thus include exon 7, so that 10 % to 30 % of functional full-length SMN proteins are produced from the *SMN2* gene.[26]

The number of *SMN2* copies varies greatly. While 50 % of individuals in the general population have two copies, others may have zero, one, three, or four. SMA is caused by the homozygous loss of the functional *SMN1* gene. In 95 % of cases, there is a deletion of exon 7 and exon 8 of the gene and point mutations are rare. The number of copies of the *SMN2* gene, determines the severity of the disease. While patients with SMA type I almost always have one or two copies of *SMN2*, individuals with type II or III typically have three or four copies of *SMN2*.[27]

23. See Peter Turnpenny & Sian Ellard, *Emery's Elements of Medical Genetics*, 14th Ed., p. 306.
24. Christian P. Schaaf, Johannes Zschocke & Lorraine Potocki, *Human Genetics: from Molecule to Medicine*, 2012, p. 297; and Michael Sendtner, 'Therapy development in spinal muscular atrophy', *Nature Neuroscience*, Vol. 13, No 7, July 2010, p. 795.
25. See Gabriela E. Oprea, et al., 'Plastin 3 is a Protective Modifier of Autosomal Recessive Spinal Muscular Atrophy', *Science* 320, 524 (25 April 2008).
26. Michael Sendtner, 'Therapy development in spinal muscular atrophy', *Nature Neuroscience*, Vol. 13, No 7, July 2010 p. 795; and Christian P. Schaaf, Johannes Zschocke & Lorraine Potocki, *Human Genetics: from Molecule to Medicine*, 2012, p. 297.
27. Michael Sendtner, 'Therapy development in spinal muscular atrophy', *Nature Neuroscience*, Vol. 13, No 7, July 2010 p. 795; and Christian P. Schaaf, Johannes Zschocke & Lorraine Potocki, *Human Genetics: from Molecule to Medicine*, 2012, p.

This genetic situation implies several directions for therapy development: (i) strategies to increase full-length SMN protein production from the *SMN2* gene, either by stimulating promoter activity or by enhancing exon 7 inclusion, (ii) gene therapy to supply extra copies of a functional *SMN* gene or the *SMN* cDNA in motor neurons, and (iii) strategies focusing on compensating the deficits that are caused by SMN deficiency leading to motor neuron dysfunction and the disease symptoms.[28]

A first step towards a potential treatment for all types of SMA was achieved in December 2011. Isis Pharmaceuticals, Inc. initiated a Phase 1 study of ISIS-SMNRx oligonucleotide drug in patients with spinal muscular atrophy (SMA). The Phase 1 study of ISIS-SMNRx drug in children with SMA between the ages of 2-14 will administer ISIS-SMNRx intrathecally as a single injection directly into the spinal fluid. However, advances in oligonucleotide technologies for correcting *SMN2* pre-mRNA splicing, so that exon 7 is efficiently included, still depend on improved technologies for drug delivery to motor neurons. If this can be achieved it is said that this could boost therapy development, not only for SMA but also for other neurodegenerative disorders (such as Alzheimer's disease).[29]

'The drug is our first drug to intervene in the splicing of RNA to increase the production of a normal protein, SMN' said C. Frank Bennett, Ph.D., Senior Vice President of Research at Isis. Together with Dr. Krainer's lab, we have validated the antisense approach to treating this disease and are now advancing this program into clinical studies'. The patent application (Cold Spring Harbor) is described in Chapter 7, section 3 to treat SMA relies on a simple, powerful antisense method that boosts SMN protein levels by fixing a genetic RNA splicing glitch. Working with Isis, we have successfully redirected splicing to increase functional SMN production. We have thoroughly validated this approach in multiple animal models, observing marked improvement in modifying the disease course in both mild and severe models of SMA,' said Adrian Krainer, Ph.D., Professor of Molecular Genetics at Cold Spring Harbor Laboratory in Long Island, NY. 'We look forward to translating this important discovery into an effective treatment for this serious disease.'[30]

297; Peter Turnpenny & Sian Ellard, *Emery's Elements of Medical Genetics*, 14th Ed., p. 306.

28. Michael Sendtner, 'Therapy development in spinal muscular atrophy', *Nature Neuroscience*, Vol. 13, No 7, July 2010 p. 795.
29. Michael Sendtner, 'Therapy development in spinal muscular atroph', *Nature Neuroscience*, Vol. 13, No 7, July 2010 pp. 795 and 799.
30. http://ir.isispharm.com/phoenix.zhtml?c=222170&p=irol-newsArticle&ID=1640792&highlight, http://www.fsma.org/LatestNews/index.cfm?ID=6573&TYPE=1150 (last visited January 2012).

As described in Chapter 2.3, stem cell therapy has also developed rapidly in the area of SMA. It is believed that human induced pluripotent stem cell (iPSC) technology can progress treatment of the disease.[31] Finally, *Nature Biotechnology's* list of recent patent applications for biomarkers includes an application claiming to predict the level of severity or disease progression of SMA in patients.[32]

### 2.3. Canavan's disease – an introduction

Canavan's disease is a devastating neurological birth disorder in which the brain deteriorates due to a defective inherited gene. The disease is inherited as an autosomal recessive disease caused by mutations in the gene for an enzyme called aspartoacylase.[33] The disease is caused by inheriting two mutated copies of the *ASPA* gene. Thus, both parents must be carriers of the defective gene and the gene must be inherited from both parents before a child becomes affected.[34] Canavan's is found in all ethnic groups, but it is more frequent among certain Semitic populations; Ashkenazi Jews from Eastern Poland, Lithuania and Western Russia, and Saudi Arabians in particular.[35]

31. Michael Sendtner, 'Tailor-made diseased neurons', *Nature*, Vol. 457/15, January 2009, p. 269; and same author 'Therapy development in spinal muscular atrophy', *Nature Neuroscience*, Vol. 13, No 7, July 2010, p. 799.
32. WO2011032109; See *Nature Biotechnology*, Vol. 29, No 5, May 2011.
33. See http://www.ninds.nih.gov/disorders/canavan/canavan.htm.
34. When both parents are found to carry the Canavan gene mutation, with each pregnancy there is a one in four (25 %) chance the child will be affected by Canavan's disease. See http://www.ninds.nih.gov/disorders/canavan/canavan.htm; and Alessandra Colaianni, Subhashini Chandrasekharan & Robert Cook-Deegan, 'Impact of Patents and Licensing Practices on Access to Genetic Testing and Carrier Screening for Tay-Sachs and Canavan Disease', Report of the Secretary's Advisory Committee on Genetics, Health and Society (SACGHS), April 2010, p. H-1. In the same report at page H-13 the authors account for the clinical utility of genetic testing for Canavan's disease. Today, testing is based on two common mutations and DNA testing is the only way to detect carrier status (one mutation in the test accounts for 97-98 % of Ashkenazi Jewish carriers, and the other mutation accounts for approximately 1 % of the Ashkenazi Jewish population and about 50 % of the non-Ashkenazi Jewish population). And see Alessandra Colaianni, Subhashini Chandrasekharan & Robert Cook-Deegan, 'Impact of gene patents and licensing practices on access to genetic testing and carrier screening for Tay-Sachs and Canavan disease', *Genetics in Medicine*, April 2010, Vol. 12, issue 4, pp. S5-S14.
35. See http://www.canavan.org.

The symptoms of Canavan's disease appear in early infancy and progress rapidly, and the clinical features are as follows.[36]

> Normally the ASPA enzyme, aspartoacylase, breaks down N-acetylaspartate acid (NAA) into the building blocks needed for myelin production. Myelin functions as a protective coating around nerve cells in the brain and the spinal cord. In patients with Canavan's Disease, mutations in the ASPA gene disrupt the function of the ASPA enzyme. NAA accumulates to dangerous levels and myelin production is disturbed, impairing the brain's normal function. 'This leaves Canavan children incapable of performing the simplest functions. Even if they live to their full life expectancy – three to ten years – they become blind, paralyzed, prone to seizures...and increasingly lost to the world around them.'[37]

The *ASPA* gene is located on chromosome 17. Canavan's disease can be identified by a simple pre-natal blood test that screens for the missing ASPA enzyme or for mutations in *ASPA* which encodes aspartoacylase.[38] However, the screening process requires the causative mutation in the family to be known in order to test whether the child carries it.[39] There is no cure, nor is

36. See further: http://www.ninds.nih.gov/disorders/canavan/canavan.htm: Symptoms of Canavan's disease which appear in early infancy and progress rapidly may include mental retardation, loss of previously acquired motor skills, feeding difficulties, abnormal muscle tone (floppiness or stiffness), and an abnormally large, poorly controlled head.
37. See http://www.canavan.org.
38. See http://www.ncbi.nlm.nih.gov/entrez/dispomim.cgi?id=271900, 'a point mutation in the *ASPA* gene (E285A; 608034.0001), several mutations in the *ASPA* gene (see, e.g., A305E, 608034.0003 and Y231X, 608034.0005), In 19 non-Jewish patients with Canavan disease, Shaag et al. (1995) identified 9 different mutations: 4 point mutations, 4 deletion mutations, and 1 exon skip. One mutation (A305E) accounted for 39.5 % of the mutated alleles and was pan-European (i.e., identified in patients of Greek, Polish, Danish, French, Spanish, Italian, and British origin) and probably the most ancient mutation. As summarized by Sistermans et al. (2000), 2 mutations account for about 98% of the alleles of Ashkenazi Jewish patients, in which population the disease is highly prevalent: E285A and Y231X. In non-Jewish patients of European origin, the A305E mutation accounts for 50% of alleles. Zeng et al. (2002) studied the ASPA gene in 22 unrelated non-Jewish patients with Canavan disease and identified 24 different mutations, of which 14 were novel. The E24G gene mutation (608034.0010) resulted in substitution of an invariable amino acid (glu) in the first esterase catalytic domain consensus sequence. Patients with the novel D249V mutation (608034.0011) manifested clinically at birth and died early'.
39. See Alessandra Colaianni, Subhashini Chandrasekharan & Robert Cook-Deegan, 'Impact of Patents and Licensing Practices on Access to Genetic Testing and Carrier Screening for Tay-Sachs and Canavan Disease', Report of the Secretary's Advisory Committee on Genetics, Health and Society (SACGHS), April 2010, p. H-12.

there any standard treatment, other than symptomatic and supportive treatment.[40]

### 2.3.1. *Canavan's – a literature case study*

#### 2.3.1.1. *State of the art*

The Reuben Matalon situation has been described in detail by *Colaianni et al.*[41] The Greenberg family – a family with two children diagnosed with Canavan's disease[42] – initially recruited Dr. Matalon to conduct research into the disease, but finally sued the Miami Children's Hospital, where Matalon was affiliated, for the exclusive rights to license the result of the research.[43]

'Matalon filed a patent application on September 29, 1993 ... and was granted two US patents [US 5679635 in 1997, and US 7217547 in 2007 both entitled 'Aspartoacylase gene, protein, and methods of screening for mutations associated with Canavan disease]. The 1997 patent covered the DNA sequence of the gene, mutated sequences associated with Canavan disease, use of the sequence in DNA testing, and test kits for Canavan disease. The 2007 patent claimed mutated versions of the aspartoacylase protein. The patents were assigned to the Miami Children's Hospital Research Institute, Inc.' where Matalon did his

40. See http://www.ninds.nih.gov/disorders/canavan/canavan.htm.
41. See http://www.ninds.nih.gov/disorders/canavan/canavan.htm; and Alessandra Colaianni, Subhashini Chandrasekharan & Robert Cook-Deegan, 'Impact of Patents and Licensing Practices on Access to Genetic Testing and Carrier Screening for Tay-Sachs and Canavan Disease', Report of the Secretary's Advisory Committee on Genetics, Health and Society (SACGHS), April 2010, p. H-4 ff.
42. Who provided tissue for the research into the disease and helped identify other affected families; see Colaianni, Chandrasekharan & Cook-Deegan, 'Impact of Patents and Licensing Practices on Access to Genetic Testing and Carrier Screening for Tay-Sachs and Canavan Disease', Report of the Secretary's Advisory Committee on Genetics, Health, and Society (SACGHS), April 2010, p. H-4 ff.; Donna M. Gitter, 'Ownership of human tissue: A proposal for Federal recognition of human research participants' property rights in their biological material, 61 *Washington and Lee Law Review*, 2005, p. 325 ff.
43. Alessandra Colaianni, Subhashini Chandrasekharan and Robert Cook-Deegan, 'Impact of Patents and Licensing Practices on Access to Genetic Testing and Carrier Screening for Tay-Sachs and Canavan Disease', Report of the Secretary's Advisory Committee on Genetics, Health and Society (SACGHS), April 2010, p. H-6. On pp. H-15 to H-16 it is said that the Canavan gene was licensed with a relative high royalty and with volume restrictions. The market was open for pre-natal screening based on a DNA test, and the gene patent was commercially significant.

research [44] Thus Reuben Matalon, isolated and patented the Canavan gene sequence and developed genetic screening tests for it.[45]

In the *Greenberg v Miami Children's Hospital Research Institute* case, a group of parents of children afflicted with Canavan's disease (including the Greenbergs) and three non-profit organisations[46] sued the Miami Children's Hospital Research Institute.

'This [the lawsuit] was due to their requirement of both an up-front payment and a per-test fee for sublicensure for Canavan disease mutation analysis (*Greenberg and co workers vs. MCH Research Institute*). The lawsuit alleges that the gene discovery was dependent upon genetic information and financial resources provided by the Canavan Foundation and that these charges limit the availability of testing.[47] Moreover, the plaintiffs claim that MCH tried to restrict access and promote a lead test center by setting a limit on the number of tests that can be performed by each licensee. Indeed, although the Canavan disease gene and the associated common Jewish mutations were first reported in 1993, relatively few laboratories have completed their negotiations with MCH, despite the ACMG (*American College of Medical Genetics*) recommendation that all Jewish couples be offered carrier testing. The confidential settlement of this suit, reached in 2003, provides for continued royalty-based genetic testing by certain licensed laboratories and royalty-free research by institutions, doctors and scientists searching for a cure'.[48]

The case was settled on confidential terms, but a press release from the Canavan Foundation, the Greenbergs and the rest of the plaintiffs reveals that clinical testing remained under licence and subject to royalty fees, but the set-

44. Colaianni, Chandrasekharan & Cook-Deegan, 'Impact of Patents and Licensing Practices on Access to Genetic Testing and Carrier Screening for Tay-Sachs and Canavan Disease', Report of the Secretary's Advisory Committee on Genetics, Health and Society (SACGHS), April 2010, p. H-6.
45. See United States District Court, S.D. Florida, Miami Division, Court decision: 264 Federal Supplement, 2d Series 1064; 2003 May 29; and Mary R. Anderlik & Mark A. Rothstein, 'Canavan Decision Favors Researchers Over Families', *Journal of Law, Medicine and Ethics*, Fall 2003, 31, 450.
46. These organisations had developed a confidential database and Canavan's disease registry; see Colaianni, Chandrasekharan & Cook-Deegan, 'Impact of Patents and Licensing Practices on Access to Genetic Testing and Carrier Screening for Tay-Sachs and Canavan Disease', Report of the Secretary's Advisory Committee on Genetics, Health, and Society (SACGHS), April 2010, p. H-4 ff.
47. See further David Magnus, Arthur Caplan & Glenn McGee, *Who own's life?* 2002, p. 104, on the Canavan Disease Screening Consortium of 1999.
48. Jean Amos & Wayne Grody, 'Development and integration of molecular genetic tests into clinical practice: the US experience', *Expert Review of Molecular Diagnostics*, 4 (4), 465-477 (2004), p. 470.

tlement allows licence-free use of the *ASPA* gene in research to cure Canavan's (gene therapy research, genetic testing in pure research, and in mice used to research Canavan's disease).[49]

In the SACGHS report the authors asked: 'What has this got to do with patents?', and they answered that the main lesson of the *Canavan* case is that exclusive property rights can be used unwisely.

'Without the property right, the problems do go away, but so also do any benefits of intellectual property. The Canavan case could easily have been a story similar to cystic fibrosis or Huntington's disease, in which the constituencies that were involved in the discovery were at the table when decisions were made about patenting and licensing. The narrative in those cases is one of scientific success leading to broad availability not only of a genetic test, but also creating new pathways for scientific advance building on the discovery of mutations in a causative gene. Patents were also part of those stories, but patenting did not cause a shift in the CF or Huntington's narrative from heroic scientific discovery to secrecy, betrayal, and greed – the way the Canavan story played out in the public media. The difference was partly about licensing strategy, but more importantly, it was about human and organizational relationship.'[50]

49. Canavan Foundation Press Release, 29 September 2003: see http://canavanfoundation.org/news/09-03_miami.php [accessed September, 2011]. See Colaianni, Chandrasekharan & Cook-Deegan, 'Impact of Patents and Licensing Practices on Access to Genetic Testing and Carrier Screening for Tay-Sachs and Canavan Disease', Report of the Secretary's Advisory Committee on Genetics, Health, and Society (SACGHS), April 2010, p. H-8. See David Magnus, Arthur Caplan & Glenn McGee, *Who own's life?* 2002, pp. 102-106, with a thorough description of the course of events (a first clause in the settlement limiting the freedom to publicly criticise the MCH licensing practice and royalty). The question of ownership of the human body has been discussed before in *Moore v Regents of the University of California*, 249 Cal Rptr 494 (1988), 793 P. 2d. 479 (1990); and Angela A. Stanton, 'Forfeited Consent: Body Parts in Eminent Domains', in Johanna Gibson (ed.), *Patenting Lives, Life patents, Culture and development*, 2008, pp. 96-99. See also Lionel Bently & Brad Sherman, 'The Ethics of patenting: Towards a Transgenic Patent System', *Medical Law Review*, 3, Autumn 1995, p. 290: 'a patient's body from being an object of care, into an information and profit yielding resource and the doctor-patient relationship, from one of altruistic carer, to genetic speculator'.
50. Colaianni, Chandrasekharan & Cook-Deegan, 'Impact of Patents and Licensing Practices on Access to Genetic Testing and Carrier Screening for Tay-Sachs and Canavan Disease', Report of the Secretary's Advisory Committee on Genetics, Health, and Society (SACGHS), April 2010, pp. H-16 and H-17.

## 2.4. Cystic fibrosis – an introduction

Cystic fibrosis (CF) is described as one of the most common autosomal recessive disorders in individuals of western European origin.[51] The disease is an inherited chronic disease that affects the lungs and pancreas. The Cystic Fibrosis Foundation describes the disease as follows:

> 'A defective gene and its protein product cause the body to produce unusually thick, sticky mucus that clogs the lungs and leads to life-threatening lung infections; and obstructs the pancreas and stops natural enzymes from helping the body break down and absorb food.'[52]

The clinical features concerning the lungs have been described as follows:

The disorder is caused by mutations in the Cystic Fibrosis Conductance Regulator gene (*CFTR*), located on chromosome 7.[53] The effect of *CFTR* mutations can be classified as pleiotropy, meaning that they cause disease symptoms in different tissues. It was first known as Cystic fibrosis of the pancreas (disruption of the exocrine function), and then further relating to intestinal glands, biliary tree, bronchial glands and sweat glands, and finally infertility that occurs in males and females.[54] Pulmonary disease, lung infections are the leading cause of morbidity and mortality in CF.[55] Approximately 90 % of CF patients die from obstructive lung disease.[56]

51. The incidence varies from 1 in 2,000 to 1 in 3,000, see Peter Turnpenny & Sian, *Ellard Emery's Elements of Medical Genetics*, 14th Ed., p. 301. The incidence is slightly lower in eastern and southern European populations, and much lower in African Americans (1 in 15,000) and Asian Americans (1 in 31,000).
52. See http://www.cff.org/AboutCF/.
53. See http://www.ncbi.nlm.nih.gov/entrez/dispomim.cgi?id=219700. Gert Matthijs writes that it was due to major international collaboration that it was shown that the gene defect in patients with CF resided on chromosome 7; Gert Matthijs, 'DNA Diagnostics in Practice', in Geertrui van Overwalle (ed.), *Gene Patents and Public Health*, 2007, p. 32. See further Subhashini Chandrasekharan et al., 'Impact of Gene Patents and Licensing Practices on Access to Genetic Testing for Cystic Fibrosis', Report of the Secretary's Advisory Committee on Genetics, Health, and Society (SACGHS), April 2010, p. C1,.
54. http://www.ncbi.nlm.nih.gov/omim/219700
55. http://www.ncbi.nlm.nih.gov/omim/219700; and Gert Matthijs, DNA 'Diagnostics in Practice', in Geertrui van Overwalle (ed.), *Gene Patents and Public Health*, 2007, p. 32.
56. Subhashini Chandrasekharan et al., 'Impact of gene patents and licensing practices on access to genetic testing for cystic fibrosis', *Genetics in Medicine*, Subhashini Chandrasekharan et al, Impact of Gene Patents and Licensing Practices on Access to Genetic Testing for Cystic Fibrosis Vol. 12, No 4, April 2010 supplement, p. S194.

'Chronic lung disease caused by recurrent infection eventually leads to fibrotic changes in the lungs with secondary cardiac failure, a condition known as cor pulmonale. When this complication occurs the only hope for long-term survival rests in a successful heart-lung transplant.'[57]

'Because CF patients cannot adequately clear their airways of the mucus buildup, they wheeze, cough, and suffer from repeated lung infections and other pulmonary pathologies ... The thick, sticky mucus found in CF patients also accumulates in the pancreas, thus preventing digestive enzymes from reaching the small intestine and leading to poor digestion, retarded growth, and persistent diarrhea.'[58]

There is no cure for CF, but there has been some progress in life expectancy. The average life expectance in the 1950s was less than 5 years and today it has increased to 36.9 years. However, CF remains a significant cause of chronic ill-health and death in childhood and early adult life.[59]

The *CFTR* gene was localized in 1987 by Lap-Chee Tsui and colleagues, and once the location of the gene was known it took several more years to explore the huge region of chromosome 7 and identify the gene.[60]

'The nearly 40-years hunt for the CF gene began in the 1950s. Using linkage analysis, researchers studied whether the CF gene was linked to blood groups but were unsuccessful ... A major difficulty in identifying the CF gene was the lack of cytologically detectable chromosome rearrangements or deletions. Such large-scale DNA changes greatly facilitated the positional cloning of some other human disease genes. In the 1980s, new technologies were applied to search for the CF gene. Researchers used restriction fragment length polymorphisms (RFLPs, which reflects sequence differences in DNA sites that can be cleaved by restriction enzymes) for linkage analysis to establish the approximate chromosomal location of genes. In 1985, Tsui et al. ... reported that an uncharacterized RFLP marker, DOCRI-917, was linked to the CF gene in 39 families with CF-affected children. It took 4 years of intensive effort by many laboratories to move from this initial linkage to find the mutated gene ... Finally, in 1989, Drs Tsui and John Riordan and coworkers from

57. See Peter Turnpenny & Sian Ellard, *Emery's Elements of Medical Genetics*, 14th Ed., p. 301.
58. Subhashini Chandrasekharan et al., 'Impact of gene patents and licensing practices on access to genetic testing for cystic fibrosis', *Genetics in Medicine*, Vol. 12, No 4, April 2010 supplement, p. S195.
59. See Peter Turnpenny & Sian Ellard, *Emery's Elements of Medical Genetics*, 14th Ed., p. 301; http://www.cff.org/AboutCF/; and Subhashini Chandrasekharan et al., 'Impact of gene patents and licensing practices on access to genetic testing for cystic fibrosis', *Genetics in Medicine*, Vol. 12, No 4, April 2010 supplement, p. S194.
60. Gert Matthijs, 'DNA Diagnostics in Practice', in Geertrui van Overwalle (ed.), *Gene Patents and Public Health*, 2007, p. 32.

HSC and Dr. Francis Collins and fellow researchers, then at the University of Michigan, identified the gene encoding the CFTR'.[61]

### 2.4.1. *Cystic fibrosis – a literature case study*

#### 2.4.1.1. *State of the art*

In 1989, when the gene was identified, the Hospital for Sick Children of Toronto and the University of Michigan filed a patent application claiming the genetic sequence for the *CFTR* gene and its use in diagnostics.[62] However, the patentees have never enforced their patent against the use of the sequence for clinical diagnostics and thus, the University of Michigan, the Hospital for Sick Children of Toronto and the John Hopkins University allow significant research. The 2010 SACGHS report suggests that the active participation of the Cystic Fibrosis Foundation in discussions about intellectual property and licensing benefitted the patient perspective and may even have influenced decisions about licensing.[63]

In his writing on DNA Diagnostics, *Gert Matthijs* uses CF as an example of an open model.[64] He writes that since its discovery, more than a thousand different mutations of the *CFTR* gene have been described, and that the information has been collated in a public Cystic Fibrosis Mutation database initiated by the Cystic Fibrosis Genetic Analysis Consortium.[65] In most cases it is sufficient to screen for a limited set of mutations in order to identify the

61. Subhashini Chandrasekharan et al., 'Impact of gene patents and licensing practices on access to genetic testing for cystic fibrosis', *Genetics in Medicine*, Vol. 12, No 4, April 2010 supplement, p. S195.
62. Gert Matthijs, 'DNA Diagnostics in Practice', in Geertrui van Overwalle (ed.), *Gene Patents and Public Health*, 2007, p. 33. See also Subhashini Chandrasekharan et al., 'Impact of Gene Patents and Licensing Practices on Access to Genetic Testing for Cystic Fibrosis', Report of the Secretary's Advisory Committee on Genetics, Health, and Society (SACGHS), April 2010, p. C-5.
63. Subhashini Chandrasekharan et al., 'Impact of Gene Patents and Licensing Practices on Access to Genetic Testing for Cystic Fibrosis', Report of the Secretary's Advisory Committee on Genetics, Health, and Society (SACGHS), April 2010, p. C-21.
64. Gert Matthijs, 'DNA Diagnostics in Practice', in Geertrui van Overwalle (ed.), *Gene Patents and Public Health*, 2007, p. 33. See further Subhashini Chandrasekharan et al., 'Impact of Gene Patents and Licensing Practices on Access to Genetic Testing for Cystic Fibrosis', Report of the Secretary's advisory Committee on Genetics, Health, and Society (SACGHS), April 2010, p. C-21; and Rochelle C. Dreyfuss, 'The patentability of (Genetic) Diagnostic in US Law and Policy', in Josef Drexl & Nari Lee (eds.), *Pharmaceutical innovation, competition and patent law – A Trilateral perspective*, September 2010, Law and Economics research paper series, Working Paper No 10-44, SSRN No 1678123, p. 14.
65. See http://www.genet.sickkids.on.ca/cftr/.

causal defect in a patient. Thus, in the Caucasian population, one specific mutation (ΔF508) is found in a large proportion (up to 70 %) of the patients.[66] Free access to the patented gene sequence for diagnostic testing has promoted the availability of the *CFTR* tests at a reasonable cost, both in terms of the possibility for diagnostic laboratories to offer test and in terms of offering licenses to companies that have developed kits for the simplified identification of the most frequent mutations.[67]

'Multiple individuals and institutions applied for patents at the same time and the discovery of the CFTR gene was characterized as a 'race'. However, academic competition more than the prospect of patents incited the intense hunt for the CFTR gene and innovation in techniques for gene mapping and positional cloning of genes, at least among the several academic groups involved. Two primary academic groups (Francis Collins and co-workers, University of Michigan, and John Riordan, Lap-Chee Tsui, and co-workers at the HSC) combined their complementary approaches to advantages and were successful in beating the competition and discovering the CFTR gene in June 1989.'[68]

66. Gert Matthijs, 'DNA Diagnostics in Practice', in Geertrui van Overwalle (ed.), *Gene Patents and Public Health*, 2007, p. 33. Other mutations account for 5 to 0.5 % each, while most of the other mutations are really rare; Subhashini Chandrasekharan et al., 'Impact of gene patents and licensing practices on access to genetic testing for cystic fibrosis', *Genetics in Medicine*, Vol. 12, No 4, April 2010 supplement, p. S194. Chandrasekharan writes that (US) carrier rates vary by ethnicity as follows: Ashkenazi Jews (1:24), Non-Hispanic Caucasians (1:25), Hispanic Americans (1:46), African Americans (1:65) and Asian Americans (1:94); see further Subhashini Chandrasekharan et al., 'Impact of Gene Patents and Licensing Practices on Access to Genetic Testing for Cystic Fibrosis', Report of the Secretary's Advisory Committee on Genetics, Health, and Society (SACGHS), April 2010, p. C-2.
67. See Gert Matthijs & Gert-Jan B. Van Ommen, 'Gene patents: from discovery to invention, A geneticist's view', in Geertrui van Overwalle (ed.), *Gene patents and collaborative licensing Models, Patent pools, Clearinghouses, Open Source Models and Liability Regimes*, p. 319; and see further Subhashini Chandrasekharan et al., 'Impact of Gene Patents and Licensing Practices on Access to Genetic Testing for Cystic Fibrosis', Report of the Secretary's Advisory Committee on Genetics, Health, and Society (SACGHS), April 2010, p. C-21.
68. Subhashini Chandrasekharan et al., 'Impact of gene patents and licensing practices on access to genetic testing for cystic fibrosis', *Genetics in Medicine*, Vol. 12, No 4, April 2010 supplement, p. S202; the authors write at p. S196: 'Dr. Francis Collins and co-workers at the University of Michigan, and Dr. Lap-Chee Tsui, John Riordan, and co-workers at the HSC in Toronto, Canada, jointly determined the nucleotide sequence of the CFTR gene. Dr. Tsui, Dr. Collins and their co-workers were the first to identify the DelF508 mutation and to then link this mutation with symptomatic CF. According to Dr. Francis Collins, all parties including the CF Foundation and the Howard Hughes Medical Institute, which partially funded their research (along with

Professor *Rochelle C. Dreyfuss* also stresses the contrast between the availability of genetic tests that are controlled by a single source provider and genetic tests that are unpatented or licensed non-exclusively. She writes that there are many patents associated with CF, but many of the underlying advances have been made by Lap-Chee Tsui of the Hospital for Sick Children in Toronto and Francis Collins.[69] She points out that there is little concern of there being 'secret' CF mutations which only one provider can interpret, and that if variations appear in only small populations, making them less commercially interesting, research interests may nonetheless lead researchers to make diagnostics available.[70] Finally, *Dreyfuss* refers to Huntington's disease, which is a rare disease, but where non-exclusive licensing practices have made testing possible in more than 50 private and non-profit laboratories.[71]

... NIH) and supported Dr. Collins as a Howard Hughes Medical Institute investigator agreed that it was important to seek patent protection for the CFTR gene and the delF508 mutation because of the implications for diagnosis and potential therapies ... given the possibility of commercial interests in both therapeutic and diagnostic applications, patent applications were eventually filed in the United States, the European Patent Office, Japan, Australia, Ireland, and Canada ... The United State and Trademark office declared a patent interference after receiving a patent application from Genzyme Corporation, with Richard Gregory as the first inventor. The Genzyme application claimed the sequence of the CFTR cDNA and rights to the CFTR-containing vector, which overlapped with claims in the Michigan HSC patent application. Subsequently, Genzyme argued that Tsui et al. failed to provide a written description of the manner and process for their invention ... The interference proceedings went on for 10 years and were resolved in part in Tsui's favor in 2002 ... The Tsui's patents covering both the wild-type CFTR cDNA sequence and delF508 mutant sequences (US 6,984,487) and the CFTR protein sequence (US 6,730,777) were granted. Genzyme was granted patent US 5,876,974 which covers methods for producing the CFTR cDNA. In 2006, Genzyme was granted US 7,118,911, which covers vectors for producing the CFTR cDNA'.

69. Rochelle C. Dreyfuss, 'The patentability of (Genetic) Diagnostic in US Law and Policy', September 2010, SSRN No 1678123, p. 13. On p. 14 she writes: 'Many of the facilities are in academic institutions, which can make testing available to patients who could otherwise not afford to pay for it. In these labs, clinicians can utilize the information acquired to better understand the disease and treatment options. And because these doctors are motivated by academic norms, their findings are disclosed through publications.'
70. Rochelle C. Dreyfuss, 'The patentability of (Genetic) Diagnostic in US Law and Policy', September 2010, SSRN No 1678123, p. 14.
71. Rochelle C. Dreyfuss, 'The patentability of (Genetic) Diagnostic in US Law and Policy', September 2010, SSRN No 1678123, p., 14 referring to the Report of the Secre-

## 3. Poly-genetic disorders

### 3.1. Alzheimer's disease – an introduction

Dementia is characterised as a syndrome that can be caused by several progressive disorders, such as Alzheimer's disease, which is the most common type of dementia, vascular dementia, dementia with Lewy bodies and frontotemporal dementia.[72] It is not possible to draw a clear line between the different types of disorders and a mixture of more than one cause is not unusual.[73]

In the 2010 World Alzheimer Report, it was estimated that in 2010 there were 35.6 million people afflicted with Alzheimer's and other dementias worldwide, and the number is expected to increase to 65.7 million by 2030 and 115.4 million by 2050.[74] Furthermore, is the report highlighted that the largest part (2/3) of all afflicted people live in low and middle income countries and in rapidly developing regions such as Latin America, China, and India.[75] The European Parliament resolution of 19 January 2011 on a European initiative on Alzheimer's disease and other dementias[76] referred to the World Alzheimer Report, where it is estimated that 9.9 million people in Europe (2010) suffer from dementia.[77]

tary's Advisory Committee on Genetics, Health, and Society (SACGHS), April 2010, p. 83.

72. See Alzheimer's Disease International, World Alzheimer Report 2010, The global Economic Impact of Dementia: http://www.alz.co.uk/research/files/WorldAlzheimerReport2010.pdf, p. 8.
73. See Alzheimer's Disease International, World Alzheimer Report 2010, The global Economic Impact of Dementia: http://www.alz.co.uk/research/files/WorldAlzheimerReport2010.pdf, p. 8.
74. Alzheimer's Disease International. World Alzheimer Report 2009; and World Alzheimer Report 2010, p. 8.
75. Alzheimer's Disease International. World Alzheimer Report 2009; and World Alzheimer Report, 2010, p. 8.
76. Available at http://www.europarl.europa.eu/sides/getDoc.do?pubRef=-//EP//TEXT+TA+P7-TA-2011-0016+0+DOC+XML+V0//EN, point C. Europe accounts for over 28 % of the total number of persons suffering from dementia, placing it second only to Asia (with 35 %), while of all the world's regions western Europe has the highest proportion of persons suffering from dementia (19 %) In the USA, over 5 million people are currently estimated to suffer from Alzheimer's disease and the number is expected to increase to 16 million by 2050; see Katie Skeehan, Christopher Heaney & Robert Cook-Deegan, 'Impact of Patents and licensing Practices on Access to genetic testing for Alzheimer's disease', Report of the Secretary's Advisory Committee on Genetics, Health, and Society (SACGHS), April 2010, p. B-1.
77. Alzheimer's Disease International, World Alzheimer Report 2010.

The total estimated worldwide costs of dementia (2010) are USD $604 billion, of which about 70 % occur in Western Europe and North America.[78]

Dementia is characterised as being an irreversible and progressive global impairment of intellect, memory, social skills and control of emotional reactions in the presence of normal consciousness.[79]

Alzheimer's has several forms but two forms are relevant in a genetic context:[80] early-onset dementia (age < 60 years, or pre-senile) and late onset (age > 60 years, or senile).[81] Familial early onset Alzheimer's is caused by an autosomal dominant mutation in one of the three genes: *PSEN1* (located on chromosome 14), *PSEN2* (located on chromosome 1), or *APP* (chromosome 21). A person with mutations in *PSEN1, PSEN2 or APP* will contract the disease if they live long enough, usually developing symptoms before age 60. However, the vast number of people afflicted with Alzheimer's have the late onset form, which is linked to the *APOE* gene (apolipoprotein E). The linkages between Alzheimer and susceptible genes (the apolipoprotein E gene, a polymorphic, with three major isoforms (ε2, ε3, ε4)) have been described as follows:

> 'Although it is clear that the *APOE* ε4 allele is the most significant risk factor for late-onset AD, found in up to 40% of cases, the strongest association is with the age of onset rather than absolute risk of developing AD. The *APOE* ε4 allele is therefore neither necessary nor sufficient for the development of AD, emphasizing the importance of other genetic and environmental etiological factors.'[82]

It is evident that *other genetic risk factors* contribute to late-onset Alzheimer disease. More than 30 putative loci have been reported, but none has been consistently replicated in other studies.[83]

78. Alzheimer's Disease International. World Alzheimer Report 2009; and World Alzheimer Report 2010, p. 4.
79. See Peter Turnpenny & Sian Ellard, *Emery's Elements of Medical Genetics*, 14th Ed., p. 243.
80. Katie Skeehan, Christopher Heaney & Robert Cook-Deegan, 'Impact of Patents and licensing Practices on Access to genetic testing for Alzheimer's disease', Report of the Secretary's Advisory Committee on Genetics, Health, and Society (SACGHS), April 2010, p. B-1.
81. See Peter Turnpenny & Sian Ellard, *Emery's Elements of Medical Genetics*, 14th Ed., p. 243.
82. See Peter Turnpenny & Sian Ellard, *Emery's Elements of Medical Genetics*, 14th Ed., p. 243.
83. See Peter Turnpenny & Sian Ellard, *Emery's Elements of Medical Genetics*, 14th Ed., p. 244. See also Katie Skeehan, Christopher Heaney & Robert Cook-Deegan, 'Impact

The European Parliament resolution points at the increased awareness of dementia.[84] It emphasises that the population of Europe is ageing, with persons aged over 80 constituting the fastest growing cohort in most European countries. In the coming decades dementia is expected to be one of the main challenges to the sustainability of national social and healthcare systems, from informal care to long-term care facilities.[85]

*3.1.1. Alzheimer's – a literature case study*

*3.1.1.1. State of the art*

The neuropathology of Alzheimer's consists of plaques of beta-amyloid protein deposited in the brain and neurofibrillary tangles of another protein called tau inside nerve cells; other abnormalities in the brain of a person with Alzheimer's can include inflammation and oxidative stress.[86]

According to the SACGHS Report's US case study of Alzheimer's late onset, three *APOE* patents were granted to Duke University, NC and then licensed to Athena Diagnostic on an exclusive basis.[87] The claims were directed to methods and testing kits based on the *APOE* genotype and observation of risk; see US 5508167 and US 6027896. As for the early onset, 'the gene hunts for *PSEN1*, *PSEN2*, and *APP* were characterized by competitive

of Patents and licensing Practices on Access to genetic testing for Alzheimer's disease', Report of the Secretary's Advisory Committee on Genetics, Health, and Society (SACGHS), April 2010, p. B-1: 'While the APOE genetic test is used in a relatively small fraction of LOAD [Late onset form] cases, the much larger number of late-onset AD cases means it is more frequently used than the genetic tests for PSEN1, PSEN2, or APP in high-risk families'. On the same page the authors write that these families are quite rare, but a 50 % risk for children of an affected member means these tests can be important for those at risk.

84. European Parliament resolution of 19 January 2011 on a European initiative on Alzheimer's disease and other dementias (2010/2084(INI))
85. http://www.europarl.europa.eu/sides/getDoc.do?pubRef=-//EP//TEXT+TA+P7-TA-2011-0016+0+DOC+XML+V0//EN, point D. See also Katie Skeehan, Christopher Heaney & Robert Cook-Deegan, 'Impact of Patents and licensing Practices on Access to genetic testing for Alzheimer's disease', Report of the Secretary's Advisory Committee on Genetics, Health, and Society (SACGHS), April 2010, p. B-2, writing that Alzheimer's accounts for 50 % to 70 % of all cases of dementia and that nearly 40 % of the population aged 85 and over has Alzheimer's.
86. See p. B-2 of the Report.
87. U.S. 5508167, U.S. 5716828, and U.S. 6027896.

races and nasty controversies, including conflicting claims of scientific priority.'[88] Five patents for the presenilin genes are mentioned in the Report.[89]

'Athena exclusively licensed the gene (sequence) patent for *PSEN2* and the methods patent for *PSEN1*. The Toronto group has another patent that appears to cover the *PSEN1* sequence, called the Alzheimer's Related Membrane Protein in their patent (U.S. 6531586), but this does not appear on Athena's list of exclusively licensed patents. ... Another patent, U.S. 6248555, was assigned to the General Hospital Corporation (Massachusetts General Hospital's holding trust for patents) in Boston. This patent covers a mutant *PSEN1* gene. Athena did not license it. Instead, another pharmaceutical licensing partner originally paid for its prosecution. When the licensing partner's interest terminated, Massachusetts General Hospital (MGH) abandoned the patent, allowing it to enter the public domain.'[90]

The range of the patent landscape has increased dramatically, as shown in the empirical case study conducted for this thesis; see Chapter 7, section 6. In 2008 *Robert Cook-Deegan* carried out a search for patents granted with 'pre-

88. See p. B-6 of the Report: Interview with Dr. Allen Roses, 5 July 2007. In the races for other Alzheimer genes, there were allegations of papers under review being delayed while reviewers geared up to claim contemporary discovery, disputes over agreements to share or not to share data and materials, use of family pedigrees and clinical materials without permission, withholding such permission arbitrarily for competitive purposes, and other shenanigans. These are recounted, in somewhat muted form, in two books: D. Pollen, *Hanah's Heirs: The Quest for the Genetic Origins of Alzheimer's Disease*, 1993; and RE Tanzi & AB Parson, *Decoding Darkness: The Search for the Genetic Causes of Alzheimer's Disease*, 2000. According to Roses, the first APP717 mutation reported in *Nature* in 1991 used two families, the larger of which was provided by Roses's laboratory, but the patenting of the discovery and the submitted publication did not acknowledge this; similarly Roses's laboratory provided approximately one-quarter of the patients for the discovery of the *PSEN1* locus, but was excluded from authorship of the publication.
89. . Four were granted to the Hospital for Sick Children (Toronto) and the University of Toronto: (U.S. 5986054 (proteins of PSEN1); U.S. 6194153 (methods claims for *PSEN1*); U.S. 6117978 (proteins of PSEN2); and U.S. 6485911 (methods of *PSEN2*). U.S. 5840540(the *PSEN2* gene and mutations) was granted only to the Hospital for Sick Children.
90. See also Katie Skeehan, Christopher Heaney & Robert Cook-Deegan, 'Impact of Patents and licensing Practices on Access to genetic testing for Alzheimer's disease', Report of the Secretary's Advisory Committee on Genetics, Health, and Society (SACGHS), April 2010, p. B-7.

senilin or *PSEN1* or *PSEN2* or Alzheimer or 'amyloid precursor'' in the claims. That search found 355 granted US patents.[91]

'Many of these are clearly for research methods, transgenic animal models, and other purposes, and do not bear directly on genetic testing. A few, however, are of clear interest. Perlegen, for example, has a patent application for 'Genetic Basis of Alzheimer's Disease and Diagnosis and Treatment Thereof' that claims a collection of polymorphic sites (US 2006/0228728 A1/WO06083854A2). Its initial claim is for an AD genotype profile, which includes APOE and APP along with many other loci associated with AD risk. Even though the patent application may not be granted, it indicates that multiplex testing for AD is being commercially pursued and is the subject of patent applications.'[92]

As for international patents, the SACGHS Report lists an application by Duke University filed with the World Intellectual Property Organization (WIPO), WO/1994/009155 for: Methods of detecting Alzheimer's disease. However, of the 60 countries listed in the application, only New Zealand, Canada, Germany, and the UK upheld the patent.[93] Finally, three applications were related to early onset Alzheimer's on the presenilin genes and were filed with the WIPO.[94]

91. See Skeehan, Heaney & Cook-Deegan, 'Impact of Patents and licensing Practices on Access to genetic testing for Alzheimer's disease', Report of the Secretary's Advisory Committee on Genetics, Health, and Society (SACGHS), April 2010, footnote 48, p. B-7.
92. See Skeehan, Heaney & Cook-Deegan, 'Impact of Patents and licensing Practices on Access to genetic testing for Alzheimer's disease', Report of the Secretary's advisory Committee on Genetics, Health, and Society (SACGHS), April 2010, p. B-1 ff.
93. See p. B-7 of the Report: 'The US patents claim increased risk assessment in individuals, while the WIPO application claims 'a method of diagnosing or prognosing Alzheimer's disease in a subject, wherein the presence of an apolipoprotein E type 4 (ApoE4) isoform indicates said subject is afflicted with Alzheimer's disease or at risk of developing Alzheimer's disease.' 49 The Canadian Intellectual Property Office (CIPO) issued patent number CA 2142300 in August 2005, 12 years after the application was filed'.
94. See p. B-7 of the Report: 'WO/1996/034099, WO/1997/027296, and WO/1998/001549. The first two were assigned to both the Hospital for Sick Children (HSC) and the University of Toronto, while the last patent was assigned only to the University. Four patents were granted in Canada. The first, CA 2200794, was assigned only to the University but the remaining three – CA 2219214, CA 2244412, CA 2259618 – were assigned to both HSC and the University of Toronto'.

## 3.2. Parkinson's disease – an introduction

Parkinson's disease (PD) is the second most common neurological disorder after Alzheimer's disease. It affects approximately 1 % of the population over the age of 50. Late onset PD begins after the age of 50, while with early onset PD the symptoms start to manifest before the age of 50.[95] It is estimated that 4 to 6 million people world-wide suffer from the disease.[96]

PD belongs to a group of motor system disorders which are the result of the loss of dopamine-producing brain cells in the substantia nigra.[97] Many symptoms occur when nerve cells in the substantia nigra die or become impaired:

> 'Normally, these cells produce a chemical messenger called dopamine, which transmits signals within the brain to produce smooth physical movements. When these dopamine-producing neurons are damaged or die, communication between the brain and muscles weakens. Eventually, the brain becomes unable to control muscle movement. In most cases of Parkinson disease, protein deposits called Lewy bodies appear in dead or dying dopamine-producing neurons.'[98]

The four primary symptoms of PD are: tremor, or trembling of the hands, arms, legs, jaw, and face; rigidity, or stiffness of the limbs and trunk; bradykinesia, or slowness of movement; and postural instability, or impaired balance and coordination. As these symptoms become more pronounced, patients may have difficulty walking, talking, or completing other simple tasks.[99]

The clinical features have been described as follows:

95. See Nicola Simola, Annalisa Pinna & Sandro Fenu, 'Pharmacological Therapy of Parkinson's disease: Current Options and New Avenues', *Recent patents on CNS Drugs Discovery*, 2010, 5, p. 221; see http://ghr.nlm.nih.gov/condition/parkinson-disease
96. See http://www.ncbi.nlm.nih.gov/entrez/dispomim.cgi?id=600116 and http://www.parkinson.org/parkinson-s-disease.aspx.
97. See http://www.ninds.nih.gov/disorders/parkinsons_disease/parkinsons_disease.htm. The disease affects the dopamine neurons in the mesencephalon, leading to a progressive and relentless motor disability and to non-motor symptoms of differing severity; See Nicola Simola, Annalisa Pinna & Sandro Fenu, 'Pharmacological Therapy of Parkinson's disease: Current Options and New Avenues', *Recent patents on CNS Drugs Discovery*, 2010, 5, p. 221.
98. See http://ghr.nlm.nih.gov/condition/parkinson-disease
99. See http://www.ncbi.nlm.nih.gov/entrez/dispomim.cgi?id=600116.

'Early symptoms of PD are subtle and occur gradually. In some people the disease progresses more quickly than in others. As the disease progresses, the shaking, or tremor, which affects the majority of PD patients may begin to interfere with daily activities. Other symptoms may include depression and other emotional changes; difficulty in swallowing, chewing, and speaking; urinary problems or constipation; skin problems; and sleep disruptions. There are currently no blood or laboratory tests that have been proven to help in diagnosing sporadic PD. Therefore the diagnosis is based on medical history and a neurological examination. The disease can be difficult to diagnose accurately. Doctors may sometimes request brain scans or laboratory tests in order to rule out other diseases'.[100]

In most cases the disease occurs in individuals with no family history of the disorder (sporadic).[101] PD is most probably the result of a complex interaction of environmental and genetic factors (idiopathic Parkinson's disease (IPD)).[102] Besides environmental factors, certain drugs may also interact. Thus, PD can have *genetic and/or environmental causes.*

The disease can have more than one genetic cause (genetic heterogeneity). Several gene loci are found to be implicated in autosomal dominant forms of PD.

'Approximately 15 percent of people with Parkinson disease have a family history of this disorder. These familial cases are caused by mutations in the *LRRK2*, *PARK2*, *PARK7*,

100. See http://www.ninds.nih.gov/disorders/parkinsons_disease/parkinsons_disease.htm.
101. See http://ghr.nlm.nih.gov/condition/parkinson-disease
102. See http://emedicine.medscape.com/article/1831191-overview. Previous studies have shown that individuals with Parkinson's disease are more than twice as likely to report being exposed to pesticides as unaffected individuals, see http://www.healthnewstrack.com/health-news-491.html and further: 'The study of related individuals who share environmental and genetic backgrounds that might contribute to Parkinson's disease enables researchers to identify specific differences in exposures between individuals with and without the disease. The research team from Duke University Medical Center (Durham, NC) and the University of Miami Miller School of Medicine Morris K. Udall Parkinson Disease Research Center of Excellence (Miami, FL, USA) recruited 319 patients and over 200 relatives. They used telephone interviews to obtain histories of pesticide exposure, living or working on a farm, and well-water drinking. The authors detected an association between pesticide use and Parkinson's disease. Among these, the strongest was between the disorder and use of herbicides and insecticides, such as organochlorides and organophosphates. No association was found between Parkinson's disease and well-water drinking or living or working on a farm, which are two commonly used proxies for pesticide exposures'. See also Christine Kelin & Michael G Schlossmacher, 'The genetics of Parkinson disease: Implications for neurological care', 136 *Nature Clinical Practice Neurology*, March 2006, Vol. 2, No 3.

*PINK1*, or *SNCA* gene, or by alterations in genes that have not been identified. Mutations in some of these genes may also play a role in cases that appear to be sporadic.'[103]

A form of autosomal recessive juvenile PD (*PARK2*, or *PDJ*) is caused by a mutation in the parkin gene (*PARK2*) on chromosome 6q. Other loci for autosomal recessive early-onset PD include *PARK6* and *PARK7*, both of which map to chromosome 1p. Another form of PD is caused by mutation in the alpha-synuclein gene (*SNCA*).[104]

At present, there is no cure for PD, but a variety of medications provide dramatic relief from the symptoms.[105] At both ends of the spectrum there are rare cases that appear to be due solely to one or other of the causes.

### *3.2.1. Parkinson's – a literature case study*

#### *3.2.1.1. State of the art*

In September 2011, the US Company Curemark LLC announced that it had been issued a US patent for methods of treating and diagnosing Parkinson's disease and related dysautonomic disorders.[106] The invention consists of a biomarker for diagnosing an individual with PD, as well as testing whether an individual may develop PD.[107]

Several patents have been issued in relation to Parkinsonism, and this literature study is not intended to be exhaustive. In 2010 a review was published of therapeutic potential for PD and recent patents; it reveals some activity which will be referred to in this section.[108] The empirical case study in Chapter 7 elaborates on this activity, but due to the immense number of patents and patent applications (both in the field of Alzheimer's disease and Parkinson's disease) this present analysis focuses solely on the identified genes involved in the diseases. Thus, in contrast to the single-gene diseases, in this study the review of Alzheimer's disease and PD only looks at applications and patents concerning a disease gene involved in contracting the dis-

103. http://ghr.nlm.nih.gov/condition/parkinson-disease.
104. http://www.ncbi.nlm.nih.gov/entrez/dispomim.cgi?id=600116.
105. See http://www.ninds.nih.gov/disorders/parkinsons_disease/parkinsons_disease.htm.
106. US 8,012,710.
107. http://checkorphan.org/grid/news/treatment/curemark-issued-patent-for-the-diagnosis-of-parkinson-s-disease (last visited January 2012).
108. See Nicola Simola, Annalisa Pinna & Sandro Fenu, 'Pharmacological Therapy of Parkinson's disease: Current Options and New Avenues', *Recent patents on CNS Drugs Discovery*, 2010, 5, pp. 221-238.

ease. Nevertheless, it is evident from the review article that much activity is directed towards addressing the symptoms rather than the disease as such.

The 2010 review article makes the following categorisation on pharmacological agents available for the treatment of PD:

*Treatment of motor symptoms of PD disease*: drugs acting on Dopamine transmission, L-3,4-Dihydroxyphenylalanine (L-DOPA): L-DOPA is an intermediate in the metabolic pathway generating Dopamine in the mammalian brain. It has a long clinical record as the most efficient antiparkinsonian medication and it is still considered the 'gold standard' in pharmacological treatment of PD. ... he importance of L-DOPA in PD management still stands, and new avenues are continuously being explored to optimize the beneficial effects of L-DOPA on PD motor impairment, limiting at the same time its adverse effects (US6746688 (issued 2004), US7309497 (issued 2007), US6548484 (issued 2003);[109]

*Direct Dopaminergic Agonists*: The use of agents which directly stimulate dopaminiergic receptors without the need for being previously converted into dopamine is a major alternative to L-DOPA in the pharmacotherapy of PD. New strategies are continuously being sought to increase the therapeutic potential of dopaminiergic agonists US20090304794, WO20090237618, EP1471892, US20090247509, US6958396 (issued 2005), US7605162 (issued 2009), WO2004000837, WO2009069828;[110]

*Inhibitors of Monoamine Oxidases*: Monoamine oxidases catalyse the oxidative deamination of both exogenous and endogenous amines, including the neurotransmitters striatal dopamine (DA), norepinephine (NE) and serotonin (5HT). The agents are acting as selective inhibitors of the MAO-B isoenzymes, which are chiefly expressed in the CNS, appear the most suited as potential therapeutics against PD. New methods for treating PD through the use of MAO-B: EP2070526, US20090312436;[111]

*Inhibitors of Catechol-O-methyl Transferases*: Different studies show that Catechol-O-methyl Transferase inhibitors increase the bioavailability of L-DOPA and stabilize its concentration in the plasma, US 20070142399;[112] *Drugs acting on Cholinergic Transmission*: Antagonists of central cholinergic receptors were the first, and for a long time the only

109. See Nicola Simola, Annalisa Pinna & Sandro Fenu, 'Pharmacological Therapy of Parkinson's disease: Current Options and New Avenues', *Recent patents on CNS Drugs Discovery*, 2010, 5, p. 223.
110. See Nicola Simola, Annalisa Pinna & Sandro Fenu, 'Pharmacological Therapy of Parkinson's disease: Current Options and New Avenues', *Recent patents on CNS Drugs Discovery*, 2010, 5, pp. 224-225.
111. See Nicola Simola, Annalisa Pinna & Sandro Fenu, 'Pharmacological Therapy of Parkinson's disease: Current Options and New Avenues', *Recent patents on CNS Drugs Discovery*, 2010, 5, p. 225.
112. See Nicola Simola, Annalisa Pinna & Sandro Fenu, Pharmacological Therapy of Parkinson's disease: Current Options and New Avenues, Recent patents on CNS Drugs Discovery, 2010, 5, p. 228.

pharmaceutical agent available to treat motor deficits accompanying PD. The clinical use of anticholinergics is now limited, being mostly narrowed to the very initial stage of PD.[113]

*Drugs acting on Glutamatergic Transmission*: Functional interaction between glutamate and dopamine receptors exists in the basal ganglia (a group of nuclei responsible for the integration of sensorimotor information and control of voluntary movement), and evidence suggests that the loss of dopamine in PD may lead to glutamatergic hyperactivity, which participates in the manifestation of motor impairment accompanying the disease. On this basis, glutamatergic antagonists have been extensively investigated as potential antiparkinsonian drugs, US20060063810, JP20011316267, WO2009054543, US7619007 (issued 2009), US20060166972, EP1295865, US7531541 (issued 2009), WO2007035823, WO2009010455;[114]

*Adenosine $A_{2a}$ receptor antagonists*: Studies in experimental animals have shown that $A_{2a}$ antagonists counteract several features of PD-like motor disability, either as monotherapy or in combination with dopaminiergic agents. Moreover, when combined to a low dose of L-DOPA, $A_{2a}$ antagonists lead to a more stable motor response than L-DOPA alone, with reduced motor fluctuations and lower dyskinetic potential. Furthermore, $A_{2a}$ antagonists have been found to afford protection against toxin-induced degeneration of SNc dopaminiergic neurons, which suggest that these agents might be able to slow down PD progression US20080242673, US7384949, US7589097;[115]

*Other pharmaceutical agents*: Pharmacological agents: histamine or cannabinoid receptors are considered as therapeutics for the management of PD-associated motor impairment and L-DOPA-induced dyskinesia, US20050245587, issued 2011 (US7935718).[116]

## 4. Designing an empirical case study

The empirical case study in Chapter 7 compares DNA patents and applications from the USPTO and the EPO for the diseases referred to above, and

113. See Nicola Simola, Annalisa Pinna & Sandro Fenu, 'Pharmacological Therapy of Parkinson's disease: Current Options and New Avenues', *Recent patents on CNS Drugs Discovery*, 2010, 5, p. 228.
114. See Nicola Simola, Annalisa Pinna & Sandro Fenu, 'Pharmacological Therapy of Parkinson's disease: Current Options and New Avenues', *Recent patents on CNS Drugs Discovery*, 2010, 5, p. 229.
115. See Nicola Simola, Annalisa Pinna & Sandro Fenu, 'Pharmacological Therapy of Parkinson's disease: Current Options and New Avenues', *Recent patents on CNS Drugs Discovery*, 2010, 5, p. 229-230.
116. See Nicola Simola, Annalisa Pinna & Sandro Fenu, 'Pharmacological Therapy of Parkinson's disease: Current Options and New Avenues', *Recent patents on CNS Drugs Discovery*, 2010, 5, p. 230.

the data for Chapter 8 looks at patents on pig-related inventions for illustration purposes, for example looking at patents for enhancements of animals and genetic markers separated from essentially biological processes, using the pig as an example. The patent search was carried out in collaboration with Patent Information Specialist and Technical Examiner, *Bodil B. Hasling*, of the Danish Patent and Trademark Office in 2004, with additional searches carried out by Bodil B. Hasling in 2005, 2007 and 2009. Finally, having found that the analysis of the four searches carried out showed very little activity on the specific genes related to the diseases, yet another search was carried out in May 2011 in order to remedy that outcome. The search history is as follows:

*SE 2004 05010 of December 2004 and SE 2005 05778 of December 2005*: *BRCA*: breast cancer (human and non-human), Canavan's, cystic fibrosis (human and non-human), Huntington's, *BDNF*, TrkB, (pig). *SE 2007 04727 of December 2007*: Breast cancer, Canavan's, Huntington's, *BDNF*, TrkB, Alzheimer's, *ADDL*, cystic fibrosis (Pig genes, minipig, Chapter 8). And *SE 2009 01735 of September 2009* (Canavan's, Huntington's, Alzheimer's, Parkinson's and more, spinal muscular atrophy (and Gottingen pig, see Chapter 8).[117]

*SE 201101296 of May 2011*. For Alzheimer's, the search was focused on familiar, early onset: Presenilin-1; Presenilin; Amyloid beta A4 protein; *PSEN1*; APP; PSEN2; late onset: APOE; SORL1. For Parkinson's the search was focused on the monogene form without multi factual in this search, autosomal dominant) *PARKIN*; *SNCA*; *UCHL1*; *LRRK2*; *PARK2*; *PARK7*; *PINK1*. For Huntington's it was focused on the huntingtin gene (*IT15* gene), *HTT*. For cystic fibrosis: cystic fibrosis transmembrane conductance regulator; *CFTR*. For spinal muscular atrophy it was focused on survival of motor neuron 2; survival of motor neuron 1; *SMN1*; *SMN2*. And for Canavan's the search was on: aspartoacylase; *ASPA*.[118]

The entire search used the Epodoc and World patents Index (WPI) databases. The 2004/2005 search profile addressed 'Nucleotide'.[119] The 2007 search profile broadened the wording from 'nucleotide' to 'sequence, 'sequid', 'nucleoti+', 'nucleic+', 'DNA', 'RNA' and 'gene'.[120] The 2009 search profile

117. See Appendix 2004 search profile, Appendix 2005 search profile, Appendix 2007 search profile, and Appendix 2009 search profile.
118. See Appendix 2011 search profile.
119. Proximity operator: 6 words space to sequence, See Appendix 2004, and 2005 search profile.
120. For the search profile references to analysis methods (G01N and C12Q1) as well as combinations of the chemical substances for the treatment of diseases (C07C and C07D) were omitted from the search; see Appendix 2007 search profile.

omitted breast cancer and included spinal muscular atrophy.[121] The 2009 search profile was further limited to activities comprising EPO documents as well as USPTO documents. Thus, the search identified and the analysis included patent families from two different jurisdictions, the EPO and the USPTO, in order to compare differences and similarities between the two systems for applications filed in both systems. Where an application has only been filed in one of the systems, it has been analysed from the patent family perspective. The patent family approach has been chosen in order to widen the patent activity perspective. The reasoning behind a similar approach used in another study is also applicable here:

> 'the fate of individual patent applications is often difficult to determine because the granting process is often lengthy, and applications have not always been made public. As patent offices have different propensities to divide submitted inventions into two or more patent applications, our analysis is based on patent families as well as individual patents. A patent family comprises all the patent applications and granted patents resulting from a specific invention.'[122]

The current study of patents related to genetic diseases was initiated on two assumptions: first, that it may be possible to predict a difference in the scope of protection for single-gene diseases and genetically heterogeneous disorders; and second, that the scope of protection or the patentable subject matter differed between US and EPO patent prosecution.[123] When it appeared that more patent applications were abandoned and withdrawn than granted, the focus of the thesis shifted. Thus, instead of highlighting the scope of protection, which was the original idea of the case study, the focus of the analysis shifted to identifying the common causes for rejecting patents.

Every EP application is publicly available in the European Patent Register, where all prosecution documents can be found. Every US patent application was looked up in the USPTO public pair system[124] where all prosecution documents are available in the file wrapper. The reasons or causes most often

121. See Appendix 2009 search profile.
122. Michael M. Hopkins, Surya Mahdi, Pari Patel & Sandy Thomas, 'DNA patenting: the end of an era?', *Nature Biotechnology*, Vol. 25, No 2, February 2007, p. 187.
123. Lori Andrews & Erin Shaughnessy Zuiker, 'Ethical, legal, and social issues in genetic testing for complex diseases, 37 *Valparaiso University Law Review*, 2003, p. 795 ff., 'The intellectual property scheme that is presumed best for disorders involving a single gene may be inappropriate for disorders in which multiple genes may each have a role'.
124. Patent Application Information Retrieval (PAIR).

referred to by the examiners and identified in the first review of the empirical study were as follows:

a) 'Real world use' and fair protection (credible, specific and substantial asserted utility and well-established utility)/Industrial applicable/functional definition/structural, see the utility guidelines in Chapter 3.
b) Obviousness, inventive step, state of the art (prior art identical with the claimed sequence) (first/second medical use type claim).
c) Enablement requirement.
d) No experimental evidence in the application to conclude, that it [e.g. the expression vector] can be used for the treatment of diseases.
e) Elect a single invention: lack of unity (claims for more than one species of the genetic invention).
f) Methods of treatment on the human body (Article 52(4) EPC/ Article 53(c) EPC 2000)[125]
g) Homology – assumption of function based on similarity is not credible (speculative use).
h) Subject matter not sufficiently defined.
i) Lack of support in the description/clarity.
j) Embryonic stem cells, Rule 23(d), Article 53(a) EPC: useful for therapeutic purposes/second medical use[126]
k) Written description requirement.
l) New matter in disclosure/re-issue.
m) Sequence list.
n) Novelty.
o) Double patenting (patentably distinct from each other). Judicially created doctrine grounded in public policy (reflected in the statute)
p) EPC Articles 83 and 84

In order to structure the causes, the categories in the case study were narrowed down to four, covering both US and EPO patent prosecution:

I. Real world use and fair protection,
II. Homology and assumption of function based on similarity,

125. One of the most frequently referred to bars to patentability in the EPO has been EPC Article 52(4) (Article 53(c), EPC 2000) concerning methods of treatment of the human or animal body.
126. Most examiners awaited the outcome of the Boards of Appeal decision in *WARF* 2/09 on embryonic stem cell patenting.

III. Obviousness (inventive step, state of the art, prior art, identity with the claimed sequence), and
IV. Written description requirement.

Real world use includes well-established utility or industrial application, and a functional and structural definition. Furthermore, in this context it includes lack of experimental evidence to conclude that the invention (e.g. an expression vector) can be used for the treatment of diseases.

For the European patent prosecution two further bars to patentability can be added, namely:

I. EPC (2000) Article 53(c), and
II. Rule 23 (d), EPC 1973 (Rule 28(c), EPC 2000))

The common causes that came out of the first review of the data were then divided into the groups shown below in Box 6.1.

*Box 6.1. Common causes*

I. Real world use and fair protection

II. Homology and assumption of function based on similarity

Huntington's, spinal muscular atrophy, Canavan's, cystic fibrosis, Parkinson's and Alzheimer

III. Obvious

IV. Written description requirement

With these common causes, the structure of the analysis was then further narrowed down. The study of patents and applications in the areas of the selected genetic diseases starts with an issue that concerns a broader perspective than the relevant disease gene. The study revealed that most patent applications relating to a particular disease seldom encompass applications relating to the specific gene causing the disorder or mutations of it; for example, a search for *cystic fibrosis* seldom encompasses applications relating to the *CFTR* gene or its mutations. The patent applications relate to genes, proteins and sequences other than the gene directly connected with diagnosing, treating or curing the disorder. Instead, the subject of an application was a treatment of a condition or complication caused by the disorder – what can be termed *the pathological process*. Given this outcome, the new search in 2011 was narrowed down to the disease genes as mentioned above.

One trend appeared early in 2008 when the author initiated this analysis. The claims in the applications often referred to a long list of diseases. Thus in a huge number of patents and patent applications this was a recurrent theme for all six diseases examined in the case study. When an invention is concerned with the pathological process of a genetic disease, it is seldom limited to only one disease. Without exception, either in the claims, in the description or in the specification, one can read that patients afflicted with *other* disorders can benefit from the methods and systems of the invention. Sometimes this applies to related diseases, e.g. cystic fibrosis and other respiratory disorders, and sometimes it applies to diseases with very little resemblance to the core disease, e.g. cystic fibrosis and cancer. This phenomenon is categorised as *a laundry list*. Extending the case study to examine the use of laundry lists in patent practice quickly became an objective of this thesis, with the aim of identifying the reason for this practice.

This phenomenon leads to a further delimitation of the analysis. By merging common causes I and II, one very important aspect of the study analysis turned out to be a reflection of a 'problem child' of patent law. In his article, *Christopher M. Holman* points out that claims relating to general molecular biology methods are a concern. In his opinion it would be better to address the wider problem of patents that make broad claims for any practical application of fundamental biological discoveries.[127] In contrast to Holman, who argues that gene patents make up only a small subset of this problematic class of patents, the present study has shown that applications claiming genes *for* fundamental biological processes (such as regulatory genes, housekeeping proteins, motor proteins, structural proteins, transport proteins, signal proteins, transcriptions factors (large class of proteins); see Chapter 2, section 2.1.d.) are a concern with a relatively big impact on the scope of protection, for example if a laundry list turns out to be relevant to future uses of the invention.[128]

127. Christopher M. Holman, 'The Impact of Human Gene Patents on Innovation an Access: A Survey of Human Gene Patent Litigation', Winter 2007, 76 *UMKC Law Review* 295, p. 30. See further Chapter 5, section 1.

128. See e.g. the 2011 case in which a US Federal Appeal Court overturned a lower-court ruling where the North Chicago-based drug giant Abbott Laboratories used Johnson & Johnson's technology to make a blockbuster rheumatoid arthritis therapy (*Humi-*

The following example below, for a patent application for *POSH interacting proteins and related methods*, is an exemplary illustration of this trend/phenomenon.[129]

The application provides novel complexes of POSH polypeptides and POSH-associated proteins. It also provides methods and compositions for treating POSH-associated diseases such as viral disorders, cancer, and neurological disorders. In the background description of the international application (WO2004078130) one can read that potential drug target validation involves determining whether a DNA, RNA or protein molecule is implicated in *a disease process* and is therefore a suitable target for development of new therapeutic drugs. Drug discovery, the process by which bioactive compounds are identified and characterized, is a critical step in the development of new treatments for human diseases. The landscape of drug discovery has changed dramatically due to the genomics revolution. DNA and protein sequences are yielding a host of new drug targets and an enormous amount of associated information. The identification of genes and proteins involved in various disease states or key biological processes, such as inflammation and immune response, is a vital part of the drug design process. Many diseases and disorders could be treated or prevented by decreasing the expression of one or more genes involved in the molecular etiology of the condition if the appropriate molecular target could be identified and appropriate antagonists developed.

In the description one example mentioned is *cancer*, in which cellular oncogenes become activated and result in the unchecked progression of cell cycle processes, could be treated by antagonizing appropriate cell cycle control genes. Another example is *Huntington's disease* and certain prion conditions, which are influenced by both genetic and epigenetic factors, result from the inappropriate activity of a polypeptide as opposed to the complete loss of its function. Accordingly, antagonizing the aberrant function of such mutant genes would provide a means of treatment. Other areas of interest mentioned in the background description include the identification of genes and proteins involved in cancer, apoptosis and neural disorders (particularly those associated with apoptotic neurons, such

*ra*). 'In a 2009 verdict, the largest patent-related damages award in U.S. history, was reversed by the U.S. Court of Appeals for the Federal Circuit in Washington ... The three-judge panel ruled the patent, co-issued to New York University and J&J's Centocor unit, is invalid. J&J, which sells the competing arthritis medicine Remicade, had convinced a jury that Abbott infringed the patent, which J&J said covered the human antibodies used in Humira. The appeals court ruled that J&J's patent application never described fully human antibodies and said at most the claims '*constitute a wish list of properties*' [emphasis added] that such an antibody would have. 'The opinion confirms that you can't claim what you didn't invent,' said William Lee of WilmerHale in Boston, who represented Abbott in the case'; See http://articles.chicagotribune.com/2011-02-23/business/ct-biz-0224-abbott-humira-20110224_1_humira-rheumatoid-arthritis-drug-knoll-pharmaceuticals-abbott-laboratories and http://www.businessweek.com/news/2011-02-23/abbott-wins-reversal-of-j-j-s-1-67-billion-patent-victory.html (last visited January 2012).

129. EP1646644 and WO2004089302, both withdrawn.

as *Alzheimer's disease*). Identifying proteins involved in one or more of these processes for use in, among other things, drug screening methods are certainly of major importance. *Additionally, once a protein involved in one or more processes of interest has been identified, it is possible to identify proteins that associate, directly or indirectly, with the initially identified protein.* Knowledge of *interactors* will provide insight into protein assemblages and pathways that participate in disease processes, and in many cases an *interacting protein* will have desirable properties for the targeting of therapeutics. In some cases, an interacting protein will already be known as a drug target, but *in a different biological context.* Thus, by identifying a suite of proteins that *interact with an initially identified protein*, it is possible to identify novel drug targets and new uses for previously known therapeutics.[130]

In the summary, it is explained that the application provides *methods of identifying an agent that inhibits trafficking of a protein through the secretory pathway*: By identifying a test agent that disrupts a complex comprising a POSH polypeptide and a POSH-AP and evaluating the effect of the test agent on the trafficking of a protein through the secretory pathway wherein an agent that disrupts localization of said POSH-AP is an agent that inhibits trafficking of a protein through the secretory pathway. The protein is specified as a myristoylated protein in certain embodiments, a viral in others and in alternative embodiments the amyloid beta precursor protein *associated* with neurological diseases. The latter embodiment is further elaborated to provide methods of identifying an agent that *inhibits* the progression of a neurological disorder, thus again identifying a test agent that disrupts a complex as identified above. POSH-AP in some aspects is HERPUD1. In other aspects the application provides methods of *treating* a neurological disorder comprising administering an agent to a subject in need thereof, wherein said agent either inhibits the Ubiquitin ligase activity of POSH or of POSH-AP.

The laundry list perspective then pops up in the claims.[131] The neurological disorders involved include *Alzheimer's disease, Parkinson's disease, Huntington's disease* (and more, but these are not included in the present case study: e.g. *schizophrenia, Niemann-Pick's disease*). POSH polypeptides modulate presenilin-mediated amyloid β generation and amyloid β being one hallmark of Alzheimer's disease. POSH polypeptides may be involved in the pathogenesis of Alzheimer's disease.[132] Furthermore, novel associations between certain disease states are described, such as POSH-AP interacting or binding directly with *SMN1* and *SMN2* (*spinal muscular atrophy (SMA)*), so that POSH polypeptides are involved in viral replication. The interaction with *SMN2* is said to mean that certain POSH polypeptides may influence the severity of SMA.

130. See WO2004089302, p. 1, line 24 ff. and pp. 2-3.
131. Claim 9. Use of an agent capable of inhibiting the ubiquitination of a POSH-AP, preferably HERPUD1, wherein the agent is selected from an siRNA construct or a small molecule, for the manufacture of a medicament for treating a neurological disorder selected from among: Alzheimer's disease, Parkinson's disease, Huntington's disease, schizophrenia, Niemann-Pick's disease, and prion-associated diseases.
132. The protein has been shown to interact with presenilin proteins and to increase the level of amyloid-beta protein following its over-expression. Alternative splicing of this gene produces multiple transcript variants, some encoding different isoforms. The full-length nature of all transcript variants has not been determined. (http://www.genecards.org/cgi-bin/carddisp.pl?gene=HERPUD1&search=herpud1.

The application was withdrawn. The 2007 supplemental search report stated that: 'The present claims 9-10 encompass compounds defined only by their *desired function*, namely the capability of inhibiting the ubiquitination of a 'POSH-AP', preferably HERPUD1 and by the very general structure of being a siRNA contrary to the requirements of clarity of Article 84 EPC, because the result-to-be-achieved type of definition does not allow the scope of the claim to be ascertained. The fact that any compound could be screened does not overcome this objection, as the skilled person would not have knowledge beforehand as to whether it would fall within the scope claimed, except for the compounds disclosed in the description ... Undue experimentation would be required to screen compounds randomly. This non-compliance with the substantive provisions is to such an extent, that a meaningful search of the whole claimed subject-matter of the claim/s could not be carried out (Rule 45 EPC)'. Thus, the communication from the European examination division concluded that the application (as filed) lacked unity (see Article 82 EPC) and asked for selection.[133]

The search design can be further narrowed down to a single heading, namely real world use and fair protection, thus including the laundry list of I and II as identified in the first design; see Box 6.1 above. In order to outline a narrower case study, it was decided to focus on a real world use from a laundry list perspective. This perspective may include some of the causes mentioned above in indents a) to i), especially indent d). Lack of unity has been shown to be a major barrier to most patent applications. Furthermore, as the analysis progressed it became evident that III and IV did not provide any guidance for classifying the patent applications, or help to make order out of the chaos of the number of patents and patent applications the search resulted in. Then, in order to structure the data so as to enable a comparable analysis to be made between the six disorders, the analysis was re-divided into four sub-parts focusing on *the disease gene* (which is mainly based on the 2011 search profile), *the pathological process* (which is mainly based on the 2004, 2005, 2007, and 2009 search profiles), and the *stem cells* and *highlights* (which are based on all search profiles). The final design for analysis can be illustrated as follows:

133. See also Jordan Paradise, Lori Andrews & Timothy Holbrook, 'Patents on Human Genes: An Analysis of Scope and Claims', *Science*, 11 March 2005, Vol. 307, p. 1566, referring to one patent where the inventor showed how a polymorphism could be used to predict asthma. The inventor additionally claimed various uses of the polymorphism to predict other conditions, without showing how the polymorphism was linked to those conditions.

*Box 6.2. Framework of analysis*

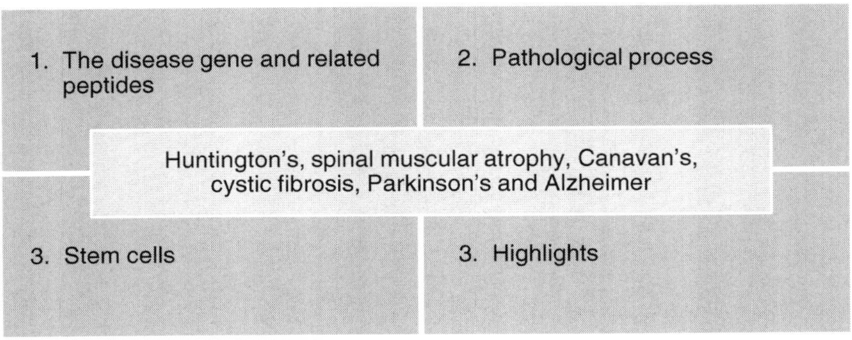


Returning to the starting point, that a laundry list approach and the patenting of biological processes may turn out to become a vulnerable point of patent laws, the decision of the European Patent Office Technical Board of Appeal in Case T-18/09 *Eli Lily*[134] adds to the analysis of the practical implication of a laundry list perspective. Elaborating on *Christopher M. Holmans* concerns about the patenting of general molecular biology methods and the *Eli Lilly* case below, it will be questioned whether laundry lists, super families (genes & proteins) [135] and protein domains fit into the current patent law regime. Thus, the frame of analysis illustrated in Box 6.2 is addressed in Chapter 7 from a laundry list perspective.

134. See Case T0018/9 *Eli Lilly and Company v Human Genome Sciences Inc.*; decision of the Technical Board of Appeal of 21 October 2009, patent EP 0939804.
135. See Point 22 in *HGS v Eli Lilly*, UKSC, 2011, 51; 'A family or superfamily of proteins is a group of proteins, all of which enjoy a significant degree of homology, i.e. they all have certain specified structural characteristics. Although the distinction is not always observed, members of a particular family will normally have close structural similarity and similar functions, whereas members of a particular superfamily, while retaining related structural characteristics, will often be more distantly related and will include members which have similar functions but also may include members with different functions. However, even that is an over-simplification, as, in some cases, proteins will have pleiotropic functions, 'that is to say a multitude of different effects on different cell types, driving multiple biological processes' – per Kitchin J at [2008] RPC 29, para 71. Accordingly, there will be cases where members of a family or superfamily have some functions which are common to all (or a majority) of the members, and other characteristics which are unique to one member (or a few members).' See further point 147.

## 5. Case T 0018/09 Eli Lilly and Company v Human Genome Sciences Inc. and laundry lists

Part III of this thesis focuses on the animal kingdom and discusses the provision in European patent law that essentially biological processes for the production of plants or animals are excluded from patentability. The assessment of the concept of 'an animal variety' and of 'essentially biological processes' concerning animals has been elaborated by a reference to the case law on plants, most recently in the joined cases tomato/broccoli G 1/07 and G 1/08.[136] As the patenting of plants has inspired the patent law approach to animals, one may question whether the principles and law behind the patenting of animals may be carried over to humans, for example the judicially created balancing test derived from the *Harvard Oncomouse* case, as discussed in Chapter 4, section 4.[137]

The dispute in Case T 0018/09 *Eli Lilly and Company v Human Genome Sciences Inc.* falls precisely within the scope of the theme of this present case study.

The case involved parallel proceedings in the UK and the EPO system.[138] The UK court of first instance revoked the patent in July 2008 on the grounds that the claims were not sus-

136. The Enlarged Boards of Appeal (review of the decisions in Case T 83/05 (OJ EPO 2007, 644) and Case T 1242/06 (OJ EPO 2008, 523). The case concerns the scope for exclusion from patentability of essentially biological processes for the production of plants pursuant to Article 53(b) EPC, see Chapter 10, section 2.1.
137. Case T-19/90 *Harvard Oncomouse* and the moral balance where the subject of balancing suffering against the benefit for humans was the oncomouse, a transgenetic non-human animal with a predisposition to develop cancer useful in cancer research. Could this balancing moral test be applied if the subject of balancing were a human embryo destroyed for a medical benefit for humans? – a balancing moral test, weighing the interests of a potential benefit for living human beings against the rights of embryos.
138. Concerning a European patent (UK) 0,939,804 filed on October, 25, 1996 and granted on August 17, 2005, revoked in the UK system in *Eli Lilly and Co v Human Genome Sciences, Inc.*, [2008] EWCA 1903 (Pat) at first instance and by Court of Appeal *Eli Lilly and Co v Human Genome Sciences, Inc.*, [2010] EWCA Civ 33, [2010] R.P.C. 14. As for the EPO system, opposition was filed by Eli Lilly and others, the Opposition Division revoked the patent in June 2008, and on 21 October 2009, in Case T 0018/09 of the Technical Board of Appeal, the Board ruled that the patent was valid. The UK Supreme Court in *Eli Lilly and Company v Human Genome Sciences Inc.*, 2011, UKSC, 51, held that the patent was validly granted, see Brian Whitehead, Stuart Jackson & Richard Kempner, 'Eli Lilly *v* Human Genome Sciences: Has the pro-patentee pendum swung too far?' Vol. 12 Issue 3 *BSLR*: case Comment, p. 100 ff.

ceptible of industrial application, and they were insufficient and obvious because of a lack of technical contribution. That decision was appealed to the UK Courts of Appeal which invalidated the patent on the grounds of lack of industrial application in February 2010. In the EPO system the appeal was conducted on the basis of the claims and was allowed by the EPO in Case T 0018/09 (the Technical Board of Appeal decision of October 2009).[139] Thus the UK Courts of Appeal and the Technical Board of Appeal came to opposite conclusions. However, in November 2011, the UK Supreme Court overruled the Court of Appeal and upheld the validity of the patent.

Reading the applicant's argument it is striking to see how processes that may have a basic value as an essential biological process that is fundamental for being a human can be patented, whereas essentially biological processes for the production of animals cannot. Admittedly, the human process in question is not for the production of a human being as in the case of animals (it is not a question of breeding people by selection and crossing).[140] The fundamental biological process is essentially the mechanism for the basic functioning of the human body, which Directive 98/44/EC Article 5(1) seemingly prohibits:

1. The human body, at the various stages of its formation and development, and the simple discovery of one of its elements, including the sequence or partial sequence of a gene, cannot constitute patentable inventions.

2. An element isolated from the human body or otherwise produced by means of a technical process, including the sequence or partial sequence of a gene, may constitute a patentable invention, even if the structure of that element is identical to that of a natural element.

The distinction between Article 5(1) and (2) of Directive 98/44/EC (Article 5(2) allowing the patenting of what is mentioned in Article 5(1), if isolated) has been discussed in Chapter 5. In Chapter 2, section 3, a description has been given of how, according to Case C-34/10 *Oliver Brüstle v Greenpeace*, human dignity can justify a ban on processes that are capable of commencing the process of development of a human being, as provided in Article 6(2)(c) of the Directive. However, the patenting of processes that are essential for the functioning of the human body are not contrary to human dignity. The concern about patenting biological processes is perhaps relevant with regard to laundry lists. The closer and more broadly that a patent claim is drafted to es-

139. Court of Appeal; *Eli Lilly and Co v Human Genome Sciences, Inc.*, [2010] EWCA Civ 33, [2010] R.P.C. 14, pp. 429 H4-H5.
140. Tine Sommer, 'Patenting the Animal Kingdom? From Cross-Breeding to Genetic Make-up and Biomedical Research', 2/2008 *IIC*, Vol. 39, p. 149. See Chapter 10.

sentially biological processes, or to gene families' involvement in the regulation of human immune cells, or similar 'block functions', the more diseases will be covered by the invention.

In a 2010 article on gene patenting, *Robert Fitt and Edward Nodder* refer to the *Eli Lilly and Company v Human Genome Sciences Inc.* case, writing that the crucial issue in that case was the evidence required to determine an industrial application when the identification itself was based on a bioinformatics technique.[141] They wrote that:

> 'Human Genome Sciences, Inc. (HGS) was one of the companies using bioinformatics techniques during the 1990s to identify new proteins of interest. One of HGS' research projects focused on identifying new members of the tumor necrosis factor (TNF) ligand superfamily. The founding member of the TNF ligand superfamily was TNF-$\alpha$ which was isolated in 1975. TNF-$\alpha$ is a cytokine, a protein that acts as an intercellular mediator in inflammation and immune responses. It was originally shown to provoke the death of tumours, hence the name 'tumor necrosis factor,' but subsequently, it was found to play a key role in inflammation and, in particular, rheumatoid arthritis. By 1996, at least a further eight members of the TNF ligand superfamily had been identified, and they had been shown to display some overlapping functions but also many individual ones in relation to a wide variety of cell types. However, they were all involved, to some degree, in the immune system and immune responses.'[142]

141. Robert Fitt & Edward Nodder, 'An Uncertain Future for Gene Patents: The view from Europe', 29, *Biotechnology Law Report*, No 6, 2010, p. 616: 'One of the direct results of the Human Genome Project was the generation of vast public and private databases of DNA and protein sequence data. These databases could be 'mined' using computer algorithms to trawl for fragments of sequence that displayed similarity (or homology) to previously characterized genes and proteins. Statistical analysis of the results obtained by homology screening enabled researchers to identify the sequences most likely to be new genes or proteins. An important distinction between the bioinformatics approach and the traditional 'wet lab' approach was that, although the activity of the gene identified using bioinformatics could be predicted on the basis of the activities of the members of the family of genes it displayed homology with, it was not possible to prove the actual activity of the gene until after it had been cloned and the protein it encoded had been assayed. The crucial issue in the present case was how much needed to be said about the proposed application, and how reliable that proposal had to be, when using bioinformatics techniques to identify a new gene encoding a protein in an established family.' See *Eli Lilly and Co v Human Genome Sciences, Inc.*, [2010] EWCA Civ 33, [2010] R.P.C. 14, p. 430, H7 and H8.
142. Robert Fitt & Edward Nodder, 'An Uncertain Future for Gene Patents: The view from Europe', 29, *Biotechnology Law Report*, No 6, 2010, p. 616; and *Eli Lilly and Co v Human Genome Sciences, Inc.*, [2010] EWCA Civ 33, [2010] R.P.C. 14, p. 430, H7.

By 1996, *Human Genome Sciences Inc.* had identified a new member of the TNF ligand superfamily (the disputed protein neutrokine-α) by the use of homology screening. The company filed a patent application claiming the full length nucleotide and amino acid sequences of the protein, therapeutic and diagnostic applications of the protein and antibodies to the protein.[143] However, there was no laboratory work to verify the biological activity and Eli Lilly stressed that at the time of filing, Human Genome Sciences had not functionally demonstrated/proved the biological activity or function of neutrokine-α.[144]

The EP neutrokine-α polypeptide patent did not refer to a laundry list in its claims, but the description includes the heading *treatment of immune system-related disorders* at [point 0120], referring to a large field of disorders.[145]

In the decision (paragraph 26), the Board stated, that 'reading the patent specification, a skilled person would distinguish the positive technical information from other allegedly contradictory and broad statements ... (e.g. the wide range of activities and conditions for which Neutrokine-α could be useful). This is because the skilled person realizes that the description of the structure of Neutrokine-α, its structural assignment to the family of TNF ligands, and the reports about its tissue distribution and activity on leukocytes, are the first essential steps at the onset of research work on the newly found TNF ligand superfamily member. In view of the known broad range of possible activities of such a molecule, the skilled person is aware of the fact that the full elucidation of all properties requires further investigations which will gradually reveal them. In this context, the skilled person regards the long listing of possible actions of Neutrokine-α and of medical conditions in which it might take part as the enumeration or generalisation of the properties of the members of the TNF ligand superfamily. This is seen as the frame in which the newly found molecule has to be placed as one could *prima facie* have a reasonable expectation that most of them could in fact be present. At any rate, none of these specific conditions and/or activities is actually claimed and the language used in the specification is in many instances quite prudent ...)'.

143. Robert Fitt & Edward Nodder, 'An Uncertain Future for Gene Patents: The view from Europe', 29, *Biotechnology Law Report*, No 6, 2010, p. 616.
144. Robert Fitt & Edward Nodder, 'An Uncertain Future for Gene Patents: The view from Europe', 29, *Biotechnology Law Report*, No 6, 2010, p. 616.
145. At the time of filing, the patent application did not fulfil the requirement of industrial application. Only later publications made it possible to determine that neutrokine-α was useful in treating diseases associated with particular immunological disorders.

Concern about the use of a laundry list can be mitigated by the requirement for industrial application (and thus real world use). It is relevant to examine the following background information on the Eli Lilly case prior to analysing the use of laundry lists in Chapter 7.

The decision of the UK case of first instance was given prior to the EPO Opposition Division deciding to revoke the patent.[146] The UK judge held that simply identifying a protein was not necessarily sufficient to confer industrial utility on it, and he invalidated the patent.[147] The patent did not identify any disease or condition which the neutrokine-α gene could be used to diagnose or treat since the prediction of the function was based on speculation. In Case T 0018/09 the Technical Board of Appeal decided contrary to the UK court and the EPO Opposition Division.[148] Finally, the UK appeal court decided contrary to the Technical Board of Appeal.

The UK judge Mr Justice Kitchin elaborated a list of principles for assessing industrial application, based on EPO and US courts and PTO case law, and these were later applied in the UK appeal decided by Lord Justice Jacob: 1. The concept of industry must be construed broadly (all industry, see *Max-Planck*, Case T 0870/04); 2. Profit is not a requirement (see *Chiron* [1996] RPC 5350, e.g. products for rare and orphan diseases may not be intended for use in trade; 3. The capability of industrial exploitation must be derivable by a skilled person from the description, read with the benefit from of the common general knowledge (*Genentech* Case T0604/04); 4. The description must disclose a practical way of exploiting the invention in at least one field of industrial activity (*Max-Planck; Salk Institute* Case T338/00); 5. Is there is a sound and concrete basis for recognising that the contribution could lead to practical application in industry? There remains a need to disclose, in definite technical terms, the purpose of the invention and how it can be used to solve a given technical problem; and there must be a real prospect of exploitation derivable directly from the specification, if not already obvious from the nature of the invention or the background art (*ZymoGenetics*; *Bayer* Case T1452/06); 6. The requirement will not be satisfied if what is described is merely an interesting research result that might yield a yet-to-be-identified industrial application (*Salk Institute*). A speculative indication of possible objectives that might or might not be achievable by carrying out research is not sufficient

146. *Eli Lilly and Co v Human Genome Sciences, Inc.*, [2008] EWCA 1903 (Pat) at first instance. It was revoked since it lacked inventive step and claimed an arbitrary member of the TNF ligand super family without known functions; see further Robert Fitt & Edward Nodder, 'An Uncertain Future for Gene Patents: The view from Europe', 29, *Biotechnology Law Report*, No 6, 2010, p. 617.
147. *Eli Lilly and Company v Human Genome Sciences, Inc.* [2008] EWHC 1903 (Pat).
148. Robert Fitt & Edward Nodder, 'An Uncertain Future for Gene Patents: The view from Europe', 29, *Biotechnology Law Report*, No 6, 2010, p. 617, write that in the lead up to the Technical Board of Appeal, both parties filed further documents, including new evidence, which had not been submitted during the UK Trial, including a new set of claims.

(*Max-Planck*). Similarly, it should not be left to the skilled reader to find out how to exploit the invention by carrying out a research program (*ZymoGenetics*); 7. The purpose of granting a patent is not to reserve an unexplored field of research for the applicant (*Max-Planck*) nor to give the patentee unjustified control over others who are investigating in that area and who might eventually find practical ways to exploit it (*ZymoGenetics*); *8. If a substance is disclosed and its function is essential for human health, then the identification of the substance having that function will immediately suggest a practical application. If, on the other hand, the function of that substance is not known or is incompletely understood, and no disease has been identified that is attributable to an excess or a deficiency of it, and no other practical use is suggested for it, then the requirement of industrial applicability is not satisfied. This will be so even though the disclosure may be a scientific achievement of considerable merit* (*Max-Planck*) [emphasis added]; *9.* Using the claimed invention to find out more about its own activities is not in itself an industrial application (*Max-Planck*); 10. As for homology, it is not a bar to patentability that the invention has been found by homologous studies using bioinformatics techniques (*ZymoGenetics*), but that may have a bearing on how the skilled person would understand the disclosure.[149]

In the *Eli Lilly* case, the appellant Human Genome Sciences, Inc. argued before the EPO Technical Board of Appeal that, based on structural features, the patent in question[150] identified neutrokine-α as a new member of the TNF ligand superfamily. Importantly, they explain that all members of that superfamily had a diagnostic and therapeutic interest since they were known to play a central role in human diseases as a result of their involvement in the regulation of human immune cells:

'Although they had pleiotropic effects, the members of that superfamily were not so diverse that a skilled person could not derive a function for a novel member. Indeed, *all members shared significant biological effects on the immune system, with* the ability to

149. See *Eli Lilly and Company v Human Genome Sciences, Inc*, UK Court of Appeal [2010] EWCA Civ 33; and Robert Fitt & Edward Nodder, 'Setting the threshold for industrial application: the UK diverges from Europe', *Journal of Intellectual Property Law & Practice*, 2010, Vol. 5, No 8, p. 564; and the same authors in, 'An Uncertain Future for Gene Patents: The view from Europe', 29, *Biotechnology Law Report*, No 6, 2010, p. 618.
150. The patent in question (EP 0939804) was granted with 27 claims. Claim 1 comprised a nucleic acid molecule comprising a polynucleotide sequence encoding a neutrokine-α polypeptide wherein the polynucleotide sequence is selected from the group consisting of: a) a polynucleotide sequence encoding the full length Neutrokine-α polypeptide having the amino sequence of residues 1 to 285 of SEQ ID NO:2; b) a polynucleotide sequence encoding the extracellular domain of the neutrokine-α polypeptide having the amino sequence of residues 73 to 285 of SEQ ID NO:2. Claim 6 comprised a recombinant vector, and claim 20 an antibody or portion thereof that binds specifically to the neutrokine-α portion of the former claimed neutrokine-α polypeptide.

both stimulate and inhibit the growth, proliferation and differentiation of immune cells. *They were known to be implicated in a number of diseases of the human immune system, which was known to be highly complex and to involve multiple classes of molecules.* Thus, it was not surprising to expect a novel member of that superfamily to have different, and even opposite, biological effects on the immune system depending on the type of immune cell and its activating stage. The list of possible Neutrokine-α activities and effects on diseases and conditions cited in the patent-in-suit reflected only the activities, diseases and conditions common to the known members of the TNF ligand superfamily and expected to be shared by a new member of that superfamily ... [emphasis added]

Therefore, the absence of information in the patent-in-suit on the relative levels of Neutrokine-α in non-lymphoma and lymphoma cells and the apparent widespread distribution of Neutrokine-α mRNA did not preclude a possible use of Neutrokine-α (or antibodies thereto) as an imaging, diagnostic or therapeutic agent.'[151]

On the other hand, the opponent Eli Lily and Company, having recognised that it could be plausibly assumed that Neutrokine-α was a member of the TNF ligand superfamily, argued that none of the claimed products occurred in nature as such.[152]

'Neutrokine-α was either *present on the cellular membrane or proteolytically cleaved* in the cell to a soluble form that was known to be the relevant biological form (namely, residues 134 to 285 of SEQ ID NO:2) No predictions could be validly made as *to the biological activities of the claimed non-natural forms of Neutrokine-α*, especially in view of the fact that the patent-in-suit did not provide any experimental data for the claimed Neutrokine-α forms.

All members of the TNF ligand superfamily (except for TNF-α and TNFβ) had their own receptor and interacted with one-to-one correspondence. Thus, a new member of that superfamily was expected to bind to an unknown receptor and to have a novel combination of properties.'[153] [Emphasis added]

The prior art clearly showed that known members of the TNF ligand superfamily differed dramatically in their properties, with different effects on different cell types and at different stages of cell activation and that their functions were complex, diverse and only incompletely understood. Therefore, the opponent claimed that the identification of an immediate concrete application for a new member was precluded without supporting data. The patent specification failed to provide an immediate concrete application (Case T

151. Case T 0018/09, para. 57, pp. 13-14.
152. See Case T 0018/09, p. 20: 'namely the isolated full-length neutrokine-α and the extracellular domain of neutrokine-α, residues 73 to 285 of SEQ ID NO:2'.
153. See Case T 0018/09, p. 21.

898/05) as its disclosure was too broad, not supported by any evidence and lacked tests for its biological properties:

'Based on these speculations, the patent specification recited *a highly implausible long list of diseases* [emphasis added], which were said to be treatable with Neutrokine-α, and a similarly implausible list of alleged uses for Neutrokine-α antagonists, which in many cases were self-contradictory'.

Having noted that only post-published evidence reported the effects (under conditions not disclosed nor mentioned in the patent application) the opponent continued:

'The appellant had done what patentees without evidence of treatable conditions do when filing patent applications, namely include as 'boiler-plate' the disclosure of a long list of possible activities and conditions (including implausible, self-contradictory ones) from which they later 'cherry-pick' the very few which have been subsequently confirmed or demonstrated'.[154]

And finally the statement of primary importance for the current case study is the laundry list, evaluated by the Board in its reasons for its decision:

'Point 27. Filing patents with such long lists of conditions and activities and subsequently relying on the few which have been confirmed or demonstrated is what the respondent criticized *as a 'boiler-plate' and 'cherry-picking' practice*. It can certainly be argued whether or not such practice is proper but it is significant that the respondent acknowledged, albeit in the form of criticism of the applicant, *that it is a practice used by patentees*, and the appellant pointed to the respondent's own US patent # 7,317,089. *Thus, the skilled person is acquainted with this practice and able to differentiate mere 'boiler-plate' from positive technical information.* In the present case, the description of the patent delivers sufficient technical information, namely the effect of Neutrokine-α on T-cells and the tissue distribution of Neutrokine-α mRNA, to satisfy the requirement of disclosing the nature and purpose of the invention and how it can be used in industrial practice.

As regards the effects on T-cells, the respondent argued that, in view of the technical difficulties involved in measuring the co-stimulation of T-cells by Neutrokine-α and in the absence of any detailed experimental information on the activities of Neutrokine-α listed in the patent-in-suit, the skilled person would not have been able to reproduce them without the undue burden of undertaking a research programme. Moreover, in its view, no industrial application can be directly derived from a mere co-stimulation ...

Point 29. The board cannot accept this line of arguments. Firstly, in the light of the great number of documents concerned with known members of the TNF ligand superfamily which – as explicitly acknowledged in the patent-in-suit ... – disclose standard assays for

154. See Case T 0018/09, p. 22.

measuring their activities and effects on (immune) cells, no particular effort would be required to verify the co-stimulation of T-cells by Neutrokine-α. Even though a few contradictory results are reported in post-published documents on file ..., there is also a convincing body of post-published evidence showing that, using standard assays, Neutrokine-α activity is indeed present on T-cells, in particular on mature T-cells at all stages of differentiation ... Secondly, the reference in the patent-in-suit to the presence of Neutrokine-α activity in lymphocytes would inevitably prompt the skilled person to look for that activity in all types of lymphocytes, not only in T-lymphocytes but also in B-lymphocytes. There is post-published evidence on file showing that Neutrokine-α activity in B-lymphocytes could be easily measured with standard assays ... Thirdly, and contrary to the respondent's view, these activities of Neutrokine-α may represent a valid basis for a possible industrial application. In particular, the inhibition of co-stimulation and/or proliferation of lymphocytes might be *prima facie* of relevance for certain immune diseases ...

Point 30. In the board's judgement, the tissue distribution of Neutrokine-α mRNA disclosed in the patent-in-suit, in particular the expression of Neutrokine-α mRNA in B-cell and T-cell lymphomas provides in itself in the context of the disclosure of a valid basis for an industrial application. The presence of Neutrokine-α in these lymphomas, which is also confirmed by post-published evidence on file ... may be used to develop appropriate means and methods for their diagnosis and treatment based on the disclosure of the patent-in-suit.'

This was thus contrary to the (later) decision of Lord Justice Jacob in the UK Appeal Court, who stated that however clever and inventive you may have been in discovering a gene sequence you cannot have a patent for it or for the protein which it encodes if you do not disclose/demonstrate how it can be used.[155] Accordingly, Lord Justice Jacob stated that what matters is that there is a sufficient specification of the function of the protein. Just describing the existence of a protein and its structure is not enough. Nor is it enough to describe the function at a high level of generality, so that the compound must have a significant function biologically and so that it (or its antibodies) may be usable to treat some disease.[156]

155. See *Eli Lilly and Company v Human Genome Sciences, Inc.*, UK Court of Appeal [2010] EWCA Civ 33, point 57. See Geertrui Van Overvalle, 'Policy levers tailoring patent law to Biotechnology. Comparing US and European Approaches', *University of California Irvine Law Review*, 2010, p. 27.
156. See *Eli Lilly and Company v Human Genome Sciences, Inc.*, UK Court of Appeal [2010] EWCA Civ 33, point 64. See further, Rochelle C. Dreyfuss & James P. Evans, 'From Bilski back to Benson: Preemption, Inventing Around, and the Case of Genetic Diagnostics', June 2011 63 *Stanford Law Rev.* 1358, referring to *Ariad Pharmaceuticals, Inc. v Eli Lilly & Co* as a question of timing: 'To be sure, other requirements for patentability can also be construed as aimed at timing. For example, *Ariad Pharmaceuticals, Inc. v Eli Lilly & Co.*, a case on exploiting the properties of the protein complex NF-κB, was decided on the ground that the applicant filed before it could

The UK Supreme Court, in a Judgment given on 2 November 2011, overruled Justice Jacobs (UK Appeal Court).[157] As an introduction, in point 2, Lord Neuberger stated, that:

'While this issue can be said to raise an important question of principle, its resolution is inevitably fact-sensitive, and therefore any answer may be of limited value in other cases. Further, the issue arises in the context of a fast-developing field, which requires a court to approach it with caution. The need for caution is reinforced by the fact that the answer may give rise to potentially far-reaching consequences for scientific research, the biotech industry, and human health. On the other hand, for those very reasons, it is particularly important that the law in this area is as clear, consistent and certain as possible.'[158]

Without further notice in point 6, the 'boiler-plate' was referred from the High Court for revocation (before Kitchin J) as follows:

'it [specifications, description] is confusingly long, diffuse, and running to over 25 closely typed pages, and nearly 200 paragraphs of descriptive text, and a further twelve pages of sequences of polypeptide amino acids and DNA nucleotides ... the specification 'contains extravagant and sometimes contradictory claims' – [2008] RPC 29, para. 134. Perhaps rather more tolerantly, the Technical Board of Appeal of the European Patent Office ('the Board') referred to the Patent as having been drafted on a 'boiler-plate' basis, which it described as 'a practice used by patentees' – T 0018/09 *Neutrokine/Human Genome Sciences*, para. 27.'

supply a written description of the substances that could be used to achieve the claimed result. The case could have been equally well argued along preemption lines – that knowledge about NF-κB, which appears to play a role in a wide variety of conditions, including memory loss and susceptibility to diseases such as cancer, is not patentable because the patent would preempt those who would follow on, elucidate the impact of NF-κB, and find ways to utilize the information to treat patients. Rebecca Eisenberg and Robert Merges have similarly argued that the utility issue is aimed at delaying the onset of patent protection to the point where more is known about the invention', See Rebecca S. Eisenberg & Robert P. Merges, 'Opinion Letter as to the Patentability of Certain Inventions Associated with the Identification of Partial cDNA Sequences', 23, *AIPLA Q.J.* 1, 20 (1995).

157. 2011, UKSC 51 (Lord Hope, Lord Walker, Lord Neuberger, Lord Clarke and Lord Collins) on appeal from 2010, EWCA Civ 33.

158. In points 96 ff. Lord Neuberger discusses consistency and policy (the wider picture) and the outcome of the former decisions as 'causing UK bioscience companies great difficulty in attracting investment at an early stage in research and development process'. See for a concern of legitimate expectation, also Chapter 5, section 3.1.2 and Chapter 10, section 2.1.

Lord Neuberger quoted from a number of EPO Board of Appeal decisions on EPC Article 57 and biological material (gene and protein sequences; points 43-82) which was summarized in general principles (point 107) headed as; (i-iv) practicable application, concrete benefit, speculative use, enable the skilled person to reproduce or exploit the claimed invention without undue burden (or having to carry out a research programme); (v-x) new proteins and its encoding gene; (xi-xv) superfamily members. Point (xv) reads: 'The position may also be different if the known members have different activities, although they need not always be 'precisely interchangeable in terms of their biological action', and it may be acceptable if 'most' of them have a common role (T 0870/04, para. 12, T 0604/04, para. 16, T 0898/05, para. 27)'. The passage on new proteins read:

'Where a patent discloses a new protein and its encoding gene:

(v) The patent, when taken with common general knowledge, must demonstrate 'a real as opposed to a purely theoretical possibility of exploitation' (T 0604/04, para. 15, T 0898/05, paras 6, 22 and 31);

(vi) Merely identifying the structure of a protein, without attributing to it a 'clear role', or 'suggest[ing]' any 'practical use' for it, or suggesting 'a vague and speculative indication of possible objectives that might be achieved', is not enough (T 0870/04, paras 6-7, 11, and 21; T 0898/05, paras 7, 10 and 31);

(vii) The absence of any experimental or wet lab evidence of activity of the claimed protein is not fatal (T 0898/05, paras 21 and 31, T 1452/06, para. 5);

(viii) A 'plausible' or 'reasonably credible' claimed use, or an 'educated guess', can suffice (T 1329/04, paras 6 and 11, T 0640/04, para. 6, T 0898/05, paras 8, 21, 27 and 31, T 1452/06, para. 6, T 1165/06 para. 25);

(ix) Such plausibility can be assisted by being confirmed by 'later evidence', although later evidence on its own will not do (T 1329/04, para. 12, T 0898/05, para. 24, T 1452/06, para. 6, T 1165/06, para. 25);

(x) The requirements of a plausible and specific possibility of exploitation can be at the biochemical, the cellular or the biological level (T 0898/05, paras 29-30);'

In point 124 Lord Neuberger stated that the former decisions did not give sufficient weight to point (x) when concentrating on the absence of firm evidence of specific therapeutic roles, as opposed to the other roles of Neutrokine-α (without taking into account the roles at other levels which could be attributed to Neutrokine-α as a result of its membership of the TNF ligand superfamily and their known activities). In points 126-128, Lord Neuberger gave his reason for the decision drawing support from T-18/09 (para. 29):

'that the disclosure of the Patent 'may represent a valid basis for a possible industrial application. In particular, the inhibition of costimulation and/or proliferation of lymphocytes might be *prima facie* of relevance for certain immune diseases.'

Neuberger further stated that if 'a patent advances an appropriately plausible function for the claimed protein, then the question of undue burden has to be considered in relation to the making of the protein, as the Board's observation at T 0604/04, para. 22 that 'the patent specification provides adequate experimental instructions for the skilled person to be able to reproduce without undue burden the [claimed] polypeptides' shows'. The necessary assays to determine the precise role and potential of the patent's disclosure would not be a 'complex task', it would simply involve 'standard assays'. The disclosure of Neutrokine-α as a member of the TNF ligand superfamily in relation to the invention's potential satisfy Article 57. As a result, the relevance of the degree of effort needed in relation to any subsequent work falls away (cf. point 128).

The Supreme Court has remitted the case back to the Court of Appeal and one can ask if the patent can still be held to be invalid?[159]

159. Brian Whitehead, Stuart Jackson & Richard Kempner, 'Eli Lilly v Human Genome Sciences: Has the pro-patentee pendum swung too far?' Vol. 12 Issue 3 *BSLR*: case Comment, p. 105. For a recent UK decision also allowing laundrylists, see *Regeneron Pharmaceuticals Inc v Genentech Inc*, 2012, EWHC 657 (Pat), Patents Courts, England and Wales, 22 March 2012. VEGF known as one of a number of angiogenic growth factors (implicated in a large number of diseases).

CHAPTER 7

# An Empirical Case Study: Huntington's disease, spinal muscular atrophy, Canavan's disease, cystic fibrosis, Alzheimer's disease and Parkinson's disease

Chapter 7 contributes to the legal literature with an empirical study of a total of 1590 patent applications spread across the selected diseases. 39 % of the patent applications studied led to the grant of a patent; 38.4 % of patent applications were rejected; and 22.6 % were still pending. The patent documentation has been produced on the basis of searches made at the Danish Patent and Trademark Office in 2004, 2005, 2007, 2009, 2011 and 2012. The analysis relating to each disease (Huntington's disease, Canavan's disease etc.) is divided into four sub-groups. The four sub-groups cover patent activity on the disease gene, patent activity on the pathological process, patent activity involving stem cells, and patent activity involving highlights which can vary from disease to disease. The most important conclusions of the empirical study are linked to the new RNAi techniques (miRNA and siRNA), which challenge the traditional gene patent and which, in the thesis, are taken as showing that the wording of Directive 98/44/EC is too static and not suited for dealing with new developments in biomedical research. Questions are also raised about the use of 'laundry lists', whereby patent applicants do not limit their applications to one specific disease, and most of the patent applications studied include a long list of diseases which can have more or less in common with the main disease.

## 1. Making a mess (out) of disorder(s)

Chapter 6 contains a literature case study of inherited disorders in a patent law context, and gives a brief introduction to the clinical features of the disorders that are the focus of this thesis: Huntington's disease (HD), spinal muscular atrophy (SMA), Canavan's disease (CD), cystic fibrosis (CF), Alzheimer's disease (AD) and Parkinson's disease (PD). It also contains a review of the patent history and legal literature relating to these diseases. HD, SMA, CD and CF are caused by the mutation of a single gene and are characterised as being *single-gene diseases* while AD and PD are characterised by being *multifactorial diseases*. With AD and PD several factors can have an influence on whether the disease is expressed, e.g. small genetic variations together with environmental factors. Chapter 6 gives the background information that is relevant to the empirical case study in the present Chapter which complements Chapter 6, and Chapter 6.4 introduces the reader to the design of the empirical case study in Chapter 7. The aim of the empirical case study in Chapter 7 is to analyse how the patent authorities on both sides of the Atlantic navigate these uncharted waters, and what challenges they meet in their work at the forefront of emerging technologies.

The following empirical case study of Huntington's disease (HD), spinal muscular atrophy (SMA), Canavan's disease (CD), cystic fibrosis (CF), Alzheimer's disease (AD) and Parkinson's disease (PD) has been narrowed down to a template scrutiny of four issues: (a) patent activity on the disease gene and its function; (b) patent activity on the pathological process; (c) patent activity involving stem cells; and (d) patent activity involving highlights defined as a focus on RNAi and ASO techniques which have been claimed in relation to the disorder, but with a general interest in the overall question of tailor-made patent law for emerging technologies. The stem cell-related patents and patent applications are also a focal point, due to the discussion of tailor-made patent law since the EU Directive 98/44/EC imposed a ban on the patenting of embryonic stem cells, as discussed in Chapter 4.4.[1] In several respects the wording of the Directive reflects an outdated approach to biomedi-

1. The present analysis is therefore a review of whether a method for generating induced pluripotent stem cells from somatic cells exists, e.g. outside the search US 20110294218 and WO 2011146607 (stable reprogramming cells: In some embodiments, the stable reprogrammed cell has a statistically significant faster rate of doubling compared with an induced pluripotent stem cell (iPSC) or an embryonic stem cell (ES). Other aspects of the invention relate to compositions comprising the reprogrammed cell, method of isolation and method of use).

cal sciences, as discussed in Chapter 2. The template comprising the four issues is to be regarded only as a starting point. For some of the diseases one or two of the headings (e.g. highlights on RNAi) have not been relevant and have thus been omitted. In order to structure heading (a), on patent activity on the disease gene and its function, the following paradigm has been used.

*Box 7.1.A. Disease genes in claims*

| | |
|---|---|
| **DNA** | Diagnostics/treatment |
| | *Production of protein*:<br>A. Treatment<br>a. Recombinant protein<br>b. Gene Therapy approaches<br>c. siRNAs/microRNAs/piRNAs/antisense/RNA silencing<br>d. Splicing correcting oligonucleotides (SMA, Duchenne muscular atrophy DMD)<br>B. Antibody production |
| **Amino-acid sequence/ protein** | Diagnostics/treatment |
| | a. Recombinant protein for treatment or antibody production<br>b. Diagnostics at the protein level (use protein as biomarker)<br>c. Produce anti-body (for diagnostics/treatment) |
| **Antibody** | Diagnostics/treatment |
| | A. Diagnostic<br>a. Protein detection in blood, cells, urine, biopsy etc.<br>B. Treatment<br>a. Neutralize protein/Bind protein<br>b. Isolate protein<br>C. Bind to protein |

According to the Danish Medical dictionary, the causes, development, and effects of diseases as well as the structural and functional change that are produced by diseases are termed 'the pathological process' and this case study uses this term to refer to an organic process occurring as a consequence of a disease.[2]

2. http://www.wordwebonline.com/en/pathologicalprocess; and Morten Pilegaard & Helge Baden, *Medicinsk ordbog*.

The heading entitled *(d) highlights* is mainly a focus on modulating gene expression either by RNA interference (RNAi) or by ASO mediated splicing correction.[3] The rapid progress in employing RNAi is also discussed in Chapter 2 section 2.1.d. and the technique can be frequently observed in the patent activity that is the subject for the present case study.[4] RNAi is used by many organisms to control gene expression at the post transciptional level. It is now also widely used as a research tool in laboratories and it is a strong candidate for use in numerous future therapies. Two types of RNAi molecules are the subject of this case study, miRNAs and siRNAs. Sophisticated ways of controlling gene expression in the complex environment of a cell include the co-ordinated regulation at different steps of the gene expression process, both by the precise targeting of a single gene and the targeting of multiple genes simultaneously. Generally speaking there are two main differences between miRNAs and siRNAs. miRNAs are encoded in the human genome and they are used by the organism/human cell for the simultaneous modulation of expression of numerous genes by imprecise binding to the target gene mRNAs (not all nucleotides of the miRNA base pair with nucleotides in the target mRNA) thereby inhibiting their translation into protein. siRNAs are typically synthetic/designed short RNA sequences, which are fully complementary (all

3. There are several different ways for RNAi which are slightly different. The recent study does not distinguish strictly between these, since it is often unclear whether an ASO will exhibit its function according to one or the other cellular mechanisms; moreover, sequences are often claimed both as an siRNA, which is synthesized chemically outside the cell, and as an shRNA, which has the same functional core sequence but is delivered via an expression vector (for instance via a virus). Other patents do not make clear what kind of RNAi they are claiming. Consequently, the present study does not distinguish between ribozymes, antisense RNA, microRNA and siRNA.
4. In 2010 a patent for the design of an siRNA was granted to the University of Massachusetts, see EP 1633890. The patent claimed novel RNAi agents with improved properties (enhanced ability of a first strand of an RNAi agent to act as a guide strand in mediating RNAi; enhanced efficacy of an siRNA duplex and e.g. promoting the entry of a desired strand of an siRNA duplex into an RISC complex, and a method for enhancing the silencing of a target mRNA). The patent title is: 'Methods and compositions for enhancing the efficacy and specificity of RNAi'. Opposition has been filed by Alnylam Pharmaceuticals, Novartis AG and Alcon Research, on the grounds of lack of an enabling disclosure, novelty, and inventive step. In 2012 a US patent for gene silencing was granted (US 8097710), which covers the use of RNA strands of 20-24 base pairs in length. It has been pointed out that the use of *3-prime overhangs on RNA strands*, which are already widely used to enable efficient silencing in human cells, is not described in the patent, so companies developing human therapeutics may have a defence against the need to license; see Jennifer Rohn, 'RNAi patent win', *Nature Biotechnology* 30, 202, 2012.

nucleotides of the siRNA base pair with nucleotides of the target gene mRNA) to the mRNA from a specific gene and their binding to the target mRNA causes degradation of the target gene mRNA by the cells' own enzymes. Thus, synthetic siRNAs can down-regulate the expression of a specific gene, whereas endogenous (coded by the human genome) miRNAs can simultaneously, and to different degrees, down-regulate the expression of multiple target genes. Finally, synthetic RNA's can function like the endogenous miRNAs if they bind without being fully complementary to an mRNA. In a patent context, the current study categorises siRNAs as being directly related to the sequence of their target gene, whereas miRNAs are a new category in patent terms. In the view of the current thesis, the wording of Article 5 of Directive 98/44/EC on the sequence or partial sequence of a gene (Article 5(2)) and the focus on industrial application (Article 5(3)) is an outdated approach to biomedical research.

Short antisense RNA oligonucleotides (ASO) used in RNAi techniques are aimed at the down-regulation of gene expression and are therefore suitable for down-regulating the expression of dominant negative mutant proteins in dominant diseases like Huntington's disease or for down-regulating the expression of secondary genes which are involved in the pathological process. In recent years another use of short antisense RNA oligonucleotides has emerged, where the ASO is instead used to correct gene expression. This type of ASO plays an important role in new therapies aimed at correcting aberrant splicing of disease genes like the *SMN2* gene in spinal muscular atrophy (SMA). In fact, the most promising therapy for SMA employs ASOs to sterically block the binding of a splicing inhibitory protein to a splicing regulatory element in *SMN2*. This approach is currently being evaluated in clinical trials. Similar approaches are also being developed, using ASOs to alter the splicing pattern of other disease genes in diseases like Duchenne muscular dystrophy and β-thalassemia.

ASOs can be delivered to the cell either by using expression vectors, which use the cells' own enzymes to produce the ASO RNA, or they can be synthesised chemically and delivered to the cell.

When the ASOs are chemically synthesised outside the cell they typically contain modified nucleotides or nucleotide analogues, which improve their functionality either by improving delivery to the target cell, by improving stability when hybridized to their target mRNA or by blocking degradation of the hybridised ASO or by improving stability. The most commonly employed

*Box 7.1.B. Antisense RNA/RNAi overview**

| | | |
|---|---|---|
| | **New developments for manipulating the expression of specific genes for experimental and therapeutic purposes** | |
| | miRNA | |
| dsRNA encoded by the human genome | miRNA occurs naturally in the human body: Human express ≈ 1,000mi RNAs | 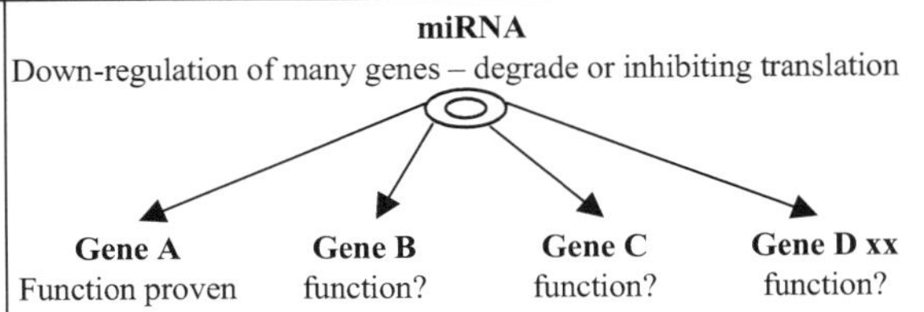<br>miRNAs are only 21 nucleotides long and are involved in many broader biological networks (controls manifold processes, mechanism contribute significantly to the regulation of gene expression), most miRNAs can target many genes (more than one mRNA).<br>*Imprecise matching*:<br>miRNAs are more complex than siRNAs: miRNAs hybridise imperfectly with their target mRNAs, repressing translation of the mRNA. Nucleotides 2 to 7 of an miRNA are most critical for targeting (out of 21 nucleotides)<br>microRNAs may down-regulate differently on each gene (have different effects) but they work in the same way on each gene. However, the gene targeted by a miRNA can have unknown functions. |
| Artificial dsRNA | artificial miRNA | Therapeutic |
| | | |
| | siRNA | Therapeutic – more similar to gene patenting |
| | not naturally occurring. Complementary to mRNA's | Produced in the cell itself or delivered into the cell experimentally.<br>Manipulate gene expression: Often 1 siRNA → 1 mRNA (exon sequences) similar to a probe, primer (PCR) *precise targeting*, perfect complementary to their target sites A full match of all 21 nucleotides involved in order to match the gene<br>Specific target mRNAs causing degradation |
| | Endo siRNA & piRNA | |
| | | piRNA: The longest siRNAs |
| | Antisense oligonucleotide | |
| | | Modifying post-transcriptional processing (splicing/polyadenylation) |

* M. Ghildiyal & P. Zamore, 'Small Silencing RNAs: An expanding universe', *Nature Review Genetics*, 2009, 10 (2), pp. 94-108; and Adam Bostanci, Jane Calvert & Pierre-Benoit Joly, 'Regulating gene regulation: patenting small RNAs', in Matthew Rimmer & Alison McLennan, *Intellectual Property and Emerging Technologies, the new Biology*, 2012, p. 214. See also Lodish et al., *Molecular Cell Biology*, 6th Ed., 2008, pp. 347-348.

*Figure 7.1. C dsRNA function*

**A. miRNA (inhibiting gene expression)**

**B. siRNA (inhibiting gene expression)**

**C. Modulation (correcting gene expression)**

**D. Modulation (correcting gene expression)**

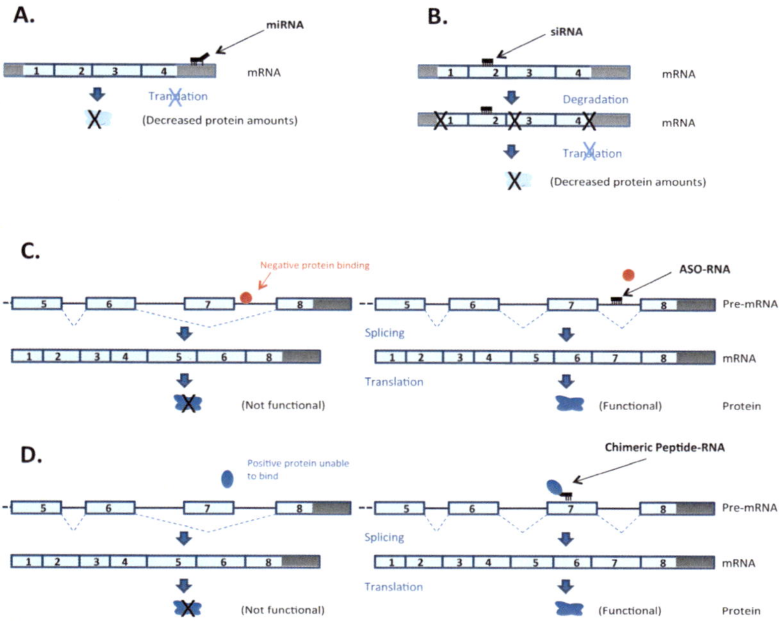


Figure A illustrates how an miRNA inhibits gene expression. Figure B illustrates how an siRNA inhibits gene expression. Both figures A and B are relevant to Huntington's disease. Binding of the antisense oligo to the target gene causes either blocking of translation of the mRNA to protein (A) or degradation of the mRNA (B). Both approaches inhibit production of a dominant negative deleterious protein and they may even be allele-specific in the sense that the ASOs can be targeted only at the mutant sequences. Figures C and D illustrate modulation or correction of gene expression by correcting splicing by an anti-sense oligonucleotide (ASO)-RNA. This can be achived by sterical blocking of the binding of a splicing inhibitory protein (C) or by recruiting a splicing stimulatory protein by means of a modified chimeric ASO containing peptides, which can recruit the stimulatory protein (D). These techniques (C and D) are relevant to SMA.

nucleotide analogues are Locked Nucleic Acids (LNA), Peptide Nucleic Acids (PNA) and Morpholino Nucleic Acids (MNA).[5]

As described in Chapter 6, the concept of 'real world use and fair protection'[6] in this thesis analysis includes a variety of reasoning such as 'well established utility/industrial application', and 'lack of experimental evidence' to conclude that the invention can be used for the treatment of diseases. When relevant, this perspective is subject to a laundry list scrutiny. As the heading of the current chapter indicates, the practice of claiming a broad list of diseases creates a mess. Almost every patent application examined has referred to one or more of the diseases that are the subject to this case study; sometimes patents are granted with the laundry list (either in the claims or in the description) and sometimes not. The concept of a 'laundry list' was coined in connection with the Huntington's applications of the present study and used by the USPTO examiners; see e.g. Huntington's, Group (6). The concept has also been called a 'wish list' or, according to Case T 0018/09 *Eli Lilly and Company v Human Genome Sciences Inc.*, a 'boiler-plate' or 'cherry-picking' practice.[7]

The boxes below show the numbers of patents and patent applications included in the case study, and their outcomes.[8] APPENDICES VII-XII contains the search outcome; appendix VII Table HD; appendix VIII: Table SMA; appendix IX: Table CD; appendix X: Table CF; appendix XI: Table AD and appendix XII: Table PD.

5. R. Kole, A.R. Krainer & S. Altman, 'RNA therapeutics: beyond RNA interference and antisense oligonucleotides', 2012, *Nat Rev Drug Discov.* 11(2), pp. 125-140.
6. On the concept of 'real world use' see Chapter 3.4.1 (Utility guidelines).
7. See Chapter 6.4.
8. See Chapter 6.4 for the design of the case study.

*Box 7.1.D. Single-gene disorder*

| **Disease** | **Huntington's (HD)** | **Spinal muscular atrophy (SMA)** | **Canavan's (CD)** | **Cystic fibrosis (CF)** |
|---|---|---|---|---|
| **Groups**<br>*[a] Disease gene<br>*[b] Pathological process<br>*[c] Stem cells<br>*[d] Highlights | **42** (6 out)<br>**0**<br>**14**<br>**3**<br>**19** | **36** (14 out)<br>**20**<br>**8**<br>**3**<br>**5** | **15** (55 out)<br>**4**<br>**6**<br>**2**<br>**3** | **313** (60 out)<br>**39**<br>**106**<br>**6**<br>**162** |
| Applications | 136 | 70 | 35 | 846 |
| Grants | 26 | 16 | 13 | 387 |
| Withdrawn/abandoned | 62 | 28 | 13 | 278 |
| Pending | 50 | 26 | 9 | 181 |

*Box 7.1.E. Heterogeneous disorders*

| **Disease** | **Parkinson's (PD)** | **Alzheimer's (AD)** |
|---|---|---|
| **Groups**<br>*[a] Disease gene<br>*[b] Pathological process<br>*[c] Stem cells<br>*[d] Highlights | 21 (14 out)<br>**4**<br>**9**<br>**4**<br>**4** | **151** (20 out)<br>**27**<br>**55**<br>**61**<br>**8** |
| Applications | 114 | 409 |
| Grants | 34 | 144 |
| Withdrawn/abandoned | 55 | 167 |
| Pending | 25 | 98 |

*Box 7.2.**

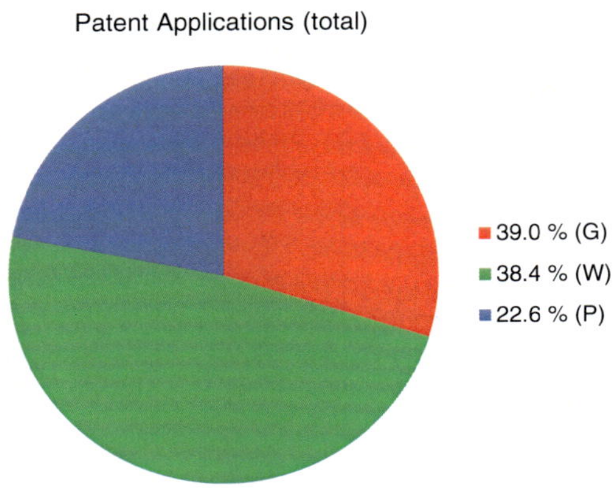


* Total patent applications: 1,590. Of which 620 (39 %) were granted (G); 611 (38.4 %) were withdrawn (W); and 359 (22.6 %) still pending (P). Status at the beginning of 2012 (AD, PD, SMA, HD, CF and Canavan's checked from February to 1 April 2012).

The following sections comprise a review of the data in the following order: first the single gene disorders: Huntington's disease, spinal muscular atrophy, Canavan's disease and cystic fibrosis; and second the genetically heterogeneous disorders: Parkinson's disease and Alzheimer's disease.

# I. SECTION
# on single gene disorders
# Huntington's disease, spinal muscular atrophy, Canavan's disease & cystic fibrosis

## 2. Huntington's chorea and juvenile Huntington's

As mentioned in Chapter 6, Huntington's disease is associated with increases in the length of a CAG triplet repeat expansion present in the coding region of the huntingtin gene *IT15*.

The CAG repeat expansion results in a long stretch of polyglutamines (Gln) close to the amino-terminus of the HD protein huntingtin (htt).[9] When above a certain length, the polyglutamine tract in the protein leads to aggregation and causes damage to neuronal cells. It is the length of the polyglutamine tract that determines the severity of disease in patients. When the disease allele is inherited from the father, the length of the repeat can increase from one generation to the next.

The literature case study in Chapter 6, section 2.1.1.1, describes the state of the art of the early patenting involving the huntingtin gene (*IT15*), the protein encoded by the gene and the protein for detecting predisposition to developing Huntington's disease (HD). The scope of protection of the first patents granted was very broad. The present case study is based on a patent search focusing on how research into the disease has been subject to patent activity thereafter.

### 2.1. Data outline

The present case study of Huntington's disease comprises 42 groups of altogether 136 patent applications, of which 26 have been granted, 62 have been withdrawn, and 50 are still pending.[10] Six groups (groups (11) (15) (21) (31) (34) and (40) in bold Appendix VII, table HD, are omitted as they do not directly concern genes as such, but e.g. chemical compounds.[11] Group (22)[12] also shows that the search on Huntington's disease came up with applications

9. Christian Landles & Gillian P Bates, 'Huntingtin and the molecular pathogenesis of Huntington's disease', *EMBO Reports*, 2004, 5, pp. 958-963, doi:10.1038/ sj.embor. 7400250.
10. See Appendix VII Table HD.
11. See Appendix VII Table HD.
12. US 2005048641, list of diseases in point [0060].

beyond the gene itself. This group claimed an infusion pump system for the delivery polynucleotides to the central nervous system, and in the description the usual laundry list of including Parkinson's, Huntington's and Alzheimer's is mentioned.

The search strategy is described in Chapter 6, section 4. The focus of the case study of HD is based on the following patent activity:

## 2.2. The empirical analysis

Box 7.2.A below shows the template of the design of the present HD case study.

*Box 7.2.A. Huntington's, analysis model*

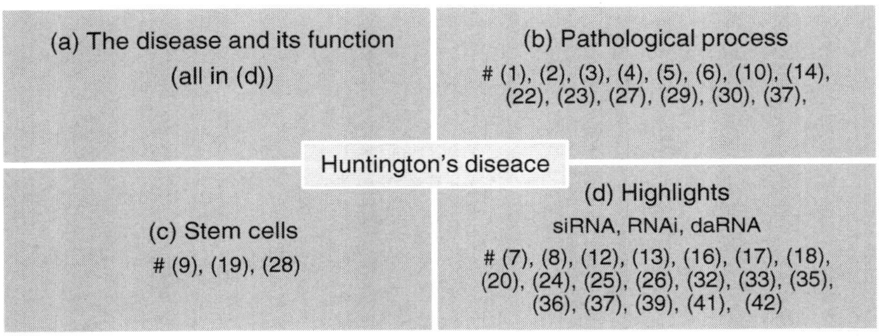


The analysis of the data on Huntington's disease can be further compressed since it was found that none of the applications claimed huntingtin, *IT15* or *HTT* as such. Thus, category (a) developed into category (d), since RNAi concerns techniques for inhibiting, decreasing or correcting gene expression of the huntingtin gene. In other words if one compares the literature case study in Chapter 6.2 with that of current chapter, the development has moved from gene patents (a) to patents for RNAi (d), and the two approaches do not seem to affect each other in a patent context.

### 2.2.b. Pathological process

The analysis of the patent activity under the heading of the HD pathological process comprises inventions that are based on chemicals or secondary genes, and such inventions do not claim the HD primary (target) gene. Instead, the inventions either enable treatment of the disease symptoms or disease pathways. The overall purpose of this section is to examine the practice which recurs throughout the case study of the drafting of patent applications with laundry lists. It is questionable and even objectionable that the examination of laundry lists is far from being subject to a strict practice by patent examiners. There seems to be no pattern to the examination of such lists and no clear conclusion can be reached as to whether there is a difference between the practices of the USPTO and the EPO on this matter. The reasons that are often repeated by the examiners are such as a 'lack of a real world use', 'assumption of function based on similarity/homology', and 'lack of experimental evidence' for claiming the treatment of diseases. Thus, in this context the heading 'real world use' is applied accordingly.[13] This is not relevant solely to Huntington's disease, but the HD data illustrate well that some applicants make claims for inventions that are far too broad, covering a list of the world's diseases. However the question remains why some applicants succeed in obtaining a grant with a comprehensive 'laundry list' while other applications are dismissed because they are considered speculative.

The category of 'the (HD) pathological process' illustrates well the laundry list aspect, and it will be supplemented by the spinal muscular atrophy (SMA), Canavan's disease (CD), cystic fibrosis (CF), Alzheimer's disease (AD) and Parkinson's disease (PD) case studies. However, the main section on this topic is in this section and it will only be briefly commented on in the analyses of the other diseases. Groups (30) (37) and (1-6) have been categorised as 'real world use and fair protection' and all relate to laundry lists or to a myriad of diseases that are claimed to benefit from the invention.[14] The listed groups have been selected as illustrative, but the overall case study found many applications relevant in this regard.[15]

13. See Chapter 6, section 4.
14. Groups (1) (2) (3) (4) (5) and (6).
15. For the HD study, see the following groups (14) (22) (23) (27) (29) and (35). *Group (22)* is mainly a curiosity. See description point [0006] stating that several studies have examined whether cannabinoid agonists can provide benefits for the treatment of HD transgenic mouse models of the disease also suggest that cannabinoids interacting with this receptor could be of therapeutic interest for HD ... *Group (27)* concerns secondary gene erythropoietin (EPO), and a method that can be used to prevent or reduce the severity of, for example, Alzheimer's disease, Parkinson's disease, Huntington's disease, epilepsy, amyotrophic lateral sclerosis, multiple sclerosis, a movement disorder, HIV-associated dementia, HIV-associated neuropathy, neuropathic pain, migraine, glaucoma, drug addiction, drug withdrawal, drug dependency, depression or anxiety. From the examination one can read: 'It is in no way predictable that any IGF peptide analogue would afford a protein having activity comparable to the one disclosed. The specification and working examples fail to teach that any IGF peptide analogue can be co-administered with EPO to give a synergistic neuroprotective

*For example group (30)* claimed the use of a histone deacetylase inhibitor (a chemical compound)[16] for the preparation of a medicament for the treatment of a polyglutamine-expansion-related neurodegenerative disease in which the disease is selected from a group consisting of: Huntington's disease,[17] Kennedy's disease, spinocerebellar ataxia diseases, DRPLA and Machado-Joseph disease.[18] The claims were *very broad*; claim 25 in the US application read: a method of treating protein-aggregation-related neurodegeneration in a patient, comprising identifying a patient at risk for protein-aggregation-related neurodegeneration, and administering to the patient a therapeutically effective amount of a histone deacetylase inhibitor. Histone deacetylases simultaneously inhibit gene expression of numerous genes and consequently a histone deacetylase inhibitor will result in increased gene expression from numerous genes.

In the final rejection in 2009 the US examiner referred to the prior art as teaching that sodium phenylbutyrate was known to be a potent inhibitor of histone deacetylase. However, the application failed to establish a reasonable correlation between histone deacetylase activity and neurodegenerative disease, and failed to lead a person of ordinary skill in the art to conclude that

effect'. *Group (29)* claimed that transition metal-cyclopentadienyl-tropane conjugates is an invention for diagnostics and mainly relevant to Parkinson's disease (dopamine transporter (DAT)) but overall a chemical that binds to monoamino transporter. *Group (14)* directs the pathological process for several diseases including Huntington's disease (cognitive impairment). The core of the invention concerns the proline transporter (an amino acid) and in part the invention addresses methods of improving cognitive performance in patients suffering from a variety of diseases (laundry list). *Group (23)* comprises a comprehensive laundry list for diseases affecting learning and memory by causing defects (in memory, orientation, learning, attention, reasoning, language, and/or the ability to perform simple tasks). The invention describes a Drosophila model for diseases affecting learning and memory and use of that model to identify compounds that are useful for treating the learning and memory-affecting components of those diseases (claim 36 lists the diseases). *Group (35)* fibroblast growth factor-2 promotes neurogenesis and neuroprotection and prolongs survival in Huntington's disease.

16. See Claim 3; Wherein the histone deacetylase inhibitor is selected from the group consisting of suberoylanilide hydroxamic acid (SAHA), butyrate, pyroxamide, depsipeptide, MS-275, and derivatives thereof.
17. A method of treating Huntington's disease in a patient, comprising identifying a patient at risk of Huntington's disease and administering to the patient a therapeutically effective amount of a deacetylase inhibitor; a method of treating Parkinson's disease in a patient, comprising identifying a patient at risk of Parkinson's disease and administering to the patient a therapeutically effective amount of a deacetylase inhibitor.
18. EP 1390491 granted, US 2004142859 pending.

the asserted utility in the treatment of neurodegenerative disease was more likely than not.

*Group (37)* on gene therapy claimed G-rich polynucleotides for the treatment of HD.[19] It related to guanidine-rich (G-rich) oligonucleotides to disrupt disease-causing protein aggregates, for example HD protein aggregates. The application US 2008051363 was abandoned. Claim 7 included a laundry list (Parkinson's disease, Alzheimer's disease, Huntington's disease, a prion disease, and a polyglutamine disease, which all share protein aggregation as disease mechanism).[20] In the final rejection in 2010, the US examiner stated that there is a difference between claiming to deliver oligonucleotides and claiming to prevent a disease. The specification did not provide guidance. Thus, the laundry list was not accepted. In the granted patent US 7807648 in the same group there did not appear to be a laundry list in the claims, since claim 16 was amended so there was a deletion of all diseases except HD. Instead, the patent as granted provided the deleted list in the summary of the invention.

*Groups (1) (2) and (3)* concerns human GPCR-like proteins. G Protein-Coupled Receptors (GPCRs) comprise a large protein family of transmembrane receptors that sense molecules outside the cell and activate signal transduction pathways inside the cell and, ultimately, cellular responses. GPCRs are involved in many diseases, and are also the target of around half of all modern medicinal drugs. The human genome encodes roughly 900 different G-protein-coupled receptors, and 360 G protein-coupled receptors concern senses, mainly the sense of smell (including receptors in the visual, olfactory (smell), and gustatory (taste) system). However, approximately 150 of the GPCRs in the human genome have unknown functions.[21]

19. EP 1940861, US 2008051363 and US 7807648.
20. See further the description point [0064]: 'Diseases where protein aggregation is causal or an associated symptom and for which the present invention may be useful for treatment and/or prevention include Down's syndrome, Alzheimer's disease (AD), Parkinson's disease (PD), amyotrophic lateral sclerosis (ALS), and prion diseases such as bovine spongiform encephalitis (BSE) and Creutzfeldt-Jakob Disease (CJD), cystic fibrosis, and the so-celled polyglutamine diseases for example, Huntington's disease (HD), dentato-rubral and pallido-luysian atrophy (DRPLA) and several forms of spino-cerebellar ataxia (SCA), also have intracellular inclusions in regions corresponding to the regions of neuronal degeneration.'
21. Lodish et al, *Molecular Cell Biology*, 6th Ed., 2008, Chapter 15 on cell signalling, p. 624.

Group (1) essentially claimed a GPCR-like protein polypeptide and methods of use.[22] The US application claimed a gene patent (based on a computer homology search) on a receptor concerning the sense of smell and usable for a wide range of disorders, including asthma, Huntington's disease to chemotherapy-induced vomiting.[23]

In a non-final rejection in April 2005 the US examiner rejected the remaining claims because they were drawn to an invention with no apparent or disclosed *specific and substantial credible utility*; see 35 USC § 101. While the application provided a description of the isolated DNA encoding a protein, it did not disclose a specific biological role for this protein *or its significance to a particular disease, disorder or physiological process which one would wish to manipulate for a desired clinical effect* (non-final rejection, page 2, emphasis added).[24] The examiner found absolutely no support for concluding that the protein of the invention was in any way involved in the claimed disorders or 'in any one or more of the plurality of causally unrelated diseases and disorders listed. The only evidence of record was that a protein of the instant invention was properly an odorant receptor. The applications failed to disclose the identity of those tissues and organs in which a protein

22. Patent No US 20040101871 and No EP 1326973. Both patent applications have now been withdrawn (the US abandonment in 2005, and the EP in 2006). As there was no reply to the 2005 examination report, the application was deemed to be withdrawn. (subject matter which falls within Article 54(2) of Directive 98/44/EC).
23. See [0051] 'particularly a human GPCR-like protein, can be used in therapeutic methods to treat disorders such as anxiety, depression, hypertension, osteoporosis, diabetes, cancer, migraine, compulsive disorders, schizophrenia, autism, neurodegenerative disorders, such as Alzheimer's disease, Parkinsonism, and Huntington's chorea, obesity, and cancer chemotherapy-induced vomiting'. See further [0204] 'as well as sleep and eating disorders, pain control, disorders involving regulation of body temperature and blood pressure'.
24. At page 2 in the non-final rejection from 12 December 2005, it is further argued: 'that it is clear from the instant specification that the receptor protein described therein as 'a human GPCR-like protein' is what is termed an 'orphan receptor' in the art. This is a protein whose cDNA has been isolated because of its similarity to known proteins. There is little doubt that, after complete characterization, this protein may be found to have a specific and substantial credible utility. This further characterization, however, is part of the act of invention and until it has been undertaken Applicant's claimed invention is incomplete'. The examiner referred to *Brenner v Madison*, 148 U.S.P.Q 689 (Sup. Ct., 1966). The invention related to a *protein of as yet undetermined function or biological significance* [emphasis added].

has shown to be expressed and the instant specification did not disclose a credible *real world use* for the protein' (35 USC § 101).[25]

The European examiner also emphasised that all the essential features must be indicated, these being the features which are *necessary to obtain the desired effect,* see EPC Article 84. The wording of a claim should not leave the reader guessing whether something falls within its terms. It follows that the claims should refer to clearly defined sequences encoding the GPCR-like protein polypeptides. Such clearly defined claims would also reflect the general principles of Article 84 EPC, namely that the scope of a granted patent should correspond to its technical contribution to the state of the art. The need for *fair protection* should be recognised.[26]

Group (2) follows the same line of argument and claimed extremely broad protection, claiming almost the whole gene family based on only 70 % homology.[27] Leucine-rich repeat-containing G protein-coupled receptor 4, also known as LGR4, is a human gene, and it belongs to the largest mammalian superfamily of proteins with seven-transmembrane domains.[28] LGR4-6 are

25. See p. 5, non-final rejection. The claims were further rejected under 35 USC § 112, first paragraph, as failing to adequately teach how to use the invention. Finally, one of the claims failed to provide the guidance needed to make claimed modifications with any reasonable expectation that the modified protein would function as required. The examiner referred to *re Fisher*, as the specification did not identify those amino acid residues in the claimed sequence that are critical to the structure and functional integrity of a protein comprising 'a GPCR activity', identify a structurally analogous protein for which this information is known and could be applied to the proteins in question, or even provide a single working example of an intentionally modified protein of the invention. Therefore, the specification left it to the practitioner of the art to engage in a substantial amount of undue experimentation (see p. 7, non-final rejection).
26. See point 3, in the annex to the communication report of 2005. Novelty and inventive step were not acknowledged. The examination in this regard is left out.
27. See US 20030139341 and EP 1290030; the inventor was Shyam Ramakrishnan *(Brighton, MA),* who was also co-inventor of (1) human gpcr-like protein. Claims 56-60 and claim 63 related to a polypeptide, an antibody, RNA, an antisense oligonucleotide a ribozyme and what may bind to the undisclosed polypeptide. The US examiner required the applicant to *elect a single invention* to which the claims must be restricted. The applicant never responded and the application was abandoned in 2006.
28. LGR4 – Wikipedia and 'Entrez Gene: LGR4 leucine-rich repeat-containing G protein-coupled receptor 4'; and Sabine Mazerbourg et al., 'Leucine-Rich Repeat-Containing, G Protein-Coupled Receptor 4 Null Mice Exhibit Intrauterine Growth Retardation Associated with Embryonic and Perinatal Lethality', 2004, *Molecular Endocrinology* 18 (9): p. 2241.

orphan receptors with unknown physiological roles.[29] The same description quoted above of the use of GPCR protein mentions the *potential* role of human LGR4-like GPCR as a remedy for several diseases and dysfunctions.[30]

The European application was closed in 2004 due to the lack of a reply to the examination report. In the 2004 examination report it was concluded that some of the claims were not patentable (Article 52(4) EPC and Boards of Appeal, Case T964/99). Furthermore the claims defined the subject matter *without giving a true technical characterization of any such compounds.* 'They describe the underlying technical problem, but neither characterizes nor discloses such a compound and therefore not allowable ... In consequence, the scope of said claims is ambiguous and vague, and the subject matter is not sufficiently disclosed and is not supported by the description (Articles 83 and 84 EPC).[31] Beside inventive step, vague phrases and lack of solution to the problem being posed, it is finally stated that one of *the claims refers to diseases, which are not further characterized ('a disease').* However, when a claim is defining a condition to be treated or diagnosed in functional terms, e.g. any condition susceptible of being improved or prevented by selective change of the activity of the disclosed protein, the claim can be regarded as clear within the meaning of Article 84 EPC and as sufficiently disclosed in the sense of Article 83 EPC only if instructions, in the form of experimental tests or any testable criteria, are available from the patent documents or from the common general knowledge allowing the skilled person to recognize which conditions fall within the functional definition and accordingly within the scope of the claim. However, this is not the case in the present application.'

Group (3), which focuses on reagents which regulate human follicles, stimulating hormone-like G protein-coupled receptors (FSH-like GPCR) and reagents which bind to human FSH-like GPCR gene products, also claimed to play a role in a huge range of dysfunctions or diseases (e.g. from obesity to

29. Sabine Mazerbourg et al., 'Leucine-Rich Repeat-Containing, G Protein-Coupled Receptor 4 Null Mice Exhibit Intrauterine Growth Retardation Associated with Embryonic and Perinatal Lethality', 2004 *Molecular Endocrinology* 18 (9): pp. 2241-2254.
30. The claimed invention was extremely broad, comprising a cell with a reagent which specifically binds to a product encoded by a polynucleotide comprising the nucleotide sequence shown in SEQ ID NO:1, whereby the activity of a human LGR4-like GPCR is reduced (and further polypeptide, antibody, RNA, ribozyme; see claims 55-60). Claim 63 also claimed what may bind to a sequence (A pharmaceutical composition, comprising: a reagent which specifically binds to a polypeptide comprising the amino acid sequence shown in SEQ ID NO:2).
31. Finally, the examiner considered the state of art (EPC Article 54(3) and (4)). An earlier application disclosed a sequence, which was 91 % identical to the sequence of the present invention, and its protein showed 85 % identity. As the earlier application also disclosed processes for the production and detection of the disclosed G-protein-coupled receptor, and e.g. methods for screening – it was prejudicial to the novelty of the subject matter of 17 claims in the subsequent application.

*neurodegenerative disorders such as Alzheimer's disease, Parkinson's disease, and Huntington's chorea,* and chemotherapy-induced vomiting).[32] Point 4 in the annex to the 2005 European examination report emphasised that the subject matter was not sufficiently defined. The wording as drafted would also include polynucleotides which were far from identical to the claimed polypeptide; it claimed a fragment not sufficiently defined and it referred to a reagent that was *not further specified by technical features.* In point 5.5 of the examination report it was stated that a speculative remedy for a certain disorder is dependent on the specific function of the polypeptide, its specific binding capability to a ligand and to its expression pattern in specific tissues relating to certain disorders. No experimental data or evidence were given that would specifically link 'the FSH GPCR-like polypeptide' to any of the mentioned disorders.[33]

Group (4) adds to the row of speculative uses and as in groups (1)-(3), group (4) had no experimental evidence, no biochemical characterization, and only blast homology search.[34] It is notable how nucleic acid and amino acid

32. See US 20040077041 and EP 1287135. From Wikipedia: 'The follicle stimulating hormone receptor or FSH-receptor (FSHR) is a transmembrane receptor that interacts with the follicle stimulating hormone (FSH) and represents a G protein-coupled receptor (GPCR). Its activation is necessary for the hormonal functioning of FSH. FSHRs are found in the ovary, testis and uterus. The gene for the FSHR is found on chromosome 2p21 in humans. The gene sequence of the FSHR consists of about 2,080 nucleotides.' The same inventor as invented (1) and (2) (Shyom Ramakrishan) filed 2 applications (one US and one European), but all applications were deemed to be withdrawn.
33. The application was closed in 2005. Point [0008]: 'Over the past 15 years, nearly 350 therapeutic agents targeting GPCRs receptors have been successfully introduced onto the market. This indicates that these receptors have a proven history as therapeutic targets. Clearly, there is an ongoing need to identify and characterise further GPCRs which can play a role in preventing, ameliorating, or correcting dysfunctions or diseases including, but not limited to, infections such as bacterial, fungal, protozoan and viral infections, particularly those caused by HIV viruses, pain, cancers, anorexia, bulimia, asthma, Parkinson's diseases, acute heart failure, hypotension, hypertension, urinary retention, osteoporosis, angina pectoris, myocardial infarction, ulcers, asthma, allergies, benign prostatic hypertrophy, and psychotic and neurological disorders, including anxiety, schizophrenia, manic depression, delirium, dementia, severe mental retardation, and dyskinesias, such as Huntington's disease and Tourette's syndrome.'
34. EP 1259618 and US 20030207430, both withdrawn. Group (5), major intrinsic protein (MIP), will not be further described but that too claimed broad gene patents (vectors, host cells, antibodies, antagonist and recombinant methods for producing human MIP-like polypeptides) useful for diagnosis, treatment, prevention and/or prognosis of various disorders in mammals, preferably humans (a laundry list).

sequences of human enzyme molecules and the use of these sequences in the diagnosis, treatment, and prevention of autoimmune/inflammation disorders, genetic disorders, neurological disorders, and cell proliferative disorders can be claimed so very broadly. What was claimed was an amino acid sequence ((b) a naturally occurring amino acid sequence having at least 90 % sequence identity, (c) a biologically active fragment, (d) an immunogenic fragment) all selected from the group consisting of SEQ ID NO:1-26, and the laundry list was almost endless, naming almost every known disease.[35]

However, *the alleged function* of the enzyme was not substantially shown. It was limited to a statement of how measurement of enzymatic activity may be carried out, without actually providing any concrete results. The European examination division considered that a mere assumption of this function on the basis of partial sequence similarity with known enzymes was not credible, and that the potential field of use was merely speculative.[36] The US application also related to nucleic acid and amino acid sequences of human enzyme molecules, and to the use of these sequences in the *diagnosis, treatment, and prevention of autoimmune/inflammation disorders, genetic disorders, neurological disorders, and cell proliferative disorders* (i.e. the invention provides a method of treating a disease or condition associated with decreased expression of functional HEM, comprising administering the composition to a patient in need of such treatment).[37]

35. US 20030207430, point [0232]; the description consisted of a one page of list of diseases in alphabetic order.
36. The generation of amino acid and DNA sequences *without any purpose* is not industrially meaningful according to Rule 23e (3). EP 1259618 was withdrawn since the alleged function of the enzyme was not substantially shown.
37. The US examiner (office communication, 2004) concluded that no claims were in a condition to be allowed. *Most of the claims lacked a substantial and specific asserted utility (no specific biological function).* For example, it was stated on page 10, concerning the claimed protein (ALEX1); that the specification provided no clue as to (1) what is an ALEX1 protein, (2) its biological function, (3) which type of enzyme is it, (4) its substrate/target, or (5) the biological processes associated with the polypeptide of SEQ ID NO: 10 or its substrate/targets. There was no experimental evidence provided by the Applicants to indicate that the polypeptide of SEQ ID NO: 10 is an enzyme. Furthermore, even if one assumed that the Applicants had asserted a utility for the polypeptide of SEQ ID NO: 10, the Applicant's asserted utility for the polypeptide of SEQ ID NO 10 (enzyme), particularly in view of a lack of knowledge as to the type of enzyme ALEX1 is, its specificity/substrate, and/or specific biological processes associated with ALEX1 or its substrates (if ALEX1 is an enzyme), was not specific or substantial since it would require or constitute carrying out further research to identify or reasonably confirm *a real world* context of use. See *e.g. Brenner*

Group (6),[38] comprising secreted proteins (or the genes that encode them), claimed to detect, treat and prevent medical diseases. It was stipulated in the description that: 'despite the great progress made in recent years, only a *small number of genes encoding human secreted proteins have been identified*'.[39] However, most of the applications in this group were withdrawn on the basis of the reasoning that Congress did not intend patents be granted for chemical compounds '*whose sole 'utility' consists of its potential role as an object of use-testing*' (*Brenner, 148 USPQ 689*); only speculative biological activities had been described thus requiring further research to be carried out to identify or reasonably confirm *a 'real world' context of use*, so the applications did not define specific and substantial utility.[40]

*v Manson, 383 U.S. 519, 148 USPQ 689 (Sup. Ct. 1966).* Furthermore the application failed to comply with the *written description requirement* (35 USC 112): page 13: 'The general knowledge in the art concerning a naturally occurring variant, such as an allele, does not provide any indication of how a naturally-occurring variant is representative of unknown naturally-occurring variants as other species clearly differ in structure in some respect and often differ drastically in function. *In the case of alleles, there is no indication in the art that would suggest that the structure and function of one provides guidance to the structure and function of the others* ... the genus of nucleic acids is a large variable genus with the potentiality of encoding many different proteins. As taught by the art, a high degree of structural homology may not reply in functional homology (page 13). Page 15: ... The art as described above, clearly teaches that even one amino acid substitution can result in a polypeptide having different function, therefore the claimed polynucleotides can potentially encode proteins of many different functions which cannot be inferred by structural homology alone. ... Page 16 the unpredictability of the art in regard to assigning function based on structural homology.'

38. See US 20020012966, EP 1471072, EP 1212342, EP 1173477, US 7368531, US 7381800, US 2007032414, and US 6420526: Craig A. Rosen and Steven M. Ruben (MD), and one grant to Shi Yanggu (and Steven M. Rugen) US 7381800.
39. See the description of US # US 20020012966 [0006]. These secreted proteins include the commercially valuable human *insulin, interferon, Factor VIII, human growth hormone, tissue plasminogen activator, and erythropoeitin.* Thus, in the light of the pervasive role of secreted proteins in human physiology, there is a need to identify and characterise novel human secreted proteins and the genes that encode them. [0011] In the invention in question, a 'secreted' protein refers to those proteins capable of being directed to the ER, secretory vesicles, or the extracellular space as a result of a signal sequence, as well as those proteins released into the extracellular space without necessarily containing a signal sequence.
40. See e.g. the EP examiner: lack of inventive step because the application does not disclose the function of the claimed protein or its gene. In US 20020012966, page 8 of the final rejection. And further in the final rejection: The specification asserts that

Shortly after the filing of the application for *18 Human secreted proteins* (US 2020012955, 23 Claims) a US patent (US64200526) was granted for *186 human secreted proteins.*[41] There is no image file wrapper published for this patent which claims *an isolated protein comprising amino acid residues 27 to 234 of SEQ ID NO:478* (and 57 other claims). Furthermore, another application (US 20040235113) within this patent family was filed in 2004. The first 24 claims were cancelled due to election and claims 24-44 were rejected for failing to comply with the written description requirement. However, this application went through and *the patent was granted in June 2008 (US7381800) with a new (and narrower) title: Antibodies to HBIMF63 polypeptide.* The application almost copies the abstract and description of the former application ('Briefly, the uses include, but are not limited to the detection, treatment, and/or prevention of *Alzheimer's Disease, Parkinson's Disease, Huntington's Disease, Tourette's Syndrome, meningitis*') Despite reducing claim from a gene patent to an antibody against polypeptide, the overall framework still comprises diagnostic and therapeutic methods useful for diagnosing and treating diseases, which can still be considered speculative.

Patent No US 7368531 was issued in May 2008. It has 40 claims (claim 1 is for an isolated polypeptide comprising amino acid residues 20 to 283 of SEQ ID NO: 5761). This patent also relates to human secreted proteins/polypeptides, and claimed a variety of polypeptide sequences from a protein. The description included a laundry list of 44 different disorders, including cystic fibrosis, and Alzheimer's.

The application (US 2007037206 claiming antibodies, gene sequences, antagonists, agonist and more) comprised 32 claims and the description of preferred diseases [point 0055] was repeated in the patent as issued. In a non-final rejection in 2004, the US examiner emphasised that DNA, protein and antibodies are related to each other, but that each product had acquired a separate status in the art.

translation products of this gene share sequence homology with ligand binding protein which is thought to be important in cell-to-cell communication and signal transduction. Further, the translation product of the gene also shares sequence homology with lymphocytoma proliferation activating peptide (LPAP) ... Based on the specification and examples, no biological activity has been set forth for the polypeptide of clone HBIMF63, nor has any use for the polypeptide itself been provided. Page 4: Other activities that the protein may exhibit are listed throughout pages 275-277 of the specification. However, these activities are speculative. In summary, the proteins claimed do not have a credible, specific or well-established or even demonstrable utility and therefore lacks utility under 35USC 101.

41. US application filed in 1998, US 6420526, 2002 (58 claims). Same inventor group.

The examiner then expressly referred to 'a laundry list' in point 8 as follows:

'The claimed invention is directed to polynucleotides and the encoded polypeptide (see page 2). Further, the specification on page 19 asserts that the polynucleotides and polypeptides can be used in assays to test for one or more biological activities and can be used to treat associated diseases for which *a laundry list* is provided on pages 11-12. The diseases and conditions which include but are not limited to, reproductive disorders, hyperproliferative disorders, cardiovascular and immune activity disorders. Further, the polypeptides and antibodies directed to these polypeptides are useful in providing immunological probes for differential identification of the tissues (s) or cell type (s). *However, this asserted use is not demonstrated in the specification.* The specification *provides no indicia of the claimed polypeptide and polynucleotide in association with any specific disease or disorder*, for example, those listed on pages 11-12 or any data via *in vitro* or *in vivo* studies. In addition, *no working examples* are provided to support the asserted use.'

### 2.2.c. Stem cells[42]

This category will only be discussed briefly. It is mainly part of the case study in order to follow the technological development as the European legislation prohibits the commercial exploitation of human embryonic stem cells (hES) (see Chapter 4). It is thus interesting to follow whether that ban has been circumvented by new technological developments. The main discussion in this case study is in the Parkinson's disease (PD) review.

Category (c) stem cells did not reveal activity on disease-specific induced pluripotent stem cells, although the scientific literature shows that this is ongoing in biomedical research.[43] The three groups studied in the present search are most likely non-exhaustive in a broader search context. The category relating to Huntington's disease comprises the following three groups which are all more traditional stem cell patent applications.

Group (19) claimed a new medical system by the administration of mesenchymal stem cells (cells that differs from ES) in amounts effective to treat the genetic disease or disorder in an animal.[44] The claims also include a laundry list (claims 8-21: Gaucher's disease, Parkinson's disease, Alzheimer's disease, Huntington's disease, Charcot-Marie-Tooth syndrome, Zellweger syndrome, autoimmune polyglandular syndrome, Marfan's syndrome, Werner syndrome, adrenoleukodystrophy, Menkes disease, malignant infantile osteoporosis, spinocerebellar ataxia, spinal muscular atrophy ...).

42. Groups (9) (19) and (28). In the latter, the applicant also described a comprehensive laundry list ([0029] cystic fibrosis, [0037]; US 2009035347.
43. See e.g. In-Hyun Park et al., 'Disease-specific induced pluripotent stem cells', *Cell* 134, 877-886, 5 September 2008.
44. US 2007253931, EP 1973553, US 2009214487.

After the final rejection in August 2009, claims 1-32 were cancelled due to lack of enablement and only cystic fibrosis remained from the laundry list (claim 33). *The claims are narrowly drawn to a well-characterized genetic disease (cystic fibrosis) and a well-defined population of cells (allogenic mesenchymal stem cells).* Group (28) broadly claimed isolated stromal cells.[45] The invention related to isolated marrow stromal cells ('MSCs') for use in treatment of central nervous system ('CNS') diseases, disorders or conditions. Most patents in this group have been granted.

**2.2.d. Highlights: RNA interference (RNAi)**

A huge number of patent applications concerning Huntington's disease involved RNA interference (RNAi). This reflects the fact that Huntington's disease is a dominant disease caused by a mutation (triplet repeat expansion) that results in production of a dominant negative protein. RNAi methods which specifically remove the mutant mRNA that encodes the dominant negative protein are therefore very promising as a future therapy. In particular, RNAi methods which are allele-specific (mutation specific) have a great potential and many of the patents therefore claim a specific RNAi sequence which only targets, for instance, a common SNP variation. This approach would leave the mRNA from the patient's normal functional allele intact and ensure that normal huntingtin (which is required for normal cell function) is still produced in the patient.

The following review of the patent activity can be divided into the following headings:

(a*) The RNAi technique as such
(b*) Targeting the *HTT* gene

Patents under the first heading which comprises claims for the RNAi technique as such, also include a comprehensive laundry list. This is not surprising as the technique can be used to down-regulate or inhibit the expression of disease genes in general. The second heading, the specific use of RNAi in regard to huntingtin (2), also comprises claims for other diseases.[46]

45. US 5716616, EP 0871457, US 2003039639, US 2003059412, EP 1304112, EP 1304113, EP 0871457, US 2003202966, US 2004166097, US 6974571, EP 1304112, EP 1304113, US 2009035347, US 2009214492.
46. See Chapter 2, section 2.1.d.1, where the patents for the fundamental technology are mentioned. See further Antonio Orlacchio, Giorgio Bernardi, Aldo Orlacchio & Sa-

*(a*). The RNAi technique as such*[47]

This heading comprises two groups claiming an invention based on dsRNA and/or siRNA techniques or even claiming the technique itself.

*Group (7)*, attenuation of gene expression using dsRNA,[48] is extremely broad, referring to and claiming the technique/method as such without a specific sequence.[49] The claims are very broad and cover from dsRNA at least 20 nucleotides long to dsRNA at least 400 nucleotides long as well as the dsRNA being an siRNA The claims cover both *in vivo* and *in vitro* and open up a host of new medical applications. Examples include the construction of anti-viral agents, anti-tumour agents, and therapeutics designed to block the expression of specific alleles of genes that have been implicated in autosomal dominant genetic diseases, such as Huntington's chorea.[50] It is remarkable a claim is made for the use of siRNAs to block the expression of specific alleles without defining which nucleotide changes (Mutations or SNPs) are targeted and in which genes. It is fundamental to the therapeutic use of the siRNA technique for dominant negative diseases that one can specifically target mRNA from only one (the disease allele) of the two alleles in the patient cells. Typically one would want to design a siRNA that will only bind to one of the two possible sequences harbouring a SNP variation. Because the disease allele harbouring the repeat expansion can be defined by the presence of one of two possible nucleotides in cis at each SNP position in the HTT gene, it would be possible to specifically target the mutant mRNA by designing allele-specific siRNAs for the various SNPs and use the SNPs where the patient's normal allele and the mutant allele are heterozygous (are different). Huntington's disease is mentioned in a laundry list as a repeat disease under the category of neurological diseases.

bata Martino, 'Patented therapeutic RNAi strategies for neurodegenerative diseases of the CNS', *Expert Opinion on Therapeutic Patents*, 2008, Vol. 18, No 10, p. 1167. The review refers to three HD applications on therapy by RNAi: US 20070167389, on intercranial delivery of siRNA; US 20070092889 (PD essentially) on siRNAs and surgically implanting a catheter in the brain to obtain a predetermined dosage infusion; and US 20050042646 on targeting siRNAs directed to mRNA for huntingtin.

47. Groups (7) and (8)
48. See US 20040147475 which is abandoned, and US 2002114784 and EP 1147204 which are still pending.
49. See also US 20090130751 on RNA interference (RNAi) molecules targeted at a nucleic acid sequence, and methods of using these RNAi molecules to reduce off-target toxicity (Markush reduction).
50. See description point [0027].

The office communication of the US 20040147475 (abandoned) stated a claim rejection (35 USC § 112), page 5: *the disclosure provides examples of attenuation of gene expression of three vertebrate species while the claims are directed to all vertebrate species, which number in the tens of thousands. ... While the zebra fish may be a model organism for studies of gene function, this does not speak to the ability of zebra fish to model human gene therapy.* Furthermore: Anticipated rejection cf. USC §102 and 103: Applicants additionally refer to recent court decisions as evidence that the disclosure of the Fire et al. reference is not enabling. Applicants refer to the decision in *Impax Laboratories, Inc. v Aventis Pharmaceuticals, Inc.* after remand from the federal circuit, noting that the District Court of Delaware determined that mention of riluzole *in a list of several compounds to treat a myriad of diseases did not anticipate claims using riluzole to treat ALS.* Applicants liken this broad disclosure to Fire et al. by noting that Fire et al. simply recites vertebrate cells as one of many cell types in which the target gene can reside. However, as noted above, the many cell types referred to by applicants are actually disclosed at column 8, lines 13-15 as including plants, animals, protozoa, bacteria, viruses, or fungi; one of ordinary skill immediately recognizes that *vertebrates are not part of a large laundry list,* but as one of the two types of animal cells. The instant specification provides an almost identical definition of the cells useful in the claimed method at page 2. The cell into which the dsRNA is introduced can be derived from or contained in any organism (e.g., plant, animal, protozoan, virus, bacterium, or fungus).

In the same category one can find *group (8)* claiming RNAi mediated treatment of polyglutamine (POLYQ) repeat expansion diseases using short interfering nucleic acid (siNA).[51] The problem to be solved by the invention is the provision of more stable interfering nucleic acids for down-regulating Huntingtin. However, the application broadly claimed the whole siRNA technique.

The description comprises a comprehensive laundry list e.g. [0226-234] proliferative disease, inflammatory disease, autoimmune disease, infectious disease, neurologic disease, respiratory disease, cardiovascular disease, ocular disease, metabolic disease with quite many diseases listed in alphabetic order (more than two hundred including Alzheimer's disease, Canavan's disease, Huntington's disease, Parkinson's disease, spinal muscular atrophy) in point [0230].

51. EP 1735443, US 2007042983. The inventor is McSwiggen, Chapter 2, section 3 who has been granted patents on siRNAs for the study, diagnosis and treatment of diseases and conditions associated with polyglutamine repeat (polyQ) allelic variants, for example Huntington's disease.

The 2006 European examination found that the subject matter of the claims: double-stranded short interfering nucleic acids targeting a huntingtin RNA and having one or more chemically modified nucleotides, is new, however the *problem to be solved* by the invention may therefore be regarded as the provision of *more stable interfering nucleic acids for the down-regulation of huntingtin*. The *solutions* proposed in the claims of the present application cannot be considered as *involving* an *inventive* step (Article 33(3) PCT) *as the same chemical modifications and structures have already been employed for the same purpose in prior art*. The 2009 US office action: Applicant is correct that the aforementioned *structural features are disclosed by the specification. However, there is not a single species actually made by the inventors of the instant case; rather, the 'claimed' invention is described in terms of a mere wish or plan for making an asymmetric double-stranded nucleic acid having at least about 3-nucleotide complementarity. The mere disclosure of structural requirements without reducing the claimed invention to actual practice does not satisfy the written description requirement set forth in the first paragraph of 35 U.S.C. 112, because short double-stranded nucleic acid molecules, especially the instantly claimed at least about 3-nucleotide complementarity between the asymmetric lengths of the two strands, were relatively unknown and highly unpredictable in the art at the time of the invention. Particularly, the instantly claimed double-stranded nucleic acid molecule inherently confers a functionality or utility such that it is 'capable of mediating RNA interference (RNAi) against target gene* expression' [*emphasis added*]. Hence, the instant invention claimed in pure structural terms is inherently inseparable from its *intended functionality or utility*.[52]

### *(b*). Targeting the HTT gene*

The patents concern the targeting of various proportions of the *IT15* gene sequence by short (typically 15-19 nt long) complementary RNA oligonucleotides in order to down-regulate *IT15* gene expression. Some patents claim several sequences (e.g. groups (12) and (18)), whereas others claim a particular sequence (e.g. group 16). As mentioned above, as HD is a dominant disease, where the aggregating protein product derived from the disease allele causes the disease, it is highly relevant to target only the disease allele in patients (allele-specific siRNA). Consequently, some of the patents (e.g. group (25) and group (38)) claim sequences with reported frequent SNP variations. The idea here is that the requirement for a siRNA to be completely complementary to the target mRNA in order to cause degradation, can be exploited in such a way that only the mutant disease-causing mRNA is degraded in patients, because the nucleotide at the SNP position will determine whether the siRNA is fully complementary. siRNAs that are not allele-specific are of little therapeutic value, since they will cause degradation of both mutant and normal *IT15* mRNA in patients, causing lack of normal IT15 protein and may therefore themselves cause neuronal disease.

52. All abandoned.

Another trend concerns how the siRNAs are delivered and the chemistry employed, for example: *group (20)*, claiming a dsRNA comprising at least one modified nucleotide; *group (32)*, siRNA-mediated gene-silencing with viral vectors (delivery); and *group (36)*, claiming a siRNA agent designed to target neural cells (comprising a lipophilic moiety, a phosphorothioate or a 2'-OMe modification). Figure 7.2 illustrates the mechanism behind the allele-specific siRNA for Huntington's disease.

*Figure 7.2.*

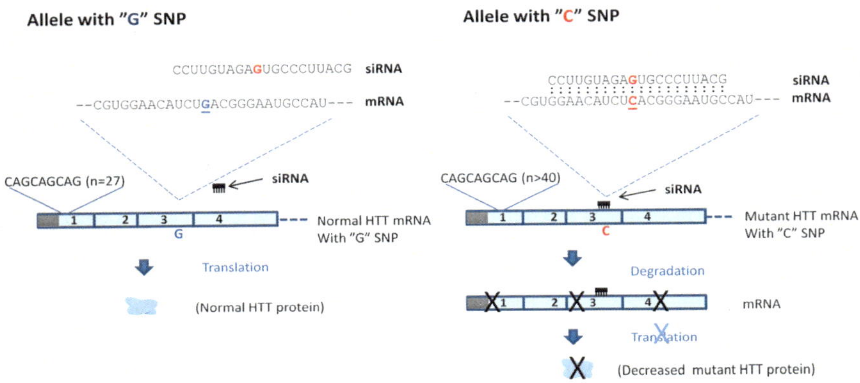


Thus, the data have been arranged in sub-groups (*several sequences, a particular sequence, sequences with reported frequent SNP variations, and siRNA delivery).* Some groups qualify for more than one sub-group but they are only referred to in one.

*Several sequences*[53]
*Group (12)* claim new isolated double-stranded short interfering nucleic acid molecule comprising complementary sense and antisense regions, useful for manufacturing a medicament for preventing or treating Huntington's disease.[54] It is targeted at a specific sequence, an expression vector (encoding

53. Groups (12) (17) (18) (26) (37) and (42).
54. EP 2014769 granted in 2010, claiming the antisense region, has from 15 to no more than 19 contiguous 5 nucleotides that are complementary to a human huntingtin transcript, these nucleotides being encoded by a sequence selected from the group consisting of SEQ ID NO: I to 3; the sense and antisense regions have at least 15 contiguous nucleotides that are complementary to each other and form a duplex,

the siRNA) and a rodent model to study HD. *Group (17)*[55] claimed siRNAs suitable for inhibiting huntingtin genes in both humans and macaca mulatta (macaques), due to the high degree of sequence identity. *Group (18)*[56] discloses many siRNA sequences targeting the huntingtin gene in primates, including macaca mulatta and homo sapiens. The siRNAs claimed are not designed to target particular SNPs. *Group (20)*[57] claimed a dsRNA for inhibiting the expression of a human huntingtin gene in a cell, where the dsRNA comprises a sense strand consisting of SEQ ID NO:792 and an antisense strand consisting of SEQ ID NO:793. This dsRNA also comprises at least one modified nucleotide. *Group (24)*[58] claimed siRNA for two genes (HD huntingtin, and spinocerebellar ataxin-1 (SCA-gene)), viral vectors (an AAV-1 expressed siRNA) and a mouse model.[59] *Group (42)* targets many siRNAs and claimed to cause allele-specific silencing of disease genes.[60] *Group (26)*[61] comprises dsRNAs, homologous to a certain sequence targeted at a huntingtin mRNA.

and the double-stranded short interfering nucleic acid molecule inhibits 10 the expression of endogenous wild-type and exogenous human mutated huntingtin genes in cells of a non-human mammal which express both these huntingtin genes.

55. US 2006257912 was abandoned and EP 1885854 is still pending.
56. US 7902352 was granted and EP 2056881 is still pending.
57. US 2007099860, US 7320965, EP 1941059. The examiner stated that although the disclosure teaches a method for inhibiting Huntington's disease in a cell in vitro, by administering a double-stranded ribonucleic acid (dsRNA) for inhibiting the expression of a human HD gene, *such a disclosure would not be considered enabling since the state of antisense-mediated gene inhibition in vivo is highly unpredictable.*
58. US 2005255086, EP 1844148, US 2008274989 are all abandoned.
59. A person of ordinary skill in the art would have been expected to be able to design any siRNA targeted to any mRNA transcript because Fosnaugh et al. detail the steps to effectively find a target site in any RNA and design and test siRNA molecules for specific RNAi activity. Fosnaugh et al. teach known target sites that have previously been targeted by antisense compounds which are useful as well as known target sites that have been shown in art to be responsible for certain diseases.
60. US 2009105169 (abandoned)) compared to group (24), each claimed a specific (but different) allele. The application embraced a huge number of siRNAs [0007] [0008] and many different genes [0006]. The alleles of the targeted gene may differ by seven or fewer base pairs out of 21 base pairs (e.g., 7, 6, 5, 4, 3, 2 or 1 base pairs). They may even differ by only one base pair out of 21 base pairs. Examples of targeted gene transcripts included transcripts encoding beta-glucuronidase, TorsinA, Ataxin-3, Tau, or huntingtin. The targeted genes and gene products (i.e., a transcript or protein) may be from different species of organisms, such as a mouse allele or a human allele of a target gene. A kind of laundry list can be identified in point [0137].
61. US 7589189 granted.

*Sequences with reported frequent SNP variations*[62]
The above three groups all comprises siRNA and they more or less claim the same.

*Group (16)* claimed silencing mRNA encoding mutant huntingtin.[63] Without defining specific sequences or SNPs, the inventor claimed all siRNAs that target SNPs with an allele frequency of more than 35 % in the IT15 gene. They also claim a specific siRNA that targets a frequent SNP (RS363125) in the *IT15* gene.[64]

The extended European Search Report (October 2010) stated that the invention lacked novelty as the prior art disclosed 19 bp siRNA targeting both variants of the particular SNP (RS363125) in the *IT15* gene identified in the application (i.e. the prior art comprises the same SNP region). Some of the siRNAs in the prior art differed from the claimed sequence at one nucleotide outside the polymorphic position and can be considered as variants. The report stated that the claims concerning SNPs in the huntingtin gene having an allelic frequency of at least 35 % in a sample population disclose a limited number of such SNPs and it is an undue burden to determine which SNPs fall within this definition because the allelic frequency can change in sample populations (EPC Articles 83 and 84). It is interesting to note that emphasis is on the fact that determining which SNPs have an allele frequency of 35 % is an undue burden, rather than pointing to the lack of the provision of specific sequences or SNP IDs for the claimed siRNAs/SNPs.

*Group (25)*[65] concerns a method for treating a subject having or at risk of having a disease characterised or caused by a gain-of-function mutant protein. Claim 1 is for the medical use of an allele-specific siRNA targeting a mutant huntingtin gene. The gene involved comprises an expanded trinucleotide repeat region (claim 2) or the mutant protein comprises an expanded polyglutamine domain (claim 3). However, the diseases in claim 4 are selected from a group consisting of *Huntington's disease, spino-cerebellar ataxia type 1, spino-cerebellar ataxia type 2, spino-cerebellar ataxia type 3, spino-cerebellar ataxia type 6, spino-cerebella ataxia type 7, spino-cerebellar ataxia type 8, spino-cerebellar ataxia type 12, fragile X syndrome, fragile XE MR, Friedreich's ataxia, myotonic dystrophy, spinal bulbar muscular disease*

62. Groups (16) (25) and (38).
63. EP 2046993, US 2009186410. The inventor claimed a specific siRNA by targeting a single nucleotide polymorphism (SNP) present at genomic site RS362331.
64. All withdrawn.
65. EP 1670518 still pending (extension of examination granted in 2010) and US 7947658 granted.

*and dentatoiubral-pallidoluysian atrophy* which do not all involve expanded polyglutamine domains. The invention was originally for 43 different SNPs (claims 7 and 8), but the claims were narrowed down to HD in the granted US patent. The non-final action stated as follows:

'enabling for direct delivery of a siRNA targeting that specific huntingtin sequence at site 2886 and a resultant inhibition or treatment effect, but is not enabling for treating a subject at risk for the disease and is not enabling for treatment of Huntington's disease or any other disease via the instantly recited broad method, which only requires targeting any polymorphic site distinct from a disease causing mutation. The instant specification does not draw an adequate nexus between inhibition of any polymorphic site distinct from a disease causing gene with a resultant treatment effect of such a broad genus of diseases and does not draw an adequate nexus between targeting any such polymorphic site even specific to a huntingtin target sequence with predictable cleavage or a treatment effect.'[66]

*siRNA delivery*[67]

It is crucial that the siRNAs reach the target cells, for instance neuronal cells in the brain. This is typically ensured by using special modifications of chemically synthesised oligonucleotides or by using viral vectors, which are designed to preferentially target the relevant cells or by employing mechanical devices such as pumps and catheters.

*Group (32)*[68] is for siRNA targeted at a nucleic acid sequence that encodes huntingtin or ataxin-1, and is thus limited to specific siRNA for the two disorders,[69] and it comprises different forms of delivery, e.g. plasmid and virus

66. See the 2010 US final rejection,
67. Groups (13) (24) (20) (32) (33) (36) (39) and (41)
68. EP 1534861, US 2005042646, US 2011111491, US 2004241854, US 2010144026, US 2005106731, US 2005255086, US 2006009408, EP 1576118, US 2010190243, US 2004023390, EP 1581635, EP 1844148, and US 2008274989 – all abandoned.
69. EP 1576118. Withdrawn in 2009, but some remarks from the EPO examiner on Article 56 EPC are as follows: 'Prior art discloses expression constructs for producing shRNAs used in gene silencing. Huntington's disease and the different ataxias are mentioned as target for these interfering RNAs (... proteins of Table I which are aberrantly produced and associated with a disease phenotype). In the view of the examining division this implicitly means the disease-causing allele of a particular gene. Hence, prior art addresses the exact same problem as does the present application, i.e. specific inhibition of the Huntington's disease gene by RNA interference. The subject matter of present claims differs from that teaching in that short interfering RNAs (i.e. with strand lengths of 15 to 30 nucleotides) are used as the inhibitory agents. The problem to be solved by the present invention may therefore be regarded as providing alternative silencing RNAs for the disease-causing allele of the HD gene. The proposed solution being interfering RNAs having strand lengths of 15 to 30 nucleotides.'

(claims 17-20). *Group (33)* claimed a treatment for neurodegenerative disease through intracranial delivery of siRNA. It comprises the steps of surgically implanting a catheter so that a discharge portion of the catheter lies adjacent to a predetermined infusion site in the brain, and discharging a predetermined dosage of at least one substance capable of inhibiting production of at least one neurodegenerative protein. The present invention also provides valuable small interfering RNA vectors, and methods for treating neurodegenerative disorders such as Alzheimer's disease, Parkinson's disease, Huntington's disease, Spinocerebellar Ataxia Type 1, Type 2, Type 3, and/or dentatorubral-pallidoluysian atrophy.[70] Claims 16-18 claimed a gene patent covering all appropriate siRNAs corresponding to the gene and the gene itself. *Group (39)* claimed *polypeptide-nucleic acid conjugates* and uses thereof.[71] The description states that the invention claimed conjugates that can be used for transporting RNAi agents, for example, siRNA agents, to cells, tissues, or organs to treat cancer, neurodegenerative diseases, or lysosomal storage diseases. According to claim 7, the claimed composition is able to cross a mammalian blood-brain barrier efficiently. [72] *Group (41)* claimed devices, systems and methods for improving memory and/or cognitive function through brain delivery of siRNA (e.g. intercranial access device).[73] *Group (36)* also claimed

70. EP 1569662, EP 2145628, US 7605249, US 7829694, US 2010063134, EP 1829547, EP 1832292. All US applications were granted whereas the EP applications were withdrawn (after oral proceedings where objections were directed to the lack of an inventive step (EPC Article 56 (intracranial delivery already known)).
71. EP 2235175 and US 2011039785, both pending. It comprises a comprehensive laundry list in claims 41 and 43: 'wherein said neurodegenerative disease is multiple sclerosis, schizophrenia, epilepsy, Alzheimer's disease, Parkinson's disease, Huntington's disease, amyotrophic lateral sclerosis (ALS), or a stroke and in claim 43 wherein said lysosomal storage disease is mucopolysaccharidosis (MPS-I; i.e., Hurler syndrome, Scheie syndrome), MPS-II (Hunter syndrome), MPS-IIIA (Sanfilippo syndrome A), MPS-IIIB (Sanfilippo syndrome B), MPS-IIIC (Sanfilippo syndrome C), MPS-IIID (Sanfilippo syndrome D), MPS-VII (Sly syndrome), Gaucher's disease, Niemann-Pick disease, Fabry disease, Farber's disease, Wolman's disease, Tay-Sachs disease, Sandhoff disease, metachromatic leukodystrophy, or Krabbe disease.'
72. In another embodiment, the polypeptide is produced by recombinant genetic technology or chemical synthesis. The polypeptide-nucleic acid conjugates of the invention can be admixed or formulated with a pharmaceutically acceptable carrier [0006-0008]. Claim 4 comprising 112 sequences.
73. US 7618948 and US 8058251, both granted and the last US application (US 2010184835) is still pending.

an allele-specific siRNA for Huntington's disease.[74] The siRNA agent is designed to target neural cells (comprising a lipophilic moiety, a phosphorothioate or a 2'-OMe modification).

## 3. Spinal muscular atrophy

The literature case study in Chapter 6, section 2.1.1, on spinal muscular atrophy (SMA) is mostly a description of the disorder and its genetics. So far, it has been difficult to find literature on SMA patents. As described in Chapter 6, section 2 there are two genes involved in SMA, namely *SMN1* and *SMN2*. A mutation in *SMN1*, usually a deletion of the entire gene, causes the disease whereas the severity of the disease is determined by the number of copies of the *SMN2* gene, which is only partly functional due to a high level of aberrant pre-mRNA splicing. One invention was referred to in the biomedical literature namely biomarkers for predicting the severity of SMA in the form of a protein, peptide, polypeptide, polynucleotide and many other forms such as a vitamin.[75]

### 3.1. Data outline

The present case study on spinal muscular atrophy comprises 50 groups of altogether 105 patent applications, of which 14 groups representing 35 applications were deleted due to an error in the search term. In the current search 'SMA' refers to 'spinal muscle atrophy', but it was found that it also refers to 'smooth muscle actin' or restriction enzyme for cloning techniques.[76] Some of the groups omitted also demonstrate the broad palette of associated diseas-

74. Almost every application withdrawn (EP 1937066, US 2010267810, US 2011027882, US 2007105803, US 2007161591 and US 2010267810); only EP 2239328 is still pending.
75. WO 2011032109 and *group (29)*, comprising a method for predicting the level of severity of spinal muscular atrophy or disease progression in a patient or population of patients comprising: (a) measuring the level of at least one marker and/or the ratio of at least two markers ... in a biological sample obtained from one or more patients; and (b) comparing the measured level or ratio to a reference value or range of reference values, wherein a change in the level of the at least one marker or the ratio of at least two markers is indicative of the level of severity of spinal muscular atrophy or disease progression in the patient or population of patients.
76. Groups (1) (5) (7) (8) (9) (10) (17) (19) (20) (23) (24) (25) (27) and (28), all in bold in appendix VIII Table SMA.

es[77] that are claimed to benefit from the invention (from *group (9)* (CXCR4), associated medical conditions are e.g. cancer, AIDS, inflammatory disease, diabetes, multiple sclerosis and autoimmune hepatitis).[78]

The latest update from March 2012 shows 70 applications, of which 16 patents have been granted, 28 applications have been withdrawn, and 26 applications are still pending, see Appendix VIII, table SMA. The search strategy has been described in Chapter 6, section 4. The focus of the case study of SMA is based on the following patent activity:

### 3.2. The empirical analysis

The case study template for the SMA analysis model is shown in box 7.3.C below.

*Box 7.3.1. SMA analysis model**

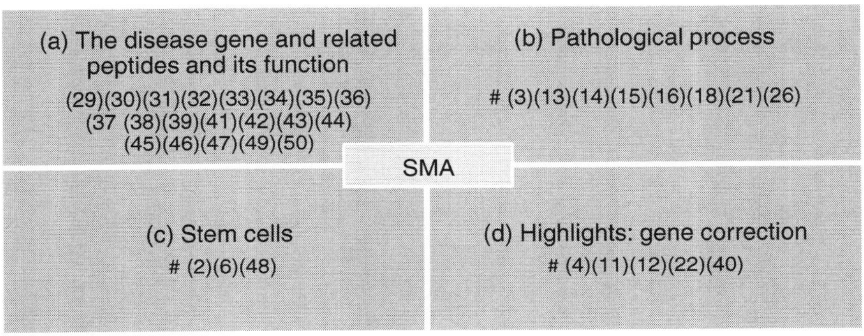


* Heading (c) is omitted from the SMA analysis. None of the groups comprised new technology that may circumvent the ban on patenting embryonic stem cells in Europe, and it is almost a repetition of the HD study since all groups claimed a laundry list, e.g. group (2) embraced almost all human diseases in point [0017] and same in group (6) point [0069-70].

### 3.2.a. The disease gene, related peptides and its function

The category 'the disease gene and related peptides and its functions' contain activities relating to the genes involved in the disease (*SMN1*, *SMN2*).

77. Group (10) in particular has a wide laundry list given that CCR5 is normally considered HIV relevant.
78. EP 1883651 and US 2009054322, both abandoned.

The category comprises the following groups:

(21) (29) (30) (31) (32) (33) (34) (35) (36) (37: 8) (38) (39) (41) (42) (43) (44) (45) (46) (47) (49) and (50).

Some of the groups claimed delivery systems in the form of vectors or nanoparticles and they can be further divided into sub-groups of delivery systems and protein replacement therapy. The major problem in gene therapy is ensuring effective gene delivery and transient expression of a gene, thus inducing a therapeutic effect. Another group claimed treatment of SMA and they also are in a separate sub-group. The last group comprises a variety of inventions relating to polypeptides, molecules and probes for diagnosis, gene correction and chemicals used to enhance gene expression and splicing efficiency. *Groups (20)* and *(40)* have to be assessed under heading (d) highlights, as they relate to gene correction; see Figure 7.1.D above.

*Delivery systems*
*Groups (21) (31) (32) (36) (38) and (45) all claim vectors for gene therapy, and groups (41) (42) (44)*[79] *and (46 claim) nanopolymers for improved delivery of nucleic acid and protein replacement therapy.* Thus these groups can be categorised as systems for delivery to the target tissue (in this case neurons or more specifically motor neurons in the spinal cord) in a broad sense, since they cover a variety of diseases of which SMA is just one disease among many.

*Group (21)* claimed to deliver recombinant adeno-associated virus (rAAV).[80] The invention relates generally to the effective delivery of viral vectors to cells of the central nervous system (CNS), which is particularly useful in the treatment of neurodegenerative disorders and motor neuron diseases. The invention describes the delivery of several genes including *SMN2*.[81] In fact, the reference to *SMN2* shows that the laundry list technique

79. US 2004266715, US 2007213290 on neurite regeneration relates to vectors capable of directing the expression of a factor, such as a vector (Retinoic Acid Receptor (RAR) .beta.2) relating to neurite growth. It included a giant laundry list. It was abandoned in 2009 since the applicant did not posses at least the EIAV vector prior to the filing date (PCT/GB), and hence priority was denied.
80. EP 1402043 and US 2005032219, both withdrawn.
81. See point [0033]: a therapeutic polypeptide comprised in said rAAV virions may be for example SMN2, a trophic factor or a polypeptide capable of decreasing glutamate toxicity. The US examiner stated that the specification did not provide reasonable enablement for the claims on the nucleic acid molecule that can effectively be expressed

can be problematic since the correction for SMA would be much better if *SMN1* and not *SMN2* were delivered. *Group (26)* [82] comprises a method for the delivery of TTC peptide to nerve cells. It is a therapeutic peptide with the same ability to bind nerve cells and be retrogradely transported through a synapse as the native toxin. Claim 22 broadly claimed compositions wherein the active molecule is selected from the group consisting of protein SMN (survival motor neuron), BDNF (brain-derived neurotrophic factor), NT-3 (Neurotrophin--2-3), NT-4/5 (Neurotrophin 4/5), GDNF (glial cell-line derived neurotrophic factor), IGF (insulin-like growth factor), PNI (protease nexin I), SPI3 (serine protease inhibitor protein), ICE (interleukin -1 converting enzyme), Bcl-2, GFP (green fluorescent protein), endonucleases, antibodies or drugs specifically directed against neurodegenerative diseases. *Group (31)* claimed methods for administering AAV vectors to synaptically connected neurons, claiming gene therapy on a giant laundry list on Alzheimer's, Parkinson's, SMA, Huntington's, paraplegia and more.[83] *Group (32)* claimed a method for treating motor neuron disease using a lentiviral vector system to transduce a target site.[84] It was originally filed for a bundle of genes (nucleotides) encoding a protein selected from the group consisting of SMN-1, GDNF, IGF-1, VEGF, XIAP, NIAP, and bcl-2, but during examination it was narrowed down due to lack of technical evidence. The claim then comprised the use of a vector system in the manufacture of a medicament to treat and/or prevent spinal muscular atrophy (SMA) by transducing a target site in a neonate, wherein the vector system is or comprises at least part of a rabies G envelope protein.

in a second connected population of neurons to treat or prevent any human motor neuron disease (e.g., spinal cord injury, amyotropic lateral sclerosis, spinal muscular atrophy, Kennedy's disease, hereditary spastic hemiplegia).

82. EP 1049712, US 2003004121, US 2004170651, US 2005060761, EP 1681300, EP 1049712 and US 7435792.
83. EP 1402043 withdrawn; claims 24-26, 47, 53-55, 60-61. The examiner stated that in absence of further evidence, the claims were considered to lack sufficient disclosure as far as the first and second population of synaptically connected neurons do not correspond to the populations of the examples (Article 83 EPC). See similarly *Group (36)*, gene therapy for spinal cord disorder, AVV vector and a giant laundry list; EP 1879623 and US 2009069261. See further *group (45)*, AAV vector again used to administer to a first population carrying the nucleic acid that could effectively be expressed in the second connected population of neurons to treat or prevent diseases within a giant laundry list; US 2005032219, abandoned.
84. EP 1551974, US 2004076613 and US 2008131400, all abandoned.

The examination stated that the wording of the claim encompassed the use of the vector system for the manufacture of a medicament to transduce an *in utero* target site, despite the fact that there was no technical evidence showing that the technical problem had been credibly solved.

*Group (38)* provides delivery of a transgene product to the spinal cord by administering an AAV vector.[85] The invention aims at correcting a subject's motor function, in particular, motor function affected by disease or injury to the brain and/or spinal cord. The bundle of genes in claim 19 comprises insulin growth factor-1 (IGF-1), calbindin D28, parvalbumin, HIF1-alpha, SIRT-2, VEGF, SMN-1, SMN-2, CNTF (ciliary neurotrophic factor), sonic hedgehog (shh), erythropoietin (EPO), lysyl oxidase (LOX), progranulin, prolactin, ghrelin, neuroserpin, angiogenin, and placenta lactogen. In a non-final action in December 2010 the US examiner emphasised the biological function of the large number of proteins claimed, and stated that the specification failed to provide adequate guidance and evidence for how to treat various diseases and disorders with various transgenes via gene therapy *in vivo*. Moreover it was questioned whether sufficient or therapeutic amounts of a transgene, including the *SMN-l* transgene, could be delivered as claimed.

*Group (46)* encompasses polymeric carriers for delivery of a class of oligonucleotides that modulate RNA splicing. The concept is for carriers for the delivery of nucleic acid agents to cells and tissues.[86] As inventing functionalised carriers such as nanopolymers and nanovesicles for improved delivery of nucleic acid agents (NAA) focuses on the delivery process, it claimed broad applicability for treating numerous diseases and disorders, including neurodegenerative and neuromuscular disorders. Inserting oligonucleotides in nanoparticles using different chemical compounds (see e.g. claim 18 and the specific inclusion of human exon 7 of the *SMN2* gene) is one way of modulating or correcting gene expression in order to increase the amount of functional SMN protein by means of the body's own machinery and thus modulate the severity of the disease or cure the disease.

*Replacement therapy*

*Group (41)* claimed therapeutic compositions for protein replacement therapy e.g. therapeutic agents and nitrogen-containing lipids, nanoparticles, microparticles, and complexes containing lipid/therapeutic agent complexes. 'Replacement therapy' in the application refers to the use of a protein to reconsti-

85. EP 2029742, still pending, whereas US 2010267812 is abandoned.
86. WO 2007133812 and US 2009011004 both withdrawn.

tute a deficiency or to increase otherwise low levels of the protein, e.g., in an individual who has a disease or disorder, or has a predisposition for a disease or disorder caused by or associated with a protein deficiency or low levels of a protein.[87] The laundry list technique is used and e.g. claim 4 claimed proteins comprising: Bruton's tyrosine kinase (BTK), ornithine transcarbamylase (OTC), survival motor neuron 1 protein (SMN1), galactocerebrosidase, N-sulfoglucosamine sulfohydrolase, N-acetylglucosaminadase, iduronate-2-sulfatase, alpha-glucosidase, sulfatase-modifying factor 1 (SUMF1), glucocerebrosidase (GCB), alpha galactosidase, alpha iduronidase, beta glucuronidase, or N-acetylgalactosamine-4-sulfatase. The claimed diseases are subject to differing compositions of lipid, micro and nanoparticles targeted at the form of delivery of the protein.

*Treatment of spinal muscular atrophy*

*Groups (30) (33)*[88] *(39) (43) (47) and (50)* all claim treatment of spinal muscular atrophy. *Group (30)* addresses *SMN2* and the splicing process.[89] It is well recognised that all viable patients with SMA have one or more copies of the *SMN2* gene, but that *SMN2* only results in low amounts of the SMN protein due to aberrant pre-mRNA splicing, where exon 7 is skipped. If the skipping of exon 7 from the *SMN2* gene can be prevented *in vivo*, *SMN2* protein production will be increased and this will be a treatment for SMA. Alternatively, if expression (transcription) of the *SMN2* gene can be up-regulated, this will also lead to increased levels of SMN protein which may also be used therapeutically. Many of the groups concern drugs/chemical substances, which are claimed either to increase transcription or improve splicing of *SMN2*. For example, *Group (39)*[90] claimed tetracycline compounds modulating *SMN2* mRNA splicing.[91] *Group (43)* claimed hydroxyurea in an amount

87. US 2011052673, point [0003] still pending, and EP 2244694 also pending.
88. Group (33), on the use of a histone deacetylase inhibitor for treating muscular dystrophies. The invention concerns in particular the treatment of dystrophies such as Duchenne's dystrophy or Becker's dystrophy, but it claimed atrophy as well; e.g. Claim 41 for a protein coded by the SMN1 gene (EP 1565175 pending, US 2006148684 abandoned and US 2008160108 pending).
89. EP 1133993 withdrawn.
90. EP 2144614 pending, and US 7935687 granted.
91. The US examiner stated that the Applicant had discovered a new property or advantage of a method made obvious by the prior art. MPEP 2145 II stated: 'The fact that Applicant has recognized another advantage which would flow naturally from following the suggestion of the prior art, cannot be the basis for patentability when the differences would otherwise be obvious'; *Ex parte Obiaya*, 227 USPQ 58, 60.

sufficient to ameliorate symptoms of SMA.[92] *Groups (47) and (50)* provide a method for modulating SMN gene expression in a cell, comprising contacting an effective amount of sodium-proton exchanger inhibitor to the cell, to increase the expression level of SMN exon 7 in the cell. [93]

*The group comprises diverse inventions: polypeptides, molecules and probes*
Groups (29) (34) (35) and (42) all claim inventions relating to *SMN1* or *SMN2*, but on a distant correlation.[94] Group (29), just entered in the European phase, very broadly claimed different compounds as biomarkers without specifying particular pivotal factors. *Group (34)*[95] comprises both HD and SMA (*POSH* interacting proteins and related methods) and *group (35)*[96] claimed the use of ubiquitin ligase inhibitors as useful for the treatment of neurological disorders, with an extensive laundry list of proteins which may be inhibited, including SMN1 and SMN2.

### 3.2.b. Pathological process

The category of 'pathological process' contains both activities relating to the gene involved in the disease (*SMN1*, *SMN2*) and activities relating to the symptoms of the disease. Once again following the structure of the overall analysis, the category of 'pathological process' contains most applications and patents relating to the symptoms or related conditions caused by the disease. The heading comprises the following groups (13) (14) (15) (16) and (18). Together, these groups claimed formula and synthetic peptides, and others claimed antibodies, vectors and peptides. *Groups (15-18)* connect to the pathological process but lack close connection to SMA. Their main features

92. US 6573330.
93. Group (47), US 2010152204 abandoned. Whereas group (50), US 6376508 granted, for histone deacetylase (the inventor differs from group (33)). Sodium-proton exchanger inhibitor induces the expression of SRp20 protein, thus modulating splicing of the SMN gene. The examiner in the final rejection 2011 stated that the invention of Group I functions modulated SMN gene expression in a cell, while the invention of Group II functions treated spinal muscular atrophy. The search fields are thus diverse, due to the inventions being patentably distinct. Group (49), concerned another chemical compound (bytyrate) up-regulating SMN2. The application included an extensive laundry list in claim 132. One US patent US 2010234402 pending, and one WO 2008150509 withdrawn.
94. WO 2011032109, applied for by the US SMA foundation.
95. EP 1646644 and WO 2004089302, both withdrawn; see group (43), appendix VI Table HD.
96. EP 1651226 granted and US 7659277 granted.

relate to muscle fibres relevant for tissue deterioration in general, skeletal muscle growth and for increasing muscle mass.

### 3.2 c. Stem cells

None of the groups (2) (6) or (48) showed activity on Induced Pluripotent Stem cells. *Group (48)* comprises the generation of brachial, thoracic and lumbar spinal motor neurons from embryonic stem cells in the absence of all-trans retinoic acid supplement.[97]

### 3.2.d. Highlights: Gene expression correction

The present heading differs from the RNAi technique described in the HD section. Where RNAi, divided into microRNA or siRNA, is used to modulate gene expression by down-regulating or inhibiting gene expression, the present techniques with antisense oligonucleotides are used to correct gene expression.

The heading comprises four groups (4) (12) (22) and (40) some of which are focused on a delivery system but at the same time are directed at a defined antisense-oligonucleotide SEQ ID no 1 (group (4)) as well as an antisense-oligonucleotide targeted to intron 7 of a nucleic acid encoding SMN2 (group 12).

*A race to seq-id*

*Group (4)*, on new polyethylene glycol-polyethylenimine-Nucleic Acid Agents (PEG-PEI-NAA) polyplex, is closely related to the technique described in groups (46) (42) and (41).[98] The applications are deemed to have been withdrawn. The US application was a continuation of application 60/755,113 of 30 December 2005 (provisional application expired: polymer-oligonucleotide RNA splicing compound). The application claimed a method for chemical modification of ASOs so that delivery to the target cells is improved. The claims had an extensive laundry list of neuromuscular diseases (e.g., dystrophies, ALS, SMA), cardiovascular disease, brain function/disorders, and cancer biology and therapeutics.

97. US 2010196332 pending.
98. *Group (4)* comprises WO 2007133812 and US 2009011004 abandoned. The US office communication of 23 December 2010 required a restriction under 35 U.S.C 121 and 372. The special technical feature was a PEG-PEl bound nucleic acid, or a synthetic vesicle encapsulating a nucleic acid. These species lack unity of invention because they are not linked so as to form a single general inventive concept under PCT Rule 13.1.

*Group (12)*[99] addresses SMA only; the technique has been illustrated above in section 1 figure 7.1.C. There are no other diseases involved and the inventors did not claim a laundry list. The technique claimed is comparable to *Group (41)* under the CF section. The patent claimed ASOs which can correct splicing (stimulate exon 7 inclusion) of *SMN2* by blocking binding of splicing inhibitory proteins to sequences located in intron 6, exon 7 and intron 7 of the SMN2 gene. The ASOs are claimed for use in the treatment of spinal muscular atrophy (SMA). Originally several ASOs located in intron 6, exon 7 and intron 7 where claimed in the same patent. The examiner suggested that the various ASOs were regarded as separate inventions because of prior art and because their effects were not identical; some of the ASOs had an inhibitory effect instead of the desired stimulatory effect.

In the 2010 European search opinion the examiner referred to the prior art (a patent application was filed in December 2005 and granted on 23 November 2010 for SMA treatment via targeting SMN2 splice inhibitory sequences).[100] However, the European examiner stated that there appeared to be no prior art disclosure of *SMN2* antisense oligonucleotides having the specific sequences and the 2'-O-methoxyethyl sugar modification on every position. Hence, the product claims and the specific use and method claims were directed to novel subject matter. The patent is still pending. One of the ASOs (ASO-10-27) claimed is currently being tested in clinical trials with SMA patients.

*New group (12a)* is a competitor (prior art) to group (12). *Group (12a)* spinal muscular atrophy (SMA) treatment targets *SMN2* splice site inhibitory sequences. The patent for the sequence of the intronic splicing silencer that is targeted by ASO-10-27 was granted in November 2010.[101]

It is remarkable that claim 16 in the granted US patent claimed ASOs where each nucleotide of the antisense compound comprises a 2'-methoxyethyl (2'-MOE) moiety. This is exactly the same modification that the European examiner stated was without prior art and therefore new subject matter in group (12), thus the priority date of the EP patent application (group 12) was prior to that of the US group (12a).

Yet another *new group (12b)*[102] has been filed by the applicants of group (12), with 107 claims covering the whole region of the gene (exon 7 and in-

99. EP 1910395 priority 23 June 2005) and US 2010216238 filed 2006; both pending.
100. The literature (Singh et al) referred to by the examiner also reveals a patent which is not mentioned by the examiner – added to the analysis as group (12a).
101. US 7838657, filed with 45 claims. The applicant differs from that of *group (12)* and it was filed on 5 December 2005.
102. EP 24442816, filed before group (12a).

tron 6 without specifying the coding region). However, the January 2012 amended application has only 16 claims and claimed a specific method for delivering ASOs covering intron 7 (including ASO-10-27) to the cerebrospinal fluid in the brain.

*Group (22)*[103] relates to figure 7.1.D above in section 1 illustrating chimeric ASO used to correct defective pre-mRNA splicing and improve gene expression. It claimed chimeric molecules that can be delivered to SMA patients in order to enhance the incorporation of exon 7 of the *SMN2* gene. The invention is based on a design of a PNA-peptide chimeric molecule, which recognises and binds to a splicing enhancer in exon 7, thereby tethering an RS domain which will stimulate exon 7 inclusion.

The description states that the chimeric molecules are preferably delivered *in utero* or at an appropriate time after birth. The SMA mouse model shows that an appropriate time is forty-eight hours after birth. An appropriate time after birth for humans is the time corresponding to forty-eight hours in the mouse model. The administration of the chimeric molecules at a significant time after birth can prevent further degeneration of motor neurons and/or partially reverse the course of a disease after its onset.[104] The European patent was granted in 2012, and the examiner stated that none of the available prior art disclosed chimeric molecules that modulate splicing of pre-mRNA. However, the US patent application was abandoned because of the prior art demonstrated with another gene. The prior art teaches modulation of splicing by simply tethering to splicing enhancers and also teaches that artificially tethering RS domains to splicing enhancers located near the splicing site is sufficient to activate splicing. Thus the prior art teaches the concept of activating splicing by artificially tethering the splicing machinery to splicing enhancers via moiety which specifically targets the splicing enhancers and tethering the splicing machinery to pre-m-RNA via base-pairing segments.[105]

*Group (40)*[106] claimed compounds and methods for modulating protein expression by altering the splicing of a pre- mRNA, by 'arresting translation and/or by interrupting poly-adenylation of a pre- mRNA'. In the written opinion from 2009, the European examiner stated that the claims therefore embrace an undefined number of unspecified conditions all allegedly capable of being treated or prevented by the modulation of the expression of any target

103. EP 1416972 granted January 2012, and US 2005054836 abandoned.
104. Description in US 2005054836, point [0126].
105. PCT filed in November 2001. The US application was filed in May 2003, whereas the European application was filed in November 2001.
106. See EP 2219680.

protein. The claims must be directed towards a clearly identifiable agent for use in the defined, real treatment of any pathological condition in order to make a technical contribution to the art, and in order that a skilled person should be able to unambiguously identify the group of patients to be treated by the therapy claimed.

*RNAi techniques and delivery*
It is obvious from the nature of the disease that RNAi methods for down-regulating the primary disease genes, *SMN1 and SMN2*, are not relevant, since that would merely worsen the lack of a functional protein. *Group (11)* claimed new iRNA agent relating to the pathological process, more precisely down-regulation of a secondary gene that may play a role in the pathological process. The aim of the invention is to knock out a gene that inhibits the growth of nerve cells, thus the pathological process is the inhibition of nerve growth or elongation as a result of nerve injury or damage.[107] The granted US patent describes the correlation between the RHO-A gene and treatment. The laundry list comprises disorders associated with *RHOA* expression resulting in a bundle of diseases, including Alzheimer's, SMA and schizophrenia.[108]

## 4. Canavan's disease (CD)

Canavan's disease is caused by mutations in the *ASPA* gene for an enzyme encoding aspartoacylase, and the disease is categorised as a demyelinating disease (for more detail see Chapter 6, section 2.3). The substrate for aspartoacylase, N-acetylaspartate acid, or NAA, a naturally-occurring compound in the brain, is thought to be instrumental in the synthesis of myelin. In those afflicted with CD, however, an inherited mutation prevents production of aspartoacylase and NAA accumulates to dangerous levels, impairing the brain's normal function.[109]

### 4.1. Data outline

The following case study of Canavan's disease comprises 19 groups of altogether 35 patent applications, of which 13 applications have been granted, 13 applications have been withdrawn, and 9 applications are still pending. The

107. Claimed in the application (claim 18); see US 2007042984, US 7772200 and EP 1909850.
108. US 7772200, detailed description. 25 claims granted October, 2010.
109. See http://www.canavan.org.

March 2011 search on *ASPA* and aspartoacylase found 53 groups (some overlapping), but only 2 groups, *(18)* and *(19)*, were relevant to CD.[110] The search term also caught fragments present in bacteria (a transcriptional terminator used for cloning) and other examples not relevant to the case study. Thus, 51 groups from the 2011 search and 4 groups from the initial search have been omitted.[111]

From the initial search *group (4)*, on methods for identifying genetic traits, was omitted from the analysis due to its general approach as a DNA-based diagnostic method for screening for the presence or absence of mutations or polymorphisms in a plurality of genes. It also included a very broad laundry list.[112] *Group (7)*[113] was omitted as it was more concerned with chemical compounds than genes. *Group (7)* also included a broad laundry list in claims 9-12.[114]

The focus of the case study of CD follows the overall structure, using the four categories. However, as CD is a recessive disease caused by lack of aspartoacylase, there are no RNAi applications, so the highlight category is omitted. The review is based on the following patent activity.

### 4.2. The empirical analysis

Heading (a) covers the *ASPA* gene with regard to the DNA sequence of the gene, mutated *ASPA* sequences associated with the disease, and delivery methods for gene therapy on *ASPA*. The pathological process more distantly comprises three sub-groups of antibodies, small peptides and demyelinating diseases in general, and processes from gene families. As such, this heading comprises a laundry list perspective since the claimed processes are suitable for several diseases. Stem cells will only be briefly analysed, as this heading did not contain any new techniques that circumvent the current ban on the patenting of embryonic stem cells.

110. The search outcome is available as Appendix IX table CD.
111. Groups (4) (7) (14) (17) (1*) (3*-13*) and (15*-53*), all in bold in Appendix IX table CD.
112. See e.g. claim 5 and claim 27 CF, HD and CD and a bundle of other diseases.
113. Use of ribonucleotide reductase inhibitor for treating e.g. Alzheimer's disease, multiple sclerosis, congenital metabolic disorder, rheumatoid arthritis, insulin dependent diabetes mellitus, lipid storage disease, stroke, myasthenia gravis (US 2009047250, EP 2030615 and EP 2030615)
114. See US 2009047250, claim 9: Alzheimer's disease, Parkinson's disease, CD and HD but not limited to these diseases.

*Box 7.4.1. Canavan's analysis model*

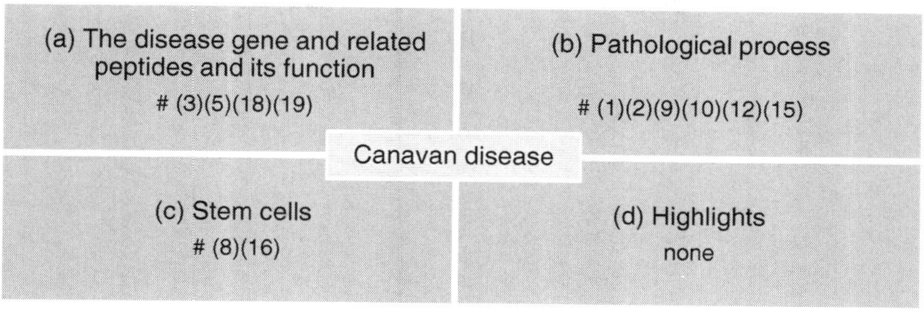


**4.2.a. The disease gene, related peptides, and its function**

The heading 'disease genes, related peptides, and its function' is divided into two sub-groups, one on the Matalon patents, groups (3) and (5), and one on delivery systems, groups (18) and (19).

*The Matalon patents on ASPA*

The sub-group comprises *group (3)* and *group (5)* which are described in Chapter 6, section 2.3.1.1. These are the Matalon patents assigned to the Miami Children's Hospital Research Institute, Inc. The groups are the starting point for the case study search for activity in the aftermath of the Matalon patents.

*Group (3)*[115] comprises one patent granted in May 2007 with only 2 claims out of 65 filed.[116] In a letter of 20 July 2006 the applicant rendered all rejection moot by cancelling most of the claims. No agreement with any aspects of the office action was implied. What was claimed was three disea-

115. US 7217547 Reuben Matalon.

116. The non-final rejections reduced the numbers of claims and in 2005 only 13 claims remained. The multitude of possibilities of biologically active fragments which meet the limitations of the claims is vast. Since the claims were drawn *to so many different combinations of potential polypeptides*, a skilled person could not necessarily envision the detailed structures of each and every different biologically active fragment of human aspartoacylase, because the specification did not describe which amino acids of the 313 of SEQ ID No: 2 are absolutely essential and critical for the peptide to maintain *both* its structure and functionality. Therefore, the conception would not be achieved until reduction to practice had occurred, regardless of the complexity or simplicity of the methods of making the claimed invention.

se-causing mutations; among them was p.A305E, the most frequent mutation found in European patients (including Danish patients).

It is interesting to note that the first Matalon patent (*Group (5))* granted in 1997 have the same title and consists of 44 claims, claiming e.g. the protein, and a process for screening a potential Canavan's disease carrier or patient for the presence of an identified mutation in an aspartoacylase gene. The new patent grants the applicants three further mutations, whereas the original patent only claimed one mutation, p.E285A, which is the most frequent mutation in Ashkenazi Jewish populations.

*Delivery systems*

The sub-group comprises *group (18)* and *group (19)*. *Group (18)*[117] claimed delivery systems for gene therapy for the brain, using an adeno-associated vial (AAV) vector which contains *ASPA* for delivery to the brain, employing lipid-entrapped polycation-condensed nucleic acid comprising lipid, which is claimed to enhance the delivery. This system has been used for gene therapy in children with CD. *Group (19)* claimed general methods and compositions for delivering enzymes and nucleic acid molecules to the brain, bone and other tissues, and methods for increasing blood-brain barrier permeability in a subject by using hyaluronidase.[118] It covers a variety of enzyme deficiency diseases, e.g. Canavan's disease, Fabry disease, Gaucher's disease, various forms of mucopolysaccharidosis (e.g., Hurler's syndrome, Scheie syndrome, Hurler-Scheie syndrome, Sanfillippo A syndrome, Morquio A syndrome, Morquio B syndrome etc.), Niemann-Pick disease, Schindler disease and Pompe disease. In the final office action (US, 2009) the examiner argued that the specification failed to provide an enabling disclosure for the treatment of any disease or disorder, but merely constituted a delivery method.

### 4.2.b. Pathological process

The heading has been divided into three sub-groups. The first covers antibodies, the second covers small peptides and the third covers demyelinating diseases and supergene families. The heading comprises the following groups (1) (2) (9) (10) (12) and (15).

117. EP 0969876 withdrawn, US 6436708.
118. EP 1521523 withdrawn, US 7807618.

*Small peptides*

*Group (9)*[119] relates to a neurotrophic peptide whose actions appear to be restricted to the nervous system where it promotes neurotransmitter synthesis and neurite outgrowth in certain neuronal populations. The invention is directed to the pathological process in many diseases with neurite damage, and it has an immense laundry list in claim 28 (including Parkinson's disease, Alzheimer's disease, Huntington's disease and Canavan's disease).

*Antibodies*

The sub-group comprises group (10) and group (15).[120] Group (10)[121] comprises antibodies to CD100 which belongs to the semaphorin family of cell surface receptors and is directly implicated in demyelination and axonal degeneration. This is why Canavan's disease is listed as one of many diseases in the laundry list in claim 70 (which also lists Alzheimer's disease and a bundle of other diseases).[122]

*Demyelinating diseases and superfamilies*

The sub-group comprises groups (1) (2) and (12). The heading 'demyelinating diseases' has been chosen since the inventions all claim broad biological processes or superfamilies involved in processes involving demyelination. Thus processes concerning nerve cells (e.g. inhibiting axonal growth) and transportation of anions (e.g. N-acetylaspartate necessary for the myelin synthesis) are relevant for demyelinating diseases.

*Group (1)* [123] claimed an isolated nucleic acid molecule (SEQ ID NO:1, SEQ ID NO:3), a peptide, antibodies and a method for detecting the peptide (probe and primer). The invention is based, in part, on the discovery of a nov-

119. EP 2121930 pending, EP 2009103 withdrawn, US 2010029572 pending.
120. EP 1993608 and US 2009220529, both withdrawn, concerned antibody-drug conjugates that are intended for the treatment and/or diagnosis of diseases such as tumours and/or inflammatory reactions. No further analysis of this group will be made as it is broadly related to cancer and comprises a broad list of medical conditions [146-148], of which Canavan's is not an obvious target.
121. EP 2069404 and US 7919594, both granted.
122. See description [0098]. There is laundry list in the granted EP patent claim 12 whereas there is no laundry list in the US patent. See Claim 12: 'An immunoglobulin according to anyone of claims 1 through 7 for use in a method for treating an autoimmune disease or inflammatory disease in a subject, wherein said disease is associated with CD 1 DO-expressing cells in said subject; wherein said autoimmune disease or inflammatory disease is optionally selected from [a list of diseases]'.
123. See US 2002193582 description point [0007]. The group also comprises EP 1270726.

el transporter family member, referred to here as '69624', which may be involved in the transport of N-acetylaspartate, which is the substrate of ASPA.

The application was abandoned under 35 U.S.C. §101, as it lacked a credible, specific and substantial asserted utility or a well-established utility. The specifications did not disclose a function for *the claimed protein* in the context of the cell or organism. No well-established utility exists for newly-isolated complex biological molecules. For example, the examiner wrote that there is no disclosure of particular disease states correlating to an alteration in levels or forms of the polypeptide such that the polypeptide could be used as a diagnostic tool. *Also the breadth of claims failed to recite particular biological activities, and undue experimentation would be required of a skilled artisan to make and/or use the claimed invention.*[124]

In *Group (2)*,[125] the invention claimed NgR proteins and biologically active Nogo protein fragments,[126] compositions and methods for modulating the expression or activity hereof, and peptides which block Nogo-mediated inhibition of axonal growth.[127] Claims 36 and 37 comprise methods relating to central nervous system disorders resulting from cranial or cerebral trauma, spinal cord injury, stroke or a demyelinating disease. Claim 7 clarifies that demyelinating disease is selected from a group consisting of multiple sclerosis, monophasic demyelination, encephalomyelitis, multifocal leukoencephalopathy, panencephalitis, Marchiafava-Bignami disease, pontine myelinolysis, adrenoleukodystrophy, Pelizaeus-Merzbacher disease, Spongy degeneration, Alexander's disease, Canavan's disease, metachromatic leukodystrophy and Krabbe's disease.

124. See Office communication from US examiner Sandra L. Wegert, non-final rejection of February 2006, p. 7.
125. US 7119165, US 20020012965, and EP 1248803: the granted US patent US 7119165 and the abandoned application US 20020012965 were both examined by patent examiner Sandra L. Wegert. This invention was filed by the inventor both as a European patent application and as two US patent applications. The European application is still pending. One US patent consisting of 50 claims was granted in 2006 and the other application was abandoned in 2002.
126. Identical to amino acids 27 309 of SEQ ID NO:2, wherein the polypeptide inhibits NOGO-receptor-mediated neurite outgrowth inhibition (complete protein).
127. The protein family Nogo, particularly Nogo-A, has been identified as an inhibitor of remyelination in the CNS. Antagonising this inhibitor results in improved remyelination, as it is involved in the RhoA pathway; see http://en.wikipedia.org/wiki/Neuroregeneration (also known as reticulon 4 receptor; see http://en.wikipedia.org/wiki/Reticulon_4_receptor (last visited, May 2011).

### 4.2.c. Stem cells[128]

The two sub-groups claim inventions that broadly target nerve cells and the connection to Canavan's disease is not obvious, but as a demyelinating disease it is claimed among several other diseases.[129]

*Group (8)*[130] comprises glial-restricted progenitors (GRPs), which are a kind of stem cell different from ES. *Group (16)* comprises stem cell therapy that can be used in a variety of diseases such as degenerative diseases, acute injury or neurological conditions e.g. Alzheimer's disease, Canavan's disease, Huntington's disease, and Parkinson's disease.[131]

## 5. Cystic fibrosis (CF)

The patent activity on CF has been characterised in the literature case study in Chapter 6, section 2.4.1.1. It has been referred to as an open model since the main patents (Francis Collins, Lap-Chee Tsui, John Riordan) on the *CFTR* gene have never been enforced against clinical diagnostics. The economic interest in this disease is reflected in the massive number of patent applications that present case study has found, even though it is not intended to be exhaustive of all CF patent activity. Research into the disease can still be measured by the patent activity which is perhaps due to the open model of the CTFR gene patents and the common Δ 508 mutation (described in Chapter 6), and as one of the most common severe recessive genetic diseases it is worth finding new methods of screening, detection of new mutations and treatments.[132]

128. Groups (8) and (16) .
129. Even Alzheimer's disease, Huntington's disease, Parkinson's Disease which are not demyelination diseases but a condition of neurodegenerative or stroke, *group (8)*.
130. EP 2190981 and US 2009155223, both abandoned.
131. Most applications are still pending, but one has been abandoned. Committed progenitor cells are made by selecting from a cultured source of a cell population which co-express CD49c (ITGA3 gene) and CD90 (THY1 gene) and modifying the cell population. The committed progenitor cells can be used to treat a human suffering from a degenerative, traumatic, acute injury, cardiac or neurological condition and to formulate pharmaceutical compositions; see http://www.genecards.org/cgi-bin/ carddisp.pl?gene=THY1&search=CD90 and http://www.ncbi.nlm.nih.gov/omim/605025 (Integrins are a family of cell surface adhesion molecules).
132. See also Subhashini Chandrasekharan et al., 'Impact of gene Patents and licensing Practice on Access to genetic testing for Cystic Fibrosis, Patents and Licensing for Cystic Fibrosis', *Genetics in Medicine*, April 2010, 12 (4 Suppl.), S194-S211.

### 5.1. Data outline

The search for patent activity covering cystic fibrosis comprises 325 groups. 49 groups were deleted as their connection to the *CFTR* gene was too remote.[133] In fact, the search found almost every aspect of biomedical patents claiming other genes, far away from the core of the study of the *CFTR* gene.[134] The remaining 276 groups were then divided into sub-groups. The main sub-groups comprised the template headings of the case study design: (a) the disease gene, related peptides and its function, (b) pathological process, (c) stem cells, and (d) highlights. Of the remaining 276 groups containing 846 patent applications, of which 387 have been granted, 278 have been withdrawn, and 181 are still pending.[135] The search strategy has been described in Chapter 6, section 4. The focus of the empirical case study is on the patent activity shown in box 7.5.1. below.

### 5.2. The empirical analysis

Table 7.5.1 below contains an overview of the data arranged in the template four groups according to the overall design of the case study.

133. Highlighted with red in Appendix X table CF, Groups (16) (24) (13*) (17*) (42*) (53*) (55*) (58*) (97*) (100*) (110*) (113*) (116*) (127*) (146*) (168-169*) (173*) (175*) (177*) (217*) (223*) (230*) (244*) (248-249*) (251*) (253-254*) e.g. macular degeneration), (258*) (262*) (264*) (266*) (268*) (277*) (282*) (285-286*) (299*) (311*) (317*) (320*) (324*) (342*) (367-368*) (386*) (428*) and (434*).
134. See e.g. *Group (24),* relating to dsRNA for inhibiting the expression of the *Aha* gene (Activator of Heat Shock Protein 90 ATPase (Aha)). It is not CFTR as such that is claimed, but the implication of HSP90 in the degradation of CFTR mutations in the CFTR gene leading to defective folding and ubiquitination of the protein as a consequence of HSP90 ATPase activity. Following ubiquitination, CFTR is degraded before it can reach its site of activity. Lack of active CFTR then leads to the development of cystic fibrosis in human subjects having such a mutation. Therefore, the inhibition of HSP90 activity may be beneficial for subjects suffering from cancer or cystic fibrosis [description point, 0005].
135. Latest status update March 2012.

*Box 7.5.1. CF analysis model*

**(a) The disease gene and related peptides and its function**

# (10) (16) (19) (1*) (2*) (5*) (9*) (10*) (16*) (18*) (19*) (25*) (28*) (47*) (52*) (79*) (99*) (100*) (104*-105*) (136*) (145*) (158*) (160*) (171*) (181*) (197*) (201*) (214*) (232*) (252*) (267*) (268*) (293*) (303*) (318*) (364*) (391*) (404*)

**(b) Pathological process**

# (1) (2) (3) (4) (8) (9) (10) (11) (12) (14) (15) (18) (19) (20) (21) (25) (26) (27) (28) (29) (30) (31) 32) (33) (34) (35) (36) (37) (38) (39) (40) (8*) (36*) (41*) (54*) (59*) (64*) (71*) (73*) (82*) (84*) (93*) (122*) (123*) (139*) (148*) (149*) (154*) (156*) (164*) (167*) (170*) (176*) (181*) (183) (185*) (187*) (190*) (193*) (194*) (195*) (198*) (199*) (200*) (202*-203*) (207-209*) (215*-216*) (219*-222*) (225*) (228*-229*) (234*) (236*) (238*) (243*) (245*) (255*-256*) (260*) (263*) (265*) (270*-271*) (273*) (275*-276*) (278*-280*) (283*-284*) (292*) (313*) (323*) (340*-341*) (376*) (378*-379*)

laundry list

**(c) Stem cells**

# (5) (6) (7) (101*) (132*) (161*)

**(d) Highlights: Trangenic animals and claims on techniques as such**

# (2) (19) (23) (41) (3*) (4*) (5*) (6*) (7*) (8*) (11*-12*) (14-15*) (20*-22*) (23*-24*) (26*-27*) (29*-32*) (34*-35*) (37*-40*) (42*-43*) (45*-46*) (48*) (51*) (56*-57*) (61*-63*) (66*-70*) (72*) (74*) (76*-77*) (83*) (85*-92*) (94*-95*) (103*) (106*-109*) (111*-112*) (114*-115*) (119*-121*) (124*-125*) (128*-130*) (131*) (133*-138*) (140*-144*) (147*)(150*-153*) (155*) (157*) (159*) (163*) (165*) (166*) (174*) (178*) (182*) (184*) (186*) (189*) (191*-192*) (196*) (205*-206*) (211*-213*) (218*) (227*) (233) (259*) (237*) (240*) (281*) (287*-288*) (290*) (294*-298*) (301*) (306*) (308*) (312*) (315*-316*) (325*) (337*) (344*) (352*) (377*) (383*-384*) (388*-389*) (400*) (410*-412*) (421*-422*) (427*) (433*) (435*)

### 5.3.a. The disease gene and related peptides

*The first CFTR patents on the sequence, protein, diagnosis etc.*

The search performed in 2011 came up with further patent applications and grants relating to the disease gene. First, it identified the patents from the CF group mentioned in Chapter 6 (Francis Collins, Lap-Chee Tsui, John Riordan, groups (391*) (2*) and (10*)) as well as the Gregory patent which interfered with the CF group (groups (1*) and (404*)).

The first patents on the *CFTR* gene, groups (391*)[136] (2*)[137] and (10*),[138] obtained protection for the *CFTR* gene and its Δ508 mutation. The claim as granted comprised the sequence of the normal cDNAs (CF gene), the Δ508 mutant cDNAs (a mutant CF gene), genetic testing, the normal and mutant *CFTR* products and mixtures thereof (proteins, vectors and cell lines expressing the normal and mutant *CFTR* genes), antibodies, and methods for screening (assay, probe). Group (10*) also broadly claimed the use of the genetic and protein information in developing DNA diagnosis, protein diagnosis, carrier and patient screening, cloning of the gene and manufacture of the protein, and development of cystic fibrosis affected animals. Thus, it would seem that most doors to further patenting were closed.

Groups (1*)[139] and (404*)[140] are claimed by another inventor (Gregory). They overlap with the former groups (Tsui et al.) as they claimed the sequence of the *CFTR* cDNA and the CFTR-containing vector, arguing that Tsui had failed to provide a written description. The USPTO declared a patent interference which went on for ten years and was resolved in part in Tsui's favour. Tsui was granted both the wild type *CFTR* cDNA sequence and the Δ508 mutant sequence, whereas Gregory was granted methods of producing *CFTR* cDNA.[141]

*Later mutation diagnosis and gene tests*

The 2011 search revealed patent activity on mutation diagnosis and gene tests. A mutation in the *CFTR* gene has an impact on protein function. Different mutations can change the function of the CFTR protein in different ways which in turn has an influence on the severity of the disease. CF is caused by a wide range of mutations but, as mentioned in Chapter 6, one mutation (Δ 508) accounts for 70 % of CF chromosomes worldwide.[142] Approximately

136. US 5776677.
137. EP 0489058 granted; EP 1418183 withdrawn; US 6201107, US 6730777, US 6902907 and US 6984487, all granted.
138. EP 0667900, US 6001588 and US 5981178 all granted.
139. EP 0446017 withdrawn.
140. US 5876974 and US 7118911 both granted.
141. Subhashini Chandrasekharan et al., 'Impact of gene Patents and licensing Practice on Access to genetic testing for Cystic Fibrosis, Patents and Licensing for Cystic Fibrosis', *Genetics in Medicine*, April 2010, 12 (4 Suppl.), S194-S211, p. 4 in NIH Public Access, pp. 5-6 and footnotes 29-31, Interference No 103882, 103933, 104228 *Gregory vs. Tsui*, 2002.
142. Subhashini Chandrasekharan et al., 'Impact of gene Patents and licensing Practice on Access to genetic testing for Cystic Fibrosis, Patents and Licensing for Cystic Fibro-

1,900 mutations[143] in the CFTR gene have been verified, and the present search identified a scientific and commercial interest in patenting a detection test. Some of the applications claimed mutation testing for several mutations in *CFTR*, some tests included the Δ 508 mutation (e.g. *groups (28*) (52*) (99*) (100*) (145*) (171*) and (267*)*), some claimed disease detection through antibodies or other genes than the *CFTR* gene (e.g. *groups (104*) (105*) (160*) (201*) (197*) (52*) (158*) (452*) and (136*)*[144]), yet others claimed biomarkers and assays (e.g. *groups (232*) (9*) (18*) (158*) (160*) (293*) and (181*)*), and some claimed a gene test that measured on chromosome imbalance (*groups (364*) and (268*)*).

*Group (18*)* provides chips of immobilised probes, and methods for using the chips, for comparing a reference polynucleotide sequence of known sequence with a target sequence showing substantial similarity with the reference sequence, but differing in the presence of e.g., mutations. *Groups (28*)*[145] and *(52*)*[146] both claimed a method for detection as such; e.g. *group (28*)* claimed a method based on polymerase chain reaction (PCR) amplification and oligonucleotide ligase assay (OLA) reaction providing for the analysis of complex genetic systems in a single reaction vessel. *Group (136*)*[147] claimed a method of diagnostics based on mass spectrometric detection aimed at a protein test rather than a gene test, since proteins can be subjected to high throughput biomarker screening, such as tandem mass spectrometry.

Two of the groups will be mentioned here as examples. Group (171*)[148] claimed detection of a bundle of mutations, including Δ 508, by a method for

sis', *Genetics in Medicine*, April 2010, 12 (4 Suppl.), S194-S211, p. 4 in NIH Public Access: 'a deletion of three nucleotides in DNA causes the protein to lack the amino acid phenylalanine (F) at position 508'.

143. http://www.genet.sickkids.on.ca/Home.html.
144. Group (424*) claimed a diagnostic method based on mass spectrometric detection, where CF was one of several diseases in a laundry list e.g. claim 75 Huntington's disease, prostate cancer, Fragile X syndrome type A, myotonic dystrophy type I, Kennedy disease, Machado-Joseph disease, dentatorubral and pallidolyusian atrophy, spinobulbar muscular atrophy, and aging. And claim 76, wherein the nucleic acid comprises at least a portion of a gene selected from the group consisting of BRCAl, BRCA2, APC, dystrophin gene, fi-globin, Factor IX, Factor Vllc, ornithine-d-aminotransferase, hypoxanthine guanine phosphoribosyl transferase, CFTR, p53, and a protooncogene.
145. EP 0777749, US 5912148 and US 6130073 all granted.
146. EP 0928832 and US 2003008281 both withdrawn.
147. EP 1296143, US 6322970, US 6207370 and EP 1010008 all granted.
148. EP 1567540 granted, and US 2006147938 withdrawn.

detecting a plurality of *CFTR* alleles, whereas *group (52*)* claimed twelve mutations. *Group (18*)*[149] provides an array in the form of chips for several diseases (HIV chip, CFTR chip, P53 Chip (cancer) and mitochondrial chip) testing several mutations at the same time.

Finally, some groups claimed screening methods for CF, for example by testing blood samples from pregnant women in order to determine whether the child/foetus is afflicted with CF (*group (214*)*).[150] *Group (105*)*[151] provides a method for using the *NDPK* gene as a biomarker since that gene is affected in CF patients (*NDPK* is essential for cell secretion and differentiation).[152] *Group (160*)*[153] claimed the *UCHL1* gene as a biomarker for CF.

*Change in CFTR genetic code enhancing activity*

One patent application (*group (252*)*[154]) claimed a change in the *CFTR* genetic code, causing enhanced protein production. Changing the DNA sequence does not alter the amino acid sequence, but enhances the protein production by using more efficient alternative amino acid codons. What is claimed is the DNA sequence that differs from the natural *CFTR* gene sequence. Group (5*) claimed amino-terminally truncated *CFTR*. 118 amino acids have been deleted from the CFTR protein (amino acid terminal deletion) claiming that the protein have the ability to restore chloride conductance in bronchial epithelial cells derived from individuals diagnosed with CF. Another patent application claimed a reduced *CFTR* protein (removed 40% from 1480 amino acids to 836 amino acids, claiming a sequence patent on *CFTR, group (25*)*[155]).

**5.3.b. The pathological process**

Cystic fibrosis often causes severe lung infections, and is thus often connected with inflammation. The present case study found that many applications focused on lung infections, for example abnormal production of mucus obstructing the airways and resulting in permanent lung damage and other ef-

149. EP 0730663, US 5837832 and US 5861242 all granted.
150. EP 1838877 and US 2009317797 both pending.
151. EP 1157274 withdrawn.
152. EP 1157274 description.
153. EP 1506217 and US 2006292562 both withdrawn.
154. EP 2076291 pending, and US 2011035819 abandoned.
155. EP 0751994 withdrawn, and both US applications granted as US 5958893 and US 5639661.

fects of the disease.[156] Thus, inventions under this heading claimed treatments with other targets than CFTR, for example other genes, proteins, chemicals or compounds (such as proteins or chemicals that increase gene expression). Most of the inventions in this category claim or describe laundry lists since the invention is not solely directed at CF patients. For example, *group (4)* claimed treatment with NF-kB oligonucleotide decoy molecules.[157] It claimed a giant laundry list and claimed to overcome the restriction to a single disease by arguing that NF-KB plays a key role in each disease.[158] One

156. See http://ghr.nlm.nih.gov/condition/cystic-fibrosis and e.g., group (17) claiming a method of inhibiting biofilm formation or reducing biofilms by administering a charged compound to a subject, (being a (claim 2) polyamino acid, (claim 8) polyamino acid between 50 and 300 amino acids in length, (claim 9) a protease inhibitor, (claim 3) a non-defined peptide (both charged and neutral amino acids), and (claim 45) a synthetic polymer). According to the background description of the US application, by adulthood, 80 % of CF patients are chronically infected with the special *P. aeruginosa* bacterium. The description also points out that chronic *P. aeruginosa* infection is associated with increased morbidity and mortality in CF patients. According to the description, persistent *P. aeruginosa* infection is associated with numerous phenotypic and genetic changes made by the bacteria within the CF airway, including the formation of biofilms. Thus the detection of a specific pattern of quorum-sensing signalling molecules in the CF airway suggests that *P. aeruginosa* exists primarily in the biofilm form, a conclusion supported by the inability of antibiotics and host defence mechanisms to eradicate the infection; point [0002], [0003].
157. EP 1691817 granted, US 7378509 granted, US 2007010474 pending, and US 2007078102 abandoned.
158. A more detailed list of diseases and pathogenic conditions which are targeted for prevention and/or treatment by the NF-.kappa.B decoys of the invention includes psoriasis, eczema and atopic dermatitis; systemic scleroderma and sclerosis; responses associated with inflammatory bowel disease (IBD) (such as Crohn's disease and ulcerative colitis); ischemic reperfusion disorders including surgical tissue reperfusion injury, myocardial ischemic conditions such as myocardial infarction, cardiac arrest, reperfusion after cardiac surgery and constriction after percutaneous transluminal coronary angioplasty, stroke, and abdominal aortic aneurysms; cerebral oedema secondary to stroke; cranial trauma, hypovolemic shock; asphyxia; adult respiratory distress syndrome; acute-lung injury; Behcet's Disease; dermatomyositis; polymyositis; multiple sclerosis (MS); meningitis; encephalitis; uveitis; osteoarthritis; lupus nephritis; autoimmune diseases such as rheumatoid arthritis (RA), Sjorgen's syndrome, vasculitis; diseases involving leukocyte diapedesis; central nervous system (CNS) inflammatory disorder, multiple organ injury syndrome secondary to septicaemia or trauma; alcoholic hepatitis; bacterial pneumonia; antigen-antibody complex mediated diseases including glomerulonephritis; sepsis; sarcoidosis; immunopathologic responses to tissue/organ transplantation; inflammations of the lung, including pleurisy, alveolitis, vasculitis, pneumonia, chronic bronchitis, bronchiectasis, diffuse panbron-

US patent was granted after the office communication on adequate written description and obviousness, and the laundry list in the description survived the prosecution. The European patent was granted with the claimed laundry list, but later withdrawn due to non-payment of the renewal fees, and the one US application was abandoned.[159] NF-kB is one out of many transcription factors that regulate numerous human genes.[160]

The heading is divided into two sub-sections: one for treatment and the other for drug screening.

*Treatment*

This sub-heading comprises the following groups: (1) (2) (3) (4) (8 ) (9) (10) (11) (12) (14) (15) (18) (19) (20) (21) (25) (26) (27) (28) (29) (30) (31) 32) (33) (34) (35) (36) (37) (38) (39) (40) (8*) (36*) (41*) (54*) (59*) (64*) (71*) (73*) (82*) (84*) (93*) (122*) (123*) (139*) (148*) (149*) (154*) (156*) (164*) (167*) (170*) (176*) (181*) (183) (185*) (187*) (190*) (193*) (194*) (195*) (198*) (199*) (200*) (202*-203*) (207-209*) (215*-216*) (219*-222*) (225*) (228*-229*) (234*) (236*) (238*) (243*) (245*) (255*-256*) (260*) (263*) (265*) (270*-271*) (273*) (275*-276*) (278*-280*) (283*-284*) (292*) (313*) (323*) (340*-341*) (376*) and (378*-379*).

*Group (10)* claimed medication for CF, asthma, chronic obstructive pulmonary disease (COPD) and more (anti-IFN-gamma Ab). *Group (11)* comprises an inhaler including a powder (medicament)[161] for the treatment and prevention of asthma and other conditions associated with reversible airways obstruction, and for treating, for example, cystic fibrosis and cancer. *Group (25)* is directed to the use of bortezomib, which is a proteasome inhibitor for the treatment of a condition selected from the group of cystic fibrosis, diabetes insipidus, hypercholesterinaemia and long QT-syndrome-2. *Group (71*)*[162] concerns a novel approach for treating cystic fibrosis using, in par-

chiolitis, hypersensitivity pneumonitis, idiopathic pulmonary fibrosis (IPF), and cystic fibrosis, and the like.

159. US 2007078102.
160. A transcription factor (sometimes called a sequence-specific DNA-binding factor) is a protein that binds to specific DNA sequence, thereby controlling the movement (or transcription) of genetic information from DNA to mRNA. See Alberts et al., *Essential Cell Biology*, 3rd Ed., p. 285.
161. US 7018657/EP 1274403 granted for powder comprising a medicament and lipids and form liposomes (smaller than 1000 mm) when dispersed in buffer.
162. EP 0984784 and US 6635627 both granted.

ticular, anti-cancer chemotherapy. The invention is also applicable to the treatment of rheumatoid polyarthritis or asthma.

*Treatment – specific CF*

The group of specific CF treatments comprises inventions where the treatment directs *CTFR* itself such as group (93*) and group (8*). *Group (8*)* comprises a method for generating CFTR function in cells containing mutant CFTR. The method comprises an agent which increases the level of mutant *CFTR* at cellular locations where wild-type *CFTR* normally functions. Group (93*) comprises a method of administering a therapeutically effective amount of docosahexaenoic acid (DHA).[163] This method is particularly useful for treating subjects suffering from a disorder characterised by a defect in the CF gene, e.g., cystic fibrosis, or a chronic inflammatory disorder, e.g., ulcerative colitis, Crohn's disease, chronic pancreatitis, asthma, rheumatoid arthritis or chronic gastritis. The application also claimed a method for ameliorating the effects of cystic fibrosis in a newborn and a method for increasing surfactant levels in a foetus.

*Drug screening: premature stop codons (PTCs)*

A significant proportion of disease causing mutations in the *CFTR* gene (including some of the most frequent disease causing mutations) generate premature stop codons (PTCs). Approximately 1,900 mutations have been identified in the *CFTR* gene, of which 160 are PTCs.[164] These mutations are severe because they encode a truncated non-functional CFTR protein. In addition, mutant mRNAs containing PTC mutations are usually degraded by the cellular process of nonsense-mediated mRNA decay (NMD). In recent years, drugs have appeared, for instance aminoglycosides (which are common antibiotics), which can overcome the deleterious effect of PTCs by fooling the ribosomes of the cell to read through the PTCs and thus rescue the protein. Identification of this type of readthrough drug is not only interesting in relation to treatment for CF, but may have enormous potential since more than 11% of all mutations that cause inherited diseases in humans are PTC mutations.[165] Although these drugs are general and may be useful for treatment of

163. EP 106191 and US 2004127567 both withdrawn, and US 6552081 granted.
164. http://www.genet.sickkids.on.ca/Home.html
165. In-frame PTCs account for approx 11% of all described disease causing mutations in human genetic diseases, M. Mort, D. Ivanov, D.N. Cooper& N.A. Chuzhanova, 'A meta-analysis of nonsense mutations causing human genetic disease', *Human Mutation*, Vol. 29, No 8, 2008, pp. 1037-1047.

all kinds of diseases caused by PTC mutations in a disease gene, CFTR plays a central role because some of the most prevalent diseases causing mutations in the *CFTR* gene in certain populations are PTC mutations (like the p.W1282X mutation in the Ashkenazi Jewish population, which is found in up to 40 % of Israeli patients with CF). Moreover, some of the initial clinical trials with aminoglycoside drugs were performed using nose droplets in patients with *CFTR* PTC mutations.

The search identified three groups within this area (122*) (220*), and (239*), all claimed by one US company: PCT Therapeutics Inc.

*Group (122*)* claimed a method of *in vivo* screening for compounds that modulate premature translation termination and nonsense-mediated mRNA decay.[166] It further claimed a kit for screening for compounds that modulate translation termination and nonsense-mediated mRNA decay and it claimed e.g. the method wherein the nucleotide sequence is an endogenous gene, and wherein the endogenous gene optionally encodes [alpha]-globin, [beta]-globin, factor VIII, factor IX, vWF, p53, dystrophin, CFTR, Rb, MSH1, MSH2, APC, Wt1, hexosaminidase A, neurofibromin 1, neurofibromin 2, or an immunoglobulin. *Group (220*)* claimed a method for treating a disease associated with a nonsense mutation, consisting of administering an effective amount of a chemical to a patient in need.[167] A disease associated with a nonsense mutation is a genetic disease, cancer, an autoimmune disease, a blood disease, a collagen disease, diabetes, a neurodegenerative disease, a proliferative disease, a cardiovascular disease, a pulmonary disease, an inflammatory disease or central nervous system disease, cystic fibrosis or Duchenne muscular dystrophy. Finally *group (239*)* claimed a method for producing an effective amount of a functional readthrough protein encoded by a nucleic acid sequence comprising a nonsense mutation. The method consisted of the oral administration of an effective amount of a nonsense codon suppressor agent to a human subject. Non-exhaustive examples of diseases that can be treated, managed and/or prevented by functional readthrough proteins encoded by genes comprising nonsense mutations include: amyloidosis, LINCL, haemophilia, Alzheimer's disease, atherosclerosis, gigantism, dwarfism, hypothyroidism, hyperthyroidism, cystic fibrosis, aging, obesity, Parkinson's disease, Niemann Pick's disease, familial hypercholesterolemia, retinitis pigmentosa,

166. EP 1238110, US 6458538 and US 7026122 all granted; EP 2036991 pending; and US 2006134681 abandoned.
167. EP 1874306 granted, and US 2009149513 pending.

muscular dystrophy (e.g., Duchenne muscular dystrophy), spinal muscular atrophy and Marfan syndrome.[168]

*Drug Screening (CFTR)*

The heading comprises the following groups (181*) (236*) (263*) (265*) (379*) and (378*). These are also categorised by the purpose of drug screening, but target CFTR directly. *Group (236*)* claimed a method for screening candidate agents for chaperone activity, the method comprising: combining a candidate biologically active agent with a cell comprising genetic sequences encoding an S-tagged conformationally defective protein of interest in which the conformationally defective protein of interest is Δ F508-CFTR.[169]

*Formula and synthetic peptides*

The heading 'formula and synthetic peptides' comprises patents and patent applications that fall under the (b) pathological process heading but where the claims have been constructed in such a way that the DNA sequence or gene claimed was presented as a formula or a synthetic peptide.

*Groups (8)* and *(14)*, categorised as CF pathological processes, concern a Kunitz domain which are the active domains of proteins that inhibit the function of protein degrading enzymes. *Group (8)* comprises modified protease inhibitors.[170] US 20070020252 and the correlated EP 1663279 claimed a synthetic peptide (DX-890). The claimed part of the peptide DX-890 is a synthetic protease inhibitor that comprises the DX-890 amino acid sequence. However, except for 4 amino acids out of 53, in its natural environment the synthetic DX-890 equals the sequence for a part of the natural gene AMBP.[171]

### 5.3.c. Stem cells

This heading comprises groups (6) (7) (101*) (132*) and (161*), none of which comprises claims for induced pluripotent stem cells. For example, *group (6)* concerned stem cell therapy administering mesenchymal stem cells for treating genetic diseases and disorders. However it directed the claims to a laundry list with 20 subjects ranging from cystic fibrosis, spinal muscular disease to Charcot-Marie-Tooth disease. Each distinct disease representing a

168. EP 2007895 and US 2010292307 both pending.
169. EP 1982189 pending, and US 7790364 granted.
170. EP 1663279, US 2007020252 and US 20050089515 abandoned, and only US 75504277 granted.
171. The protein inhibits human neutrophil elastase (hNE).

distinct technical solution to the problem of finding therapeutic applications of MSC and a distinct invention.[172] The first invention following claim 1 comprises a method of treating cystic fibroses by e.g. limiting the progression of fibrosis and scar expansion.

In the office communication it was stated that no technical feature has been identified which could be regarded as a special technical feature unifying all or part of the subjects. It was observed that the application did not provide any example directed to any of the listed diseases. The first part of example 1 described an animal model for studying MSC homing in the lungs of irradiated rats; no results are presented. The last part of example 1 and examples 2 and 3 then described protocols for clinical trials of MSC in the treatment of cystic fibrosis, Wilson's disease and amyotrophic lateral sclerosis, all of which begin with the statement 'depending upon the outcome of the intravenous and intraosseous administrations of MSCs to rats as hereinbefore described', which implies that no study had been actually carried out.[173]

*Group (7)* concerned methods of producing human stem cells by parthenogenetically activating human oocytes by manipulation of oxygen tension, including manipulation of Ca2+ under high $O^2$ tension using anionophore and contacting oocytes with serine threonine kinase inhibitors (DMAP) under low $O^2$ tension, isolating inner cell masses (ICMs) from the activated oocytes, and culturing the cells of the isolated ICMs under high $O^2$ tension. Moreover, methods were described for the production of stems cells from activated oocytes in the absence of non-human animal products, including the use of human feeder cells/products for culturing ICM/stem cells. Stem cells produced by the disclosed methods were also described.

172. See EP 1973553, November 2010 communication.
173. From the EP examination it follows that the prior art teaches the isolation and expansion of MSC and proposed their use in the treatment of various acquired or genetic diseases, including diseases of the lung such as cystic fibrosis. *The actual examples are not directed to any specific disease but study the homing of the cells after injection and the stability of expression in transgenic MSC.* For genetic diseases, in addition to transgenic cells there is also consideration of the use of unmodified cells from healthy donors (claim 1). In view of the prior art, as far as it is directed to the treatment of cystic fibrosis, the subject matter of claims 1-3, 5-10 and 12-14 is not novel (Article 54(2) EPC), and the alternate embodiment of claims 4 and 11 (intraosseous administration, not supported by any actual example) does not involve an inventive step (Article 56 EPC).

**5.3.d. Highlights: Transgenic animals and claims for techniques as such**

This heading has been divided into three sub-groups: transgenic animals, delivery and gene therapeutic vectors and proteins (some of which target *CFTR* and some of which do not specifically target CFTR), and claims for general techniques. Transgenic animals accounted for four groups[174] and they concern techniques for animal cloning (*group (106*)*) and a ('laundry') rabbit that expresses apo(A) and apo(B)[175] where the invention uses for example therapeutic gene promoters such as CFTR and factor IIIV. One group *(297*)* claimed the production of genetically modified animals, in which the genetic modifications are engineered in somatic cells cultured *in vitro* by gene targeting. Genetically modified cells may then be used as nuclear donors' loci in animal chromosomes which are suitable sites for transgene addition to cells. *Group (165*)* concerned transgenic animals and the production of e.g. *CFTR* and a laundry list, whereas *group (259*)* claimed a transgenic CF pig.

The inventions of general techniques such as gene therapy were in (*group (131*)*) comprising methods and compositions for targeted chromosomal genomic alterations using modified single-stranded oligonucleotides, and a competing method to that described in the SMA section, and group *(147*)* claiming methods for spliceosome-mediated RNA trans-splicing (and e.g. repair of an exogenously supplied *CFTR* target molecule carrying an Δ F508 deletion in exon 10).[176] *Group (41)* addressed antisense oligonucleotides which combat aberrant splicing and methods of using the same.[177] The group has been separated from the rest of the groups since the method using antisense oligonucleotides (which can bind to RNA) is close to the method used in relation to spinal muscle atrophy (SMA).[178] The claims in the present patents are broad and are primarily directed at splicing correction in beta globin (haemophilia). However, they also claimed use of an antisense oligonucleotide for the manufacture of a medicament for the treatment of disease involving aberrant splicing, e.g. Tay-Sachs syndrome, phenylketonuria and cystic fibrosis.[179]

174. Groups (90*) (165*) (259*) (106*) and (297*).
175. Apolipoprotein, group (90*).
176. US 2002115207, description point [0039]. The groups within this category were (41) (131*) (147*) and (233*).
177. EP 0698092, US 5627274, US 5665593, US 5916808 and US 5976879, all granted to the same inventor.
178. See section 7.3.2.d.
179. See EP 0698092 claims 20 and 21.

*Delivery and gene therapeutic vectors and proteins*

A large group comprised delivery methods, for example gene therapeutic vectors or proteins for substituting for the defective CFTR protein in target cells – typically the lung epithelium.[180]

The inventions based on adeno-associated virus (AVV) for delivery include the following groups: (2) (18*) (19*) (21*) (23*) (38*) (74*) (89*) (95*) (150*) and (186*). AAV vectors may have utility for gene therapy, but a significant obstacle has been the inability to generate such recombinant vectors in sufficient quantities to be clinically useful for human gene therapy. *Group (19*)* claimed packaging systems and processes for packaging AAV vectors that effectively circumvent these problems.

*Group (2)* comprises modified polypeptides.[181] The applicants claimed a general method of gene therapy using the CD46 receptor directed at cancer, and CF was only part of a disease laundry list. Adenovirus can be directed to a cell by this invention comprising a fusion polypeptide derived from the extracellular region of CD46 as a specific binding site for adenovirus and a component capable of binding to a cell surface molecule. *Group (11*)* provides gene therapy vectors, which are especially useful for cystic fibrosis, and methods for using the vectors.

180. Those not targeting CFTR specifically (laundry list approach): (2) (19) (3*) (5*) (6*) (7*) (11*-12*) (14-15*) (20*-24*) (26*-27*) (29*-32*) (34*-35*) (37*-40*) (43*-46*) (48*) (51*) (56*-57*) (61*-63*) (66*-70*) (72*) (74*) (76*-77*) (83*) (85*-89*) (91*-92*) (94*-96*) (103*) (106*-109*) (111*-112*) (114*-115*) (119*-121*) (124*-125*) (128*-130*) (133*-138*) (140*-144*) (150*-153*) (155*) (157*) (159*) (163*) (166*) (174*) (178*) (182*) (184*) (186*) (189*) (191*-192*) (196*) (205*-206*) (211*-213*) (218*) (227*) (237*) (240*) (281*) (287*-290*) (294*-298*) (301*) (306*) (308*) (312*) (315*-316*) (325*) (344*) (352*) (377*) (383*-384*) (388*-389*) (400*) (411*-412*) (421*) (427*) (433*) and (435*); and those more specifically targeting CFTR: 4*) (8*) (16*) (19*) (79*) (157*) and (240*).

181. EP 1482052, EP 1627067 and 0076540, all applications have been withdrawn (PCT Rule 13.1). The inventions listed in Groups I-IX do not relate to a single general inventive concept under PCT Rule 13.1 because, under PCT Rule 13.2, they lack the same or corresponding special technical features for the following reasons: The special technical feature linking groups I-IX appears to be that they all relate to a polypeptide derived from the extracellular domain of CD46. However, Seya et al. teach that the extracellular and transmembrane domains of CD46 are essential to measles-virus-mediated syncytium (1997, *Biochem J. Vol.* 322, pp. 135-144). Therefore, the technical feature linking the inventions of groups I-IX does not constitute a special technical feature as defined by PCT Rule 13.2, as it does not define a contribution over the prior art.

*Group (23)* concerns inventions of swellable particles for drug delivery. The delivery of therapeutic agents to the pulmonary system has been used for the treatment of local lung diseases such as asthma, cystic fibrosis, and chronic obstructive pulmonary disease. Group (26*) claimed nanoclusters for delivery of therapeutics and group (29) claimed gene delivery comprising a therapeutic agent to deliver to the cell by magnetic means. Group (157)* claimed a method of treatment *in situ* in the lungs of mammals using a biological device (microcapsule) for the treatment of CF. Thus the group headed 'delivery' can consist of every form of delivery from AAV to the use of nanoclusters. Sometimes only the method of delivery is claimed, but often it is accompanied by a laundry list.

# II. SECTION on Genetically Heterogeneous Diseases Alzheimer's disease & Parkinson's disease

## 6. Alzheimer's disease

Biomedical research into Alzheimer's disease (AD) is still undergoing considerable developments but, as mentioned in Chapter 6, there is still no successful treatment. The research has been focused on two essential proteins: beta amyloid protein and tau protein.[182] The association of the *APOE* E4 allele with AD is significant but the *APOE* genotype is not only associated with AD. The present search into *presenilin 1*, *presenilin*, amyloid beta A4 protein, *PSEN1*, *PSEN 2*, *APP*, *APOE*, and *SORL1* has resulted in a lot of inventions that are not relevant to AD. Addressing strictly the primary genes was very important for enabling the inquiry to remain closely related to AD, leaving out inventions for such as chemotherapeutics which only *might* have an impact on Alzheimer's, and *APOE* which is targeted at atherosclerosis which is completely different from AD and thus outside the scope of this case

182. Both Alzheimer's disease and Parkinson's disease occur in two different forms. There is a very frequent sporadic form of AD, which is multifactorial with a genetic component, and a rarer familial form following an autosomal dominant inheritance pattern with several known disease genes. The search comprised mainly the autosomal dominant inheritance pattern as a consequence of the search outcome. The late onset gene searched for AD (APOE, SORL1) did not find any applications.

study.[183] Inventions claiming genes as AD primary genes but which do not find support in the biomedical literature, e.g. *group (89)* claiming PAMP mutation as a cause of AD, have also been omitted. Furthermore inventions such as a talking toilet paper holder also fall outside the scope of gene patents.[184]

Most cases of early-onset Alzheimer's disease are caused by mutations in one of three genes: *APP*, *PSEN1*, or *PSEN2*. [185] These three genes all follow an autosomal dominant inheritance pattern. When altered, large amounts of a dominant negative toxic protein fragment called amyloid β-peptide are produced in the brain. [186] The build up of this peptide in the brain results in aggregate formation into clumps called β-amyloid plaques, which are toxic. Inhibiting the formation of β-amyloid plaques by diminishing amyloid β-peptide aggregation is a major area of patent activity.

The main focus of this case study is the amyloid beta A4 protein and other primary disease genes, while secondary genes, delivery methods or drugs are only discussed briefly. Section II of the case study is not intended to be a full review, but is an inquiry into activity in the area where there is no single disease gene. The study bears out the observation in Section I, that biological targets and processes are claimed for treatments for multiple diseases (laundry list approach).[187] In particular, the diseases covered by this case study have all been claimed in laundry lists (sometimes all at the same time) relating to treatment drugs, delivery (AVV) and e.g. vaccines, despite the wide differences between the diseases. Parkinson's, Alzheimer's and Huntington's are often claimed together, but SMA, CF and Canavan's can also appear in the same laundry list. Furthermore, it seems that AD always can be claimed in connection with oligonucleotide techniques that inhibit gene expression (e.g. *group (6*)* in Appendix XI table AD. This is because the disease is caused by a dominant negative mutant protein so that, if initiated early, down-regulation of the expression of the mutant protein would constitute a potential therapy.

Finally, as with the single-gene disease case study, the review of heterogeneous diseases is not intended to be a full-scale inquiry. For example, the search based on the primary genes for Alzheimer's disease found 409 appli-

183. The groups that are concerned with atherosclerosis as a primary disease are (53*) (55*) (68*) (77*) (78*) (83*) (108*) (109*) (114*) (126*) and (132*).
184. US 2005117760.
185. See http://www.ncbi.nlm.nih.gov/books/NBK1161/ (last visited March 2012).
186. See http://ghr.nlm.nih.gov/condition/alzheimer-disease (last visited March 2012).
187. See e.g. Groups (5) (6) (7) (16) (48) (52) (53) (17*) (51*) (58*) (71*) (75*) (79*) (80*) (88*) and (98*).

cations. A study published in February 2012 identified 48,000 patents for Alzheimer's, grouped into 23 clusters. The 23 clusters fell into two groups. One group comprised drugs targeted at beta amyloid protein (amyloid grouping) and the other group comprised drugs targeted at tau proteins and serotonin and other alternative pathways (tau grouping).[188]

### 6.1. Data outline

The focus of the case study of Alzheimer's is based on the patent activity listed in appendix XI Table AD, 2011. It comprises 171 groups representing patent families of altogether 409 patent applications, of which 144 applications have been granted, 167 have been withdrawn, and 98 are still pending.[189]

### 6.2. The empirical analysis

Box 7.6.1 below shows the patent activity for Alzheimer's disease, categorised under four headings that differ from the template for the case study for single disease genes. The review of AD and PD is less intensive and it is solely aimed at indicating any observable differences between patent activity when the invention relates to single-gene diseases and genetically heterogeneous diseases. Heading b) is not discussed as it relates to secondary disease genes which were not the target of present search, for example in the form of secondary modifier genes that can help degradation of amyloid 4 betaprotein.

### 6.2.a. Primary genes

In the section headed 'primary genes', most applications concerned beta amyloid plaques as described in the introduction to this section on AD disease. The prevention of the toxic protein fragment (amyloid beta peptide) has been

188. Mike Lloyd et al., 'Clearing the fog: Patenting trends for the treatment of Alzheimer's disease. What can Network Patent Analysis tell us about developments in the treatment of this crippling condition?', February 2012, p. 3, available at http://www.patents4life.com/wp-content/uploads/2012/02/Whitepaper_Clearing the fog _February2012.pdf. The study was conducted using Network Patent Analysis, a patent analysis method that maps the 'collective intelligence' of patent applicants to categorise and rank patents, technologies and patent applications.
189. Left out: (13) (26) (7*) (34*) (37*) (47*) (65*) (72*) (94*) (97*) (99*) (104*) (107*) (111*) (117*) (118*) (119*) (122*) and (141*).

*Box 7.6.1. Alzheimer's analysis model*

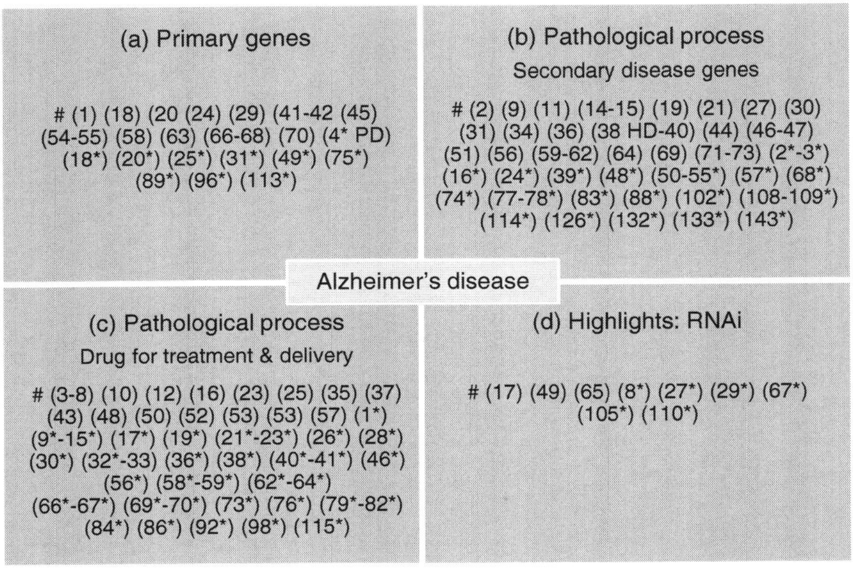


targeted in many applications and vaccines against its formation being developed. Mutations of the primary disease genes *APP*, *PSEN1* and *PSEN2* are involved in the production of amyloid beta peptide in the brain.[190] Some of the inventions (e.g. groups (66) and (70)) concern treatments such as drugs relating to the primary gene; these include APP peptides which are useful in the treatment of Alzheimer's disease and as cognitive enhancers. Thus *group (66)* provides compounds derived from APP peptides[191] and conjugates which can cross the blood-brain barrier as well as pharmaceutical compositions containing such conjugates (pharmaceutically acceptable carriers). *Group (70)* claimed recombinant antibodies that are specific for beta-amyloid in the form of a method for preventing or inhibiting the accumulation of amy-

190. See http://www.ncbi.nlm.nih.gov/books/NBK1161/ (last visited March 2012).

191. See e.g. EP 0954524 claiming formulae comprising up to 30 amino acid residues with the formula representing a reverse-order sequence corresponding to amino acid residues 332 to 328 of human APP and from zero to 30 successive amino acid residues of human APP extending in each direction therefrom. *Group (22)* also claimed peptide formulas.

loid-p peptides and the aggregation of peptides which form amyloid deposits in the brain.[192]

Both Alzheimer's disease and Parkinson's disease occur in two different forms. There is a very frequent sporadic form, which is multifactorial with a genetic component; and there is a rarer familial form following an autosomal dominant inheritance pattern with several known disease genes. The search result comprised mainly the autosomal dominant inheritance pattern. A search associated with the late onset gene for AD (*APOE*, *SORL1*) did not find any applications.

The 'primary genes' heading can be divided into several sub-headings. The first sub-heading claimed '*vaccines and antibodies*' and it shows a considerable development of therapeutic vaccinations against amyloid peptides covering a broad range of approaches.[193] One claim is for administering an antibody to beta-amyloid to a patient (*group (18)*), one claim is for a method for producing beta-amyloid peptides and antibodies for use in a vaccine against amyloid peptides (*group (20)*). *Group (27)* provides a vaccine via gene therapy (antibodies and a laundry list where one can find secondary as well as primary genes).[194] *Group (29)* also claimed anti-A beta peptide antibodies prophylaxis by inducing the production of the antibody against this antigen, using a recombinant adeno-associated virus (rAAV). *Group (55)* claimed a method for improving the immune effect (antibodies: APOE amyloid beta peptides).[195]

192. Drugs/delivery for primary genes are claimed in *groups (30*) (32*) (49*) (67*) (62*) (59*) (80*) (81*)* and *(84*). Group (80*)* claimed a delivery vaccine in the context of a laundry list. Group (81*) claimed a chemical compound (drug) enabling a regulation of the APOE level in cells. [Group (84*) comprises an anti-plaque denture cleansing tablet and a treatment for different subtypes of AD and claim 9 (EP 1585520) has a genomic mutation in an amyloid precursor protein gene, an ApoE gene, or a presenilin gene. Group (62*) claimed sequestration of amyloid ss (Ass) into a plasma administered to a person. *Group (67*)* claimed a method for reducing A[beta] production in a neuronal cell expressing an ApoER2 receptor providing the cell with an agent that inhibits the binding of ApoE to ApoER22.
193. Groups (18) (20) (27) (29) (41) (45) (54) (55) (58) (63) (67) and (20*).
194. Claiming e.g. adoptive cellular gene therapy. A method for treating or ameliorating a neurodegenerative disease, comprising the administration of a composition comprising nucleic acid that induces a cellular immune response.
195. The highest ranked patent of the 48,000 mentioned in the introduction to the AD section was US 7189819, which appeared to protect the stage III (trial) Pfizer/Elan/Johnson & Johnson drug bapineuzumab, and which was successfully litigated up to the US Federal Circuit by the patent applicant to obtain an increase in its patent term. See similarly US 7195761 for the Eli Lilly stage III drug solaneuzumab; see Mike

The next sub-heading is '*transgenic animal disease models*'.[196] One of the groups is more directed at Parkinson's than AD, thus *group (4*)* claimed a PD mouse model comprising *tau* mutations. *Group (25)* comprises a transgenic animal model whose genome comprises the first transgenic DNA sequence encoding a human APP Swedish or a human APP London protein (mutations), and a second transgenic DNA sequence encoding a human presenilin 2 protein comprising a N141I substitution (mutation).[197] Finally, there is the model of the 'Aarhus pigs' (*group (31)*[198] and a supplementary search done by the author). The pig model holds a mutant human APP gene with the Swedish mutation (APPsw695) or the *PS1* gene harbouring a p.P117L mutation.

### 6.2.b. Group for delivery methods – e.g. gene therapeutic vectors or modification of peptides and proteins

There are differences in the claimed delivery methods. One method involves the use of an AAV virus for gene therapy, targeting the disease gene by e.g. delivering a siRNA to down-regulate gene expression or to deliver the gene. The other method involves the delivery of a drug in order to affect gene expression, e.g. the patient's disease gene *in situ* or a secondary gene involved in the pathological process.

For example *group (1*)* covers a gene therapy chimeric AAV vector, granted with a claimed laundry list where the subject has or is at risk of a disorder selected from a muscular dystrophy including Duchenne or Becker muscular dystrophy, haemophilia A, haemophilia B, multiple sclerosis, diabetes mellitus, Gaucher's disease, Fabry disease, Pompe disease, cancer, arthritis, muscle wasting, heart disease including congenital heart failure or peripheral artery disease, intimal hyperplasia, a neurological disorder including epilepsy, Huntington's disease, Parkinson's disease or Alzheimer's disease, an autoimmune disease, cystic fibrosis, thalassemia, Hurler's disease, Krabbe's disease, phenylketonuria, Batten's disease, spinal cerebral ataxia, LDL recep-

Lloyd et al., 'Clearing the fog: patenting trends for the treatment of Alzheimer's disease. What can network patent analysis tell us about developments in the treatment of this crippling condition?', February 2012, p. 3, available at http://www.patents4life.com/wp-content/uploads/2012/02/WhitepaperClearingthefog February2012.pdf.

196. Groups (24) (42) (68) (4*PD, tau) (18*) (25*) (31*, Aarhus pig) and (113*).
197. E.g. US 2010306868. See also Group (68) claiming an animal model for APP mutations (670, 717 mutations).
198. The Pig model for atherosclerosis (group (31*)) is less interesting than the added search.

tor deficiency, hyperammonemia, anaemia, arthritis, a retinal degenerative disorder including macular degeneration, adenosine deaminase deficiency, and cancer including tumour-forming cancers. Surprisingly many applications have been granted within this category.

### 6.2.c. Highlights RNAi

The last group, headed 'highlights', concerns RNAi and anti-sense oligonucleotides. As shown in the previous reviews in this chapter, especially in the sections on Huntington's disease (section 2) and spinal muscular atrophy (section 3), new techniques may question the traditional gene patent both in terms of why e.g. patents for a disease gene can be granted and at the same time siRNA patents are still possible (scope of protection), and in terms of lawmaking (detailed provisions or case law approaches). In the 2008 article, also mentioned under the HD section on therapeutic RNAi patents, it is stated that AD is one of the most preferred targets for the development of RNAi therapeutics, generating siRNAs to develop useful methods for the down-regulation of beta-secretase, APP, Pin-1, presenilin 1 (PS-1) and PS-2 gene expression, as well as other genes involved in these pathways.[199]

The group comprises groups (17) (49) (65) (8*) (27*) (29*) (67*) (105*) and (110*). A search relating to one disease is seldom confined only to siRNAs for that disease. Not only can one find laundry lists with several diseases, but inventions related to one of the other searched diseases were often identified. A few examples were also found in connection with the AD search e.g. *group (17)*, which identifies siRNAs targeting *SNCA*, which is a primary disease gene for PD. However, in this patent there were also claims for sequences which were claimed to target PD, AD, multiple system atrophy and

199. See Antonio Orlacchio, Giorgio Bernardi, Aldo Orlacchio & Sabata Martino, 'Patented therapeutic RNAi strategies for neurodegenerative diseases of the CNS', *Informa Healthcare, Expert Opinion, Ther.Patents*, 2008, 18 (10), p. 1166. The review refers to Tsai (US 20070185042), Tanzi and Xie (US 20060172964, reducing A-beta in AD brains) and Mcswiggens (US 20070185043). The application based on the McSwiggen invention is in the present case study *group (9)* and categorized as a secondary gene under heading b) regarding siRNA against Nogo and Nogo receptor. Activating nogo (a receptor that is a glycosylphosphatidylinositol-linked cell surface molecule forming a heteromeric receptor complex with either LINGO-1 and p75, or LINGO-1 and TROY (a tumour necrosis factor (TNF) receptor family member) initiates a cascade that leads to the activation of Rhoa and, ultimately, to the inhibition of axonal growth. AD is part of a laundry list in that application. Recent studies indicates that NGR is also expressed in non-neuronal cells and modulates macrophage responses connected to AD.

Lewy body dementia. Similarly, *group (21)* claimed RNAi molecules targeted at a nucleic acid sequence that encodes polyglutamine repeat diseases. As mentioned previously, Huntington's disease (HD) is caused by the expansion of a polyglutamine repeat, but spinocerebellar ataxia type 1 (*SCA1*), dentato-rubral-pallidoluysian atrophy (DRPLA), Macado-Joseph disease (MJD), and spinobulbar muscular atrophy (SBMA) are also strongly associated with an expanded polygluatmine repeat. Despite this, the patent contains a laundry list that includes many diseases where expanded polyglutamine tracts have not been identified or indicated as playing a role. The list includes, but is not limited to, amyotrophic lateral sclerosis (ALS), hereditary spastic hemiplegia, primary lateral sclerosis, spinal muscular atrophy, Kennedy's disease, Alzheimer's disease, Parkinson's disease, multiple sclerosis, and repeat expansion neurodegenerative diseases, e.g., diseases associated with expansions of trinucleotide repeats such as polyglutamine (polyQ) repeat diseases, e.g., Huntington's disease (HD), spinocerebellar ataxia (*SCA1*, *SCA2*, *SCA3*, *SCA6*, *SCA7*, and *SCA17*), spinal and bulbar muscular atrophy (*SBMA*), dentato-rubropallidoluysian atrophy (*DRPLA*). Group (49) claimed very broadly several different types of ASO RNAs (short interfering nucleic acid (siNA), short interfering RNA (siRNA), double-stranded RNA (dsRNA), micro-RNA (miRNA), and short hairpin RNA (shRNA)) with lengths of 'about 18 to about 23 nucleotides' for modulation of the expression of beta-secretase (*BACE*), amyloid precursor protein (*APP*), *PIN-1*, presenilin 1 (*PS-1*) and/or presenilin 2 (*PS-2*) genes in Alzheimer's disease. Group (65) provides small interfering RNA targeted at BACE 1, and different pump delivery systems. The original application comprised 152 claims with siRNAs targeted at neurodegenerative disorders such as Alzheimer's disease, Parkinson's disease, Huntington's disease, Spinocerebellar Ataxia Type 1, Type 2, Type 3, and/or dentatorubral-pallidoluysian atrophy. The granted patent has only retained the siRNAs targeted at BACE-1, which is specific for AD, but the abstract and descriptions still contain all the other diseases. *Group (105*)* focuses primarily on the use of ribozyme ASOs, with an immense list of target sequences (BACE, MetAP2, *PSEN2* etc.), but it also claimed all other types of RNAi ASOs targeted at these targets. The application has been withdrawn. *Group (110*)* also focuses on the use of ribozyme ASOs, with an immense list of target sequences from the BACE gene. The application is pending.

It is remarkable that none of the siRNAs (or other types of inhibiting ASO RNAs) seems to be targeted at SNPs in the disease genes. As mentioned previously, it is vital that only the mutant mRNA from one disease allele in the patient is targeted, and as described for HD this can be achieved by designing the siRNA (also shown in the figure 7.2) so that it binds to a frequent SNP (the lack of perfect match to one of the alleles will then result in lower or no down-regulation of the normal allele).

## 7. Parkinson's disease

As described in Chapter 6, there are several mutations in the PD primary genes that cause Parkinson's disease. In particular, in this empirical case study it seems that mutations in the parkin *PARK2* gene, which causes the disease, are the most researched. The parkin gene is part of the cellular protein degradation system (ubiquitin-proteasome system) and mutations in the parkin gene cause loss of function of this system. Parkinson's disease caused by *PARK2* mutations is inherited in an autosomal recessive fashion and results in juvenile disease.

### 7.1. Data outline

The following case study of Parkinson's disease (PD) comprises 21 groups representing patent families of altogether 114 patent applications, of which 34 applications have been granted, 55 withdrawn, and 25 are still pending.[200]

200. Appendix XII Table PD. Omitted: (4) (5) (11) (1*) (2*) (6*) (24*) (27*-30*) and (33*). E.g. group (4) claimed a chemical composition comprising a cytokine inhibitory drug and a major laundry list of AD, PD, HD and more. However, one claim (US 20070190070, claim 11) claimed the second active ingredient as a dopamine agonist, a monoamine oxidase inhibitor (MAO), which is the core focus of the Buffalo University application. The applications within this group have all been abandoned. The US examiner stated that: 'It does not define any structural features commonly possessed by members of the genus that distinguish from others. One skilled in the art therefore cannot, as one can do with a fully described genus, visualize or recognize the identity of the members of the genus. A definition by function, as we have previously indicated, does not suffice to define the genus'; at 1406 (emphases added). 'In the instant case 'a dopamine agonist', 'a monoamine oxidase inhibitor', 'a catechol-O-methyltransferase inhibitor', 'an acetylcholinesterase inhibitor', 'an antiemetic', 'an anti-inflammatory agent', recited in the instant claims are purely functional distinction. Hence, this functional recitation read on any compounds that might have the recited functions. However, the specification merely provides those particular compounds for functional compounds for the method herein (see pages 53-54 of the specification herein). Thus, Applicants functional language at the points of novelty fails to meet the requirements set forth under 35 U.S.C. 112, first paragraph. Claims employing functional language at the exact point of novelty, such as Applicants', neither provide those elements required to practice the inventions, nor 'inform the public dur-

## 7.2. The empirical analysis

Box 7.7.1 below shows the patent activity for Parkinson's disease. The PD template differs from the case study for single disease and from the AD template. It will focus in particular on stem cell activity and siRNA. The 2011 search on Parkin, *SNCA, UCHL1, LRRK2, PARK2, PARK7, PINK1* only found activities on Parkin, and new search on the other genes was carried out in April 2012. The purpose of this search was to verify whether only Parkin has been the subject of activity or whether there were activities on the other genes which the first search had missed, so it did not expand the number of applications that form part of the analysis. Heading b) on the pathological process is mainly omitted from the analysis, but it does have one relevant issue since it comprises secondary genes and laundry lists where, for example, AD, PD, HD and others are claimed. For example, nogo receptor binding proteins (*group (3)*) have been claimed in respect of almost every disease in the present case study, despite the fact that an invention e.g. regarding Canavan's disease differs from that of an invention for PD, even though its function of decreasing inhibition of axonal growth can be relevant in many cases.

*Box 7.7.1. Parkinson's disease analysis model*

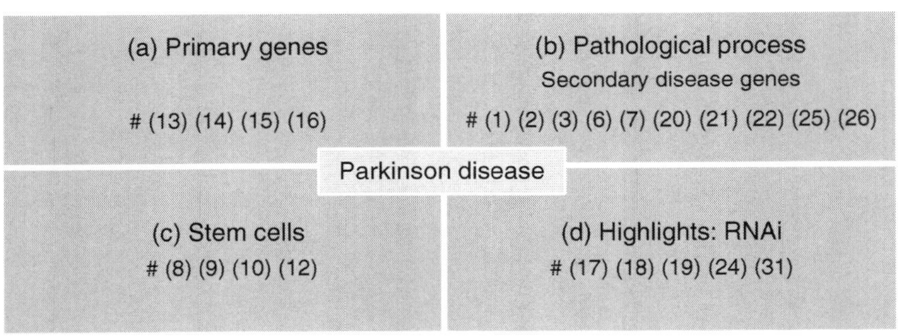


### 7.2.a. Primary genes

The first patent (Shimizu and Mizuno, Boehringer Ingelheim International GmbH) that fits into the heading 'primary genes' is *group (13)*, where four

ing the life of the patent of the limited of monopoly asserted' (Genera Electric Company v Wabash Appliance Corporation et al. 37 USPQ at 468 (US Supreme Court 1938)). The predictability or unpredictability: The instant claimed invention is highly unpredictable'.

granted patents claimed an isolated DNA or gene that is responsible for Parkinson's disease and is useful in diagnosing and treating the disease. According to this invention, the isolated DNA or gene comprises the full-length parkin sequence or partial sequence, or a sequence hybridisable thereto or hybridisable with a complementary strand thereof. What is claimed is the parkin gene in a very broad sense, for diagnosis, prevention and treatment of the disease. It also claim and was granted 'encoding gene products a protein having a substantially equivalent function to a protein'. Its scope of protection embraced the genomic sequence of Parkin, recombinant vectors, antibodies to the protein, gene therapy, production of the protein and primers and probes. Altogether one wonders how subsequent inventors can claim anything more. A recent patent application (Aventis Pharma S.A, March 2012) claimed an isolated nucleic acid molecule encoding a human parkin with at least one genetic alteration comprising: a) a deletion of one or more exons, in combination or otherwise; b) a multiplication of exons; c) a point mutation; d) a deletion of 1 or more contiguous base pairs; e) an insertion of 1 or more contiguous base pairs; or f) a combination thereof.[201] The same applicant was granted a patent in August 2011 claiming nucleic acids coding for mutated or truncated forms of the human parkin gene, or forms comprising multiplications of exons, and the corresponding proteins and antibodies. The invention also concerns methods and kits for identifying mutations of the *parkin* gene, and for studying compounds for therapeutic purposes.[202]

*Group (15*)* claimed C. elegans (animal model) for the investigation of neurodegenerative diseases, in particular Parkinson's disease, uses and methods for the discovery of substances and genes which can be used in the treat-

201. See US 2012064598, Aventis Pharma S.A. (outside the 2011 search).
202. See US 7998667, Aventis Pharma S.A, (outside the 2011 search). Claim 1: 'A method for detecting an autosomal recessive parkinsonian syndrome in a symptomatic subject comprising analyzing a nucleic acid sample obtained from said subject for the presence of one or more heterozygous genetic alterations in a parkin gene comprising SEQ ID NO:1, wherein the presence of one or more heterozygous genetic alterations in said parkin gene indicates that said subject has an autosomal recessive parkinsonian syndrome, wherein said genetic alteration in said parkin gene comprises: a) a heterozygous deletion of any one of the following single exons or exon combinations: exons 2 and 3; exon 2; exons 2, 3 and 4; exon 3; exon 5; exons 3-6; exon 6; exons 6-7; exon 8; exons 7, 8, and 9; or exons 3-9; b) a heterozygous duplication of any one of exons 3, 6, 7 or 11, c) a heterozygous point mutation resulting in any one of the following amino acid changes: Gln34Arg; Asn52Met; Lys161Asn; Lys211Asn; Arg256Cys; Arg275Trp; Asp280Asn; Cys268Stop; Cys289Gly; Arg348Glu; Gly328Glu; or Gly430Asp, wherein said amino acid is encoded by SEQ ID NO:1, d) a heterozygous triplication of exon 2, or e) a combination of a)-d) thereof.'

ment and identification of a nematode gene from C. elegans, which is homologous to the human parkin or alpha-synuclein genes.[203]

Recently, an article from the University of Buffalo[204] stated that a team from the University had applied for patent protection for a screening platform for the discovery of a new treatment that might mimic the protective functions of parkin. Once parkin is mutated, they write, it can no longer precisely control the action of dopamine which supports the neural computation required for movement. The team also found that parkin mutations prevent it from tightly controlling the production of monoamine oxidase (MAO), which catalyzes dopamine oxidation. Parkin normally ensures low levels of MAO, but mutated parkin lacks this quality resulting in high levels of MAO that are toxic. The team found that nerve cells from Parkinson's patients expressed MAO at a high level and directed a new design of drugs to reduce the expression of MAO.[205]

### 7.2.b. Stem cells

All the diseases of the present case study have had patent activity that comprised stem cell research and all the European applications have been evaluated with a *WARF*-awaiting clause, so the review of the activity is less interesting since the use embryonic stem cell techniques has been *de facto* outside the scope of patent protection since 2000 when the Biotech Directive 98/44/EC entered into force. Thus, the purpose of the current case study is to identify developments that can circumvent the European position on embryonic stem cells. Both Parkinson's disease and spinal muscle atrophy seems to direct research to this area. The PD study recognises four *groups (8) (9) (10)*[206] and *(12)*, of which *group (8)*[207] can be used as an example of long discussions in the examination of embryonic stem cells. The invention provides neural precursor cells and stem cells which can give rise all three major cell

203. *Group (16*)* withdrawn claim ASP26 mutation claiming 'an isolated nucleic acid sequence which is at least 70 % identical to a nucleic acid sequence such as shown in SEQ ID NO: 3, coding for a polypeptide functionally homologous to a mutated Parkin polypeptide or a fragment of said polypeptide, the amino acid sequence of said polypeptide differing from the wild type amino acid sequence such as shown in SEQ ID NO: 2 at least in including a mutation at Asp 126.'
204. http://www.buffalo.edu/news/13170 last visited April 2012.
205. Jiang H, et al., 'Parkin controls dopamine utilization in human midbrain dopaminergic neurons derived from induced pluripotent stem cells', 2012, *Nat Commun.* 3:668.
206. The invention is present in CF group (6) laundry list: CF, AD, HD, PD and more.
207. EP 1228195, US 6878543, US 2005191745, US 7303912, US 2008138895 and US 7651853.

sub-types of the CNS, i.e. neurons, astrocytes, and oligodendrocytes. From these studies, there is a clear need for cells capable of proliferating to make large numbers of cells as well as a capacity for neural differentiation in order to make the appropriate 'adult' cells capable of integrating and restoring function to a diseased area in the CNS. This was examined by the EPO examiner, pursuant to EPC Article 53(a) and Rule 28(c) EPC, as follows:

> In the communication dated 29 October 2009 the Examiner states that the present invention does not exclude that neural tissues such as NS4 cells are obtained from a human source between weeks 9 and 12 of the first trimester.
>
> Thus, in the absence of an indication in the application as filed, in the light of the state of the art at the time of filing, of a way to carry out the invention whereby the claimed cells are not derived exclusively by the use involving the destruction of human embryos, the claimed invention is considered to fall under the exclusion of Rule 28(c) EPC in accordance with decision G2/06.

Group (9)[208] claimed an isolated human stem cell line derived from parthenogeneic blastocysts. Parthenogeneic activation of mammalian oocytes may be used as an alternative to fertilisation by sperm/NT to prepare oocytes for embryonic stem cell generation. The invention also comprises a laundry list, see EP 2155861, claim 7: a disease selected from the group consisting of Parkinson's disease, Huntington's disease, Alzheimer's disease, ALS, spinal cord defects or injuries, multiple sclerosis, muscular dystrophy, cystic fibrosis, liver disease, diabetes, heart disease, macular degeneration, cartilage defects or injuries, burns, foot ulcers, vascular disease, urinary tract disease, AIDS, and cancer.

The article mentioned above describing the new findings of the Buffalo University School of Medicine and Biomedical Sciences is highly relevant to the present case study, since it reveals patent activity on induced pluripotent stem cells (iPSCs).[209] The article states that it is the first time that human dopamine neurons have ever been generated from Parkinson's disease patients with parkin mutations. Using iPSCs allowed the research team to 'reverse engineer' human neurons from human skin cells taken from four subjects: two with a rare type of Parkinson's disease in which the parkin mutation is the

208. See EP 2188861 and US 2008299091 (Patient-specific stem cell lines derived from human parthenogenetic blastocysts).
209. It converted human cells into induced pluripotent stem cells (iPSCs) that could then be converted into nearly any cell in the body, mimicking embryonic stem cells. See Jiang H, et al., 'Parkin controls dopamine utilization in human midbrain dopaminergic neurons derived from induced pluripotent stem cells', 2012, *Nat Commun.* 3:668.

cause of their disease, and two healthy subjects who served as controls. The material available for the study thus circumvented the ban on embryonic stem cell patenting but it did not make research any easier.[210] A literature search reveals that the techniques have been developed for many diseases (including CF, SMA, HD and PD) and the overall assumption is that human induced pluripotent stem cells have the potential to revolutionise personalised medicine by allowing immuno-compatible stem cell therapies to be developed.[211] Patent applications found outside the search (e.g. US 8058065 and US 8048999, both issued 2011) and an application from 2011 (US 2011039332) comprise four genes Oct3/4, KLF4, c-Myc and Sox2 that are relevant for reprogramming a differentiated somatic cell to derive an induced pluripotent stem (iPS) cell. However promising this technique may seem, it is not considered safe and new scientific data shows that safety is a major issue since genetic material is altered during the reprogramming.[212] Today, in the US and Europe iPS is commercialised in non-patented forms by stem cell biobanks.[213]

### 7.2.d. Highlights: RNAi

The last sub-heading refers to RNAi techniques that are also reviewed in relation to HD and AD, and which are also relevant to PD.[214] *Group (17)* essentially relates to AD, providing methods and reagents useful in modulating gene expression in a variety of applications, including use in therapeutic, diagnostic, target validation, and genomic discovery applications associated with Alzheimer's disease. Specifically, the invention relates to small interfer-

210. Jiang H, et al., 'Parkin controls dopamine utilization in human midbrain dopaminergic neurons derived from induced pluripotent stem cells', 2012, *Nat Commun.* 3:668. The problem has been that human neurons live in a complex network in the brain and are thus off-limits to invasive studies. Feng explains: 'Before this, we didn't even think about being able to study the disease in human neurons ... The brain is so fully integrated. It's impossible to obtain live human neurons to study'.
211. A.D. Ebert et al., 'Induced pluripotent stem cells from a spinal muscular atrophy patient', 2009, *Nature* 457(7227), pp. 277-80; I.H. Park et al., 'Disease-specific induced pluripotent stem cells', 2008, *Cell.*, 134(5), pp. 877-886; Athurva Gore et al., 'Somatic coding mutations in human induced pluripotent stem cells', *Nature*, 3 March 2011, Vol. 471, p. 63; and Houbo Jiang et al., 'Parkin controls dopamine utilization in human midbrain dopaminergic neurons derived from induced pluripotent stem cells', *Nature Communications*, February 2012 (DOI:10.1038/ncomms1669).
212. Scientific Daily, 2 March 2011, 'Mutations found in Human Induced Pluripotent stem cells'.
213. See Coriell: http://ccr.coriell.org/ and ATCC: http://www.lgcstandards-atcc.org/
214. Comprising groups (17) (18) (19) (24) and (31).

ing RNA (siRNA) molecules capable of mediating RNA interference (RNAi) against beta-secretase (BACE), PIN-1, presenilin-1 (PS-1) and presenilin-2 (PS-2) polypeptide and polynucleotide targets.[215] The McSwiggens patent was granted to Sirna Therapeutics, Inc., and together with *group (18)* it is part of a bundle of patents on the technique itself since they have invented the chemical linkage of 2'-O-methyl nucleotides and the company has around 350 patents on the technique. The US patent US 7989612 granted in 2011 comprises methods and reagents useful in modulating gene expression in a variety of applications, including use in therapeutic, diagnostic, target validation, and genomic discovery applications. The invention relates to synthetic chemically modified small nucleic acid molecules, such as short interfering nucleic acid (siNA), short interfering RNA (siRNA), double-stranded RNA (dsRNA), micro-RNA (miRNA), and short hairpin RNA (shRNA) molecules capable of mediating RNA interference (RNAi) against target nucleic acid sequences. The small nucleic acid molecules are useful in the treatment of any disease or condition that responds to modulation of gene expression or activity in a cell, tissue, or organism. Where the above mentioned patent is general and granted for the technique as such, *group (18)* comprises RNAi against secondary genes (contrary to PD primary genes). A step specifically towards PD can be seen in *group (31)* where McSwiggen claimed *SNCA*.[216]

It is remarkable that once again the abstract and description embraces several PD genes, *PARK1* (*SNCA*), *PARK2*, *PARK7*, and/or *PARK5* whereas the claims are only for *PARK1* (*SNCA*). The application was abandoned, and as for the enablement requirement, the final rejection in 2007 emphasised that the post-filing art indicated that the art of *in vivo* delivery of double-stranded interfering nucleic acids for selectively inhibiting gene expression in animals and humans is unpredictable. Although the applicants clearly discussed possible routes of delivery and methods for administering nucleic acids, these teachings were general in nature, and did not teach the ordinary artisan how to effectively deliver siRNA or any other double-stranded nucleic acids, or

215. In Group (17), EP 1423404 has been granted.

216. See further Antonio Orlacchio, Giorgio Bernardi, Aldo Orlacchio & Sabata Martino, 'Patented therapeutic RNAi strategies for neurodegenerative diseases of the CNS', *Informa Healthcare, Expert Opinion, Ther.Patents*, 2008, 18 (10), p. 1167. The review refers to PD applications from Sirna Therapeutics: US 20050137155 on methods and reagents useful in modulating PARK1 (SNCA), PARK2, PARK7 and PARK5 gene expression relating to the full scale of siRNA, dsRNA, miRNA and shRNA molecules capable of mediating RNAI against the expression of the SNCA gene (and for other diseases that correspond to the modulation of PARK1 and PARK2.

combinations thereof, to any target tissues and cells *in vivo* so as to effectively reduce *gene expression of any SNCA gene* to effectively treat and/or prevent a neurological disorder in any subject. The examiner stated that the amount of disclosure was insufficient, given the level of unpredictability in the art. For example, the application did not appear to teach a person skilled in the art how *to effectively target tissues and cells in the brain*. Similarly, while the application was enabling *for the use of double-stranded nucleic acids to transfect cells in culture, it did not enable the use of these molecules* in vivo *in a way that would reasonably enable a person skilled in the art to use the invention so as to obtain the desired result, e.g., phenotype or outcome in an individual*. Most of the examples given in the application, such as working and prophetic examples, were directed to the cellular uptake and inhibition of *SNCA* (*PARK*) by siRNAs *in vitro* (in cell culture), and the one example directed at an animal model was prophetic in nature and did not adequately represent the delivery of siRNA to animals and humans for the treatment or prevention of disease. The discussion of the prior art is very interesting, as it shows how new developments can be approached by patent examiners even before they are fully developed. The following communication is from 2007, and the examiner referred to the prior art as follows:

> The prior art teaches, then, that siRNAs, antisense oligonucleotides, and ribozymes can be used to produce the same effect, albeit with different potencies and by different biochemical mechanisms. siRNAs, antisense oligos, and ribozymes can both be used to inhibit gene expression *in vivo* or *in vitro,* via mRNA degradation or translation attenuation, and, thus, both types of nucleic acids may be used to prevent the expression of a gene in a cell. For example, Bass teaches that antisense RNA is another technique to prevent the expression of particular genes (page 429). Thus, in this sense, siRNAs, antisense oligos, and ribozymes are recognized equivalents that may be used for the same purpose: reducing or inhibiting gene expression. (See for example MPEP §2144.06, substituting equivalents known for the same purpose.)[217]
>
> Nevertheless, as explained above, siRNAs possess certain advantages over antisense oligos and ribozymes which would motivate one of ordinary skill in the art to select siRNAs over antisense oligos to more efficiently block and/or reduce the expression of any given target gene, particularly a gene known to be involved in a neurodegenerative disease such as Alzheimer's. One of skill would have been further motivated to make and use chemically modified siRNAs for sequence specific inhibition of amyloid precursor protein, given that the prior art teaches and suggests using chemical modification to prevent nuclease-catalyzed degradation of RNA and DNA-based drugs in cells and tissues in vitro and in vivo.

217. The technology was developed and used in the following order: (1) ribozymes, (2) dsRNA, and finally (3) siRNA/microRNA.

*Group (24)* contains an RNAi laundry list claiming HD, AD and PD. In the group only application EP 2239328 remains (Alnylam Pharmaceuticals Inc.) for neurological diseases in general. Originally claim 9 claimed a siRNA where the human carries a genetic variation in a parkin gene. This would allow an allele-specific siRNA, but this is not relevant in the case of parkin, since that is a recessive disorder. The patent application is still pending, however, due to the requirement for unity of the invention (EPC Article 82 and Rule 164(1)) claims for AD and PD were deleted.

## 8. Hands-on patent activities relating to genetic diseases

This empirical case study has comprised 1,590 patent applications spread across the selected diseases (HD, SMA, CD, CF, AD and PD). 39 % of the patent applications studied resulted in the grant of a patent; 38.4 % of patent applications were rejected; and 22.6 % were still pending. The analysis relating to each disease is divided into four sub-groups: patent activity on the disease gene, patent activity on the pathological process, patent activity involving stem cells, and patent activity involving highlights which varied from disease to disease. The outcome of the case study with regard to emerging technologies (most importantly RNAi techniques) will be the subject of further examination in Chapter 10. In order to give a brief summary of how research into the diseases has developed, this section provides a broad overview of the results. The case study comprises two sections: Section I concerns single-gene disorders and Section II concerns genetically heterogeneous diseases.

*Section I – single-gene disorders*
Huntington's disease is associated with increases in the length of a CAG triplet repeat expansion present in the coding region of the huntingtin gene *IT15*. On developments relating to *Huntington's disease*, according to the state-of-the-art described in Chapter 6, comprising patents on huntingtin, IT15 or HTT as such, it can be concluded that the focus has moved from gene patents to patents for RNAi, without these affecting each other in a patent context. A huge number of the patent applications concerning Huntington's disease involved RNA interference (RNAi). This reflects the fact that Huntington's disease is a dominant disease caused by a mutation (triplet repeat expansion) that results in the production of a dominant negative protein. RNAi methods which specifically remove the mutant mRNA which encodes the dominant negative protein are therefore a very promising future therapy. In particular, RNAi methods that are allele-specific (mutation specific) have a

great potential and many of the patents therefore claim a specific RNAi sequence which only targets a common SNP variation, for example. This approach would leave the mRNA from the patient's normally functioning allele intact and ensure that normal huntingtin (which is required for normal cell function) is still produced in the patient. siRNAs that are not allele-specific are of little therapeutic value, since they will cause degradation of both mutant and normal *IT15* mRNA in patients, leading to a lack of normal IT15 protein and they may therefore themselves cause neuronal disease.

The HD case study showed examples of the extensive use of 'laundry lists'. There appeared to be no pattern to the examination of such lists and no clear conclusion can be reached as to whether there is a difference between the practices of the USPTO and the EPO on this matter. This is not solely relevant to Huntington's disease, but the HD data illustrate well that some applicants make claims for inventions that are far too broad, covering a list of the world's diseases. It is unclear why some applicants succeed in obtaining a grant with a comprehensive 'laundry list', while other applications are dismissed because they are considered speculative.

Two genes are involved in SMA, namely *SMN1* and *SMN2*. A mutation in *SMN1*, usually a deletion of the entire gene, causes the disease while the severity of the disease is determined by the number of copies of the *SMN2* gene, which is only partly functional due to a high level of aberrant pre-mRNA splicing. References to *spinal muscular atrophy* are less common in the patent law literature. However, the disease is highly relevant in patent law contexts, especially with regard to gene replacement therapy and gene expression correction. Gene expression correction differs from the RNAi technique described in the HD empirical case study. Where RNAi, divided into microRNA or siRNA, is used to modulate gene expression by down-regulating or inhibiting gene expression, the present techniques with antisense oligonucleotides are used to correct gene expression. It is obvious from the nature of the disease that RNAi methods for down-regulating the primary disease genes, *SMN1 and SMN2*, are not relevant, since that would merely worsen the lack of a functional protein.

*Canavan's disease* is caused by mutations in the *ASPA* gene for an enzyme encoding aspartoacylase, and the disease is categorised as a demyelinating disease. CD is a recessive disease caused by lack of aspartoacylase, and there are no RNAi applications, so the highlight category is omitted. The Matalon patents on *ASPA,* which are described in Chapter 6, section 2.3.1.1, are the starting point for the case study search for activity in the aftermath of these 'Matalon' patents. The new patent grants the applicants three further mutations, whereas the original patent only claimed one mutation, p.E285A, which is the most frequent mutation in Ashkenazi Jewish populations.

The Laundrylist aspects are also present in regard to CD. The techniques claimed were such as delivery systems for gene therapy for the brain, using an adeno-associated vial (AAV) vector which contains *ASPA* for delivery to the brain, employing lipid-entrapped polycation-condensed nucleic acid comprising a lipid which is claimed to enhance delivery. This system has been used for gene therapy in children with CD. Furthermore the section on CD comprises a heading on 'demyelinating diseases', chosen because the inventions all claim broad biological processes or superfamilies involved in processes involving demyelination. Thus processes concerning nerve cells (e.g. inhibiting axonal growth) and transportation of anions (e.g. N-acetylaspartate necessary for the myelin synthesis) are relevant for demyelinating diseases.

*Cystic fibrosis* is caused by mutations in the *CFTR* gene located on chromosome 7. In the literature case study in Chapter 6, section 2.4.1.1, the patent activity on CF is characterised as an open model since the main patents (Francis Collins, Lap-Chee Tsui, John Riordan) on the *CFTR* gene have never been enforced against clinical diagnostics. The economic interest in this disease is reflected in the massive number of patent applications that the empirical case study has found, even though the study is not intended to be exhaustive of all CF patent activity. That research into the disease can be measured by the patent activity is perhaps due to the open model of the CTFR gene patents and the common Δ 508 mutation (described in Chapter 6), and as it is one of the most common severe recessive genetic diseases it is worth finding new methods for screening and detecting new mutations and treatments. The 2011 search revealed patent activity on mutation diagnosis and gene tests. A mutation in the *CFTR* gene has an impact on protein function. Different mutations can change the function of the CFTR protein in different ways which in turn have an influence on the severity of the disease. CF is caused by a wide range of mutations but, as mentioned in Chapter 6, one mutation (Δ 508) accounts for 70% of CF chromosomes worldwide. Approximately 1,900 mutations in the CFTR gene have been verified, and the present search identified a scientific and commercial interest in patenting a detection test. Some of the applications claimed mutation testing for several mutations in *CFTR*, some tests included the Δ 508 mutation, some claimed disease detection through antibodies or genes other than the *CFTR* gene, yet others claimed biomarkers and assays and some claimed a gene test that measured chromosome imbalance.

Finally, some groups claimed screening methods for CF, for example by testing blood samples from pregnant women in order to determine whether the child/foetus is afflicted with CF. One patent application claimed a change in the *CFTR* genetic code, causing enhanced protein production. Further patents claimed treatment of CF and drug screening. One of the highlights of the CF case study concerned transgenic animals.

*Section II – genetically heterogeneous diseases*

Section II of the case study is an inquiry into activity in the area where the disease is not related to a single gene. The study bears out the observation in Section I, that biological targets and processes are claimed for treatments for multiple diseases (laundry list approach). In particular, the diseases covered by this case study have all been claimed in laundry lists (sometimes all at the same time) relating to treatment drugs, delivery (AVV) and e.g. vaccines, despite the wide differences between the diseases. Parkinson's, Alzheimer's and Huntington's are often claimed together, but SMA, CF and Canavan's can also appear in the same laundry list. Finally, as with the single-gene disease case study, the review of heterogeneous diseases is not intended to be a full-scale inquiry. For example, the search based on the primary genes for Alzheimer's disease found 409 applications. A study published in February 2012 identified 48,000 patents for Alzheimer's, grouped into 23 clusters.

Most cases of early-onset *Alzheimer's disease* are caused by mutations in one of three genes: *APP*, *PSEN1*, or *PSEN2*. These three genes all follow an autosomal dominant inheritance pattern. When altered, large amounts of a dominant negative toxic protein fragment called amyloid β-peptide are produced in the brain. The build up of this peptide in the brain results in aggregate formation into clumps called β-amyloid plaques, which are toxic. Inhibiting the formation of β-amyloid plaques by diminishing amyloid β-peptide aggregation is a major area of patent activity. The main focus of the AD case study is the amyloid beta A4 protein and other primary disease genes, while secondary genes, delivery methods and drugs are only discussed briefly.

The last group, headed 'highlights', concerns RNAi and anti-sense oligonucleotides. As shown especially in the sections on Huntington's disease (section 2) and spinal muscular atrophy (section 3), new techniques may put in question the traditional gene patent, for example in terms of why patents for a disease gene can be granted while at the same time siRNA patents are still possible (scope of protection). AD is one of the most preferred targets for the development of RNAi therapeutics, generating siRNAs to develop useful methods for the down-regulation of beta-secretase, APP, Pin-1, presenilin 1

(PS-1) and PS-2 gene expression, as well as other genes involved in these pathways.

There are several mutations in the PD primary genes that cause *Parkinson's disease*. In this empirical case study, mutations in the parkin *PARK2* gene, which cause the disease, appear to be the most researched. The PD template differs from the case study for single-gene disorders and from the AD template. It focuses in particular on stem cell activity and siRNA since it reveals patent activity on induced pluripotent stem cells (iPSCs). Using iPSCs allowed researchers to 'reverse engineer' human neurons from human skin cells taken from four subjects: two with a rare type of Parkinson's disease in which the parkin mutation is the cause of their disease, and two healthy subjects who served as controls. The material available for the study thus circumvented the ban on embryonic stem cell patenting but it did not make the research any easier. A literature search reveals that the techniques have been developed for many diseases (including CF, SMA, HD and PD) and the overall assumption is that human induced pluripotent stem cells have the potential to revolutionise personalised medicine by allowing immunocompatible stem cell therapies to be developed.

RNAi molecules are useful in the treatment of any disease or condition that responds to modulation of gene expression or activity in a cell, tissue or organism. The PD case study included patents granted for the technique as such, and for RNAi directed at secondary genes (contrary to PD primary genes). A step specifically aimed at PD claimed the full scale of siRNA, dsRNA, miRNA and shRNA molecules capable of mediating RNAI against the expression of the SNCA gene.

PART III

# The Animal Kingdom

CHAPTER 8

# Patenting the Animal Kingdom? From Cross-Breeding to Genetic Make-Up and Biomedical Research

Part III of the thesis deals with the animal kingdom. It consists of one chapter, Chapter 8, which reviews the legal position of patents relating to animals and which, like Chapter 7, contains an empirical analysis, dealing with patent activity relating to pigs. The discussion of the parallels of the prohibition of patents for essentially biological processes for the breeding of plants or animals is taken further in Chapter 10. A process is essentially biological when, taken as a whole, it depends on natural phenomena such as crossing and selection. In contrast, natural biological processes in relation to humans are not, in themselves, non-patentable.

There are two main purposes of the present study. The first objective (Part II) is to examine the substantive patent law on animal patenting, and the second objective (Part III), based on empirical data, is to examine patent law in the light of the patents issued by and applications made to both the United States Patent and Trademark Office (USPTO) and the European Patent Office (EPO),[1] in an effort to identify the current challenges to animal patenting. The legal protection of technological developments, from crossbreeding to genetically based breeding, as well as the use of animal models in biomedical research, calls for legal certainty. It is time to turn the spotlight on the issue of animal patents.

* This chapter was first published in 2/2008 *IIC* p. 139-178. It will be elaborated in Chapter 10.

1. The patent search was carried out in collaboration with Patent Information Specialist and Technical Examiner, Bodil B. Hasling, Danish Patent arnd Trademark Office, in December 2005.

# I. Introduction and Case Studies

## 1. Overview

Commercial exploitation of biological material can be divided into three major categories, namely biological material derived from humans, from plants and from animals. From a global perspective it can be argued that, so far, the particular focus on DNA sequence patenting has been the major concern with regard to humans. Thus opposition has mostly been directed towards safeguarding human dignity and integrity.[2] Plant breeders' rights to genetic resources are well known and therefore the step from recognising plant breeders' rights to plant patents is not difficult. The plant variety monopoly was created and accepted long ago.[3] As yet, animal patents as such have remained more or less out of the limelight.[4] Admittedly, the *Harvard Oncomouse* case has been on every patent lawyer's lips since 1988,[5] the *Edinburgh* case since 1999,[6] and in 2005 Greenpeace drew the public's attention to the Monsanto patent applications on pig breeding.

Today's patent literature in the area of biotechnology focuses on humans and plants rather than on animals. To a certain extent farmers buy, sell, eat and own farm animals, and people in general buy pets and sports animals. As long as animals are dealt with without offending *ordre public*, morality and welfare standards, this is generally accepted by society. Legal protection of inventions regarding animal materials is acceptable, as long as any new and

2. See e.g. Case C-377/98 *The Netherlands v European Parliament and Council of the European* Union.
3. Cornish & Llewelyn; '*Intellectual Property*', 5th Ed. 2003, p. 220. In the 1950s and 1960s various countries already introduced a special plant variety right. The UPOV Convention from 1961 established the International Union for the Protection of New Varieties of Plants, and the EU introduced the Community Plant Variety Right in 1994.
4. See also the Opinion of Advocate General Jacobs delivered on 14 June 2001, in Case C-377/98, para. 60: 'It appears for example that some Member States already grant patents for biotechnological inventions involving animals: In France for example, a patent was granted in 1991 for a process for producing a transgenic mouse, and in Italy the first patent concerning a transgenic mammal was granted in 1996'.
5. Cf. patent EP 169672 and case T-315/03. See Andrew Sharples and Duncan Curley, 'Death of the Oncomouse' in *BSLR*, Vol. 8 issue 3 (2005/2006), p. 122: 'on 14 March 2006 the Oncomouse patent was finally revoked, not on any substantive grounds based on morality or prohibitions on the patenting of animals, but on a technicality'.
6. Cf. patent No EP 695351.

exclusive rights do not interfere with the traditional breeding processes and animal ownership.[7]

In the US and Europe, there have been several patent applications with claims for animal substances, as for example, the above mentioned Harvard Onco mouse, Leland Stanford (mouse), Tracy and Dolly (sheep), and the Edinburgh patent (animal stem cells). All of the patents mentioned have been subject to opposition before the European Patent Office.[8] In Europe, it seems that concerns relating to animal patents are developing on a *case-by-case* approach, based on opposition raised by Greenpeace and other special interest groups or organisations.[9] And very often the opposition has succeeded either in obtaining the withdrawal of the patent or the deletion of original claims, thereby limiting the scope of protection.[10] Limitation was often unrelated to the issue of patenting an animal breed, but concerned the inclusion of humans.

The US *Harvard Oncomouse* patent (1988) was the first animal patent (for a laboratory research specimen).[11] The patent was claimed for transgenic, non-human mammals, which had been genetically modified so as to predispose them to cancer. The related and contested European patent application for the Oncomouse was filed in 1985. The patent expired on 24 June 2005, having been subject to opposition and appeals (the claim was subsequently restricted from transgenic animals in general to cover rodents and finally mice only). The proprietor never paid the printing fees required by the EPO for publishing the patent in its amended form, and the patent never succeed-

7. The first proposal for Directive 98/44/EC (OJ 1989 C 10/3) Article 3(1) referred to the possibility of animal variety protection.
8. There is no corresponding opposition procedure in the US. On the other hand, infringement actions in Europe must be brought in each Member State, whereas the US has achieved uniformity by the establishment of the Court of Appeal of the Federal Circuit in 1982.
9. EPC, Article 99. The Article makes provision for opposition proceedings to take place after the publication of the grant to any person – the post-grant opposition which does not (yet) exists in the USA. In *Animal Legal Defence Fund (ALDF) v Quigg*, (932 F.2d 920) decided April 1991, the CAFC affirmed that individual farmers and groups of animal husbandmen, or non-profit organisations whose goal is the protection of animals, lacked standing. The suit was filed against Chakrabarty. The post-grant opposition system is included in the draft of the Substantive Patent Law Treaty, currently being negotiated.
10. Furthermore, EPC 2000 inserted a new system in Art. 105a. At the request of the patentee, a European patent may be revoked or limited by an amendment of the claim.
11. US4736866. Max F. Rothschild and Scott Newman (eds.), *Intellectual Property Rights in Animal Breeding and Genetics*, 2002, p. 9.

ed.[12] In the *Leland Stanford* patent (1988) the issue again was a genetically modified animal, this time involving a test animal for AIDS research. Following the opposition procedure, the patent was revoked on formal grounds in 2002.[13] The patent for a sheep (*Tracy*) relates to protein production by a mammal. The patent claim was for the animal cell and the animal itself whose genetic material includes the genetic construct. The invention may be used to produce, for example, blood coagulation factors (particularly factors VIII and IX) and proteins for foodstuffs, including natural or altered milk proteins of the host mammal, or enzymes. It was granted in 1997 and an appeal was rejected in 2002.[14]

As early as 1997, nuclear transfer technology gave rise to another patent application on a sheep, known as '*Dolly* the sheep'.[15] Both applications originally covering human beings (embraced by the general term 'animal') has been granted. The international publication pamphlet related the invention to the generation of animals including, but not limited to, genetically selected and/or modified animals, and cells useful in their generation. The technique (reproductive cloning) allows the production of genetically identical individuals. It was argued in the application that the invention has clear advantages for both research (e.g. as biological controls) and commercial applications (e.g. the multiplication of genetically valuable livestock, uniformity of meat products, animal management). According to the examination report from July 2006, objections to the application are still outstanding, e.g. the method of preparing a non-human mammal includes an invasive step practised on a non-human animal body (violation of former Article 52(4) EPC, now Article 53(c)), and that when adapting the description to the new set of claims the applicant is requested to make sure that the whole application is restricted to non-human animals. A reply was given by the applicant in October 2006, and the intention to grant communicated in September 2007.

The *Edinburgh* Patent comprised a method of using genetic engineering to isolate, select and propagate stem cells (also including embryonic stem cells) from more differentiated cells in a cell culture, in order to obtain pure stem

12. See Sharples and Curley, 'Death of the Oncomouse' in *BSLR*, 2005/2006, p. 122, 'The invention was never a commercial success. It was more a fight about the relationship between biotech inventions and the moral dimension of patent law'.
13. EP0322240, The Board of Trustees of the Leland Stanford Junior University.
14. EP0274489, Clark Anthony John, see Sigrid Sterckx (ed.), *Biotechnology, Patents and Morality*, 2000, p. 210.
15. See Patent EP0847237 (not yet granted) and EP0849990 (granted in 2001, producing live offspring by nuclear transfer).

cell cultures as well as a transgenic animals generated from them. The patent claim's use of the term 'animals', without being limited to non-human animals, triggered public interest. The patent was limited after an opposition hearing and was thereafter effective (2002).[16] The *Seabright Patent,* which was granted in 2001, consisted of claims based on the creation of a transgenic fish characterised by the incorporation of a chimeric gene of non-human origin. This patent also covers the tests for determining transgenic fish.[17]

The rapid development of technology and the increase in genetic knowledge in society has continued to challenge human creativity to enhance the genetic make-up of farm animals, pets, sporting animals, transgenic animals, safeguarding endangered species and so forth. Today's technologies can intervene at germ cell level in order to change heredity and thereby secure the desired traits in the next generation. Cloning technologies are being used both in relation to farm animals and biomedical research (Polly, Dolly, Tracy) to satisfy human needs, and further developments are being made.[18] In 2003 the first US *Copy Cat* was cloned for AIDS medical purposes, and it could interest people wanting to resurrect a lost pet by cloning.[19] Since animals often eaten as food as well as being used for human amusement, e.g. betting on horses and watching dog races, it is obvious that animals often serve human purposes, and humans and animals are far from being treated equally within or outside the patent system.

16. EP0695351, Edinburgh University.
17. Patent EP0578653 B. See the Report from the Commission to the European Parliament and the Council, Development and implications of patent law in the field of biotechnology and genetic engineering, COM (2002) 545, point 3.2.2. See also the Norwegian patent on transgenic salmon, NO321650.
18. J. Suk et al, 'Dolly for dinner? Assessing commercial and regulatory trends in cloned livestock' in *Nature Biotechnology*, Vol. 25, Number 1, January 2007. Polly was the first transgenic clone containing a randomly integrated transgene for human factor IX. The article contains a timeline of key events in mammalian cloning comprising sheep, cattle, mice, goats, pigs (first gene replacement cloned animal), rats, rabbits, mules, horses, cats, dogs and ferrets. It is expected that products derived from the offspring of cloned cattle and pigs are likely to enter the food chain before 2010, and furthermore that milk and meat from the progeny of cloned animals are the products most likely to enter human food chain in the near future.
19. See Charbonneau, 'A cat fight with a twist', http://news.bbc.co.uk/1/hi/sci/tech/1820749.stm. 'What is Warm and Fuzzy Forever? With Cloning, Kitty', *New York Times*, February 15, 2002.

## 2. Historical context: From Red Dove to Chakrabarty

### a) Patentability of traditional breeding

A very early attempt to patent a traditional method of breeding came before the German Supreme Court in 1969. The patent in suit claimed a method for breeding a dove with red plumage, which is considerably larger with respect to other doves of the same colour, has a considerably larger wing spread, the colour of the plumage of the wings being considerably more beautiful and more intense, and having a craw which is extremely large in relation to the size of the body. The Supreme Court held that '... in principle no obvious reason exist for excluding methodical exploitation of natural biological forces and phenomena from patent protection. Thus, a method for breeding animals can represent a teaching to utilize natural biological forces and phenomena and therefore be patentable.' The process was ultimately held unpatentable as it could not be repeated:

> '... the position of the Patent Office is supported in the present case by the fact that the disclosure of the breeding method in the patent specification and its characterisation in the patent claim does not ensure a genetically identical repetition of such breeding method, and under no circumstances can it be assumed that the same genetic results would be obtained with a high degree of certainty. The initial animal species are characterised in a general manner, according to their kind and without indicating individual hereditary characteristics. Therefore, in attempting to repeat the breeding method, a person skilled in the art would be required to use such animals as might correspond to the kind described with respect to their looks. The 2-step selection is directed to two characteristics which, however, are defined only in general terms (size and color). Carrying out the steps of selection thus still leaves room for many variations because the phenotypical approach does not promise a sufficiently predetermined result.
>
> The above reasons do not permit the conclusion with certainty that the breeding method can be repeated, particularly since we are here concerned with breeding an animal in the upper range of the evolution scale, and having complex hereditary characteristics.'[20]

Because of the requirement for repeatability, patent protection is not suitable for the traditional breeding of plants and animals and the results thereof.[21] Yet *Peace and Christie* argue that animal breeders do not usually attempt to develop new breeds, and as animals can only be reproduced sexually, it is not possible to keep the traits of a breed perfectly stable. On the contrary, a herd

20. 1 IIC 142 (1970).
21. See *Rote Taube* (Red Dove), 1 IIC 141-142 (1970) and Singer & Stauder, *The European Patent Convention, A Commentary*, 3rd ed. Thomson, Sweet & Maxwell, Vol. 1, Article 53, note 46 ff.

will evolve from generation to generation.[22] The authors also emphasise that sufficient genetic variation must be kept in an animal population in order to avoid inbreeding. Prior to the development of genetic engineering, farmers were only able to cross-breed and select desirable traits as a means for enhancing livestock. As such, the implication of essentially biological breeding processes relates to the reproduction. As long as farmers breed animals by essentially biological breeding processes, identical offspring are more or less out of reach.

**b) Patentability of breeding via inserting exogenous genes (DNA breeding)**

The first attempt to patent an organism that was genetically modified by the insertion of an exogenous gene came before the US Supreme Court in 1990, *Diamond v Chakrabarty.*[23] Unlike the Red Dove decision, the issue in the *Chakrabarty* case concerned genetic breeding.

In this case it was decided that whether the invention was alive or dead was irrelevant to its patentability.[24] The Supreme Court held that the statutory provisions in 35 USC § 101 were intended by Congress to be broad enough to '*include anything under the sun that is made by man*'.[25] The essential *product-of-nature doctrine* (1889), as formulated in *Ex parte Latimer*, survived the new development as a limitation on patentable subject matter.[26] Conse-

22. See Peace and Christie, 'Intellectual Property Protection for the Products of Animal Breeding' in (1996) *EIPR*, p. 215.
23. 447 U.S. 303 (1980). The inventor Ananda Chakrabarty, a microbiologist, filed a patent application for a strain of genetically altered Pseudonomas bacterium capable of digesting multiple components of crude oil, a characteristic which makes them useful in controlling oil spills. The patent claimed a bioengineered bacterium to consume oil slicks – a patent on the living altered bacterium. The landmark decision has been commented on ever since see e.g. Elisabeth Joy Hecht, 41 *AM. U.L. Rev*. 1023 (1992).
24. The bacterium was not a product of nature; it was a product of the inventor. See Kevles, 'The Advent of Animal Patents: Innovation and Controversy in Engineering and Ownership of Life' in Max F. Rothschild and Scott Newman (eds.) *Intellectual Property Rights in Animal Breeding and Genetics*, 2002, p. 19.
25. See Burchfield, *Biotechnology and the Federal Circuit*, (1995), p. 38. Before this decision, the CCPA (formerly the United States Patent and Trademark Office, Board of Patent Appeals and Interferences) in *re Bergy* established that living micro-organisms can be 'manufactures' or 'compositions of matter' and thus are statutory subject matter.
26. *Ex parte Latimer*, 1889 C.D. 123, 46 O.G 1638, disallowed a claim to cellular tissue that exists in the leaves and needles of pine tree. According to Oliver Mills the basis

quently, the relevant distinction is that between *products of nature* (nature's handiwork) and man-made inventions that are the result of human activity.[27] Subsequent to *Diamond v Chakrabarty*, in *Ex parte Allen* (1989), concerning a triploid oyster, the USPTO declared that patents could in principle be granted for non-human animals. The issue was not whether the subject matter was a product of nature, but whether it was made by man.[28]

Following the development in *Ex parte Allen*, a general resistance emerged to the patenting of living matter. In mid-May 1987, the US Commissioner of Patents and Trademarks was urged to impose an eight-month moratorium on the processing of animal patent claims until Congress could develop a policy on the issue.[29] According to *Elisabeth Joy Hecht*, the Harvard Onco mouse patent was issued in the US on the same day that the moratorium expired.[30] By that time, legislative proposals had been put forward to remove higher animals from the ambit of 35 USC § 101.[31] As a consequence, the US patent authority soon after announced that 35 US § 101 covered non-naturally occurring non-human multi-cellular living organisms, including animals. But to be patentable, an animal must be given a new form, quality,

of the decision was that fibre that nature produces is intended for the use of all men, and consequently does not exclusively belong to any one man, see Mills, *Biotechnological Inventions, Moral restraints and Patent law*, 2005, p. 108.

27. See Burchfield, 1995, op. cit. footnote 25, pp. 40-41: a claim to a plant product, the fibre of the needle of the *Pinus australis* tree, was rejected as directed to a product of nature. For Europe see Directive 98/44/EC Article 3(2) and the European Patent Convention, Implementing Regulation 23 c (EPC 2000 Rule 27(a)). See Singer & Stauder, *The European Patent Convention, A Commentary*, 3rd Ed., Vol. 1, Article 53, note 28.2.
28. See Rothschild and Newman (eds.), op. cit. footnote 11, pp. 22 and 68.
29. See Rothschild and Newman (eds.), op. cit. footnote 11, p. 23. See also Hecht, *41 AM. U.L. Rev. 1071 (1992). recommending Congress to amend Section 101 to include transgenic animals;* and McGovern, 'Biotechnology and the Patenting of Living Organisms', 3 *Animal Law 221*, 1997, concerning Congressional Inaction, Section C.
30. See Hecht, 41 *AM. U.L. Rev.* 1023 (1992).
31. See Burchfield, Biotechnology and the Federal Circuit, 1995, op. cit. footnote 26, p. 42. on proposals for a moratorium on animal patents and an outright ban on patents for vertebrate or invertebrate animals, modified, altered, or in any way changed through genetic engineering (footnote 41); and Sease, 'From Microbes, To Corn Seeds, To Oysters, To Mice: Patentability of New Life Forms' in 38 *Drake Law Review* 1988/89, p. 565 in regard to a Transgenic Animal Patent Reform Act of 1988 which was never adopted.

property or combination not present in the original article existing in nature.[32] The Onco Mouse was considered such an article not present in nature.

## II. The Legal Framework and Analysis of the Legal Concepts

### 1. Relevant international and european statutes[33]

#### a) The Strasbourg Convention and the European Patent Convention

Article 2(b) of the 1963 Strasbourg Convention already excluded plant or animal varieties or essentially biological processes for the production of plants or animals from patent protection per se. The 1973 version of the European Patent Convention (EPC) (and EPC 2000)[34] addresses the question of patentable subject matter at different levels. Part II, concerning patentability, defines a patentable invention in Article 52(1) (EPC 2000) as 'any inventions, in all fields of technology, provided that they are new, involve an inventive step and are susceptible of industrial application'. Article 52(4) (EPC 1973) included a provision whereby methods for the treatment of the human or animal body by surgery or therapy and diagnostic methods practised on the human or animal body were outside the scope of an invention.[35] Today, EPC 2000 expresses part of this provision in Article 53(c) as an exemption to patentability,[36] and under Article 53(b) plant or animal varieties or essentially

32. See Blattman et al. in Rothschild and Newman (eds.), op. cit. footnote 11, p. 69.
33. As the relevant legal concepts are present in diverse European and international regulations (EPO, EC, TRIPS), the analysis of the concepts will be dealt with together below. The word *diverse* refers to the fact that the TRIPS Agreement is a global agreement, and the EPC is a European Convention covering more countries than the EU Member States. The legal power derived from the different systems is far from equal.
34. EPC 2000 enters into force on 13 December 2007 at the latest. As a consequence, Implementing Regulations 23b-e becomes Regulations 26-29. The Implementing Regulations for EPC 2000 were adopted by the Administrative Council on 7 December 2006.
35. See Directive 98/44/EC, Recital 35. 'Whereas this Directive shall be without prejudice to the provisions of national patent law whereby processes for treatment of the human or animal body by surgery or therapy and diagnostic methods practiced on the human or animal body are excluded from patentability.'
36. See the Explanatory Remarks. While these surgical or therapeutic methods constitute inventions, they have been excluded from patentability by the fiction of their lack of industrial applicability. It is undesirable to uphold this fiction since methods of treatment and diagnostic methods are excluded from patentability in the interests of public

biological processes for the production of plants and animals are also excluded. The latter provision was already included in the 1973 version. Article 53 (EPC 2000) reads as follows:

> 'European patents shall not be granted in respect of:
> (a) inventions the commercial exploitation of which would be contrary to 'ordre public' or morality; such exploitation shall not be deemed to be so contrary merely because it is prohibited by law or regulation in some or all of the Contracting States;
> (b) plant or animal varieties or essentially biological processes for the production of plants or animals; this provision shall not apply to microbiological processes or the products thereof.
> (c) methods for treatment of the human or animal body by surgery or therapy and diagnostic methods practised on the human or animal body; this provision shall not apply to products, in particular substances or compositions, for use in any of these methods.'

Article 53 EPC is complemented by the new Implementing Regulations 23b-23e (Rule 26-29, EPC 2000) that are an implementation of European Directive 98/44/EC on Biotechnological Inventions (see below c) into the framework of the EPC.

**b) The TRIPS agreement**

In 1994, before the adoption of Directive 98/44/EC, the WTO Agreement on Trade-Related Aspects of Intellectual Property Rights (TRIPS) entered into force. The Agreement frames minimum standards, and its Part II, Section 5 contains provisions concerning patents. TRIPS Article 27(1) reflects the traditional patentability requirements, and apart from the ban on distinctions relating to the field of technology, its wording causes no surprise.[37] However, TRIPS Article 27(2) take into account the EPO case law, thus confirming the concepts of *ordre public* and morality in the legal text so as *to protect human, animal or plant life or health or to avoid serious prejudice to the environment.*[38] The exemption could be elaborated in conjunction with the concerns outlined, such as the protection of human life or the avoidance of serious prejudice to the environment, since damage to animals (and plants) could

health. It is therefore preferable to include these inventions in the exceptions to patentability in order to group the categories of exceptions together in Article 53 (which has, in general, been brought into line with the wording of TRIPS).

37. See Pires de Carvalho, *The TRIPS Regime of Patent Rights*, 2nd Ed., 2005, p. 216.
38. Case law from the EPO e.g. case T 356/93 reasons, paragraphs 5 and 6 interpreting Article 53(a) EPC as operate with the same 'modern' interpretation.

have consequences for the health of humans as well as the environment in which animals are considered an integral part.[39]

Under TRIPS Article 27(3)(a), Members may also exclude diagnostic, therapeutic and surgical methods for the treatment of humans or animals from patentability. The provision replicate the 1973 EPC as mentioned above.[40] TRIPS Article 27(3)(b) allows Members to exclude plants and animals other than micro-organisms, and essentially biological processes for the production of plants or animals other than non-biological and microbiological processes from patentability.[41]

### c) Directive 98/44/EC

The first draft of the European biotechnology directive, from 1988, was rejected by the European Parliament due to its lack of an ethical dimension.[42] Ethical considerations were later added, as well as the substantial expansion of the scope and aim of the Directive.[43] However animals were not the trust of this controversy.[44] Thus the patenting of animal substances is subject to the

39. See UNCTAD-ICTSD Project on IPRs (intellectual property rights) and Sustainable Development, *Resource Book on TRIPS and Development*, 2005, p. 376. The concept of health may be deemed to cover the satisfaction of basic needs such as food.
40. EPC 2000 contains a minor change, as the wording of Article 53(c) has been brought into line with TRIPS.
41. See UNCTAD-ICTSD Project on IPRs and Sustainable Development, *Resource Book on TRIPS and Development*, 2005, pp. 388-389 and 396. The clause was subject to review already in 1999; the review has not yet been completed.
42. See COM(92)589 final of December 16 1992. Parliament requested the traditional framework to be supplemented by more precise guidelines. Patent law should contain certain impenetrable barriers so as to provide guidance for those interpreting the concepts of public policy and morality.
43. The drafts are the following: COM(88)496; COM(92)589; COM(93) 570, COM(94) 245, COM(95) 661 and COM(97) 446. See Kamstra et al, Special report, *Patents on Biotechnological Inventions*: The E.C. Directive 2002, p. 3 and 7: 'In December 1995 the Commission published a new draft of the Directive which took into account the concerns of the European Parliament. In all, over 450 amendments were considered by various committees and in the end the Committee on Legal Affairs and Citizens Rights approved 65 amendments. In the run up to the First Reading in July 1997, an additional 47 amendments were tabled by various groups. Given that the Commission's text contained only 33 Recitals and 20 Articles, this is an incredible figure.' Transposition of the Directive had to take place before July 2000. See further Tine Sommer, 38 *IIC* 30 (2007).
44. See Kamstra et al, op cit. footnote 43, p. 4: '... On top of the questions of farmers' privilege, the European Parliament showed strong concerns over the patentability of nucleotide sequences derived from human genetic research, the role of ethical and

traditional patent law requirements as stipulated in Art. 3(1). The attempt to reshape these requirements and make them stricter in relation to human genes in Art. 5(3) is evidently not addressed to animal genes.[45] In contrast to both the EPC and TRIPS, there is no provision in the Directive relating to methods for treatment by surgery or therapy or diagnostic methods.[46]

The Directive introduced statutory definitions for biological material, microbiological processes and essentially biological processes (cf. Arts. 2(2), 4(2) and (3)). These concepts will be addressed below, taking into account the diverse European regulation.

## 2. Legal foundations for limiting animal patents in Europe

### a) Introduction

The concepts of both 'essentially biological processes' (see below b) and 'animal varieties' (see below c) have been used to limit patentable subject matter at least under European law. However, unlike the protection of plant varieties, there is no corresponding sui generis protection of animal varieties.[47]

moral criteria in the patenting process, and a desire to extend the farmers' privilege to transgenic animals. In particular, three amendments were approved by the European Parliament at its second reading in May 1994'. One of the three amendments (c) concerned 'The exclusion from patentability of transgenic animals where certain conditions are not met, and the incorporation of the criterion of proportionality in order to assess the acceptability of 'the suffering or physical handicaps inflicted on the animals'.'

45. On the other hand, in Recital 27 the guidance on prior informed consent is directed to both plant and animal material: the patent application should, where appropriate, include information on the geographical origin of such material, if known; whereas this is without prejudice to the processing of patent applications or the validity of rights arising from granted patents.
46. Besides the general reference in Recital 35, in contrast to the final Directive the 1988 draft Directive Article 18 followed the wording of EPC Article 53(4), the 1992 draft listed the exclusion under *ordre public* and morality in Article 2(3)(b) in regard to human processes for modifying the genetic identity for a non-therapeutic purpose contrary to the dignity of man, as well as Article 8 with a wording similar to EPC Article 52(4). See COM(92) 589, p. 9 for further comments on therapeutic purpose.
47. See Cornish & Llewelyn, *Intellectual Property*, 2003, p. 220: '... the exclusion of 'animal varieties' from patenting seems merely to have reflected traditional expectations. The expression 'animal variety' is an artificial one in any EPC language; and indeed the term in German may well refer to an entire animal species, but in French and English only to sub-species. The exclusion was adopted without animal breeders having

**b) The definition of 'essentially biological processes'**

Essentially biological processes for the production of plants or animals are not patentable. The patent-law concept of *essentially biological processes* for the production of animals arose before the possibilities of biotechnological animal breeding. It was probably intended to separate natural and artificial breeding processes, and to exclude from patentability the ordinary workings of the animal's biological system.[48] Conversely, this means that essentially *non*-biological processes are patentable. If an animal can be obtained through genetic engineering without any natural breeding, the invention relating to such an animal may be patentable.[49] This concept is enshrined in the European patent system (EPC Article 53(b)), Implementing Regulation 23b(3) (EPO 2000, Regulation 26(5)). Directive 98/44/EC, Art. 2(2) and Recital 33 define this as follows: 'A process for the production of plants or animals is essentially biological if it consists entirely of natural phenomena such as crossing or selection'.

The previous drafts of the Directive used the following definitions and wording:

- 1988, Article 7: A process in which human intervention consists in more than selecting an available biological material and letting it perform an inherent biological function under natural conditions shall be considered patentable subject-matter.
- 1992, Article 6: Essential biological processes shall not be patentable. In determining whether this exclusion applies, human intervention and its impact on the result achieved shall be taken into account. A process which, taken as a whole, does not exist in nature and is more than a mere production process shall be patentable.
- 1995, Article 6: Essentially biological processes for the production of plants or animals shall not be patentable.
- Article 4(3): Essentially biological process for the production of plants or animals means any process which, taken as a whole, exists in nature or is not more than a natural plant-breeding or animal-breeding process.

The concept of essential biological processes has been defined in EPO case law e.g. case T-356/93 and case T-320/87. Case T-356/93 concerning plants reads as follows:

experience of exclusive rights of any kind. This may help to explain the curious drafting.' The EPO's official linguistic usage refers to *animal variety*, *Tierrasse*, and *race animale* – similar to the usage in Directive 98/44/EC of *animal varieties*, *Tierrassen* and *les races animales* (in English, German and French along with the 19 other official language versions).

48. UNCTAD-ICTSD, op. cit. footnote 39, p. 223.
49. COM(2002) 545, p. 14.

'40.1. Claim 7 does not relate to a process which is 'essentially biological' within the meaning of this provision [Article 53(b), EPC] because the step of transforming the plant cells or tissue with a recombinant DNA, regardless of whether its performance is dependent on change or not, is an essential technical step which *has a decisive impact on the desired final result*. If it is not successfully performed, plants or seeds could most probably still be regenerated from the plant cells or tissue and they could replicate, but they would not display the desired distinctive characteristic of having the heterologous DNA integrated in their *genome in a stable manner. Therefore, although the subsequent steps of regenerating and replicating the plants or seeds make use of the 'natural' machinery, the decisive step, namely the insertion of the relevant DNA sequence into the genome of the plant, could not occur without human intervention*. In this respect, it is also noted that the regeneration step is not entirely biological, but rather agrotechnical, since some degree of technical intervention is required in the selection of the proper working conditions. Therefore, the process of claim 7 as a whole is not 'essential biological' and, thus, not excluded from patentability under Article 53 (b) EPC, first half-sentence *[emphasis added]*.'

In case T-320/87 point 6, it was stated that:

Like any exception to a general rule of this kind the exclusion of 'essentially biological' processes for the production of plants (or animals) has to be narrowly construed. This is underscored by the fact that this exclusion does not apply to microbiological processes or the products thereof, as also stated in Article 53(b) EPC. The Board takes the view that whether or not a (non-microbiological) process is to be considered as 'essentially biological' within the meaning of Article 53(b) EPC has to be judged on the basis of the essence of the invention taking into account the totality of human intervention and its impact on the result achieved. *It is the opinion of the Board that the necessity for human intervention alone is not yet a sufficient criterion for its not being 'essentially biological'. Human interference may only mean that the process is not a 'purely biological' process, without contributing anything beyond a trivial level*. It is further not a matter simply of whether such intervention is of a quantitative or qualitative character *[emphasis added]*.'[50]

Due to the exclusion of essentially biological processes, conventional breeding methods are generally not patentable. In contrast thereto, methods based on modern biotechnology, where the technical intervention is significant, would be patentable.[51] Thus in the OncoMouse case T-315/03 point 13.3.5 it was expressly pointed out as self-evident that a process that includes genetic manipulation does not consist entirely of natural phenomena.

50. Also Case T-356/93: (in regard to plants) a patent can be granted if the production of a genetically modified plant comprises at least one essentially technical stage that cannot be carried out without human (technical) involvement and has a decisive influence on the final result.
51. See UNCTAD-ICTSD, op. cit. footnote 39, p. 393.

The legislative history of Article 53(b) EPC has been outlined in the pending decision case T 83/05 concerning plants (broccoli), where two questions have been referred to the Enlarged Board of Appeal (G 2/07). In paragraph 38 and further, the Board of Appeal examines the preliminary Draft Strasbourg Convention, and points 40 and 42 read as follows:

40. Article 2 of the Preliminary Draft Convention of the Council of Europe (cf. Doc EXP/Brev (61) 2 rev., p. 26) read:
'The words 'susceptible of industrial application' shall be understood in the widest sense. Nevertheless, the Contracting States shall not be bound to provide for the grant of patents, in respect of new plant or animal species or of purely biological, horticultural or agricultural (agronomic) processes.'

After thorough discussions in a committee meeting of 7 to 10 November 1961, this provision was substantially amended in that the words 'horticultural or agricultural (agronomic)' were deleted and the remaining phase ('purely biological processes') was replaced by the current wording 'essentially biological processes for the production of plants and animals'. A memorandum of the secretariat of the committee (cf. Doc.EXP/Brev 61(8), p. 4-5) contains the following explanations:

'The processes for the 'production of plants or animals' referred to in the new text include those which may produce *known* varieties as well as those which may produce *new* ones, it being understood that only new varieties can eventually qualify for protection *in themselves*. Selection or hybridisation of existing varieties may be mentioned as examples of such processes (in the vegetable kingdom). The new text specifies that the processes which may be ineligible for patents are *essentially* (and no longer *purely*) biological. It was evident that the exclusion should be extended to cover processes which were fundamentally of this type if, as a *secondary* feature, 'technical' devices were involved (use of a particular type of instrument in a grafting process, or of a special greenhouse in growing a plant), it being understood that such technical devices may perfectly well be patented themselves, but not the biological process in which they are used...

42. The above review of the Travaux Préparatoires shows that the drafters of the provision regarded 'biological' as being in opposition to 'technical', that they deliberately chose the adverb 'essentially' to replace the narrower term 'purely' and that they considered plant breeding processes based on selection and hybridisation to fall under the exclusionary provision even if secondary features of the processes were characterised by the use of technical devices'.

Furthermore in T-83/05, points 51 and 52, the Board of Appeal set out the legislative history of Rule 23b(5) EPC and Article 2(2) Biotech Directive. It is the viewpoint of the board that Rule 23b(5) EPC is somewhat difficult to understand and that the Rule appears to be self-contradictory to some extent (processes which consist entirely of natural phenomena are considered to be essentially biological processes for the production of plants), since the systematic crossing and selection as carried out in traditional plant breeding would not occur in nature without the intervention of man (point 53). In points 54 and 55, it is held that there are two approaches (the EPO case law and the new Rule 23b(5)):

'... particular when taking into account the adverb 'entirely', the wording of Rule 23b(5) EPC aims at a very narrow construction of the process exclusion contained in Article 53(b) EPC. The board interprets Rule 23b(5) EPC as meaning that a process which, apart from 'natural phenomena' (which appears to cover crossing and selection by way of a legal fiction), contains an additional feature of a technical nature would be outside the ambit of the process exclusion. As already pointed out in decision T 1054/96 (above point 46), this has not been the approach adopted by the boards of appeal before the introduction of Rule 23b(5) EPC. The board thus finds it difficult to concur with the statement contained in the EPO's Notice cited above (point 50), according to which the interpretation developed by the boards falls within the framework of the definition given in the new rule.'

Finally, in point 55 the board emphasises that Rule 23b(5) EPC was meant to be a full definition. Having said that, the board then examines any doubts as to the applicability of Rule 23b(5). It is the view of the board that the introduction of Rule 23b(5) EPC does not finally settle the question as to what the correct approach is when interpreting the process exclusion of Article 53(b) EPC. Did Rule 23b(5) EPC change the law by narrowing the scope of the process exclusion contained in Article 53(b) EPC and thus expanding the area of patentable subject matter?[52] In points 57-60 three lines of arguments are elaborated, e.g. that the provisions of the Convention shall prevail in the case of conflicts with provisions of the Implementing Regulations. EPO case law reflecting the true meaning of Article 53(b) EPC cannot be overturned by an amendment to the Implementing Regulations (point 57). Thus it is stated that the correct approach for interpretation of Article 53(b) EPC is still to be determined (points 60-61).

The examination in T 83/05 is very interesting since the patent in suit claims a method for the production of the plant (Brassica oleracea) with elevated levels of certain glucosinolates. The method contains several steps of crossing and selection, and of utmost importance are especially the steps that specifies that molecular markers are used to select hybrids with the desired genetic combination encoding expression of elevated levels of the glucosinolates. Thus points 65-66 read:

65. The approach adopted by Rule 23b(5) EPC (above, point 54) would lead to the conclusion that at least the first feature relied upon by the respondent would be sufficient to bring the claimed process outside Article 53(b) EPC. While there has been some dispute between

52. Point 59, decision T-83/05.

the parties how broadly the term 'molecular markers' should be understood, in the board's view the use of such markers involves subjecting plant material to an analytical laboratory process. This means that a technical step requiring human invention is performed before or during selection.

66. If, however, the approach adopted by the present board in its previous decisions T 320/87 and T 356/93 were still the correct one, none of the features relied upon by the respondent would make the claimed method escape the process exclusion of Article 53(b) EPC.

- The use of molecular markers such as DNA markers is, on a general level, a well-known step in the selection of plants with desired characteristics. *Methods to discover and produce molecular markers that segregate with a desired trait were commonly known from the prior art and had already been used in the context of Brassica species (above, point 6). This has been acknowledged by the respondent (above, section VI). The board does therefore not consider that this feature is able to contribute anything beyond a trivial level to the claimed invention [emphasis added].*[53]

From the EPO Examination Guidelines it follows:

'... To take some examples, a method of crossing, inter-breeding, or selectively breeding, say, horses involving merely selecting for breeding and bringing together those animals having certain characteristics would be essentially biological and therefore unpatentable. On the other hand, a process of treating a plant or animal to improve its properties or yield or to promote or suppress its growth e.g. a method for pruning a tree, would not be essentially biological since although a biological process is involved *the essence of the invention is technical [emphasis added].*'[54]

The guidance from the examination guidelines so far provided is directed more to the examination of a plant patent application than to an application relating to animals.[55] As *Singer & Stauder* point out in their Commentary on the concept, this was concluded by the EPO Board of Appeal case law long before the adoption of Directive 98/44/EC.[56]

53. Two questions are thus referred to the Enlarged Board of Appeal for decision. The questions relates to the interpretation of the concept of (non-micro) biological processes for the production of plants.
54. Guidelines for substantive examination, draft, status March 2007, Part C, Chapter IV-8 on Patentability, point 4.6.2. Processes for the production of plants or animal.
55. The illustration given of soil (as yet another biological substance) cannot be applied to animal foodstuff for promoting growth.
56. See Singer & Stauder, op. cit. footnote 21, Article 53, note 46 ff. The German wording: *Züchtung* (breeding) and the French *Obtention* (points to breeding methods) and Kamstra *et al*, op. cit. footnote 43, p. 30.

The concept of *essentially biological processes* can be supplemented by the concept of *microbiological processes.* A microbiological process means any process involving or performed upon or resulting in microbiological material. Whereas essentially biological processes are excluded from patentability *per se*, this is not the case when an invention concerns a microbiological or other technical process or a product obtained by means of such a process, see EPC Art. 53(b), Implementing Regulation 23c (EPO 2000, Regulation 27(c)) and Directive 98/44/EC Article 2(1)(b) and Article 4(3), stating that biotechnological inventions are also patentable if they concern: 'a microbiological or other technical process or a product obtained by means of such a process other than a plant or animal variety'.

It follows from case T-356/93 that 'according to the current practice of the EPO, the term 'microorgamism' includes not only bacteria and yeasts, but also fungi, algae, protozoa and human, animal and plant cells, i.e. all generally unicellular organisms with dimensions beneath the limits of vision which can be propagated and manipulated in a laboratory'.[57] Furthermore, the Examination Guidelines define microbiological processes as covering not only processes performed upon microbiological material or resulting in such, but also processes which, as claimed, include both microbiological and non-microbiological steps.[58] The products of a microbiological process may also be patentable *per se*.[59] In G 1/98 (point 5.2) the Enlarged Board of Appeal clarified that processes of genetic engineering are not identical with microbiological processes.

### c) The definition of 'animal varieties'

The concept of 'animal variety' from the Strasbourg Convention survived the technological developments which were incorporated into the European Patent Convention (EPC) in 1973, into the TRIPS Agreement in 1994, as well

57. Point 34.
58. Draft March 2007, Part C, Chapter IV point 4.7.1.
59. The interpretation of the concept can be illustrated by reference to Case T-356/93 para. 40.9 and 40.10, concerning a *multi-step process for the generation of a transgenic plant:* '... regardless of the decisive impact that the microbiological process step has on the final result, the multi-step process whereby the plant according to Claim 21 is produced is not a microbiological process within the meaning of Article 53 (b), EPC, second half-sentence (cf. points 38 and 39 supra). Accordingly, such a plant may not be considered to be 'the product of a microbiological process',' Decision of 21 February 1995.

as into Directive 98/44/EC in 1998, without defining 'animal variety'.[60] From a historical point of view, the somewhat empty standard of 'variety' was introduced in order to distinguish breeders' rights from patentable rights and in order to avoid double protection.[61]

Thus it follows from the European patent system (EPC Art. 53(b), Implementing Regulation 23c (b) (EPO 2000, Regulation 27(1)(b))) as well as Directive 98/44/EC, Art. 4(2) and Recital 29), that a patentable biotechnological invention may concern animals if the technical feasibility of the invention is not confined to a particular animal variety.[62] The EPO Examination Guidelines provide no further information. However, it follows from case G 1/98 and case T 315/03 that the principles concerning plants and 'plant varieties' should be applied to animals. Thus, a patent should not be granted for a single animal variety, but can be granted if varieties may fall within the scope of its claim. Further in the OncoMouse case, T 315/03, point 11.8, it was emphasised that:

> 'The definition of animal variety (or species or race) by reference to taxonomical rank would be consistent with the position in relation to plant varieties and in the interest of legal certainty, allowing assessment under Article 53 (b) EPC as interpreted by Rule 23c (b) EPC to be made by considering whether, in a case concerning animals, the technical feasibility of the invention is not confined to a particular animal variety (or species or race).'[63]

60. The definition of a plant variety was introduced in UPOV 1961 Article 2. There is no legal definition of an animal variety. This can be defined as a taxonomical grouping ranking next below a sub-species (where present) or species, whose members differ from others of the same species or sub-species in minor but permanent or heritable characters (Definition taken from the Shorter Oxford English Dictionary, cf. the conclusions of Advocate General Jacobs, in Case C-377/98 para. 131 and COM(2002) 545, p. 13.
61. See case G 1/98 concerning the historical reason for the exclusion of plant varieties.
62. In relation to plant varieties, see G 1/98 NOVARTIS II/Transgenic Plant, where the Enlarged Board decided that claims to transgenic plants were not to be disallowed merely because they embraced plant varieties. See Recital 32 of Directive 98/44/EC.
63. However, The Technical Board of Appeal emphasised that, contrary to a plant variety there is no single term relating to animals in the EPC. As pointed out in case T 315/03 (paragraphs 11.5-11.8) the three different terms used in each of the three official languages would lead to the absurd result that the outcome of an Article 53(b) EPC objection would depend on the language of a case (due to Article 177(1) EPC) – with German having the highest taxonomic order 'species' ('Tierarten') and thereby offering the widest objection).

The Enlarged Board of Appeal in decision G 1/98 stated that it does not make any difference for the requirements under the UPOV Convention or under the Regulation on Plant Variety Rights, how a variety was obtained, point 5.3:

'Whether a plant variety is the result of traditional breeding techniques, or whether genetic engineering was used to obtain a distinct plant grouping, does not matter for the criteria of distinctness, homogeneity and stability and the examination thereof. This means that the term 'plant variety' is appropriate for defining the borderline between patent protection and plant breeders' rights protection irrespective of the origin of the variety. The argument that the legislator of the EPC did not envisage the possibility of genetically-modified plant varieties and for this reason could not have had the intention of excluding them from patentability cannot be accepted. Laws are not restricted in their application to situations known to the legislator. Since plant varieties are excluded, the only question is the conditions under which they are excluded. *The Enlarged Board of Appeal supports the view of the referring Board (Reasons, point 92) that the mere fact of being obtained by means of genetic engineering does not give the producers of such plant varieties a privileged position relative to breeders of plant varieties resulting from traditional breeding only* [emphasis added]. Given the purpose of Article 53(b) EPC, question 4 [Plant varieties as products of processes using recombinant gene technology] has to be answered in the negative. Article 4(1)b and (3) of the Biotechnology Directive, using language corresponding to Article 53(b) EPC, is intended to be interpreted in the sense outlined above, since Recital 32 of the Directive postulates that a new plant variety bred as a result of genetically modifying a particular plant variety is still excluded from patent protection, even if the genetic modification is the result of a biotechnological process.'

In case T-315/03 concerning the Oncomouse, the claimed method[64] was considered an invention for the following reason:

'The claimed method pertains to the field of genetic engineering and the transgenic animals produced by that method may be used as appropriate models for studying different aspects of cancer, such as the development of tissue-specific tumours and the effect of suspected carcinogens. [...]. They represent technical tools just as other means – such as bacterial tests and culture cell lines – are also technical tools. Thus, both the claimed method and the transgenic animal directly derived there from have a technical character. Similarly, the technical feature characterising these transgenic animals – the presence of an activated oncogene which confers an increased probability of developing neoplasm – is also found in their offspring. Therefore, this progeny would also be used as an appropriate technical

64. 'Claim 1. A method for producing a transgenic rodent having an increased probability of developing neoplasms, said method comprising chromosomally incorporating an activated oncogene sequence into the genome of a rodent. Claim 19. A transgenic rodent whose germ cells and somatic cells contain an activated oncogene sequence as a result of chromosomal incorporation into the animal genome, or into the genome of an ancestor of said animal, said oncogene optionally being further defined according to any one of claim 3 to 10.'

tool. Thus, the subject matter of claim 1 and 19 is an invention within the meaning of Article 52 EPC.'[65]

The Oncomouse was not considered a modified variety (mouse) but rather a genetic modification applicable to mice.

## 3. Issues of morality in respect of animal suffering

### a) Genetically modified animals and suffering

Apart from the concepts derived from the Strasbourg Convention, another exception has been developed with regard to modifying the genetic identity of animals. This concept has been developed from EPO case law in the earlier mentioned T-19/90 *Oncomouse* case to a statutory concept in Directive 98/44/EC Article 6(2)(d). Thus the balancing test concerning transgenic animals was developed in the decision T 19/90, paragraph 5, as mainly depending on a careful weighing up of the suffering of animals and possible risks to the environment on the one hand, and the invention's usefulness to mankind on the other.

Concern for animal welfare was included in the 1992 draft of the Directive as an explicit requirement of the European Parliament. The modernisation of the traditional morality clause in patent law originated from this request, as is shown by the explanatory memorandum to the 1992 draft Directive.[66] Article 2(3)(c) of the 1992 draft Directive was as follows: (c) 'processes for modify-

65. See paragraph 12.1.2. See also Singer & Stauder op. cit. footnote 21, Article 53, note 32-43. Consequently, it is the animal variety which results from the breeding methods which is excluded from patentability, and not animals *per se*. And the authors continue by saying that variety protection is a special form of protection with its own conditions (distinctiveness, unity and stability). In contrast, patent law is a technical industrial property right which requires the repeatability of the invention, and usually this does not apply to breeding methods and their results. Genetic engineering has now overcome the obstacle of repeatability in the plant and animal world. Consequently, there is an area of overlap between the two forms of protection that requires clarification. See also Cornish & Llewelyn, *Intellectual Property*, 2003, p. 222.
66. See COM(92) 589, Explanatory Memorandum, p. 1 and footnote 36. The topic was subject to assessment by the European Group on Ethics in Science and New Technology resulting in Opinion No 7 on 21 May 1996 on the ethical aspects of the genetic modification of animals. The group called for extreme care in view of the consequences for human and animal health, the environment and society.

ing the genetic identity of animals which are likely to inflict suffering or physical handicaps upon them without any benefit to man or animal.'[67]

In the explanatory memorandum it is said that the techniques for producing transgenic animals are well known and have been described in detail, and they seek, among other things, to create laboratory animals (e.g. Harvard University's Oncomouse) or animals that produce, by means of a tumour cell that develops as a result of gene therapy, effective secretions of foreign proteins of therapeutic value.

'In view of the usefulness of this type of invention to man's well-being, in this instance his health, the Commission considers it only right and proper that investment in research thereof should be capable of being duly protected. The Commission also considers that the borderline between what is acceptable and what is not acceptable must take account of the criterion of animal suffering. The Commission acknowledges that this criterion may be difficult to evaluate, but it believes that its inclusion is necessary in order to avoid the gratuitousness of certain experiments which may be inflicted on animals in so far as they are out of all proportion to the objective pursued. While it is concerned about certain extreme situations in which animals may find themselves in some laboratories, the Commission must nevertheless point out that patent law is not the appropriate field in which to legislate thereon.'[68]

The Commission referred to the European Convention for the Protection of Vertebrate Animals used for Experimental and Other Scientific Purposes, signed in Strasbourg in 1988, which recognised that man has a moral obligation to respect all animals and to have due consideration for their capacity for suffering and memory.

67. See COM(92) 589, p. 13.
68. See COM(92) 589, p. 13. See also the opinion of the Economic and Social Committee on the 'Proposal for a European Parliament and Council Directive on the legal protection of biotechnological inventions', OJ 96/C295/03 p. 13, point 3.3.3. Concerning ethical questions the Committee approves in particular the Parliaments intention to provide clear guidance on the legitimate ethical questions. However, the Committee emphasises that patent law does not interfere in procedures for examining compatibility with existing rules on ethics, such as procedures applying in particular to upstream research activities and to downstream marketing of the products. A closer view at the guidance provided by Art. 6(2) discloses that the list merely concerns the field of research, whereas Art. 6(1) concern commercial exploitation, See Geertrui Van Overwalle, Legal and Ethical Aspects of Bio-Patenting, in *Death of Patents*, Peter Drahos (ed), 2005, p. 222.

The 1995 draft of the Directive, read as follows:

'(b) processes for modifying the genetic identity of animals which are likely to cause them suffering or physical handicaps without any substantial benefit to man or animals, and also animals resulting from such processes, whenever the suffering or physical handicaps inflicted on the animals concerned are disproportionate to the objectives pursued.'

Under the heading of exceptions to patentability, Directive 98/44/EC Article 6(2) and Implementing Regulation 23d(d)/EPO 2000 Rule 28(d) exclude inventions that fall under specific listed headings.[69] The list, which was formulated by the European Commission, refers to animals in lit. (d), while (a) to (c) refers to humans only. The Article 6(2) exception reads as follows: (d) processes for modifying the genetic identity of animals which are likely to cause them suffering without any substantial medical benefit to man or animal, and also animals resulting from such processes.[70]

The EPC Examination Guideline (draft 2007) Part C, Chapter IV-7, point 4.1, regarding exceptions to patentability (EPC Article 53(a)) emphasises that in general the provision is likely to be invoked only in rare and extreme cases. Addressing the list in Implementing Regulation 23d(d)/Rule 28 (EPO 2000), and Recital 45, it is further explained that 'substantial medical benefit' includes any benefit in terms of research, prevention, diagnosis or therapy.[71]

As the T-19/90 case balances suffering of animals not against substantial medical benefit to man or animal, but against usefulness to mankind, the relationship between Rule 23d(d) and the balancing test in case T 19/90 was the issue of case T 315/03. Benefit to mankind may:

'embrace a wider range of benefits than the 'substantial medical benefit' found in the Rule 23 d(d) EPC test and accordingly the T 19/90 test is broader and it is also manifestly clear

69. The ECJ stated in both case C-377/98 (para. 37-39) and case C-456/03 (para 78-79) that Art. 6(1) allows the administrative authorities and courts of the member states a wide (but far from discretionary) scope for manoeuvre whereas Art. 6(2) allows the Member States no discretion with regard to the unpatentability of the processes and uses which it sets out, since the very purpose of this provision is to give definition to the exclusion laid down in Article 6(1).
70. Although processes for cloning human beings are included in the list (Art. 6(2)(a)), there are no expressed reference for cloning animals.
71. That the list is far from exhaustive is reflected by Recital 38.

that the reverse is true, namely that if 'substantial medical benefit' is established so as to satisfy Rule 23d(d) EPC, then 'usefulness to mankind' is necessarily established.'[72]

The relationship between Rule 23d(d) and the earlier case law is summarised as follows:

'The introductory provision of the new Rules – Rule 23b(1) EPC – states those Rules are interpretative. [...] The established and authoritative statement of the case law is to the same effect. [...] None of the arguments presented to the contrary are accepted by the Board for the reasons in this section. As regards cases such as the present which fall within Rule 23d(d) EPC, the effect of this interpretation is to insert a test which, depending on the facts and thus on the outcome of the test, may be either additional or alternative to that previously established by the case law.'[73]

Since the *OncoMouse* decision (T-315/03) concerned a process for modifying the genetic identity of animals and suffering was an inevitable consequence of the very purpose of the patent,[74] Article 53(a) and Rule 23 d(d) EPC were considered and required a balancing of benefits.[75]

Furthermore in the *Oncomouse* case (T-315/03, points 13.2.7-13.2.9), the Board looked at moral issues concerning both the environment and animal welfare. It is interesting to note that the scrutiny initially covered concerns for: (a) the possibility of escape into the environment,[76] (b) the idea that evolution was being manipulated, (c) the use of animals purely as a means to an

72. Case T 315/03, paragraph 7.1. The wording of Rule 23 d (d) is similar to Directive 98/44/EC Article 6(2)(d).
73. Paragraph 7.7. Further case T 83/05, para. 57.
74. Decision T 315/03, paragraph 12.2.1 in fine.
75. Paragraph 13.2.6.
76. See also Suk et al, 'Dolly for Dinner? Assessing commercial and regulatory trends in cloned livestock' in *Nature Biotechnology*, Vol. 25, Number 1, January 2007, p. 49. The authors emphasise that the guidelines issued under Directive 2001/18, on issues more specifically related to GM/cloned animals, such as animal welfare and how the safety of animal-derived products for human health might be assessed, have not been developed and furthermore a major question remains unanswered: Can animals cloned by somatic cell nuclear transfer be considered GMOs, as defined by Directive 2001/18? See also Yang et al, Risk Assessment of meat and milk from cloned animals, op. cit, p. 77. See Manspeizer, 'The Cheshire Cat, The March Hare, and The Harvard Mouse: Animal Patents open up a New, Genetically-Engineered Wonderland' in *Rutgers Law Review*, Winter 1991(43 RULR 417).

end, (d) the detrimental effects of the tumours on the animals themselves, and (e) whether such research could be conducted without the use of animals.[77]

### b) Animal welfare outside patent law

As patents granted on animals are evidently subject to additional regulation in regard to both the exploitation of the invention and the research on which the invention is based, it is obvious that animal welfare will influence the process of invention.

At the European level there have been several attempts to reduce the number of test animals, and this could have an impact on the efforts to create transgenic animals in order to reduce the numbers.[78] Under European legislation, animal tests are only to be performed as a last resort, and the promotion of alternative testing methods has become important since, in 2003, Europe banned the testing of cosmetics and cosmetic ingredients on animals if validated alternatives are available, and, in 2009, the testing of cosmetics on animals will be banned even if no alternative testing methods exist.[79] REACH (the Registration, Evaluation and Authorisation of Chemicals) also includes specific provisions aimed at promoting alternative approaches to reduce the number of animal tests.[80]

The issues of animal welfare, escape into the environment and cloning techniques can be questioned from a patent lawyer's point of view. Patent regimes should be supplemented by other legislation on aspects of animal treatment. Using the words of Jacobs:

77. Animal welfare protection does not form part of the general principles of Community law as such. See Camm and Bowles, 'Animal Welfare and the Treaty of Rome – A legal Analysis of the protocol on Animal Welfare and Welfare Standards in The European Union' in *Journal of Environmental Law*, 2000, Vol. 12, No 2 and *Common Market Law Review*, 2002, 39: 1159-1170; Case C-189/01, *H. Jippe*. In the draft Treaty establishing a Constitution for Europe (Art. III-121) however animal welfare is one of the objectives of the Union.
78. See COM(95) 195; and Krämer, *EC Environmental law*, 5th Ed., p. 193.
79. The Commission published a report in 1994 on the number of animals used for experimental purposes. In 1991 the number of test animals was more than 12 million (including 8,409 primates). The Community fixed a target to halve the number by 2000, See IP/06/814 Reducing animal testing: Industry commits to action programme, Commission tables guidelines for labelling cosmetics as 'not tested on animals'. See also *Nature*/Vol. 418/11 July 2002, p. 116. The race is on to find alternatives to animal tests.
80. See also Nuffield Council on Bioethics, The Ethics of research involving animals, 2005 on Alternatives: Reduction, Refinement and Replacement, p. 187.

'substantive patent law cannot serve to replace or render superfluous national, European or international law which may impose restrictions or prohibitions or which concerns the monitoring of research and the use or commercialisation of its results, notably from the point of view of the requirements of public health, safety, environmental protection, animal welfare, the preservation of genetic diversity and compliance with certain ethical standards.'[81]

Furthermore as pointed out by *J. Suk et al.*: 'The emergence of food products derived from animal cloning raises numerous ethical and policy issues, but a consistent international system for the regulation and trade of such products does not yet exist. This poses a potential barrier to animal biotech industry and also increases the likelihood of public resistance.'[82]

If by granting patents in this area, the patent system is far ahead of the regulatory system, then the public resistance mentioned above will be directed towards patent law, ignoring the fact that patent law does not regulate commercial exploitation.[83] Such concerns would thus be addressed to the wrong regulatory system.[84] Since an inventor's starting point in respect of commercial exploitation will be applications for patents and marketing authorisation, it is important to recognise that the patent authorities are not a 'watchdog' agency.[85]

81. See the Opinion of Advocate General Jacobs delivered on 14 June 2001, in Case C-377/98, para. 14.
82. See Suk et al., op. cit. footnote 76, p. 47.
83. See e.g. Moufang, in Geertrui Van Overwalle (ed.) *Patent Law, Ethics and Biotechnology*, 1998, p. 70.
84. See Phillips and. Kerr, 'Frustrating Competition Through Regulatory Uncertainty, International Trade in the Products of Biotechnology' in *World Competition* 25 (1): pp. 81-99, 2002; and Kevin W. O'Connor, 'Patenting Animals and other Living Things' in 1991 *Southern California Law Review*, 65 S. Cal. L. Rev. 597: Because current statutes regulate various uses and protections for animals, no single federal policy governs all uses of genetically altered animals. In the absence of a single policy, federal agencies will rely on existing statutes, regulations, and guidelines to regulate transgenic animal research and product development. Current federally funded research efforts could lead to patents on animals. The patentability of an animal however, does not affect the manner in which the animal would be regulated by any federal agency.
85. See e.g. Wasowski, 'The Evolution of Patentable Compositions of matter: The United States Patent Office accepts Genetically Altered Animals as Patentable Subject Matter under 35 U.S.C. Section 101' in 2 *ADMLJ* 309 (1988).

## 4. Scope of protection

Chapter II of Directive 98/44/EC introduced provisions (Arts 8-11) regarding the scope of protection at four levels, namely as protection of biological material, protection of a product containing genetic information, exhaustion of rights and farmers' privilege.

In order to examine the scope of protection in Part II of this article it is useful to pay attention once again to the draft directive from 1988, 1992 and 1995. The draft provisions and the final text concerning the scope of protection on biological material and on biological material containing genetic information had the following wording:

*Table 8.A. Protection of Biological Material*

| |
|---|
| ***1988 draft Art. 12:*** 1. If the subject-matter of a patent is a process for the production of living matter or other matter containing *genetic information permitting its multiplication in identical or differentiated form*, the rights conferred by the patent shall not only extend to the product initially obtained by the patented process but also the identical or differentiated products of the first or subsequent generations thereof, said products being deemed also directly obtained by the patent process.<br>2. Any extension of the protection conferred by the patent to a process as indicated under paragraph 1 to a product obtained thereby shall not be affected by an exclusion of plant or animal varieties from patentability. |
| ***1992 draft, Art. 10:*** 1. The protection conferred by a patent on a biological material possessing, *as a result of the invention, specific characteristics* shall extend to all biological materials derived from that biological material through multiplication or propagation and possessing the same characteristics.<br>2. The protection conferred by a patent on a process that enables the production of a biological material *possessing, as a result of the invention, specific characteristics* shall extend to biological material directly obtained using that process and to any other biological material derived from such biological material through multiplication or propagation and possessing the same characteristics. *This extension of protection shall not be affected by the exclusion from patentability of plant and animal varieties provided for in Article 3 of this Directive.* |

| |
|---|
| ***1995 draft, Art. 10:*** 1. The protection conferred by a patent on a biological material possessing specific characteristics *as a result of the invention* shall extend to *any* biological materials derived from that biological material through multiplication or propagation *in an identical or divergent form and possessing those same characteristics.*<br>2. The protection conferred by a patent on a process that enables a biological material to be produced *possessing specific characteristics as a result of the invention* shall extend to biological material directly obtained through that process and to any other biological material derived from the biological material directly obtained through multiplication or propagation in an identical or divergent form and possessing the same characteristics. That protection shall not affect the exclusion from patentability of plant and animal varieties as such, pursuant to Article 4 (2). |
| **Directive 98/44/EC,** ***Art. 8***: 1. The protection conferred by a patent on a biological material possessing specific characteristics as a result of the invention shall extend to any biological material derived from that biological material through propagation or multiplication in an identical or divergent form and possessing those same characteristics.<br>2. The protection conferred by a patent on a process that enables a biological material to be produced possessing specific characteristics as a result of the invention shall extend to biological material directly obtained through that process and to any other biological material derived from the directly obtained biological material through propagation or multiplication in an identical or divergent form and possessing those same characteristics. |

If the invention contains *genetic information*, the scope of protection is tailored as follows:

*Table 8.B. Protection of a Product containing Genetic Information*

| |
|---|
| ***1988 draft, Art. 13:*** The protection for a product consisting of or containing *particular genetic information as an essential characteristic* of the invention shall extend to any products in which said genetic information has been incorporated *and is of essential importance for its industrial applicability or utility* |
| ***1992 draft, Art. 12:*** The protection conferred by a patent on a product containing or consisting of genetic information shall extend to all material in which the product is incorporated and in which the genetic information is contained and expressed. |
| ***1995 draft, Art. 11:*** The protection conferred by a patent on a product containing or consisting of genetic information shall extend to all material, save as provided for in Art. 3(1), in which the product is incorporated and in which the genetic information is contained and expressed. |
| **Directive 98/44/EC,** ***Art.9***: The protection conferred by a patent on a product containing or consisting of genetic information shall extend to all material, save as provided in Article 5(1), in which the product in incorporated and in which the genetic information is contained and performs its function. |

It is interesting to note that the 1988 draft (Table A) distinguishes biological material containing genetic information (Art. 12) and biological material containing genetic information as an essential characteristic (Table B), whereas the other drafts distinguish between biological material possessing specific characteristics and biological material containing genetic information (Table A and B). Furthermore the 1988 draft (Table A) refers to a process for the production of living matter or other matter containing *genetic information* permitting its multiplication *in identical or differentiated* form (the 1995 draft: identical or divergent). The drafts in Table A all state that the extension of protection shall not be affected by the exclusion from patentability of animal varieties. This is not explicitly referred to in the Directive 98/44/EC Art. 8 and 9.

## 5. US Patent Law vis-à-vis Animals

### a) In general

The constitutional foundation of the US patent system as it appears today is without any concerns regarding biological material, thus biotechnological inventions will be subject to the same rules of patentability as conventional biological and chemical inventions.[86]

According to *Kenneth J. Burchfield*, the uncertainty relating to the application of the traditional legal principles on obviousness and written descriptions to complex living systems has led to proposals for modifying the patentability standards in the field of biotechnology. However, he points out that proposals for radical reforms may have lost their attraction as the Federal Circuit clarified the application of traditional patentability principles to biotechnology inventions and unified this field of law.

Although US patent law embraces all life-forms, US patents are not denied on the grounds of moral concerns. Oliver Mills has emphasised that, in

86. See Burchfield, *Biotechnology and the Federal Circuit*, 2001, op. cit. footnote 43, p. 3. See also Miller et al, *Intellectual Property, Patents, Trademarks and Copyright*, pp. 8-9. A major reason why there have been so few changes in the Patent Act is that, until recently when technological change and globalisation have become politically and scientifically important, patent law had not been a politically attractive or volatile issue. Thus, given a rapidly changing society and a relatively stable statute, changes in patent law tended to come, if at all, from the courts. See Maskus, in a Council special report, Council on Foreign Relation, November 2006, Reforming U.S. Patent Policy, Getting the incentives Right.

the *Chakrabarty* decision, the Supreme Court refused to consider the 'grave risk' associated with genetic research (preferring the view that grants or denials of patents are not likely to put an end to research, or its attendant risks): 'We are without competence to entertain these arguments ... [t]he choice we are urged to make is of high policy for resolution within the legislative process after the kind of investigation, examination, and study that legislative bodies can provide and courts cannot (206 USPQ 193 at Point 11 of the Opinion).'[87]

US patent law links morality to the utility requirement under the *moral utility doctrine*. In 2001, the USPTO adopted new utility examination guidelines and guidelines for examination of patent applications under 35 USC §101 and §112.1, operating with the concepts of specific, substantial and credible use. These guidelines specifically emphasise that they do not alter the substantive requirements of 35 USC §101 and §112, nor are they designed to obviate the examiner's review of applications for compliance with all other statutory requirements for patentability. However the guidelines make no mention of morality or public policy issues. However, the guidelines make no mention of morality or public policy issues. It is, according to Oliver Mills, uncertain whether or not moral considerations will continue to be enforced by means of the utility requirement.[88]

In his examination of moral issues and US patent law, *Oliver Mills* compares the US concept of utility with the EPO concept of industrial application. The US concept is broader in some respects, in embracing elements such as morality and illegality, whereas it is narrower in other respects, since pure research is not held to be equivalent to practical utility. In the decision in *Lowell v Lewis* in 1817 the court already held that mischievous or immoral inventions were considered unuseful.[89] *Mills* argues that in the past, the courts have gradually rejected a 'utility' based on public mores for inventions of the

87. See Mills, *Biotechnological Inventions, Moral Restraints and Patent Law*, 2005, p. 112.
88. See Burchfield, *Biotechnology and the Federal Circuit*, 2005 Cumulative Supplement, p. 20. The reshaped 2001 USPTO Utility Guidelines addresses DNA sequences in general. As such the USPTO Utility Requirement demand that the application discloses *a specific, substantial, and credible utility* for the claimed isolated and purified gene whether it is a human or animal DNA sequence. However is seems most likely that the examiners in the USA and Europe scrutinise the functionality requirement alike. The specific utility requirement (or industrial application) is strictly addressed by both the US and the European examiner.
89. 3 F Cas 37 (No 1217), Circuit Court, Massachusetts 1817. See Mills, op. cit. footnote 87, p. 46, footnote 24 and p. 45.

traditional mechanical type, whereas the link between 'utility' and 'morality' is especially relevant for modern biotechnological inventions involving gene sequences not identified in terms of structure or function. Considering the US patent system as a model Mills further argues:

'Hence in respect of life-form protection a single regime operates in the US. Additionally, in the US the judiciary has embarked on judicial legislation and the courts there have laid down the conditions for patentability. In this regard, it was never intended morality should extend to validity criteria. Because of this, courts in the US do not have to consider whether or to what extent, moral norms have changed with time. Neither do US courts have to deal with the limits of the morality test. The PTO functions as a patent-granting authority and not a determiner of patent policy.'[90]

### b) Animal variety and breeding processes in US patent law

The US Supreme Court has explicitly stated that a bacterium designed to clean up oil (*Chakrabarty*), an Oyster without the unpleasant bitter taste due to the reproduction cycle (*Ex parte Allen*), and a sub-human mammal (the *Harvard Oncomouse*) are all patentable. Following the *Chakrabarty* decision, under 35 USC §101, only man-made life forms are patentable in contrast to subjects that occur naturally (the doctrine of nature).

The opposition to animal patenting in the US may have influenced the USPTO practice in this area. *Burchfield* emphasised that the scope of utility patents based on claims relating to animals had been extremely limited:

'In the six years following the Allen decision, only six patents were claiming animals, including the original *Harvard Oncomouse* patent, which contrary to its reputation contains broader claims to transgenic subhuman mammals.[91] The scope of subsequent patents to multicellular animals has been more restricted [transgenic mice (the only other patented vertebrate to date is a 'rabbit infected with HIV-1 virus].[92]

[...] By the end of 1991, about 157 animal patent applications were pending at the PTO, about 80 percent of which were directed to animals useful in biomedical applications, and most of the rest being agricultural animals with characteristics superior to conventional cows, chickens, sheep, and pigs. Under the judicial and administrative standards of patentable subject matter, if such domestic animals are not naturally occurring but are the 'artificial'

90. See Mills, op. cit. footnote 87, p. 173.
91. See Burchfield, *Biotechnology and the Federal Circuit*, op.cit. footnote 43, 1995, p. 32-33.
92. See Burchfield, *Biotechnology and the Federal Circuit*, op.cit. footnote 43, 1995, p. 42, five mice, one rabbit, and a worm.

product of human effort (with characteristics superior to conventional cows, chickens, sheep, and pigs), they are clearly patentable subject matter, whether produced by genetic engineering techniques or developed through conventional breeding procedures'.[93]

In contrast to European patent law, the Federal Circuit declined to adopt a rule that required 'genetically identical repetition'.[94] In *Amgen v Chugai* concerning Chinese hamster ovary cells producing at a superior rate, the Federal Circuit emphasised that the statute does not require the patentee to 'guarantee that every aspect of the specification be precisely and universally reproducible.'[95] A description of standard biological materials is sufficient:

> 'Particularly with respect to classical breeding methods, where the parent variety is well known in the art and traits for selection may be readily recognised by those of ordinary skill, no more should be required for enablement than a written description of the parent and the method. Similarly, where those skilled in the art are directed to standard plants or animals, such as the Chinese hamster ovary cells in *Amgen v Chugai*, enablement of the invention is provided despite the certainty that the individual genetic characteristics of particular animals will vary within a predictable range.'[96]

In regard to breeding processes, the decision in *Ex parte Allen,*[97] concerning a claim to a method for increasing growth by inducing polyploidy in Pacific oysters, elaborated on 'the breeding process concept' since the examiner refused the applicant's product-by-process claims on grounds that the claimed subject-matter was directed to 'living entities', and therefore not patentable under 35 USC §101. The US Board of Appeals 'disagreed and reversed the decision. The Board took the view that although 'laws of nature' control breeding processes, this did not address the issue of whether the subject-matter is a non-naturally occurring manufacture or composition of matter.'[98]

93. See Burchfield, *Biotechnology and the Federal Circuit*, op.cit. footnote 43, 1995, pp. 43-44, footnote 52.
94. See Burchfield, *Biotechnology and the Federal Circuit*, op.cit. footnote 43, 1995, p. 220.
95. 927 F.2d at 1212, 18 USPQ2d at 1025-26; and Burchfield, *Biotechnology and the Federal Circuit*, op.cit. footnote 43, 1995, p. 220.
96. See Burchfield, *Biotechnology and the Federal Circuit*, op.cit. footnote 43, 1995, p. 221. Deposit is not the only way to provide enablement of a plant or animal, or its tissues, in order to permit others to replicate its genetic material exactly.
97. 2 USPQ 1426-1427, 3 April 1987.
98. See Mills, op. cit. footnote 87, 2005, p. 113. In the same direction *Re Merat and Cochez,* 186 USPQ 471, decided August 7 1975. See Mills, p. 109; and for the opposite

# III. Case Studies

## 1. Introduction

The purpose of the present case study is to use selected illustrative cases to shed light on how patents and patent applications have progressed in areas determined by the core concepts of animal varieties, essential biological processes and genetically modified animals. Since this is not an exhaustive presentation the intention is mainly to expose obstacles (or their absence) regarding the patenting of animals currently present to patent law and patent authorities.[99] Special emphasis is placed on DNA claims concerning pigs, with respect to breeding processes. The pig has been chosen due to its economic and medical importance.[100]

conclusion see Edmund J. Sease, 'From Microbes, to Corn Seeds, to Oysters, To Mice: Patentability of New Life Forms' in *Drake Law Review*, 1988-9 Vol. 38, p. 558: 'The lesson of the chicken case was clear. Applications for patents which relate to living organisms, whether simple or complex, would find it difficult to pass through the United States Patent Office. Because of the variability which is inevitable in the reproduction of organisms, attempts to patent the results of reproduction would be doomed to failure under 35 U.S.C. § 112'. See Wasowski, 'The Evolution of Patentable Compositions of matter: The United States Patent Office accepts Genetically Altered Animals as Patentable Subject Matter under 35 U.S.C. Section 101' in *Administrative Law Journal*, Fall, 1988, (2ADMLJ 309).

99. As the US and European patent searches are made for equivalent patents (patent family or a WO patent application continued as US-PCT or EPC-PCT applications made under the terms of the Patent Cooperation Treaty (PCT)) the aim is to assess whether the scope of protection granted in the USA and in Europe is alike.

100. Pigs and humans are close in sequence space, and that the pig is in fact genetically closer to man than the commonly used laboratory animals. See 'A Chinese-Danish sequencing consortium' http://www.biomedcentral.com/1471-2164/6/70. Genome mapping has covered several species, including the pig. The Sino-Danish Pig, Cf. Wernersson, Schierup et al, 'Pigs in sequence space: A 0.66X coverage pig genome survey based on shotgun sequencing' in BMC Genomics 2005, 6:70, http://www.biomedcentral.com/1471-2164/6/70. Since the inner organs of pigs and humans are physiologically very much alike, the pig genome can be used as a model for the study of e.g. neurodegenerative diseases like Alzheimer's. The purpose of the Sino-Danish Pig Genome Project is to improve breeding processes as well as to solve some of the riddles about genetically transmitted diseases in humans, cf. First it was humans, now it's the pigs' turn, Public release of pig genomic sequences, http://www.eurekalert.org/pub_releases/2005-06/bgi-pro060605.php.

The case study is based on patent applications as well as granted patents.[101] As obviously not all patent applications are successful, they are only included in order to look into all activities (research and development) that are actually subject to applications for legal protection.

## 2. Animal variety and essential biological processes for the production of animals

Both the pig and other animals have been the subject of several patent applications regarding breeding processes. It is essential to define the borderline between the essential biological process (which are excluded from patent protection) and the process dominated by technical (human) interference (which can be protected). *The Red Dove decision* concluded that if the phenotypical approach does not promise a sufficient predetermined result (repeatability), then the selection step leaves room for many variations. One question is, however, whether DNA-marker-based selections sufficiently improve the phenotypical approach? Another question derived from case T 83/05: are molecular markers as a technical step that requires human intervention a well known step in the selection of swine with the desired traits (prior art already used in the context of pig species)? The focus of the study of breeding processes could essentially be narrowed down to *genetic marker assisted breeding* within the following areas:

(1) Genetic resistance to F 18 *E. coli* associated diseases (EP985052, EP1310570, US6596923, US656922, US20020129395),
(2) Methods for determining the coat colour genotype of a pig (EP842296, US6183955),
(3) DNA markers/receptor genes as genetic markers for increased litter size in pigs (EP879296, US6410227, EP958376, US5935784, US7081335),
(4) Genetic markers for fat content, weight gain, and/or for feed consumption of animals (US6803190, EP1100970),
(5) Receptor genes to enhance pork production (US6458531, WO2005/017204, EP1651777, US20030493158, US20040553582),
(6) Methods for genetic improvement of terminal boars (WO2005/015989, EP1651030, US20030492395),

101. See Supra note 2. The patent search covered 48 patents and applications and it was made in December 2005, using Epadoc and WPI (World Patent Index). New areas from 2005 up to the present are not covered by the search.

(7) DNA sequences involved in increasing or decreasing mammalian ovulation rate (US 20040092007, EP 1292674).[102]

## 3. Genetic marker assisted selection and breeding strategies

Genetic-marker-assisted selection (MAS) can be classified as a selection tool for breeding purposes. As to F 18 *E. coli associated diseases* (above (1)), the DNA marker assist the work of differentiating swine that are genetically susceptible to diseases associated with F 18 *E. Coli* infection from resistant swine. The aim is to control the breeding process by the use of a method for identifying a swine that are resistant by determining whether the swine has a mutation in the *FUT1* gene.[103] The patents claim the *FUT1* sequence itself, the specified mutation and the methods for identifying the swine. The granted patents concerning F18 *E. coli* are both US and EP patents.[104]

The 2005 EP claim differs from the 2003 EP claim by further claiming a method for identifying resistance caused by a micro-organism capable of binding to ECF18R in swine, and providing a diagnostic test useful for swine breeders. When analysing the claims it is notable that in the US the patents claim a method for *breeding the selected swine*. The claim was deleted from the granted EP patent. The International preliminary examination report as well as the initial European 2002 Annex to the Communication refers to such breeding methods as representing an essential modification of known biological and breeders' processes. It is said that the technical invention in the process by man plays a significant part in controlling the desired result. The effi-

102. See also: Administration of DNA sequences to female animal to enhance growth in offspring (EP 1364004, US 20030074679); Allelic variants of bovine somatotropin gene: a genetic marker for superior milk production in bovine (US5374523). Outside DNA marker assisted breeding: Production of mammals which produce progeny of a single sex and transgenic pigs.
103. *E. coli* (bacterial) infection is the leading cause of sickness and death in newborn pigs, See *Alfredo Flores*, 'New way to control E. coli in weaned pigs', Agricultural Research, March 2004.
104. The first US patent was granted in 2003 and the second in 2005 (respectively Patent Nos US 6596923 and US 6596622). Application US 20020129395 was granted in 2007. The corresponding EP patents are from 2003 and 2005, respectively Patent EP 985052 (2003) and EP 1310570 (2005). Both patent applications were subject to amendments during the examination procedure as to sufficiency of disclosure (e.g. the term *E. coli* was not an allowable generalisation of the examples given for *E. coli* F 18), novelty, and inventive step.

ciency and high yield associated with the method show the important technological character. Yet the claim on the animal is not repeated in the European patent, but deleted in the last minute due to an oral agreement (without further explanation than a telephone conversation). Furthermore, it is also notable that the analysis showed that the examination of the applications as to the disclosure of function both in the USA and in Europe were strict.

When determining the *coat colour in pigs* (above (2)), this can involve testing *the KIT genotype*.[105] As a test-tool, the patents claim a screening method to distinguish the coat colour genotype of a (genomic) DNA sample.[106] In the abstract of the US patent application it is stated that the traditional test mating programmes are time-consuming and costly, so it is desirable to determine the coat colour genotype of individual pigs.[107]

> 'In pigs, four alleles have been reported at the *KIT* locus, the recessive *i* allele for color, the $I^P$ allele for black spots, the dominant *I* allele for white phenotype, and $I^{Be}$ for the Belt phenotype. Beside the *KIT* locus, strong evidence that the classical extension coat color locus is identical to the melanocortin receptor 1 (*MC1R*) was found by Kijas et al. (1998). So far, four *MC1R* alleles have been reported: the $E^+$ allele for wild type, the $E^d$ allele for the dominant black phenotype, the recessive e allele for the red color phenotype, and the $E^P$ allele for the spotted phenotype (black spotting). Results by Marklund et al. (1998) indicate that there are interactions between the *KIT* and the *MC1R* loci.'[108]

The present study examines inventions concerning *genetic markers for litter size* (above (3)) as well. Thus the study cover examples of genetic typing of pigs for DNA markers or the prolactin receptor genes, thereby identifying individual males and females that carry a gene for larger litters.[109] In the patent

105. The granted patents, concerning coat colour, are both US and EP patents: US 6183955 and EP 842296.
106. Coat colour preferences have been explained in the literature, although coat colour is not a functionally important trait, pork sells for a higher price if it is sold on markets where its coat colour is preferred by consumers e.g. by *Hirooka et al*, 'Genome Scan Reveals New Coat Color Loci in Exotic Pig Cross' in *The Journal of Heredity*, 2002 (93), p. 1. On page 1 it is stated that: 'Most of the European breeds such as Large White and the different Landraces have white coat colour. On the other hand, the uniform black coat is the most common type in breeds from China.' See EP 842296 as granted (pp. 1-2).
107. US 6183955, p. 1, abstract.
108. Jf. Genome Scan Reveals New Coat Color Loci in Exotic Pig Cross, *The Journal of Heredity*, 2002 (93): 1.
109. The study includes three granted US patents and two granted EP patents: US 6410227, 5935784, 7081335 and EP 879296, 958376. The 1999 US patent claim relates to pigs, whereas the 2006 US patent claim relate to animals. See BPAI decision

applications it is assumed that, although many genes are involved in complex traits, a few of the genes involved (major genes) have major effects on the trait.[110] As litter size is very difficult to select conventionally, one approach to improvement might be to introduce beneficial genes into production lines from breeds which have significantly higher litter sizes.[111] Hence, the inventions concern a selection tool used in breeding programs in order to screen pigs to determine those which are more likely to produce larger litters (to develop desired lines and breeds).[112]

High consumption of animal fat is a significant health hazard associated with cardiovascular disease and high cholesterol. Consequently, there is an increased demand for meat products with *low fat content* (above (4)). As genetic differences exists between individual meat-producing animals as well as between breeds a method for genetically evaluating animals by assaying for the presence of at least one genetic marker – *the MC4R gene* – which is in-

(examination affirmed in part). The recitation of 'pig' appear to lack antecedent basis in the claim, as the rest of the claim refers to 'animal'. Upon return of the application, the examiner should address this discrepancy in the claim. See in general M.E. Wilson et al, 'Novel insight into the control of litter size in pigs, using placental efficiency as a selection tool' in *Journal of Animal Science*, 1999 Vol. 77 Issue 7, pp. 1654-1658 and Montgomery et al, *Journal of Reproduction and Fertility supplement*, 1995 Vol. 49, pp. 113-121; EP 879296, p. 3.

110. Cf. for example the oestrogen receptor gene (ESR) as marker to improve litter size, G. Horogh et al, 'Oestrogen receptor genotypes and litter size in Hungarian Large White pigs' in *Journal of Animal Breeding and Genetics*, vol. 122 (2005), issue 1, p. 56. See also Wilson et al, *Journal of Animal Science*, 1999 Vol. 77 Issue 7, pp. 1654-1658, 'differences in placental size and efficiency would affect litter size.'
111. Cf. US patent 641022.
112. In the EPO examination report, issued on 28 April 2003, the examiner stated that the use of a specific polymorphism detected in the PRLR gene as an indicator of increased litter size is highly dependent on the genetic background of the PIC line used. In response, it was submitted that the genotype associated with larger litter size alternates in different breeds. Other objections were raised by the examiner, e.g. under EPC Article 52(4) regarding the steps for obtaining a sample from a pig, and further novelty and the inventive steps of the product claims. See also EP 1364004 (US patent application No 20030074679) concerning the administration of DNA sequences to female animal to enhance growth in offspring. This application relates to the improvement of growth and performance where prior art discloses treating animals in general, without any distinction between males and females. The applicant replied to European examination report in February 2007 stating that prior art does not disclose the use of GHRH and mutants in a context of gene therapy by muscular injection in gestating females for the amelioration of growth of the offspring or for increasing milk production, see examination report of February 2006, point 3.

dicative of one or more of *the traits of fat content, growth rate, and feed consumption* is thus desirable.[113] The EP patent claim 1 concerned a method of identifying a non-human animal with a polymorphism at nucleotide position 678 of the MC4R gene, wherein an animal with a guanine in position 678 is indicative of said animal more likely to have less back fat, lower daily gain, and lower feed intake than an animal with an adenine at position 678, which an adenine is indicative of higher feed intake. The method comprising: (a) detecting the nucleotide present at nucleotide position 678; and (b) relating the polymorphism to the phenotype, thus enabling breeders to select more accurately those animals which phenotypically express desirable traits. The *leptin receptor* gene (above (5)) to enhance pork production was already patented as a genetic marker for leanness in pigs in the USA (2002).[114] The claims cover a breeding plan with the aim of selecting and breeding until an altered allelic frequency is achieved. A pig population produced by the claimed methods, and the offspring produced are also covered by the claims.

It is notable that the European Examination Report focuses on novelty, as it is concluded that it was not apparent which part of the application could serve as a basis for a new, allowable claim, and further that there is no inventive step, whereas no references are directed towards breeding processes. The applicant replied in January and July 2007, but currently the amended claim no. 4 still covers methods for enhancing meat production from a swine herd, using a breeding plan to produce offspring with an increased frequency of the desired allele. Claim 4 read:

113. See US patent (US 6803190) and the European patent application (EP1100970) now comprising 16 claims. The European patent was filed in July 1999 and an intention to grant the patent was announced in June 2007. The examiner stated that the invention did not meet the requirements of Article 52(4) EPC, that some of the claims were not supported by the description, and that e.g. it would be obvious for the skilled person to investigate, with a high expectation of success, the use of a method for determining polymorphisms in the MC4R gene of the type disclosed in D1 in humans or other animals when seeking to identify individuals having an obese (high fat content) metabolic trait.

114. US 6458531. One patent application concerning the leptin receptor gene to enhance pork production has so far been filed (July 2004) as a WO patent application, Cf. WO 2005/017204 (PCT/US2004023050) and EP 1651777. In December 2005 Greenpeace submitted third party observations requiring the application to be withdrawn: 'Die Patentschrift ist technisch hinsichtlich ihrer Reichweite nicht ausreichend definiert. Sie beruht nicht auf einer Erfindung (Gensequenz war vorbekannt). Es gibt keine nachvollziehbare technische Lehre, aus der eindeutig hervorgehen würde, zu welchem Zweck die beanspruchten Gen-Sequenzen verwendet werden sollen.'

'A method of enhancing meat production from a swine herd comprising:

a) screening a plurality of pigs to identify the nature of an allelic variant in the porcine leptin receptor (pLERP) gene, wherein said allelic variant produces a threonine or methionine polymorphism at amino acid number 69 of the prepro-pLEPR protein:
b) selecting those pigs having a desired allele;
c) using the selected pigs as sires/dams in a breeding plan to produce offspring, wherein the offspring have an increased frequency of the desired allele; and
d) repeating steps a) through c) until an increased allelic frequency for the desired allele is achieved'

Furthermore this study examines inventions concerning a method for the *genetic improvement of terminal boars* (above (6)).[115] The International publication broadly claims a method for producing terminal swine parent animals having improved germplasm.[116] The International search report addressed the application as relating to the subject-matter for which no search is required by the International Search Authority: Since 'breeding' in animal husbandry and in horticulture has been defined as 'the selection of stock for propagation and the act of insemination by natural or artificial means' (Webster's Dictionary), claims 1-65 are considered to refer to animal varieties and essentially biological processes for the production of said animals.[117]

Regarding the concepts of breeding of animals other than the pigs, only one illustrative example was included, namely *DNA sequences involved in increasing or decreasing mammalian ovulation rate* (above (7)), broadly directed to a mutated gene sequence GDF-9B and its corresponding encoded protein.[118] The USPTO and EPO examiners stated that the instant specification failed to describe essential core structure(s) possessed by other mutated DNA sequences and that the claimed function of the mutated GDF-9B sequences does not distinguish the invention from wild-type GDF-9B.[119] Once

115. So far only the EPC-PCT application can be traced, and the latest activity, from March 2006, only consists of a notification of the European publication number and information on the application of Article 67(3) EPC, application EP1651030/WO2005/015989.
116. And further developments based thereon e.g. claim 31. The method of claim 1 comprising making rapid genetic improvements of commercially valuable traits in the target herd and in the terminal sires for the SP herd through use of a breeding plan designed to provide an improvement selected from the group.
117. See International search report, p. 2. PCT Rule 39 and Article 17.2(a)(i) PCT.
118. See US patent application No 20040092007 and EP patent application No 1292674.
119. See the applicants reply to the examination report 26 January 2007. The fact that wild-type or mutated GDF-9B could be useful as a modulator of ovulation was not known at the date of the invention. The applicant further argued that a skilled worker

again the examination of the function of the animal DNA sequence is strictly scrutinised.

## 4. Genetically modified animals and breeding

As an example of claims regarding transgenic pigs, the present study examine a method for preparing a transgenic pig which over expresses porcine growth hormone.[120] The now-abandoned US patent related in particular to methods for creating new breeds of animals with desired characteristics, such as increased weight gain, feed efficiency, milk production or disease resistance.[121] The European patent was subject to opposition, arguing that exploitation of the claimed methods for producing transgenic pigs was contrary to morality and thus excluded by virtue of Article 53(a), EPC. One opponent claimed that the suffering of pigs which are genetically engineered for increased growth outweighed the usefulness of the invention to humanity. It was argued that the genetic engineering of farm animals for increased or faster growth had, in a number of cases, led to serious health problems. The serious risk of animal suffering provides no significant benefit to humanity – the 'usefulness' side of the balancing test was not fulfilled – and the increased demand for meat should not be satisfied by traditional animal husbandry.[122] The other opponent also focused on the balancing test. The patent in the present case only benefited humans by the increased production of pork or the production of leaner pork. There is no shortage of pork in the world market, and although leaner meat is desirable for human health, it is not comparable to the benefit of the aim of cancer research. In other words, the opponent argued that the animal suffering caused by the invention outweighed its usefulness to humanity.[123]

could not predict the effects of this gene from one species to another or the effects of any mutation. A new main request was submitted in November 2007.

120. The US patent (US 5573933) has expired due to non-payment of maintenance fees under 37 CFR 1.362. The European patent (EP309559) was revoked in May 1998.
121. See EP 309559, as granted, pp. 5-6 concerning the technique used. The process has been successfully used in relation to mice but never in attempts to enhance desired characteristics in farm animals.
122. Cf. Compassion in World Farming Limited, 16 May 1997.
123. The British Union for the Abolition of Vivisection (BUAV).

A search for transgenic pigs, as a concept, shows that it has been developed through many years for various reasons.[124] Some of these reasons are mentioned in the ISU Bioethics Outreach, e.g. the Enviropig, helping the environment by producing less phosphorus in their faeces than non-transgenic pigs. So far the *Enviro pig* has been subject to one US patent, one US and EP patent application.[125] This invention has been described as a pig which has a phytase gene placed in its salivary glands to allow better utilisation of phosphorus in feedstuffs. The genetically modified pig may prove to be part of a solution to animal waste issues (by reducing phosphate pollution).[126] The US patent was issued in March 2006.[127] The International Preliminary Examination Report (PCT, 2001) already stated that it was highly questionable whether all animals specified in the claims (originally including fish and poultry as well) were actually suitable to successfully express functional polypeptides. The European examination objected in the same way.[128]

124. A transgenic animal is defined by The Federation of European Laboratory Animal Associations as an animal, in which there has been a deliberate modification of its genome, the genetic makeup of an organism responsible for inherited characteristics, FELASA (Federation of European Laboratory Animal Science Associations September 1982, revised February 1995. *Endang Tri Margawati* 'Transgenic Animals – Derivation, Welfare, Use and Protection, in biotechnology: students speak out' in Transgenic Animals: Their Benefits To Human Welfare. http://www.actionbioscience.org/biotech/margawati.html.
125. Respectively patent US 7115795 and patent applications US 2007006334 and EP 1170994. The latter is deemed to be redrawn, because the invitation to file observations on the communication from the Examining Division was not complied with (further processing may be requested within two month of 1.June 2007). The application was closed in September 2007.
126. See Animal Pharming: The Industrialisation of Transgenic Animals, The Center for Emerging Issues December 1999 (http://www.aphis.usda.gov).
127. Cf. US Official Communication December 2005. The US patent granted in March 2006, filed new claims in July 2006 replacing all prior versions, and the applicant already in the 2005 Communication limited the animal species to mammalian species only (pigs, goats, sheep, cow and horse) excluding non-mammalian species, such as fish and birds. The 2007 US application incorporate all of the prior applications in their entirety. The new application refers to non-human animals as well as selection from the group consisting of pigs, goats, sheep, cows, horses, fish and poultry. The term mammals have been replaced by animals' and fish and poultry reinserted in the claims.
128. Cf. Communication from the inventor, University of Guelph, February 2006, a reply to the Official Communication claiming that the Official Communication merely reiterates the objections raised in the International Preliminary Examination Report. The Examination Division (EPO) maintains the objection in a written Communication

There are also transgenic pigs in the area of biomedical research, e.g. producing milk that contains human proteins that promote blood clotting. And finally there are the Nextran transgenic pigs which are intended to serve as organ donors for human beings.[129] Organ farms for hearts, livers and other spare parts have been discussed in the light of the ability to alter the pig's genetic makeup, making it less likely that organs will be rejected by the immune systems of human recipients.[130] There have been other programmes, e.g. scientists in Taiwan have bred transgenic pigs, created by adding genetic material from jellyfish into a normal pig embryo.[131]

Genetic alterations were also the issue in the invention concerning the *production of mammals which produce progeny of a single sex* involve genetically modifying the heterogametic sex[132] such that the genetically modified gamete is marked or disabled.[133]

'First, the USPTO Office Action indicates as a fact that the art teaches that integration of a transgene construct is unpredictable and an artisan cannot predict the number of transgene copies that integrate into the genome. An artisan also cannot predict the site of integration ... this is an issue because genomic DNA comprises regulatory regions that effect transgene expression. Subsequently, an artisan cannot predict whether a transgene construct has any effects on the claimed transgenic non-human mammal. Second, the specification indicates the intended use of the product made by the claimed method for mammals such as cows and pigs ... While it is understood that a dairy farmer would prefer to have more female cows and a beef farmer would prefer bulls, the specification does not provide guidance for

(December 2006) since reliable predictions concerning successful expression of heterologous proteins in the gastrointestinal tract of larger animals cannot be made.

129. Cf. http://www.bioethics.iastate.edu/classroom/threelittlepigs.html. Three Little pigs, 2003.
130. Cf. http://www.pbs.org/saf/1104/features/substitute4.htm.
131. The scientists will use the transgenic pigs to study human disease. Because the pig's genetic material encodes a protein that shows up as green, it is easy to spot, cf. http://news.bbc.co.uk/2/hi/asia-pacific/4605202.stm. Taiwan breeds green-glowing pigs.
132. The sex that carries two different sex chromosomes and therefore determines the sex of the offspring.
133. So far, two patent applications have been filed – one in US and one in Europe, See US patent application 20040088745 and EP1257167. The EP examination report given in June 2007 objected to the present claims as far as they cover human beings (Art. 53(a) and Rule 23(d) as well as Art. 53(b) since claims 20, 25-27 and 29 apparently only contain conventional breeding techniques.A US non-final rejection was mailed in January 2006 stating that claim 18 and 19 are claims drawn to transgenic mammals. This encompasses transgenic humans, which are non-statutory matter. Amending the claims to non-human mammals would obviate the rejection.

an artisan to use the other claimed non-human mammals such as a mouse, whale and dog. The use of these claimed non-human mammals is not readily apparent ...'[134]

Finally, it can be pointed out that the US examiner addressed the interpretation of wording 'substantially male' as imprecise. One interpretation of the term could be that the claims encompass hermaphroditic or mosaic non-human mammals.[135]

## 5. Genetically modified animals and biomedical research

Pigs have, as mentioned above, also been subject to patent applications in the sphere of biomedical research. Only one of the hits from the search was actually an issued patent.[136] The field of biomedical research relates to xenotransplantation and, more precisely, a process for producing miniature swine, free of the human-tropic PERV.[137] The European patent application was rejected on ground as being an essentially biological process (EPC, Article 53(b)). The examiner cited the Enlarged Board of Appeals decision in case G 1/98, referring to three approaches in this regard:

'a) only processes comprising exclusively non-biological process steps could be considered as non-essentially biological (Art. 53(b) EPC;

134. Cf. US patent application NO 20040088745, Final rejection mailed 19 September 2006. In the final rejection, the examiner maintain the rejection as the applicants' amendment does not address the issue that the natural breeding is a step used to arrive at the claimed founder non-human mammal.
135. The rejection from the US examiner is final cf. 37 CFR 1.136 (a) – shortened statutory period for reply is set to expire three month, and any extension fee pursuant to 37 1.136(a) will be calculated from the mailing date of the advisory action. The correlating patent application in Europe has been subject to a supplementary European Search report (2004), but since 2005 no activity in the patent examination has been reported. See patent application No EP 1257167. The International preliminary examination report (2003) observes that the specification provides general guidelines for the generation of a transgenic mammal, however does not provide the detailed guidance to practice the invention as claimed.
136. US 20040142449 and EP 1366171 concerned the generation of PERV-free strains of pigs for xenotransplantation. Both applications were withdrawn.
137. Expression and possible replication of PERV (porcine endogenous retroviruses), even at low levels, has a major implication on the use of pig organs and tissues in the course of xenotransplantation. The US patent was issued in 2005, but the European patent application was withdrawn in June 2006, US 6867347 and EP1250594, with the same inventor.

b) the essence of the invention taking into account the totality of human intervention, and its impact on the results achieved;[138]
c) present of at least one clearly identified non-biological process step, but would allow any number of additional essentially biological steps.'[139]

However, the Board did not state a preference for any one of the three approaches outlined. The examiner concluded that especially the step that lacks any technical features and simply refers to 'selection of those offspring free of human-tropic ERV', falls under methods excluded under Art. 53(b), EPC. It is notable that exactly the same wording was used in the corresponding US patent.

This case study analysed claims on *a transgenic non-human animal* in which the DNA has been artificially inserted into its chromosome, or has been knocked out as being expected to promote a dramatic advance, mainly in diagnoses, treatments, or studies relating to neurological diseases. The patent description describes the use of a transgenic animal and the examiner has not objected to the transgenic animal.[140] Animal models have also been examined in regard to a *gene for identifying individuals with familial dysautonomia.*[141]

## IV. Conclusion

Patent law does not compete or interfere with animal breeders' right. Thus, this article is an attempt to turn the spotlight on the issue of animal patents only, from cross-breeding to genetic make-up and biomedical research. The

138. Cf. case T-320/87. As discussed in Case T-356/93, this would have the consequence that a process containing at least one essential technical step, which could not be carried out without human intervention and which had a decisive impact on the final result, did not fall under the exclusion.
139. In order to escape the prohibition of Article 53(b) EPC, the approach adopted in Article 2(2) of the draft EC biotechnology Directive (Directive 98/44/EC) has to be addressed.
140. Cf. US patent 6436669, US patent application 20020146775 and EP patent application 0960888.
141. US patent application NO 2005204409/patentfamily and US patent application NO 2002169299, EP patent application 1373568. FD is an autosomal recessive disorder present in 1 in 3,600 live births in the Ashkenazi Jewish population. The method of producing a transgenic animal have been deleted (2006)/withdrawn by the applicant even though a disclaimer on humans would have fulfilled the request from the examiner.

case study aims at disclosing areas of present patent activity in order to clarify what the drafters could not predict in 1963.

The patent search on the pig has shown a huge patent activity relating to the identification of mutations as indicators of certain traits. Obviously, the genetic marker as a test tool can be patented, as long as it is not a well-known step, but the major question is to determine whether the enhanced offspring qualifies within the scope of protection. As animal varieties and traditional breeding are excluded from patentability, the question must be answered in a negative way. Still, case T 83/05 considers the use of DNA marker to be a technical step, although questioning whether this additional feature of a technical nature contributes something beyond a trivial level.

The genetic information is more likely to be inherited but it cannot be fully controlled by human. Thus the requirement of repeatability is not fulfilled. Referring to Directive 98/44/EC Article 8(2) – it is not a question of a pig possessing specific characteristics (*only*) as a result of the invention (and the offspring possessing those same characteristics). Thus providing a transparent breeding guidance to farmers who buy the test cannot extend protection to the pig itself, since that would affect the exclusion from patentability of animal varieties as such (cf. section 6 and the 1995 drafts 10(2)). Furthermore, human invention still consists in identifying an available biological material and letting it perform an inherent biological function under natural conditions (cf. 1988-draft, Article 7, section 4). If one looks at current patent activity, there is a huge number of patent applications which are concerned with the efforts of reducing or preventing boar taint in pigs by the use of a genetic marker, having a statistically significant association with boar taint.[142] One question can be raised in this connection: If the selection is based on a statistically significance only – will the phenotypical approach then satisfy the requirement of repeatability?

As an example of genetically modified animals, the present case study of the transgenic pig obviously did not constitute a significant benefit to mankind in a patentable manner (increased demand for meat versus animal suffering). Compared to the 1995 draft Directive, Art. 9(2) (cf. Part II, section 4, table A, it is obvious that the physical handicaps inflicted on the animal con-

142. Due to boar taints male pigs are castrated very early in life in order to avoid an unpleasant release of odour during cooking. See e.g. application No WO 2005123922, to be published as EP 1766025 in 28 March 2007 (disclosing a porcine SULT2A1 sultfotransferase), US 2005216965, EP0966683. So far the PCT International Preliminary Report states that the invention is novel, contains an inventive step and is industrially applicable.

cerned (e.g. lameness, kidney and liver damage) are disproportionate to the objectives concerned. The Enviro-pig on the other hand, if patentable, achieves a benefit to the environment in waste issues – a concern which is presented in the TRIPS-agreement (in order to avoid serious prejudice to the environment) as exclusion to patentability. However, the issue of environmental progresses are not part of Directive 98/44/EC Article 6 (2)(d), but they are part of Art. 53(a) EPC.

Spotlighted this way, it would be worth considering if only genetically modified animal models or animals (e.g. the Canadian Enviro-pig) were subject to the scope of protection conferred by Directive 98/44/EC, Arts. 8(2) and 9 as products containing genetic information. As to the genetic marker-assisted breeding, the enhanced offspring should not be claimed since the trait is occurs naturally and humans can simply predict the possibility of enhanced offspring. The implication of genetic-marker-assisted breeding still relates to the reproduction performance, including concerns of inbreeding and evolution from generation to generation; thus humans can never fully and predictably control the outcome of the offspring.

PART IV

# Patent Law: A Bellwether for Global Concerns

CHAPTER 9

# Designing Substantive Patent Law: From Life Sciences to Climate Change?

Part IV of the thesis looks at the institution of patents as a bellwether for global problems. This part contains two chapters, Chapters 9 and 10. Chapter 9 uses the metaphor of waves to consider the development of the law. Law-making repeats itself in small waves, and it is concluded that the legislator must bear in mind the past before intervening with new and possibly ill-considered legislative measures. Specifically, there is consideration of how climate change influences patent law. Is it necessary to adopt 'climate-friendly' incentives by changing patent legislation, or is the law able to absorb new challenges?

The purpose of this chapter is to emphasise that we have to learn from previous experience when addressing the issues of patent law and climate change in order to suggest appropriate solutions. If we address the problem while turning a blind eye to former failures we will never succeed. Especially in this area, where traditional environmental law and traditional patent law have had their glory days, there are lessons to be learnt by future lawmakers and policymakers working in the field of climate change.

## 1. Human health, climate change and patent law

A World Health Organization poster states that climate change hurts, not only in terms of environmental and economic damage, but also in terms of human lives.[1] The poster pictures a black foot on a cracked soil background, symbol-

* This Chapter has been published in Nordisk Miljörättslig Tidskrift, 2010, Volume: 2010; Issue: 2; page: 161. It will be further elaborated in Chapter 10.

ising that drought hits harder in the undeveloped world where people are most likely to suffer from climate change.

Human lives are often at stake. That is why we have a huge pharmaceutical and biotech industry serving human needs as well as economic purposes. The history of safeguarding human lives worldwide is nothing to be proud of. In this regard, the patent law system has been identified as a major contributor to continuing health problems in the least developed countries. Our efforts to transfer patent laws to less and least developed countries have diminished the legitimacy of exclusive rights. What will happen when we add food and energy scarcity as well as climate change issues to our present experience of safeguarding humans? To be more specific, will issues such as climate change demand special IPR treatment?

The lifecycle of patents can broadly be divided into two phases. Phase one, the *pre-grant* phase, covers the period from the filing of the patent application to the grant of the patent. Phase two, the *post-grant* phase, covers the period from the grant until the patent term expires (or lapses due to non-payment of fees). This article mainly addresses the *pre-grant* phase, the requirements for patentability and the balancing of public and private interests. However, *post-grant* measures are also important and will be briefly discussed.

## 2. Waves of law

As people get older, most discover that in some ways life repeats itself. Simple things, like the clothes they wore decades ago, come back into fashion and are reused by the new generation. Perhaps as lawyers get older, they too may wonder if law and policymaking repeat themselves and follow such *waves of life*, or '*waves of law*'. Some things never or seldom change, and one may ask if that is a sign of quality. Design classics, such as the Arne Jacobsen Egg Chair, seldom go out of date. Is patent law comparable to designer furniture? If that were the case, patent law would still be useful in a changed context, surviving the challenges of changing society without being swept away.

1. The WHO poster says: 'Climate change hurts, besides environmental and economic damage, the ultimate impact of climate change represents a toll on our most precious resource – human lives and health', see http://www.who.int/globalchange/multimedia/photos_posters/foot_corr.pdf.

Although, the patent regime has expanded rapidly into parts of the world that do not have the infrastructure, capacity, or inventors to benefit from the regime, its design has been absorbed in its traditional form, without fundamental changes. Furthermore, it has expanded to include new subject matter and embrace some of the emerging technologies. This expansion has triggered some changes in the wording of the European patent laws, but not a rethinking of the system. Business has continued as usual.

To elaborate on the *waves of law* metaphor which is the theme of this article, in the following there will be a focus on five developments in the law: the first wave is traditional international and EU environmental law; the second wave is traditional patent law and environmental law; the third wave is patent law and life sciences; the fourth wave is patent law and the biotech experience; and the fifth and final wave is patent law and climate change.

## 2.1. Traditional international and EU environmental law

By the early 1970s environmental law and policy had become a major issue. The UN Conference on the Human Environment, held in Stockholm in 1972, encouraged national and international societies to protect and improve the human environment.[2] The Stockholm Declaration (1972), and later the Brundtland Commission report (1987) on Environment and Development, *Our Common Future*, directed the world's attention to environmental law in its more modern form.[3] By emphasising that pollution is not absorbed by air and water (and is not solved, for example, by building high chimneys or piping waste water into the ocean), and that it is likely to have cross-border effects, the focus on shared responsibility became a priority on the international agenda.[4] In its different forms shared responsibility became an important issue in international environmental policy in the following years. Thus, the 2002 Johannesburg Declaration on Sustainable Development assumed the ex-

2. See New Challenge for the United Nations, OPI/433/-02726, February 1971.
3. The Stockholm Declaration was adopted on 16 June 1972; see Philippe Sand: Principles of international environmental law, p. 40.
4. See Philippe Sand: Principles of international environmental law, Ed. 2, 2003, p. 3; and Rudolf E. Blum, 'The Treat to our Environment and the Protection of Intellectual Property', in Industrial property and copyright: monthly review of the World Intellectual Property Organization, Geneva 1973, p. 243, for a description of the problems to be faced, p. 244. See Staffan Westerlund, 'Theory for Sustainable Development; Towards or Against?', in Hans Chr. Bugge & Christina Voigt (eds.): Sustainable Development in International and National Law, pp. 54-56.

istence of collective responsibility[5] and most recently point 1 of the Copenhagen Accord 2009 emphasised the 'strong political will to urgently combat climate change in accordance with the principle of common but differentiated responsibilities and respective capabilities'.[6]

At the regional level, awareness of transboundary pollution and of the interplay between environmental policies and a competitive European industry was developed in the first place by the European Court of Justice and then by the adoption of the Single European Act in 1987. As a dynamic lawmaker, the European Court of Justice has contributed to the change of priority of environmental interests which was enshrined by the inclusion of Treaty provisions designed to protect the Environment, see the former EC Treaty, Articles 130r, 130s, 130t and Articles 100a(3) and (4).

This development continued. The Maastricht Treaty (1992) was also an important phase, with the Treaty *expressly* incorporating powers aimed at safeguarding the environment as part of its general principles in Articles 2 and 3 (referring to sustainable growth, for example). Furthermore, the Amsterdam Treaty (1997) contributed *by* changing the wording of the sustainability principle (Article 2: promoting a harmonious, balanced and sustainable development of economic activities), and *especially by* the promotion of the integration principle in Article 6.[7] The latest stage in this agenda is the Treaty of Lisbon (2009) where not only has the Charter of Fundamental Rights of the European Union (2000)[8] been replaced from the date of entry into force of the Treaty of Lisbon, but new provisions relating to climate change and energy have been adopted. See Article 191 in the consolidated version of the Treaty on the functioning of the European Union where Union policy on the

5. See The Johannesburg declaration at point 5: 'Accordingly, we assume a collective responsibility to advance and strengthen the interdependent and mutually reinforcing pillars of sustainable development – economic development, social development and environmental protection – at the local, national, regional and global levels', http://www.un.org/esa/sustdev/documents/WSSD_POI_PD/English/POI_PD.htm; and Marie-Claire Cordonier Segger, 'Sustainable development in International Law', in Hans Chr. Bugge & Christina Voigt (eds.): Sustainable Development in International and National Law, p. 87.
6. See Decision -/CP15 at http://www.um.dk/NR/rdonlyres/AA2E8765-4158-4BD1-B29E-1CEDC5BF2CCE/0/COP15CopenhagenAccord.pdf.
7. In its original version the integration principle was inserted in Article 130r(2). See also Jan H. Jans and Hans H.B. Vedder: European Environmental Law, Ed. 3, Ch. 1, development and principles.
8. The principles of a high level of environmental protection, integration and sustainability are in Article 37 of the Charter.

environment shall contribute to promoting measures at international level to deal with regional or worldwide environmental problems, and in particular combating climate change. In the context of Union energy policy Article 194, there is a reference to the need to preserve and improve the environment in a spirit of solidarity between Member States.

The Single European Act introduced Treaty-based environmental principles, such as the principle of a high level of protection, the principle of sustainable development, the integration principle, the precautionary principle, the prevention principle, the source principle, and the polluter pays principle. Since its introduction, scholars of European environmental law have debated the legal quality of the environmental principles of the Single European Act. The Act has not been followed by an immense body of case-law, but the European Court of Justice (ECJ) accepts that some of the principles are legally binding.[9] In particular, Case C-284/95 *Safety Hi-Tech* has been highlighted as interpreting the former Article 130r. In paragraph 37 of its judgment, the ECJ stated that:

> 'in view of the need to strike a balance between certain of the objectives and principles mentioned in Article 130r and of the complexity of the implementation of those criteria, review by the Court must necessarily be limited to the question whether the Council, by adopting the Regulation, committed a manifest error of appraisal regarding the conditions for the application of Article 130r of the Treaty'.

The discussion of environmental principles as legally binding is still ongoing, and it is important to remember this when considering the transfer of environmental principles to patent law.[10] As Treaty-based principles they guide lawmakers, courts and, one can argue, authorities granting authorisations and exclusive rights.[11] In her legal writing *Astrid Epiney* has argued that all scholars, whether or not they support the binding effect of principles, tend to focus on the legal significance of the principles. The purpose of an open-ended

9. See Astrid Epiney, in Richard Macrory (ed.): Environmental Principles, in Reflections on 30 Years of EU Environmental Law – A High Level of Protection?, pp. 20 and 21.
10. See Estelle Derclaye, 'Should patent law help cool the planet? An inquiry from the point of view of environmental law: Part 1', in EIPR. 2009, 31(4), pp. 168-184.
11. Derclaye (n. 10) pp. 227-235; and Derclaye, 'Patent Law's role in the protection of the environment – reassessing patent law and its justifications in the 21st Century', in IIC 2009, 40(3), pp. 249-273.

principle can require only the balancing of different interests.[12] On the basis of the environmental principles we may consider whether the requirement to integrate environmental protection as a general principle of EU law or the precautionary principle can be enforced at the level of the European Patent Organisation. This issue will be dealt with below.

The protection of the environment became subject to multi-level regulation already in 1972, and one can find interplay on environmental protection between lawmakers at the global, regional and national levels. However, in general terms the environmental focus peaked after *the Amsterdam wave*, emphasising the importance of environmental protection by promoting the integration principle as a general Community principle. Since then environmental protection has lived a more retired life, where compliance with and enforcement of European environmental law are issues that remain to be solved.[13] However, climate change has brought the environment more into focus.

### 2.2. Traditional patent law and environmental law

Patent laws are not a new invention. National laws for reward schemes have been an integral part of industrial development since at least 1500.[14] The overall idea of patents has been to balance society's interests in new developments on the one hand and reward for inventors on the other hand. This is known as the reward doctrine.[15] Thus the disclosure of how an invention works is an essential requirement for granting a patent for an invention. Substantive patent law has long been harmonised at the regional and international levels. Since 1972, the European Patent Convention (EPC) has provided a common system of law for the grant of patents, and since 1994 the WTO Agreement on Trade-Related Aspects of Intellectual Property Rights (TRIPS) has provided detailed rules on substantive patent law and the enforcement of IPRs. Thus, as with environmental law, patent law is subject to multi-level

12. Epiney (n. 9) p. 21. For a more deterrent binding effect, see Estelle Derclaye, 'Not only Innovation but also Collaboration, Funding, Goodwill and Commitment: Which role for Patent Laws in Post-Copenhagen Climate Change Action', in 9 J. Marshall Rev. Intell. Prop.L 657 (2010), p. 659.
13. See Peter Pagh, 'The Battle on Environmental Policy Competence', in Richard Macrory (ed.): Reflections on 30 Years of EU Environmental Law. A High Level of Protection?, p. 5.
14. For justifications for patents, see Lionel Bentley and Brad Sherman, Intellectual Property Law, Ed. 3, p. 339; and Nuno Pires de Carvalho: The TRIPS Regime of Patent Rights, Ed, 2, p. 9.
15. Nuno Pires de Carvalho: The TRIPS Regime of Patent Rights, Ed. 3, p. 31.

regulation. Furthermore, the elements of balancing the interests in patent law and the interests of less and least developed countries can be recognised as issues in environmental law which can be classified as issues of social responsibility.

Turning to what may be termed 'traditional' patent law, one will find an attempt to reflect an express environmental focus. In this regard TRIPS contain more modern wording of patent law's traditional clause on exclusions from patentability on ethical grounds. The '*ordre public* and morality' clause, expressing environmental concern, is found as a minimum standard in TRIPS Article 27(2), which refers to the avoidance of 'serious prejudice to the environment'. As such, patent law supports an ethic norm that a law that rewards a polluting inventor is not good law.

The wording of TRIPS Article 27(2) suggests broader protection than Article 53(a) EPC. The wording is as follows:

*TRIPS Art. 27(2):* 'Members may exclude from patentability inventions, the prevention within their territory of the commercial exploitation of which is necessary to protect *ordre public* or morality, including to protect human, animal and plant life or health or to avoid serious prejudice to the environment ...'

*EPC Art. 53(a):* European patents shall not be granted in respect of 'inventions the commercial exploitation of which would be contrary to 'ordre public' or morality; such exploitation shall not be deemed to be so contrary merely because it is prohibited by law or regulation in some or all of the Contracting States.'

Despite the differences in wording, it must be pointed out that EPO case law, through the interpretation of the *ordre public* standard, does make references to the environment.[16] Thus, the *ordre public* standard of Article 53(a) EPC does constitute a bar to patentability for the exploitation of inventions that are *likely to seriously prejudice the environment*.

A relevant question in this regard is the extent of environmental protection taken into account by patent authorities. Do they consider sustainability and the precautionary principle and even make a risk assessment before granting a patent? An answer from the EPO Boards of Appeal in T 356/93, paragraph 18.5 hints at future decisions. The Board held that the *ordre public* standard presupposes that the threat to the environment is sufficiently substantiated at

16. PGS/'Basta T356/93, February 1995, Official Journal of the European Patent Office, 1995, 545, the concept of *ordre public* also encompasses the protection of the environment; see also Ulrich Schatz, 'Patents and morality', in Sigrid Sterckx (ed.): Biotechnology, Patents and Morality, p. 222.

the time when the decision to revoke is taken by the EPO. The Board further emphasised that most of the applicant's arguments were based on the *possible* occurrence of harmful events (e.g. the transformation of crops into weeds, the spread of herbicide-resistant genes to other plants, damage to the ecosystem etc.). However, the documentary evidence must sufficiently substantiate the existence of a threat to the environment such as to represent a bar to patentability under Article 53(a).[17]

Thus it seems that the *precautionary principle* is far from being recognised by the EPO Boards of Appeal. The attitude is as follows:

> 'patent offices are placed at the crossroads between science and public policy. However, at this crossroads patent offices are not alone, but find themselves side-by-side with increasing number of other authorities and bodies, in particular regulatory authorities and bodies, whose function is inter alia to ensure that the exploitation of a given technology, *regardless of whether it is protected by a patent or not*, takes place within the regulatory framework provided by laws, international treaties, administrative provisions'.[18]

The exploitation of a patent has to be assessed within the framework of national laws and regulations on the use of an invention. The Board pointed out that potential risks cannot be anticipated merely on the basis of the disclosure of the invention in the patent specification. The balancing test elaborated in the *Harvard Onco Mouse* case[19] is only applicable when there is an actual risk of harm in the form of 'conclusively documented hazards'.[20]

Is has been suggested that a patent examiner should require an environmental impact assessment from specialised authorities before granting a patent.[21] As the results of such tests are usually not available to patent offices during the processing of a patent application, this would doubtless extend the period of patent application proceedings. Today, the actual approval by the authorities is often obtained *after* the grant of the patent. If we start requiring environmental impact assessments prior to the granting of a patent, we start mixing the exploitation of a patent with the grant, and too many grey areas

17. See T356/93 Plant cells/PGS; and Tine Sommer, 'Interpreting ordre public and morality in a patent law context: Which is the correct approach?', in Bio-Science Law Review, 2006/2007, 2 at p. 69.
18. See T 356/93 at points 18.2 to 18.4 of the decision; and Sommer (n. 17) p. 69.
19. Harvard/Oncomouse [1990] O.J. EPO 476.
20. See Ulrich Schatz, 'Patents and morality', in Sigrid Sterckx (ed.): Biotechnology, Patents and Morality, p. 222.
21. See Estelle Derclaye, 'Patent Law's role in the protection of the environment – reassessing patent law and its justifications in the 21st Century', in IIC 2009, 40(3), p. 258.

will distort the objective of the patent system, subjecting it to even further problems of loss of legitimacy. Article 53(a) does not provide a legal basis for controlling and preventing technological hazards.[22]

Following the discussion of environmental principles, one can hardly blame patent examiners or the Boards of Appeal if they do not rely on EU environmental principles which still are associated with legal uncertainty in their own legal system. As to the EPO, the EU has no legal power or authority over it.[23] Yet, the principle of sustainability in particular is not only a general principle of the EU, it is recognised in international conventions. This principle has the global aim ensuring that technological development does not bring about the end of the world and/or that the legal protection of developments takes social responsibility into account. The author's opinion is that these principles are very useful in the post-grant phase, whereas the pre-grant phase must keep patent law to the straight and narrow, separating grant from exploitation.

## 2.3. Patent law and life sciences

The TRIPS Agreement's explicit reference to serious prejudice to the environment has been less controversial than its relation *to life saving medicine*.[24] We all know by now that the patent system *as such* is not transferable to less and least developed countries. The obvious question that scholars ask in this regard is whether the Doha process on pharmaceuticals has lessons for climate change. The focus is now on *post-grant measures* and there is an issue of accessibility, ensuring that IPRs do not unjustly prevent access to technologies.[25] One could say that post-grant measures are all about to rebalancing the system and achieving global social goals.[26]

22. Schatz (n. 20) p. 222.
23. See Tine Sommer, 'Interpreting ordre public and morality in a patent law context: Which is the correct approach?', in Bio-Science Law Review, 2006/2007, (2).
24. See Frederick M. Abbott & Graham Dukes: Global Pharmaceutical Policy, Ensuring Medicines for Tomorrow's World, 2009, EE; Thomas Cottier, 'The Doha Waiver and Its Effects on the Nature of the TRIPS System and on Competition Law, The Impact on Human Rights', in Inge Govaere & Hans Ullrich (eds): Intellectual Property, Public Policy and International Trade, at p. 175: 'a permanent regime of *lex specialis* on access to essential drugs in the TRIPS system of intellectual property protection'.
25. See Meir Perez Pugatch, 'The process of intellectual property policy-making in the 21st century – shifting from general welfare models to a multidimensional one', in EIPR 2009 31 (6), p. 311.
26. See e.g. Brad Sherman, 'Intellectual property and environmental protection', in EIPR 1991, 13(5), pp. 165-170 at p. 172.

Facilitating access to essential medicines in developing countries is an obvious starting point for assessing access by developing countries to environmentally sound technologies (ESTs). The Doha waiver enables a pharmaceutical product produced or imported under a compulsory licence in a member country to be exported to the markets of those other less or least developed countries that share the same health problem.[27] Thus, finding solutions to the barrier of *insufficient capacity* for developing countries to make effective use of compulsory licences, as in the case of HIV and malaria, can now be considered for environmentally sound technologies.[28]

Comparing the UN Framework Convention on Climate Change (UNFCCC, 1992) to the Doha agenda appears to be straightforward. The distinction of less and least developed countries in Doha terms and UN Annex II Parties and non-Annex I Parties guides us in the same direction, whether we talk about pharmaceuticals or combating climate change. The principle of common but differentiated responsibility is behind both systems in terms of technology transfer and financial resources. However, as emphasised by several scholars, it is questionable whether the feasibility of a waiver for the transfer of environmentally sound technologies is similar to that for public health. And unfortunately, the Doha process has only influenced the global pharmaceutical industry at the margin, so its structure and behaviour has been largely unaffected.[29] According to *Frederick Abbott*, at its best the Doha Declaration on the TRIPS Agreement and public health process has positively influenced governments and multilateral organisations towards taking greater responsibility for ensuring that populations and developing countries have adequate access to medicines.[30]

*Frederick M. Abbott* points to one positive outcome of the WTO public health negotiations, which is that a number of Public Private Partnerships

27. See Implementation of paragraph 6 of the Doha Declaration on the TRIPS Agreement and public health, Decision of the General Council of 30 August 2003, cf. Article 6 (i). Article 7, first sentence, adds that Members recognise the desirability of promoting the transfer of technology and capacity building in the pharmaceutical sector in order to overcome the problem identified in paragraph 6 of the Declaration.
28. See Frederick M. Abbott, 'Innovation and Technology Transfer to Address Climate Change: Lessons from the Global Debate on Intellectual Property and Public Health', ICTSD, Issue paper No 24.
29. Abbott (n. 28) p. 14.
30. The waiver is still on the agenda. The TRIPS Council Meeting in June 2010 is set to review the little-used waiver, and boost availability of affordable medicines in developing countries, see http://www.ip-watch.org/weblog/2010/05/31/drug-access-waiver-debate-looms-for-june-trips-council-meeting/.

(PPPs) have been formed on drugs for neglected diseases, e.g. the Drug for Neglected Diseases initiative (DNDi), the US President's Emergency Program for African Relief (PEPFAR), and the WHO UNITAID scaling up access to treatment for HIV/AIDS, malaria and tuberculosis.[31] There are now also orphan drugs arrangements in the EU for rarer diseases, such as Gaucher's disease, Fabry's disease.[32]

Perhaps one of the lessons we can learn from the experience with *life saving medicines* for neglected diseases is to use Public Private Partnerships for *adaptation technologies* directed to the least developed countries.

## 2.4. Patent law and the biotech experience

Biotechnology has had an influence on several disciplines of law. As examined by *Han Somsen*, there are lessons to be learnt from environmental law, such as issues of risk regulation, where the precautionary principle has been acknowledged outside EU environmental law, as in the example of food safety regulation.[33] Risk assessment and the precautionary principle are appropriate as bases for voicing opposition on behalf of the environment and future generations when these are at stake.[34] This is also the approach of legitimacy.

Turning to patent law, we may consider Han Somsen's concept of 'genetic governance'.[35] The European Community had severe difficulties with the adoption of Directive 98/44/EC (the Biotech Directive). There was a 10 year struggle before the Directive was finally adopted in 1998. Patenting life forms caused much debate, mainly by establishing a regime without field restrictions. The biotech revolution has left its own footprint on substantive patent law, and this has led to the European Community making changes to the established EPO law. The author's view is that this is where we have learnt our lesson: Patent law flexibility and dynamic is not best safeguarded by detailed rules.

How can we cope with new patentable subject matter with patentability requirements that have been established at a time when the emerging technology (in this case biotech) was immature and when the political environment is reluctant and resistant to change, while the industrial lobby is positive

31. Abbott (n. 28) p. 13.
32. Abbott & Dukes (n. 24) p. 133 f.
33. See Han Somsen, 'Some reflections on EU Biotechnology Regulation', in Richard Macrory (ed.): Reflections on 30 Years of EU Environmental Law. A High Level of Protection?, p. 329.
34. Somsen (n. 33) p. 333.
35. Somsen (n. 33) p. 335.

about change? Now, as result of *a political compromise* we are locked into outdated rules. For example, there are rules that have created a phoney doctrine of isolation. Thus, Directive 98/44/EC Article 3(2), states that:

'Biological material which is isolated from its natural environment or produced by means of a technical process may be the subject of an invention even if it previously occurred in nature'.

This doctrine governs biological material in general, as well as human genes; see Article 5(2):

'An element isolated from the human body ... may constitute a patentable invention, even if the structure of that element is identical to that of a natural element.'

Comparing the patentability requirements to those of traditional patent law, the requirement for industrial application is the only criterion that has adopted a kind of field restriction, if it is to be interpreted as a strict requirement for industrial application.[36] Directive 98/44/EC, Article 5(3) states:

'The industrial application of a sequence or a partial sequence of a gene must be disclosed in the patent application'

While the focus of the debate was on human genes, the Directive was adopted without changes to the patentability of plants and animals. Plant and animal varieties are excluded from patentability if the technical feasibility of the invention is confined to a particular plant or animal variety (Directive 98/44/EC, Article 4(2)). Essentially biological processes for the production of plants or animals are excluded. Essentially biological processes are defined in the Directive, which is not the case in traditional patent law. However this definition is more open-ended than precise.[37]

Finally, the *worst outcome of law making* is the EU system creating its own detailed moral standard for Europe. Article 6(1) of the Directive repeats EPC Article 53(a), and Article 6(2) is as follows:

'On the basis of paragraph 1, the following, in particular, shall be considered unpatentable:

36. For this discussion see, Tine Sommer, 'Scope of Gene Patent Protection and the TRIPS-Agreement – An Exclusively Nondiscriminatory Approach?', in in 38, IIC 1/2007.
37. See Tine Sommer, 'Patenting the Animal Kingdom? From Cross-Breeding to Genetic Make-up and Biomedical Research', in 38, *IIC* 2/2008, p. 149 ff.

(a) processes for **cloning** human beings;
(b) processes for **modifying** the germ line genetic identity of human beings;
(c) uses of **human embryos** for industrial or commercial purposes;
(d) processes for **modifying the genetic identity of animals** which are likely to cause them suffering without any substantial medical benefit to man or animal, and also animals resulting from such processes' (emphasis added).

Given the heading of Article 6(2), it is obvious that one or more areas are missing. The non-exhaustive list provides examples concerning humans or animals used for human purposes (mainly) but there is not a word on environmental protection.[38] Maybe the EC legislator found this was already covered by the traditional clause (Article 6(1)) or perhaps the focus was more on human genes – 'genetic governance' in the term of Han Somsen.

## 2.5. Patent law and climate change

Climate change has been on the EU agenda for more than 20 years, and the EU is a major player on the international scene.[39] However, as pointed out by *Streck and Freestone*, the 'implementation gap', which is always a big problem of European environmental policy, is particularly obvious in the context of climate change.[40]

With this background to patent law, the question is whether a change in the regulatory environment is necessary. Can patent law repair what environmental regulation cannot? Are we explicitly to make our patent law system ensure the reduction of green house gas emissions? Or should we emphasise that patent law favours environmentally sound technologies, or at least add environmental damage to our detailed moral standard? One could ask whether patent law is neglecting its social responsibility if it does not act to deal with climate change.

Patent law must develop in line with developments in society, and the surrounding legal society often collides with patent law, requiring the outdated

38. The recital 10 in the preamble to the Directive states that regard should be had to the potential of the development of biotechnology for the environment and in particular the utility of this technology for the development of methods of cultivation which are less polluting and more economical in their use of ground; whereas the patent system should be used to encourage research into, and the application of, such processes. Furthermore recital 11 addresses developing country issues, and recital 36 refers to TRIPS Article 27(2).
39. See Charlotte Streck & David Freestone, 'The EU and Climate Change', in Richard Macrory (ed.): *Reflections on 30 Years of EU Environmental Law. A High Level of Protection*?, p. 87.
40. Streck & Freestone (n. 39) p. 88.

regime to hang on for dear life. But if *real life* requires the adoption of new rules, the experience from the biotech adventure warns us that EC interference with the current state of patent law is not to be wished for. The provisions on isolated DNA sequences as patentable subject matter have revealed legal uncertainty and the different approaches of the EU Member States.[41] France, Italy and Germany have a strict requirement for industrial application of gene sequences, and France also prohibits patents on sequences themselves. The EU Commission has been very reluctant to adopt a position and as a result it has been left to national courts to interpret. The legal approach to this by the EU has been a failure, and the EPO and the EU are stuck with the outdated and mistaken creation of patent law. The USPTO 2000 Utility Guidelines as a supplement to the US statute, and the case law approach in the USA seems better able to adapt smoothly to shifting directions as to the patentability of gene technology on different grounds (35 US §§ 101, 102 or 103). Lately, in a case before the United States District Court – Southern District of New York[42] in March 2010, the claimed isolated DNA was not markedly different from DNA existing in nature, and was therefore not patentable under 35 U.S.C § 101. Detailed regulation of the patentability of emerging technologies is not an appropriate solution, but the case law approach adapts smoothly to developments. If pre-grant conditions are concise, this may also affect blocking patents as a narrower scope of protection will diminish the blocking effect of a patent.

So far patent law has collided with trade aspects of IPRs, environmental concerns, access to pharmaceuticals for least developed countries, biotech and ethical issues. What has been learnt so far is that when global concerns are at stake patent law often lacks legitimacy. These collisions have had the major impact on the patent regime that it has lost its legitimacy, stuck between efforts to adopt new approaches and failing to do so. The outcome has been devastating to the scope of protection and quality of the patents issued. *This lesson is worrisome* when we now return to climate change. It seems that exactly the same components are present – law and policymaking repeats itself. And one can ask, do we need new rules favouring green inventions and creating uncertainty? The answer depends on which phase we address: the pre-grant phase or the post-grant phase.

41. For this discussion see, Tine Sommer, 'Scope of Gene Patent Protection and the TRIPS-Agreement – An Exclusively Nondiscriminatory Approach?', in 38, IIC 1/2007.
42. See Case 1:09-cv-04515-RWS, document 255, filed 3/29/2010: American Civil Liberties Union Foundation and Public Patent Foundation.

To answer the question regarding the *pre-grant phase* it is necessary to focus on the nature of the environmentally sound technologies which ask for legal protection.[43] When referring to environmentally sound technologies, environmental lawyers think of adaptation and mitigation technologies. The former includes regional and local climate modelling, early warning, coastal zone management, water resources, agriculture and public health. The latter, mitigation technologies, includes energy supply, carbon dioxide capture and storage, and technologies for the reduction of green house gas emissions.[44] Today *feasible technologies are available*, but not all are commercially competitive without government or other support.[45] We have to identify the key barriers to more rapid development and deployment of *state-of-the-art* technologies and mechanisms that can accelerate these processes.[46]

Patent law today already embraces improvements and new technologies on turbine blades,[47] different sorts of clean technologies,[48] climate-tolerant crops for developing countries such as drought resistant crops, flood resistance, salt resistance, sea grass and other crops for bio fuels can be protected. Clean farm animals are being developed, such as pigs (not a contradiction but a difference in genetic make-up); enviro-pigs expel up to 60 per cent less phosphorus than their non-transgenic counterparts.[49] A naked or featherless chicken is being genetically engineered, saving poultry farmers large amounts of money on ventilation to prevent their chickens from overheating.[50] Isolated and clean human genes may also be patented if we find solutions for enhancing the human metabolism in order to make humans eat less or perhaps dislike meat.

43. See The new 'tagging clean energy' database created as part of a joint study on IP and climate change mitigation technologies carried out by the United Nations Environment Programme (UNEP), the EPO and the International Centre for Trade and Sustainable Development (ICTSD):
http://www.epo.org/topics/news/2010/20100611.html.
44. See Background paper: Climate Change: Technology Development and Technology Transfer, Beijing High-level Conference on Climate Change: Technology Development and Technology Transfer, Beijing, China, 7-8 November 2008, UN Department of Economic and Social Affairs, pp. vii, viii, and 2.
45. Background paper (n. 44) p. 2 f.
46. Background paper (n. 44)
47. See for example US 2010/0135806 A1: a wind turbine blade includes a plurality fins, each fin rotatably-joined to a tip of the blade.
48. Cf. http://greenenergyreporter.com/2010/06/8696/.
49. Sommer (n. 37).
50. http://www.nytimes.com/2002/05/24/world/rehovot-journal-cluck-cluck-chickens-in-their-birthday-suits.html. Although due to prior art it is not patentable.

Against this background is it really necessary to regulate patent law in detail, given that patentable subject matter today generally covers environmentally sound technologies? There are several options for amending patent law, however taking previous experience into account, patent law must be kept to the straight and narrow.

One option is to insert environmental principles and objectives in patent law conventions, either as mission statements or terms in the preamble. The *mission statement option* is a preamble text in European and international patent conventions referring to internationally recognised environmental principles which, as a measure of global legitimacy, patent law has to respect.[51] The preambles are to be respected by signatory states as parties to the conventions. Starting with environmental principles, the most suitable principles are the principle of sustainability and the precautionary principle. The sustainability principle is globally recognised and it seems to be behind almost every piece of environmental legislation or climate change regulation today. As such, the principle is very suitable for being inserted in a preamble giving a political signal that patent law aims to promote technologies that can preserve the world for future generations. The precautionary principle is more connected to risk assessments, and perhaps this is most useful as a guiding principle for granting authorisations. However, one way to use the principle and that is for regulating the scope of protection.[52] If patent examiners are faced with applications on new emerging technologies, it might be appropriate to limit the scope of protection, and not to grant a product patent. Climate change has its own principles. Apart from the principles of sustainability and the precautionary principle, the most commonly referred to principle is that of common but differentiated responsibility. In his opening speech at the conference on 'Regulation Global Concern' in Sandbjerg, May 2010, *Thomas Cottier* addressed climate change as a *common concern of mankind.*[53] He

51. See in general a proposal for a mission statement as a clear statement of the purpose of the patent system that might facilitate decisions on whether to include new subject matter under the patent system, and might be a valuable source of reference in court cases on patent related matters, STOA, Policy Options for the Improvement of the European Patent System (IP/A/STOA/FWC/2005-28/SC16 at page 33.
52. See also 'The future of the patent system', The Danish Board of Technology, page 22 http://www.tekno.dk/subpage.php3?article=1132&toppic=kategori11&language=uk.
53. UN GA Resolution 43/53, 1988: The UN Framework Convention on Climate Change, UNFCCC, in its preamble stated that changes in the Earth's Climate are a common concern of humankind, and the Convention on Biological Diversity, CBD, stated that the conservation of biological diversity is a common concern of humankind. See Thomas Cottier and Sofya Matteotti-Berkutova, International environmen-

pointed out that the principle is treaty-based and defined. Referring to the concept of a common concern of mankind in patent convention preambles may prompt awareness of the importance of a regulated *post-grant phase* such as a waiver-option for environmentally sound adaptation technologies for least developed countries.

Thomas Cottier has also written about the Doha Waiver and the amendment of TRIPS as introducing 'a new dimension to intellectual property and competition law: they reflect concerns for human rights, in particular the right to health and right to life ...The right to health and life entail components of distributional justice which intellectual property and competition laws per se should support and assist in realization'.[54]

The *option of changing the wording* can work in several ways. Patent law provisions on patentability can favour inventions of environmentally sound technologies or can fast-track processing of such patent applications.[55] The fast track option has gained ground since 2009. The United Kingdom and the USA have introduced a fast track system for green technology. The former has an accelerated procedure called the 'Green Channel' and the latter has a more detailed system.[56] This is a burden on the patent examiner but nothing indicates that a petty patent is the outcome. Some also argue that the term of protection for environmentally sound technologies ought to be extended or the fees reduced.[57]

The open question is how to make environmental impact assessments of a technology during the processing of a patent application without leaving patent examiners behind. *Estelle Derclaye* has argued that patentees ought to

tal law and the evolving concept of 'common concern of mankind' in Thomas Cottier, Olga Nartova & Sadeq Z. Bigdeli (eds.): International Trade Regulation and the Mitigation of Climate Change, p. 38.

54. Thomas Cottier, 'The Doha Waiver and Its Effects on the Nature of the TRIPS System and on Competition Law, The Impact on Human Rights', in Inge Govaere & Hans Ullrich (eds.): Intellectual Property, Public Policy and International Trade, p. 198.
55. See the USPTO pilot project, http://www.uspto.gov/patents/init_events/green_tech.jsp. For the UK, see http://www.ipo.gov.uk/pro-types/pro-patent/p-law/p-pn/p-pn-green.htm.
56. See Estelle Derclaye, 'Not only Innovation but also Collaboration, Funding, Goodwill and Commitment: Which role for Patent Laws in Post-Copenhagen Climate Change Action', in 9 J. Marshall Rev. Intell. Prop.L 657 (2010), pp. 662-661.
57. Derclaye (n. 10) pp. 227-235; and Derclaye, 'Patent Law's role in the protection of the environment – reassessing patent law and its justifications in the 21st Century', in IIC 2009, 40(3), pp. 249-273.

calculate the carbon footprint of their products.[58] Legislators just have to decide a standard, as in environmental law, and make it applicable to patents. She argues that for products it will be relatively simple, such as a reduction of emissions below a threshold in order to get favoured treatment.[59] One very promising option is to ask the patent applicant to disclose information about the environmental impact voluntarily. The only effect of this requirement would be at the infringement stage or in opposition proceedings.[60] This is a familiar approach, recalling the disclosure of origin of genetic resources.[61]

The *moral standard option* merely requires the addition of a new subparagraph (e) to the list in Directive 98/44/EC, Article 6(2), covering processes that are likely to cause harm to the environment. Adding the likelihood of harm relates to the precautionary principle.

## 3. Conclusion

As emphasised in the first section of this article, it is necessary to learn the lessons from environmental and biotechnology regulations, as well as from patent law when designing regulations on climate change. It is useless to repeat environmental principles that no one relies on and think that this is how to solve climate change problems. It is also useless simply to amend patent law with highly specific and detailed rules on environmentally sound technologies. As the biotech experience must have taught us, this may only lead to confusion and disharmony. The biotech lesson in particular, with its mix of competences between the EU and the EPO, is not the way forward. Until we adopt an EU patent, with its own patent Court of Justice, this area must be settled at the level of the EPO or the WTO.

To return to the starting point, it can be concluded that the shape of patent law is still good law. We can have a dynamic patent law system by keeping patent law to the straight and narrow. Patent law can adapt to the changing

58. EIPR/IIC. See also Derclaye (n. 56) p. 660.
59. Derclaye (n. 56) p. 660.
60. See Carolyn Abbott & David Booton, 'Using Patent Law's teaching Function to introduce an Environmental Ethic into the Process of Technological Innovation', in 21 Geo. Int'l Envtl. L. Rev. 219 (2009), pp. 248-249, cited in Estelle Derclaye, 'Not only Innovation but also Collaboration, Funding, Goodwill and Commitment: Which role for Patent Laws in Post-Copenhagen Climate Change Action', in 9 J. Marshall Rev. Intell. Prop.L 657 (2010), p. 660.
61. See e.g. the preamble to Directive 98/44/EC, recital 27.

nature and pace of technology if we keep patentability standards straight and thereby restore its global legitimacy. The case law approach can be guided by terms in preambles, but to suggest that patent examiners should ask for technical advice or should address assessments long before the authorisation authorities are capable of doing so, is perhaps not the most appropriate solution. Patent law can consider the *common concern of mankind.* But amending pre-grant patent law cannot solve the emerging problems of lack of capacity and more urgent concerns in the least developed countries. In other words, patent law cannot save us from climate change.

We have to be open to new methods, other market-based instruments and guidance by means of taxation, fines and standards. Environmental rules are in themselves indirect incentives to promote clean technologies, and adding economic incentives may be more effective than amending *pre-grant* patent law. Given market forces, environmental regulation has an impact on the kinds of inventions seeking protection, thereby reducing the global carbon footprint without changing the wording of patent law.

Why not consider a system awarding firms with prizes for technology transfer arrangements? The emphasis on climate change must put severe pressure on solving emerging problems in the least developed countries. To add common concern principles to patent convention preambles would reflect the social responsibility of patent law. Inventions benefiting a cleaner environment also benefit future generations and must be disseminated throughout the world. Efforts must be made to focus on adaptation technologies for the less developed world where people are most likely to suffer from.

CHAPTER 10

# Patent Law: A Bellwether for Global Concerns

The thesis concludes with Chapter 10, which has a summary of the findings and a proposal to protect the dynamism and flexibility of patent law. In particular, it is suggested that multilevel regulation and the pluralist approach to the sources of law should be united through the establishment of a *global public order* for patent law, based on the principle known in international environmental law of *the common concern of mankind.*

## 1. Facing backwards

This thesis has focused on patent law, and has to some extent touched on environmental law, trade law and EU law in a diverse fashion. The overall inquiry has been based on the title of the thesis: 'Can law make life (too) simple?'. It addresses human gene patent protection, and patenting within the animal kingdom and in the area of climate change, with an initial agenda of examining the expansion of the patentable subject matter in these fields. Several questions have guided the inquiry, such as: *whether* the ban on patenting 'essentially biological processes for the production of animals' can apply to human gene patenting; *whether* it is possible to regulate biological complexity in a detailed and static legal framework without compromising future developments beyond protein therapeutics and even beyond the multi-functionality of RNAi molecules; and finally, *whether* dealing with climate change may require a redesign of patent law.

The purpose of the thesis has been to use patent law as an example of multilevel regulation for dealing with the common concerns of mankind – concerns that ought to be separated from the economic values of trade and patent laws and reflected in a global norm for patent law governance, as an obligation *erga omnes*. Under this heading the thesis also aims to illustrate how the EU is expanding its powers, and to question the benefits of mixing legal powers at the regional level, with particular regard to the proposal for enhanced cooperation on the EU patent, which may further divide Europe. At the regional level the EU has complicated the multilevel approach by continuing to interfere with the competences of the European Patent Organisation (EPO) by establishing an EU patent order. The thesis discusses the disadvantages of tailor-made patent law within the confusing regional regulatory framework established by Directive 98/44/EC on the legal protection of biotechnological inventions. The question 'Can law makes life (too) simple?' thus also refers to the mixing of competences, which may have seemed unproblematic in principle from the perspective of the EU legislators, but life is seldom that simple in practice. The thesis has identified several conflicts, such as over the European definition of an embryo, the interpretation of Article 6 of Directive 98/44/EC (Chapter 4), purpose-bound protection and the implementation of the Directive in the Member States (Chapter 5), and from the EPO perspective the mix of interpretations of the Implementing Rules of the European Patent Convention (EPC) and EPC Article 164 (Chapters 4 and 8), and finally the premature proposal for a two-speed EU Patent in this confusing legal environment (Chapter 3).

The thesis is divided into four parts. Part I looks at the multilevel regulation and new technologies, and sets the overall framework; Part II considers the human genome; Part III looks at the animal kingdom; and Part IV reviews patent law which is a prime global concern, with the aim of giving concrete examples of Europe's mistaken approach to tailor-making patent law. Parts II, III and IV all deal with patents, from human genes, to animal and plant genes. However, on the issue of climate change, the question of the stimulation of green technologies has prompted thoughts of tailor-made patent law for green technologies, which are not necessarily gene based.

The analysis of the US patent system was initially chosen in order to determine the scope of protection of gene patents and to compare the scopes of protection of the US Patent and Trademark Office (USPTO) and the EPO. However, in carrying out the review it became clear that it would be more than difficult and extremely time-consuming to assess the general differences, and the analysis was redirected to a focus on the technology claimed. At the start of the case study, Europe and the USA had patent practices that were very much alike (allowing the patenting of DNA sequences) and the intention with the study was to focus on the differences between the systems. In the legal literature it has always been claimed that the USA is less restrictive (e.g. patentable subject matter) in its patent practice than Europe. However, this picture has changed. Not only has the analysis of patent practice turned out to be similar for the two systems, but the EU has started to provide Europe with tailor-made static provisions, whereas the US Supreme Court has started to be sensitive in redefining the scope of patentable subject matter. This has taken place over decades, and it is leaving Europe behind. It is time to change the current thinking of the EU legislator, and to demand a rethink of the European patent system and to push it back to where it started more than three decades ago.[1] Enhancing the role of the European Patent Organisation will give Europe greater patent legitimacy covering more countries than the EU Member States. Thus the agreement of 3 May 2012 between the EPO and WIPO on enhanced cooperation is more likely to ensure a coherent international patent system in support of innovation around the globe.[2]

### 1.1. Lessons from the US Supreme Court and the Court of Justice of the European Union (CJEU)

The pending *AMP v Myriad* case is an example showing current status of gene patents in the USA. As mentioned in Chapter 5.3.1.2, the current view of isolated DNA claims is that they are patentable subject matter. The majority opinion (Judge Lourie and Judge Moore) stated that the isolated DNA in the case are drawn to patentable subject matter, being molecules that are markedly different (with distinctive characteristics) from naturally occurring DNA (separate chemical entities).[3]

One of the features that Judge Lourie addressed was the difference between isolated and naturally occurring DNA molecules, since the former has been cleaved or synthesised so as to consist of just a fraction of a naturally occurring DNA molecule. The argument was that the chemical bonds of an

1. In a letter to the Members of the European Council of 26 April 2012, the President of the European Council, Herman van Rompuy, urged the delegations in the Council to show a greater spirit of compromise on several matters. As for the Unitary Patent, he emphasised that this deal is needed now due to its crucial importance for innovation and growth (see EUCO 72/12, Presse 180, PR PCE 64). It is an open question if the patent package will be accepted by the Parliament.
2. See press release (PR/2012/709) of 3 May 2012. Alexander Stack, '*International Patent Law*, Cooperation, Harmonisation and Institutional Analysis of WIPO and the WTO, 2011, p. 110.
3. The claims at issue fall into three categories: 1) claims that cover the isolated BRCA genes (claim 1 of the '282 patent, claim 1 of the '473 patent, and claims 1 and 6 of the '492 patent); 2) claims that cover only the BRCA cDNA (claims 2 and 7 of the '282 patent and claim 7 of the '492 patent); and 3) claims that cover portions of the BRCA genes and cDNA as small as 15 nucleotides long (claims 5 and 6 of the '282 patent), see Document 2010-1406, Bryson, Circuit Judge, concurring-in-part and dissenting in part I, p. 4.

isolated DNA molecule differ[4] from those of a non-isolated sequence in the chromosome (without introns and with different ends).[5] The isolation thus creates a distinctly non-natural molecule. In his argumentation, Judge Lourie emphasises that DNA is a chemical polymer, comparing it with nylon. He characterises the bases (A, C, G and T as described in Chapter 2.2.1.a) as chemically bonded to each other and states that, whether carried out by chemical or biological means, the process of polymerisation of the monomer units (bases working together as a code) results in a new molecule.

Moore (and Lourie) also argues for a scale perspective. Instead of a string of four [or five] nucleotides, the chromosome is millions of nucleotides; instead of a two-monomer molecule, the isolated molecules claimed in this case range from 15 nucleotides to thousands (or tens of thousands) of nucleotides. His argument is that just because the same series of letters appears in both the chromosome and in an isolated DNA sequence, this does not mean they are the same molecule. While the isolated DNA molecules claimed in this case are undoubtedly inspired by the corresponding naturally occurring sequence on the chromosome, man must create these isolated DNA molecules.[6]

Thus, the isolation makes a difference due to the change in the chemical bonds in the isolated DNA, as well as other differences such as being a part of a chromosome or as short as fifteen nucleotides (the *BRCA1* cDNA sequence has approximately six thousand nucleotides).[7]

Judges Lourie, Moore and Bryson all stated that cDNA is eligible for patenting. cDNA, has a unique sequence of DNA bases (A, C, G, T) which is not present in nature. The naturally occurring gene sequence includes both introns (which are removed) and exons (which are included in the mature RNA). The cDNA sequences, which are complementary to the mature RNA, do not include the introns.

Judge Moore did not exactly support Judge Lourie's testing of covalent bonds and she did express doubts about patenting the isolated DNA. How-

4. Different properties from the parent molecule from which it is derived, e.g. the bonds connected to a phosphate, or a hydroxyl or from sugar via a phosphodiester bond, to a phosphate. See Document 2010-1406, Lourie, Circuit Judge, A, ii) p. 42 ff.
5. This argument disregards the fact that some applicants claim the genomic sequence (comprising both exon and intron) and some cDNA (~mRNA).
6. See Document 2010-1406, Moore, Circuit Judge, concurring-in-part, II, p. 11 ff.; and Lourie, Circuit Judge, A, ii) p. 42 ff.
7. See *Association for Molecular Pathology et al. v United States and Trademark Office et al.*, 702 F. Suppl.2d. 181, appeal docketed No 2010-1406 (Fed. Cir. 22 June 2010).

ever, she seems to have argued that the only option was to vote in favour of patentability since a change of practice had to come from Congress.[8]

Judge Bryson did not have policy concerns nor did he support the covalent bond test. According to Judge Bryson it was clear that the claims comprised the genes themselves, which appear in nature on the chromosomes of living human beings. Bryson stated that the only material change made to those genes from their natural state is the change that is necessarily incidental to the extraction of the genes from the environment in which they are found in nature. Thus the isolated genes are not materially different from the native genes. The magic of the chemical bond did not convince Bryson. On the contrary, isolated genes are not new molecules. Native *BRCA* genes are chemically bonded to other genes and histone proteins, but cleaving those bonds to isolate the *BRCA* genes does not turn the isolated genes into 'different materials.' Bryson stated that a chemical bond is merely a force between two atoms or groups of atoms strong enough 'to make it convenient for the chemist to consider [the aggregate] as an independent molecular species'.[9]

Judge Bryson found that breaking the chemical bonds that hold a gene in place in the body in order to isolate it does not change the genetic coding sequence. The '282 patent, claim 1, comprising all isolated DNAs coding for the BRCA1 protein, covers a truly immense range of substances from the cDNA that is 5,914 nucleotides long to the isolated gene that contains more than 120,000 nucleotides. And the patent does not define the upper end of that range because it does not identify a unique nucleotide sequence for the 120,000 nucleotide-long isolated *BRCA1* gene.[10] The isolation, Bryson argued, takes place according to nature's predefined boundaries, i.e., at points that preserve the ability of the gene to express the protein for which it codes.

The issue of utility was also discussed by Bryson, who stated that the identity of the function in the isolated gene is the key to its value, and Myriad could have claimed more narrowly to achieve the utility attached to the segments of cDNA (e.g. probes and primers). He then stated that many of the

8. See Document 2010-1406, Moore, Circuit Judge, concurring-in-part, IV, p. 20 ff. She concluded her reasoning at p. 31, as follows: 'The patents in this case might well deserve to be excluded from the patent system, but that is a debate for Congress to resolve. I therefore decline to extend the 'laws of nature' exception to include isolated DNA sequences'. See also Lourie, Circuit Judge, pp. 47-48.
9. See Document 2010-1406, Bryson, Circuit Judge, concurring-in-part and dissenting in part, I, B, p. 7 ff. (*AMP v USTPO*, 702 F. Suppl.2d. 181).
10. See Document 2010-1406, Bryson, Circuit Judge, concurring-in-part and dissenting in part, I, B, p. 9 ff. (*AMP v USTPO*, 702 F. Suppl.2d. 181).

cDNA segments will not work as primers since some will be too long or too short, and some palindromic and fold in on themselves. Patentability is thus connected to an identified subset of the segments that work as primers, with 'markedly different characteristics from any found in nature and having the potential for significant utility'.[11]

The decision has been vacated by the Supreme Court, and the Federal Circuit has been ordered to reconsider its decision in the light of the recent Supreme Court decision in *Mayo v Prometheus*, as described in Chapter 5.3.1.[12] The outcome of that case might be that the isolated DNA is not patentable either because the information in the DNA represents a law of nature, or the DNA is a natural phenomenon, and or the isolation process is already known.[13] With such an outcome, in addressing the patentable subject matter of DNA molecules in new ways it will be relevant to consider Bryson's comments on utility, which will (still) be the important parameter for future assessments of the patentability of genetic products.

*Lessons from the Court of Justice of the European Union (CJEU)*

The lessons from the CJEU are of very little value to the discussion of DNA molecules as patentable subject matter. Furthermore, the wording of Directive 98/44/EC is a product of misleading legislation. Not only can gene sequences be patented, but the isolation doctrine has been qualified and upgraded to a criterion for the assessment of patentability. Indeed, for better or for worse there is a specific EU definition for embryonic stem cells as described in Chapter 4 (Case C-34/10 *Oliver Brüstle*), but the main CJEU contribution has been on the scope of protection, limiting DNA patents to perhaps purpose-bound protection; see Chapter 5 (Case C-428/08 *Monsanto*). The *Monsanto* outcome may have an impact on the ongoing debate in Europe, together with Bryson's argumentation referred to above, especially in the case of a renewed Federal Circuit assessment of DNA molecules as patentable subject matter. Once there is a new decision on genes as patentable subject matter from the other side of the Atlantic Ocean, the European assessment of purpose-bound protection and the US utility requirement may

11. See Document 2010-1406, Bryson, Circuit Judge, concurring-in-part and dissenting in part, II, p. 15 ff. (*AMP v USTPO*, 702 F. Suppl.2d. 181).
12. http://www.patentlyo.com/patent/2012/03/gene-patents-amp-v-myriad-genetics.html
13. http://www.patentlyo.com/patent/2012/03/gene-patents-amp-v-myriad-genetics.html. Natural Process and known Elements are not patentable cf. http://www.patentlyo.com/patent/2012/03/mayo-v-prometheus-natural-process-known-elements-normally-no-patent.html.

change accordingly. *Merges and Duffy* describe the situation as follows: 'in the area of cDNA sequences, the class of researchers capable of discovering sequences may be large and relatively homogeneous. Finding a substantial use for sequences may be the really difficult step, and the ability to take that step may be what really differentiates a few researchers from the many'.[14]

### 1.2. Lessons from tailor-made patent law provision and human gene patents

The question of whether the law can make life (too) simple arose when the author was conducting her preliminary analysis of the empirical study in 2008. Reading the correspondence from patent examiners, applicants, and patent documentation in general prompted the question of whether the EU Directive provisions relating to genes and gene sequences make genetics too simple. The wording of Article 5(2) of Directive 98/44/EC that 'an element isolated from the human body or otherwise produced by means of a technical process, including the sequence or partial sequence of a gene, may constitute a patentable invention' seemed inadequate for addressing the biological processes and pathways that actually find their way into patent documents. In the author's earlier writing the question was raised as to whether genes are too complex to be treated in the legislative framework as a chemical compound. The EPO, EU and the USPTO all address DNA molecules as chemical entities that are patentable, and the concept of isolation has been embodied with a legal quality. In the USA this development took place via case law whereas in Europe the development was enacted in statutes; see Chapter 5.2.1.[15] It has further been argued by the author that claiming general molecular biological processes is a concern. Indeed a ridiculous example of this was found in the PD 2012 search group (24**), claiming a new signalling pathway. By inserting a chemical, the inventor claimed to be changing the pathway.

In a way, it is difficult to measure the case study of the thesis with the current state in court practice. One can ask again whether the approach to gene sequences only as worded in Directive 98/44/EC, Article 5, represents an outdated approach to biomedical research or whether the design of Article 5

14. *Robert Patrick Mergers & John Fitzgerald Duffy,* Patent Law and Policy: Cases and Materials, *5th Ed., 2011, p.* 258.
15. In Chapter 1, section 4.3, it was said that the case study aimed at finding differences in patent prosecution due to this difference. This task turned out to be almost impossible mainly because the requirements were assessed within the two systems in almost the same way. The general outcome did not the impression of very different patent prosecutions.

is unaffected by time? If the isolation process is a known process, DNA as such is a natural phenomenon and if the information it contains represent laws of nature, what is then the substance of Article 5? The focus on genes, as currently being investigated as patentable subject matter, can be supplemented by a focus on RNAi molecules that in fact challenges standard gene approaches with the utility question directed to proteins, primers and probes. Are RNAi molecules to be considered as genes?[16] The idea behind Article 5(3) of Directive 98/44/EC seems to be of immense importance for evaluating new technologies in biomedical research, such as the recognition of the RNAi molecule as being crucial for complex biological processes.

As for single-disease disorders, the empirical case study in Chapter 7, section I, addressed new techniques that challenge the traditional gene patent, namely modulating gene expression either by RNA interference (RNAi) or by ASO mediated splicing correction (sASO). RNAi is used by many organisms to control gene expression at the post-transciptional level, it is widely used as a research tool in laboratories, and importantly it is also a strong candidate for use in numerous future therapies.

Section II, comprising genetically heterogeneous diseases, did not differ from the first study on single-gene disorders. The aim of the different focus was an assumption that patent activity ought to differ if the causes were multifactorial rather than being caused by a single gene, if they were autosomal recessive and not dominant and so on. The assumption was not confirmed, mainly due to the design of the searches on Alzheimer's disease (AD) and Parkinson's disease (PD) patents. The subject of the patent search was the primary disease genes, but in section II there can be several for the same disease. Operating with primary disease genes and secondary disease genes helps prevent the inquiry from becoming too complicated and broad, but narrowing the inquiry in this way diminishes the differences between patent activities in sections I and II. Finally, both AD and PD patent activity did comprise RNAi techniques.

There are two questions with regard to the 'new' techniques. As in the case of the doctrine of isolation now being too trivial to be a subject for patenting (known process), the first question concerns whether the general RNAi technique is also considered a known process. The background information in Chapter 2, section 2.1.d.1, refers to: the first patent application, the Carnegie patent on cancer genes or pathogen genes; the German scientist

16. See Adam Bostanci, Jane Calvert & Pierre-Benoit Joly, 'Regulating gene regulation: patenting small RNAs', in Matthew Rimmer & Alison McLennan (eds.), *Intellectual property and emerging technologies, the new biology*, p. 222.

Thomas Tuschl, who has several basic patents on the technique; and the US Scientist McSwiggen.[17] The HD case study also contained patent applications on the techniques as such; see Chapter 7.2.2.d. Thus, the techniques as such are not inventive unless of course new ways are invented (perhaps an elaboration on the (3-prime) overhang approach).[18] The second question concerns the different forms of RNAi and sASO. How are genes and RNAi molecules to be compared in a patent perspective? These questions can be addressed in three subclasses: miRNA, siRNA and sASO.

The two types of RNAi molecules, *miRNAs* and *siRNAs*, can be distinguished by some important features regarding their occurrence and functions. One of the important differences is that *microRNAs* occur naturally and are encoded in the human genome by miRNA genes, whereas *siRNAs* are typically short synthetic antisense oligonucleotides which are fully complementary to a single naturally occurring gene. Another important difference is how they function; microRNAs are only partly complementary to their target genes and can therefore bind to and simultaneously down-regulate the expression of numerous genes and typically not all their target genes are known. In contrast, all nucleotides of an siRNA must base-pair with nucleotides of their target gene, making them specific for a single gene, and they may even be specific for a single variant of a gene if they are designed to target the subsequence with the nucleotide variant (SNP or disease causing mutation). siRNAs may therefore be suitable for down-regulating the expression of dominant negative mutant proteins in dominant diseases like HD.

Another type of synthetic short antisense RNA oligonucleotide has emerged in recent years, namely the splicing of modulating antisense oligonucleotides (sASOs). Like siRNAs, sASOs are also fully complementary to a single gene, but they are designed with chemical modifications, so that their

17. The Carnegie patent (US 6506559) on methods for inhibiting target genes with double-stranded RNA; Tuschl patents (US 7056704 and US 7078196 and application US 20050059005) on methods for preparing double-stranded RNA molecules from 19-25 nucleotides in length with 3' overhang and miRNA; Sirna Therapeutics (US 7022828) and Crooke (US 6107094 and US 5898031). One company, Alnylam, has licensed key patent positions in both general and platform technologies; see Steven L Highlander & Travis M Wohlers, 'Patent strategies for therapeutic RNAi', *Expert Opin. Ther. Patents*, 2008, 18(11), p. 1292; and Adam Bostanci, Jane Calvert & Pierre-Benoit Joly, 'Regulating gene regulation: patenting small RNAs', in Matthew Rimmer and Alison McLennan (eds.), *Intellectual property and emerging technologies, the new biology*, p. 218; and Chapter 7.1.
18. Chapter 7.1, referring to Jennifer Rohn, 'RNAi patent win', *Nature Biotechnology*, 30, 202, March 2012.

binding does not cause degradation of their target gene. Instead, sASOs block the binding of cellular factors to the pre-mRNA of the targeted gene in order to modulate or correct splicing. sASOs play an important role in new therapies aiming at correcting aberrant splicing of disease genes like the *SMN2* gene in spinal muscular atrophy (SMA).

In a patent context, the current study categorises siRNAs (and sASOs) as being directly related to the sequence of their target gene, whereas miRNAs are a new category. How should miRNAs be approached when they may influence an undefined number of genes and entire disease pathways when only few of their functions may be realized? Should miRNA patents be approached in the same way as regular gene patents, since they are encoded by the genome? If so, how should gene function and utility be defined? A solution could be to claim that miRNAs are not patentable as such, but that specific and documented functions of microRNAs on specified target genes (in relation to a disease process or treatment) are patentable. In this way other unknown functions of miRNAs are not blocked.

How is it possible to account for other effects of siRNAs than those on their completely matched target genes? What if a siRNA has another (unintended) function in a 'miRNA-like' way by binding without full complementarity to genes other than the target gene they are designed for? Maybe the same solution as referred to above could be used, namely to ensure that claims are always restricted to a specific target gene.

In the study of *Huntington's disease (HD)* the highlights on RNAi illustrate the rapid pace of technological development. The literature case study in Chapter 6.2.1.1.1 describes the first HD gene patents (1987, 1997 and 2002) on the huntingtin gene (*IT15*), as well as the huntingtin gene encoded by the protein. As mentioned in the chapter, no rights have been enforced or licences issued for financial gain. Returning to the Chapter 7 study, this reveals the further developments in the efforts to treat HD. There is a huge number of patent applications for RNAi. As mentioned in Chapter 7, section 2.2.d, RNAi methods which specifically remove the mutant mRNA that encodes the dominant negative protein are very promising as a future therapy. In particular, RNAi methods which are allele-specific (mutation specific) have great potential and many of the patents therefore claim a specific RNAi sequence which only targets a common SNP variation. This approach would leave the mRNA from the normal functional allele in the patient intact, and ensure that normal huntingtin (which is required for normal cell function) is still produced in the patient.

Using this technique, some patents have targeted various parts of the *IT15* gene sequence by using short (typically 15-19 nt long) complementary RNA oligonucleotides to down-regulate *IT15* gene expression; see Chapter 7, section 2.2.d.b. Some patents claim several sequences, others claim a particular sequence, and others claim sequences with reported frequent SNP variations. The idea here is that the requirement for a siRNA to be completely complementary to the target mRNA in order to cause degradation, can be exploited in such a way that only the mutant disease-causing mRNA is degraded in patients, because the nucleotide at the SNP position will determine whether the siRNA is fully complementary. siRNAs that are not allele-specific are of less therapeutic value, since they will cause degradation of both mutant and normal *IT15* mRNA in patients, causing lack of normal IT15 protein, and they may therefore themselves cause neuronal disease. Another trend concerns how siRNAs are delivered and their chemistry exploited.

The case study on *spinal muscular atrophy (SMA*) found activities relating to the genes involved in the disease (*SMN1 and SMN2*). The literature case study in Chapter 6.2.2.1.1 did not reveal immense patent activity, since the disease has not previously been elaborated in patent literature.

Some of the SMA patent groups claimed delivery systems in the form of vectors or nanoparticles and protein replacement therapy. The major problem with gene therapy is gene delivery so as to ensure effective delivery and expression of a gene to induce a therapeutic effect.

The SMA highlight addressed *gene expression correction*, which is distinguished from the RNAi technique described above for HD. While RNAi is used to modulate gene expression by *down-regulating* or *inhibiting* gene expression, the present techniques with antisense oligonucleotides are used to *correct* gene expression (splicing).

Some patents claim a method for chemical modification of sASOs (see figure 7.1.C), so that delivery to target cells is improved. They include extensive laundry lists (neuromuscular diseases (e.g., dystrophies, ALS, SMA), cardiovascular disease, brain function/disorders; and cancer biology and therapeutics. Others address only SMA, claiming sASOs which can correct splicing (stimulate exon 7 inclusion) of *SMN2* by blocking the binding of splicing inhibitory proteins to sequences located in intron 6, exon 7 and intron 7 of the *SMN2* gene. The sASOs are claimed for use in the treatment of spinal muscular atrophy (SMA). Originally several ASOs located in intron 6, exon 7 and intron 7 where claimed in the same patent.

It is interesting to compare the study of *cystic fibrosis* in Chapter 7 with the literature study in Chapter 6.2.4.1.1, as it is based on an open model and many *CFTR* mutations have been identified since the identification of the *CFTR* gene in 1989. The empirical study came up with new mutations (today, around 1,900 mutations have been verified) and easy ways of identifying mutations, e.g. in blood samples from pregnant women. The study also found inventions related to delivery and gene therapeutic vectors and proteins seeking to substitute the defective CFTR protein in target cells, and it found inventions claiming to change the CFTR genetic code to enhance protein production.

As for *Alzheimer's disease*, the first sub-heading claims '*vaccines and antibodies*' and it shows a considerable development of therapeutic vaccinations against amyloid peptides covering a broad range of approaches. One kind of delivery process used an AAV virus for gene therapy directed at the disease gene, e.g. by delivering a siRNA to down-regulate gene expression or to deliver the gene. Another kind addressed the delivery of a drug in order to affect gene expression, e.g. the patient's disease gene *in situ* or a secondary gene delivery of genes involved in the pathological process. The AD study also comprised RNAi and anti-sense oligonucleotides techniques.

*One group* (group 24) contains an RNAi laundry list claiming HD, AD and PD. In the group, only one EP application remains (EP 2239328, Alnylam Pharmaceuticals Inc.) which is for neurological diseases in general. Originally, claim 9 claims a siRNA where the human carries a genetic variation in the parkin (*PARK2* gene). This would allow allele specific siRNA, but this is not relevant in the case of parkin, since when caused by parkin mutations Parkinson's disease is inherited recessively.

The PD search also found broad claims for an isolated DNA or gene that is responsible for Parkinson's disease and useful for diagnosis, prevention and treatment. According to this invention the isolated DNA or gene comprises the full-length DNA sequence or a partial sequence, or a sequence hybridisable thereto or hybridisable with a complementary strand thereof (probes, primers, gene tests). Why does that not block for siRNA design? In the summary of the HD empirical study above, claims were identified that comprised a specific RNAi sequence, which only targets a common SNP variation.

It is remarkable that none of the siRNAs (or other types of inhibiting ASO RNAs) seems to be targeted at SNPs in the AD/PD disease genes. As mentioned previously, it is vital that only the mutant mRNA from one disease allele in the patient is targeted and, as described for HD, this can be achieved by designing the siRNA so that it binds across a frequent SNP (the lack of a perfect match to one of the alleles will then result in less or no down-regulation of the normal allele). However, in the 2012 search, one PD group (group 16**) claimed SNCA specifically inhibiting the SNCA (*PARK1* gene, being autosomal dominant) by siRNA. The 2012 PD search shows one application on siRNA targeted at synuclein mRNA. The siRNA may be aimed at down-regulating expression of the frequent A30P mutant, and/or the A53T mutant. This allele-specific knockout (down-regulation) by siRNAs (SNPs) can be considered therapeutic and the relationship between a DNA sequence (probe) patent and that of the correlated siRNA (with a SNP variant) is a new aspect of the *evaluation of patentable subject matter* in the USA. Is the evaluation of the scope of protection for DNA sequence patents (e.g. a mutation-specific probe) facing a new future competing with a siRNA targeted at an SNP variant?

The PD search was also interesting with regard to the prospect of stem cell activity. Both PD and SMA seem to attract research into alternatives to embryonic stem cell research, such as induced pluripotent stem cells (iPSCs).

Comparing the current dispute about genes as patentable subject matter with Chapter 7, Box 7.1.A, on *Disease Genes Claims*, and Box 7.1.B, on *Antisense RNAi overview*, the current dispute concerns DNA sequences, proteins for treatment or antibody production, protein as biomarkers, antibodies and the like. However, RNAi is yet to be evaluated. This thesis points to RNAi being considered a research tool with a distinction being made between miRNA and siRNA which provides a solution under the headings of 'artificially made' or 'naturally occurring'. However, new questions may arise when inventors claim inventions covering many siRNAs while claiming to be allele-specific silencing of disease genes. What will be the blocking effect of such a patent?

The traditional patentability requirements have to be assessed in the light of RNAi techniques. The antisense technique has been subject to scrutiny in *Enzo Biochem, Inc. v Calgene, Inc.* applying antisense technology to regulate three genes in *E. coli*.[19] The patents were invalid because the breadth of enablement was not commensurate with the scope of the claims.

19. *Enzo Biochem, Inc. v Calgene, Inc.*, 188 F.3d 1362, 52 USPQ2d 1129 (Fed. Cir. 1999). The two patents in this case were targeted at genetic antisense technology (which aims to control gene expression in a particular organism); see http://www.uspto.gov/web/offices/pac/mpep/documents/2100_2164_06_b.htm. (last visited May 2012). See further John LeGuyader, 'Decision does not necessarily determine the outcome for examination of antisense patent applications recently filed because current knowledge and level of skill in the art is high (antisense has progressed as a technology since 1989)', at http://www.docstoc.com/docs/56788347/Patenting-Interfering-RNA; and see Robert A. Schwartzman, 'Elaborating on the unpredictable factor for antisense', at http://www.docstoc.com/docs/103623546/Patenting-Antisense-Oligonucleotides-and-Methods (last visited May 2012).

Since RNAi is still in its infancy (first reported dsRNA in 1998[20]), it will still 'benefit' from the status of being an unpredictable field. This has explicitly been the case in several of the applications of new techniques in the case study in Chapter 7. *McLeod et al.* compare the situation for miRNA with that of the genomic patent landscape of the late 1990s, when high-throughput sequencing genes and ESTs pushed the legal limits for patentable utility. They ask whether miRNA patent strategies will face a similar fate.[21]

First, it is interesting to consider how prior art can be assessed in the light of both siRNAs and miRNAs. As for siRNAs, it was asked above in regard to PD whether a patent for the gene (probes and primers) would anticipate a later siRNA. Since siRNAs have the sequence of the target gene itself, it will almost inevitably be anticipated by prior art in this situation. However, in the case of miRNAs the situation is different. MiRNAs raise two questions. First, does a gene patent also anticipate intergenic miRNA genes? The majority of miRNA genes are in fact intergenic, meaning that they are located within other genes and often (around 40 %) they are located in an intron of that gene. Knowing the gene subject for the patent does not give an answer to how the intergenic miRNA works, and the functioning of the miRNA may be completely unrelated to the overlapping gene. The second question concerns the gene where the miRNA has one of its many unknown functions. If a patent is granted for a target gene and the miRNA happens to have a function in the same gene, it seems unlikely that such a gene patent (e.g. on a large sequence) should include a yet unknown microRNA gene which is e.g. 80 nucleotides long in its unprocessed form, and where the processed microRNA is only partly complementary and not described (nor anticipated) in relation to the gene patent as granted.[22]

Second, the utility requirement/requirement of industrial application (35 U.S.C. § 101 and EPC Article 57) is both interesting and of utmost importance. Some authors find that the utility requirement constitutes a minimum barrier to patentability (the more focused the better, directed to a specif-

20. A. Fire et al., *Nature* 391, 806-811, 1998. See Bonnie W McLeod et al., 'The 'real world' utility of miRNA patents: lessons learned from expressed sequence tags', *Nature Biotechnology*, Vol. 29, No 2, February 2011, p. 129.
21. See Bonnie W McLeod et al., 'The 'real world' utility of miRNA patents: lessons learned from expressed sequence tags', *Nature Biotechnology*, Vol. 29, No 2, February 2011, p. 129.
22. Steven L Highlander & Travis M Wohlers, 'Patent strategies for therapeutic RNAi', *Expert Opin. Ther. Patents*, 2008, 18(11), p. 1293, also discuss prior art for microRNAs. However they do not separate the utility of microRNA from microRNA as a 'gene' in itself, so their discussion does not sufficiently describe the problem.

ic disease, RNAs and gene targets) and that the hurdle may be embodied in the enablement and description requirements which will become easier to satisfy as technologies mature.[23] But the view in this thesis is that the utility requirement/requirement of industrial application cannot be that easy to satisfy. The focus on miRNA ought also to be subject to particular scrutiny. The questions about the one-to-one relationship between gene and function raised in this thesis with regard to biological pathways and complex systems are also applicable to miRNA. As pointed out in Chapter 7, Box 7.1.B, downregulation may respond differently on each gene and have unknown functions. One miRNA can target hundreds of different genes, and several miRNAs can target the same gene, which is even more striking than the multifunctionality of protein-coding genes.[24] As also discussed by *McLeod et al.*, if interpreted as a requirement for specific, substantial and credible use, the US utility requirement can be difficult to satisfy and a good deal of research will often be required to conform to a predicted utility. However as they write, 'recent reports suggest that the most accurate target-prediction algorithm achieves only 66 % precision in predicting true targets, suggesting that even patents and applications that predict potential targets having an established utility using known algorithms might be argued to still lack real world utility where the association between the miRNA and the predicted target has not been confirmed experimentally.'[25] Thus, the two steps discussed above for the assessment of miRNA both have to be viewed in the context of utility. Characterising miRNA without knowledge of its utility would result in lack of patentability, whereas finding one specific, substantial and credible function does not provide information and evidence about associated functions. This being said, miRNA leads to greater uncertainty about blocking effects than do genomic DNA sequences.

As shown in the sections in Chapter 7 on Huntington's disease, spinal muscular atrophy, Alzheimer's disease and Parkinson's disease, the discovery of 'new' or previous unknown biological processes may put in question the traditional gene patent. First, there is a question why, for example, e.g. pa-

23. Steven L Highlander & Travis M Wohlers, 'Patent strategies for therapeutic RNAi', *Expert Opin. Ther. Patents*, 2008, 18(11), p. 1295.
24. Adam Bostanci, Jane Calvert & Pierre-Benoit Joly, 'Regulating gene regulation: patenting small RNAs', in Matthew Rimmer & Alison McLennan (eds.), *Intellectual property and emerging technologies, the new biology*, p. 223.
25. Bonnie W McLeod et al., 'The 'real world' utility of miRNA patents: lessons learned from expressed sequence tags', *Nature Biotechnology*, Vol. 29, No 2, February 2011, p. 129.

tents for a disease gene can be granted at the same time as RNAi patents are still possible (scope of protection). Next there is a question of a shift in focus from gene patents (proteins, antibodies and more) to RNAi such as miRNA and siRNA, including methods of delivery. Will the current practice on industrial application for DNA patents be transferred, and if so will it work for small RNA molecules? So far, the impact of the US reshaping of DNA as patentable subject matter has focused more on the traditional DNA sequence of a gene (genomic sequence, cDNA sequence, primers and probes) which is the subject of the current *Myriad* dispute and less (if at all) on the consequences for RNAi molecules and other yet-to-be discovered applications.[26] The

26. Isabell Huys, Gert Matthijs & Geertrui van Overwalle, 'The fate and future of patents on human genes and genetic diagnostic methods', *Nature reviews, Genetics*, Vol. 13, June 2012, p. 441 ff., distinguishing between patent claims for human genes and patent claims for genetic diagnostic methods; describing broad claims for genetic sequences (isolated, purified genes, encoding proteins, oligonucleotide primers, location of gene on a chromosome). Christopher M. Holman writes about gene patents in the light of whole genome sequences (WGS) based on Judge Bryson's question to Myriad's attorney, as to whether the company's patent would be infringed by whole genome sequencing; see Christopher M. Holman, 'Debunking the myth that whole-genome sequencing infringes thousands of gene patents', *Nature biotechnology*, Vol. 30, No 3, March 2012; IP theory, Vol. 2: issue 1, 2011, p. 1. Holman writes that no one can own a gene and that many unfortunately incorrectly believe that genes can be patented per se. Correcting this view, Holman emphasises that the only right conferred by a patent is the right to exclude others from making, using, selling, or exporting a non-naturally occurring man-made product or process defined in the claims. Holman also distinguishes between the isolated non-naturally occurring man-made DNA molecule being patentable and the corresponding naturally occurring gene. Patents are also available, he writes, for inventive methods of using isolated DNA molecules (recombinant protein expression, methods for analysing genes and genetic variations (e.g. genetic diagnostic test). The overall message of his article is that whole-genome sequencing will not infringe granted gene patents, first because there are not that many and there is a distinction between a DNA or amino acid sequence being mentioned in a patent claim and a gene being claimed. Holman find that most patent claims are directed towards proteins, not genes or DNA. It also covers claims that refer to DNA sequences but only cover inventions such as fusion genes, vaccines or methods of DNA processing and analysis that are totally unrelated to sequencing or genetic analysis. Holman did find patents claiming DNA molecules that correspond in sequence to at least some portion of a human gene, and claims for the analysis or detection of specific gene sequences or genetic variations. However, they would not prevent whole gene sequencing (WGS) as WGS does not rely on DNA amplification. In his discussion, he also mentions the issue of introns (present in human genomic DNA) contrary to cDNA and he discusses the length of isolated DNA sequences, and short fragments of genomic DNA used for large-scale DNA sequencing, and most of-

Bryson approach seems to be the best solution for addressing future techniques for processes and methods.

*1.2.1. Laundry lists and the dissemination of knowledge*

The empirical case study found that, for many applications within the pathological process, and regarding RNAi techniques it appeared that the use of a *laundry list* in the description did not qualify as speculative use, whereas a laundry list in the claims was often met with rejection on the ground of speculative use (claiming all the world's diseases). Despite a rejection of a claim, the laundry list has continued to stand in the description and sometimes a US patent has been granted with a claimed laundry list while the European patent claiming the same laundry list has been rejected and vice versa. It has been impossible to determine the existence of a fixed practice on laundry lists. The laundry lists in patent applications relating to Canavan's disease comprised inventions claiming broad biological processes and gene superfamilies involved in processes for demyelination, and immense laundry lists were claimed. In section II on Alzheimer's disease and Parkinson's disease,

ten patent claims are for full-length protein-encoding sequences. Finally, the discussion of smaller fragments of isolated DNA, usually probes or primers but also fragments capable of encoding a peptide (antigene response), and others with even smaller fragments, and with reference to Cook-Deegan he writes that 'that fragment claims of this type could be declared invalid based on anticipation by a large number of prior-art DNA sequences falling within the scope of the claim'. In Thomas B. Kepler, Colin Crossman & Robert Cook-Deegan, 'Metastasizing patent claims on *BRCA1*', *Genomics*, May 2010, 95 (5), pp. 312-314, the authors' analysis was directed at the claim of the *Myriad* patents which seemed particularly broad, and the patent extends to portions of most genes in the human genome and probably most genes in nature as well. There is more literature concerning what is patented; see e.g. Philippe G. Ducor, *Patenting the Recombinant products of Biotechnology and other Molecules*, 1998; Amanda Warren-Jones, *Patenting rDNA, human and animal biotechnology in the United Kingdom and Europe*, 2001, writing about antisense technology pp. 16-18; Eddy Ventose, *Medical Patent Law – The challenges of medical Treatment*, 2011, p. 218 ff.; Goldbach, Vogelsang-Wenke & Zimmer, *Protection of Biotechnological matter under European and German Law*, 1996, especially on proteins, p. 134 ff.; Paul England (ed.), *Intellectual Property in the Life Sciences, A global Guide to rights and their application*, 2011, p. 87 ff. See further *Robert Patrick Mergers & John Fitzgerald Duffy,* Patent Law and Policy: Cases and Materials, 5th Ed., 2011, p. 255, on EST (*In re Fisher* 421 F 3.d 1365 (Fed. Cir. 2005); and the Trilateral Project WM4 concerning SNPs (Report on comparative study on examination practice relating to single nucleotide polymorphisms (SNPs) and haplotypes available at http://www.trilateral.net/projects/biotechnology/Haplotypes.pdf. (last visited may 2012).

searches for one disease are seldom confined to siRNAs for the searched disease, secondary or primary disease genes. Not only can laundry lists be found with several diseases, but the claimed inventions frequently identified several of the other searched diseases in current case study.

As mentioned previously in Chapter 7 regarding AD, *group (21)* claimed RNAi molecules targeted at a nucleic acid sequence encoding polyglutamine repeat diseases. Huntington's disease (HD) is caused by expansion of a polyglutamine repeat, and spinocerebellar ataxia type 1 (*SCA1*), dentatorubral-pallidoluysian atrophy (DRPLA), Macado-Joseph disease (MJD), and spinobulbar muscular atrophy (SBMA) are also strongly associated with an expanded polyglutamine repeat. Despite this, the patent application contains a laundry list that includes many diseases in which expanded polyglutamine tracts have not been identified or been indicated to play a role such as amyotrophic lateral sclerosis (ALS), hereditary spastic hemiplegia, primary lateral sclerosis, spinal muscular atrophy, Kennedy's disease, Alzheimer's disease, Parkinson's disease, and multiple sclerosis.[27]

Even though many functions depend on the biological context, the widespread use of laundry lists in patent practice puts in question one of the very purposes of patent documentation; disclosure must be considered as being made in exchange for the monopoly granted, thus providing sources of information to the public. The descriptions in the majority of patent applications are loaded with general and irrelevant information, sloppy writing and more, and are not consistent with EPC Articles 83 and 84 and 35 USC § 112. This leaves the reader with a concern that, in the main, patents are constructed in order to support future speculative uses, and are less an information tool addressed to peers.[28] Patent authorities could usefully tighten up the requirements for acceptable practice and thus direct their resources to examining strict applications.

27. The author has found that examiners often refer to the requirement of unity in the context of the '*Markush* claim' (see EPC Article 82 and PCT Rule 13.1). However, this did not appear to be a barrier to all laundry lists.
28. EPC Article 69 and its protocol, EPC examination guidelines, June 2012, Part. H, Chapter IV, point 4, and especially Part F-IV 4.3 on inconsistency. See further Singer & Stauder, *EPC – A commentary*, 3rd Ed. Vol. I, Article 69, point 5. Sven J.R. Bostyn, '*Enabling Biotechnological Inventions in Europe and the United States, A study of the patentability of proteins and DNA sequences with special emphasis on the disclosure requirement*', 2001. p. 145 ff.

# 2. Safeguarding patent law – facing the future

## 2.1. Patenting the animal kingdom – an update

Chapter 8 comprises an article written in 2007 and the focus on essentially biological processes has developed since then. At the time of writing section 2.b, Case T 83/05 had been referred to the Enlarged Board of Appeal (22 May 2007 – Case G 2/07), whereas Case T 1242/06 had been referred to the Enlarged Board of Appeal in April 2008 (Case G 1/08) by the time the article was published. The Enlarged Board of Appeal decided to consider the cases in consolidated proceedings in 21 April 2008.

Case G 2/07 *Tomatoes/State of Israel* concerned methods for breeding tomato plants that produce tomatoes with reduced fruit water content. According to the case document, the claimed method contained steps of interspecies crossing (a Lycopersicon esculentum plant with a Lycopersicon spp. plant to produce hybrid seed), collecting, growing, pollinating and selecting as follows:

> 'Plants are grown from the first generation of hybrid seeds, the plants of the most recent hybrid generation are pollinated and the seeds produced are collected. Plants are again grown from the seeds of this most recent hybrid generation. This is followed by a final step, i.e. allowing the fruit to remain on the vine past the point of normal ripening and screening for reduced fruit water content as indicated by extended preservation of the ripe fruit and wrinkling of the fruit skin.'[29]

It was argued by appellant I that plants belonging to separate species are generally not capable of interbreeding and that the selection criterion of capability of dehydration was not a natural one. Tomatoes expressing this phenotype had no evolutionary advantage in any given environment over corresponding plants not expressing it. Furthermore, selection of plants with tomato fruits having an increased dry weight percentage implied an additional technical step. Fruit samples are first weighed fresh, then dried in an oven and weighed again in their dried state.[30]

Case T 83/05 *Broccoli/Plant Biosciences*[31] has been described in Chapter 8, section 2.b. The patent in question claimed a method for the production of broccoli (*brassica oleracea*) with elevated levels of certain glucosinolates (potentially anti-carcinogenic substances). The method contained several

29. Case T 1242/06 (3.3.04), 4 April 2008, (EP 1211926), reasons for decision, points 5 and 6.
30. Case T 1242/06 (3.3.04), 4 April 2008, the submissions by appellant I, Point XI.
31. Case T 83/05 (3.3.04), May 2007, concerning EP 1069819.

steps of crossing and selection. However, selection was assisted by the use of molecular markers to select hybrids with the desired genetic combination encoding expressions of elevated levels of glucosinolates. The Enlarged Board of Appeal decided the Joined Cases G 1/07 and G 1/08 *Tomato/Broccoli* on 9 December 9 2010, after having received comments from the EPO President in 2009 and numerous *amicus curiae* briefs.

The decision is interesting in two ways. *First* it considers the relationship between the Directive and its implementation in the EPC Implementation Rules. It sends several signals about the relationship, adding to the number of situations in which the perspective of the EU has been contrasted with that of the EPO. In point VI of the decision (the President's submission) the EPO President emphasised that the Implementing Regulations must be interpreted in the light of Article 164(2) EPC, thus emphasising a dynamic and harmonised interpretation of the process exclusion under Article 53(b) EPC. The Enlarged Board of Appeal initiated the analysis in point 2.2 by clarifying that the function of the Implementing Regulations is to determine in more detail how the EPC should be applied, whether or not governing issues of substantive patent law. In this regard the law-making powers of the EPO Administrative Council can be inferred from Article 164(2) EPC.

The verbatim transposition of the Directive in the Implementing Rules in Rule 23b(5) (1973) (Rule 26(5) (2006/2010)) was based on the assumption that the Board's practice already reflected the true meaning of Article 53(b) EPC.[32] The Enlarged Board of Appeal interpreted Article 53(b) EPC by focusing on the presence of a technical step which cannot be carried out without human intervention and which has a decisive impact on the final result (Case T 356/93). In point 3.2.3, the Board referred to the judicial definition of 'essentially biological processes' and identified key elements in particular in decision T 320/87. The key elements, in no particular order, are as follows:

1. The totality of human intervention and its impact on the result achieved is to be determined.
2. This has to be judged on the basis of the essence of the invention.

32. See point 5. The wording of Rule 26(5) of the Implementing Regulations is a definition of essentially biological processes derived from Directive 98/44/EC, Article 2, as follows: 'A process for the production of plants or animals is essentially biological if it consists entirely of natural phenomena such as crossing or selection'. The wording of Directive 98/44/EC, Article 2(2), has the following definition: 'A process for the production of plants or animals is essentially biological if it consists entirely of natural phenomena such as crossing or selection'.

3. The impact must be decisive.
4. The contribution must go beyond a trivial level.
5. The totality and the sequence of the specified operations must neither occur in nature nor correspond to the classical breeders' processes.
6. The required fundamental alteration of the character of a known process for the production of plants may lie either in the features of the process, i.e. in its constituent parts, or in the special sequence of the process steps, if a multistep process is claimed. In some cases the effect of this can be seen in the result.

As for Rule 26(5) EPC, elaborating on the definition of essentially biological processes, the Enlarged Board emphasised that the Rule had to be interpreted in the light of Directive 98/44/EC. Surprisingly, it went further in point 4.2 of the decision by emphasising that Directive 98/44/EC contains a complete definition so that Rule 26(5) EPC can only be interpreted as being intended to give an exhaustive definition.[33] Referring to the Vienna Convention, Article 31(1), which is applicable to the EU-EPO relationship, it concluded that the meaning of a term of a treaty cannot be established purely semantically but must be interpreted in good faith, in accordance with the ordinary meaning to be given to the terms of the treaty in their context (in this case, the context of the terms 'crossing' and 'selection' in the relevant provisions is the processes for the production of plants). The terms must be placed in the context of the acts of a breeder intervening in order to obtain a desired result. Within this context, crossing and selection are not natural phenomena but generally involve human intervention.[34]

After analysing Article 2(2) of the Directive, the Enlarged Board found no indication as to how the legislator intended the Article to be interpreted. The consequence of the self-contradictory wording of Article 2(2) (combining two elements of different concepts into a single definition, and citing one of these concepts as an example of the other)[35] and its literal transposition into Rule 26(5) EPC is that the term must be interpreted on its own authority by the Enlarged Board of Appeal.[36]

33. See further in point 4.5, where the Enlarged Board interpreted the EPO Rule so as to conform to the EU interpretation by verifying that it is the text of the Directive and its legal history which have to be considered when looking for further clarification of the meaning to be given to Rule 25(5) of the Implementing Regulations.
34. Point 4.4.
35. Essentially biological: natural phenomena means crossing or selection.
36. Point 5.

*Second*, the case is important since it defines the term 'essentially biological processes' and in principle this definition and the reasoning for it can be measured against the initial question asked in Chapter 1, section 2.2, on the differences between the approaches to animals and plants. See Directive 98/44/EC Article 4(1) with Article 2(2), and the corresponding Article 5(1) for humans which states that: 'The human body, at the various stages of its formation and development, and the simple discovery of one of its elements, including the sequence or partial sequence of a gene, cannot constitute patentable inventions.' However, there is no prohibition of essentially biological processes in relation to humans. On the contrary, it follows from Article 5(2) that: 'An element isolated from the human body or otherwise produced by means of a technical process, including the sequence or partial sequence of a gene, may constitute a patentable invention, even if the structure of that element is identical to that of a natural element'.

Returning to the definition of *essentially biological processes* for plants and humans (Rule 26(5), of the Implementing Regulations) which was the subject of the case, in point 6.4.2.3 the Enlarged Board of Appeal reached a conclusion that was based on a historical context.[37] The legislator intended to exclude from patentability the conventional methods for the breeding of plant varieties at that time (sexual crossing and subsequent selection of the plants having the desired traits). With reference to the replacement in the legislation of the words 'purely biological' by 'essentially biological', it was obviously the intention of the legislator to emphasise that the mere fact of using a technical device (e.g. a greenhouse) in the breeding process should not be sufficient to make an invention patentable.[38] The use of technical means to obtain the product (crossing, growing or selection in greenhouse surroundings) was not a relevant issue, but the technical devices might be patentable in themselves.

The above-mentioned consideration is still important today. As emphasised by the Enlarged Board of Appeal, the picture today is much more so-

37. The EPC *travaux préparatories*, including the Strasburg Convention *travaux préparatories*, and the Comments on the first Preliminary Draft Convention relating to a European Patent Law, 14 March 1961.
38. The Enlarged Board found that this approach indicated that: 'the legislator did not wish breeding processes to be patented in which the technical measures used are only means serving to bring about processes for the production of plants which are otherwise based on biological forces. This is made abundantly clear by the additional remark in the explanatory notes that such technical devices may perfectly well be patented in themselves but not so the biological process in which they are used'; see point 6.4.2.3.

phisticated. The replacement of word 'purely' by 'essentially' cannot be ignored today as the wording has remained over time. The very important teaching from the Board is as follows:

*'Be that as it may, the Enlarged Board is unable to see why the legislator's decision to provide appropriate patent protection for 'secondary' features such as technical devices or means (today e.g. markers) by allowing them to be patented in themselves but not to extend protection to the biological process in which they are used, would no longer be justified today, merely because today many more such technical possibilities exist. On the contrary, given that there is a certain tendency to ever broaden the technical field covered by a patent by drafting claims directed to all envisaged contexts in which the invention might potentially be used, the fact that the legislator did not want such an extension of protection in the field considered here is still a valid consideration to be respected.* Hence, it must be concluded that the provision of a technical step, be it explicit or implicit, in a process which is based on the sexual crossing of plants and on subsequent selection does not cause the claimed invention to escape the exclusion if that technical step only serves to perform the process steps of the breeding process.' [Emphasis added][39]

In fact the italicised text could also be applied to human gene patents. In the light of the ongoing debate about genes as patentable subject matter, precisely the same can be argued. Article 5(1) of Directive 98/44/EC ought to be the only provision in a context of tailor-made patent law (though overall the US doctrine of products of nature is to be preferred). Article 5(2) is in fact contrary to the approach of the Enlarged Board of Appeal highlighted in italics. Furthermore, it addresses the same issues as have been addressed by Judge Bryson in *AMP v Myriad*, merely using different words or being set in a different context.[40]

If an additional step of a technical nature is needed to bring the natural phenomena of crossing and selection outside the scope of the exclusion, it

39. Point 5.
40. See above section 1.1: 'the only material change made to those genes from their natural state is the change that is necessarily incidental to the extraction of genes from the environment in which they are found in nature'. In her *amicus brief* of 24 June 2010, point 3.4, pp. 14-15, Sigrid Sterckx compared the exclusion with that of medical treatment. Her focus is on the direct process and direct product which cannot be patented. She argued that a claim should only be allowed if the characterising technical step is the final step of the claimed process, and the plants or reproductively-capable plant components (e.g. seeds) should only be patented as the direct product of such a process.

may introduce a trait into the genome or modify a trait in the genome of the plant produced that is not the result of the mixing of the genes of the plants chosen for sexual crossing. Finally, the Board stated that the evaluation of 'essentially biological processes' does not have regard for whether a step of a technical nature is a new or known measure, whether it is a trivial or a fundamental alteration of a known process, whether it does or could occur in nature, or whether the it is essence of the invention.

This ruling set aside Case T 320/87 and Case T 356/93.[41] If a method comprises a single step of crossing entire genomes of two breeding partners, it is considered an essentially biological process regardless any technical measure involved.[42] In May 2011, the EPO Guidelines for Examination was amended accordingly. It now follows from the 2012 Guidelines Part G-ch.II-16, 5.4.2, that essentially biological processes (based on the sexual crossing of whole genomes and subsequent selection) are excluded from patentability, even if there are other technical steps relating to the preparation of the plant or animal or its further treatment in the claims before or after crossing and selection. The use of an additional feature of a technical nature, such as a genetic marker to select parent or progeny, does not change this assessment. However, inserting a gene or trait into the plant by genetic engineering does not rely on a recombination of whole genomes and the natural mixing of plant genes, and hence this approach to enhancing the plant is patentable (or the application of growth-stimulating substances or radiation).

*Animal models*

The use of animal models in biomedical research has been further illustrated by the case study. There are often claim for an animal model (rodent, zebra fish, fruit flies and so on) among the other claims, and sometimes the animal model itself is the invention. The study of CF in Chapter 7, section 5.3.d, found different models and the analysis of applications relating to Parkinson's disease and Alzheimer's disease found claims for a PD mouse model comprising *tau* mutations, a transgenic animal model whose genome com-

41. According an interlocutory decision of 31 May 2012, in the appeal proceedings of T 1242/06, further questions of law has been referred to the Enlarged Board of Appeal (maintenance of EP 1211926 in amended form). The questions concern the issue as to whether the exclusion of essentially biologically processes for the production of plants may affect the allowability of product claims directed to plants or plant material, such as a fruit.
42. Franz-Josef Zimmer & Benjamin Quest, 'When is a Process for Production of Plants 'Essentially Biological?'', 30 *Biotechnology Law Report*, No 4, 2011, p. 462.

prises two APP mutations, and the 'Aarhus pig' containing a mutant APP gene or a mutant PS1 gene.[43]

## 3. In Conclusion: A global principle of common concern of mankind

In Chapter 9 there is an elaboration on the waves of law and the question is asked whether the lack of change is a sign of quality. Is the fact that the patent regime has undergone very little change compared with some of the other IPR disciplines a sign of solid legislation? Patent law design reflects a rather static approach, but it is by no means intended to blindfold patent authorities and courts. The message has been fully illustrated by the main subject of the thesis, DNA molecules (as well as embryonic stem cells and other areas discussed in the thesis), where for example the US Supreme Court has sufficient room for manoeuvre in its case law approach to doctrine, whereas the European legislator has tied the hands of the courts of its Member States and perhaps the hands of the EPO as well, by tailor-making patent law for biotechnology. A legal basis of doctrines such as laws of nature, products of nature etc., as developed by the US courts, is much more dynamic and flexible than more specific and static rules. The development of patent law and policy is in fact better safeguarded and balanced by courts navigating on the basis of doctrines than by legislation. Legislators will always find it hard to keep pace with technological developments, with the concomitant risk of patent law being stuck with outdated provisions for new technologies.[44]

The issue can be illuminated by examining the legal expectations connected with these two approaches. Legal expectations can be attached to tailor-made rules to a much greater extent than to case law which has never been evaluated by a superior court. This must be the outcome of the *Myriad* case (if the patentability of DNA molecules is re-defined) as well as the Joined

43. E.g. US 2010306868. See also group (68), claiming an animal model with APP mutations (670, 717 mutations).
44. See e.g. Dan L. Burk & Mark A. Lemley, *The Patent crisis and how the Courts can solve it*, 2009. See the same authors in 'Courts and the patent system', *Regulation*, Summer, 2009, p. 18 ff., arguing that instead of legislation, Congress should let the courts continue to resolve the patent crisis; and 'Don't tailor make a patent act', May 2009, *The National Law Journal* (http://www.nlj.com); Arti K. Rai, 'Intellectual Property Rights in Biotechnology: Addressing New Technology', *Wake Forest Law Review*, 1999, Vol. 34, p. 827 ff.; William Cornish, *Intellectual Property, Omnipresent, Distracting, Irrelevant*, 2004, Chapter 1, p. 33, 'Who determines Patent Policy?'.

Cases G 2/07 and G 1/08 *Tomatoes/Broccoli* from the Enlarged Board of Appeal. Judge Bryson did not have any concerns about the precedents of former USPTO practice, and the Enlarged Board of Appeal stated that there is no 'legitimate expectation' that a former interpretation will not be overruled in the future. Recognising such an expectation as legitimate would undermine the function of the Enlarged Board of Appeal.[45]

Chapter 9 discusses the application of environmental principles, requiring the balancing of different interests and it points to the principle of sustainability as a recognised global principle.[46] It can be said that the balancing of interests is a matter of social responsibility. One of the purposes of the approach of this thesis to multi-level regulation, and in particular multilevel governance of patent law, has been to shed light on the different social interests that challenge the overall economic standards of patent law. Table 1.1, in Chapter 1, illustrates the fragmentation of international and regional laws in some areas of the law, and this may play a role in safeguarding patent legitimacy. The table is not intended to be exhaustive, and it can be added to or supplemented with several other legal disciplines, such as human rights. However, the overall aim has been to propose an international norm aiming at covering the global exchange of values represented in Table 1.1, but separated from its original context and elevated to a kind of *global public order*. Such a global norm can already be said to represent the deep structure of law inherited from its original context and transmitted to a global context. While aimed at restoring the balance between economic standards and social values, it also mirrors a departure from the overall economic dimension of international patent law.[47] The principle discussed in Chapter 9, section 2.5, which is

45. Point 2.5 also states that this holds particularly true for issues on which there is no coherent body of decisions, all to the same effect, but where the case law consists only of a limited number of individual decisions, as is presently the case.
46. See e.g. Hans Chr. Bugge & Christina Voigt, *Sustainable Development in International and National Law*, 2008; Ricardo Meléndez-Ortic & Pedro Roffe, *Intellectual Property and Sustainable Development, Development Agendas in a Changing World*, 2009; Henning Grosse Ruse-Khan, 'A Real Partnership for development? Sustainable development as a Treaty Objective in European Economic Partnership Agreements and Beyond', *Journal of International Economic Law*, 2010, Vol. 13, No 1, pp. 139-180; Winfriend Lang (ed.), *Sustainable Development and International Law*, 1995.
47. See Chapter 1, section 4.2; and Kaarlo Tuori, *Critical Legal Positivism*, 2002; Harold D. Lasswell & Myres S. McDougal, 'Jurisprudence for a free society', *Studies in law, Science and Policy*, 2006, Vol. I, p. 167 ff.; Ulla Neergaard, Ruth Nielsen & Lynn Roseberry, *European Legal method, Paradoxes and Revitalisation*, 2011, p. 101. See e.g. by way of analogy, Robert Howse & Michael J. Trebilcock, 'The free trade-fair

built on the UN principle of the *common concern of mankind*, seems able to accommodate the imbalance between developed and less developed countries, with its asymmetries in health care, nutrition, biological resources, traditional knowledge and climate change, just to name a few. However, it can also incorporate concerns related to the patenting of essentially biological resources and human DNA molecules and future developments.

As discussed in Chapter 1, the TRIPS Agreement cannot be considered a frozen document, and amendments to TRIPS have been discussed for a long time. *Annette Kur* and *Marianne Levin* have proposed making profound amendments to reform TRIPS. Of special interest is the proposal to amend Article 7, on objectives, and Article 8, on public policy concerns, which serve to inform the interpretation of the TRIPS Agreement. They suggest including a new provision as Article 7(b), stating that the protection and enforcement of intellectual property rights should ensure, to the advantage of society as a whole, a balance of rights and obligations so that, in particular, the scope of the protection conferred by an intellectual property right corresponds to the contribution made to creativity and innovation. It follows from their explanatory memorandum that the provision provides for a principled guideline – an overarching concept towards which the IP system should be geared.[48] Article 8(1) in the proposed amended form comprises the TRIPS principles, which address public policy concerns such as public health and nutrition, and it is more formulated as an obligation for Members to protect these values.[49] The proposal also suggests a new Article 8(a) on the balancing of interests, as follows:

'Members shall take due account of the objectives and principles set out in Articles 7 and 8 when formulating or amending their regulations. In doing so, they shall ensure that the protection granted reflects a fair balance between private economic interests and the larger public interests as well as the interests of third parties.'

trade debate: Trade, labor, and the environment', Chapter 5 in Jagdeep S. Bhandari & Alan O. Sykes, *Economic Dimensions in International law, Comparative and Empirical Perspectives*, 1997, p. 193.

48. Annette Kur & Marianne Levin (eds.), *Intellectual Property Rights in a fair world trade system, proposals for reform of TRIPS*, 2011, p. 533, where it is also stated that its impact will primarily be felt when decisions on the legislative or adjudicative level are allowed a margin of discretion, as for instance when the threshold for protection is determined, or when limitations are formed, or their ambit is explored in a given case.
49. Annette Kur & Marianne Levin (eds.), *Intellectual Property Rights in a fair world trade system, proposals for reform of TRIPS*, 2011, p. 536, where the legal qualities of the values are further elaborated.

The proposed Article 8(2) states that users may, without the consent of the right holder, use protected subject matter, provided that such use does not unreasonably prejudice the legitimate interests of the right holder, taking into due consideration the normal exploitation of the right. The provision lists factors that must be taken into account and according to the umbrella provision in Article 8(2)(a)(v), unauthorised use may be justified by serious cultural or social considerations, thus providing for a certain amount of built-in flexibility.[50]

The proposal for amendments to TRIPS goes into great detail on how to update and restore the TRIPS framework in order to improve and balance the trade-related aspects of the IPR system. There have been many suggestions in the literature on how to find the proper balance for the IPR system and its inherent *one-size-fits-all* approach.[51] This thesis also proposes the establishment of an overall balancing principle, recognising the *common concerns of mankind* as a global patent norm aimed at *erga omnes* unification.[52] Capturing the global concerns of today and tomorrow may be made possible by means of including a mandatory norm in the preambles of international patent conventions embracing both *pre-grant* and *post-grant* patent legislation and prosecution, addressed to patent legislators, authorities and courts at all levels. Thus it is not only a TRIPS principle, but a principle that ought to be in-

50. Annette Kur & Marianne Levin (eds.), *Intellectual Property Rights in a fair world trade system, proposals for reform of TRIPS*, 2011, p. 545.
51. See e.g. Annette Kur & Vytautas Mizaras, *The Structure of Intellectual property law, Can one size Fit All? A TRIP Intellectual Property*, 2011; Rochelle C. Dreyfuss, Harry First, & Danielle L. Zimmerman, *Working within the Boundaries of Intellectual Property, Innovation Policy for the Knowledge Society*, 2010; Geertrui Van Overwalle & Ester van Zimmerren, 'Functions and Limits of Patent Law', in Eric Claes, Wouter Devroe, & Bert Keirsbilck (eds.) *Facing the limits of the law*, p. 415. See further on global approaches to IPRs: Keith E. Maskus & Jerome H. Reichman, *International Public Good and transfer of Technology, Under a Globalized Intellectual Property regime*, 2005; and Robert C. Bird & Subhash C. Jain, *The Global challenge of Intellectual Property Rights*, 2008. European Parliament, *STOA*, 'Policy options for the improvement of the European patent system', 2006.
52. Thomas Cottier & Sofya Matteotti-Berkutova, 'International environmental law and the evolving concept of 'common concern of mankind'', in Thomas Cottier, Olga Nartova & Sadeq Z. Bigdeli, *International trade regulation and the mitigation of climate change, world trade forum*, p. 21 ff. For an overall approach to a global norm, see Konrad Zweigert & Hein Kötz, *Introduction to comparative law*, 3rd Ed., 1998, p. 3: 'the development of nothing less than a common law of mankind (*droit commun del'humanite*). A world law must be created – not today, perhaps not even tomorrow – but created it must be, and comparative law must create it'.

serted in a wide range of international patent conventions. In the light of the non-exhaustive Table 1.1 in Chapter 1, the principle of the *common concern of mankind* is a dynamic principle and it encompasses a wide range of concerns.

The first meeting of the UNEP Group of Legal Experts to examine the concept of the *common concern of mankind* in relation to global environmental issues was held in 1990, and the second meeting took place in 1991.[53] In a report from the second meeting it was stated that there was a growing interest among states in the concept of the common concern of mankind, especially in relation to climate change and the sustainable use of biological diversity. The report (point 3) pointed out key aspects of the concept such as: (a) possible implications of the concept for specific obligations in the relevant international treaties; (b) implications for the human right to a healthy environment; and (c) implications for equitable burden-sharing and fair compensation. The legal quality of the concept was discussed in point 4, stating that the concept has no legal consequences in terms of rights and duties. Thus, it does not infringe the sovereign rights of states. Furthermore, it is distinct from the concept of the common heritage of mankind.[54]

The principle has been discussed in the literature, and treated together with '*common heritage and concern of humankind'*, though the common concern principle is newer than 'common heritage'.[55] The common concern principle implies that 'the whole humanity is

53. UNEP Secretariat, Report, II Meeting of the Group of Legal Experts to examine the concept of 'Common Concern of Mankind' in relation to Global Environmental Issues, Geneva, 20-22 March 1991, Revista/documents, II DH, Vol. 13, 1991, pp. 253-258, at p. 253. Already in a 1988 resolution (A/RES/43/53).
54. See e.g. A.A. Cançado Trinidade Antônio, 'Environment and Development: Formulation and Implementation of the Right to Development as a Human Right', available at www.bibliojuridica.org/libros/4/1985/10.pdf (last visited May 2012).
55. Kemal Baslar, *The Concept of the Common Heritage of mankind in International Law*, 1998, especially the discussion at p. 110: 'in order to eliminate this chronic problem [common heritage of mankind without legal effect/power] we advance the principle of common concern according to which only those natural and cultural resources which globally affect the survival and welfare of mankind can be exploited, conserved or protected under the common heritage regime'; Tuula Honkonen, *The Common but Differentiated Responsibility Principle in Multilateral Environmental Agreements, Regulatory and Policy Aspects*, 2009, p. 67; Patricia Birnie, Allan Boyle & Catherine Redgwell, *International Law and the Environment*, 2009, p. 129 ff.; and Alexander Kiss & Dinah Shelton, *International Environmental Law*, 2004, pp. 31-36, at p. 35 the authors write: 'The common heritage of mankind is a concept that emerged at the end of the 1960s to challenge older concepts of *res nullius* and *res communis* in legal approaches to common resources. *Res nullius*, which in most sys-

entitled to express its concern' and the main impact of common concern appears to be that it gives the international community of states both a legitimate interest in resources of global significance and common responsibility to assist in their sustainable development.[56] In the words of *Kemal Baslar* 'the hidden key point is the collective and vital interest of all mankind'.[57]

The concept of the *common concern of mankind* serves as a guiding principle and in its origin it is confined to *global* environmental issues capable of causing significant adverse effects on the environment. While the current global environmental agenda has been tailor-made to fit the interests of the developed countries, it was stated in point 5 of the report that more attention by the international community would be required with respect to the environmental protection of global commons. It was emphasised that the provision of a life of dignity to all, in a clean, safe and healthy environment, should be a matter of common concern of mankind.

Although the principle was originally meant to justify action to combat global environmental problems, implying the existence of a common interest of humankind to protect resources, it covers concerns that should be managed by the international community as a whole.[58] Further, the common concern principle may have given birth to 'the common but differentiated responsibility principle'. This principle states that all parties in the international community (in all parts of the world) should participate and address environmental

tems included wild animals and plants, belong to no one and can be freely used and appropriated when taken and captured. The concept of *res communis* implies the reverse, common ownership that precludes individual appropriation but allows common use of resources e.g. navigation on the high seas. The concept of common heritage of mankind is distinct from both earlier concepts, in part because of its inclusion of the word 'heritage', connoting a temporal aspect in the communal safeguarding of areas incapable of national appropriation. Special legal regimes have been created for deep seabed and its subsoil, Antarctica, the Moon, the geostationary orbit of satellites, and areas, sites and monuments that form essential parts of the cultural heritages of humanity. The nature of the common heritage is a form of trust whose principal aims are exclusive use for peaceful purposes, rational utilization in a spirit of conservation, good management or wise use, and transmission to future generations'.

56. Tuula Honkonen, *The Common but Differentiated Responsibility Principle in Multilateral Environmental Agreements, Regulatory and Policy Aspects*, 2009, p. 67, footnote 184.
57. Kemal Baslar, *The Concept of the Common Heritage of mankind in International Law*, 1998, p. 110.
58. Tuula Honkonen, *The Common but Differentiated Responsibility Principle in Multilateral Environmental Agreements, Regulatory and Policy Aspects*, 2009, p. 68.

problems. Common responsibility includes considering the needs of others.[59] The principle involves a direct response to differences between countries and their capacities to take action at national level, so there are the two dimensions of the allocation of rights or obligations and the re-distribution of resources.[60] As the principle of common but differentiated responsibility is derived from the principle of the common concern of mankind, these concerns can be absorbed by *the common concern of mankind principle* as well as being placed in a patent law context.

The connection between the concept of the *common concern of mankind* and patent law can seen in the later adopted 1992 UN Convention on Biological Diversity, Article 16, on access and transfer of technology. Point 7, of the UN report referred to above stated that the experts considered the principles relating to the common concern of mankind which could be reflected in a possible Convention on Biological Diversity (prior to the adoption of the CBD). It was stressed in the report that the benefits of biotechnological research should be equitably shared and made available to the developing countries which, in most cases, were the original sources of the gene pools on which the research was based. The issues of conservation, and access to biomaterials were also emphasised. In the conclusion of the discussion on benefit-sharing it was stated that there should be a legal reflection of a full scale partnership which simultaneously seeks to protect the environment while looking to raise up the developing countries. Thus, in modern global policymaking, the common concern principle is connected to the only relevant international environmental convention (the CBD) so far that has explicitly related to IPRs. However, the final text of the CBD did not refer to the principle by name.[61] In this regard it is also interesting to note the discussion in point 9 of the report of a possible new code of international environmental principles to be elaborated by the 1992 UN Conference on Environment and Development, to address the issues of the environment and development coherently. It was stated a proper solution could be an authoritative statement of the world community, in the form of a code on the issue of the environment and development.

As a conclusion, it is *first* and foremost important to stress the conclusion from Chapter 5 that TRIPS is not a barrier to a reassessment of patentable

59. Tuula Honkonen, *The Common but Differentiated Responsibility Principle in Multilateral Environmental Agreements, Regulatory and Policy Aspects*, 2009, p. 2.
60. Tuula Honkonen, *The Common but Differentiated Responsibility Principle in Multilateral Environmental Agreements, Regulatory and Policy Aspects*, 2009, pp. 2-3.
61. See Alexander Kiss & Dinah Shelton, *International Environmental Law*, 2004, p. 358, addressing biological diversity, for the first time recognised in a global treaty as a common concern of humankind.

subject matter. From the doctrine of nature perspective, the reshaping of the patentability of DNA molecules can be absorbed by the principle of the common concern of mankind.[62] This would be true for any other value-based evaluation of patentable subject matter, such as embryonic stem cells or essentially biological processes. *Second*, the CBD concern for a fair share of the world's genetic resources and disclosure of origin can also be absorbed by the principle.[63] *Third*, the concerns for global health equity and food security, as well as climate change concerns, can also be addressed as a *common concern of mankind*.[64]

62. See Chapter 5, sections 2 and 3 with literature references. And see Andrew Auchincloss Lundgren, *The Eligibility of DNA Sequence Discoveries for Federal Patent protection: A legal research guide*, Vol. 45, 2004; and Andrew W. Torreance, 'Patenting Human Evolution', *Kansas Law Review*, 2008, Vol. 56, p. 1075 ff.
63. Concerns of this nature have been discussed by many scholars for example: Ulrich Brand, Christoph Görg, Joachim Hirsh & Markus Wissen, *Conflicts in Environmental regulation and the Internationalisation of the State, contested terrains*, 2008, p. 144 and Chapter two on the value of nature: the convention on biological diversity and the commercialization of genetic resources, p. 53 ff; Charles Lawson, 'The role of patents in biodiversity conservation', *Nature Biotechnology*, Vol. 27, No 11, 2009, p. 994 ff.; Klaus Bosselmann, 'Plant and Politics: The International Legal Regime concerning Biotechnology and Biodiversity', *Colorado Journal of International Environmental Law and Policy*, 1996, Vol. 7, p. 111 ff.; I. Serageld in & G.J. Persey, *Biotechnology and Sustainable development, Voices of the South and North*, 2003; Bruce Harper, 'TRIPS Article 27.2: An Argument for Caution', *William and Mary Environmental Law and Policy Review*, 1997, Vol. 21, p. 381 ff.; Philippe G. Le Prestre, *Governing Global Biodiversity, The evolution and implementation of the Convention on Biological Diversity*, 2002; Krishna R. Dronamraju, *Biological and Social Issues in Biotechnology Sharing*, 1998; Nordic Council of Ministers, 'Access and Rights to Genetic Resources, A Nordic approach', Nord 2003:16; Martin Khor, *Intellectual Property, Biodiversity and sustainable development, resolving the difficult Issues*, 2002; Michael Jeffery, Jeremy Firestone & Karen Bubna-Litic, *Biodiversity, Conservation, law + Livelihoods, Bridging the North-South divide*, 2008; Vandana Shiva, *Protect or Plunder?, understanding Intellectual Property Rights, global issues*, 2001; same author *Biopiracy, The Plunder of nature and Knowledge*, 1997; Jonathan Curci, 'The protection of Biodiversity and Traditional Knowledge', in] *International Law of Intellectual Property*, 2011; Francesco Francioni & Tulli Scovazzi, *Biotechnology and International law*, 2006.
64. See UN General Assembly, 'Keeping the promise: a forward-looking review to promote an agreed action agenda to achieve the Millennium Development Goals by 2015', Report of the Secretary-General, 12 February 2010 (A764/665). Concerns of this nature have been discussed in the literature, see e.g. Tzen Wong & Graham Dutfield, *Intellectual Property and human development, Current trends and future scenarios*, 2011, Chapters 2 and 3; Matthew Rimmer, *Intellectual Property and climate Change, Inventing Clean Technologies*, 2011; International Centre for Trade and Sus-

The approach to environmental principles described in Chapter 9 was critical of their transfer to patent law; see Chapter 9, section 2.1. Why would the international environmental principle of the common concern of mankind be any different? In her PhD thesis, the author discussed the environmental principles in an EU context, arguing that in a way these principles have acted to promote the integration of environmental concerns beyond trade and economic interests. The criticisms in Chapter 9 are directed at patent authorities' use of the precautionary principle for guidance in the pre-grant phase, due to its practical implications and the uncertainty about a harmonised and global consensus about its content. However, the *common concern principle* is broader and more flexible and addresses several concerns covered by this umbrella principle as a *global common principle.* As has been argued in Chapter 9 and above, it is possible to transfer the principle to a broader context, thereby re-

tainable Development, 'The Climate Technology Mechanism: Issues and challenges', *ICTSD*, 18 March 2011; K, Maskus, 'Differentiated Intellectual Property Regimes for Environmental and Climate Technologies', 2010, No 17, OECD publishing; and the same author, 'Intellectual Property and the transfer of green technologies: An Essay on economic perspectives', 2009, *W.I.P.O.J.*, No 1, p. 133 ff.; and in the same journal Peter Drahos, 'The China-US relationship on Climate change, Intellectual property and CCS: Requiem for a Species?', p. 125 ff.; OECD Studies on Environmental Innovation, Environmental Policy, Technological Innovation and Patents, 2008; F. Scott Kieff, 'Sustainable Agriculture: Food for the Future, Patents for environmentalists', 2002, Washington University Journal of Law and Policy, p. 307 ff.; Russell Thomsen & Elisabeth Webster, 'The Role of Intellectual property Rights in Addressing Climate Change: The case of Agriculture', 2010, 2 *W.I.P.O.J.* Issue 1, p. 133 ff.; Michael Blakeney, 'Climate Change and gene patents', *Queen Mary Journal of Intellectual Property*, Vol. 2, No 1, 2012, pp. 2-13; Andrew W. Torrance, Patent law, 'Hippo and the Biodiversity Crisis', 9, *The John Marshall Review of Intellectual Property Law*, 624 (210) (Patent law might seem an unlikely remedy for the mass extinction of biodiversity); same author in: An Extinction Bar to Patentability, *The Georgetown Int'L Envtl. Law Review*, 2008, Vol. 20, p. 237 ff.; Christopher E. James, 'The impact on Agricultural research by genetic material patents and the need for clarity and reform in patent law for genetic material', 11 *Drake Journal of Agricultural law*, 2006, p. 253 ff; Duncan Matthews; 'When framing meets law: Using human rights as a practical instrument to facilitate access to medicine in developing countries', (2011), 3 *W.I.P.O.J.* issue 1, p. 113 ff.; Chuan-Feng Wu, 'Raising the right to health concerns within the framework of international property law', 5 *Asian J. WTO & Int'l Health & Policy*, 2010, p. 141 ff.; Stephen M. Mauer, Arti Rai & Andrej Sali, Neglected diseases, Finding Cures for Tropical Diseases: Is open Source the Answer?, PLoS Medicine, Vol. 1, issue 3, December 2004, p. 56; Meir Perez Pugatch, Rachel Chu & David Torstensson, 'Assembling the pharmaceutical R&D puzzle for needs in the developing world', *Pugatch Consilium*, May 2012; DNDi Policy Brief, 'Why an essential Health R&D Convention is needed', April 2012.

flecting the issues of access to medicine, environmentally sound technologies and so on, but it is also possible to take it into account at the national level, thus supporting the social responsibility of the patent regime. Viewing patent law as a bellwether for global concerns may legitimate the adoption of an overall principle of guidance, such as the proposed 'common concern of mankind'. Building a bridge between 'the future we want' in RIO +20, bio-economies, and/or future needs to be addressed *and* patent law may express the very nature of patent law as being a 'lightening rod' for new developments and concerns.[65]

The principle of the common concern of mankind rises above discussions about securing the future sustainability of the patent regime. Perhaps the principle would be able to put an end to arguments about the tailor-making patent law for nanotechnologies or new emerging technologies, and other yet to be discovered technologies. Unfortunately this exists in the imagination, but life is never that simple!

65. Jean-Pierre Beurier, Alexandre Kiss and Said Mahmoudi, '*New Technologies and Law of the Marine Environment*', 2000, p. 215 ff.; A.C. Tridade and David J. Attard, The Implication of the 'Common Concern of Mankind' concept, in Toru Iwama, '*Policies and Laws on Global Warming: International and Comparative Analysis*', 1991, p. 33 ff.; Philippe sands & Jacqueline Peel, *Principles of International Environmental Law*, third ed., 2012, p. 210 ff. and 233 ff.

Sandrine Maljean-Dubois and Lavanya Rajamani, *Implementation of International Environmental Law*, 2011, p. 134 ff. See also the UN RIO +20 Conference on Sustainable Development, reaffirming the three Rio Conventions; UNFCCC, CBD and the UN Convention to Combat Desertification in Those Countries Experiencing Serious Drought and/or Desertification, Particularly in Africa (CCD), paragraph II, A, A/CONF.216/L.1; '*The Future we want*': final document of the Rio+20 Conference; UN, '*Our common vision*', June 2012; COM(2012) 60 final, Communication from the Commission to the European Parliament, the Council, the European Economic and Social Committee and the Committee of the Regions: 'Innovating for Sustainable Growth: A Bioeconomy for Europe', p. 6: Investment in knowledge, innovation and skills on key enabling technologies; 'The Challenges of the 21st Century and perspectives of a Bioeconomy', in National Research Strategy Bioeconomy 2030; 'Our Route towards a biobased economy', Federal Ministry of Education and Research, Berlin 2011, p. 4; 'Bio-economy', Council Report 2010, Bio-Economy Innovation, Berlin 2010; Ronnie Hall, 'Bio-economy versus Biodiversity', Global Forest Coalition, April 2012; The Climate Technology Mechanism: Issues and challenges, ICTSD, Information note number 18. March 2011; Technologies for Climate Change and Intellectual Property: Issues for Small Developing Countries, ICTSD, Information note number 12. October 2009.

# Bibliography


**Books**

*Abbott, Frederick M. & Dukes, Graham:* Global Pharmaceutical Policy, Ensuring Medicines for Tomorrow's World, Edward Elgar, 2009.

*Adelman, Martin J.*: Patent Law Perspectives, second edition, LexisNexis, 2009.

*Andersen, Per, Letto-Venamo, Pia, Modeér, Kjell Åke & Vogt, Helle:* Liber Amicorum Ditlev Tamm, Law, History and Culture, Djøff Publishing.

*Anker, Helle Tegner & Olsen, Birgitte Egelund:* Miljøretlige emner, Festskrift til Ellen Margrethe Basse, Jurist- og Økonomforbundets Forlag, 2008.

*Arnull, Anthony, Dashwood, Alan, Dougan, Michael, Ross, Malcolm, Spaventa, Eleanor & Wyatt, Derrick:* European Union Law, London, Sweet & Maxwell, 2006.

*Atkinson, Rob, Terizakis, Georgios & Zimmerman, And Karsten:* Sustainability in European Environmental Policy, Challenges of Governance and Knowledge, Routledge, 2011.

*Bently, Lionel & Sherman, Brad:* Intellectual Property Law, 3rd edition, Oxford University Press, 2009.

*Bernier, Louise:* Justice in Genetics, Intellectual Property and Human Rights from a Cosmopolitan Liberal Perspective, Edward Elgar publishing limited, 2010.

*Bessen, James and Meurer*, Michael J., Patent failure, how Judges, Bureaucrats and Lawyers put innovation at risk, Princeton University Press, 2008.

*Bird, Robert C. & Jain, Subhash C.*: The Global Challenge of Intellectual Property Rights, Edward Elgar, 2008.

*Birnie, Patricia, Boyle, Alan & Redgewell, Catherine:* International Law & the Environment, 3rd Edition, Oxford University Press, 2009.

*Blakeney, Michael*: Trade Related Aspects of Intellectual Property Rights: A Concise Guide to the TRIPS Agreement, (Intellectual Property in Practice), Sweet & Maxwell, 1996.

*Bostyn, Sven J.R.*: Enabling Biotechnological Inventions in Europe and the United States, A study of the patentability of proteins and DNA sequences with special emphasis on the disclosure requirement, European Patent Office, 4th edition, 2001.

*Brand, Ulrich, Görg, Christoph, Hirsch, Joachim & Wissen, Markus:* Conflicts in Environmental Regulation and the Internationalisation of the State, Contested terrains, Routledge, 2008.

*Bray, Alberts, Johnson, Hopkin, Raff, Lewis & Walter, Roberts:* Essential Cell Biology, 3rd edition, Garland Science, Taylor & Francis Group, 2010.

*Brownsword, Roger & Yeung, Karen:* Regulating Technologies, Legal Futures, Regulatory Frames and Technology Fixes, Hart Publishing, 2008.

*Bruun, Niklas:* Intellectual Property Beyond Rights, WSOY, 2005.

*Bucknell, Duncan:* Pharmaceutical, Biotechnology, and Chemical Inventions, Volume I, Oxford University Press, 2011.

*Bucknell, Duncan:* Pharmaceutical, Biotechnology, and Chemical Inventions, Volume II, Oxford University Press, 2011.
*Bugge, Hans Christian & Voigt, Christina:* Sustainable Development in International and National Law, Europa Law Publishing, 2008.
*Búrca, Gráinne de & Scott, Joanne:* The EU and the WTO, Legal and Constitutional Issues, Hart Publishing, 2001.
*Burchfiel, Kenneth J.:* Biotechnology and the Federal Circuit, Cumulative supplement, The Bureau of National Affairs, 2000.
*Burchfiel, Kenneth J.:* Biotechnology and the Federal Circuit, Cumulative supplement, The Bureau of National Affairs, 2005.
*Burchfiel, Kenneth J.:* Biotechnology and the Federal Circuit, The Bureau of National Affairs, 2nd. Ed. 2010.
*Burk, Dan L. & Lemley, Mark A.:* The Patent Crisis and How the Courts Can Solve It, University of Chicago, 2009.
*Byers, Michael:* The Role of Law in International Politics, Oxford University Press, 2001.
*Carolan, Michael S.:* Decentering Biotechnology, Assemblages Built and assemblages Masked, Ashgate, 2010.
*Carrier, Michael A.:* Innovation for the 21st Century, Harsenessing the Power of Intellectual Property and Antitrust Law, Oxford University Press, 2009.
*Castle, David:* The Role of intellectual Property rights in Biotechnology Innovation, Edward Elgar, 2009.
*Chalmers, Damain, Hadjieemmanuil, Christos, Monti, Giorgio & Tomkins, Adam:* European Union Law, Cambridge University Press, 2006.
*Claes, Erik, Devroe, Wouter & Keirsbilck, Bert:* Facing the Limits of the Law, Springer-Verlag Berlin Heidelberg, 2009.
*Claybourne, Anna*: The Usborne Introduction to Genes & DNA, EDC publishing, 2006.
*Cook, Trevor:* EU Intellectual Property Law, Oxford University Press, 2010.
*Correa, Carlos M. & Yusuf, Abdulqawi A.:* Intellectual Property and International Trade, The TRIPS Agreement, Kluwer Law International, 2008.
*Correa, Carlos M.:* Intellectual Property Rights, the WTO and Developing Countries, The TRIPS Agreement and Policy Options, Zed Books, 2000.
*Correa, Carolos M.:* Research Handbook on the Interpretation and Enforcement of Intellectual Property under WTO Rules, Intellectual Property in the WTO Volume II, Edward Elgar Publishing, 2010.
*Correa, Carolos M.:* Research Handbook on the Protection of Intellectual Property under WTO Rules, Intellectual Property in the WTO Volume I, Edward Elgar Publishing, 2010.
*Craig, Paul & Gráinne de Búrca:* EU Law, Text, Cases and Materials, 5th Edition, Oxford University Press, 2011.
*Craig, Paul & de Búrca, Gráinne* (ed.), The Evolution of EU Law, second ed. 2011.
*Crespi, R.S.:* Patents a Basic Guide to Patenting in Biotechnology, Cambridge Studies in Biotechnology 6, 1988.
*Cullis, Roger:* Patents, Inventions and The Dynamics of Innovation, A multidisciplinary study, Edward Elgar, 2007.
*Curci, Jonathan:* The Protection of Biodiversity and Traditional Knowledge in International Law of Intellectual Property, Cambridge University Press, 2010.

*Daniel, Bugge Thorbjørn, Ekholm, Thomas, Starup, Peter & Steinicke Michael:* Grundlæggende EU-ret, EU Efter Lissabontraktaten, 2nd edition, Jurist- og Økonomforbundets Forlag, 2011.

*Danielsen, Jens Hartig*: Parallelhandel og varernes frie bevægelighed, Djøf forlag, 2005.

*Damm, Leo A.*: Opfinderrettens genstand, 1937.

*Dashwood, Alan, Dougan, Michael, Rodger, Barry, Spaventa, Eleanor and Wyatt, Derrick*: Wyatt and Dashwood's, European Union Law, sixth edition, 2011.

*Deller, Anthony W.:* Patents Claims, 1, 2nd Edition, Sections, 1-323, CBC, 1971.

*Deller, Anthony W.:* Patents Claims, 2, 2nd Edition, Sections, 324-455, CBC, 1971.

*Deller, Anthony W.:* Patents Claims, 3, 2nd Edition, Sections, 456-583, CBC, 1971.

*Dinwoodie, Graeme B., Hennessey, William O., Perlmutter, Shira & Austin, Graeme W.:* Intellectual Property Law and Policy, 2nd edition, Lexis Nexis, 2008.

*Drahos, Peter & Mayne, Ruth:* Global Intellectual Property Rights, Knowledge Access and Development, Palgrave Macmillan, 2002.

*Drahos, Peter:* The Global Governance of Knowledge, Patent Offices and their Clients, Cambridge University Press, 2010.

*Dronamraju, Krishna R.:* Biological and Social Issues in Biotechnology Sharing, Ashgate, 1998.

*Ducor, Phillippe G.:* Patenting the Recombinant Products of Biotechnology and Other Molecules, Kluwer Law International, 1998.

*Dutfield, Graham & Suthersanen, Uma:* Global Intellectual Property Law, Edward Elgar, 2008.

*Dutfield, Graham:* Intellectual Property Rights and the Life Science Industries, A Twentieth Century History, Ashgate Publishing, 2003.

*Dutfield, Graham:* Intellectual Property, Biogenetic Resources and Traditional Knowledge, EarthScan 2004.

*Eeckhout, Piet:* External Relations of the European Union, Legal and Constitutional Foundations, Oxford University Press, 2004.

*Engelbrekt, Antonina Bakardjieva, Bernitz, Ulf, Domeij, Bengt, Kur, Anette & Nordell, Per Jonas:* Festskrift till Marianne Levin, Norstedts Juridik, 2008.

*England, Paul* (Consulting Editor): Intellectual Property in the Life Sciences, A Global Guide to Rights and Their application, Globe Law and Business, 2011.

*Evans, Malcolm D.:* International Law, 2nd Edition, Oxford University Press, 2006.

*Fairhurst, John:* Law of the European Union, 8th Edition, Pearson Educated Limited, 2010.

*First, Harry, Zimmerman, Diane L. & Dreyfuss, Rochelle C.:* Working within the Boundaries of Intellectual Property, Innovation Policy for the Knowledge Society, Oxford University Press, 2010.

*Flanagan, Anne & Montagnani:* Intellectual Property Law, Economic and Social Justice Perspectives, Edward Elgar Publishing, 2010.

*Follesdal, Andreas, Wessel, Ramses A. & Wouters, Jan:* Multilevel Regulation and the EU, The Interplay between Global, European and National Normative Processes, Martinus Nijhoff Publishers, 2008.

*Francioni, Francesco and Scovazzi, Tulli*: Biotechnology and International law, 2006, Hart Publishing.

*Francioni, Francesco:* Environment, Human Rights & International Trade, Hart Publishing, 2001.

*Gebreselassie, Abeba Tadesse:* The Sustainability of Plants and Plant Intellectual Property, Djøf forlag, 2011.

*Gendreau, Ysolde:* An Emerging Intellectual Property Paradigm, Perspectives from Canada, Edward Elgar, 2008.
*Germer, Peter:* Indledning til Folkeretten, 4th Edition, Jurist og Økonomforbundets Forlag, 2010.
*Gervais, Daniel:* Intellectual Property, Trade and Development, Strategies to Optimize Economic Development in a TRIPS-Plus Era, Oxford University Press, 2007.
*Gervais, Daniel:* The TRIPS Agreement, Drafting History and Analysis, Thomson, Sweet & Maxwell publishing, third ed., 2008.
*Ghidini, Gustavo & Genovesi, Luis Mariano:* Intellectual Property and Market Power, ATRIP Papers 2006-2007, Eudeba, 2008.
*Gibson, Johanna:* Patenting Lives, Life Patents, Culture and Development, Ashgate Publishing Limited, 2008.
*Ginsburg, Jane C. & Dreyfuss, Rochelle Cooper:* Intellectual Property Stories, Foundation Press, 2006.
*Goldbach, Vogelsang-Wenke & Zimmer:* Protection of Biotechnology Matter under European and German Law, A Handbook for Applications, VCH-Law Books, 1997.
*Greenhalgh, Christine & Rogers, Mark:* Innovation, Intellectual Property, and Economic Growth, Princeton University Press, 2010.
*Griller, Stefan & Ziller, Jacques (Eds.):* The Lisbon Treaty, EU Constitutionalism without a Constitutional Treaty? SpringerWienNewYork, 2008.
*Grubb, Philip W. & Thomsen, Peter R.:* Patents for Chemicals, Pharmaceuticals, and Biotechnology, Fundamentals of Global Law, Practice, and Strategy, 5th Edition, Oxford University Press, 2010.
*Hacon, Richard and Pagenberg, Jochen*: Concise European Patent law, second ed. (eds), 2010.
*Harmon, Robert L., Homan, Cynthia A. and McMahon, Charles M.*: Patents and the Federal Circuit, 2010 Supplement, Ninth Edition, BNA Books.
*Hartley, Trevor C.:* European Union Law in a Global Context, Texts, Cases and Materials, Cambridge University Press, 2004.
*Health, Christopher & Sanders, Anselm Kamperman:* Intellectual Property & Free Trade Agreements, International Intellectual Property Law Series, Hart Publishing, 2007.
*Heath, Christopher & Sanders, Anselm Kamperman*: IIC studies, Studies in Industial Policy and Copyright Law, Hart Publishing, 2005.
*Heller, Michael:* The Grid Lock Economy, How Too Much Ownership Wrecks Markets, Stops Innovation and Costs Lives, Basic Books, 2008.
*Hesselink, Martijn W.:* The New European Private Law, Essays on the Future of Private Law in Europe, Kluwer Law International, 2002.
*Holyoak & Torremans:* Intellectual Property Law, sixth edition, (2010).
*Honkonen, Tuula*: The Common but Differentiated Responsibility Principle in Multilateral Environmental Agreements, Regulatory and Policy Aspects, Kluwer law International, 2009.
*Jaenichen/McDonell/Haley, Jr.:* From Clones to Claims, Heymanns Intellectual Property, 3rd edition, 2002.
*Jans, Jans H. & Vedder, Hans H.B.:* European Environmental Law, 3rd Edition, Europa Law Publishing, 2008.
*Jeffery, Michael L., Firestone, Jeremy & Bubna-Litic, Karen*: Biodiversity, Conservation, Law + Livelihoods, Bridging the North-South Divide, Cambridge University Press, 2008.

*Jorde, Lynn B., John C. Carey, Michael J. Bamshad*, and *White, Raymond L.*, Medical Genetics, third edition, Elsevier, 2006.

*Kamstra, Gerald et al*: Special report, Patents on Biotechnological Inventions: The E.C. Directive 2002, Sweet & Maxwell, 2002.

*Kelsen, Hans:* General Theory of Norms, Translated by Michael Hartney, 1991.

*Kieff, F. Scott & Nack, Ralph:* International, United States and European, Intellectual Property, Selected Source Material, 2007-2008, Aspen Publishers, 2006.

*Kiss, Alexander*: International environmental Law, third ed., 2004.

*Kohr, Martin:* Intellectual Property: Biodiversity and Sustainable Development, Zed Books, 2002.

*Kronenberger, Vincent:* The European Union and the International Legal Order: Discord or Harmony?, T.M.C. Asser Press, 2001.

*Kur, Annette & Levin, Marianne:* Intellectual Property Rights in a Fair World Trade System, Proposals for Reform of TRIPS, Edward Elgar Publishing 2011.

*Lang, Winfried:* Sustainable Development and International Law, Graham Trotman & Martinus Nijhof, 1995.

*Le Preste, Philippe G.:* Governing Globsl Biodiversity, Ashgate, 2002.

*Ledendahl, Jonas:* Copyright Protection of Software under the TRIPS Agreement, Software reengineering and reverse engineering in the context of international trade law, Doctoral Thesis, Lund University, School of Economics and management, 2010.

*Lidgard, Hans Henrik:* Protecting and Transferring Biotech Inventions, Áuthors and Studentlitteratur, 2004.

*Lipscomb III, Ernest Bainbridge:* Patent Claims, 3rd Edition, Volume I-IV, Thomsen/West, 2006.

*Llewelyn, Margaret & Adcock, Mike:* European Plant Intellectual Property, Hart publishing, 2006.

*LLP, Kaye Scholer:* Pharmaceutical and Biotech Patent Law, Practising Law Institute, 2008.

*Lasswell, Harold D. and McDougal, Myres S.*: Jurisprudence for a free society, Studies in law, Science and Policy, New Haven, 2006, Vol. I-II.

*Lodish, Berk, Kaiser, Krieger, Scott, Bretscher, Ploegh and Matsudaira*: Molecular Cell Biology, Sixth edition, 2008.

*Luginbuehl, Stefan:* European Patent Law, Towards a Uniform Interpretation, Edward Elgar, 2011.

*Lundgren, Andrew Auchincloss:* The eligibility of DNA Sequence Discoveries for Federal Patent Protection, A Legal Research Guide, William S. Hein & Co., Inc., 2004.

*Macklin, Ruth:* Double Standards in Medical Research in Developing Countries, Cambridge University Press, 2004.

*MacQueen, Hector, Waelde, Charlotte & Laurie, Graeme*: Contemporary Intellectual Property, Law and Policy, Oxford University Press, 2008.

*Macrory, Richard:* Reflections on 30 Years of EU Environmental Law, A High Level of Protection, The Avosetta Series 7, Europa Law Publishing, 2006.

*Magnus, David, Caplan, Arthur & McGee, Glenn:* Who Owns Life?, Prometheus Books 2002.

*Mahop, Marcelin Tonye:* Intellectual Property, Community Rights and Human Rights, The biological and genetic resources of developing countries, Routledge, 2010.

*Malanczuk, Peter:* Akehurst's, Modern Introduction to International Law, 7th Edition, Routledge, 1987.

*Malbon, Justin & Lawson, Charles:* Interpreting and Implementing the TRIPS Agreement, is it Fair? Edward Elgar, 2008.

*Malinowski, Michael J., Foley, Hoag & Eliot LLP*: Biotechnology, Law, Business, and Regulation, Aspen Law & Business, 2002.

*Maniatis, Spyros & Bently, Lionel:* Perspectives on Intellectual Property, Volume 4, Intellectual Property and Ethics, Sweet and Maxwell Limited, 1998.

*Maskus, Keith E. & Reichman, Jerome H.:* International Public Goods and Transfer of Technology, Under a Globalized Intellectual Property Regime, Cambridge University Press, 2005.

*Maskus, Keith E.:* The WTO, Intellectual Property Rights and the Knowledge Economy, Critical Perspectives on the Global Trade System and the WTO, Edward Elgar, 2004.

*Matthews, Duncan:* Intellectual Property, Human Rights and Development, The Role of NGOs and Social Movements, Edward Elgar, 2011.

*McLean, Sheila A.M.:* Genetics and Gene Therapy, The International Library of Medicine Ethics and Law, Ashgate, 2005.

*Meléndez-Ortiz, Ricardo & Roffe, Pedro:* Intellectual Property and Sustainable Development, Development Agendas in a Changing World, Edward Elgar, 2009.

*Merges, Menell & Lemley:* Intellectual Property in the New Age, 4th Edition, Wolters Kluwer Publishing, 2007.

*Mergers, Mendell & Lemley*: Intellectual property in the new technology age, Fourth ed. 2007.

*Merges, Robert Patrick & Duffy, John Fitzgerald:* Patent Law and Policy: Case and Materials, 5th Edition, Lexis Nexis, 2011.

*Micklos, David A. and Freyer Greg A.*: DNA Sciences, A first Course, second edition, 2003.

*Miller, Joseph Scott:* Patents, Critical Concepts in Intellectual Property Law, Edward Elgar, 2010.

*Mylly, Thomas:* Intellectual Property and European Economic Constitutional Law: The Trouble with Private Information power, Publications of IPR University Center, Vaajakoski, Gummerus Kirjapaino Oy, 2009.

*Nard/Barnes/Madison:* The Law of Intellectual Property, Aspen Publishers, 2006.

*National Research Council:* A Patent System, For the 21st Century, National Academy of Sciences, 2004.

*National Research Council:* Global Dimensions of, Intellectual Property Rights, In Science and Technology, National Academy Press, 1993.

*National Research Council:* Reaping the Benefits of: Genomic and Proteomic Research, Intellectual Property Rights, Innovation and Public Health, National Academy of Sciences, 2006.

*Neergaard, Ulla & Nielsen, Ruth:* EU Ret, 6th Edition, Thomson Reuters, 2010.

*Neergaard, Ulla, Nielsen, Ruth & Roseberry, Lynn:* European Legal Method, Paradoxes and Revitalisation, Djøf publishing, 2011.

*Neergaard, Ulla, Nielsen, Ruth & Roseberry, Lynn:* The Role of Courts in Developing a European Social Model, Theoretical and Mehthodological Perspectives, Djøf Publishing, 2010.

*Nielsen, Ruth, and Tvarnø, Christina D.*, Retskilder & Retsteorier, 3. Ed. 2011.

*Nordic Council of Ministers:* Access and Rights to Genetic Resources, NORD, 2003.

*Nussbaum, Robert L., Roderick R. McInnes, Willard, Huntington F.*, Thompsons & Thompsons, Genetics in medicine, Seventh Ed., 2007.

*OECD Studies on Environmental Innovation:* Environmental Policy, Technological Innovation and Patents, OECD, 2008.

*Olsen, Birgitte Egelund & Sørensen, Karsten Engsig:* Europæiseringen af Dansk Ret, Jurist- og Økonomforbundets Forlag, 2008.

*Olsen, Birgitte Egelund (ed.), Steinicke, Michael (ed.), Sørensen, Karsten Engsig:* WTO Law – From a European perspective, Forlaget Thomsen, 2006.

*Overwalle, Geertrui Van:* Gene Patents and Collaborative Licensing Models, Patent Pools, Clearinghouses, Open Source Models and Liability Regimes, Cambridge University Press, 2009.

*Overwalle, Geertrui Van:* Gene Patents and Public Health, Establissements Emile Bruylant, 2007.

*Ovey, Clare & White, Robin C.A.:* Jacobs & White, The European Convention on Human Rights, 4th Edition, Oxford University Press, 2006.

*Pagenberg/Conish:* Interpretation of Patents in Europe, Application of Article 69 EPC, Heymanns Intellectual Property, 2006.

*Palombi, Luigi:* Gene Cartels, Biotech Patents in the Age of Free Trade, Edward Elgar, 2009.

*Palombi, Luigi*: The Patenting of Biological Materials In the Context of the Agreement on Trade-Related Aspects of Intellectual Property Rights, PhD Thesis, University of New South Wales, Sydney (2004).

*Paterson, Gerald*: 'The European Patent System' (2001).

*Prestre, Philippe G. Le*: Governing Global Biodiversity, The evolution and implementation of the Convention on Biological Diversity, Ashgate, 2002.

*Pedersen, Hanne & Zahle, Henrik (eds.):* Legal Polycentricity: Consequences of Pluralism in Law, Dartmounth Publishing Company, 1995.

*Pila, Justine:* The Requirement for an Invention in Parent Law, Oxford University Press, 2010.

*Pires de Carvalho, Nuno:* The Trips Regime of Patent Rights, 3rd edition, Kluwer Law International, 2010.

*Piris, Jean-Claude:* The Lisbon treaty, A legal and political analysis, 2010.

*Pogge, Thomas, Rimmer, Matthew & Rubenstein, Kim:* Incentives for Global Public Health, Patent Law and Access to Essential Medicines, Cambridge University Press, 2010.

*Pollen, Daniel*: Hannah's Heirs: The Quest for the Genetic Origins of Alzheimer's Disease (New York: Oxford University Press, 1993).

*Pottage, Alain & Sherman, Brad:* Figures of Invention, A history of Modern Patent Law, Oxford University Press, 2010.

*Prime, Terence:* European Intellectual Property Law, Ashgate publishing, 2000.

Prinz zu Waldeck und Pyrmont, W., Adelman, M.J., Brauneis, R., Drexl, J., Nack, R. (Eds.): Patents and Technological Progress in a Globalized World, Liber Amicorum Joseph Straus, Springer, 2008.

*Pugatch, Meir Perez:* The Intellectual Property Debate, Perspectives from law, Economics and Political Economy, Edward Elgar, 2006.

*Rasmussen, Hjalte:* On law and Policy in The European Court of Justice, A comparative study in judicial policymaking, 1986.

*Rasmussen, Hjalte:* EU-Ret I Kontekst, 5th Edition, Forlaget Thomsen, 2003.

*Resnik, David B.*: Owning the Genome, A Moral Analysis of DNA Patenting, State University of New York 2004.

*Revesz, Richard, Philippe Sands and Richard B. Stewart: Environmental Law, the Economy, and Sustainable development, 2000*

*Rimmer, Matthew:* Intellectual Property and Biotechnology, Biological Inventions, Edward Elgar, 2008.

*Rimmer, Matthew*: Intellectual Property and climate Change, Inventing Clean Technologies, EE, 2011.

*Rothschild, Max F., & Newmann, Scott:* Intellectual Property Rights in Animal Breeding an Genetics, CABI publishing, 2002.

*Ross, Alf,* On law and Justice, 1958.

*Sand, Philippe*: Principles of international environmental law, Ed. 2, 2003.

*Sandholtz, Wayne & Sweet, Alec Stone:* European Integration and Supranational Governance, Oxford University Press, 2004.

*Schaaf, Christian, Johannes Zschocke, Lorraine Potocki: Human genetics:* From Molecules to medicine, Wolters Kluwer, 2012.

*Schovsbo, Jens & Rosenmeier, Morten:* Immaterialret, 2. udgave, Jurist- og Økonomforbundet, 2011.

*Serageldin and Persey, G.J.*: Biotechnology and Sustainable development, Voices of the South and North, 2003, CABI publishing.

*Seville, Catherine:* EU Intellectual Property Law and Policy, Edward Elgar, 2009.

*Shiva, Vandana:* Biopiracy: The Plunder of Nature and Knowledge, Library of Congress, 1997.

*Shiva, Vandana:* Protect or Plundor?, Understanding the Intellectual Property Rights, Zed books ltd. 2001.

*Shiva, Verdana:* Monocultures of the Mind, Perspectives on Biodiversity and Biotechnology, Zed books, 2000, 5th edition.

Singer & Stauder, Europäisches Patentübereinkommen, kommentar, 5. Auflage, 2010, page 20 ff. Carl Heymanns verlag.

*Singer/Stauder:* The European Patent Convention – a Commentary, 3rd edition, Volume I, Preamble to Article 89, Thomsen, Sweet & Maxwell, 2003.

*Singer/Stauder:* The European Patent Convention – a Commentary, 3rd edition, Volume II, Preamble to Article 90-178, Thomsen, Sweet & Maxwell, 2003.

*Sollie, Paul & Düwell, Marcus:* Evaluating New Technologies, Methodological Problems for the Ethical Assesment of Technology Developments, Springer, 2009.

*Sommer, Tine:* Affaldsreguleringen I EU-retlig belysning, E.M. Basse ApS og forfatteren, 1997.

*Stac, Alexander*; International Patent Law, Cooperation, harmonization and Institutional analysis of WIPO and the WTO, EE 2011.

*Steiner, Josephine Steiner & Woods, Lorna:* EU Law, 10th Edition, Oxford University Press, 2009.

*Sterckx, Sigrid:* Biotechnology, Patents and Morality, Ashgate Publishing Limited, 2000.

*Suthersanen, Uma, Dutfield, Graham & Chow, Kit Boey:* Innovation without Patents, Harnessing the creative spirit in a Diverse World, Edward Elgar, 2007.

*Sørensen, Karsten Engsig & Nielsen, Poul Runge:* EU-retten, 5th edition, Jurist- og Økonomforbundets Forlag, 2010.

*Teló, Mario:* The European Union and Global Governance, Routledge/GARNET series: Europe in the World, 2009.

*The University of Copenhagen & The Danish Council of Ethics:* Conference Report and Summaries The Ethics of Patenting Human Genes and Stem Cells, Det Etiske Råd, 2005.

*Thorley, Simon, Miller, Richard, Burkill, Guy, Birss, Colin & Campbell, Douglas*: Terrell on the Law of Patents, Intellectual Property Library, London, Sweet & Maxwell, 2006.

*Torremans, Paul & Plomer, Aurora:* Embryonic Stem Cell Patents, European Law and Ethics, Oxford University Press, 2008.

*Torremans, Paul:* Holyoak & Torremans, Intellectual Property Law, 6th edition, Oxford University Press, 2010.

*Tuori, Kaarlo & Sankari, Suvi:* The Many Constitutions of Europe, Edinburgh/Glasgow Law and Society Series, Ashgate, 2010.

*Tuori, Kaarlo:* Critical Legal Positivsm, Ashgate Publishing Limited, 2002.

*Turnpenny, Peter and Ellard, Sian*, EMERY's Elements of Medical Genetics, 13 TH edition, ELSEVIER.

*Ullrich, Hanns:* Technology and Competition Technologie Et Concurrence, Groupe De Boeck, 2009.

*UNCTAD-ICTSD Project on IPRs and Sustainable Development:* Resource Book on TRIPS and Development, Cambridge University Press, 2005.

*Veronese, Andrea & Watchorn, Peter:* Procedural law under the EPC-2000: A practical guide for patent professionals and candidates for the European qualifying examination, 2nd Edition, Medienhaus Kastner, 2011.

*Ventose, Eddy*: Medical Patent Law – The challenges of medical Treatment, EE, 2011.

*Wacks, Raymond*, Understanding Jurisprudence, An Introduction to Legal Theory, (2. ed), 2009.

*W. Cornish & D. Llewelyn:* Intellectual Property: Patents, Copyright, Trade, Marks and Allied Rights, 6th Edition, London, Sweet and Maxwell, 2007.

*Waelde, Charlotte & MacQueen, Hector:* Intellectual Property, The Many Faces of the Public Domain, Edward Elgar, 2007.

*Warshofsky, Fred*, The Patent wars, the battle to own the world's technology, *John Wiley & Sons,* 1994.

*Warren-Jones, Amanda*: Patenting rDNA, Human and Animal Biotechnology in the United Kingdom and Europe, Lawtext publishing, 2001.

*Watson, James D.*, The Double Helix, A personal Account of the Discovery of THE STRUCTURE OF DNA, A touchstone book, Simon & Schuster, 1968.

*Westerlund, Li:* Biotech Patents, Equivalency and Exclusions under European and US Patent Law, Kluwer Law International, 2002.

*Westkamp, Guido:* Emerging Issues in Intellectual Property, Trade, Technology and Market Freedom: Essays in Honour of Herchel Smith, Edward Elgar, 2007.

*Wolfrum, Anne Sophie:* Patentschuts für medizinische Verfahrenserfindungen im Europäischen Patentsystem und im UD-Recht, Mohr Siebeck, 2008.

*Wouters, Jan, André Nollkaemper and Wet, Erika de (editors)*, The Europeanisation of International law, The Status of International law in the EU and its Member States 2008.

*Wong, Tzen & Dutfield, Graham:* Intellectual Property and Human Development, Current Trends and Future Scenarios, Cambridge University Press, Public Interest Intellectual Property Advisers, 2011.

*Wunderlich, Jens-Uwe & Bailey, David J.:* The European Union and Global Governance, A Handbook, 1st edition, Routledge International Handbooks, 2011.

*Zander, Joachim:* The Application of the Precautionary Principle in Practice, Comparative Dimensions, Cambridge University Press, 2010.

**Journal articles**

*Abbott, Carolyn & Booton, David*: 'Using Patent Law's teaching Function to introduce an Environmental Ethic into the Process of Technological Innovation', 21 Geo. *Int'l Envtl. L. Rev.* 219, 2009.

*Abbott, Frederick M.*: 'Innovation and Technology Transfer to Address Climate Change: Lessons from the Global Debate on Intellectual Property and Public Health', *ICTSD*, Issue paper No 24.

*Abbott, Frederick M.*: 'Seizure of Generic Pharmaceuticals in Transit Based on Allegations of Patent Infringement: A Threat to International Trade, Development and Public Welfare', 43 *W.I.P.O.J.* No 1, 2009.

*Abbott, Frederick M.*: Worst Fears Realised: 'The Dutch Confiscation of Medicines Bound from India to Brazil', Comment – Bridges, No. 1, February – March 2009, www.ictsd.net.

*Abbott, Frederick M. and Reichman, Jerome H.*: 'The DOHA round's public health legacy: strategies for the production and diffusion of patented medicines under the amended TRIPS provisions', *Journal of International Economic Law* 10(4), p. 921-987, 2007.

*Aerts, Rob J.:* 'The legitimacy of Patent Law-Making in Europe and the U.S. – A tentative Comparison',Vol. 38 *IIC* 165-182 (2/2007)

*Aerts, Rob J.*: 'Biological patents in Europe – Functions of Recombinant DNA and expressed proteins and satisfaction of the Industrial Applicability Requirement', 39 *IIC* 282-306 (3/2008).

*Agovic, Amina:* 'Stem cell patents on a knife edge', *Journal of Intellectual Property law and Practice*, 2008, vol. 3, no. 11.

*Allarakhia, Minna and Wensley, Anthony*: 'Innovation and Intellectual property rights in systems biology, A framework for characterizing system-based knowledge is needed to understand the new complexities arising from the assignment of IP rights to biological information', *Nature Biotechnology*, December 2005, Volume 23, Number 12.

*Allott, Philip*: 'The concept of international law', in Michael Byers (ed), *The role of law in international politics, essays in international relations and international law*, Oxford University Press, 2001.

*Amos, Jean and Grody, Wayne*: 'Development and integration of molecular genetic tests into clinical practice: the US experience', *Expert Rev. Mol. Diagn*, 4 (4), p. 465-477 (2004).

*Andrews, Lori B.*: 'The gene patent dilemma: balancing commercial incentives with health needs', *Houston Journal of Health law and policy*, 2002, 2HOUJHLP65, pp. 65-106.

*Andrews, Lori B.*: 'Genes and patent policy: rethinking intellectual property rights', *Nature Review*, Genetics, 2002, vol. 3, p. 803-808.

*Andrews, Lori, Paradise, Jordan, Holbrook, Timothy and Bochneak, Danielle*: 'When patents threaten science', *Science*, vol. 314, 2006, p. 1395-1396.

*Arnold, Beth E. and Ogielska-Zei, Eva*: 'Patenting genes and genetic research tools: Good or bad for innovation?' *Annu. Rev. Genomics Hum. Genet*, 2002, 3 p. 415-32.

*Aymé, S., Matthijs, G., and Soini, S.*: 'Patenting and licensing in genetic testing, Recommendations of the European Society of Human genetics', *European Journal of Human genetics*, 2008, 16, p. S3-S9.

*Baird, Patricia*: 'Will genetics be used wisely', *ISUMA*, Canadian Policy Research Journal Vol. 2 (1), Spring 2001, pp. 94-101.

*Bassil, Nick:* 'Developments in patentability of inventions relating to Human Embryonic Stem cells', *Bioscience Law Review*, Vol. 12, Issue 1.

*Beneyto, José Maria*: 'From Nice to the Constitutional Treaty: Eight theses on the (Future) Constitutionalisation of Europe', in Stefan Griller, Jacques Ziller (eds.), *The Lisbon Treaty, EU Constitutionalism without a Constitutional Treaty*, 2008.

*Bently, Lionel and Sherman, Brad*: 'The Ethics of patenting: Towards a Transgenic Patent System', 3 *Medical Law Review*, pp. 275-291, Autumn 1995.

*Berman, Helen M. and Dreyfuss, Rochelle C.*: 'Reflections on the Science and Law of Structural biology, Genomics, and Drug development', 53 *UCLA L. Rev.* 871 (2005-2006).

*Bernhardt, Jeffery and Fox, James*: 'Gene sequences are still patentable: the Federal Circuit decides the Myriad genetics case', *Bioscience Law Review*, vol. 12, issue 2.

*Blakeney, Michael*: 'Climate Change and Gene Patents', *Queen Mary Journal of Intellectual Property*, Vol. 2, No. 1, 2012, p. 2-13.

*Blakeney, Michael*: 'International Intellectual Property Jurisprudence after TRIPS', in *Intellectual Property in the new Millennium*, Essays in Honour of William R. Cornish, D. Vaver and L. Bently, 2004

*Bosselmann, Klaus*: 'Plant and Politics: The International Legal Regime concerning Biotechnology and Biodiversity', *Colorado Journal of International Environmental Law and Policy*, 1996, vol. 7.

*Bostanci, Adam, Calvert, Jane and Joly, Pierre-Benoit*: 'Regulating gene regulation: patenting small RNAs', in Matthew Rimmer and Alison Mclennan (eds), *Intellectual Property and Emerging technologies, The New Biology*, 2012.

*Bostyn, Sven J.R.:* 'A decade after the birth of the biotech directive: was it worth the trouble?', in Emanuela Arezzo & Gustavo Ghidini, *Biotechnology and Software Patent Law*, A Comparative Review of New Developments, 2011.

*Bradshaw, James:* 'Gene Patent Policy: Does issuing gene patents accord with the purposes of the U.S. Patent System', 37 *Willamette Law Review*, 2001.

*Brownsword, Roger*: 'So what does the world need now, reflections on regulating technology', in Roger Brownsword and Kate Yeung (eds), *Regulation technology*, Hart publishing, 2008.

*Búrca, Gráinne de*: 'The EU on the Road from the Constitutional Treaty to the Lisbon Treaty', Jean Monnet Working Paper 03/08.

*Búrca, Gráinne de*: 'The Evolution of EU Human Rights Law', in Paul Craig and Gráinne de Búrca (eds), *The Evolution of EU Law*, 2011.

*Burgess, Paul, Hutt, Peter Barton, Farokhzad, Omid C., Langer, Robert, Minick, Scott and Zale, Stephen*: 'On firm ground: IP protection of therapeutic nanoparticles', *Nature Biotechnology*, vol. 28, number 12, December 2010, p. 1267.

*Burk, Dan L.*: 'Copyrightability of recombinant DNA sequences', 1989, *Jurimetrics J*, Section of Science & Technology, American Bar Association, p. 490-491.

*Burk, Dan L. & Lemley, Mark A.*: 'Is patent law Technology Specific?' 17 *Berkeley Tech. L. J.* 1155 (2002)

*Burk, Dan L. & Lemley, Mark A.*: 'Policy Levers in Patent law', 2003, *University of Minnesota Law School public law & Legal Theory Research paper series*, Research paper no. 03-11.

*Burk, Dan L. & Lemley, Mark A.*: 'Courts and the patent system', *Regulation*, Summer, 2009, page 18 ff.

*Camm, Tara and Bowles, David:* 'Animal Welfare and the Treaty of Rome – A legal Analysis of the protocol on Animal Welfare and Welfare Standards in The European Union', *Journal of Environmental Law*, 2000, Vol. 12, No 2 and *Common Market Law Review*, 2002, 39: pp.1159-1170.

*Callaway, Ewen:* 'European ban on stem-cell patents has silver lining', *Nature* 478, p. 441, October 2011.

*Carbone, Julia, Gold, E. Richard, Sampat, Bhaven, Chandrasekharan, Sushashini, Knowles, Lori, Angrist, Misha & Deegan, Robert Cook*: 'DNA patents and diagnostics: not a pretty picture', *Nature Biotechnology*, vol. 28, number 8, 2010, p. 784 ff.

*Caulfield, Timothy*: 'Underwhelmed: hyperbole, Regulatory Policy, and the genetic Revolution', 45 *McGill Law Journal* 2000, p. 437 f.

*Chahine, Kenneth G.*: 'Anchoring gene patent eligibility to its constitutional mooring', Commentary, *Nature biotechnology*, vol. 28, number 12, December 2012, p. 1253 ff.

*Chan, Sarah*: 'Patents in synthetic biology, May hinder future research and restrict access to innovation', *British Medical Journal* (BMJ) 2010; vol. 340; c1315 editorials.

*Chandrasekharan, Subhashini et al*: 'Impact of gene patents and licensing practices on access to genetic testing for cystic fibrosis', *Genetics in Medicine*, vol. 12, Number 4, April 2010 supplement, p. S194.

*Charbonneau, Nicolle:* 'A cat fight with a twist', http://news.bbc.co.uk/1/hi/sci/tech/1820749.stm

*Chisholm, Daniel P.*: Note: 'The effect of the USPTO's written Description Guideline on Gene Patent applications', 35 *Suffolk University Law review*, p. 543.

*Cockbain, Julian and Sterckx, Sigrid*: 'Is the Enlarged Board of Appeal of the European Patent Office authorized to extend the bounds of the patentable? The G-3/85 second medical indication/EISAI and G-2/08 Dosage Regime/Abbott respiratory cases', 3/2011 *IIC*, p. 257 ff.

*Collins, Francis S., Eric D. Green, Alan E. Guttmacher and Guyer, Mark S.*: 'A vision for future genomics research', *Nature* 422, 835-847 (24 April 2003).

*Conley, John M.*: 'Gene patents and the product of nature', 2010, 84 *Chi.-Kent L. Rev.*109.

*Conley, John M. and Makowski, Roberte*: 'Back to the future: rethinking the Product of Nature Doctrine as a barrier to Biotechnology Patents', 85 *Journal of the Patent and Trademark Office Society* 301 (2003),
http://www.whoownsyourbody.org/conley_article.pdf.

*Cook-Deegan, Robert and Heaney, Christopher*: 'Patents in genomics and human genetics', *Annu. Rev. Genom. Human Genet.* 2010, 11: p.383-425.

*Correa, Carlos M.*: 'Patenting Human DNA', *The Journal of World Intellectual Property* (2007), vol. 10, no. 6, p. *419-437*.

*Cottier, Thomas*: 'The Doha Waiver and Its Effects on the Nature of the TRIPS System and on Competition Law, The Impact on Human Rights', in Inge Govaere & Hans Ullrich (eds.): *Intellectual Property, Public Policy and International Trade* and in NCCR Trade Regulation Working Paper No. 2006/21.

*Cottier, Thomas and Matteotti-Berkutova, Sofya*: 'International environmental law and the evolving concept of 'common concern of mankind' in Thomas Cottier, Olga Nartova and sadeq Z. Bigdeli, *International trade regulation and the mitigation of climate change*, World Trade Forum, p. 21 ff.

*Craig, Paul*: 'Integration, Democracy and Legitimacy', in Paul Craig and Gráinne de Búrca, *The Evolution of EU law*, 2011.
*Craig, Paul*: 'The treaty of Lisbon, process, architecture and substance', *E.L. Rev.* 2008, 33 (2), p. 137-166.
*Cremona, Marise*: 'The Union as a global actor: Roles, Models and Identity', *Common Market Law* Review 41, 2004.
*Demissie, Hailemichael Teshome:* 'Taming matter for the welfare of humanity: Regulating Nanotechnology', in Roger Brownsword and Karen Yeung (eds), *Regulating Technologies, Legal Futures, Regulatory Frames and Technology Fixes*, 2008.
*Derclaye, Estelle*: 'Should patent law help cool the planet? An inquiry from the point of view of environmental law: Part 1', in *EIPR* 2009, 31(4), pp. 168-184.
*Derclaye, Estelle:* 'Patent Law's role in the protection of the environment – reassessing patent law and its justifications in the 21st Century', in *IIC* 2009, 40(3), pp. 249-273.
*Derclaye, Estelle:* 'Not only Innovation but also Collaboration, Funding, Goodwill and Commitment: Which role for Patent Laws in Post-Copenhagen Climate Change Action', 9 J. *Marshall Rev. Intell. Prop.L* 657 (2010).
*Derzko, Natalie M.*: 'In search of a compromised solution to the problem arising from patenting biomedical research tools', 2004, 20 *Santa Clara Computer and High Technology Law Journal*, p. 347 ff.
*Dolder, F.:* 'Schranken der Patentierbarkeit biotechnologischer Erfindungen nach dem Europäischen Patentübereinkommen', 1984 *Mitteilungen der deutschen Patentanwält* p. 1-7.
*Drahos, Peter:* 'Biotech Patents, Markets and Morality', *EIPR* 1999 p. 441.
*Drahos, Peter*: 'The China-US relationship on Climate change, Intellectual property and CCS: Requiem for a Species?', *W.I.P.O.J.* No. 1, p. 125 ff.
*Dreyfuss, Rochelle Cooper*: 'Nonobviousness: a comment on three learned papers', 12 *Lewis & Clark Law Review* 435 (2008).
*Dreyfuss, Rochelle*: 'TRIPS and Essential Medicine', in Thomas Pogge, Matthew Rimmer and Kim Rubenstein (ed.) *Incentives for Global Public Health, Patent Law and access to Essential Medicines*, 2010.
*Dreyfuss, Rochelle C. and Evans, James P.*: 'From Bilsky back to Benson: Preemption, Inventing Around, and the Case of Genetic Diagnostics', *Stanford Law Review*, Vol. 63, issue 6, June 2011, p. 1349.
*Dreyfuss, Rochelle C.*: 'The patentability of (Genetic) Diagnostic in US Law and Policy', published September 2010, Law and Economics research paper series working paper no. 10-44, SSRN no. 1678123; Josef Drexl and Nari Lee (eds), in Pharmaceutical Innovation, Competition and Patent law – A Trilateral Perspective.
*Dutfield, Graham*: 'Is the world ready for Substantive Patent Law Harmonization?', in Peter Draho (ed) *Death of Patents*, 2005.
*Dykxhoorn, Derek M.*: 'RNA interference as an anticancer therapy: a patent perspective', *Informa*, Expert opinion, Expert opin.Ther. Patents (2009) 19(4): p. 475-491.
*Ebert, Allisson D. et al.*: 'Induced pluripotent stem cells from a spinal muscular atrophy patient', *Nature*, Vol. 457, January 2009, p. 277.
*Eisenberg, Rebecca S.*: 'Bargaining over the transfer of property research tools: Is this market failing or emerging?', In Rochelle C. Dreyfuss, Diane Leenheer Zimmerman and Harry First (eds), *Expanding the Boundaries of Intellectual Property, Innovation Policy for the Knowledge Society*, 2001.

*Eisenberg, Rebecca S.*: 'Biotech patents: looking backward while moving forward', *Nature biotechnology*, Vol. 24, number 3, March 2006.

*Eisenberg, Rebecca S.*: 'Molecules vs. Information: should patents protect both?' 8 *B.U.J.Sci. & Tech. L.* 190, 196 (2002).

*Eisenberg, Rebecca S.*: 'Re-examining the role of patents in appropriating the value of DNA sequences', 2000, *Emory University School of Law*, p. 783 ff

*Eisenberg, Rebecca S.*: 'Wisdom of the ages or dead-hand control? Patentable subject matter for diagnostic methods after In re Bilsky', 3 *Case W. RES. J.L. Tech. & Internet 1*(2012), (forthcoming).

*Eisenberg, Rebecca S. & Merges, Robert P.*: Opinion Letter as to the Patentability of Certain Inventions Associated with the Identification of Partial cDNA Sequences, 23, *AIPLA Q.J.* 1, 20 (1995).

*Esoy, Cherylyn A. P.*: 'The PTO's 2001 revised utility examination guidelines for gene patent applications: has the PTO exceeded the scope of authority delineated by the Court's interpretation of a 'useful' invention?' 33 *Seton Hall Law Review*, 2002, p.127.

*Farmer, Stacey J. and Grund, Dr. Martin*: 'An Overview of the New European Convention (EPC 2000) and its potential impact on European Patent Practices', 2006/2007 Vol. 2, *BSLR*.

*Fellmeth, Aaron Xavier* and *Demaine, Linda J.*: 'Reinventing the Double helix: A novel and nonobvious reconceptualization of the Biotechnology patent', *Stanford Law Review*, November 2002 55 STNLR 306.

*Fennell, Lee Anne*: 'Common Interest tragedies', *Northwestern University Law Review, Vol. 98, p. 907, 2004.*

*Fitt, Robert and Nodder, Edward*: 'Specific, Substantial and Credible? A New Test for Gene Patents', *Bio-Science Law Review,* Vol. 9, issue 5.

*Fitt, Robert and Nodder, Edward*: 'An uncertain future for gene patents: the view from Europe', 29 *Biotechnology Law Report*, Number 6, 2010.

*Flanagan, John K.*: 'Gene Therapy and patents', *Journal of the Patent and Trademark Office Society* 80 JPTOS 739, 1998.

*Flores, Alfredo:* 'New way to control E. coli in weaned pigs', Agricultural Research, March 2004.

*Geiger, Christophe*: 'Intellectual Property shall be protected? Article 17(2) of the Charter of Fundamental Rights of the European Union: A mysterious provision with an Unclear Scope', European Intellectual Property Review (*EIPR*) 2009, 31 (3).

*Gerards, M.*, B.J. *van den Bosch* , K. *Danhauser* , V. *Serre* , M. *van Weeghel* , R.J. *Wanders* , G.A. *Nicolaes* , W. *Sluiter* , K. *Schoonderwoerd* , H.R. *Scholte* , H. *Prokisch* , A. *Rötig* , I.F. *de Coo* , *Smeets, H.J.*: 'Riboflavin-responsive oxidative phosphorylation complex I deficiency caused by defective ACAD9: new function for an old gene', *Brain*. 2011 Jan;134(Pt 1): 210-9.

*Gervais, Daniel J.*: 'The changing landscape of international intellectual property', *in* Christoffer Heath & Anselm Kamperman Sanders (eds.) *Intellectual property & Free trade Agreements*, 2007.

*Gervais, Daniel*: 'TRIPS and Development', in Daniel Gervais (ed) *Intellectual Property, Trade and Development, Strategies to Optimize Economic Development in the TRIPS-Plus Era*, 2007.

*Gitter, Donna M.*: 'International conflicts over patenting human DNA sequences in the United states and the European Union: An argument for compulsory licensing and a fair use excemption', 76 *New York University Law Review*, 1623, 2001.

*Ghildiyal, M. and Zamore, P.*: 'Small Silencing RNAs: An expanding universe', *Nature Review Genetics*, 10 (2), 2009, p. 94-108.

*Gold, E. Richard & Caulfield, Timothy A:* 'The moral tollbooth: a method that makes use of patent system to address ethical concerns in biotechnology', *The Lancet*, Vol. 359, June 29, 2002.

*Gore, Athurva, Li, Zhe, Fung, Ho-Lim, Young, Jessica E., Agarwal, Suneet, Antosiewicz-Bourget, Jessica, Canto, Isabel, Giorgetti, Alessandra, Israel, Mason A., Kiskinis, Evangelos, Lee, Je-Hyuk, Loh, Yuin-Han, Manos, Philip D., Montserrat, Nuria, Panopoulos, Athanasia D., Ruiz, Sergio, Wilbert, Melissa L., Yu, Junying, Kirkness, Ewen F., Belmonte, Juan Carlos Izpisua, Rossi, Derrick J., Thomson, James A., Eggan, Kevin, Daley, George Q., Goldstein, Lawrence S. B. and Zhang, Kun*: 'Somatic coding mutations in human induced pluripotent stem cells', *Nature*, 3 March 2011, Vol. 471, p. 63.

*Green, Jeremy B.A.:* 'Patenting: European stem-cell ruling is misleading', *Nature* 479, 41, November 2011.

*Grinberg, Daniel*: 'Recent Patents on DNA & Gene Sequences', 2008, 2, p. 40, *Bentham Science Publisher, Ltd.*

*Grund, Martin & Burda, Martin R.*: 'The New German Patent Law Fails to Transpose EC Directive *98/44/EC* Correctly', 3 *BSLR* 129 (2004/2005).

*Gulliford, Michael John*: 'Much ado about gene patents: The Role of Foreseeability', *Seton Hall Law Review*, 2004 (34 SHLR 711).

*Handler, Michael & Bryan Christopher Mercurio*: 'Intellectual Property in Bilateral and Regional trade Agreements: Commentary and analysis', 2009, pp. 308-341.

*Harper, Bruce*: TRIPS Article 27.2: 'An Argument for Caution', *William and Mary Environmental Law and Policy Review*, 1997, vol. 21, p. 381 ff.

*Hawkins, Jeffrey B.*: 'Review: The Patent Wars: The Battle to Own the World's Technology', by Fred Warshofsky, *Harvard Journal of Law & Technology*, vol. 9, no. 1 1996.

*Helfner, Laurence R.*: 'Regime shifting: The TRIPS Agreement and New Dynamics of International Intellectual Property Lawmaking', 2004, 29 *Yale Journal of International Law.*

*Helfner, Laurence R.*: 'Tragedy of the Anticommons. Can Patents Deter Innovation?' in Biomedical Research, *Science,* 1 May 1998, Vol. 280. no. 5364, p. 624-701 and *Harvard Law Rev.* 111, p. 621-688, 1998.

*Helwegen, Wim*: 'The research exemption from a nanotechnology perspective', *E.I.P.*R. 2010, issue 7, 2010, p. 342.

*Highlander, Steven L. & Wohlers, Travis M.*: 'Patent strategies for therapeutic RNAi', *Expert Opin. Ther. Patents* (2008) 18(11), p. 1292.

*Hilf, Meinhard*: 'The ECJ's Opinion 1/94 on the WTO – No Surprise, but Wise?' 6 *EJIL* (1995) p. 245-259.

*Hirooka et al*: 'Genome Scan Reveals New Coat Color Loci in Exotic Pig Cross', *The Journal of Heredity 2002* (93), p. 1.

*Hoffman, David C.*: 'A modest proposal: towards inproved access to biotechnology research tools by implementing a broad experimental use exception', 89 *Cornell L. Rev.*, 2004, p. 993 ff.

*Hoffmeister, Frank:* 'Outsider or frontrunner? Recent developments under international and European law on the status of European Union in international organizations and treaty bodies', 44 *CML Rev,* 2007, p. 41-68.

*Holman, Christopher M.*: '*Bilski*: assessing the Impact of a Newly Invigorated Patent-Eligilility Doctrine on the Pharmaceutical Industry and the Future of Personalized Medicine', available at http://ssrn.com/abstract=1424493

*Holman, Christopher M.*: 'Debunking the myth that whole-genome sequencing infringes thousands of gene patents', *Nature Biotechnology*, *Vol. 30 (3), Mar 7,* 2012.

*Holman, Christopher M.*: 'Patent border wars: defining the boundary between scientific discoveries and patentable inventions', Opinion, *Trends in Biotechnology*, Vol. 25, No. 12, 2007, p. 541.

*Holman, Christopher M.*: 'Will gene patents impede whole genome sequencing? Deconstructing the myth that 20% of the human genome is patented, *IP Theory*: Vol. 2: Iss. 1. 2011.

*Holman, Christopher M.*: 'The Impact of Human Gene Patents on Innovation an Access: A Survey of Human Gene Patent Litigation', Winter 2007, 76 *UMKC Law Review,* p. 295.

*Hopkins, Michael M., Surya Mahdi, Pari Patel & Thomas, Sandy*: 'DNA patenting: the end of an era?' *Nature biotechnology*, Vol. 25, number 2, February 2007, p. 187

*Horogh, G. et al:* 'Oestrogen receptor genotypes and litter size in Hungarian Large White pigs', *Journal of Animal Breeding and Genetics*, vol. 122 (2005), issue 1, p. 56.

*Huys, Isabelle, Overwalle, Geertrui Van and Matthijs, Gert*: Gene and genetic diagnostic method patent claims: a comparison under current European and US patent law, *European Journal of Human genetics*, 2011, 19, p. 1104-1107.

*Huys, Isabelle, Berthels, Nele, Matthijs, Gert and Overwalle, Geertrui Van:* 'Legal uncertainty in the area of genetic diagnostic testing', *Nature biotechnology*, volume 27, number 10, October 2009, p. 903 ff.

*Huys, Isabelle, Matthijs, Gert and Overwalle, Geertrui Van:* 'The fate and future of patents on human genes and genetic diagnostic methods', *Nature reviews, genetics*, Vol. 13, June 2012, p. 441 ff.

*Hydén, Håkan*: 'Rättslig pluralism ur ett normvetenskapligt perspektiv', in Bettina Lemann Kristiansen & Anette Storgaard (eds), *Nordisk retssociologi, status – aktuelle udfordringer* – visioner, 2010.

*Jacob, sir Robin*: 'Creating the community patent and its courts', in *Intellectual Property in the new millennium*, Essays in Honour of William R. Cornish, D. Vaver and L. Bently, 2004.

*Jaeger, Thomas*: 'The EU Patent: CUI BONO ET QUO VADIT?', *Common Market Law Review* 47: 63-115, 2010.

*Jaenichen, Hans-Rainer*: 'Reduction to practice: All inventions are created equal: refinement of T 1329/04 in favor of the completeness of DNA inventions without wet biology expiriments', *Biotechnology Law Report*, vol. 26, number 1, 2007.

*Jain, Nada:* 'To patent or not to patent: Gene Therapy in the European Union and the United States', *Cardozo Journal of International and Comparative Law*, Winther 1996, 4 CDZJICL p.103.

*James, Christopher E.*: 'The impact on Agricultural research by genetic material patents and the need for clarity and reform in patent law for genetic material', 11 *Drake Journal of Agricultural Law*, 2006, p. 253 ff.

*Jensen, Kyle* and *Murray, Fiona:* 'Intellectual property landscape of the human genome', *Science* 310, No. 5746 (2005), p. 239-40.

*Jiang, H., Ren, Y., Yuen, E. Y., Zhong, P., Ghaedi, M., Hu, Z., Azabdaftari, G., Nakaso, K., Yan, Z., Feng, J.*: 'Parkin controls dopamine utilization in human midbrain dopaminergic neurons derived from induced pluripotent stem cells', *Nat Commun.* 3:668, 2012.

*Kane, Eileen M.*: Symposium: The third annual IP/gender: The unmapped connections symposium: Molecules and conflict: 'Cancer, Patents, and woman's health', *American University Journal of Gender, Social Policy & the Law*, 2007, (15 AM. U.J. Gender Soc. Pol'y & L. 305).

*Kane, Eileen M.*:'Patent-Mediated Standards in Genetic testing', 2008 *Utah L. Rev.* 837, number 3.

*Kane, Eileen M.*: 'Patenting Genes and Genetic Methods: What's at Stake?' *Journal of Business & Technology Law*, 2011, vol. 6, no. 1.

*Kane, Eileen M.*: 'Splitting the gene: DNA Patents and the Genetic Code', *Tennessee law review* 71 Tenn.L. Rev. 707, Summer, 2004.

*Kane, Eileen M.*: 'Patenting Genes and Genetic Methods: What's at Stake?' *Journal of Business & Technology Law*, 2011, vol. 6, no. 1.

*Kane, Eileen M.*: 'Patent ineligibility: maintaining a scientific public domain', 80 *St. John's L. Rev.* 519.

*Karczewski, Lisa A.*: 'Biotechnological gene patent applications: the implications of the USPTO written description requirement guidelines on the biotechnology industry', 31 *McGeorge Law Review*, 2000, p.1043.

*Kass, Leon R.*: 'Ageless bodies and happy souls', in *The New Atlantis*, a Journal of Technology and Society, number 1, Spring 2003, p. 9.

*Keiper, Adam*: 'Nanethics as a Discipline?' *The New Atlantis*, Spring 2007, p. 56.

*Kelin, Christine and Schlossmacher, Michael G.*: 'The genetics of Parkinson disease: Implications for neurological care', 136 *Nature clinical practice neurology*, March 2006 vol. 2 no 3.

*Kelsen, Han:* 'The concept of the legal order', Translated by Stanley L. Paulson, *The American journal of jurisprudence*, (27 Am. J. Juris. 64, 1982).

*Kepler, Thomas B., Crossman, Colin and Cook-Degan, Robert*: 'Metastasizing patent claims on *BRCA1*', *Genomics*, 2010 May, 95 (5): p. 312-314.

*Kesselheim, Aaron S. and Mello, Michelle M.*: 'Gene patenting – Is the Pendulum swinging back?' *The New England Journal of Medicine*, 362; 20 May 2010, p. 1855.

*Kevles, Daniel J:* 'The Advent of Animal Patents: Innovation and Controversy in Engineering and Ownership of Life', *in Max F. Rothschild and Scott Newman* (eds.) *Intellectual Property Rights in Animal Breeding and Genetics*, CABI, 2002.

*Kieff, F. Scott*: 'Perspectives on properties of the human genome project', Academic Press, 2003.

*Kieff, F. Scott*: 'Sustainable Agriculture: Food for Future, Patents for environmentalists', 2002, Washington University Law and policy, p. 307 ff.

*Kingston, William*: *'Intellectual Property in the Lisbon Treaty', European Intellectual Property Review, 30, (11), 2008, pp. 439-443.*

*Klabbers, Jan*: 'Safeguarding the Organizational *Acquis*: The EU's External Practice', *International Organizations Law Review, Volume 4, Number 1* (2007).

*Kock, Michael A.*: 'Court of Justice of the European Union limits patents on DNA sequences: Much ado about nothing or the beginning of erosion for biotech patents?', Vol. 11, *BSLR*, issue 1, p. 3 ff.

*Kock, Michael A.*: 'Purpose-bound protection for DNA sequences: in through the back door?', *Journal of Intellectual property Law and Practice*, 2010, vol. 5, no. 7, p. 495 ff.

*Kohlhepp, Peter M.*: 'When the invention is an invention: Revitalizing Patentable Subject Matter to exclude unpredictable processes', 93 *Minnesota Law Review*, 2008, p. 808.

*Kohler, Stefan*: 'Getting the legal system ready for Biotechnology', Country Report', I *JIBL* Vol. 2 (2005).

*Kolata, Gina*: 'What is Warm and Fuzzy Forever? With Cloning, Kitty', *New York Times*, February 15, 2002.

*Kole, R., Krainer, A.R. and Altman, S.*: 'RNA therapeutics: beyond RNA interference and antisense oligonucleotides', *Nat Rev Drug Discov.*, 2012, 11(2), p. 125-140.

*Koss, Cara*: 'Oysters & Oligonucleotides: Concerns and proposals for patenting research tools', 25 *Cardozo Arts & Ent LJ* 747, 2007.

*Krauss, Jan B. and Takenaka, Toshiko*: 'A special rule for compound protection for DNA-sequences – impact of the ECJ 'Monsanto' decision on Patent Practice', 93 *J. Pat. & Trademark Off. Soc'y* 189, 2011.

*Kuhn, Jeffrey M.*: 'Patentable subject matter matters: new uses for an old doctrine', 22 *Berkeley Tech. L. J.* 2007.

*Kuo, P., Doudeva, L.G., Wang, Y., Shen, C.J., Yuan, H.S.*: 'Structural insights into TDP-43 in nucleic-acid binding and domain interactions', Nucleic Acids Research 37, 2009, (6), pp. 1799-1808 and http://en.wikipedia.org/wiki/TARDBP

*Kwok, Roberta*: 'DNA's master craftsmen', 22, *Nature*, VOL 468, 4 November 2010.

*Landles, Christian & Bates, Gillian P.*: 'Huntingtin and the molecular pathogenesis of Huntington's disease', *EMBO reports* (2004) 5, p. 958.

*Lavenex, Sandra*: 'EU external governance in 'wider Europe'', *Journal of European Public Policy*, 2004, 11:4.

*Lavenex; Sandra & Schimmelfenning, Frank*: 'EU rules beyond EU borders: theorizing external governance in European Politics', *Journal of European Public Policy*, 2009, 16:6.

*Lawson, Charles*: 'The role of patents in biodiversity conservation', *Nature Biotechnology*, Vol. 27, number 11, 2009, p. 994 ff.

*Lawson, Gavin*: 'The ECJ says 'no' to the Community patent Court', Opinion 1/09 Of the ECJ (Full Court), *Journal of Intellectual Property law & Practice*, 2011, Vol. 6, no. 8, p. 512.

*Leber, Thomas M.:* 'Unity of Chemical and Biotechnological Markush Claims under the PCT and EPC – Consistency of the PCT and EPC Guidelines with the Law', *IIC* 2009, 40(2), p. 206-212.

*Sweet, Alec Stone*: 'The European Court of Justice, Living Reviews in European Governance', Vol. 5, (2010), No. 2.

*Lee, Peter Yun-hyoung*: 'Inverting the logic of scientific discovery: applying common law patentable subject matter doctrine to constrain patents on biotechnology research tools', 19 Harvard journal of law and technology, 2005, page 79 ff. (stem cells).

*Lee, R. C., R. L. Feinbaum and Ambros, V.*: 'The C. elegans heterochronic gene lin-4 encodes small RNAs with antisense complementarity to lin-14', Cell 75 (5): December 1993, p.843-54.

*Leibovitz, John S.*: 'Inventing a nonexclusive Patent System', 111 *Yale Law Journal*, 2002.

*Macer, Darryl*: 'Whose Genome Project', *in* Sheila A. M. Mclean, *Genetics and Gene Therapy*, Ashgate, 2005.

*Manspeizer, David:* 'The Cheshire Cat, The March Hare, and The Harvard Mouse: Animal Patents open up a New, Genetically-Engineered Wonderland', *Rutgers Law Review*, Winter 1991(43 RULR 417).

*Maskus, K.*: 'Differentiated Intellectual Property Regimes for Environmental and Climate Technologies', no. 17, *OECD publishing,* 2010.

*Maskus, K.*: 'Intellectual Property and the transfer of green technologies: An Essay on economic Perspectives', 2009, *W.I.P.O.J.* No. 1, p. 133 ff.

*Matthews, Duncan*: 'Lessons from negotiating an Amendment to the TRIPS Agreement: Compulsory licensing and access to medicine', in Guido Weatkamp (ed), *Emerging Issues in Intellectual Property, Trade, Technology and Market Freedom*: Essays in Honour of Herchel Smith, 2007.

*Matthews, Duncan*: 'When framing meets law: Using human rights as a practical instrument to facilitate access to medicine in developing countries', (2011), 3 *W.I.P.O.J.* issue 1, p. 113 ff.

*Matthijs, Gert*: 'DNA diagnostics in practice', in Geertrui Van Overwalle (ed), *Gene Patents and Public Health*, 2007.

*Matthijs Gert*: 'The European opposition against the BRCA gene Patents', *Familial Cancer*, 2006, 5:95-102.

*Matthijs, Gert and Gert-Jan Van Ommen*: 'Gene Patents: from discovery to invention, A geneticist's view', in Geertrui Van Overwalle (ed), *Gene Patents and Collaborative Licensing Models, Patent pools, Clearinghouses, Open Source Models and Liability Regimes*, 2009 and Gert Matthijs, DNA diagnostics in practice, in Geertrui Van Overwalle (ed.), *Gene Patents and Public Health*, 2007.

*Mauer, Stephen M., Rai, Arti and Sali, Andrej*: 'Neglected diseases, Finding Cures for Tropical Diseases: 'Is open Source the Answer?', *Plos Medicine*, Vol. 1, issue 3, December 2004, p. 56.

*Mcbride, M. Scott*: 'Patentability of Human genes: our patent system can address the issue without modification', 85 *Marquette Law Review*, 2001, p. 511 ff

*McGee, Glenn and Banger, Elisabeth:* 'Ethical Issues in the patenting and control of stem cell research', in David Magnus, Arthur Caplan, Glenn McGee (eds), *Who Owns Life*? 2002.

*McGovern, Matthew:* 'Biotechnology and the Patenting of Living Organisms', 3 *Animal Law* 221, 1997.

*McLennan, Alison:* 'Building on BioBricks: constructing a commons for synthetic biology research', in Matthew Rimmer and Alison Mclennan (eds), *Intellectual Property and Emerging Technologies, The new biology*, 2012.

*McLennan, Alison and Rimmer, Matthew*: 'Cosmo, Cosmolino: Patent law and nanotechnology', in Matthew Rimmer and Alison Mclennan (eds), *Intellectual Property and Emerging technologies, The new biology*, 2012.

*Menell, Peter S.*: 'Forty years of wondering in the wilderness and no closer to the promised land: Bilsky's superficial textualism and the missed opportunity to return patent law to its technology mooring', 2011, 63 *Stanford Law Review*, 1289 ff.

*Micklitz, Hans-W.:* 'Judicial Activism of The European Court of Justice and the Development of the European Social Model in Anti-discrimonation and Consumer Law', *in* Ulla Neergaard et al, *The role of Courts in Developing a European Social Model*, Theoretical and Methodological Perspectives, 2010.

*Minssen, Timo and Schwartz, Robert M.*: 'US patent Eligibility in the wake of *Bilsky v Kappos*: 'Business as Usual' in the age of New technologies?', 30 *Biotechnology Law Report* 3, number 1, 2011 (sequel in 30 Biotechnology law report 57, number 1, 2011).

*Morgan, Gareth & Haile, Lisa A.*: 'A shadow falls over gene patents in the United States and Europe', *Nature Biotechnology*, vol. 28, number 11, 2012, p. 1172.

*Moses, Lydia Bennett*, Recurring Dilemmas: 'The law's race to keep up with technological chance', *Journal of Law, Technology & Policy*, Vol. 2007 No.2, p. 240.

*Moufang, R.:* 'Patenting of Human Genes, Cells and Parts of the Body: the Ethical Dimensions of Patent Law', (1994) 25 *IIC* p. 487-515.

*Mowzoon, May*: 'Access versus Incentives: Balancing policies in genetic patents', *Arizona State Law Journal*, Fall, 2003 (35 AZSLJ 1077), p. 1088.

*Murray, John*: 'Owning genes: disputes involving DNA sequence patents', 1999, *Chicago-Kent Law Review*, p. 231 ff.

*Murphy, Brian P. and Murphy, Daniel P.*: 'Bilsky's maschine-or-transformation' test: Uncertain prognosis for diagnostic methods and personalized medicine patents', 20 *Fortham Intell. Prop. Media & Ent.* L.J., 755, 2009-2010.

*Musungu, Sisule F.*: 'The TRIPS Agreement and Public Health', in Carlos M. Correa & Abdulqawi A. Yusuf (eds), *Intellectual Property and International Trade, The TRIPS Agreement*, 2. ed., 2008.

*N., Joseph D.*: 'DNA to proteins and perspectives on protecting the fruits of structural genomic research', *JIBL* Vol 01, 2004.

*Neergaard, Ulla and Nielsen, Ruth*: 'Where did the spirit and its friends go? On the European Legal Method(s) and the Interpretation Style of the Court of Justice of the European Union', in Ulla Neergaard, Ruth Nielsen and Lynn Roseberry (eds.), *European Legal Method, Paradoxes and Revitalisation*, 2011.

*Neumann, M., Sampathu, D.M., Kwong, L.K., Truax, A.C., Micsenyi, M.C., Chou, T.T., Bruce, J., Schuck, T., Grossman, M., Clark, C.M., McCluskey, L.F., Miller, B.L., Masliah, E., Mackenzie, I.R., Feldman H., Feiden, W., Kretzschmar, H.A., Trojanowski, J.Q. and Lee, V.M.:* 'Ubiquitinated TDP-43 in frontotemporal lobar degeneration and amyotrophic lateral sclerosis'. Science, 314, (5796), 130-3, October 2006, doi:10.1126/science.1134108 and http://en.wikipedia.org/wiki/TARDBP

*Newman, Stuart A.:* 'Averting the clone age. Prospects and perils of human developmental manipulation, Journal of Contemporary health law and policy', 19 *JCHLP* 431, 2003.

*Nielsen, Rasmus*: 'Genomics: in search of rare human variants', *Nature*, 467, 1050-51, October 2010.

*O'Connor, Kevin W.:* 'Patenting Animals and other Living Things', *Southern California Law Review*, 1991, 65 S. Cal. L. Rev. 597.

*Odell-West, Amanda*: 'The legacy of Myriad for gene-based diagnostics: a new policy and regulatory option', *Journal of Intellectual Property Law and Practice*, 2009, Vol. 4, number 4, p. 267-277.

*Olsen, Byron V.:* 'The biotechnology balancing act: Patents for gene fragments, and licensing the 'useful arts', *Albany Law Journal of Science and Technology*, 7 ALBLJST 295, 1997.

*O'Malley, Maureen A., Adam Bostanci and Calvert, Jane*: 'Whole-genome patenting', *Nature reviews genetics*, June 2005, vol. 6, p. 506.

*Oprea, Gabriela E. et al*: 'Plastin 3 is a Protective Modifier of Autosomal Recessive Spinal Muscular Atrophy', *Science* 320, 524, 25 April 2008.

*Orlacchio, Antonio, Bernardi, Giorgio, Orlacchio, Aldo and Martino, Sabata*: 'Patented therapeutic RNAi strategies for neurodegenerative diseases of the CNS', *Informa Healthcare*, Expert Opinion, Ther.Patents (2008) 18 (10): p. 1166.

*Oser, Andreas*: 'Patenting (partial) gene sequences taking partial account of the EST issue', *IIC*, vol. 30 no. 1/1999, p. 1 ff.

*Ott, Andrea:* 'The ECJ and International law, Thirty years of case-law by the European Court of Justice on International law: A pragmatic approach towards its integration', *in* Vincent Kronenberger (ed.): *The European Union and the International Legal Order, Discord or Harmony*? 2001

*Overwalle, Geertrui van*: Editorial: 'The CJEU's Monsanto soybean decision and patent Scope – as clear as mud?', *IIC* Vol. 42, No. 1/2011 p. 2.

*Owerwalle, Geertrui Van*: 'Policy Levers Tailoring Patent Law to Biotechnology. Comparing US and European Approaches', *University of California Irvine Law Review*, *1*(2), 2011, pp. 435-517.

*Overwalle*, *Geertrui Van:* 'Legal and Ethical Aspects of Bio-Patenting', in Peter Drahos, *Death of Patents,* Queen Mary Intellectual Property Institute and Lawtext Publishing Limited, 2005.

*Overwalle, Geertrui Van*: 'Study on the patenting of inventions related to human stem cell research', European Communities, p. 218, http://papers.ssrn.com/sol3/papers.cfm?abstract_id=17, 2002.

*Overwalle, Geertrui Van*: 'Turning patent swords into shares', *Science*, 2010, vol 330, p. 1630.

*Overwalle, Geertrui Van and Zimmerren, Ester van*: 'Functions and Limits of Patent Law', in Eric Claes, Wouter Devroe, and Bert Keirsbilck (eds), *Facing the limits of the law*, Springer, 2009.

*Pagenberg, Jochen:* 'Another Year of Debates on Patent Jurisdiction in Europe and No End in Sight?', 38 *IIC* 805 (2007).

*Palombi, Luigi*: 'The Impact of TRIPS on the Validity of the European Biotechnology Directive', 1 *JIBL* Vol. 2, 63 *et seq.* (2005).

*Paradise, Jordan*: 'European opposition to exclusive control over predictive breast cancer testing and inherent implications for U.S. patent law and public policy: A case study of the Myriad genetics BRCA patent controversy', *Food and Drug Law Journal*, 2004, p. 133 ff.

*Paradise, J., Andrews, L. and Holbrook, T.:* 'Patents on Human Genes: An analysis of Scope and Claims', *Science* 307, no. 5715 (2005): pp. 1566-67.

*Paradise, Jordan, Andrews, Lori and Holbrook, Timothy*: 'Patents on Human Genes: An Analysis of Scope and Claims', *Science*, 11 March 2005, Vol. 307, p. 1566.

*Park, In-Hyun, Arora, Natasha, Huo, Hongguang, Maherali, Nimet, Ahfeldt, Tim, Shimamura, Akiki, Lensch, M. William, Cowan, Chad, Hochedlinger, Konrad and Daley, George Q.*: 'Disease-Specific Induced Pluripotent Stem Cells', *Cell* 134, 877-886, September 5, 2008.

*Pagh, Peter*: The Battle on Environmental Policy Competence *in* Richard Macrory (ed.): *Reflections on 30 Years of EU Environmental Law. A High Level of Protection*? 2006.

Peace, Nicholas and Christie, Andrew: 'Intellectual Property Protection for the Products of Animal Breeding', (1996) *EIPR*, 4 European Intellectual Property Review, p. 213-233.

*Pernice, Ingolf*: 'Salvaging the Constitution for Europe, A Reform Treaty for the EU', 2007, WHI, Paper 4/07.

*Pessango, Stephen*: 'Prometheus and Bilsky: Pushing the bounds of patentable subject matter in medical diagnostic techniques with the mashine-or-transformation test', *American Journal of Law and Medicine*, 36 (2010) p. 650.

*Phillips, Jeremy*: 'The English Patent as a Reward for Invention, The Importation of an Idea', 1983, 2 *EIPR* 41.

*Phillips, Peter W.B. and Kerr, William A.:* 'Frustrating Competition Through Regulatory Uncertainty, International Trade in the Products of Biotechnology' in *World Competition* 25 (1): pp. 81-99, 2002.

*Pila, Justine*: 'Chemical Products and Proportionate patents, before and after *Generics v Lundbech'*, available at http://ssrn.com/abstract=1355524, *King's Law Journal,* Vol. 20, p. 489-526, 2009.

*Pitcher, Amanda S.:* 'Contrary to first impression, genes are patentable: should there be limitations?' 6 JHTHCLP 284, *Journal of Health Care Law and Policy*, 2003.

*Plummer, Aurora:* Towards systemic legal conflict: Article 6(2)(c) of The EU Directive on Biotechnological Inventions *in Embryonic Stem Cell Patents, European Law and Ethics*, Aurora Plummer and Paul Torremans (eds), 2009.

*Pugatch, Meir Perez*: 'The process of intellectual property policy-making in the 21st century – shifting from general welfare models to a multidimensional one', *EIPR* 2009 31 (6), p. 311.

*Rai, Arti:* 'An Audience with', in *Nature*, April 2007, Vol. 6, p. 264.

*Rai, Arti K.*: 'Intellectual Property Rights in Biotechnology: Adressing New technology', *Wake Forest Law Review*, 1999, vol. 34, p. 827 ff.

*Raith, Raimund*: 'The European community, The WTO – TRIPS agreement and the WIPO Conventions', in Vincent Kronenberger (ed.), *The European Union and the International Legal Order: Discord or Harmony*, 2001

*Reed, C.*: 'Taking Sides on Technology Neutrality', (2007) 4:3 *SCRIPTed* 263 <http://www.law.ed.ac.uk/ahrc/script-ed/vol4-3/reed.asp>, referring to Bert-Jaap Koops, 'Should ICT Regulation be Technology-Neutral' in Bert-Jaap Koops, Miriam Lips, Corien Prins & Maurice Schellekens, *Starting Points for ICT Regulation: deconstructing prevalent policy one-liners* (The Hague: TMC Asser Press 2006), p. 83-90.

*Reichman, Jerome H and Dreyfuss, Rochelle Cooper*: 'Harmonisation without consensus: Critical reflections on Drafting a Substantive patent law treaty', in Daniel Gervais (ed) *Intellectual Property, Trade and Development, Strategies to Optimize Economic Development in the TRIPS-Plus Era*, 2007.

*Risch, Michael*: 'Everything is patentable', *Tennessee Law Review*, vol. 75, 2008, p. 616.

*Rogers, Eric J.*: 'Can you patent genes? Yes and No', 93 *J. Pat. & Trademark Off. Soc'y*, 2001, p. 39 ff.

*Rohn, Jennifer*: 'RNAi patent win', *Nature Biotechnology* 30, p. 202, March 2012.

*Rosas, Allan*: 'The European Court of Justice: Sources of Law and Methods of interpretation', in *The WTO at Ten, The Contribution of the Dispute Settlement System*, Edited by Giorgio Sacerdoti, Alan Yanovich, Jan Bohanes, 2006.

*Rowlandson, Malene:* 'WARF/Stem cells (G2/06): the ordre public and morality exception and its impact on the patentability of human embryonic stem cells', *European Intellectual Property Review*, 2010 (E.I.P.R) 32(2).

*Ruse-Khan, Henning Grosse*: 'A Real Partnership for development? Sustainable development as a Treaty Objective in European Economic Partnership Agreements and Beyond', 13 *Journal of International Economic Law*, 81, p. 139-180.

*Rutz, Berthold*: 'Synthetic biology and patents, A European perspective', *EMBO reports*, vol. 10, special issue, 2009, p. S15.

*Rutz, Berthold*: 'Patentability of synthetic biology inventions in Europe', *Biotechnology journal*, 2010, 5 p. 11.

*Ryan, Antony L. and Brooks, Roger G.*: 'Innovation vs. Evasion: Clarifying Patent rights in second-generation genes and proteins', 17 *Berkeley Technology Law Journal*, 2002, p. 1265 ff.

*Samuelson, Pamela*: 'Enriching discourse on public domains', 55 *Duke L. J.*, 2006.

*Sand, Inger-Johanne*: 'From hierarchical national constitutions to combinations of constitutional texts and institutions', *in* Kaarlo Tuori and Suvi Sankari, *The Many Constitutions of Europe; The Heroic Saga of Juridical Constitutionalization*, 2010.

*Sand, Inger-Johanne*: (R'e)Constructing the Boundaries of the market: EU Law and Institutions Analysed through the Lens of Discontinuity', in Hanne Petersen, Anne Lise Kjær, Helle Krunke og Michael Rask Madsen (eds), *Paradoxes of European Legal Integration*, 2008.

*Sand, Inger-Johanne*: 'Teoretiske utfordringer for rettsociologien i globaliseringens tid' *in* Bettina Lemann Kristiansen & Anette Storgaard (eds), *Nordisk retssociologi, status – aktuelle udfordringer – visioner*, 2010.

*Sand, Inger-Johanne:* 'Constitutionalism and the Multi-Coded Treaties of the EU: Changing the concepts of Constitutionality' in Kaarlo Tuori and Suvi Sankari, *The Many Constitutions of Europe*, 2010.

*Sant David Paul & Coster, Christoph de*: 'The Unitary patent and the European patent court: A reality check', *BSLR* Vol 12, Issue 1.

*Schade, Jürgen*: 'Is the Community (EU) patent behind the times? – Globalization urges Multilateral Cooperation', *IIC* 7/2010, p. 806 ff.

*Schade, Jürgen*: 'Synergies created by international cooperation in the patent area?', in Wolrad Prinz zu Waldeck und Pyrmont, Martin J. Adelmann, Robert Brauneis, Josef Drexl, Ralph Nack (eds), *Patents and technological Progress in a Globalized World*, Liber Amicorum Joseph Straus, 2009.

*Schatz, Ulrich:* 'Patentability of Genetic Engineering in European Patent Office Practice', 1 *IIC* 1998, p. 2 ff.

*Schatz, Ulrich:* 'Patents and Morality', in Sigrid Sterckx (ed), *Biotechnology, Patents and Morality*, 2[nd] ed. 2000.

*Schmidt, Charles*: 'Negotiating the RNAi patent thicket', *Nature biotechnology*, Vol. 25, Number 3 March 2007.

*Schneider, Michael*: 'Patents in Europe and their Courts – is there light at the end of the tunnel?' *in* Wolrad Prinz zu Waldeck und Pyrmont, Martin J. Adelmann, Robert Brauneis, Josef Drexl, Ralph Nack (eds), *Patents and technological Progress in a Globalized World*, Liber Amicorum Joseph Straus, 2009.

*Schovsbo, Jens*: 'Om pandaens tommeltot, gamle dage, Kina men – især – patentretten', in Per Andersen, Pia Letto-Vanamo, Kjell Åke Modéer & Helle Vogt (eds.), *Liber Amicorum Ditlev Tamm, Law, History and Culture,* 2011.

*Schuster, Jennifer L.*: 'Combining the Components of life: The application of Patent Extraterritoriality Doctrine to Biotechnology', *Indiana Law Journal*, Winter 2008, 83 Ind. L. J. 363.

*Sease, Edmund J.:* 'From Microbes, to Corn Seeds, to Oysters, To Mice: Patentability of New Life Forms, *Drake Law Review'*, 1988-99, Vol. 38, p. 558.

*Sell, Susan K.*: 'Industry strategies for intellectual property and trade: the quest for TRIPS, and post TRIPS strategies', 10 *Cardozo J. Int'l & Comp. L.* 74, 2002.

*Sendtner, Michae:* 'Tailor-made diseased neurons,' *Nature*, Vol 457/15, January 2009, p. 269.

*Sendtner, Michael*: 'Therapy development in spinal muscular atrophy', *Nature Neuroscience*, vol. 13, number 7, July 2010, p. 795.

*Sharples, Andrew and Curley, Duncan:* 'Death of the Oncomouse', *BSLR*, Vol. 8 issue 3, 2005/2006, p. 122.

*Sherman, Brad*: 'Intellectual property and environmental protection', *EIPR* 1991, 13(5), pp. 165-170.

*Simola, Nicola, Pinna, Annalisa and Fenu, Sandro*: 'Pharmacological Therapy of Parkinson's disease: Current Options and New Anenues, Recent patents on CNS Drugs Discovery', 2010, 5, p.221.

*Sinha, Surya Prakash*: 'Legal Polycentricity', *in* Hanne Petersen and Henrik Zahle, *Legal Polycentricity: Consequences of Pluralism in Law.* Dartmounth Publishing Company, 1995.

*Singer, Joseph William*: 'The reliance interest in property revisited', *Stanford Law Review*, Vol. 40, No 3 (Feb., 1988), p. 611-751.

*Smith, Mary Breen*: 'An end to Gene Patents? The Human Genome Project versus the United States Patent and Trademark Office's 1999 Utility Guidelines',73 *University of Colorado Law Review*, 2002.

*Snead, O. Carter*: 'Neuroimaging and Capital Punishment', *The New Atlantis*, Winter 2008.

*Sollie, Paul and Düwell, Marcus*: 'Evaluating New Technologies: An Introduction', in Paul Sollie, Marcus Düwell (eds), *Evaluating New technologies, Methodological Problems for the Ethical Assessment of Technology Developments*, 2009.

*Sommer, Tine*: 'Patenting the Animal Kingdom? From Cross-Breeding to Genetic Make-Up and Biomedical Research', 2/2008 IIC Vol. 39, p. 139 f.

*Sommer, Tine*: 'Interpreting Ordre Public and Morality in a Patent Law Context: Which is the correct Approach', *Bio-Science Law Review*, vol. 9, issue 2, 2006/2007.

*Sommer, Tine:* 'Genpatentet og patentetik i EU – en harmoniseret retsvildfarelse?' in Birgitte Egelund Olsen & Karsten Engsig Sørensen (ed) Europæiseringen af dansk ret, 2008.

*Sommer, Tine*: 'Om avlsmarkører, Miljøhensyn og social ansvarligheds betydning for enerettens legitimitet', Helle Tegner Anker og Birgitte Egelund Olsen (eds), *Miljøretlige emner*, Festskrift til Ellen Margrethe Basse, 2008.

*Sommer, Tine*: 'Patentmyndigheder som bioteknologiens moralske vogter', *Nordisk Immateriellt Rättsskydd*, 2005 (74), hæfte 4, p. 333-344.

*Sommer, Tine*: 'Patentering af det humane genom', *UfR* 2004B 214.

*Sommer, Tine*: 'Retlig beskyttelse af bioteknologi, den danske lovgivning vedrørende implementeringen af direktiv 98/44/EF af 6. juli 1998 om retlig beskyttelse af bioteknologiske opfindelser', *UfR* 2000 B 457.

*Sommer, Tine*: 'Sædelighedsbegrebet i relation til bioteknologi', *Juristen* (10), december 2001.

*Sommer, Tine*: 'The Scope of Gene Patent Protection and the TRIPS agreement – An Exclusively Nondiscriminatory Approach?' 38 *IIC* 49 (1/2007).

*Sommer, Tine*: 'Udførselsrestriktioner og miljøhensyn', *Juristen* nr. 10, 1998.

Somsen, Han: 'Some reflections on EU Biotechnology Regulation', in Richard Macrory (ed.): *Reflections on 30 Years of EU Environmental Law. A High Level of Protection?* 2006.

*Stanton, Angela A.*: 'Forfeited Consent: Body Parts in Eminent Domains', in Johanna Gibson (ed), *Patenting Lives, Life patents, Culture and Development*, 2008.

*Sterckx, S. and. Cockbain, J:* 'Assessing the Morality of the Commercial Exploitation of Inventions Concerning Uses of Human Embryos and the Relevance of Moral Complicity: Comments on the EPO's WARF Decision', (2010) 7:1 *SCRIPTed* 83, http://www.law.ed.ac.uk/ahrc/script-ed/vol7-1/sterckx.asp

*Stolzenburg, Friederikke, Ruskin, Barbara A. and Jaenichen, Hans-Rainer*: 'Of incomplete inventions', *EPL information*, 1/2006.

*Straus, Joseph:* 'Patenting Human Genes in Europe – Past developments and prospects for the future', 26 (6) *IIC* 920 1995.

*Strauss, Joseph*: 'Genomics and the food industry', in D. Vaver and L. Bently, (eds) *Intellectual Property in the New Millennium*, Essays in Hounour of William R. Cornish, 2004.

*Strauss, Joseph*: 'The scope of protection conferred by European Patents on transgenic plants and on methods for their protection', in *Festskrift til Marianne Levin*, 2008.

*Streck, Charlotte & Freestone, David*: 'The EU and Climate Change' in Richard Macrory (ed.*): Reflections on 30 Years of EU Environmental Law. A High Level of Protection?* 2006.

*Suk, J. et al:* ' for dinner? Assessing commercial and regulatory trends in cloned livestock', *Nature Biotechnology*, Vol. 25, Number 1, January 2007.

*Summers, Teresa M.*: 'The Scope of Utility in the Twenty-first Century: New Guidance for Gene-related Patents', 91 *Georgetown Law Journal*, January 2003.

*Tahan, Charles*: 'The Nanotechnology R(evolution)' in Fritz Allhoff, Patrick Lin, James Moor and John Weckert (ed), *Nanoethics, The Ethical and Social Implications of nanotechnology*, 2007.

*Takenaka, Toshiko*: 'Harmony with the rest of the world? The America invents Act', *Journal of Intellectual Property Law and Practice Advance*, November 11, 2011, available on http://ssrn.com/abstract=1940008 .

*Taubman, Antony*: 'The public domain and international intellectual property law treaties', in Charlotte Waelde, Hector Macqueen, *Intellectual Property, The many faces of the public Domain*, 2007.

*Thomsen, Russel and Webster, Elisabeth*: 'The Role of Intellectual property Rights in Addressing Climate Change: The case of Agriculture', 2010, 2 *W.I.P.O.J.* Issue 1, p. 133 ff.

*Tilmann, Winfriend*: 'Scope of protection for gene sequence patents', *IIC* Vol. 36, 8/2005, p. 909.

*Torremans, Paul L.C:*. 'Patentability of human stem cell or synthetic biology based inventions', in Emanuela Arezzo & Gustavo Ghidini, *Biotechnology and Software Patent Law,* 2011.

*Torrance, Andrew W.*: 'An extinction bar to patentability', *The Georgetown Int'L Envtl. Law review*, 2008, Vol. 20, p. 237 ff.

*Torrance, Andrew W.*: 'Metaphysics and patenting life', *UMKC Law Review*, 2007, vol. 76: 2, p. 364 ff.

*Torrance, Andrew W.*: 'Patent law, Hippo and the Biodiversity Crisis', *The John Marshall Review of Intellectual Property Law*, 9 J. Marshall Rev. Intell. Prop.L. 624, 2010.

*Torrance, Andrew W.*: 'Patenting human evolution', *Kansas Law Review*; 2008, vol. 56, p. 1075 ff.

*Tritel, Marc: 'Obtaining useful patent protection for nucleotide and amino acid sequences in the USA', Journal of Intellectual property law & practice, July, 2009,* 4(9), p. 623-628.

*Tuori, Kaarlo*: 'Towards a multi-layered view of modern law', in *Justice, Morality and Society*, A Tribute to Aleksander Peczenik on the occasion of his 60th birthday 16 November 1997.

*Tuori, Kaarlo*: 'The many Constitutions of Europe' in Kaarlo Tuori and Suvi Sankari, *The many constitutions of Europe;* The Heroic Saga of Juridical Constitutionalization, Edinburgh/Glasgow Law and Society Series, Ashgate, 2010.

*Tvedt, Morten Walløe*: 'One Worldwide Patent System: What's in it for developing countries?' *Third World Quarterly*, Vol. 31, No. 2, 2010.

*Tvedt, Morten Walløe*: 'The Path to One Universal Patent', *Environmental Policy and Law*, Vol 37, No 4, 2007, pp. 297-305.

*Ullrich, Hanns*: 'European Intellectual Property Protection', in *Intellectual Property in the new Millennium*, D. Vaver and L. Bently (eds.), 2004.

*Ullrich, Hann:s:* 'Harmony and unity of European Intellectual property protection', in *Intellectual Property in the new Millennium*, Essays in Honour of William R. Cornish, D. Vaver and L. Bently, 2004.

*Verbeure, Birgit, Matthijs, Gert and Overwalle, Geertrui Van*: 'Analysing DNA patents in relation with diagnostic genetic testing', *European Journal of Human genetics* (2006), 14, p. 26-33.

*Wadlow, Christopher*: 'Strasbourg, the Forgotten Patent Convention, and the origins of the European Patent Jurisdiction', *IIC* Vol. 41, 2/2010, p. 123.

*Walterscheid, Edward C.*: 'Patents and the Jeffersonian Mythology', 29, *J. Marshall L. Rev*. 269 (1995).

*Warnock, Mary:* 'The Ethical Regulation of Science', *Nature*, Vol. 450/29 November 2007.

*Warren-Jones, Amanda:* 'Finding a 'Common Morality Codex' for Biotech – A question of Substance, 'morality equated to the 'ethical values' whereas public order equated to the 'legal values', 6/2008, *IIC* Vol. 39.

*Wasowski, Robert S.:* 'The Evolution of Patentable Compositions of matter: The United States Patent Office accepts Genetically Altered Animals as Patentable Subject Matter under 35 U.S.C. Section 101', *Administrative Law Journal*, Fall, 1988, (2ADMLJ 309).

*Wells, Angus J.*: 'Patenting new life forms: an ecological perspective', 1994, *European Intellectual Property Review*, E.I.P.R. 1994, 16 (3), p. 113.

*Wernersson, Schierup et al.*: 'Pigs in sequence space: A 0.66X coverage pig genome survey based on shotgun sequencing', in *BMC Genomics* 2005, 6:70.

*Westerlund, Staffan*: 'Theory for Sustainable Development; Towards or Against?' in Hans Chr. Bugge & Christina Voigt (eds.): *Sustainable Development in International and National Law*, 2008.

*Westin, Lorelei Perez*: 'Genetic Patents: Gatekeeper to the promised cures', *Thomas Jefferson law Review*, Fall 2002, 25 TJLR 271.

*Whiteside, Stephen G.*: 'Patents claiming genetically engineered inventions: A few thoughts on obtaining broad property rights', 1996, *New England Law Review*, p. 1019 ff.

*Wiener, Antje*: 'Enacting meaning-in-use: qualitative research on norms and international relations', *Review of international studies*, vol. 35, p. 176.

*Willoughby, Dr. Kelvin W.*: 'How much does technology really matter in patent law? A comparative analysis of Doctrines of appropriate patentable subject matter in American and European Patent law', 18, Fed. Cir. B.J. 63 2009, p. 63 ff.

*Wilson et al:* 'Novel insight into the control of litter size in pigs, using placental efficiency as a selection tool', *Journal of Animal Science*, 1999 Vol. 77 Issue 7, pp. 1654-1658.

*Wouters, Jan*: 'Dominic Coppens and Bart de Meester: The European Union's external relations after the Lisbon Treaty', in Stefan Griller, Jacques Ziller (eds.): *The Lisbon Treaty, EU Constitutionalism without a Constitutional Treaty,* 2008.

*Wolrad Prinz zu Waldeck und Pyrmont*: 'Special legislation for Genetic Inventions – A violation of Article 27(1)?' in Wolrad Prinz zu Waldeck und Pyrmont, Martin J. Adelmann, Robert Brauneis, Josef Drexl, Ralph Nack (Editors) *Patents and Technological Progress in a Globalized World, Liber Amicorum Joseph Straus*, 2009.

*Wu, Chuan-Feng*: 'Raising the right to health concerns within the framework of international property law', 5 *Asian J. WTO & Int'l Health & Policy*, 2010, p. 141 ff.

*Yang, Xiangzhong et al:* 'Risk Assessment of meat and milk from cloned animals', *Nature Biotechnology* 25, 77-83 (2007), p. 77.

*Yu, Peter K.*: 'The objectives and principles of the TRIPS Agreement', in Carlos M. Correa, *Research Handbook on the Protection of Intellectual Property under WTO Rules, Intellectual Property in the WTO*, Vol. I, 2010.

*Yu, Sa*: 'Political Privilege, Legal Rights, or Public Policy Tool? A History of the Patent System', ATRIP ESSAY COMPETITION 2009: http://www.atrip.org/26.htm.

*Zakem, Aaron J.*: 'Rethinking patentable subject matter: Are Statutory categories useful?' 30 *Cardozo L. Rev. 2009*, p. 2990.

*Zimmer, Dr. Franz-Josef and Quest, Dr. Benjamin*: 'EPO Enlarged Board of Appeal decide on dosage regimes (G2/08) and Treatment by Surgery (G 1/07)', 29 *Biotechnology Law Report* 485, number 5, 2010, p. 485 f.

*Zimmer, Franz-Josef and Quest, Benjamin*: 'When is a process for Production of Plants 'Essentially Biological?', 30 *Biotechnology Law Report*, number 4, 2011, p. 462 f.

*Zimmeren, Ester van, Verbeure, Birgit, Matthijs, Gert and Overwalle, Geertrui Van*: 'A clearing house for diagnostic testing: the solution to ensure access to and use of patented genetic inventions', *Bulletin of the world health organization*, May 2006, 84 (5), p. 352.

*Zimmeren, Ester Van and Overwalle, Geertrui Van*: 'A Paper Tiger? Compolsory License Regimes for Public Health in Europe', *IIC* 1/2011, vol. 42, p. 4-40.

*Zimmeren, Ester van, Verbeure, Birgit, Matthijs, Gert and Overwalle, Geertrui Van*: 'Model for facilitating access to patents on genetic inventions', *Nature Review*, 2006, vol. 7, p. 143 ff.

*Zimmeren, Ester van, Verbeure, Birgit, Matthijs, Gert and Overwalle, Geertrui Van*: 'Patent pools and diagnostic testing, Trends in Biotechnology', vol. 24, no. 3 March 2006, p. 115 ff.

*Zuhn Jr., Donald L.*: 'DNA patentability: shutting the door to the utility requirement', 34 *John Marshall Law Review*, 2000-2001, p. 973-999.

*Reports, websites and more*

'Reaping the benefits of Genomic and Proteomic Research, Intellectual Property Rights, Innovation and Public Health', *National Research Council of the National Academies*, 2006.

'Stem Cell Patents', European Patent Law and Ethics Report, Nottingham July 2006, project co-ordinator Aurora Plomer.

'The Climate Technology Mechanism: Issues and challenges', *ICTSD*, March 18, 2011; http://ictsd.org/i/publications/103789/.

Alzheimer's Disease International, World Alzheimer Report 2010, 'The global Economic Impact of Dementia':
http://www.alz.co.uk/research/files/WorldAlzheimerReport2010.pdf.

Canavan Foundation Press Release. September 29, 2003.
See http://canavanfoundation.org/news/09-03_miami.php.

Coriell: http://ccr.coriell.org/ and ATCC: http://www.lgcstandards-atcc.org/.

*Crouch, Dennis:* 'Reviewing the New USPTO Post Grant Review System with Reference to EPO Oppositions', January 6, 2012,
http://www.patentlyo.com/patent/2012/01/uspto-post-grant.html?utm_source=feedburner&utm_medium=email&utm_campaign=Feed%3A+PatentlyO+%28Dennis+Crouch%27s+Patently-O%29.

EPO: 'Scenarios for the future, how might IP regimes evolve by 2025?' What global legitimacy might such regimes have? 2007.

EU, Commissions report on 'human embryonic stem cell research', SEC (2003) 441 April 2003.

European Group on Ethics in Science and New Technology resulting in Opinion No 7 on 21 May 1996 on 'the ethical aspects of the genetic modification of animals'.

*Georgieva, Bilyana P. and Love, Jane M.*, 'Human Induced pluripotent stem cell: a review of the US patent landscape', special report, *Regen. Med.* (2010) 5(4), p. 581-591.

http://stemcells.nih.gov/info/basics/basics10.asp (Stemcells)

http://www.epo.org/news-issues/issues.html (Nanotechnology and Patents, 2011 (brochure).

http://www.epo.org/news-issues/issues/classification.html. (Patent Classification)

http://www.ornl.gov/sci/techresources/Human_Genome/project/about.shtml (HUGO)

http://www.wipo.int/treaties/en/ (WIPO)

http://www.wto.org/english/thewto_e/minist_e/min01_e/mindecl_e.htm. (DOHA)

*Intellectual Property watch*: 'Echoes of global patent war at WIPO Annual General Assembly, where interfaces are mentioned as the new buzzword', see the Annual Meeting http://www.wipo.int/pressroom/en/articles/2010/article_0038.html.

Interview with Chief Judge Paul R. Michel, US Patent Reform of July, 14, 2011 in *Intellectual Property Watch.*

*Jerome* REICHMAN, 'Patent Law Harmonization and the Draft SPLT', Paper presented to the World Intellectual Property Organization's (WIPO) Open Forum on the Draft Substantive Patent Law Treaty (SPLT), International Conference Center (ICC), Geneva, Switzerland, 1-3 March 2006.

*Johnston, Josephine* and *Wasunna, Angela A:.* Patents, 'Biomedical Research, and treatments: Examining Concerns, Canvassing Solutions', Hasting Center Report, January-February, 2007.

*Keith E. Maskus*: 'Reforming U.S. Patent Policy, Getting the incentives Right', Council special report, Council on Foreign Relation, November 2006.

*Lloyd, Mike, Mokdsi, Georg, Spielthenner, Doris, Beadle, Debbie and Stark, Amanda*: 'Clearing the fog: Patenting trends for the treatment of Alzheimer's disease. What can Network Patent Analysis tell us about developments in the treatment of this crippling condition?' February 2012.

Marks & Clerk's 'Life Sciences Report 2011'.

*New, William*: US Patent Reform signed into law, *Intellectual Property Watch* on 16/09/2011, http://www.ip-watch.org/.

Nordic Council of Ministers, 'Access and Rights to Genetic Resources, A Nordic approach', NORD 2003:16.

Nuffield Council on Bioethics, 'The ethics of patenting DNA', 2002.

Nuffield Council on Bioethics, 'The Ethics of research involving animals', Latimer trend & company, 2005.

OECD Studies on 'Environmental innovation, Environmental Policy, Technological Innovation and Patents', 2008.

Opinion of the European Group on Ethics in Science and New Technologies to the European Commission, Opinion No 16, 'Ethical Aspects of Patenting Inventions Involving Human Stem Cells', 7 May 2002.

Opinion of the Group of Advisors on the Ethical Implications of Biotechnology in Science and New Technologies to the European Commission, Opinion No. 8, 'Ethical Aspects of patenting inventions involving elements of human origin', 25 September 1996.

'Patenting Human genes and stem cells', A Report, The Danish Council of Ethics, 2004.

'Recommendations for a Patent System of the Future', Report by a working group under the Danish Board of Technology, 2007:
http://www.tekno.dk/pdf/projekter/p05_recommendations_for_a_patent_system_of_the_future.pdf

Report of the Secretary's advisory Committee on Genetics, Health, and Society (SACGHS), 'Gene Patents and Licensing Practices and their Impact on Patient Access to Genetic Tests', April 2010.

Report presented to the Norwegian Advisory Board on 'Ethical Aspects of Patenting', October 2007 held in Brussels, see the report 'Patentnemnd uten portefølje? En analyse av etiske utfordringer ved patentering, Oslo 2008, available at
http://etikkom.episerverhotell.net/Documents/Publikasjoner-som-PDF/Patentnemnd%20uten%20portef%C3%B8lje%20%282008%29.pdf (accessed July 2011).

*Saez, Catherine*: *Intellectual Property Watch*, WIPO, 'The 'Sleeping Beauty' of Climate Change Policy, Urged to Awaken', *IP Watch*, July 14, 2011.

*Saez, Catherine*: *Intellectual Property Watch*: 'Patentable Subject Matter, IP Waiver for health Discussed at WTO', March 2011, http://www.ip-watch.org/

*STOA*, 'Policy Options for the Improvement of the European Patent System' (IP/A/STOA/FWC/2005-28/SC16.

'Synthetic Biology: Applying Engineering to Biology', Report of a NEST High-level Expert Group, 2005, page 5, available at
ftp://ftp.cordis.europa.eu/pub/nest/docs/syntheticbiology_b5_eur21796_en.pdf (last visited January 2012).

'The fate of Gene patents under the New Utility Guidelines', 2001, *Duke Law & Technology Review*, February 28, 2001, p. 8.

Trilateral Project WM4 concerning SNPs 'Report on comparative study on examination practice relating to single nucleotide polymorphisms (SNPs) and haplotypes' available at http://www.trilateral.net/projects/biotechnology/Haplotypes.pdf. (last visited may 2012).

UK, pilot project http://www.ipo.gov.uk/pro-types/pro-patent/p-law/p-pn/p-pn-green.htm.

'Understanding our genetic inheritance' the U.S. Human Genome Project, The first five years: Fiscal years 1991-1995, published April 1990, *NIH Publication* No. 90-1590: *http://www.ornl.gov/sci/techresources/Human_Genome/project/5yrplan/summary.shtml*

USPTO pilot project, http://www.uspto.gov/patents/init_events/green_tech.jsp.

WIPO Standing Committee on the law of patents, Twelfth session, Geneva, June 23 to 27, 2008, 'Report on the International Patent System'.
WTO Public Forum e.g. September 2011, 'Seeking answers to global trade challenges': http://www.wto.org/english/forums_e/public_forum11_e/public_forum11_e.htm

**Table of cases**

*European Union Court of Justice*

Case 6/64 *Flamingo Costa v Enel,* 1964 ECR 1964 585
Case 22/70 *European Agreement on Road Transport (ERTA)*, 1971 ECR 41
Case 15/74 *Centrafarm v Sterling Drug*, 1974 ECR 1147
Case 120/78 *Rewe-Zentral AG v Bundesmonopolverwaltung für Branntwein (Cassis de Dijon),* 1979 ECR 649
Case 104/81 *Hauptzollamt Mainz v CA Kupferberg*, 1982 ECR 3641
Case 240/83 *Procecureur de la Republique v ADBHU,* 1985 ECR 532
Case 302/86 *Kommission v Denmark,* 1988 ECR 4607
Case C-30/90 *Commission v Great Britain,* 1992 ECR I-829
Cases C-92 and 326/92 *Phil Collins*, 1993 ECR I-5145
Case C-53/96 *Hermes v FHT*, 1998 ECR I-3603
Joined Cases C-300/98 and 392/98, *Dior*, 2000 ECR I-11307
Case C-377/98, *The Netherlands v European Parliament and Council,* 2001 ECR I-7079
Case C-360/00 *Land Hessen*, 2002 ECR I-5089
Case C-36/02 *Omega Spielhallen- und Automatenaufstellungs-GmbH,* 2004 ECR I-6515
Joined cases C-402/05 P and C-415/05 P, *Kadi and Al Barakaat*, ECR 2008, I-6351
Case C-431/05 *Merck Genéricos – Produtos Farmacêuticos Ld v Merck & Co. Inc. and Merck Sharp & Dohme Ld,* 2007 ECR I-7001
Case C-465/07 *Elgafaji v Staatssecretaris van Justitie*, 2009 ECR I-921
Case C-428/08 *Monsanto*, 2010 ECR I-6765
Case C-34/10 *Oliver Brüstle*, October 18, 2011, OJ C 362/5, 10 December 2011
Opinion of the Court (Full Court) of 8 March 2011. Opinion delivered pursuant to Article 218(11) TFEU – Draft agreement – Creation of a unified patent litigation system – European and Community Patents Court – Compatibility of the draft agreement with the Treaties. Opinion 1/09 available at http://eur-lex.europa.eu/Lex UriServ/LexUriServ.do?uri=CELEX:62009V0001:EN:NOT (last visited January 2012).

*US Table of Cases*

*American Fruit Growers*, 282 U.S. 1931
*Arrhythmia Research Technology, Inc. v Corazonix Corp.* 958 F 2.d 1053 (Fed. Cir. 1992)
*AMP v USPTO and Myriad Genetics*, Case No. 2010-1406 in appeal from The United States District Court for The Southern District of New York, in case no. 09-CV-4515; 702 F. Supp. 181 (S.D.N.Y. 2010), AMP v USPTO, Fed. Cir. 2011, 2010-1406, decided July, 2011, 653 F.3d 1329 (Fed. Cir. 2011)
*Brenner v Manson*, 383 U.S. 519, 148 U.S.P.Q. (BNA) 689 (1966)
*Classen Immunotherapies, Inc v Biogen* IDEC, 381 F. Supp. 2d 452 (D. Md. 2005); No. 2006-1634, 2006 U.S. App. LEXIS 10808 (Fed. Cir. Dec. 19, 2008)
*Cochrane v Badische Anilin & Soda Fabrik*, 111 U.S. 293 (1884)
*Cochrane v Deener*, 94 U.S 780 (1876)
*Diamond v Chakrabarty* 447 U.S. 303 (1980)
*Diamond v Diehr*, 450 U.S. 175, 209 USPQ 1 (1981)

*Enzo Biochem, Inc. v Calgene, Inc.*, 188 F.3d 1362, 52 USPQ2d 1129 (Fed. Cir. 1999)
*Ex Parte Latimer*, 1889 C.D. 123, 46 O.G. 1638 (Com. Pat. 1889)
*Funk Brothers Seed Co v Kalo Inoculant* Co., 333 U.S. 127, 76 USPQ 280 (1948)
*Gottschalk v Benson*, 409 U.S. 63, 175 USPQ 673 (1972)
*In re Abele*, 684 F. 2d 902, 214 USPQ 682 (C.C.P.A. 1982)
*In re Bell*, 991 F.2d 781, 26 USPQ2d 1529 (Fed. Cir. 1993)
*In re Bilski*, 545 F.3d 943, 88 (2008).USPQ2d 1385 (Fed. Cir. 2008); Cert. granted sub nom., *Bilsky v Doll*, 129 S. Ct 2735 (2009). *Bilsky v Kappos*, 561 U.S., 130 S.Ct. 3218 (2010)
*In re Deuel*, 51 F.3d 1552, 1559, 34 USPQ2d 1210, 1215 (Fed. Cir. 1995)
*In re Fisher*, 427 F.2d 833, 166 USPQ 18 (CCPA 1970); 72 USPQ2d 1020 (Bd. Pat. App. & Interf. 2004); 421 F 3.d 1365, 76 USPQ2D 1225 (Fed. Cir. 2005)
*In re Merz*, 97 F.2d 599, 601, 25 C.C. P.A 1314, 1938 Dec. Comm'r Pat 728 (C.C.P.A. 1935)
*KSR Int'l Co v Teleflex* Inc., 550 U.S. 398, 82 USPQ2d 1385 (2007).
*Moore v Regents of the University of California*, 249 Cal Rptr 494 (1988), 793 P. 2d. 479 (1990)
*Morton v New York Eye Infirmary*, 17 F. Cas. 879 (S.D.N.Y 1862) (No. 9,865).
*O'Reilly v Morse*, 56 U.S. 62 (1854)
*Palin v Singer*, 36 U.S.P.Q.2d 1050 (D.Vt 1995)
*Parke-Davis & Co. v H.K. Mulford* Co., 189 F.2d 95 (S.D.N.Y. 1911), modified, 196 F. 496 (2d Cir. 1912)
*Parker v Flook*, 437 U.S. 584, 198 USPQ 193 (1978)
*Prometheus Labs. Inc. v Mayo Collaborative Servs.* 581 F.3d 1336, 92 USPQ2d 1075, 1082-83 (Fed. Cir.2009); *Prometheus Labs. v Mayo Collab. Servs.*, 628 F.3d 1347 (Fed. Cir. 2010), *cert. granted sub nom.* Mayo Collab. Servs. v Prometheus Labs, 131 S.Ct. 3027 (2011); *Mayo Collaborative Services DBA Mayo Medical Laboratories, et al, Petitioners v Prometheus Laboratories,* Inc. 566 US____(2012), no. 10-1150.
*In re Comiskey,* 499 F.3d 1365, 84 USPQ2d 1670 (Fed. Cir. Sep. 20, 2007)
*In re Grams*, 888 F.2d 835, 12 USPQ2d 1824 (Fed. Cir. 1989)
*In re Nuijten* (500 F. 3 d 1346 (Fed. Cir. Sept. 20, 2007)
*Rubber-Tip Pencil Co. v Howard*, 87 U.S. (20 Wall) 498, 507 (1874)
*The American Wood-Paper Co v The Fibre Disintegrating Co*, 90 U.S. (23 Wall.) 566 (1874)
*The telephone cases* 126 U.S. 1, 531 (1888)
*Union Carbide Co v American Carbine Co*, 181 Fed. 106 (2d Cir. 1910)

*EPO Boards of Appeal*
Case G 1/98 *NOVARTIS II/Transgenic Plant, OJ EPO 2000, 511*
Case G 2/02 *India/AstraZeneca*, OJ EPO 2004, 483
Case G 2/06 *WARF* dated 25 Nov. 2008 (306 OJEPO 5/2009); Comments by the President of the European Patent Office, September 2006, http://www.cipa.org.uk/download/EPO_WARF.pdf (last visited January 2012). T-1374/04 *Stem cells/WARF, OJ EPO 2007*
Case G 1/04 *Diagnostic Methods*, OJ EPO 2006 331
Case G 2/07 *Broccoli* and G-1/08 *Tomato*, decision of 10 December 2010 (review of decision T 83/05 (OJ EPO 2007, 644 and T 1242/06 (OJ EPO 2008, 523). Interlocutory decision of 31 May 2012, Tomatoes II.

Case G 1/07 *Treatment by surgery/Medi-Physics*, decision of 15 February 2010
Case G 2/08 *Dosage regime/Abbott respiratory*, decision of 19 February 2010
Case T 0018/09 *Eli Lilly and Company v Human Genome Sciences*, Decision of October 21, 2009 (UK: Eli Lilly and CV v Human Genome Sciences, INC, [2008] EWCA 1903 (Pat); Court of Appeal Eli, Lilly and CV v Human Genome Sciences, INC, [2010] EWCA Civ 33, [2010] R.P.C. 14, 429 H4-H5.
Case T 320/87 *Hybrid Plants*, OJ EPO 1991 71
*Case T 19/90 Harvard, OJ EPO 1990 476, T 315/03 Harvard/Transgenic animals, OJ EPO 1/2006 15*
Case T 356/93 *Plant cells/Plant Genetic Systems, OJ EPO 1995 545*
Case T 272/95 Howard Florey/Relaxin, *OJ EPO* 1995 *388*
Case T 552/00 *Smithkline Beecham Biologicals v Wyeth Holdings,* decision of 30 October 2003
Case T 1191/01 *ICOS Corp,* (not reported)
Case T 1079/03, *Edinburgh University* (not reported).
Case T 80/05, *Method of diagnosis/University of Utah*, decision of 19.11.2008; *Myriad/breast cancer cases; T 666/05* November 13, 2008, case T 902/07, 7 September 2010 (interlocutory decision EP0858467); *T 1213/05 and T 0156/08 (BRCA2/University of Utah) of 14.1.2011*
*Case T 83/05, Plant bioscience/Broccoli, OJ EPO 2007, 644*
Case T 1242/06, *Tomato,* OJ EPO 2008, 523
Case T 1819/07, *Nanosized catalyst/General Electric*) of 15.3.2011, decision of 3 March 2011
Case T 0136/09, *Salmon Brands A/S*, of 14.1.2010

**Table of instruments and legislation**
*Treaties and Secondary Legislation*
Treaty of Lisbon amending the Treaty on European Union and the Treaty establishing the European Community, signed at Lisbon, 13 December 2007, *Official Journal C 306 of 17 December 2007.*
Treaty of the Functioning of the European Union, *OJ. C 83 30 March 2010.*
Charter of Fundamental Rights of the European Union, *Official Journal C 83 of 30.3.2010.*
Paris Convention for the Protection of Industrial Property, 1883.
The law of Treaties, the Vienna Convention 1969, entering into force in 1980.
Convention for the Grant of European patents (Munich, October 5, 1973) as revised by the Act revising Article 63 EPC of 17 December 1991 and the Act revising the EPC of 29 November 2000.
Implementing Regulation to the European Patent Convention of 5 October 1973 as adopted by decision of the Administrative Council of the European Patent Organisation of 7 December 2006 and as last amended by decision of the Administrative Council of the European Patent Organisation of 26 October 2010.
Protocol of interpretation of Art. 69 EPC of 5 October 1973 as revised by the Act revising the EPC of 29 November 2000.
EPO Guideline for substantive examination, available at http://www.epo.org/law-practice/legal-texts/guidelines.html
The Strasbourg Convention, 1963, Convention on the Unification of Certain Points of Substantive law on Patents for Inventions of 27 November 1963.

Budapest Treaty on the International Recognition of the Deposit of Microorganisms for the Purposes of Patent Procedure, April 28, 1977, and amended on September 26, 1980.

Patent Law Treaty (PTL), adopted at Geneva on June 1, 2000.

Patent Cooperation Treaty (PCT), done at Washington on June 19, 1970 (amended September, 28, 1979; modified on February 3, 1984, and October 3, 2001 (as in force from April 1, 2002.

Directive 98/44/EC of the European Parliament and of the Council of 6 July 1998 on the legal protection of biotechnological inventions, OJ L 213, 30.7.1998, p. 13-21.

Directive 2004/48/EC of the European Parliament and of the Council on the enforcement of intellectual property rights.

Agreement on Trade Related Aspects of Intellectual Property Rights, 1994 (TRIPS Agreement) 33 ILM 1197.

Declaration on the TRIPS agreement and public health http://www.wto.org/english/thewto_e/minist_e/min01_e/mindecl_trips_e.htm

Decision of the General Council of 30 August 2003, Document number WT/1/540 available at http://www.wto.org/english/tratop_e/trips_e/implem_para6_e.htm.

European Parliament resolution 04/10/2001 on the patenting of BRCA1 and BRCA2 ('breast cancer') genes.

European Parliament resolution of 19 January 2011 on a European initiative on Alzheimer's disease and other dementias (2010/2084(INI))

Fellesnordisk betænkning NU 1963:6.

Regulation (EC) No 141/2000 of the European Parliament and of the Council of 16 December 1999 on orphan medicinal products.

Implementation of paragraph 6 of the Doha Declaration on the TRIPS Agreement and public health, Decision of the General Council of 30 August 2003.

Regulation (EC) No 816/2006 on compulsory licensing of patents relating to the manufacture of pharmaceutical products for export to countries with public health problems.

Regulation (EC) No 2100/94 as amended by Council Regulation (EC) No 15/2008 of 20 December 2007.

European Parliament resolution of 12 July 2007 on the TRIPS Agreement and access to medicines, PA_TA(2007)0353.

Italian implementation of the Directive *(Legge n. 78 del* 22 *febbraio 2006)* Art. 3, published on 10 March 2006.

Intellectual Property Code, Arts. L. 611-18 and L. 613-2-1, both inserted by Art. 17 (a)( III ) of Law No. 2004-800 of 6 August 2004.

Australian Law Reform Commission, Reform Issue 91, Summer 2007-08 – Animals, http://www.austlii.edu.au/au/other/alrc/publications/reform/reform91/

*US*

U.S. Patent Law, 35 U.S.C. §§ 1 et seq., 1953.

1999 USPTO Revised Interim Utility Guidelines which were later promulgated in January 2001.

53644, Federal Register, Vol. 75, No. 169, Wednesday, September 1, 2010, Notices, see further Manual of Patent Examining Procedure (MPEP), MPEP http://www.uspto.gov/web/offices/pac/mpep/mpep.htm

Revised utility examination guidelines, request for comments, 64 Fed. Reg. 71440 (Dec. 21, 1999) and discussion of Public comments (1092, Federal Register, vol. 66, no. 4,

January 5, 2001, Notices and the Manual of Patent Examining Procedure (MPEP), Eighth Edition, August 2001, Revision 8, July 2010.
http://www.uspto.gov/web/offices/pac/mpep/documents/2100_2107_01.htm.
The America Invents Act (AIA) The Leahy Smith America Invents Act (House bill (HR. 1249, enacted September 16, 2011).
NIH guidelines for human stem cells research.
http://stemcells.nih.gov/policy/2009guidelines.htm (last visited January, 2012).
NIH Principles and Guidelines for Recipients of NIH Research Grants and Contracts on Obtaining and Disseminating Biomedical Research Ressources: Final Notice. Federal Register 64 (246). December 23. Notices. P. 72090,
http://ott.od.nih.gov/pdfs/64FR72090.pdf and
http://ott.od.nih.gov/policy/genomic_invention.html.

*Environmental instruments and papers*

The Stockholm Declaration on the human environment, 1972.
Convention on Biological Diversity, 1992.
Rio-declaration, The United Nations Conference on Environment and Development, June 1992, Rio + 20, June 2012.
The Millennium Declaration on Development Goals, 2000.
*Cartegena protocol (2000) The Cartagena Protocol on Biosafety*
Nagoya Protocol, The *Nagoya Protocol on Access to Genetic Resources and the Fair and Equitable Sharing of Benefits Arising from their Utilization to the Convention on Biological Diversity* (2010).
Brundtland-Commission, Report on Our Common Future, 1987.
United Nations Framework Convention on Climate Change (UNFCCC), 1992. UN GA Resolution 43/53, 1988: The UN Framework Convention on Climate Change, UNFCCC.
Kyoto Protocol on climate change, 1997, The Bali Roadmap 2007, Cancun Agreement 2011.
The Johannesburg declaration, Earth Summit 2002.
http://www.un.org/esa/sustdev/documents/WSSD_POI_PD/English/POI_PD.htm;
Background paper: Climate Change: Technology Development and Technology Transfer, Beijing High-level Conference on Climate Change: Technology Development and Technology Transfer, Beijing, China, 7-8 November 2008, UN Department of Economic and Social Affairs.
Bonn Guidelines on Access to Genetic Resources and Fair and Equitable Sharing of the Benefits Arising out of their Utilization (2002)
http:// www.cbd.int/programmes/socio-eco/benefit/bonn.aspx#.
EPO, UNEP patent law study, Patents and clean technology bridging the gap between evidence and policy, Report from sep. 2010, http://www.epo.org/news-issues/issues/clean-energy/study.html (last visited, December 2011).

*EU Proposals*

First Community Patent Convention, O.J. 1976, L 17/1.
COM(1988)496 PROPOSAL FOR A COUNCIL DIRECTIVE ON THE LEGAL PROTECTION OF BIOTECHNOLOGICAL INVENTIONS, OJ C 10, 13.1.1989, pp. 3-8.

COM(1992)589 final of December 16 1992. Amended proposal for a COUNCIL DIRECTIVE on the legal protection of biotechnological inventions, OJ C 44, 16.2.1993, pp. 36-50.
COM(1993) 570, LIST OF PROPOSALS PENDING BEFORE THE COUNCIL ON 31 OCTOBER 1993 for which entry into force of the Treaty on European Union will require a change in the legal base and/or a change in procedure.
COM(1994) 245, Opinion of the Commission pursuant to Article 189 b (2) (d) of the EC Treaty, on the European Parliament's amendments to the Council's common position regarding the proposal for a European Parliament and Council Directive on the legal protection of biotechnological inventions amending the proposal of the Commission pursuant to Article 189 a (2) of the EC Treaty – COD 1994/0159.
COM(1995) 312: Economic and Social Committee on the 'Proposal for a European Parliament and Council Directive on the legal protection of biotechnological inventions', OJ 96/C295/03.
COM(1995) 661 final, Proposal for a European Parliament and Council Directive on the Legal Protection of Biotechnological Inventions Proposal for a EUROPEAN PARLIAMENT AND COUNCIL DIRECTIVE on the legal protection of biotechnological inventions, OJ C 296, 8.10.1996, pp. 4-10.
COM(1995) 688, Greenpapers on Innovation.
The First Action Plan for Innovation in Europe – Innovation for growth and employment, COM (1996) 589.
COM (2010) 2020 final, Commission Communication from the Council, Europe 2012, A strategy for smart, sustainable and inclusive growth.
COM(1997) 446. Amended proposal for a European Parliament and Council Directive on the legal protection of biotechnological inventions, OJ C 311, 11.10.1997, pp. 12-30.
COM(1997) 314 Promoting innovation through patents – Green Paper on the Community patent and the patent system in Europe.
COM(2000) 1, EU Commission communication on the precautionary principle 1 of February 2, 2000.
COM(2000) 412 final Proposal for a Council regulation on the Community Patent, O.J. 2000, C 377E/278.
Proposal for a Council Regulation on the Community Patent, revised text of October 2009, Interinstitutional file 2000/0177 (CNS)(15149/09), and December Council document 16113/09, Add 1. Second Community Patent Convention (Agreement relating to the Community patent, including a protocol on the settlement of litigation), O.J. 1989, L 401/1.
COM(2002) 545, Report from the Commission to the European Parliament and the Council, Development and implications of patent law in the field of biotechnology and genetic engineering.
COM(2003) 827 & 828 final, proposal for a Council decision conferring jurisdiction on the Court of Justice in disputes relating to the Community patent, and concerning appeals before the Court of first instance.
COM(2005) 312, The Second Report of the Commission (2005) given in accordance with Article 16 (c) of Directive 98/44/EC. Report from the Commission to the Council and the European Parliament – Development and implications of patent law in the field of biotechnology and genetic engineering.
COM(2007) 165 final, Communication Enhancing the Patent System in Europe. Council document 16113/09, Add 1. revising the text from 2000 (COM (2000) 412.

COM(2008) 394, a unified patent litigation system, first suggested in a Commission Communication of April 2007 (IP/07/463).

COM(2010) 350 Proposal for a Council Regulation (EU) on the translation arrangements for the European Union patent.

COM(2008) 465 final of July, 16, 2008 Communication from the Commission to the European Parliament, the Council and the European Economic and Social Committee – An Industrial Property Rights Strategy for Europe.

COM(2010) 790 Proposal for a COUNCIL DECISION authorising enhanced cooperation in the area of the creation of unitary patent protection.

COM(2011) 215 Proposal for a REGULATION OF THE EUROPEAN PARLIAMENT AND OF THE COUNCIL implementing enhanced cooperation in the area of the creation of unitary patent protection.

COM(2011) 216-C7-0145/2011-2011/0094(CNS)), and documents from October, 26, 2011 (revised Presidency text, 15539/11 PI 133 COUR 59, and 15472/11 PI 131 CODEC 1676) *and* October 19 and October 26, 2011 (draft agreement on a Unified Patent Court and draft statute).

Documents from October, 26, 2011 (revised Presidency text, 15539/11 PI 133 COUR 59, and 15472/11 PI 131 CODEC 1676) *and* October 19 and October 26, 2011 (draft agreement on a Unified Patent Court and draft statute).

COM(2011) 809 final, Brussels November 30, 2011, proposal for a regulation of the European Parliament and of the Council establishing Horizon 2020 – The framework Programme for Research and Innovation (2014-2020).

Working document 15539/11 PI 133 COUR 59, Draft agreement on a Unified Patent Court and draft Statute. Revised presidency text of October 26, 2011. See press release, 3133 rd Council meeting, December 5, 2011. The Polish Presidency is committed to reach an agreement before 2011.

COM(2012) 790, Enhanced cooperation, COM (2011) 215, and the Council decision authorising enhanced cooperation in the area of the creation of unitary patent protection 2011/167/EU (O.J. 2011 L 76/53).

Draft report on the proposal for a Council regulation implementing enhanced cooperation in the area of the creation of unitary patent protection with regard to applicable translation arrangements COM (2011) 216, (2011/0094 (CNS), COM (2010) 350 final (translation arrangements for the European Union Patent).

Strategy papers see COM (2002) 480, Commission working document on the planned Community patent jurisdiction, COM (2007) 167 Communication from the Commission: Enhancing the Patent System in Europe.

Council Decision 2011/167/EU (OJ L 76/53.

**APPENDIX I**

# 2004 Search Profile

Databases : EPODOC, WPI

| SS | Results | |
|---|---|---|
| 1 | 13246 | BREAST+ 6D (CANCER+ OR CARCINOM+ OR TUMOUR+) |
| 2 | 994 | 1 AND (NUCLEOTID+ 6D SEQUENC+) |
| 3 | 172 | 1 AND BRCA+ |
| 4 | 106 | 3 AND (US/PN OR EP/PN) |
| 5 | 791 | 2 AND (US/PN OR EP/PN) |
| 6 | 21 | 4 AND (US/PN AND EP/PN) |
| 7 | 185 | 5 AND (US/PN AND EP/PN) |
| 8 | 181 | 7 NOT |
| 9 | 13 | 6 AND HUMAN+ |
| **10** | **12** | **9 NOT (MOUSE+ OR RAT OR RATS+ OR RODEN+ OR WISTAR+ OR PIG OR BOVIN+ OR PIGS)** |
| 11 | 115 | 8 AND HUMAN+ |
| 12 | 95 | 11 NOT (MOUSE+ OR RAT OR RATS+ OR RODEN+ OR WISTAR+ OR PIG OR BOVIN+ OR PIGS) |
| 13 | 95 | 12 NOT |
| **14** | **10** | **13 NOT (G01N+ OR C12Q1+)/IC** |
| 15 | 9 | 6 NOT 10 |
| 16 | 0 | 15 NOT (G01N+ OR C12Q1+)/IC |
| 17 | 0 | 6 AND TRANSGEN+ |
| 18 | 13428 | 1 OR BRCA+ |
| 19 | 375 | 18 AND TRANSGEN+ |
| 20 | 138 | 19 NOT HUMAN+ |
| **21** | **23** | **13 NOT G01N33+/IC** |
| 22 | 91 | 20 NOT G01N33+/IC |
| 23 | 12 | 22 AND (US/PN AND EP/PN) |
| **24** | **12** | **23 NOT (21 OR 10)** |
| 25 | 13278 | 1 OR BRCA |
| 26 | 1815 | 25 AND (ANIMAL+ OR RAT OR RATS+ OR MOUSE+ OR PIG OR PIGS OR SWINE+ OR PORCIN+) |
| 27 | 427 | 26 AND (MODEL+ OR TRANSGEN+) |
| 28 | 291 | 27 NOT (G01N33+)/IC |
| 29 | 67 | 28 AND (US/PN AND EP/PN) |
| **30** | **61** | **29 NOT (10 OR 24)** |
| 31 | 5 | CYCTIC+ 2D FIBROS+ |
| 32 | 3989 | CYSTIC+ 2D FIBROS+ |
| 33 | 3990 | 31 OR 32 |

| | | |
|---|---|---|
| 34 | 1789 | 33 AND (GENE+ OR +DNA+ OR SEQUEN+) |
| 35 | 661 | 34 AND HUMAN+ |
| 36 | 417 | 35 NOT (ANIMAL+ OR MOUSE+ OR MURIN+ OR RAT OR RATS OR WISTA R+ OR PIG OR PIGS OR SWIN+ OR PORCIN+) |
| 37 | 106 | 36 AND (US/PN AND EP/PN) |
| **38** | **73** | **37 NOT G01N33+/IC** |
| 39 | 1716 | 34 NOT 38 |
| 40 | 428 | 39 AND (US/PN AND EP/PN) |
| 41 | 304 | 40 NOT G01N33+/IC |
| 42 | 210 | 41 AND (NUCLEO+ OR RECOMBIN+ OR SEQUENC+ OR NUCLEI+) |
| **43** | **47** | **42 AND (TRANSGEN+ OR MODEL+)** |
| 44 | 76 | CANAVAN+ |
| 45 | 29 | 44 AND (GENE+ OR DNA+ OR SEQUENC+ OR NUCLEO+ OR NUCLEI+ OR RECOMBIN+) |
| **46** | **4** | **45 AND (US/PN AND EP/PN)** |
| 47 | 1522 | HUNTINGTON+ 6D CHOREA+ |
| 48 | 411 | 47 AND (GENE+ OR +DNA OR SEQUENC+ OR NUCLEO+ OR NUCLEI+ OR RECOMBIN+) |
| 49 | 9 | 44 AND +DNA |
| 50 | 0 | 49 NOT 45 |
| **51** | **109** | **48 AND (EP/PN AND US/PN)** |
| 52 | 306494 | C12N15+/IC |
| 53 | 2239 | 52 AND (PIG? OR PIGL+ OR BOAR? OR GILT? OR SOW? OR PIGGY+ OR SWINE+ OR PORC+ OR PORKER+ OR HOG?) |
| 54 | 169 | 53 AND CDNA |
| 55 | 44 | 54 AND (EP/PN AND US/PN) |
| **56** | **4** | **55 AND TRANSGEN+** |
| 57 | 448 | 53 AND (EP/PN AND US/PN) |
| 58 | 19 | 57 AND BREED+ |
| **59** | **22** | **56 OR 58** |

Databases : EPODOC, WPI

| SS | Results | |
|---|---|---|
| 1 | 317 | BDNF |
| 2 | 92 | TRKB |
| 3 | 383 | 1 OR 2 |
| **4** | **47** | **3 AND (EP/PN AND US/PN)** |
| 5 | 158 | 3 AND PD<1999 |
| **6** | **17** | **5 AND SCLEROS+** |
| 7 | 43 | ADDL |
| **8** | **5** | **7 AND (EP/PN AND US/PN)** |

**Resultat:**

# 2005 SEARCH PROFILE

```
Databases : WPI, EPODOC

  SS Results
   1  425143 (EP S 200410)/PN OR (EP S 200411)/PN OR (EP S
             200412)/PN OR (EP S 2005)/PN
   2  959216 (US S 200410)/PN OR (US S 200411)/PN OR (US S
             200412)/PN OR (US S 2005)/PN
   3   66656 1 AND
   4   66656 ..LIM
   5     325 BREAST+ 6D (CANCER+ OR CARCINOM+ OR TUMOUR+)
   6      14 5 AND (NUCLEOTID+ 6D SEQUENC+)
   7       0 5 AND BRCA+
   8       5 6 AND HUMAN+
   9       4 8 NOT (MOUCE+ OR MICE+ OR RAT OR RATS+ OR RO-
             DEN+ OR WISTAR+ OR PIG OR PIGS OR BOVIN+)
  10       1 9 NOT MOUSE+
  11       0 10 NOT ( G01N+ OR C12Q1/+)/IC
  12     325 5 OR BRCA+
  13       7 12 AND TRANSGEN+
  14       3 13 NOT HUMAN+
  15       2 14 NOT G01N33/+/IC
  16       1 10 NOT G01N33/+/IC
  17      74 12 AND (ANIMAL+ OR RAT OR RATS OR MOUSE+ OR
             MICE+ OR PIG OR PIGS OR SWINE OR PORCIN+)
  18      13 17 AND (MODEL+ OR TRANSGEN+)
  19       8 18 NOT G01N33/+/IC
  20       6 19 NOT (15 OR 16)
  21       0 CYCTIC+ 2D FIBROS+
  22     128 CYSTIC 2D FIBROS+
  23      55 22 AND (GENE+ OR +DNA+ OR SEQUEN+)
  24      15 23 AND HUMAN+
  25      11 24 NOT (ANIMAL+ OR MOUSE+ OR MICE OR MURIN+ OR
             RAT OR RATS OR WISTAR+ OR PIG OR PIGS OR SWIN+
             OR PORCIN+)
  26       9 25 NOT G01N33/+/IC
  27      46 23 NOT 26
  28      32 27 NOT G01N33/+/IC
  29      17 28 AND (NUCLEO+ OR RECOMBIN+ OR SEQUENC+ OR NU-
             CLEI+)
  30       4 29 AND (TRANSGEN+ OR MODEL+)
```

| | | |
|---|---|---|
| 31 | 0 | CANAVAN+ |
| 32 | 1 | HUNTINGDON+ 6D CHOREA+ |
| 33 | 69 | HUNTINGTON+ 6D CHOREA+ |
| 34 | 19 | 33 AND (GENE+ OR +DNA+ OR SEQUENC+ OR NUCLEO+ OR NUCLEI+ OR RECOMBIN+) |
| 35 | 1451 | C12N15/+/IC |
| 36 | 29 | 35 AND (PIG? OR PIGL+ OR BOAR+ OR GILT? OR SOW? OR PIGGY+ OR SWINE+ OR PORC+ OR PORKER+ OR HOG?) |
| 37 | 1 | 36 AND CDNA |
| 38 | 0 | 37 AND TRANSGEN+ |
| 39 | 2 | 36 AND BREED+ |
| 40 | 2 | BDNF OR TRKB |
| 41 | 0 | ADDL |

# 2007 SEARCH PROFILE

File : WPI
SS Results
1 349057 (EP S 200512)/PN OR (EP S 2006)/PN OR (EP S 2007)/PN
2 791651 (US S 200512)/PN OR (US S 2006)/PN OR (US S 2007)/PN
3 149817 1 AND 2
4 149817 ..LIM 3
5 115 HER_2
6 1236 BREAST+ 6D (CANCER+ OR CARCINOM+ OR TUMOR+ OR TUMOUR+)
7 6 CANAVAN+
8 21 HUNTINGDON+
9 2064 ALZHEIMER+
10 709 HUNTINGTON+
11 34 BRAIN_DERIVED_NEUROTROPHIC
12 27 BDNF
13 7 TRKB
14 2 ADDL?
15 7 TRKB?
16 27 BDNF?
17 774 OR 7, 8,10,11,12,13,14,15,16
18 356 CYSTIC+ 2D FIBROS+
19 1294 5 OR 6
..LIM ALL
20 3832 *M4/PR/ALL
21 0 *M4/PR/ALL
22 0 *M4/PR/ALL
23 0 *M4/PR/ALL
24 0 *M4/PR/ALL
25 1 *M4/PR/ALL
26 3832 19:25
27 2744 26 AND (SEQID OR (SEQ W ID) OR SEQUENC+ OR DNA OR RNA OR +NUCLE
OT+ OR +NUCLEIC+ OR GENE? OR GENETIC+)
28 149817 3
29 524 27 AND 28
30 339 29 AND HUMAN+
**31 190 30 NOT (G01N OR C12Q1)/ICA/EC**
32 774 17
33 1116 *M4/PR/ALL
34 0 *M4/PR/ALL
35 0 *M4/PR/ALL

36 0 *M4/PR/ALL
37 0 *M4/PR/ALL
38 1 *M4/PR/ALL
39 1116 32:38
40 793 39 AND 3
41 170 40 AND (SEQID OR (SEQ W ID) OR SEQUENC+ OR DNA OR RNA OR +NUCLE
OT+ OR +NUCLEIC+ OR GENE? OR GENETIC+)
**42 72 41 NOT (G01N OR C12Q1 OR C07C OR C07D)/ICA/IC**
43 356 18
44 852 *M4/PR/ALL
45 0 *M4/PR/ALL
46 0 *M4/PR/ALL
112/112
47 0 *M4/PR/ALL
48 0 *M4/PR/ALL
49 1 *M4/PR/ALL
50 852 43:49
51 384 50 AND 3
52 123 51 AND (SEQID OR (SEQ W ID) OR SEQUENC+ OR DNA OR RNA OR +NUCLE
OT+ OR +NUCLEIC+ OR GENE? OR GENETIC+)
**53 63 52 NOT (G01N OR C12Q1 OR C07C OR C07D)/ICA/EC**
54 21 ..INDEX /IN
ELLEGAARD
55 0 54 AND (PIG? OR SWINE+)
56 0 54 AND (SEQID OR (SEQ W ID) OR SEQUENC+ OR DNA OR RNA OR +NUCLE
OT+ OR +NUCLEIC+ OR GENE? OR GENETIC+)
57 5 ..INDEX /IN
ELLEGARD
58 24 54 OR 57
59 0 58 AND (PIG OR PIGS OR BOVIN+ OR SWIN+)
60 3 58 AND GENE+
61 581850 (PIG? OR PIGL+ OR MINIPIG+ OR (MINI W PIG?) OR BOAR+ OR GILT+ O
R SOW? OR PIGGY+ OR SWINE+ OR PORC+ OR PORKER+ OR HOG?)
62 4275 61 AND 3
63 3348 *M4/PR/ALL
64 1066 *M4/PR/ALL
65 0 *M4/PR/ALL
66 0 *M4/PR/ALL
67 0 *M4/PR/ALL
68 2 *M4/PR/ALL
69 5025 62:68
70 4356 69 AND 3
71 40 70 AND TRANSGEN+
72 114 70 AND ?DNA

73 215 70 AND (+GENETIC+ OR GENE? OR DNA OR +NUCLEOT+ OR +NUCLEIC+)
74 220 71 OR 72 OR 73
75 122 74 NOT (G01N OR C12Q1)/EC/ICA
**76 58 75 AND (A01 OR C12N15)/ICA/EC**
**77 64 75 NOT 76**
**Fuldtekst US og EP**
**14 ((MINI W PIG?) OR MINIPIG+)/CLMS**

# 2009 SEARCH PROFILE

File : EPODOC
SS Results
1 932260 (US S 200710)/PN OR (US S 200711)/PN OR (US S 200712)/PN OR (US S 2008)/PN OR (US S 2009)/PN
2 413041 (EP S 200710)/PN OR (EP S 200711)/PN OR (EP S 200712)/PN OR (EP S 2008)/PN OR (EP S 2009)/PN
3 1345301 1 OR 2 (burde være: (1 or 2) and (us and ep)/pn
4 1345301 ..LIM 3
5 1524 CANAVAN+ OR HUNTINGDON+ OR HUNTINGTON+ OR ALZHEIMER+ OR
(BRAI
N_DERIVED_NEUROTROPH+) OR BDNF OR TRKB OR ADDL? OR TRKB? OR BDNF?
6 204 5 AND (SEQID OR (SEQ W ID) OR SEQUENC+ OR DNA OR RNA OR +NUCLEO
T+ OR +NUCLEIC+ OR GENE? OR GENETIC+)
7 87 6 NOT (G01N OR C12Q1 OR C07C OR C07D)/ICA/EC
8 165 CYSTIC+ 2D FIBROS+
9 36 8 AND (SEQID OR (SEQ W ID) OR SEQUENC+ OR DNA OR RNA OR +NUCLEO
T+ OR +NUCLEIC+ OR GENE? OR GENETIC+)
10 21 9 NOT (G01N OR C12Q1 OR C07C OR C07D)/ICA/EC
11 20192 (PIG? OR PIGL+ OR MINIPIG+ OR (MINI W PIG?) OR BOAR+ OR GILT+ O
R SOW? OR PIGGY+ OR SWINE+ OR PORC+ OR PORKER+ OR HOG?)
12 21 11 AND TRANSGEN+
13 33 11 AND ?DNA
14 247 11 AND (SEQID OR (SEQ W ID) OR SEQUENC+ OR DNA OR RNA OR +NUCLE
OT+ OR +NUCLEIC+ OR GENE? OR GENETIC+)
15 258 12:14
16 192 15 NOT (G01N OR C12Q1 OR C07C OR C07D)/ICA/EC
File : WPI (DW=200957)
SS Results
1 835736 (US S 200710)/PN OR (US S 200711)/PN OR (US S 200712)/PN OR (US S 2008)/PN OR (US S 2009)/PN
2 366213 (EP S 200710)/PN OR (EP S 200711)/PN OR (EP S 200712)/PN OR (EP S 2008)/PN OR (EP S 2009)/PN
3 1053744 1 OR 2 (burde væe : (1 or 2) and (us and ep)/pn
4 1053744 ..LIM 3

..LIM ALL
5 0 TRANSFER_FROM SS 5 OF EPODOC:
6 1315 *XT
7 0 TRANSFER_FROM SS 8 OF EPODOC:
8 147 *XT
9 1053744 ..LIM 3
10 9278 CANAVAN+ OR HUNTINGDON+ OR HUNTINGTON+ OR ALZHEIMER+ OR
(BRAI
N?_DERIVED_NEUROTROPH+) OR BDNF OR TRKB OR ADDL? OR TRKB? O
R BDNF?
11 9328 10 OR 6
12 2636 11 AND (SEQID OR (SEQ W ID) OR SEQUENC+ OR DNA OR RNA OR +NUCLE
90/91
OT+ OR +NUCLEIC+ OR GENE? OR GENETIC+)
13 1106 12 NOT (G01N OR C12Q1 OR C07C OR C07D)/ICA/EC
14 1095 12 NOT (G01N OR C12Q1 OR C07C OR C07D)/ICA/IC.FLD/IC
15 75 14 NOT (HUNTINGTON+ OR ALZHEIMER+)
16 1020 14 NOT 15. Disse hits overføres efter aftale til EPODOC og søges fuldtekst I TXTEP,
TXTUS2 og TXTUS3 med søgeprofilen: (Huntingdon+ or huntington+ or alzheim+) 6D GEN+ . Disse
hits overførs til WPI:
**25 250 20 OR 24. 35 Hertil er lagt de nyeste referencer, der ikke er med i WPI-søgningen,**
**fra samme søgning minus fuldtekst i Epodoc. Tidligere sete referencer er manuelt sorteret fra.**
27 147 *XT
28 6022 27 OR (CYSTIC+ 2D FIBROS+)
29 1053744 ..LIM 3
30 679 28 AND (SEQID OR (SEQ W ID) OR SEQUENC+ OR DNA OR RNA OR +NUCLE
OT+ OR +NUCLEIC+ OR GENE? OR GENETIC+)
31 322 30 NOT (G01N OR C12Q1 OR C07C OR C07D)/ICA/IC.FLD/IC . Disse hits overføres
overføres efter aftale til EPODOC og søges fuldtekst I TXTEP, TXTUS2 og TXTUS3 med søgeprofilen:
FIBRO+ 6D (GEN+ OR DNA). Disse hits overførs til WPI:
**36 123 OR 32, 33, 34, 35 Hertil er lagt de nyeste referencer, der ikke er med i WPIsøgningen,**
**fra samme søgning minus fuldtekst i Epodoc. Tidligere sete referencer er manuelt sorteret fra.**
11 20192 (PIG? OR PIGL+ OR MINIPIG+ OR (MINI W PIG?) OR BOAR+ OR GILT+ O
R SOW? OR PIGGY+ OR SWINE+ OR PORC+ OR PORKER+ OR HOG?)
12 21 11 AND TRANSGEN+
13 33 11 AND ?DNA

14 247 11 AND (SEQID OR (SEQ W ID) OR SEQUENC+ OR DNA OR RNA OR +NUCLE
OT+ OR +NUCLEIC+ OR GENE? OR GENETIC+)
15 258 12:14
16 192 15 NOT (G01N OR C12Q1 OR C07C OR C07D)/ICA/EC
38 35808 PIG? OR PIGL+ OR MINIPIG+ OR (MINI W PIG?) OR BOAR+ OR GILT+ OR
SOW? OR PIGGY+ OR SWINE+ OR PORC+ OR PORKER+ OR HOG?
39 348 38 AND TRANSGEN+
40 575 38 AND ?DNA
41 1717 38 AND (SEQID OR (SEQ W ID) OR SEQUENC+ OR DNA OR RNA OR +NUCLE
OT+ OR +NUCLEIC+ OR GENE? OR GENETIC+)
42 1749 39:41
43 1180 42 NOT (G01N OR C12Q1 OR C07C OR C07D)/ICA/IC.FLD/IC
44 1145 43 AND ((US S A?)/PN OR (EP S A?)/PN). . Disse hits overføres overføres efter aftale
til EPODOC og søges fuldtekst I TXTEP, TXTUS2 og TXTUS3 med søgeprofilen: (pig.......) 6D
(GEN+ OR DNA). Disse hits overførs til WPI:
51 246 OR 46, 47, 48, 49, 50
52 160 51 AND ((US S A?) S (2007 OR 2008 OR 2009))/PN
53 68 51 AND ((EP S A?) S (2007 OR 2008 OR 2009))/PN
**54 185 52 OR 53. Hertil er lagt de nyeste referencer, der ikke er med i WPI-søgningen, fra**
**samme søgning minus fuldtekst i Epodoc. Tidligere sete referencer er manuelt sorteret fra.**
File : EPODOC
SS Results
24 0 .LIM ALL
33 1441 (SPINAL W MUSKEL W ATROPH+) OR SMA OR SMN1 OR SMN2
34 76 33 AND (SEQID OR (SEQ W ID) OR SEQUENC+ OR DNA OR RNA OR +NUCLE
OT+ OR +NUCLEIC+ OR GENE? OR GENETIC+)
91/91
35 39 34 NOT (G01N OR C12Q1 OR C07C OR C07D)/ICA/EC
File : WPI
SS Results
..LIM ALL
58 1302 (SPINAL W MUSKEL W ATROPH+) OR SMA OR SMN1 OR SMN2
59 231 58 AND (SEQID OR (SEQ W ID) OR SEQUENC+ OR DNA OR RNA OR +NUCLE
OT+ OR +NUCLEIC+ OR GENE? OR GENETIC+)
60 146 59 NOT (G01N OR C12Q1 OR C07C OR C07D)/ICA/IC/IC.FLD
**61 79 60 AND (US OR EP)/PN**
File : EPODOC
SS Results
**36 14 35 AND (EP OR US)/PN**

English fulltext query, CA1, EP1, US2, US3
(GOETTINGEN OR GOTTINGEN) AND ((PIG? OR MINIPIG+) s (dna or gen+))
154 referencer overført til WPI:
67 49 OR 65, 66
68 24 67 NOT (G01N OR C12Q1 OR C07C OR C07D)/ICA/IC/IC.FLD
69 16 68 AND (US S A? S (2007 OR 2008 OR 2009))/PN
70 11 68 AND (EP S A? S (2007 OR 2008 OR 2009))/PN
**71 21 69 OR 70**

**APPENDIX V**

# 2011 SEARCH PROFILE

**Spinal muskel atrophy**

```
WPI:
14 (motor_neuron_2) or (motor_neuron_1) or SMN_1 or SMN_2
15 14 and (ep or us)/pn

Fuldtekst EP, US
((MOTOR_NEURON_2) OR (MOTOR_NEURON_1) OR SMN_1 OR
SMN_2)/CLMS
```

**Cystisk fibrose**

File : WPI

| SS | Results | |
|---|---|---|
| 1 | 0 | CYSTIC FIBROSIS TRANSMEMBRANE CONDUCTANCE REGU-LATOR CFTR |
| 2 | 641 | ((CYSTIC W FIBROSIS) 3D REGULATOR) OR CFTR |
| 3 | 254 | 2 AND (EP AND US)/PN not (g01n or c12q1 or c07c or c07d)/ec/ica |

```
((CYSTIC W FIBROSIS W TRANSMEMBRANE W CONDUCTANCE W REGULA-
TOR) OR CFTR)/CLMS i EP og US fuldtekst.
```

**Canavan**

```
WPI
 SS Results
 34   70 ASPARTOACYLASE+ OR ASPA
 35   30 34 AND (EP AND US)/PN not (g01n or c12q1 or c07c or
         c07d)/ec/ica
16.05.11 12:57 (Patent search)

English fulltext query
(aspartoacylase+ or ASPA)/clms
```

**Huntington**

```
16.05.11 12:23 (Patent search)

English fulltext query
(IT15 or HTT)/clms
```

| | | |
|---|---|---|
| 23 | 92 | IT15 OR HTT |
| 24 | 21 | 23 AND (EP AND US)/PN |
| 25 | 8 | 24 NOT (G01N OR C12Q1 OR C07C OR C07D)/ICA |

**Alzheimers**
16.05.11 11:20 (Patent search)

English fulltext query
(PRESENILIN or (AMYLOID_BETA_A4) or PSEN_1 or PSEN_2 or APOE or SORL_1)/clms sammen med AND (US and EP)/pn og NOT (g01n or c12q1 or c07c or c07d)/ica/ec

Søgningen er kun kørt i WPI:
File : WPI

| SS | Results | |
|---|---|---|
| 1 | 236 | PRESENILIN |
| 2 | 0 | 1 AND (EP AND UD)/PN |
| 3 | 90 | 1 AND (EP AND US)/PN |
| 4 | 39 | AMYLOID BETA A4 |
| 5 | 10 | 4 AND (US AND EP)/PN |
| 6 | 100 | 3 OR 5 |
| 7 | 177 | ( OR PSEN_1, PSEN_2, APOE, SORL_1) AND (US AND EP)/PN |
| 8 | 1918 | APP |
| 9 | 753 | 8 AND ALZHEIM+ |
| 10 | 398 | 8 AND (GENE OR SEQUENS OR SEQ) |
| 11 | 177 | 7 AND (US AND EP)/PN |
| 12 | 266 | 6 OR 11 |
| 13 | 85 | 12 NOT (G01N OR C12Q1 OR C07C OR C07D)/ICA/IC.FLD |

**Parkinsson**
16.05.11 10:59 (Patent search)
File : WPI

| SS | Results | |
|---|---|---|
| 14 | 215 | ( OR PAR-KIN,SNCA,UCHL 1,LRRK 2,PARK 2,PARK 7,PINK 1) |
| 15 | 49 | 14 AND (US AND EP)/PN |
| 16 | 6 | 15 NOT (G01N OR C12Q1 OR C07C OR C07D)/ICA/IC.FLD |

English fulltext query
( OR PARKIN,SNCA,UCHL_1,LRRK_2,PARK_2,PARK_7,PINK_1)/ clms (ord skal forekomme i kravene)og begrænset med AND (US and EP)/pn i WPI og not (G01N OR C12Q1 OR C07C OR C07D)/ICA/EC i Epodoc

# 2012 SEARCH PROFILE

13.04.12 15:55 (Patent search)

English fulltext query
(or BRPK, PARK_6)/clms

German fulltext query
(or BRPK, PARK_6)/clms

---

13.04.12 15:43 (Patent search)

English fulltext query
dj_1/clms

German fulltext query
dj_1/clms

---

13.04.12 15:33 (Patent search)

English fulltext query
(or AR_JP, PDJ, PRKN, PRKN2, (ubiquitin w E3 w ligase))/clms

German fulltext query
(or AR_JP, PDJ, PRKN, PRKN2, (ubiquitin w E3 w ligase))/clms

---

13.04.12 15:28 (Patent search)

English fulltext query
(or DRDN, PARK_8, RIPK_7, ROCO2)/clms

German fulltext query
(or DRDN, PARK_8, RIPK_7, ROCO2)/clms

---

13.04.12 15:28 (Patent search)

English fulltext query
(or DRDN, PARK_8, RIPK_7, ROCO2)/clms

---

13.04.12 15:21 (Patent search)

English fulltext query
(or MSY1,( neuron w cytoplastic W protein), (protein w 9_5), PARK_5, PGP9_5, (ubiquitin_thiolesterase), UBL_1)/clms

---

13.04.12 15:11 (Patent search)

English fulltext query
(or alpha_synuclein, NACP, PARK_1, PARK_4, PD1, SYUA_HUMAN)/clms

# Huntington's disease Table

**Table 7.2.A. HD data**[1]

| **Huntington (W/A: Withdrawn/Abandonment, P: Pending, G: Granted)** | |
|---|---|
| **1)** | Regulation of human gpcr-like protein (EP1326973 (W), US20040101871(A)) |
| **2)** | Regulation of human lgr4-like g protein-coupled receptor (US20030139341 (A), EP12900030 (W)) |
| **3)** | Regulation of human follicle stimulating hormone-like g protein—coupled receptor (US20040077041 (A), EP1287135 (W)) |
| **4)** | Human enzyme molecules (US20030207430(A), EP1259618 (W)) |
| **5)** | Major intrinsic protein (MIP)-like polynucleotides, polypeptides and antibodies (US20020055142 (A), US20030148947 (A), EP1233777 (W)) |
| **6)** | Human secreted protein (US20020012966 (A), EP1471072 (W), EP1212342 (W), EP1173477 (W), US7368531 (G), US7381800 (G), US7026447 (G), US2007032414 (A), US6420526 (G)) |
| **7)** | Composition and methods for in vivo and in vitro attenuation of gene expression using double stranded RNA (US20040147475 (A), US2002114784 (P), EP1147204 (P)) |
| **8)** | RNA interference mediated treatment of polyglutamine (POLYQ) repeat expansion diseases using short interfering nucleic acid (siNA) (EP1735443 (W), US2007042983 (A)) |
| **9)** | Method of generating stem cells and embryonic bodies carrying disease-causing mutations and methods of using same for studying genetic disorders (EP1694825 (P), US2007269790 (P)) |
| **10)** | Use of IL-6/IL-6 chimera in Huntington's disease (US2007172454 (A), EP1317273 (G), US2004057928 (A)) |
| **11)** | **Introducing nucleic acids e.g. 2-amino adenosine into Drosophila embryos for inducing mutation and developing Drosophila model for human disease e.g. Huntington (US2009133140: US7732659 (G), WO2009067634: EP2227082 (P))** |
| **12)** | New isolated double-stranded short interfering nucleic acid molecule comprising complementary sense and antisense regions, useful for manufacturing a medicament for preventing or treating Huntington's disease (EP2014769 (G), WO2009007855 (W)) |
| **13)** | New isolated RNA duplex, useful for suppressing the accumulation of huntingtin (US2009130751 (P), WO2008134646: EP2152874 (P)) |

1. Status on the applications last checked on February 29, 2012.

| | |
|---|---|
| **14)** | Treating/managing cognitive impairment associated with e.g. attention-deficit/hyperactivity disorder, Down syndrome and Huntington's disease in a patient (WO2008067121: EP2089018 (W), US2008153811 (A)) |
| **15)** | **New salicyl alcohol-creatine phosphate derivatives useful for treating e.g. Huntington disease (US2007281910 (A), WO2007146088 (W))** |
| **16)<br>15*<br>38** | Silencing mRNA encoding mutant Huntington (WO 2008005562: EP2046993 (W), US2009186410 (A)) |
| **17)** | New nucleic acid molecule (US2006257912 (A), EP1885854 (P)) |
| **18)** | Nucleic acid duplex (US2007261126: US7902352 (G), EP2056881(P)) |
| **19)** | New medical system (US2007253931 (A), EP1973553 (W), US2009214487(A)) |
| **20)** | New double-stranded ribonucleic acid (US2007099860: US7320965 (G), EP1941059 (W)) |
| **21)** | **Hydrogenated pyrido (4,3-b)indole for the treatment of Huntington disease (US2007117834 (A), EP1937263 (P))** |
| **22)** | Use of a composition comprising a cannabinoid-2 selective agonist (EP1758566 (W), US2007248702 (A)) |
| **23)** | Evaluating a compound for treating a mental defect (US2008187492 (P)) |
| **24)** | New small interfering RNA useful for treating or preventing Huntington disease (US2005255086 (A), EP1844148 (W), US2008274989 (A)) |
| **25)** | Treating a subject having or at risk for a disease (EP1670518 (P), US2009118206: US7947658 (G)) |
| **26)** | Small interference double-stranded RNA which suppresses expression of Huntington gene (WO2004101787 (W), US2008015158: US7589189 (G)) |
| **27)** | Providing acute neuroprotection (US2004092444: US2005197284: US7439063 (G), EP1535065 (P), US2009093409 (P)) |
| **28)** | Isolated stromal cells (US5716616 (G), EP0871457 (G), US2003039639: US6653134 (G), US2003059412: US6974571 (G), EP1304112 (G), EP1304113 (G), US2003202966 (A), US2004166097 (A), US2009035347 (A), US2009214492 (P)) |

**Table 7.2.B. HD data (2011*, huntingtin, IT15, HTT)**

| Huntington (W/A: Withdrawn/Abandonment, P: Pending, G: Granted) | |
|---|---|
| **29):1*/18*** | Transition metal-cyclopentadienyl-tropane conjugates (WO 0140239: EP 1233968 (W), US 2002111486: US6605717 (G)) |
| **30): 2*/20*** | Method for treating neurodegenerative, psychiatric and other disorders with deacetylase inhibitors (WO 02090534: EP 1390491 (G), US 2004142859 (P)) |
| **31): 3*/8*** | **Synthetic antibody phage libraries (WO 03102157: EP 1513879 (P), US 2011112279 (P), US2008050377: US7799899 (G), EP 2083018 (W), WO 2004094476: EP 1613660 (P), US 2005119455: US7985840** |

| | |
|---|---|
| | **(G)/Compositions and methods relating to stop-1 (EP 2305710 (P))** |
| **32):4*/6*/7*/11*/19*/21*** | SiRNA-mediated gene silencing with viral vectors (EP 1534861 (W), US 2005042646 (A), US 2011111491 (P), US 2004241854 (A), US 2010144026 (A), US 2005106731 (A), US 2005255086 (A), US 2006009408 (A), WO 2004013280: EP 1576118 (W), US 2010190243 (A), US 2004023390 (A), WO 2004058940: EP 1581635 (W), WO 2006083800: EP 1844148 (W)), US 2008274989 (A)) |
| **33):5*/10*/26*** | Treatment of neurodegenerative disease through intracranial delivery of short interfering RNA (siRNA) (WO 2004047872: EP 1569662 (W), EP 2145628 (G), US 2004162255: US 7605249 (G), US 2004220132: US 7829694 (G), US 2010063134: US8119611 (G), EP 1829547 (W), EP1832292 (W)) |
| **34: 9*** | **Method for assessing behavioral predisposition (WO 2005006971: EP 1644537 (W), US 2005037405 (G))** |
| **35:12*/25*** | Fibroblast growth factor-2 promotes neurogenesis and neuroprotection and prolongs survival in huntington's disease (WO 2007014156: EP 1904092 (P), US2009111748 (P)) |
| **36):13*/17*** | Methods and compositions for treating neurological disease (WO 2007022470: EP 1937066 (W), WO 2007022506 (W), US 2010267810 (A), US 2011027882 (A), US 2007105803 (A), US 2007161591 (A), EP 2239328 (P), US 2010267810 (A)) |
| **37): 14*** | G-rich polynucleotides for the treatment of huntington's disease (WO 2007044851: EP 1940861 (W), US 2008051363 (A), US 2007105805: US 7807648 (G)) |
| **39): 16*** | Polypeptide-nucleic acid conjugates and uses thereof (WO 2009079790: EP 2235175 (P), US 2011039785 (P)) |
| **40): 22*** | **System and method for delivering polynucleotides to the central nervous system (US 2005048641 (A))** |
| **41): 23*** | Devices, systems and methods for improving memory and/or cognitive function through brain delivery of siRNA (US 2006178328: US 7618948 (G), US 2009060987: US 8058251 (G), US 2010184835 (P)) |
| **42): 24*** | Allele-specific silencing of disease genes (US 2009105169 (A)) |
| **43) SMA** | **Posh interacting proteins and related methods (EP1646644 (W), WO2004089302 (W))/** |

# Spinal muscular atrophy table

**Table 7.3.A. SMA data**[1]

| SMA (W/A: Withdrawn/abandonment, P: Pending, G: Granted) | |
|---|---|
| **1)** | **Use of angiotensin converting enzyme 2 protein or nucleic acid (EP2077119 (W), WO2009086572: EP2227346 (P))** |
| 2) | New human pluripotent stem cell (WO2008148105 (W), US2009053182 (P)) |
| 3) | Use of SCO-Spondin polypeptide (EP2027868 (W), WO2009027350: EP2195007 (P)) |
| 4) | New polyethylene glycol-polyethylenimine-Nucleic Acid Agents (PEG-PEI-NAA) polyplex (WO2007133812 (W), US2009011004 (A)) |
| **5)** | **Isolating endothehelial stem/progenitor cells and mesenchymal stem cells (WO2008060037 (W), US2009124007 (A), US2009053182 (P))** |
| 6) | Generating neural and glial cells (WO2008026198: EP2064319 (P), US2010003751 (P)) |
| **7)** | **Method useful for treating and/or limiting fibrotic disorders (WO2008008772: EP2051727 (P))** |
| **8)** | **Preparing of cardiomyocyte-linage cells (US2007010012 (P), EP1910518 (P))** |
| **9)** | **New isolated polynucleotide (EP1883651 (P), US2009054322 (A))** |
| **10)** | **New soluble molecule (WO2007094005: EP1991578 (P), US2010227805 (P))** |
| 11) | New iRNA agent (US2007042984: US7772200 (G), EP1909850 (W)) |
| 12) | Antisense oligonucleotide (WO2007002390: EP1910395 (P), US2010216238 (P)) |
| 13) | Method for treating a lower motor neuron disease (US2007031418, EP1877445, both W/A) |
| 14) | Treating degenerative disease (US2006269520 (A), EP1843791 (W)) |
| 15) | Treating, preventing, inhibiting or reducing tissue deterioration, injury or damage (EP1853293 (P), US2008274098 (P)) |
| 16) | New pharmaceutical preparation (US2006025340: US7316998 (G), EP1755648 (P), US2009192080: US 7981857: (G)) |
| **17)** | **Stimulating stem cell recruitment, proliferation, or differentiation (US2004248796 (A), EP1594527 (W))** |
| 18) | Pharmaceutical preparation for inhibiting myostatin, treating muscular dystrophy (US2006030522: US7790910 (G), EP1729793 (P), US7456149(G), US2009098114: US7807631(G) **BRAGE** |
| **19)** | **New composition (US2004009591: US7498167 (G), US2005013807: US** |

1. Status on the applications last checked March 2012.

| | |
|---|---|
| | **7807148 (G), EP1549738 (G), US2009169528: US7915042 (G))** |
| **20)** | **Plasmid vector for producing eukaryotic protein/DNA in nematodes (EP1451332 (G), US2005014251: US 7741098 (G))** |
| 21) | Delivering recombinant adeno-associated virus (rAAV) (EP1402043 (W), US2005032219 (A)) |
| 22) | Novel chimeric molecules (EP1416972 (G), US2005054836 (A)) |
| **23)** | **New transgenic nonhuman mammal (US2002199213: US7041870 (G), US2006026698: US7576258 (G), US2006037093: US 7816578 (G), EP1354034 (G), EP1916303 (G))** |
| **24)** | **Making of tissue-engineered vascular construct (EP1339356 (G), US2004044403 (A))** |
| **25)** | **New expression vectors for use in animal cells (US2004038394: US7422874 (G), EP1392835 (G))** |
| 26) | Delivery of a composition to the central nervous system or spinal cord (US2003004121: US7435792 (G), US2004170651: US7923216 (G), US2005060761: US7923015 (G), EP1681300 (W), EP1049712 (G)) |
| **27)** | **Plasmid derived from Micromonospora sp. (EP 0064350 (G), US4478937 (G))** |
| **28)** | **Plasmid PATM3 from Streptomyces castaneoglobisporous IFO 13669 (EP0067071 (G), US4518698 (G))** |
| 29) | Biomarkers for SMA (WO2011032109 (P)) |

**Table 7.3.B. SMA, data (2011*, SMN1, SMN2, survival of motor neuron 1, survival of motor neuron 2)**

| SMA (W/A: Withdrawn/abandonment, P: Pending, G: Granted) | |
|---|---|
| 30): 1* | Substances for the treatment of spinal muscular atrophy (WO0166129: EP 1133993 (W)) |
| 31): 2* | Methods of administering vectors to synaptically connected neurons (WO 03004660: EP1402043 (W)) |
| 32): 3* | Vector system (WO 2004031390: EP1551974 (W); US 2004076613 (A), US 2008131400 (A)) |
| 33): 4* | Use of a histone deacetylase inhibitor for treating muscular dystrophies (WO2004050076: EP1565175 (P), US2006148684 (A), US2008160108 (P)) |
| 34): 5* | Group Posh interacting proteins and related methods (EP1646644 (W), WO2004089302 (W))/See HD |
| 35): 6* | Ubiquitin ligase inhibitors and methods related thereto (WO 2005007141: EP 1651226 (G), US 2007015768: US 7659277 (G)) |
| 36): 7* | Gene therapy for spinal cord disorders (WO 2006119341: EP1879623 (P), US 2009069261 (P)) |
| 37): 8* | Compositions and methods for modulation of SMN2 splicing (WO 2007002390: EP1910395 (P), US2010216238 (P)) see group 12. |
| 38): 9* | Gene therapy for amyotrophic lateral sclerosis and other spinal cord disorders (WO 2007146046: EP 2029742 (P), US 2010267812 (A)) |

| | |
|---|---|
| 39):10* | Methods for treating spinal muscular atrophy using tetracycline compounds (WO 2008127722: EP 2144614 (P), US 2009118269: US 7935687 (G)) |
| 40):11* | Compounds and methods for modulating protein expression (WO 2009064920: EP 2219680 (P)) |
| 41):12* | Therapeutic compositions comprising an amine-containing lipid and a protein (WO 2009097332: EP 2244694 (P), US 2011052673 (P)) |
| 42):13* | Therapeutic compositions (WO 2009097349: EP 2244697 (P), US 2011059172 (P), US 2011052673 (P) |
| 43):4* | Hydroxyurea treatment for spinal muscular atrophy (US 2003040543: US 6573300 (G)) |
| 44):15* | Neurite regeneration (US 2004266715 (W), US 2007213290 (A)) |
| 45):16* | Methods of administering vectors to synaptically connected neurons (US 2005032219 (A)) |
| 46): 17* | Improved carriers for delivery of nucleic acid agents to cells and tissues (WO 2007133812 (W), US 2009011004 (A)) |
| 47):18* | Treatment for spinal muscular atrophy (US 2010152204 (A)) |
| 48):19* | Generation of brachial, thoracic and lumbar spinal motor neurons from embryonic stem cells in the absence of all-trans retinoic acid supplement (US 2010196332 (P)) |
| 49):20* | Methods and compositions for treating spinal muscular atrophy (US 2010234402 (P), WO 2008150509 (W)) |
| 50):21* | Treatments for spinal muscular atrophy (US 6376508 (G)) |

# APPENDIX IX

## Canavan's disease

**Table 7.4.A. Canavan's data[1] and data 2011* (Aspartoacylase, ASPA)**

| Canavan (W/A: Withdrawn/Abandonment, P: Pending, G: Granted) | |
|---|---|
| 1) | 69624, a novel human transporter family and uses thereof (US 2002193582 (A), EP 1270726 (W)) |
| 2) | Nogo receptor-mediated blockade of axonal growth (US 7119165 (G), US 2002012965(A), EP 1248803(G)) |
| 3) | Aspartoacylase gene, protein, and methods of screening for mutations associa-ted with Canavan disease (US 2003017473: US7217547 (G)) |
| **4)** | **Approaches to identify genetic traits (US2002119442 (A), US 2003039996 (A), EP 1230393(W))** |
| 5) | Aspartoacylase gene, protein and methods of screening for mutations associated with Canavan disease (US5679635 (G)) |
| 6) | Conjugate used for treating e.g. diabetes comprises fusion peptide containing glucagon-like peptide 1 sequence having specified amino acids and nonapeptide sequence, conjugated to synthetic/natural polymer (EP1972349 (W), WO2008113601: EP2121032 (P)) |
| **7)** | **Use of ribonucleotide reductase inhibitor for treating e.g. Alzheimer's dise-ase, multiple sclerosis, congenital metabolic disorder, rheumatoid arthritis, insulin dependent diabetes mellitus, lipid storage disease, stroke, myasthe-nia gravis (US2009047250 (P), EP2030615 (W))** |
| 8) | Treating a demyelinating condition, e.g. neurodegenerative disease, multiple sclerosis or transverse myelitis, in a subject comprises administering transplan-tation-competent glial-restricted progenitors (GRPs) to the patient (WO2009023795: EP2190981 (W), US2009155223(A)) |
| 9) | New neurotrophic peptide, useful for manufacturing a medicament for treating and/or preventing neurodegenerative diseases, for improving learning-memory capacities in an individual, and for treating or enhancing motor deficiencies (WO2008113536: EP2121930 (P), EP2009103 (W), US2010029572 (P)) |
| 10) | New immunoglobulin (Ig) that specifically binds CD100 comprises optimized complementarity-determining region (CDR) region and optimized framework region (FWR), useful for treating autoimmune or inflammatory disease and cancer (WO2008100995: EP2069404 (G), US2008219971: US7919594 (G)) |
| 11) | New neuroprotective compound comprising specific amino acid sequence use-ful for treating neurodegenerative disease e.g. Alzheimer disease, amyotrophic |

1. Last check on the status March 2012.

| | |
|---|---|
| | lateral sclerosis, diabetic neuropathy, stroke (WO2007128493: EP2032596 (G), EP1857463 (W), US2009105159: US8034778 (G)) |
| 12) | Treating demyelinating disorders comprises administering to a human a dose of a soluble lymphotoxin beta receptor to promote remyelination (WO2008049053: EP2073833 (P), US2010099608: US8067375 (G)) |
| 13) | Spherical or non-spherical microcapsule for treating e.g. diabetes comprises surface coatings comprising crosslinked polymers, and core comprising crosslinked polymers and cells capable of expressing and secreting a glucagon-like peptide-1 (WO2007128443: EP2015736 (G), EP1854455 (G), US2010068289 (P)) |
| **14)** | **Novel human artificial chromosome vector comprising human chromosome fragment having telomere and subtelomere sequences, centromere and foreign DNA, and deletion of long/short arm distal ends, for treating diseases e.g. diabetes (WO2008013067: EP2048229 (P), US2010011454 (P))** |
| 15) | New drugs conjugated to antibody via an enzyme cleavable linker, useful for treating cancers or infectious diseases (WO2007105027: EP1993608 (W), US2009220529 (A)) |
| 16) | New substantially homogenous cell population that co-express CD49c, CD90 and telomerase, useful for treating degenerative, traumatic, acute injury, cardiac or neurological conditions in humans (WO03025149: EP1438392 (P), US2003059414 (P), US2005233452 (P), US2007231309 (A), US2007264232 (P)) |
| **17)** | **New nucleotide concatemers comprising a cassette of nucleotide sequence, useful for controlling and coordinating expression of large numbers of heterologous genes in a single host cell (WO02059296: EP1360308 (G), US2004110174: US8008459 (G))** |
| 18): 2* | Delivery system for gene therapy to the brain (WO 9846273: EP 0969876 (W), US 2002076394: US 6436708 (G)) |
| 19): 14* | Methods and compositions for delivering enzymes and nucleic acid molecules to brain, bone, and other tissues (WO 03100031: EP 1521523 (W), US2003215432: US7807618 (G)) |

**Table 7.4.B. Data (2011*: ASPA, ASPARTOACYLASE)**

| **Canavan (W/A: Withdrawn/Abandonment, P: Pending, G: Granted)** | |
|---|---|
| 1)* | **Novel linear derivatives of arginine vasopressin antagonists, EP 0296892, US 5055448)** |
| 2)* | Delivery system for gene therapy to the brain (EP 0969876, WO 9846273, US 2002076394, US 6436708 ) |
| 3)* | **Method for producing threonine and isoleucine (EP 1179597, US 2002110876, )** |
| 4)* | **Process for producing l-lysine (EP 1253195, EP 1253195, US 2010173368, US 7723081** |
| 5)* | **Glucose dependent release of insulin from glucose sensing insulin derivatives (EP1290024, WO 0192334, US 2002028767:  US 7316999)** |
| 6)* | **Process for the fermentative preparation of L-amino acids using strains of** |

| | |
|---|---|
| | **the enterobacteriaceae family with overexpressed rseb gene (EP 1382685, US 2004191885, US 7638313, WO 2004067757, EP 1587938)** |
| 7)* | **Process for the preparation of L-amino acids using strains of the entero-bacteriaceae family which contain an attenuated ugpb gene (EP 1407021, WO 03008616, WO 03008616, US 2005124047, US 2004115780, US 7256021, EP 1407035, EP 1407035, WO 03008604, WO 03008604, WO 03008603, WO 03008603, WO 03008602, WO 03008602, WO 03008600, WO 03008600)** |
| 8)* | **Process for the preparation of L-amino acids using strains of the entero-bacteriaceae family which contain an attenuated aspa gene (EP 1407035, WO 03008616, WO 3008616, US 2005124047, US 2004115780)** |
| 9)* | **A process for the production of L-amino acids using strains of the entero-bacteriaceae family which contain an enhanced iclr gene (EP 1446493, WO 03038106, WO 03038106, US 2005064561, US 7442530)** |
| 10)* | **Amino acid-producing bacteria and a process for preparing l-amino acids (EP 1481075, WO 03074719, US 2004082040, US 7319026)** |
| 11)* | **Process for the preparation of L-threonine using strains of the family en-terobacteriaceae (EP 1483387, WO 03076627)** |
| 12)* | **Process for the preparation of l-amino acids using strains of the family en-terobacteriaceae (EP 1483393, WO 03076635, US 2005095688, 2004214294, WO 2004087937, EP 1608764)** |
| 13)* | **Promoter and plasmid system for genetic engineering (EP 1499709, WO 03089621, WO 03089621, US 2004235099, US 7132527, US 2005147968, US 2011046014, US 2009156430, US 2007065867, US 7510869)** |
| 14)* | Methods and compositions for delivering enzymes and nucleic acid molecules to brain, bone, and other tissues (EP 1521523, WO 03100031, US2003215432: US7807618) |
| 15)* | **Vaccine composition comprising lipooligosaccharide with reduced phase variability (EP 1524990, EP 2258387, EP 2258386, EP 2255826, EP 2258385, EP 2258384, EP 1961427, EP 1961427, EP 1524993, EP 1524992, EP 1524991, WO 2004015099, WO 2004015099, WO 2004014417, WO 2004014417, WO 2004014419, WO 2004014418, WO 2004014418, US 2011033500, US 2006240045, US 2006034854, US 2006057160, US 2006051379, US 7838014)** |
| 16)* | **Vaccine composition comprising transferring binding protein and hsf from gram negative bacteria (EP 1524991, EP 2258387, EP 2258386, EP 2255826, EP 2258385, EP 2258384, EP 1961427, EP 1961427, EP 1524993, EP 1524992, EP 1524990, WO 2004015099, WO 2004015099, WO 2004014417, WO 2004014417,WO 2004014419, WO 2004014418, WO 2004014418, US 2011033500, US 2006240045, US 2006034854, US 2006057160, US 2006051379, US 7838014)** |
| 17)* | **Vaccine compositions comprising l2 and/or l3 immunotype lipooligosaccha-rides from lgtb- neisseria minigitidis (EP 1524992, EP 2258387, EP 2258386, EP 2255826, EP 2258385, EP 2258387, EP 2258386, EP 2258384, EP 1961427, EP 1961427, EP 1524993, EP 1524991, EP 1524990, WO 2004015099, WO 2004015099, WO 2004014417, WO 2004014417, WO** |

| | |
|---|---|
| | 2004014419, WO 2004014418, WO 2004014418, US 2011033500, US 2006240045, US 2006034854, US 2006057160, US 2006051379, US 7838014) |
| 18)* | Neisserial vaccine compositions comprising a combination of antigens (EP 1524993, EP 2258387, EP 2258386, EP 2255826, EP 2258385, EP 2258384, EP 1961427, EP 1961427, EP 152499, EP 1524991, EP 1524990, WO 2004015099, WO 2004015099, WO 2004014417,WO 2004014417, WO 2004014419, WO 2004014418, WO 2004014418, US 2011033500, US 2006240045, US 2006034854, US 2006057160, US 2006051379, US 7838014) |
| 19)* | Method for refolding neisserial NSPA protein (EP 1532168, WO 2004020452, WO 2004020452, US 2010291137, US 2008025993) |
| 20)* | A process for the production of L-amino acids using strains of the entero-bacteriaceae family which contain an enhanced fadR gene (EP 1548120) |
| 21)* | Microorganism producing L-threonine having inactivated GALR gene, method of producing the same and method of producing L-threonine using the microorganism (EP 1561811, US 2005176114, US 7229794, WO 2005075625) |
| 22)* | Process for the fermentative preparation of L-threonine, L-lysine, L-valine and L-homoserine using strains of the enterobacteriaceae family which contain an attenuated YJGF ORF sequence (EP 1587938, WO 2004067757, US 2004191885, US 7638313) |
| 23)* | A process for the production of L-amino acids using strains of the entero-bacteriaceae family which overexpress the GALP gene, coding for a galactose proton symporter (WO 2004087937, DK 1608764, US 2004214294) |
| 24)* | Process for the production of L-amino acids using strains of the entero-bacteriaceae family which contain an enhance YFIDD ORF and/or PFLD gene (EP 1611245, US 2004235122, US 7211415, US 2007141681, WO 2004090149) |
| 25)* | Biocatalyst for producing d-lactic acid (EP 1669460, US 2007065930, WO 2005033324) |
| 26)* | Process for preparing L-amino acids using strains of the enterobacteriaceae family (EP 1697530, WO 2005064000, DK 1697530, US 2005153403, US 2005164356, US 2005164357, US 2005176112, US 2007141681, US 2011053227, EP 1719814, US 2006257979) |
| 27)* | Process for the preparation of L-amino acids using strains of the entero-bacteriaceae family (EP1706482, US 2005164356, WO 2005071062, US 2005095687, US2005095688) |
| 28)* | Method for producing L-threonine using recombinant enterobacteriaceae with increased enolase activity (EP 1706493, US 2005164357, WO 2005071094, WO 2005071094) |
| 29)* | Process for the preparation of L-amino acids using strains of the family enterobacteriaceae (EP 1711593, US 2005176112, US 7575905, WO 2005075627, DK 1711593) |
| 30)* | Process for the preparation of L-amino acids using strains of the family Enterobacteriaceae (EP 1719814, US 2011053227, US 2006257979) |
| 31)* | Method for the preparation of L-amino acids using strains of the entero-bacteriaceae family with increased expression of YTFQ-ORF (EP 1719818, |

| | |
|---|---|
| | **US 2006286644, US 7553645)** |
| **32)*** | **Method for the production of L-amino acids using strains from the entero-bacteraceae family (EP 1776451, US 2009068711, US 7759094, WO 2006015669, US 2009068711, US 7759094, WO 2006015669, EP 1776451)** |
| **33)*** | **Industrially useful microorganism (EP 1829965, US 2008009041, WO 2006057341)** |
| **34)*** | **Methods and organisms for the growth-coupled production of 1,4-BUTANEDIOL (EP 2185708, WO 2009023493, US 2009047719)** |
| **35)*** | **Neisserial vaccine compositions comprising a combination of antigens (EP 2255826, EP 2258387, EP 2258386, EP 2258385, EP 2258384, EP 1961427, EP 1961427, EP 1524993, EP 1524992, EP 1524991, EP 1524990, WO 2004015099, WO 2004015099, WO 2004014417, WO 2004014417, WO 2004014419, WO 2004014418, WO 2004014418, US 2011033500, US 2006240045, US 2006034854, US 2006057160, US 2006051379, US 7838014)** |
| **36)*** | **Neisseria vaccine composition (EP 2258387, EP 2258386, EP 2255826, EP 2258385, EP 2258384, EP 1961427, EP 1961427, EP 1524993, EP 1524992, EP 1524991, EP 1524990, WO 2004015099, WO 2004015099, WO 2004014417, WO 2004014417, WO 2004014419, WO 2004014418, WO 2004014418, US 2011033500, US 2006240045, US 2006034854, US 2006057160, US 2006051379, US 7838014)** |
| **37)*** | **Bacterium belonging to the genus escherichia and method for producing L-threonine or L-isoleucine (EP 1179597, EP 1179597, US 2002110876)** |
| **38)*** | **Process for the fermentative preparation of L-amino acids using strains of the enterobacteriaceae family (US 2005032178)** |
| **39)*** | **Process for the production of l-amino acids using strains of the entero-bacteriaceae family which contain an enhanced FADR or ICLR gene (US 2005064561, US 7442530, WO 03038106, EP 1446493)** |
| **40)*** | **Process for the preparation of l-amino acids using strains of the entero-bacteriaceae family which contain an attenuated acek gene (US 2005124047, WO 03008616, US 2004115780, US 7256021, EP 1407035, EP 1407021, EP 1407021, WO 03008604, WO 03008603, WO 03008602, WO 03008600)** |
| **41)*** | **Vaccine composition (US 2006057160, EP 2258387, EP 2258386, EP 2255826, EP 2258385, EP 2258384, EP 1961427, EP 1524993, EP 1524992, EP 1524991, EP 1524990, WO 2004015099, WO 2004014417, WO 2004014419, WO 2004014418, US 2011033500, US 2006240045, US 2006034854, US 2006051379, US 7838014.)** |
| **42)*** | **Mutant protein and refolding method (US 2006141563, WO 2004020463, EP 1572729)** |
| **43)*** | **Process for preparing L-amino acids using improved strains of the entero-bacteriaceae family (US 2006257979, EP 1719814, US 2011053227, US 2006286644)** |
| **44)*** | **Biocatalyst for production of d-lactic acid (as amended) ( US 2007065930, EP 1669460, WO 2005033324)** |
| **45)*** | **Methods and organisms for the growth-coupled production of 1,4-butanedio (US 794748, US 2009047719, WO 2009023493, EP 2185708)** |

| | |
|---|---|
| **46)*** | **Promoter and plasmid system for genetic engineering (US 2009156430, US 200514796, US 2011046014, US 2004235099, US 7132527)** |
| **47)*** | **Gene Expression Cassette and a Transformant, and a Method for Manufacturing 2-Deoxy-Scyllo-Inosose and a Method for Purifying 2-Deoxy-Scyllo-Inosose Using Said Transformant (US 2010015672, EP 1865056, WO 2006109479, WO 2006112000)** |
| **48)*** | **Method for producing L-lysine (US 2010173368, EP 1253195, US 7723081, WO 0153459)** |
| **50)*** | **Promoter and plasmid system for genetic engineering (US 2011046014, US 2005147968, US 2004235099, US 7132527, US 2009156430)** |
| **51)*** | **Linear derivatives of arginine vasopressin antagonists (US 5055448, EP 0296892)** |
| **52)*** | **Imidazolidine derivatives (US 5554594)** |
| **53)*** | **Bacteria containing aspartate-semialdehyde dehydrogenase, phosphoenolpyruvate carboxylase, and transhydrogenase to produce L-lysine in escherichia, and methods of using same (US 7723081, EP 1253195, US 2010173368)** |

# Cystic fibrosis table

**Table 7.5.A. CF data**

| Cystic fibrosis (W/A: Withdrawn/Abandonment, P: Pending, G: Granted) | |
|---|---|
| **1)** | Methods to produce lung surfactant formulations via lyophilization and formulations and uses thereof (EP1811966 (W), US2006205663: US7582312 (G)) |
| **2)** | Modified polypeptides for targeting cell-entry of the adenoviruses of subtype B (EP1482052 (W), EP1627067 (W), US2007065408 (A)) |
| **3)** | Delivery peptides, their constructs with active agents and use (EP1694696 (A), US2007086981(A)) |
| **4)** | NF-kB oligonucleotide decoy molecules (EP1691817 (W), US2005182012: US7378509 (G), US2007010474 (P), US2007078102 (A)) |
| **5)** | Methods of generating stem cells and embryonic bodies carrying disease-causing mutations and methods of using same for studying genetic disorders (EP 1625156 (G), US2007027074: US7528104 (G)) |
| **6)/231*/ 372*** | Treating a genetic disorder comprising administering mesenchymal stem cells to the animal (US2007253931 (A), US2008286249 (A), US2009214487 (A), WO 2007084354: EP 1973553 (W), US 2010098672 (A), WO 2009111030: EP 2262513 (P), US 2010330052 (P), US 2010291047 (A)) |
| **7)** | Producing human stem cells (US2007141702: US 7732202 (G), EP1948791 (P)) |
| **8)** | Modified protease inhibitors (EP1663279 (W), US2007020252 (A), US20050089515 (A), US20070041959: US7550427: (G)) |
| **9)** | Surfactant protein C esters (US 20060148701: US 7173009: (G), EP1192184 (G), US7022671 (G), US5876970 (G), US1994278189 (A)) |
| **10)** | Treatment of type 1 immune response-mediated inflammatory lung disease by modulation of IFN-gamma activity (EP1611154 (W), US2007031377 (A)) |
| **11)** | Particulate composition (EP1274403 (G), US2003161886: US7018657 (G), US2006093557 (A)) |
| **12)** | Methods of preventing and treating RVS infections and related conditions (US2006115485 (A), EP1812068 (P)) |
| **13)** | Pegylated interferon Alpha-1B (US2006029573 (A), EP1771197 (W)) |

| | |
|---|---|
| **14)** | Genetically engineered human-derived kunitz domains that inhibit human neutrophil elastase (EP0797666 (G), US 5663143 (G), US2003223977 (A), EP1734121 (W), US2003175919: US 7078383 (G), US 2006134087 (A)) |
| **15)** | Use of secretin-receptor ligands in treatment of cystic fibrosis (CF) and chronic obstructive pulmonary disease (COPD) (US 2002142956: US6780839 (G), EP1296708 (W), US2004191238 (A), US2006147384 (A), EP1736153(G), US2011039759 (P)) |
| **16)/247*)** | New human peptide for reducing chloride transport through chloride channel and for treating disorder related to aberrant ATP-binding cassette transporter-related disorders (WO 2007137163: EP 2029157 (W), US 2009203598: US 8106013 (G) ) |
| **17)** | Biofilm formation (e.g. associated with cystic fibrosis) (US 2008199509 (P), WO2008101084: EP2114460 (P)) |
| **18)** | Treating an infection (US2008221029 (A), WO 2009052411: EP2217259 (P)) |
| **19)** | Target analyte (US2008117416 (P), EP2076172 (W)) |
| **20)** | Promoting mucus clearance (US2008199410 (P), EP2063903 (P)) |
| **21) 22)** | Administering substituted phenyl compound (US2007281906 (P), EP2053915 (P), US2009203620 (P)) |
| **23)** | Swellable particles for drug delivery (US2008031824 (P), EP2007358 (P)) |
| **24)**<br>**(246*)**<br>**(375*)** | **RNAI modulation of AHA and therapeutic uses thereof (US2008113931: US7812150 (G), WO2007137156: EP2023937 (G), US201103882: US8114985 (G))** |
| **25)** | Bortezomib (EP1912664 (P), US2009017006 (A)) |
| **26)** | New nanocluster (US2007172653: US7651770 (G), US2008075667: US 7993749 (G), EP1968550 (P), US2009053316 (A), US2009081295 (P)) |
| **27)** | Charge-Modified Lysozyme Antimicrobial Compositions, Surfac-tants, and Methods for Infections and Cystic Fibrosis (US2008118489 (A), WO2007053800 (W)) |
| **28)** | Leukotriene B4 hydroxylase (US2006247200: US7341990 (G), WO2006119071 (W)) |
| **29)** | Gene delivery (EP1885450 (G), US2008286361 (P)) |
| **30)/210*)** | New virus-like particle (WO 2006059141: EP 1827394 (P), US 2009041724 (P)) |
| **31)** | Hybrid Recombination for genome manipulation (EP1755640 (P), US2008241116: US7842503 (G)) |
| **32)** | Use of androstane compounds and e.g. cystic fibrosis, autism, sickle cell disease (US2004138187 (A), US2005101581 (A), EP1539183 (W), US2007203107 (A), US2007213309: US7935839 (G), US2007265236(A), US2007275937: US7482334 (G), US2007275938: US7547687 (G), US2008085873 (A)) |
| **33)** | Product and process for liquefaction of mucus or sputum |

| | |
|---|---|
| | (US2004131606: US7195766 (G), EP1551455 (P), US2008031865: US7534438 (G)) |
| **34)** | Recombinant intact minicells containing therapeutic nucleid acid molecules, useful for preparing a medicament for treating e.g. cystic fibrosis (EP1446489 (G), US2005222057 (A), US2008020441 (A), US2008188436 (A)) |
| **35)** | A non-inflammatory vector composition useful for treating e.g. cystic fibrosis (EP1242608 (W), US2003148975 (A), US2009069257 (A)) |
| **36)** | Nucleic acid delivery complex (EP1244805 (G), US2003153081 (A), EP1566445 (W), US2009209037 (A)) |
| **37)** | New complex of nucleic acid (EP1229941 (G), US7276594 (G), US2009203768 (A)) |
| **38)/98*/117*)/ 302*)** | Adeno-associated viruses and uses thereof (WO 0125465: EP1224313 (W), US 7241447 (G), US 2008166758 (A), US 2008226600 (A), EP 1849872 (W), WO 9960146: EP 1082444 (G), US 6436392 (G), US 2002197237: US 6897045 (G), US 2005255087: US 7803622 (G), US 2009265796 (A)) |
| **39) 80*** | Sense mRNA therapy (EP1021549 (W[1]), US2003083272 (A), EP1892299 (P), US2009093433 (A) WO 9914346: EP 1021549 (W), EP 2292771 (P)) |
| **40)** | Human DNase II (EP0914417 (G), US6265195 (G), US6569429 (G), US2003219428 (A), US2005130209 (A), US2009041742 (A)) |
| **41)** | Antisenseoligo-nucleotide (EP0698092 (G), US5627274 (G), US5665593 (G), US5916808 (G), US5976879 (G) |

**Table 7.5.B. CF data (2011*, Cystic Fibrosis Transmembrane Conductance Regulator; CFTR)**

| **Cystic fibrosis (W/A: Withdrawn/Abandonment, P: Pending, G: Granted)** | |
|---|---|
| **1*)** | New diagnostic and treatment methods involving the cystic fibrosis transmembrane regulator (EP 0446017 (W) ) |
| **2*)** | Cystic fibrosis gene (WO 9102796: EP 0489058 (G), EP 1418183 (W), US 6201107 (G), US 6730777 (G), US 6902907 (G), US 6984487 (G)) |
| **3*)/ 436*** | Transfection of lung via aerosolized transgene delivery (WO 9312756: EP 0619742 (G), US 5641662 (G), US 6468798 (G)) |
| **4*)/417*** | Gene therapy for cystic fibrosis transmembrane conductance regulator activity (CFTR) (WO9312240: EP 0625207 (W), US 6620795 (G), US 6001644 (G)) |

1. The examination division refused the application on grounds that the requirement of art. 123(3) and 84 EPC were not fulfilled, cf. appeal T 652/06 (later redrawn).

| | |
|---|---|
| **5*)/ 381*/414*)** | Adeno-associated virus with inverted terminal repeat sequences as promoter (WO 9324641: EP 0644944 (G), US 5990279 (G), US 5587308 (G), US 5866696 (G), US 5989540 (G), EP 1164195 (G), US 6165781 (G), US4797368 (G))) |
| **6*)/394*** | Expression cassette with regularoty regions functional in the mammalian host (WO 9324640: EP 0646178: (W), WO 9325673 (W), US 5827703 (G) |
| **7*)** | Defective adenoviruses and corresponding complementation lines (WO 9428152: EP 0652968 (G), US 6040174 (G), US 2001049136: US 7005277 (G), US 2003170885: US 7067309 (G), US 6133028 (G)) |
| **8*)** | Methods and therapeutic compositions for treating cystic fibrosis (WO 9404671: EP 0659211 (W), US 5750571 (G)) |
| **9*)/385*)** | Method of discriminating nucleic acid and testing set for discriminating nucleic acid (WO 9502068: EP 0664339 (W), US 5688643 (G), US 5674898(G)) |
| **10*) 413*) 415*)** | Introns and exons of the cystic fibrosis gene and mutations at various positions of the gene (WO 9110734: EP 0667900 (G), US 6001588(G), US 5981178 (G)) |
| **11*)/ 406*** | Gene therapy for cystic fibrosis (WO 9412649: EP 0673431 (W), US 6093567 (G), US 5882877 (G)) |
| **12*) 431*** | Method for preparing a viral aerosol (WO 9511664: EP 0680315 (G), US 5952220 (G), US 6344194 (G), US 2002106349:US 6582694 (G)) |
| **13*)** | **Expression of heterologous polypeptides in halobacteria (WO 9421789: EP 0690913 (W), US 5641650 (G), US 6010885 (G))** |
| **14*) 331*)** | Gene transfer into pancreatic and biliary epithelial cells (WO 9426915: EP 0698109 (G), US 2005048043 (A), EP 1211319 (W)) |
| **15*)** | Adenovirus vectors for gene therapy sponsorship (WO 9428938: EP 0710123 (W)) |
| **16*)** | Deoxyribonucleic acids containing inactivated hormone responsive elements (WO9510606: EP 0723587 (W), US 5688677 (G)) |
| **17*)/ 396*)** | **Stable cell lines capable of expressing the adeno-associated virus replication gene (WO 9513392: EP 0728214 (G), US 5837484 (G))** |
| **18*) 401*)** | Arrays of nucleic acid probes on biological chips (WO 9511995: EP 0730663 (G), US 5837832 (G), US 5861242 (G)) |
| **19*)** | Generation of high titers of recombinant AAV vectors (WO 9513365: EP 0733103(G), US 5658776 (G)) |
| **20*)** | Packaging cell lines for generation of high titers of recombinant AAV vectors (EP 0796339 (W), US 2003175974: US 6924128 (G), US 2005148076 (A)) |
| **21*)/ 416*)** | Recombinant protein production in bovine adenovirus expression |

| | |
|---|---|
| | vector system (WO 9516048: EP 0736100 (G), US 6001591 (G), US 6086890 (G), US 5820868 (G), US 2002192185: US 7025967 (G)) |
| **22*)/420*** | Method of preparing a viral vector by homologous intermolecular recombination (WO 9617070: EP 0742834 maintained as amended (G), US 6110735 (G), US 6281000 (G)) |
| **23*)** | Integrative recombinant adenoviruses, preparation thereof and therapeutical uses thereof (WO 9523867: EP 0748385 (W), US 6033885 (G)) |
| **24*)/395*** | Combined use of two expression cassettes for the production of a useful protein (WO 9524491: EP 0749489 (G), US 5834237 (G)) |
| **25*)/382*** | C-terminally truncated CFTR genes and proteins (WO 9525796 : EP 0751994 (W), US 5958893 (G), US 5639661 (G)) |
| **26*)/ 50*)/ 408*** | Novel internal ribosome entry site, vector containing same and therapeutical use thereof (WO 9601324: EP 0769062 (W), US 5925565(G), WO 9849334: EP 0918875 (W), US 6783977 (G)) |
| **27*)/ 426*/444*** | Method of gene therapy using DNA vectors without selection marker gene (WO 9605297: EP 0775203 (G), US 6573100 (G), US 6265218 (G)) |
| **28*)** | Coupled amplification and ligation method (WO 9606190: EP 0777749 (G), US 5912148 (G), US 6130073 (G) ) |
| **29*)/ 380*)/ 387*** | Recombinant adenovirus comprising a chimeric penton base protein (WO 9607734: EP 0778889 (G), EP 1679376 (W), US 5712136 (G), US 5731190 (G), US 5559099 (G)) |
| **30*)** | Method for the production of recombinant proteins, plasmids and modified cells (WO 9608572: EP 0781341 (G), US 6143518 (G)) |
| **31*)** | Viral vectors and line for gene therapy (WO 9704119: EP 0784693 (G), US 6204060 (G), US 2003203488 (A)) |
| **32*)/ 33*)/ 60*) / 403*)** | Use of a non-mammalian DNA virus to express an exogenous gene in a mammalian cell (WO 9609074: EP 0785803 (W), US 6238914 (G), US 5731182 (G), US 7192933(G), US 5871986 (G)) |
| **34*)/ 423*** | Improved adenovirus and methods of use thereof (WO 9613597: EP 0787200 (G), US 6203975 (G), US 6001557 (G)) |
| **35*)** | Container with multi-phase composition for use in diagnostic and therapeutic applications (WO 9608234: EP 0788348 (G), EP 1588699 (G)) |
| **36*)** | Xanthine derivatives compounds useful for activating chloride conductance in animal cells (WO 9616084: EP 0794963 (G)) |
| **37*)/ 399*/ 402*** | Hybrid adenovirus-AAV virus and method of use thereof (WO 9613598: EP 0797678 (G), US 5871982 (G), US 5856152 (G), US 6387368 (G), US 6251677 (G)) |
| **38*)/392)*** | Direct gene transfer into the ruminant mammary gland (WO 9622379: EP 0805869 (W), US 5780009 (G)) |
| **39*)/ 65*)** | Chimeric adenoviral fiber protein and methods of using same |

| | |
|---|---|
| | (WO 9626281: EP 0811069 (G), US 6127525 (G), US 5770442 (G), WO 9840509: EP 0973927 (W)) |
| **40*)** | Homogeneous water-in-perfluorochemical stable liquid dispersion for administration of a drug to the lung of an animal (EP 0824343:WO 9635411 (W), US 5770585 (G)) |
| **41*)** | S-adenosyl methionine regulation of metabolic pathways and its use in diagnosis and therapy (WO 9633703: EP 0824345 (W), EP 1221615 (W), US 6596701 (G), US 6020139 (G)) |
| **42*)/307*)/ 393)*/ 418)*** | **Pulmonary delivery of aerosolized medicaments (WO 9632149: EP 0825885 (revoked after opposition), US 6019968 (G), US 6258341 (G), US 5780014 (G), US 2007042048 (A), US 2003068279:US 7097827 ((G), but court proceedings), EP 1428524 (W), US 2003086877 (A), US 2007122418 (A), US 2003198601 (A), US 6673335 (G))** |
| **43*)/409*)432*** | Cationic lipid: DNA complexes for gene targeting (WO 9640962: EP 0832270 (G), US 5830878 (G), US 6344446 (G), US 5932241 (G), WO 9640963:EP 0832272 (W)) |
| **44*)** | Coarse spray delivery of functional biologic materials (WO 9635467: EP 0843565 (W), US 6227195 (G)) |
| **45*)** | Adenovirus vectors for gene therapy (WO 9630534: EP 0843731 (G), EP 1923467 (W), US 2002164782 (A), US 2003147854: US 7318919 (G), US 5585362 (G), US 5707618 (G), US 5824544 (G)) |
| **46*)/447*)** | Compositions useful for transferring therapeutically active substances into a target cell, and their use in gene therapy (EP 0890362 (G), US 6656734 (G)) |
| **47*)** | CFTR gene regulator (WO 9735005: EP 0896621 (W), US 2002086836: US 6573073 (G), US 2003229045 (A)) |
| **48*)/49*)/81*)** | Adenoviral vector deleted of all E4-ORF except ORF6 (EP 0905253 (W), EP 0911413 (W), EP 1024198 (W)) |
| **51*)** | Defective Adenoviruses and corresponding transcomplementant cell lines (EP 0919624 (W), EP 0919627 (maintained as amended (G), EP 0919626 (W), EP 0919625 (G), EP 1149916 (P)) |
| **52*)/303*)** | Cystic fibrosis test based on the detection of mutations in the CFRE gene by ARMS (EP 0928832 (W), US 2003008281 (A)) |
| **53*)/430*)** | **Cellular internalization of PIGR stalk and associated ligands (WO 9746588: EP 0934338 (G), US 6042833 (G), US 6340743 (G))** |
| **54*)** | CFTR channel activator compounds and pharmaceutical compositions containing same (WO 9805642: EP 0937044 (G), US 6630482 (G)) |
| **55*)** | **Dispersible macromolecule compositions and methods for their preparation and use (WO 9741833: EP 0948317 (G), US 2007020199 (P), US 2002122827:US 6592904 (G), US 2003215514: US7138141 (G), US 6423344 (G), US 6051256** |

| | |
|---|---|
| | **(G))** |
| **56*)** | AAV split-packaging genes and cell lines comprising such genes for use in the production of recombinant AAV vectors (WO 9827204: EP 0950091 (W), US 2002081721: US 6541258 (G)) |
| **57*)** | EHV-1 vectors (WO 9827216: EP 0953054 (G), US 6387685 (G), US 2002090716: US 6706515 (G)) |
| **58*)** | **Induction of immunological tolerance (WO 9715243: EP 0955948 (G), EP 1586284 (G), US 2002192190 (A), US 2010260816 (A), US 2008317770 (A), US 2004047848: US 7361333 (G), US 2004191227: US7361334 (G))** |
| **59*)** | Correction of genetic defects using chemical chaperones (WO 9737645: EP 0959877 (W), US 2001021500: US 6541195 (G), US 5900360 (G), US 6270954 (G)) |
| **61*)** | Improvements in gene therapy using compositions comprising a polynucleotide and a nuclease inhibitor (EP 0967289 (W)) |
| **62*)/346*)** | An improved method for the production and purification of adenoviral vectors (WO 9822588: EP 0968284 (P), EP 1707631 (W), EP 1760151 (G), WO 0032754: EP 1135469 (G), US 2002182723: US 7125706 (G), US 7732129 (G), US 2005089999: US 7445930 (G), US 2007155008 (A), US 6194191 (G), US 2008050770 (A), WO 2006052302: EP 1815029 (W), US 2010105124 (P), US 2006275781 (A)) |
| **63*)** | Butyrate prodrugs derived from lactic acid (WO 9840064: EP 0969828 (G), US 6225350 (G)) |
| **64*)/241*)/ 314*)/356*)** | Modified proteins which bind extracellular matrix components (WO 9844938: EP 0973538 (G), US 2003118551:US 6864082 (G), US 2011065179 (P), US 2008241905: US 7820157 (G), EP 2008664 (G), US 2005244376: US 7347998 (G)) |
| **66*)** | Use of adenoviral E4 reading frames to improve expression of a gene of interest (EP 0974668 (G), EP 1199368 (G), US 6475480 (G) ) |
| **67*)** | Novel adenoviral vectors capable of facilitating increased persistence of transgene expression (WO 9846780: EP 0975782 (W), US 6020191 (G)) |
| **68*)/ 291)*** | Adenoviral vectors comprising a modified e4 region but retaining E4ORF3 (WO 9846779: EP 0975785 (W), US 6100086 (G), US 6358507 (G), US 2002028194: US 6485720 (G)) |
| **69*)** | Novel transgene expression system for increased persistence (WO 9846781: EP 0975786 (W), US 5981275 (G)) |
| **70*)** | Sequence alterations using homologous recombination (WO 9842727: EP 0977771 (W), US 6200812 (G), US 6074853 (G), US 5948653 (G)) |
| **71*)** | Anti-cancer products for treating cystic fibrosis (WO 9853839: EP 0984784 (G), US 6635627 (G) ) |
| **72*)/ 398*)** | Method for the production of non-group C adenoviral vectors |

| | |
|---|---|
| | (WO 9853087: EP 0988389 (G), US 5849561 (G)) |
| **73*)** | Therapeutic nanospheres comprising 4-PHENYL BUTYRATE (WO 9856370: EP 0989849 (W), US 6207195 (G)) |
| **74*)/ 442*** | Modified adenoviral fiber and target adenoviruses (WO 9844121: EP 0991763 (G), US 6569677 (G), US 2003175243: US7256036 (G)) |
| **75*)** | Nucleic acid analysis using complete N-mer arrays (EP 0995804 (W), US 2002012913 (A)) |
| **76*)/126*)/437*)** | Improvements in gene therapy using compositions comprising a polynucleotide and a nuclease inhibitor or interleukin-10 (EP 1254961 (W), US 6528312 (G), WO 9956784: EP0996739 (W)) |
| **77*)/289*)** | Use of magnesium ($Mg^{2+}$) for the enhancement of gene delivery in gene therapy (EP 0998945 (W), US 2002019357 (A) EP 1004320 (G), US 2002156045: US 6881724 (G)) |
| **78*)** | Chemical modification of DNA using peptide nucleic acid conjugates (WO 9913719: EP 1014790 (W), US 6165720 (G), US 6746868 (G)) |
| **79*)** | Chimeric polypeptide comprising the fragment B of shiga toxin and peptides of therapeutic interest (WO 9903881: EP 1017715 (G), US 6613882 (G), US 2004047883:US 7488809 (G), US 2010322913 (P)) |
| **82*)/ 310*)** | Flavonoids for cystic fibrosis therapy (WO 9918953: EP 1024803 (P), US 2003096762: US 7335683 (G), US 5972995 (G), US 6329422 (G)) |
| **83*)** | Transcriptionally-activated AAV inverted terminal repeats (ITRs) for use with recombinant AAV vectors (WO 9920773: EP 1025243 (G), US 2002164783: US 6936466 (G), US 6346415 (G), US 2006019382 (A)) |
| **84*)** | Activating C1- secretion (WO 9920294: EP 1032410 (W)) |
| **85*)/438*)** | Pseudo-type retroviral vectors with modified surface capsid proteins and method for the production thereof for selective gene transfer (WO 9928488: EP 1034291 (G), US 6544779 (G)) |
| **86*)** | Cell-specific retroviral vectors with antibody domains and method for the production thereof for selective gene transfer (WO 9928489: EP 1034292 (G), US 2005191740: US 7531322 (G)) |
| **87*)/ 359*)** | Method and means for producing high titer, safe, recombinant lentivirus vectors (WO 9931251: EP 1036182 (G), US 6165782 (G), US 5994136 (G), US 2008286836 (P), US 2002173030: US 6924144 (G), US 2005255597: US 7083981 (G), US 6428953 (G)) |
| **88*)** | Feline immunodeficiency virus gene therapy vectors (WO 9936511: EP 1045921 (W), US 2003104611 (A), US 2002068354 (A), US 2002048805 (A)) |
| **89*)** | Recombinant adeno-associated virus (EP 1046711 (G)) |
| **90*)** | A transgenic rabbit that expresses a functional human lipoprotein |

| | |
|---|---|
| | (A) (WO 9935241: EP 1049767 (G), US 6512161 (G)) |
| **91*)** | Nucleic acid delivery vehicles (WO 9940214: EP 1053342 (W), US 6287857 (G)) |
| **92*)** | Methods for purified AAV vector production (WO 9941399: EP 1056880 (W), US 2002045250 (A)) |
| **93*)/439*)** | Novel methods for treating disorders in which docosahexaenoic acid (DHA) levels are affected (WO 9945916: EP 106191 (W), US 6552081 (G), US 2004127567 (A)) |
| **94*)** | Production methods and uses of a retrovirus (WO 9950285: EP 1066319: (W)) |
| **95*)** | Preparation of recombinant adenovirus carrying a REP gene of adeno-associated virus (WO 9953084: EP 1068343 (W), US 6521426 (G), WO9613598: EP079678 (G)) |
| **96*)** | Adenovirus for the induction of APOPTIC gene expression (WO 9946371: EP 1070122 (W)) |
| **97*)/261*)** | **Porcine adenovirus vector (WO 9953047: EP 1071758 (G), EP 2145961 (P), US 2003130187: US 7335353 (G), US 6492343 (G))** |
| **99*)** | Probes used for genetic profiling (WO 9964626: EP 1084273 (W) |
| **100*)** | **Allelic series of genomic modifications in cells (WO 9967361: EP 1088055 (W), US 2005186675:US 7554004 (G), US 2005191658 (A), US 6835867 (G))** |
| **101*)/102*)** | Non-embryonic ependymal neural stem cells and method for their isolation (WO 9967363: EP 1090105 (G), US 2003092176: US 7320872 (G)) |
| **103*)** | Complementation cells comprising an inducer and/or a repressor (EP 1096018 (W)) |
| **104*)/452*)** | Immunoassays to detect diseases or disease susceptibility traits (WO 0042436: EP 1141726 (G), US 2007031881: US 7867717 (G), US 7148016 (G), US 2011091895: US7981692 (G)) |
| **105*)** | Methods of determining altered NDPK functions (WO 0052467: EP 1157274 (W)) |
| **106*)/272*)** | Genetic modification of somatic cells and uses thereof (WO 0051424: EP 1159415 (G), EP 2199392 (W)) |
| **107*)** | Complex for transferring an anionic substance of interest into a cell (EP 1161957 (W), US 2002055174 (A)) |
| **108*)** | Methods, compositions, and cells for encapsidating recombinant vectors in AAV particles (WO 0065038: EP 1173596 (W), US 6893865 (G), US 2006166318 (A)) |
| **109*)/350*)** | A combined growth factor-deleted and thymidine kinase-deleted vaccinia virus vector (WO 0073479: EP 1180157 (P), US 2003031681: US 7208313 (G), US 2007154458 (P)) |
| **110*)** | **Method of measuring plasma membrane targeting of GLUT4 (WO 0075188: EP 1189943 (W))** |

| | |
|---|---|
| **111*)/319*)/446*)** | Type V PDE-inhibitors for the treatment of cystic fibrosis (WO 0064424: EP 1194148 (G), US 6599907 (G), Cystic fibrosis medicaments (US 2003229117: US 6936618 (G)) |
| **112*)** | Enhanced delivery via serpin enzyme complex receptor ligands (WO 0108708: EP 1200616 (W), US 2006228407 (A), US 2004152653 (A)) |
| **113*)/332*)/450*** | **Identification of peptides that facilitate uptake and cytoplasmic and/or nuclear transport of proteins, DNA and viruses (WO 0115511: EP 1210362 (W), US 2003104622 (A), US 6881825 (G), US 2003219826 (A), US 2005074884 (P), US 2004152653 (A))** |
| **114*)** | Adenoviral vectors having nucleic acids encoding immunomodulatory molecules (WO 0063406: EP 1210447 (W), US 2004023389 (A)) |
| **115*)** | Sequence-specific DNA recombination in eukaryotic cells (WO 0116345: EP 1214440 (G), US 2003027337 (P), EP 1681355 (G)) |
| **116*)/ 118*) /250*)** | **TAR RNA binding peptides (WO 0125486: EP 1218544 (G), WO 0125188: EP 1226115 (W), EP 2071036 (W))** |
| **119*)** | Recombinant sendai virus vector for introducing exogenous genes to airway epithelia (WO 0132898: EP 1228232 (G), US 7314614 (G) ) |
| **120*)** | Permanent amniocyte cell line, the production thereof and its use for producing gene transfer vectors (WO 0136615: EP 1230354 (G)) |
| **121*)** | Use of cationic lipids for intracellular protein delivery (WO 0143778: EP 1237581 (W)) |
| **122*)** | Methods of assaying for compounds that inhibit premature translation termination and nonsense-mediated RNA decay (WO 0144516: EP 1238110 (G), US 6458538 (G), US 2003008317:US 7026122 (G), EP 2036991 (P), US 2006134681 (A)) |
| **123*)** | Histidine copolymer and methods for using same (WO 0147496: EP 1242052 (G), US 2003045465: US 7163695 (G), US 2003165567: US 7070807 (G)) |
| **124*)/ 440*)** | Translentiviral vector system (WO 0144481: EP 1244797 (G), US 6555342 (G)) |
| **125*)** | Methods of preparing gas and gaseous precursor-filled microspheres (EP 1252885 (W), WO 9428780: EP 0712293 (G)) |
| **127*)/321*)** | **Complementation assay assessing protein solubility and folding (EP 1257829 (G), US 2004002065: US 6727070 (G), US 2005019829: US 7455978 (G), US 2007178498: US 7635569 (G))** |
| **128*)** | P-glycoprotein modifier-containing medicinal compositions to be delivered to the large intestine (WO 016425: EP 1260233 (W), US 2003158097 (A)) |
| **129*)/ 443*)** | Methods for homologous recombination (WO 0168882: EP |

| | |
|---|---|
| | 1263972 (G), US 6569681 (G)) |
| **130*)** | Ligands directed to the non-secretory component, non-stalk region of PIGR and methods of use thereof (WO 0172846: EP 1268555 (W)) |
| **131*)** | Targeted chromosomal genomic alterations with modified single stranded oligonucleotides (WO 0173002: EP 1268768 (W), US 2003051270: US 6936467 (G), US 2003217377: US 7258854 (G), US 2004014057: US 7226785 (G)) |
| **132*)** | Human pancreatic epithelial progenitor cells and methods of isolation and use thereof (WO 0177300: EP 1272616 (G), US 6436704 (G), US 2002192816: US 7078231 (G)) |
| **133*)** | Compositions and methods for (in vivo) delivery of polynucleotide-based therapeutics (WO 0180897: EP 1278551 (W)) |
| **134*)** | Detection of variations in the DNA methylation profile (WO 0177373: EP 1278892 (W), US 2007026393 (A)) |
| **135*)** | Ubiquitin promoter in vectors for gene therapy in respiratory tract (WO 0191800: EP 1289566 (W), US 2004047846 (A)) |
| **136*)/424*)** | Mass spectrometric detection of polypeptides (EP 1296143 (G), US 6322970 (G), US 6207370 (G), WO 9912040: EP 1010008 (G)) |
| **137*)** | Isolation of a cell-specific internalizing peptide that infiltrates tumor tissue for targeted drug delivery (WO 0202147: EP 1297002 (W), US 2002102265: US 6919425 (G), US 2006188437 (A)) |
| **138*)** | Expression vectors (WO 0202783: EP 1297168 (W), US 2002106729 (A), US 2005060762: US 7378273 (G)) |
| **139*)/322*)/336*)** | Use of fluorescein derivatives for the treatment of diseases responsive to the activation of the cystic fibrosis transmembrane conductance regulator chloride channel (WO 0205794: EP 1299092 (W), US 2004092578: US 7304090 (G), WO 0205793: EP 1299098 (W), US 2004006127 (A)) |
| **140*)** | Multifunctional nanodevice platform (WO 0187348: EP 1301211 (P), US 2002165179 (A), US 6471968 (G)) |
| **141*)** | New non-phospholipid lipid vesicles (npLV) and their use in cosmetic, therapeutic and prophylactic applications (EP 1304103 (G), US 2005037200 (A)) |
| **142*)** | Expression vectors containing hybrid ubiquitin promoters (WO 0224932: EP 1319082 (G), EP 1624067 (W), US 2002090719: US 6667174 (G), US 2007003521: US 7452716 (G)) |
| **143*)/335*)** | Therapeutic use of aerosolized S-nitrosoglutathione in cystic fibrosis (WO 0232418: EP 1333823 (G), US 2003212004: US 6723703 (G), US 2005131064 (A)) |
| **144*)** | Method for detecting methylation states for a toxicological diagnostic (WO 0240710: EP 1337668 (W), US 2004048279 (A)) |
| **145*)** | Microarray method of genotyping multiple samples at multiple loci (WO 0203849: EP 1343911 (G), US 2005153318 (A), US |

| | |
|---|---|
| | 6913879 (G)) |
| **146*)/300*)** | **Method for identifying compounds to treat medical pathologies associated with molecular crystallization (WO 02057796: EP 1354209 (W), US 2002132758 (A))** |
| **147*) 309*)** | Methods and compositions for use in spliceosome mediated RNA transsplicing (WO 02053581: EP 1358203 (W), US 2002115207 (A), US 2003027250 (A), US 2003077754 (A), US 2006088938 (A), US 2003148937 (A)) |
| **148*)** | Accessory cholera enterotoxin and analogs thereof as activators of calcium dependent chloride channel (WO 0209642: EP 1365786 (W)) |
| **149*)/330*)** | Use of a peptide which activates guanylate-cyclase c for the treatment of respiratory airway problems via the airways, medicament, inhalation devices and method of diagnosis (WO 02098912: EP 1392729 (W), US 2005032684 (A)) |
| **150*)/ 328*** | Pseudotyped adeno-associated viruses and uses thereof (WO 03006616: EP 1419245 (W), US 2003103939 (A), US 2004248301 (P)) |
| **151*)** | Yeast membrane protein expression system and its application in drug screening (WO 03018817: EP 1427831 (P), US 2005054108 (A), US 2009143308 (P)) |
| **152*)** | Porcine adenovirus E1 and E4 regions (WO 03040305: EP 1430137 (W), US 2010008947 (A), US 2003099615: US 7569217 (G)) |
| **153*)** | Paramyxoviruses as gene transfer vectors to lung cells (WO 03029274: EP 1438399 (W), US 2005048030 (A)) |
| **154*)** | Conductance of improperly folded proteins through the secretory pathway and related methods for treating disease (WO 03049717: EP 1441697 (P), US 2003149113 (A), US 2008025921: US8058314 (G)<br>US 2003236300 (A)) |
| **155*)** | Chimeric viral vectors for gene therapy (WO 03029433: EP 1442125 (W)) |
| **156*)** | Pharmaceutical formulations comprising substituted xanthine compounds (WO 03037345: EP 1448203 (W), US 2006052404 (A)) |
| **157*)/334*/ 360*** | Methods of treatment in situ in the lungs of mammals (WO 03039566: EP 1450831(W), US 2005106128 (A), US 2008292611 (A)) |
| **158*)/343*)** | Method for detecting diseases that are associated with defects of cystic fibrosis transmembrane conductance regulator (CFTR) protein (WO 03073103: EP 1478929 (G), US 2005221273: US 7160729 (G)) |
| **159*)** | Stabilization of polynucleotide complexes (EP 1491217 (W)) |
| **160*)** | Methods of identifying genomic and proteomic biomarkers for |

| | |
|---|---|
| | cystic fibrosis, arrays comprising the biomarkers and methods of using the arrays (WO 03102142: EP 1506217 (W), US 2006292562 (A)) |
| **161*)162*)172*)/ 347*)** | Expansion and transdifferentiation of human acinar cells (WO 03100038: EP 1507849 (W), US 2006122104 (A), WO 03102171: EP 1507848 (W), WO 03102134: EP 1578925 (W), US 2004127406 (A), US 2006275900 (A)) |
| **163*)** | Biodegradable macromers for the controlled release of biologically active substances (US 2004197369 (A), US 6153211 (G), US 6703037 (G)) |
| **164*)** | CFTR modifier genes and expressed polypeptides useful in treating cystic fibrosis and methods and products for detecting and/or identifying same (WO 03102140: EP 1513870 (W), US 2005158747 (A)) |
| **165*)** | Method of selecting cells for somatic cell nuclear transfer (WO 2004012499: EP 1534065 (W)) |
| **166*)** | High titer recombinant AAV production (WO 2004027048: EP 1546343 (G), US 2004058439: US 7091029 (G)) |
| **167*)** | Cystic fibrosis transmembrane conductance regulator protein inhibitors and uses thereof (WO 2004028480: EP 1549321 (P), US 2004063695 (P)) |
| **168*)/349*)358*)** | **Method of isolation and self-assembly of small protein particles from blood and other biological materials (WO 2004030622: EP 1556694 (G), US 2004137523: US 7138255 (G), US 2007128642: US 7871772 (G), EP 2068147 (W), US 2005142611 (A), US 2008262206: US 7872108 (G), US 2007122799 (P))** |
| **169*)/ 345*)** | **Stable aerosol formulation of peptides and proteins with non-CFC propellants (WO 2004043442: EP 1562561 (W), US 2006140874 (A))** |
| **170*)** | Methods for treatment of thiol-containing compound deficient conditions (WO 2004042020: EP 1562563 (G), US 2004087527 (P), US 2006058245: US 7498047 (G) ) |
| **171*)** | CFTR allele detection assays (WO 2004046688: EP 1567540 (G), US 2006147938 (A)) |
| **173*)/326*)/327*)** | **Methods and compositions for diagnosing dysplasia (WO 2004044178: EP 1578940 (W) US 2004146907 (A))** |
| **174*)** | Chromatographic methods for adenovirus purification (WO 2004020971: EP 1585964 (W), US 2004106184 (A)) |
| **175*)** | **Restoring function to gonadotropin releasing hormone receptor mutants (WO 2004069859: EP 1599494 (P))** |
| **176*)** | Pyrazole modulators of ATP-binding cassette transporters (EP 1601657 (P)) |
| **177*)/180*** | **Specific inhibition of allograft rejection (WO 2004083244: EP 1606313 (W), WO 2004083404:EP 1616016 (P), US** |

| | |
|---|---|
| | **2005118676 (A)))** |
| **178*)/179*/333*)** | Pharmico-gene theraphy of epithelial sodium channel associated disoders (WO 2004089423: EP 1608410 (W), US 2008206198 (A), US 2008249050 (A), US 2005095225 (A), US 2005037497: US 7749491 (G), WO 2004090145: EP 1608763 (W)) |
| **181*)** | Regulation of cell surface proteins (WO 2004082690: EP 1622626 (W), US 2006263781 (A)) |
| **182*)** | Methods and compositions for the production of adenoviral vectors (WO 2004104190: EP 1627057 (W), US 2004229335 (A), US 2005158283: US 7419808 (G)) |
| **183*)** | Use of N-alkanols as activators of the CFTR channel (WO 2005011659: EP 163854 (P), US 2007191493 (P)) |
| **184*)/348*)** | Minus strand RNA viral vectors carrying a gene with altered hypermutable regions (US 2007105208 (A), WO 2005001082: EP 1642966 (G)) |
| **185*)** | Pyrimidine derivatives as modulators of ATP-binding cassette transporters (WO 2004111014: EP 1646615: (G)) |
| **186*)** | Mutant adeno-associated virus virions and methods of use thereof (WO 2005005610: EP 1660678 (G), US 2005053922 (P), US 2009202490 (P)) |
| **187*)/188*)/ 242)*/257*) 274)*** | Modulators of ATP-binding cassette transporters (WO 2005026137: EP 1664006 (W), US 2005176789: US7973169 (G), WO 2005075435:EP 1716122 (P),WO 2006002421: EP 1773816 (P), US 2008071095 (P), US 2009298876 (P), US 2009227797: US8101767 (G), US 2006074075: US 7495103 (G), WO 2006127588: EP 1891018 (G), WO 2007117715: EP 2007756 (P), WO 2007044560: EP 1933831 (W), US 2005059687 (P), US 2005113423 (A), US 2010144798 (A), US 2005148648: US 7598412 (G), US 2010125090 (A), US 2006052358 (P), US 2007105833 (P), US 2007238775 (P), US 2007244159: US7776905 (G), US 2008009524: US 7671221 (G), WO 2009006315 (W), US 2010105739: US8039491 (G), WO 2006099256: EP 1865949 (P), US 2008019915: US 7659268 (G), US 2010210638: US7973038 (G), US 2010087435: US7956052 (G), US 2008113985: US 7741321 (G), US 2008306062 (P), WO 2007056341: EP 1945632 (P), US 2008044355: US 7691902 (G), WO 2007087066: EP 2016065 (G), US 2010249113 (P), US 2008176899: US7977322 (G), US 2009143381 (P), US 2010113555 (P), US 2009131492: US 7645789 (G), US 2011071206 (P), US 2011060024 (P), US 2010184739 (P), US 2010331344 (P), WO 2009064959: EP 2217572 (P)) |
| **189*)** | Polysaccharides for pulmonary delivery of active agents (WO 2005032483: EP 1667633 (W)) |
| **190*)** | Flavonoids for cystic fibrosis therapy (EP 1669067 (P)) |
| **191*)** | Method for the production of fusion proteins in transgenic mam- |

| | |
|---|---|
| | mal milk (WO 2005024044: EP 1670931 (W), US 2006105347: US 7550263 (G), US 2010021612 (A)) |
| **192*)** | Expression of dominant negative transmembrane receptors in the milk of transgenic animals (WO 2005037191: EP 1678298 (W)) |
| **193*)/351*)** | Use of glucosidase inhibitors for therapy of mucoviscidosis (WO 2005046672: EP 1680105 (G), US 2007213357: US7973054 (G)) |
| **194*)** | Modulators of ATP-binding cassette transporters containing cycloalkyl or pyranyl groups (WO 2005035514: EP 1680411 (W)) |
| **195*)** | Thiazoles and oxazoles useful as modulators of ATP-binding cassette transporters (WO 2005049018: EP 1682127 (G), EP 2140865 (P), US 2005130970: US 7407976 (G), US 2009018140: US 7846951 (G)) |
| **196*)** | Biomolecule partition motifs and uses thereof (WO 2005053811: EP 1701772 (P), US 2007031924: US 7732165 (G)) |
| **197*)/339*)** | Methods of designing, synthesizing, and propagating reference nucleic acids (WO 2005056768: EP 1709059 (G), US 2005202441: US 7425413 (G)) |
| **198*)** | Artificial mutation controls for diagnostic testing (WO 2005086938: EP 1730162 (P), US 2007072189 (P)) |
| **199*)** | Hydrazide-containing CFTR inhibitor compounds and uses thereof (WO 2005094374: EP 1740532 (P), US 2005239740: US 7414037 (G), US 2009048207: US 7888332 (G), US 2011119775 (P)) |
| **200*)** | Compounds having activity in increasing ion transport by mutant-CFTR and uses thereof (WO 2005120497: EP 1765347 (W), US 2008319008: US 7939558 (G)) |
| **201*)** | Methods for measuring chloride channel conductivity (WO 2006009986: EP 1766415 (P), US 2006008849 (A)) |
| **202*)/204*)** | Muscarinic antagonists with parp and sir modulating activity as cytoprotective agents (WO 2006008118: EP 1768674 (P), US 2007265251 (P), WO 2006008119: EP 1781300 (P), US 2007207998: US8039464 (G)) |
| **203*)/224*)** | Use of a deubiquitinating compound for enhancing the expression of membrane proteins on the cell surface (WO 2006002453: EP 1771194 (P), US 2010129343 (P), US 2007218043 (A), WO 2007002972: EP 1912664 (P), US 2009017006 (A)) |
| **205*)** | Gene or drug delivery system (WO 2007008220: EP 1793865 (P), US 2007207194 (P)) |
| **206*)** | Dendrimer based compositions and methods of using the same (WO 2006033766: EP 1796537 (P), US 2009104119 (P)) |
| **207*)/353*)** | Use of paullone derivates for the production of medicaments for the treatment of mucoviscidosis and diseases related to protein addressing errors in cells (WO 2006042948: EP 1799225 (G), US 2008076756: US 7605153 (G) |

| | |
|---|---|
| **208*)** | Use of purine derivatives for the production of medicaments for the treatment of mucoviscidosis and diseases related to protein addressing errors in cells (WO 2006042949: EP 1802310 (P), US 2007275986 (P)) |
| **209*)/363*** | Use of pyrrolopyrazine derivatives for the production of medicaments for the treatment of mucoviscidosis (WO 2006042950: EP 1807084 (G), US 2009048260 (P)) |
| **211*)** | Chimeric vectors (WO 2006066066: EP 1828390 (G), US 2011104119 (P), US 2008269149: US 7892809 (G)) |
| **212*)** | Topical administration permitting prolonged exposure of target cells to therapeutic and prophylactic nucleic acids (WO 2006079014: EP 1838353 (W), US 2007066552 (A)) |
| **213*)** | Compositions and methods for modulating gene expression using self-protected oligonucleotides (WO 2006074108: EP 1838875 (W), US 2009005332 (P)) |
| **214*)** | Non-invasive, prenatal, in-vitro method for detecting the normal healthy condition, the condition of a healthy carrier or the condition of a carrier inflicted with cystic fibrosis (WO 2006077322: EP 1838877 (P), US 2009317797 (P)) |
| **215*)** | Medicaments (WO 2006096398: EP 1853258 (P), US 2010280080 (A), US 2008113996: US 7767682 (G)) |
| **216*)** | Compounds having activity in correcting mutant-CFTR processing and uses thereof (WO 2006101740: EP 1863522 (P), US 2008318984: US8143295 (G)) |
| **217*)** | **Rodent pest control (WO 2006095128: EP 1863849 (G), US 2010158892 (P))** |
| **218*)** | Novel hTMC promoter and vectors for the tumor-selective and high-efficient expression of cancer therapeutic genes (WO 2006096815: EP 1871887 (W), US 2007092968 (P)) |
| **219*)** | Materials and methods for enhanced degradation of mutant proteins associated with human disease (WO 2006116716: EP 1874118 (P), US 2010087474 (P)) |
| **220*)** | Compositions of an orally active 1,2,4-oxadiazole for nonsense mutation suppression therapy (WO 2006110483: EP 1874306 (G), US 2009149513 (P)) |
| **221*)226*)** | Method for determining transport activity of a transport protein (EP 1892530 (W), US 2010015727 (P), WO 2008022956: EP 2064548 (G), US 2010016955 (P), WO 2007025089: EP 1934237 (W), WO 2007025098: EP 1928893 (W), WO 2007025284: EP 1934238, (P), US 2010016956 (P)) |
| **222*)** | LPA2 receptor agonist inhibitors of CFTR (WO 2007011905: EP 1906972 (P), US 2007078111 (A)) |
| **223*)** | **In vitro diagnostic kit for identification of human papillomavirus in clinical samples (WO 2007017699: EP 1910576 (W), US 2011070576 (A))** |

| | |
|---|---|
| **225*)** | Modulators of cystic fibrosis transmembrane conductance regulator (WO 2007021982: EP 1912983 (G), US 2008161371: US 7999113 (G), WO 2008147952: EP 2170901 (P), US 2010168158 (P), WO 2009036412: EP 2201010 (P), US 2010261750 (P), WO 2009076593: EP 2231671 (P), US 2011008259 (P)) |
| **227*)** | Gene transfer into airway epithelial stem cell by using lentiviral vector pseudotyped with RNA virus or DNA virus spike protein (WO 2007049752: EP 1950307 (P), US 2009162320 (P) ) |
| **228*)** | Use of guanabenz and its derivatives for making drugs for treating cystic fibrosis and diseases related to a protein addressing deficiency in the cells (WO 2007060342: EP 1954289 (P), US 2009306430 (P)) |
| **229*)/376*** | Compounds and methods for Thiol-containing compound efflux and cancer treatment (WO 2007073518: EP 1954681 (P), US 2006135585: US 7790762 (G), US 2011104305 (P)) |
| **230*)** | **Method for correcting a lipid imbalance in a subject (WO 2007068116: EP 1959945 (P), US 2008269341 (P))** |
| **232*)** | Compounds useful in CFTR assays and methods therewith (WO 2007075946: EP 1974212 (W), US 2009246820 (P)) |
| **233*)** | RNA containing modified nucleosides and methods of use thereof (WO 2007024708: EP 1979364 (P), US 2009286852 (P)) |
| **234*)/365*** | Quinolin-4-one derivatives as modulators of ABC transporters (WO 2007075901: EP 1979367 (P), US 2009105272 (P)) |
| **235*)** | Non-invasive fetal genetic screening by digital analysis (WO 2007092473: EP 1981995 (P), US 2007202525: US 7888017 (G), US 2010255493 (P), US 2010255492 (P), US 2010256013 (P), US 2010124752 (P), US 2010124751 (P), US 2009170114 (P), US 2009170113: US 800818 (G), US 2009170114 (P)) |
| **236*)** | Compositions and methods for high throughput screening of pharmacological chaperones (WO 2007089591: EP 1982189 (P), US 2007270486: US 7790364 (G)) |
| **237*)** | Acryloyloxyethylphosphorylcholine containing polymer conjugates and their preparation (WO 2007100902: EP 1988910 (P), WO 2007100905 (W), US 2010166700 (P)) |
| **238*)** | Solid forms of N-[2,4-BIS(1,1-dimethylethyl)-5-hydroxyphenyl]-1,4-dihydro-4-oxoquinoline-3-carboxamide (WO 2007079139: EP 1993360 (P), US 2011064811 (A), WO 2009038683: EP 2222304 (P), US 2009099230 (P)) |
| **239*)** | Methods for the production of functional protein from DNA having a nonsense mutation and the treatment of disorders associated therewith (WO 2007117438: EP 2007365 (P), US 2011046136 (P)) |
| **240*)/373*** | Compositions and methods for therapy of cystic fibrosis (WO 2007110628: EP 2007895 (P), US 2010292307 (P)) |

| | |
|---|---|
| **243*)** | Treatment of protein misfolding (WO 2007137237: EP 2021352 (W), WO 2007137239 (P), US 2008014191 (A)) |
| **244*)** | **Identification of tumor suppressor genes in an acute myeloid leukaemia model (WO2007139373: EP 2021498 (W), US 2010003677 (A))** |
| **245*)/354*)** | Compositions of N- 2,4-BIS(1,1-dimethylethyl)-5-hydroxyphenyl¨-1,4-dihydro-4-oxoquinoline-3-carboxamide (WO 2007134279: EP 2021797 (G), US 2008090864: US 7553855 (G), US 2010069434: US 8076357 (G)) |
| **248*)** | **Method for treatment of diarrhea-predominant irritable bowel syndrome (WO 2007130882:EP 2034844, (P) US 2007254050 (A))** |
| **249*)** | **Methods for in vivo identification of endogenous mRNA targets of microRNAs (WO 2008024499: EP 2054529 (P), US 2010267573 (P))** |
| **251*)** | **Assays for detecting native-state proteins and identifying compounds that modulate the stability of native-state proteins (WO 2008034016: EP 2074223 (W), US 2008131907 (A))** |
| **252*)** | Codon optimized CFTR (WO 2008045548: EP 2076291 (P), US 2011035819 (A)) |
| **253*)** | **Quinoline derivatives for modulating DNA methylation (WO 2008046085: EP 2078008 (W), EP 2174938 (G), US 2008175814 (A), US 2010129320 (A), WO 2009049132: EP 2205586 (P), US 2009099106: US 7790746 (G))** |
| **254*)** | **Method for treatment of macular degeneration (WO 2008042399: EP 2089016 (P), US 2009247483 (P))** |
| **255*)/355*)** | Macromolecular conjugates of cystic fibrosis transmembrane conductance regulator protein inhibitors and uses thereof (WO 2008079897: EP 2117601 (P), US 2008171793 (P)) |
| **256*)** | Azaindole derivatives as CFTR modulators (WO 2008127399: EP 2118103 (P), US 2009253736 (P)) |
| **258*) 374*** | **A method for detection of liver cancer, risk of liver cancer, risk of recurrence of liver cancer, malignancy of liver cancer and progression of liver cancer with time based on methylated cytosine in the BASP1 gene (EP 2126121 (G), US 2010304372 (P))** |
| **259*)** | Transgenic animal models of disease (WO 2008121199: EP 2141985 (P), US 2009235368: US 7989675 (G), US 2009241203 (P)) |
| **260*)** | CFTR inhibitor compounds and uses thereof (WO 2008121877: EP 2142498 (P), US 2008269206 (P)) |
| **262*)** | **Compositions and methods for treating disorders associated with salt or fluid retention (WO 2008137318: EP 2152266 (P), US 2010215779 (P))** |

| | |
|---|---|
| **263*)** | Screening assay to identify correctors of protein trafficking defects (WO 2008138123: EP 2152276 (P), US 2011009351 (P)) |
| **264*)/369*** | **Methods and compositions for diagnosis and/or prognosis of ovarian cancer and lung cancer (WO 2008157383: EP 2156187 (W), US 2010272635 (P))** |
| **265*)** | A screening method for compounds that reduce ER stress (WO 2008150186: EP 2162541 (P), US 2010215641 (P)) |
| **266*) 361*)** | **Biomarkers predictive of the responsiveness to TNF-alpha inhibitors in autoimmune disorders (WO 2008150491: EP 2165194 (P), US 2009017472: US 8092998)** |
| **267*)** | Analysis of nucleic acids by digital PCR (WO 2009019455: EP 2183387 (P), US 2009053719 (A)) |
| **268*)/ 269*/362*/ 364*** | Determining a nucleic acid sequence imbalance (WO 2009013492: EP 2183692 (P), US 2009029377 (P), US 2009087847 (P), WO 2009013496: EP 2183693 (P)) |
| **270*)** | Isothiazolopyridinones useful for the treatment of (inter alia) cystic fibrosis (WO 2009038913: EP 2187884 (P), US 2010249180 (P)) |
| **271*)/370*)** | Small molecule correctors of delta F508 CFTR trafficking (WO 2009003289: EP 2190850 (P), US 2010280004 (P)) |
| **273*)** | Modulation of protein trafficking (WO 2009062118: EP 2217239 (P), US 2010331297 (P)) |
| **275*)/371*)** | Pharmaceutical compound for the prophylaxis and/or symptomatic treatment of cystic fibrosis (WO 2009083211: EP 2231113 (P), US 2010285139 (P)) |
| **276*)** | Formulations of 3-(6-(1-(2,2-difluorobenzo[D][1,3]DIOXOL-5-YL) cyclopropanecarboxamido)-3-methylpyridin-2-YL)benzoic acid (WO 2009076141: EP 2237768 (P), US 2009176839 (P)) |
| **277*)** | **Polymer derived from linear polyethylenimine for gene transfer (WO 2009112402: EP 2257587 (P), US 2011091407 (P))** |
| **278*)** | Methods for treating a condition characterized by dysfunction in protein homeostasis (WO 2009100037: EP 2271213 (P), US 2009203605 (P)) |
| **279*)** | Pyridyl derivatives as CFTR modulators (WO 2009123896: EP 2271621 (P), US 2009246137 (P)) |
| **280*)** | Heteroaryl derivatives as CFTR modulators (WO 2009108657: EP 2271622 (P), US 2009221597 (P)) |
| **281*)** | Recombinant production of authentic human proteins using human cell expression systems (WO 2009114702: EP 2271762 (P), US 2011053221 (P)) |
| **282*)** | **TDP-43-storing cell model (WO 2009125646: EP 2272955** (P), **US 2011034447 (P))** |
| **283*)** | Compounds, compositions and methods comprising oxadiazole derivatives (WO 2009131957: EP 2278879 (P), US 2009264441 (P)) |

| | |
|---|---|
| **284*)** | Divalent hydrazide compound conjugates for inhibiting cystic fibrosis transmembrane conductance regulator (WO 2009146144: EP 2288592 (P), US 2009253799 (P)) |
| **285*)** | **Humanization of rabbit antibodies using a universal antibody framework (EP 2307458 (P), US 2011117091 (P))** |
| **286*)/305*)/357*)** | **Method of preparing a somatic cells for nuclear transfer (US 2002012660 (A), US 2008250517 (A), US 2003024002 (A))** |
| **287*)** | Cationic amphipile compositions for interacelluar delivery of therapeutic molecules (US 2002013282 (A)) |
| **288*)** | Chimeric adenoviral vectors (US 2002019051 (A), US 6479290 (G), WO 9961638: EP 1002120 (G)) |
| **290*)** | Compositions and methods for in vivo delivery of polynucleotide-based therapeutics (US 2002019358: US 6875748 (G)) |
| **292*)/425*)** | Activating Cl- secretion (US 2002035138 (A), US 6251930 (G)) |
| **293*)** | Methods of determining altered NDPK functions and the diagnosis of cystic fibrosis (US 2002065225 (A)) |
| **294*)** | Method for the preparation of a viral vector by inter-molecular homologous recombination (US 2002090715: US 7067310 (G)) |
| **295*)** | Material and methods relating to a novel retrovirus (US 2002094576 (A)) |
| **296*)/338*)** | Ligands directed to the non-secretory component, non-stalk region of plgR and methods of use thereof (US 2002102657: US 6855810 (G), US 2005201932: US 7404954 (G)) |
| **297*)** | Transgenic animals produced by homologous sequence targeting (US 2002108136 (A)) |
| **298*)** | Compositions and methods for the pulmonary delivery of aerosolized macromolecules (US 2002117170: US 6797258 (G), US 6372258 (G)) |
| **299*)** | **Composition for pulmonary administration comprising a drug and a hydrophobic amino acid (US 2002127188 (A), US 2003129141: US 6921527 (G), US 2003185765 (A))** |
| **301*/304*/ 366*** | Methods for producing purified adenoviral vectors (US 2002177215: US 7510875 (G), US 2010055763 (A), US 2003008375 (A)) |
| **306*)** | Intracellular protein delivery compositions and methods of use (US 2003054007 (A)) |
| **308*)/329*** | Trans-lentiviral vector particles and transduction of eukaryotic cells therewith (US 2003072938: US 7622300 (G), US 2008233639 (A), US 2005014260: US 7259014 (G)) |
| **311*)** | **In vivo homologous sequence targeting in cells (US 2003105039 (A), US 2005214944 (A))** |
| **312*)** | Methods and compositions for in vivo gene therapy (US 2003109475 (A), US 6627615 (G)) |
| **313*)** | Method for treating cystic fibrosis (US 2003114375 (A)) |

| | |
|---|---|
| **315*)** | Use of a nuclease inhibitor or interleukin-10 (IL-10) for the preparation of a therapeutic composition for improving transfection of a polynucleotide into a cell and compositions useful in gene therapy (US 2003124723 (A)) |
| **316*)** | Porcine adenovirus E1 region (US 2003143200 (A)) |
| **317*)** | **Methods for identifying RNA binding compounds (US 2003153523: US 6875736 (G), US 6503713 (G), US 2005221368: (A))** |
| **318*)** | Genostics (US 2003198970 (A)) |
| **320*)** | **Targeted chromosomal genomic alterations in plants using modified single stranded oligonucleotides (US 2003236208 (A), WO 0192512: EP 1297122 (W))** |
| **323*)** | Anticancer products for treating cystic fibrosis (US 2004014675 (A), US 2004009924 (A)) |
| **324*)** | **Method for the rapid selection of homozygous primary cell lines for the production of transgenic animals by somatic cell nuclear transfer (US 2004025193 (A), US 2006191025 (A))** |
| **325*)** | Chemeric viral vectors for gene therapy (US 2004086485 (A)) |
| **337*)** | Methods of preparing gaseous precursor-filled microspheres (US 2005163716 (A)) |
| **340*)** | Pharmacoperones for correcting disease states involving protein misfolding (US 2005203019: US 7842470 (G), US 2011065661 (A)) |
| **341*)** | Polysaccharides for pulmonary delivery of active agents (US 2005207988: US 7560444 (G)) |
| **342*)** | **Local regional chemotherapy and radiotherapy using in situ hydrogel (US 2005208138 (P), EP 1355566 (P))** |
| **344*)** | Induction of apoptic or cytotoxic gene expression by adenoviral mediated gene codelivery<br>(US 2006002895 (A), US 6899870 (G)) |
| **352*)** | Identification of peptides that facilitate uptake and cytoplasmic and/or nuclear transport of proteins, DNA and viruses (US 2007253968 (A)) |
| **367*)** | **Diagnosing fetal chromosomal aneuploidy using genomic sequencing with enrichment (US 2010112590 P))** |
| **368*)** | **Methods for identifying immunobinders of cell-surface antigens (US 2010216170 (P), WO 2010096941: EP 2321354 (P))** |
| **377*)** | Gene therapy vector for cystic fibrosis (US 5240846 (G)) |
| **378*)/405*** | Method of treating cystic fibrosis using 8-cyclopentyl-1,3-dipropylxanthine or xanthine amino congeners (US 5366977 (G), US 5877179 (G)) |
| **379*)** | Method of treating cystic fibrosis (US 6083954 (G)) |
| **383*)** | Cationic amphiphiles for intracellular delivery of therapeutic molecules (US 5650096 (G), US 6383814 (G)) |

| | |
|---|---|
| **384*)** | Adenovirus vector for gene therapy (US 5670488 (G)) |
| **386*)** | **Compositions of lipids and stabilizing materials (US 5705187 (G), US 5733572 (G))** |
| **388*)/390*/ 397*/407*/ 419*** | Cationic amphiphiles containing dialkylamine lipophilic groups for intracellular delivery of therapeutic molecules (US 5719131 (G), US 5840710 (G), US 5910487 (G), US 5767099 (G), US 6071890 (G), US 5756353 (G)) US 5767099 (G)) |
| **389*)** | Cationic amphiphiles containing steroid lipophilic groups for intracellular delivery of therapeutic molecules (US 5747471 (G)) |
| **391*)** | Methods of detecting cystic fibrosis gene by nucleic acid hybridization (US 5776677 (G)) |
| **400*)** | Expression of cloned genes in the lung by aerosol- and liposome-based delivery (US 5858784 (G)) |
| **404*)(451*)** | Method for producing DNA encoding cystic fibrosis transmembrane conductance regulator (CFTR) protein in E. coli (US 5876974 (G), US 7118911 (G)) |
| **410*)** | Methods of preparing gas-filled liposomes (US 5935553 (G)) |
| **411*)** | Cationic amphiphile compositions for intracellular delivery of therapeutic molecules (US 5939401 (G)) |
| **412*)/429*)** | Cationic amphiphile/DNA complexes (US 5948767 (G), US 6331524 (G)) |
| **421*)** | Nucleic acid that encodes a chimeric adenoviral coat protein (US 6153435 (G)) |
| **422*)** | Methods for treating disorders in which docosahexaenoic acid (DHA) levels are affected (US 6180671 (G)) |
| **427*)** | Method of purified rAAV vector production in non-human cell line transfected with cocksackie and adenovirus receptor (US 6303371 (G)) |
| **428*)** | **Method of measuring plasma membrane targeting of GLUT4 (US 6303373 (G))** |
| **433*)** | Mammalian cell lines expressing bovine adenovirus functions (US 6379944 (G)) |
| **434*)/445*)** | **Carbamates and compositions thereof, and methods for their use for treating cancer, inflammation, or a viral infection (US 6420591 (G), US 6583309 (G))** |
| **435*)** | Therapeutic delivery systems (US 6443898 (G)) |
| **441*)** | Permanent amniocytic cell line, its production and use for the production of gene transfer vectors (US 6558948 (G)) |
| **448*)** | Purified adenoviral compositions (US 6726907 (G), US 6194191 (G)) |
| **449*)** | Methods for compositions for in vivo gene delivery (US 6806084 (G)) |

APPENDIX XI

# Alzheimer's disease

**Table 7.6.A. Data: Alzheimer's**

| **Alzheimer's disease (W/A: Withdrawn/Abandonment, P: Pending, G: Granted)** | |
|---|---|
| **1)** | ADDL: (US6218506 (G), EP998495 (G)) |
| **2)** | BDNF, TrkB: (US5986070 (G), EP1233075 (W), US5622862 (G), EP1246938 (W), US2003232365 (A), US2003157493 (A), US5235043 (G), EP1100900 (G), US6780639(G), US6171821(G)) |
| **3)/ 28)/ 136*)/ 125*)** | Use of selenium compounds, especially selenium yeast in the treatment of Alzheimer, neurogenerative diseases and diabetes (EP 1779856 (W), US2007122491 (P), US 2007155687 (A), US20080004255 (A), US20080008692 (P), EP1779858 (P), EP1774972 (G), EP2087895 (W), EP1779857 (W), (US2008107755 (A), EP2012806 (P)) |
| **4)** | Compositions and methods for treatment of protein misfolding and protein aggregation diseases (EP 1814578 (W), US 2007185028 (A)) |
| **5)** | Carbamate compounds for use in treating neurodegenerative disorders (EP1802294 (P), US2007021500 (A)) |
| **6)** | Ubiquitin ligase inhibitors and methods related thereto (EP1651226 (G), US2007015768 (G)) |
| **7)** | Methods and compositions for modulating serum cortisol levels (EP1628660 (W), US2006194815 (A)) |
| **8)** | Peptides antibodies directed thereagainst and methods using same for diagnosing and treating amyloid-associated diseases (EP1638517 (P), US2005020809 (A), US2006234947: US 7781396 (G), US2007021345 (A)) |
| **9)** | RNA interference mediated inhibition of NOGO and NOGO receptor gene expression using short interfering Nucleid Acid (SiRNA)(EP1675949 (W), US2007185043 (A)) |
| **10)** | Methods for reducing oxidative damage. Damage of these molecules has been implicated in numerous clinical disorders, such as atherosclerosis, preeclampsia, Alzheimer's disease, Parkinson's disease and arthritis (EP1722807 (P), US2006084606 (G): US7550439 (G), US2007015711: US 7781405 (G)) |
| **11)** | NOGO-A neutralising immunoglobulin's for treatment of neurological diseases (EP1711530 (G), US2006029603: US 7780964 (G), US2007065430: US 7988964 (G), EP1706426 (W)) |
| **12)** | Mitoquinone derivatives used as mitochondrially targeted antioxidants (EP1664069 (P), US2006229278 (A), EP1047701 (G)) |
| **13)** | **Pharmaceutical composition comprising proteins and/or polypeptides and colloidal particles (EP1664069 (P), US2006229278 (A))** |
| **14)** | Alzheimer/Cancer: New methods and compositions for protecting and treating at least one muscarinic receptor from dysfunction not resulting from oxidative |

| | |
|---|---|
| | stress, toxic actions of metals or infectious agents by administering a pyrophosphate analog (EP1278525 (G), US2002028786: US7084126 (G)), US2005272642: US7972595 (G), |
| **15)** | Compounds that bind HER2 (EP1189931 (G), US2003171278:US6987088 (G)) |
| **An additional search (group 16-73) performed September 2009 finds that the activity within Alzheimer and Parkinson has increased** | |
| **16)** | Identification of a novel cysteine-rich cell penetrating peptide (EP2090584 (W)) |
| **17)** | Double-stranded RNA, useful for treating, preventing or managing a neurodegenerative disorder e.g. Parkinsons or Alzheimer disease (US2009176729 (A)) |
| **18)** | Treating Alzheimers disease (US2009155256 (A)) |
| **19)** | Treating abnorm blood brain barriers associated disease e.g. Alzheimer's disease (US2008279846 (A)) |
| **20)** | Useful as compositions and vaccines for treating Alzheimer's disease (EP2020240 (W)) |
| **21)** | miRNA and siRNA – useful for treating a neurodegenerative disease (US2009036395 (A)) |
| **22)** | Novel peptide and vaccines for preventing/treating neurodegenerative diseases (EP1992639 (P)) |
| **23)** | New fusion protein useful for prevention or treatment of conditions e.g. Alzheimer's disease (US2008242590 (P)) |
| **24)** | Model animal of Alzheimer's disease and uses thereof (US2007294779: US 7745688 (G), EP1870472 (G)) |
| **25)** | New linear or cyclic peptide useful for treating e.g. Alzheimer's disease (US2008139472 (A)) |
| **26)** | **Diagnostic agents (as contrast media for diagnosis of Alzheimer's disease (US2007269384 (P))** |
| **27)** | Treating or ameliorating a neurodegenerative disease e.g. Alzheimer's disease (US2008292603 (P)) |
| **29)** | Compositions and methods for the treatment and prophylaxis of Alzheimer's disease (US2009123488 (A)) |
| **30)** | PEGylated IGF-I fusion protein useful for the treatment of Alzheimers (EP2059261 (G), US2008119409: US 7625996 (G)) |
| **31)/33)** | New polypeptide useful for preparing a pharmaceutical composition for treating a neurogenerative disease, e.g. Parkinson, Alzheimers (US 2009098631 (A), EP2049660 (P), (EP1893754 (P), US2006275264: US7527785 (G)) |
| **32)** | Negative strand RNA viral vector: treating, preventing Alzheimers disease (EP2034014 (W), US2009246170 (A)) |
| **34)** | New isolated ErbB receptor-derived peptides, useful for treating disease, e.g. Parkinson's and Alzheimer's disease (US2009092617: US 7897723 (G), EP2049568 (W)) |

| | |
|---|---|
| 35) | Use of a compound which causes RNA interference, for treating CNS disorder, e.g. Alzheimer, Huntington and/or Parkinson's disease (EP2004823 (P), US2009176728 (P)) |
| 36) | E2 (PGE2) receptor; Alzheimer's disease or Parkinson's disease(US2007232556 (A)) |
| 37) | AD-associated biomarkers (US 2009181008 (A)) |
| 38) | New polynucleotide composition useful in treating or preventing a condition related to the aggregation of proteins, e.g. Parkinsson's disease or Alzheimer's disease (US2007105805: US 7807648 (G), US2008051363 (A), EP1940861 (W)) |
| 39) | Truncated memapsin 2 compositions and treatments (US2007092517 (A), EP1922083 (W), US2009214554 (A)) |
| 40) | Target peptidase/Alzheimer's disease (US2009217402 (P)) |
| 41) | Monoclonal antibody useful for treating Alzheimer's disease (US2006292152: US7763250 (G), EP1877442 (P)) |
| 42) | Gene coding a protein containing C-terminal fragment of mutant human amyloid beta precursor protein – Transgenic mouse showing Alzheimer (EP1730285 (W), US 2008060090 (A)) |
| 43) | Alzheimers, gene therapy for spinal cord disorder (EP1879624 (G), US2008292593 (P)) |
| 44) | Minus strand RNA viral vector, i.e. paramyxoviral vector, e.g. Sendai virus vector for use in treating or preventing Alzheimer's disease (EP1888629 (P), US 2009197801 (P)) |
| 45) | Specific binding agent that binds amyloid-beta, Alzheimer's/Parkinson's disease (US2008292639: US 7906625 (G), EP1841455 (W)) |
| 46) | Il-17F for the treatment and/or prevention of Neurologic diseases (EP1799248 (G), US2008286227 (A)) |
| 47) | RAGE polypeptides useful for treating diabetes, Alzheimer's disease and more (US2006084145 (A), EP1799700 (W), US2009098122: US 8043616 (G)) |
| 48) | Treating neurological disorder e.g. Alzheimer's disease (US2005037992 (A), US2007254835 (A)) |
| 49) | siNA molecule that directs cleavage of an amyloid precursor protein (APP) useful for treating Alzheimer's disease (US2005032733 (A), US2005191618 (A), US2005233329 (A), EP1644498 (W), US2005176024 (A), US2008032942 (A)) |
| 50) | Tripartitle raftophilic structure and their uses (EP1732612 (G), US2008317767 (P)) |
| 51) | cPLA2, for treating e.g. Alzheimer's disease (EP1758996 (P), US2008287380 (P)) |
| 52) | Method for gene delivery to neuronal cells, Alzheimer's, Parkinson, Huntington, (EP1747277 (G), US2009041665 (P)) |
| 53) | Therapeutic Agent for neurodegenerative diseases, Alzheimer's, Parkinson (EP1754485 (W), US2008227699: US8076449 (G)) |
| 54) | Novel adeno-associated virus vector comprising DNA useful for treating Alzheimer's disease (US2009004144 (P), EP1634956 (G)) |

| | |
|---|---|
| **55)** | Supramolecular antigenic constructs, useful for treating or preventing disorders, e.g. Alzheimer's (US2006073158 (P), EP1763364 (G), US2007281006 (P)) |
| **56)** | New BChE peptide – Alzheimer, Parkinson, Huntington (US2009169520 (A)) |
| **57)** | Alpha, beta-fibrils associated with Alzheimer's disease (US2005244334: US7384910 (G), EP1699418 (W), US2008227721: US 7759311 (G)) |
| **58)** | Misfolded and/or aggregated protein and treatment of Parkinson, Alzheimer's disease, Huntington (EP1684797 (W), US2008085269 (A)) |
| **59)** | Suppressing beta or gamma secretase activity (EP1683527 (W), US2007275887 (A)) |
| **60)** | Compound comprising lpa (EP1660539 (G), US 2009105149 (A)) |
| **61)** | Compositions, useful to treat e.g. Parkinson's and Alzheimer's disease (US2005042753 : US7329638 (G)) |
| **62)** | Providing acute neuroprotection – Alzheimer's, Parkinson, Huntington (EP1535065 (P), US2004092444: US2005197284: US7439063 (G), US2009093409 (P)) |
| **63)** | Amyloid deposits of Abeta; Alzheimer's disease (US2004213800 (A), EP1594969 (P), US2006188512 (P), US2008279873 (P)) |
| **64)** | Extracellular signal regulated kinase sequence; Alzheimer (US2003229209: US7371828 (G)) |
| **65)** | New medical system; Alzheimers, Parkinson, Huntington (US2004162255: US 7605249 (G), US2004220132: US 7829694 (G), US2005048641 (A), EP1569662 (W), EP1829547 (W), EP1832292 (W), US2009060987: US 8058251 (G)) |
| **66)** | Derivatives of polypeptides, useful for treating neurodegenerative diseases e.g. Alzheimer's (US2003166529:US7491702 (G), EP1381627 (G), US2004106552: US 7622446 (G), US2008153759 (A)) |
| **67)** | New synthetic immunogenic but non-deposit forming peptide, particular for treating e.g. Alzheimer's disease (US 2003166558:US 7479482 (G), EP1572894 (P), US2009163420 (P)) |
| **68)** | TgCRND8 for discovery and development of compounds for treating Alzheimer's disease (US2003093822: US7371920 (G), EP1292187 (G), EP1728423 (G), US 2008301827: US 7767879 (G)) |
| **69)** | Ly6 family protein expressed in the brain tissue; treatment and prevention of nervous system degenerative diseases such as Alzheimer's (EP1114861 (G), US2004254340: US7432363 (G), US2008275218: US7582732 (G), US2009017479 (A)) |
| **70)** | Preventing or inhibiting progression of Alzheimer's disease (EP0994728 (G), US2002086847 (A), US2003073655 (P), EP2006303 (P), US2009069258 (A)) |
| **71)** | Neuronal cell death associated with e.g. HIV infection, excite-toxicity or Alzheimer's disease (US6613740 (G), US2004053313: US7264947 (G),US2009124543 (P)) |
| **72)** | DSAS and ALZAS – used for detecting and treating Down's syndrome and Alzheimer's (EP0970203 (G), US2005064439 (A), EP0970203 (G), US2009192295 (P)) |

| | |
|---|---|
| 73) | New small molecule inhibitor (EP1651226 (G), US2007015768: US7659277 (G)) |

**Table 7.6.B. Data 2011* (Presenilin 1, Preselinin, Amyoid beta A4 protein, PSEN1, PSEN 2, APP (late onset: APOE, SORL1)**

| | |
|---|---|
| 1* | Chimeric Vectors (US2011104119 (P), WO2006066066: EP1828390 (G), US2008269149: US7892809 (G)) |
| 2*/ 101* | Presenilin associated membrane protein and uses thereof (US6812337 (G), US2004143860: US7439326 (G), US2011212459: US8067224 (G)) |
| 3*/ 100* | Identification of sel-12 and uses thereof (US6787641 (G), US2004163136: US6958320 (G), WO9711956 (W): EP0854881 (W)) |
| 4*/ 13*/ 120* | Pathogenic Tau mutations in transgenic mice (US6664443 (G), US2002026651:US6475723 (G), WO9957129: EP1077990 (W)) |
| 5* | Obligoribonucleotides with enzymatic activity (US19990476387: US6617438 (G), US19990474432 : US 6528640 (G), US19990301511: US 6482932 (G), US19980186675: US 6127535(G)) |
| 6*/8*/ 93*/ 106* | Processes for the synthesis of oligomers using phosphoramidite compositions (US6610842 (G), US6528640 (G), WO0068241: EP1175432 (W), US2003229220: US6858722 (G)) |
| 7* | **Process for purifying a protein (US 1996026740)** |
| 9*/ 45* | Purified Apo A and Apo E compounds and methods for using them (US6506879 (G), US5990081(G), WO9627608: EP0813541 (G), US5834596 (G)) |
| 10* | Peptide/lipid complex formation by co-lyophilization (US6455088 (G); US6287590 (G), US2007167351 (A), US2003099714: US7189411 (G), WO9917740: EP1019025 (G), EP1358874 (P)) |
| 11* 71* | Modulation of nitric oxide production (US6436996 (G), WO9916464: EP1019079 (W)) |
| 12* | Process for purifying apolipoprotein A or apolipoprotein E (US6423830 (G), US6090921 (G)) |
| 14* | Methods for increasing ApoE levels for the treatment of neurodegenerative disease (US6274603 (G), US2001051602 (A), WO9915159: EP1017375 (W)) |
| 15* | Process for purifying a compound (US6107467 (G), US6767994 (G)) |
| 16* | Sel-10 and uses thereof (US6087153 (G), WO9905307: EP0998578 (W)) |
| 17* | Method for expressing a gene in the absence of an immune response (US5866552 (G)) |
| 18* 54* | Creation of human apolipoprotein E isoform specific transgenic mice in apolipoprotein deficient 'knockout' mice (US5767337 (G), WO9705247: EP0842271 (W)) |
| 19*/35* | Method for production and purification or recombinant Apolipoprotein E from bacteria (US5525472 (G), EP0600875 (W)) |

| | |
|---|---|
| **20*** | Vaccine for alzheimer's disease (US2011076323 (P), WO2009090650: EP2231180 (P)) |
| **21*** | Methods and compositions for altering cell function (US 2011038889 (P), US2008004255 (A), US2007155687 (A), US2007122491 (P)) |
| **22*/ 95*** | Methods for Alzheimer's Disease Treatment and Cognitive Enhancement (US2011021508 (P), WO03075850: EP1490050 (P), US2005065205 (A), US2005037984 (P), US2010029744 (P), EP2172246 (P), US2003171356: US6825229 (G)) |
| **23*/ 124*** | Use of S-adenosylmethionine (SAM) and superoxide dismutase (SOD) for the preparation of medicaments for the treatment of alzheimer's disease (EP2249868 (P), EP2095828 (W), US2011020301 (P)) |
| **24*/ 129*** | IL-1alpha immunization induces autoantibodies protective against atherosclerosis (US2011008282 (P), US2009191149 (A), WO2007132338: EP2021033 (G)) |
| **25*** | Quadruple transgenic non-human animal (US2010306868 (P), WO2009056257: EP2207885 (P)) |
| **26*** | Methods of inhibiting calcineurin with APOE analogs (US2010305024 (P), WO2009018477: EP2190868 (P)) |
| **27*/ 51*/ 123*** | Micromirs (US2010298410 (P), WO2009043353: EP2205737 (P), US2010280099 (P), US2009143326 (P), US2008292603 (P), WO2008097927: EP2114153 (P)) |
| **28*** | Synthetic apolipoprotein e mimicking polypeptides and methods of use (US2010286025 (P), WO2009032702: EP2195340 (P)) |
| **29*** | Combination Treatment For The Treatment of Hepatitis C Virus Infection (US2010280099 (P), US2010298410 (P), US2009143326 (P), EP2203559 (P)) |
| **30*** | Treatment of viral infections (US2010221273 (P), US2007117746 (G), EP2277911 (P), US2007117746: US7691382 (G), WO2005058959: US1709075(G)) |
| **31*** | Pig model for atherosclerosis (US2010138939 (P), WO2008106982: EP2134847 (P)) |
| **32*** | Polyhydroquinoline compounds and dihydropyridine compounds for inhibiting beta-amyloid production (US2010119599 (P), WO2008070875: EP2120557 (W)) |
| **33*/ 60*/ 61*/128*** | Role of limonoid compounds as neuroprotective agents (US2006079446: US7309606 (G), US2008064652 (A), EP1796712:WO2006029125 (W), US2008090897 (A), WO2008021368: EP2056818 (G), US2010056617 (P)) |
| **34*/ 127*** | **Base-modified RNA for increasing the expression of a protein (US2010047261 (A), WO2008052770: EP2083851 (W))** |
| **35*** | |
| **36*** | Apolipoprotein A1 Mimetics and Uses Thereof (US2010009919: US8124580 (G), US2006069030: US7569546 (G), US2006069030: US7569546 (G), US2010009919: US8124580 (G), WO2006020040: EP1771473 : (P)) |
| **37*/ 129*** | **Blocking the migration or metastasis of cancer cells by affecting adhesion proteins and the uses of new compounds thereof (US2010004190 (P),** |

| | |
|---|---|
| | **US20050117745 (A), US20080195112 (A), US20050289142 (P), US20050267523:US7871419 (G), US20050117760 (A))** |
| **38*/ 42-44*/ 87*** | Methods for targeting modified raav vectors to mammalian cells (US2009286321 (A), US2006093589 (A), US2009149414 (A), US2009148949 (A), US2009148411 (A)) |
| **39*** | Method for treating dementia or alzheimer's disease (US2009226439 (A), US2006240007 (A)) |
| **40*** | APOE mimetic agents (US2009221713: US8063106 (G), WO2007020386: EP1917273 (P), EP2279743 (P), EP2279744 (W)) |
| **41*** | Salicylanilides enhance oral delivery of therapeutic peptides (US2009163408 (P), WO2008021088: EP2049137 (P)) |
| **46*** | Synthetic Peptide Copolymers for Treatment and Prevention of Cardiovascular Disorders (US2009130121 (A), WO2006082581: EP1853287 (W)) |
| **47*** | **T-Cadherin antigen arrays and uses thereof (US2009123471 (A), WO2006045849: EP1804829 (W))** |
| **48*/126*** | Use of a fusion protein targeting the ed-b fibronectin domain for treatment of atherosclerosis (US2009117073 (A), EP1917980 (W), WO2008052679: EP2086587 (W)) |
| **49*/135*** | APO E analogs and methods for their use (WO2006029028: EP1796708 (P), US2011306550 (P), US2009042783: US7947645 (G)) |
| **50*** | PKC activation as a means for enhancing sAPPALPHA secretion and improving cognition using bryostatin type compounds PKC activation as a means for enhancing sAPPALPHA secretion and improving cognition using bryostatin type compounds (US2009030055 (P)) |
| **52*** | Antagonists Against Interaction of Pf4 and Rantes (US2008287652: US8110552 (G), WO2007042263: EP1934249 (P)) |
| **53*** | Methods for Treating Vascular Disease (US2008286285: US7959918 (G), WO2005117975: EP1755670 (W)) |
| **55*/ 103*** | Method of treating dyslipidemic disorder (US2008269111: US7994120(G), US2004067873 (A), WO03096983: EP1511508 (P)) |
| **56*** | Cannabinoid Compositions and Methods of Use Thereof (US2008262079 (A), WO2006024958: EP1789054 (W)) |
| **57*/137*** | Lipoprotein-Based Nanoplatforms (US2008253960 (A), WO2006073419: EP1778298 (W)) |
| **58*** | Methods and Compositions for Treatment of Diseases Associated with Aberrant Microsatellite Expansion (US2008213182: US7964570 (G), WO2005086825: EP1734827 (W), US7964570 (G)) |
| **59*** | Artificial low-density lipoprotein carriers for transport of substances across the blood-brain barrier (US2008160094: US7803400 (G), US2007134313: US7682627 (G), US7220833(G), US2007264351: US7514402 (G), US2007292413: US7576055 (G), US7576055 (G), US2004204354: US7220833(G), EP1581186:WO2004050062 (P), EP1581203 (W)) |
| **62*** | Combination Therapy for Preventing or Treating Alzheimer's Disease, and Kit Therefor (US2008051690: US7935348 (G), WO2006005706: EP1765388 (P)) |
| **63*** | Methods and compositions for altering cell function (US2008008692 (P)) |

| | |
|---|---|
| **64*** | A process for purifying apolipoprotein a or apolipoprotein e from human plasma (WO9807751: EP0934337 (G)) |
| **65*** | **Promoters exhibiting endothelial cell specificity and methods of using same for regulation of angiogenesis (US2007286845 (P), WO2008132729: EP2152317 (P))** |
| **66*** | A process for purifying apolipoproteins and a composition for use in the process (WO9811127: EP0954524 (G), WO9811140: EP0942935 (W)) |
| **67*** | Treatment of alzheimer's disease with inhibitors of APOE binding to APOE receptor (US2007248599 (P), WO2007098417: EP1991252 (P)) |
| **68*** | Prevention And Treatment Of Vascular Disease With Recombinant Adeno-Associated Virus Vectors Encoding Apolipoprotein A-I And Apolipoprotein A-I Milano (US2007202081 (P), WO2005097206: EP1732383 (W)) |
| **69*** | Autologous upregulation mechanism allowing optimized cell type-specific and regulated gene expression cells (US2007166366 (A)) |
| **70*** | Self-rearranging DNA vectors (US2007166286 (A), US2004028653:US7259009 (G), US2004028653:US7259009 (G), US2007166286 (A), WO0220814: EP1317560 (W)) |
| **72*** | **Medicinal composition comprising ACAT inhibitor and insuline resistance improving agent (WO03084572: EP1493448 (W))** |
| **73*/131*** | Bioconjugates comprising sulfated polysaccharides and their uses (US2007110814: US7642240 (G), US2007081976: US7517856 (G), WO2007043050: EP1933879 (P)) |
| **74*/139*** | Medical treatment using an RNAi agent targeting a human notch signaling pathway member (WO2005073250: EP1713914 (W), US2007093440 (A)) |
| **75*** | Compositions and Methods for Producing Apolipoprotein Gene Products in Lactic Acid Bacteria (US2007048823 (A), WO2007023476: EP1917278 (W)) |
| **76*** | PKC activation as a means for enhancing sappalpha secretion and improving cognition using bryostatin type compounds (US2007037871 (A)) |
| **77*** | Use of Inhibitors of the Ubiquitin Proteasome Pathway as a Method of Increasing Contractility of the Heart (US2007015777 (A)) |
| **78*** | Composition for preventing atherosclerosis (US2007009618: US7485328 (G), US2005136137: US7357951 (G), WO03086437: EP1495765 (W)) |
| **79*/134*** | Charged lipoprotein complexes and their uses (US2006217312 (P), WO2006100567: EP1871341 (G), EP2289490 (P)) |
| **80*** | Immunogenic cd91 ligand-antigenic molecule complexes and fusion proteins (US2006217298 (A), WO2004069206: EP1594437 (W)) |
| **81*/85*/ 90*/91*** | 1,1-and 1,2-Bisphosphonates as apoliprotein e modulators (US2006205692 (A),WO03094851: EP1506208 (W), US2006128667 (A), WO03094852: EP1504013 (W), WO03070179: EP1485106 (W), (US2006009423 (A)) |
| **82*** | G-type peptides and other agents to ameliorate atherosclerosis and other pathologies (US2006205669 (A), US2012004720 (P), WO2006034056: EP1799242 (P)) |

| | |
|---|---|
| **83*/116*/130*** | Compounds and methods for lowering cholesterol levels without inducing hypertrigylceridemia (US2006142200 (A), WO0177136: EP1274717 (G), EP1972635 (W), US2002123093 (A)) |
| **84*** | Therapeutic formulations for the treatment of beta-amyloid related diseases (US2006135403 (A), WO2004058239: EP1581203(G), US2005031651 (A), EP1585520 (W)) |
| **86*/142*** | Dendrimer conjugates for selective of protein aggregates (US2006127350 (P), WO2004047869: EP1567195(G)) |
| **88*/ 144*** | Raav expression systems for genetic modification of specific capsid proteins (US2006088936 (P), WO2004027019: EP1496944(G)) |
| **89*** | A presenilin associated membrane protein and uses thereof (WO0060069: EP1165779 (W) |
| **92*** | Methods and compositions for the production of adenoviral vectors (US2005158283: US7419808 (G), US2004229335 (A), WO2004104190: EP1627057 (W)) |
| **94*** | **Pharmaceutical composition comprising an ACAT inhibitor and an insulin resistance reducing agent (US2005119314 (A))** |
| **96*** | Medicinal compositions (US2005065102 (A), WO03002146: EP1413314 (W)) |
| **97*** | Progenitor cells and methods of using same (US2004228847 (A), WO2004078927: EP1596659 (W)) |
| **98*** | Process for purifying a compound (US2004225120: US7193056 (G)) |
| **99*** | **Factor ix gene expression cassettes (WO0198482: EP1297120 (W))** |
| **102*** | Method and reagent for the inhibition of telomerase enzyme (US2004102413 (A)) |
| **104*** | **Methods and compositions for the treatment of ischemic reperfusion (US2004038891 (A), WO03097696: EP1556413 (W))** |
| **105*** | Nucleic acid based modulators of gene expression (WO0116312: EP1212416 (W)) |
| **107*** | **Labeled oligonucleotides, methods for making same, and compounds useful therefor (US2003208061: US6825338 (G), WO02079216: EP1379541 (W))** |
| **108*** | Compositions and methods for treating cardiovascular disorders (US2003176511 (A), WO0204031: EP1303264 (W)) |
| **109*** | Vaccine for the treatment of atherosclerosis (US2003165503 (A), WO0164008: EP1267908 (G)) |
| **110*/ 112*** | Method and reagent for the treatment of alzheimer's disease (US2003143708 (P), US2003092003 (A), (US2003143708 (P))) |
| **111*** | **Gene expression in monocytes and macrophages (US2003138411 (A), WO9742337: EP0914456 (G))** |
| **113*/ 121*** | Presenilin-deficient multipotent cell lines and screening methods for intramembrane-regulated proteolytic activities using these lines (US2003059938 (A), EP1257633 (W)) |
| **114*** | Method and reagent for the treatment of cardiac disease (US2003050259 (A)) |
| **115*** | Therapeutic agents and methods of use thereof for treating an amyloidogenic disease (US2002133001 (A), WO0242462: EP1346041 (G)) |

| | |
|---|---|
| **117*** | **Liver-specific gene expression cassettes, and methods of use (US2002076798: US7351813(G))** |
| **118*** | **Methods and compositions for treating IgE-related disease using NNT-1 inhibitors (US2002041873: US6849260 (G), WO0215977: EP1309374 (W))** |
| **119*/ 140*** | **Method for recombinant adeno-associated virus-directed gene therapy (US2002037867(A), US2001006955 (A), WO9809657: EP0932418 (G), EP1696036 (G))** |
| **122*** | **Blocking the metastasis of cancer cells and the uses of new compounds (WO2009117196: EP2254899 (P))** |
| **126*** | Use of a fusion protein between a cytokine and an antibody targeting the ed-b fibronectin domain for the treatment of atherosclerosis (WO2008052679: EP2086587 (W), EP1917980 (W), US2009117073 (A), EP1917980 (W)) |
| **132*** | Use of inhibitors of the ubiquitin proteasome pathway for increasing contractility of the heart (WO2007011757: EP1901764 (W)) |
| **133*** | Method for treating dementia or alzheimer's disease with a cd20 antibody (WO2006116369: EP1874350 (W)) |
| **141*** | **An autologous upregulation mechanism allowing optimized cell type-specific and regulated gene expression cells (WO2004099377: EP1625220 (W))** |
| **143*** | PKC activation as a means for enhancing sapp-a secretion and improving cognition using bryostatin type compounds (WO2004004641: EP1551387 (W), US2011196028 (P)) |

## APPENDIX XII

# Parkinson's disease table

**Table 7.7.A. Data**

| **Parkinsson (W/A: Withdrawn/Abandonment, P: Pending, G: Granted)** | |
|---|---|
| **1)** | Use of human growth hormone in multiple system atrophy ( EP1651250 (W), US2007066519 (A)) |
| **2)** | In vivo gene therapy of Parkinson's disease (EP1677833 (G), US2006239966 (P)) |
| **3)** | NOGO receptor binding protein (EP1606409 (G), US2007059793: US7785829 (G)) |
| **4)** | **Methods of using and compositions comprising selective cytokine inhibitory drugs for the treatment and management of disorders of the central nervous system (EP1605935 (W), US2004175382 (A), EP1729764 (W), US2007190070 (A))** |
| **5)** | **Isolated neuronal cell material free of glial cells, useful for transplantation, e.g. Treat to parkinson's disease (EP1185625 (G), US2006099192 (A))** |
| **6)** | New isolated contiguous peptide sequence of 6-13 amino acids (EP1765862 (W), US2008226639 (A)) |
| **7)** | Use of osteopontin or an agonist activity (EP1389130 (G), US2004235720: US7217687 (G), US2005176639: 7297099 (G), US2007225214 (A), US2008213234 (A)) |
| **8)** | New in vitro adhesion culture of glial fibrillary acidic protein immunoreactive (GFAP positive) cells (EP1228195 (P), US6878543 (G), US2005191745: US7303912 (G), US2008138895: US7651853 (G)) |
| **9)** | New isolated human stem cell line (WO2008124142: EP2155861 (P), US2008299091 (P)) |
| **10)** | Mesenchymal stem cells (US2007253931 (A), EP1973553 (W), US2008286249 (A), US2009214487 (A)) |
| **11)** | **Treating nitric oxide or cytokine mediated disorder (WO2005123055 (W), US2009111812 (A))** |
| **12)** | Producing differentiated progenitor cells for treating Parkinson's or Alzheimer's disease (EP1226239 (P), US2005265976 (A), US2009104697 (P)) |

**Table 7.7.B. Data (2011*: Parkin, SNCA, UCHL1, LRRK2, PARK2, PARK7, PINK1)**

| Parkinson | |
|---|---|
| 1* | **Suture dispenser and method of winding a suture in a dispenser (EP 0676172 (G), US 5788062 (G), US 5667155 (G))** |
| 2* | **Suture dispenser winding fixture (EP 0755657 (G))** |
| **13) 3*/32*/40*** | DNAs or genes participating in parkinson's disease (WO 9940191: EP 1063294 (G), US 6716621 (G), US 2005003385 (A), US 6716621 (G), WO 9940191: EP 1063294 G)) |
| **14):** 4*/33*/ | Compositions useful for regulating parkin gene activity (WO 0160857: EP 1259606 (G), US 2002155577: US 7132396 (G), US 7132396 (G), US 2006287230 (P)) |
| **15):** 5*/32*/35* | Nematodes as model organisms for the investigation of neurodegenera-tive diseases, in particular parkinsons disease, uses and methods for the discovery of substances and genes which can be used in the treatment of the above disease states and identification of a nematode gene (WO 0170944: EP1268535 (P), US 2003177507: US 7399900 (G), US 2008254475 (P), EP 2319864 (P)), EP 1268535 (P)) |
| 6*) | **Two-tiered multistrand suture folder (EP 1346693 (G), US 2003178325: US 6739450: (G))** |
| **16):** 7* | Means for inhibiting proteolytical processing of parkin (WO 02079459: EP 1373515 (W), US 2004198650 (A)) |
| **17) :** 8*/23*/36* | RNA interference mediated treatment of alzheimer's disease using short interfering nucleic acid (siNA)(US 2003190635 (A), WO 03070895: EP1423404 (G), EP 1741781 (W) |
| **18):** 9*/10*/11*/ 12*/13*/14*/15*/ 17*/18*/20*/27*/ 38*/42*/43*/45*/ 46*/48*/49*/ 50*/51*/52*/54*/ 55* /56*/58*/59*/61* | RNA interference mediated inhibition of gene expression using short interfering nucleic acid (siRNA) EP 2278004 (P), EP 2287305 (P), EP 2042510 (W), WO 03070970: EP 1470257 (W), WO 03070903: EP 1549660 (W), WO 03070968: EP 1470256 (W), WO 03070744: EP 1472269 (W), WO 03070918: EP 1458741 (P), WO 03070886: EP 1472267 (W), WO 03072704: EP 1476458 (W), EP 1476574 (W), EP 1432724 (W), WO 03074654: EP 1423406 (G), US 2010311812: US 7943757 (G), US 2010228018 (P), US 2010305191 (P), US 2010184825: US 7897757 (G), US 2009281164 (P), US 2009023676 (A), US 2009137513 (A), US 2009137512: US 8067575 (G), US 2009247606 (A), US 2009247613 (P), US 2009192105 (A), US 2009137511: US7928219 (G), US 2009137509 (A), US 2009137508: US 7928218 (G), US 2009023675 (P), WO 03070914 (W), WO 03070911 (W), US 2004192626 (P), US 2006275903 (A), US 2007167393 (P), US 2006281175 (A), US 2007004667 (A), US 2006276635: US 7956176 (G), US 2007004665 (A), US 2006293272 (A), US 2007004664: US7923547 (G), US 7923547 (G), US 2006293271: US7989612 (G), US 2007004663 (A), US 2005014172 (A), US 2011065778: US 8076472 (G), US 2006160757 (A), US 2005048529 (A), US 2010099750: US 7858771 (G), US 2009099120: |

| | |
|---|---|
| | US 7659390 (G) |
| **19): 16*** | RNA interference mediated target discovery and target validation using short interfering nucleic acid (siRNA) (WO 03070966: EP 1474433 (W)) |
| **20): 19*/34*** | Phage displayed PDZ domain ligands (WO 03004604: EP 1493028 (W), US2003148264 (W)) |
| **21): 21*** | **Binding constructs and methods for use thereof (WO 2005017148: EP 1654358 (G), US 2010279932 (P), US 2005136049: US 7829084 (G), US 2007237779: US 7754209 (G), US 2009196870 (P))** |
| **22): 22*** | Method for production of a compound in a eukaryotic cell (WO9424289 (W), EP 1741721 (W), EP 1799829 (G), US 2008026421: US 8129143 (G), WO 2006040340: EP1799829 (G))) |
| **23): 24*** | **Amidases from aspergillus niger and their use in a food production process (EP 1799822 (W))** |
| **24): 25*** | **Methods and compositions for treating neurological disease (WO 2007022470: EP 1937066 (W), WO 2007022506 (W), US 2010267810 (A), US 2011027882 (A), US 2007105803 (A), US 2007161591 (A), US 2007161591 (A), EP 2239328 (P), US 2011027882 (A))** |
| **25): 26*/44*** | Method for the treatment of neurological disorders by enhancing the activity of beta –glucocerebrosidase (WO 2007150064: EP 2040548 (G), US 2011052613 (P), US 2008009516: US7829579 (G) |
| **26): 28*/60*** | Polypeptides, antibody variable domains & antagonists (WO 2009074634: EP 2220115 (P), US 2010260853 (P)) |
| **27): 29*** | **Lytic domain fusion constructs and methods of making and using same (WO 2009094634: EP 2252627 (P), US 2009233861 (P), US 2009233860 (P), US 2009269341 (P))** |
| **28): 30*** | **Blocking the metastasis of cancer cells and the uses of new compounds thereof (WO 2009117196: EP 2254899 (P), )** |
| **29): 31*/53*** | **Recombinant bacteria with e. Coli hemolysin secretion system and increased expression and/or secretion of hlya, process of manufacturing and uses thereof (WO 2009098246: EP 2254902 (P), US 2009208461 (P))** |
| **30): 37*** | **Oligonucleotide mediated inhibition of hepatitis B virus and hepatitis C virus replication (US 2004127446 (A))** |
| **31): 39*** | RNA interference mediated treatment of parkinson disease using short interfering nucleic acid (siRNA) (US 2004219671 (A)) |
| **33): 57*** | **Blocking the migration or metastasis of cancer cells by affecting adhesion proteins and the uses of new compounds thereof (US 2010004190 (P))** |

**Table 7.7.C. Data (2012**: Synonyms for Parkin, SNCA, UCHL1 (PARK5), LRRK2 (PARK8), PARK2 (AR-JP), PARK7 (DJ1), PINK1.[1] See APPENDIX 2012 SEARCH PROFILE)**

| **Parkinson** | |
|---|---|
| 1** | New nucleic acid conjugate comprising endosomal release-promoting moiety useful for treating e.g. cancer (WO2010057154, EP2362729, US2011230420) |
| **2**** | **New transgenic plant transformed with an expression cassette comprising an isolated polynucleotide encoding a full-length hosphatidate cytidylyltransferase polypeptide, useful for producing plants with increased yield (WO2010020552, EP2329027, US2011145948)** |
| 3** | Use of compound comprising specific amino acid sequence for preventing or treating synucleinopathies e.g. Lewy body disorders, Parkinson's disease, Parkinson's disease with dementia or dementia with Lewy bodies (WO2009103105, US2011092434, EP2334326) |
| 4** | Introduction of foreign gene into early embryo of non-human primate animal, by placing embryo in sucrose solution to increase volume of perivitelline space, and injecting viral vector having human foreign gene into perivitelline space (WO2009096101, EP2246423, US2011055939) |
| 5** | Producing recombinant adeno-associated virus (AAV) viral particles in a mammalian cell by co-infecting a mammalian cell capable of growing in suspension with recombinant herpes virus and allowing the virus to infect the mammalian cell (WO2009097129, EP2242840, US2011229971) |
| **6**** | **New composition comprises natural or synthetic beta-glycolipid and immunoglobulin molecule, useful for treating, preventing, ameliorating or delaying the onset of an immune-related disorder, e.g. hyperlipidemia and atherosclerosis (WO2009090656, EP2242513, US2011200617)** |
| 7** | New conjugate comprising angio-type polypeptide conjugated to nucleic acid molecule, useful for treating e.g. multiple sclerosis, schizophrenia, epilepsy, stroke, Fabry disease (WO2009079790, EP2235175, US2011039785) |
| **8**** | **Treating synucleinopathy e.g Alzheimer's disease, in a subject that are not clinically diagnosed lysosomal storage disease involves administering e.g. acid-beta-glucocerebrosidase polypeptide or cathepsin D polypeptide (WO2008144591, EP2154969, US2011038851)** |
| 9** | Treating or preventing a disease characterized by Lewy bodies or alpha-synuclein aggregation in the brain by administering to a patient having or at risk of the disease an antibody that binds to an epitope of human alpha-synuclein (WO2008103472, EP2118300, US2010086545, US2010278814, US8092801) |

1. BRPK (PINK, PARK6). See http://ghr.nlm.nih.gov/gene.

| | |
|---|---|
| 10** | **Use of binding molecule producing microorganism, for treating immune mediated diseases, preferably T-cell mediated diseases (WO2008071751, EP2097452, US2010303777)** |
| 11** | **New composition comprising a molecular blockade agent to a costimulatory molecule, useful for regulating immune response against infection or trauma (WO2007111931, EP2010207, US2010015143)** |
| 12** | Increasing tissue resident macrophage activity in an organ or tissue of an animal afflicted with amyloidosis by systemically administering a composition comprising the colony stimulating factor polypeptide or polynucleotide (WO2008060610, EP2089049, US2010266528) |
| 13** | New peptide consisting of D-amino acids, useful for treating alpha-synucleinopathy, e.g. Parkinson's disease or multiple system atrophy (WO2008003943, EP2044102, US2009286745) |
| 14** | **Inhibiting immune complex formation in a subject comprises administering to the subject a composition comprising a purified polypeptide that inhibits Fc-mediated immune complex formation (WO2007030475, EP1940436, US2008187490, US2008200392, US2008207498, US2009105138)** |
| 15** | Preparation of peptidic folding precursor by generating switch element(s) in amino acid sequence of peptide having potential for self-assembling and fibrillogenesis, and subsequently reacting modified peptide with specific enzyme (WO2006136906, EP1907416, US2010152418) |
| 16** | New small interfering RNA (siRNA) that down regulates expression of a synuclein gene, useful for treating neurodegenerative disease, e.g. synucleinopathy, Alzheimer's disease, Parkinson's disease, or multiple sclerosis (WO2006039253, EP1799826, US2007172462, US2010286242) |
| 17** | **New antibody program dead modulate immune respond regulate receptor treat disease inflammation disorder allergic transplant reject cancer (WO2004056875, US2004213795, EP1576014, US2006210567, US2008311117, US7488802, US7521051, US2010028330, EP1576014, US8088905)** |
| 18** | New medical system comprise intracranial access device map interference RNA vector encode deliver useful treat disorder (WO2004047872, US2004162255, US2004220132, US2005048641, EP1569662, EP1829547, EP1832292, US2009060987, US7605249, EP2145628, US2010063134, US2010184835, US7829694, EP2371959, US8058251, US8119611) |
| 19** | New oligomer useful as therapeutic agent comprise several monomers comprising an active moiety and a backbone moiety (WO03068799, EP1483397, US2005080237) |

| | |
|---|---|
| 20** | Remedies and/or preventives containing neuropeptides with nerve-protecting effect for treatment of conformational diseases including Creutzfeldt-Jacob disease, Gerstmann-Strassler syndrome and Alzheimer' disease (WO03051387, US2004132648, EP1457210) |
| 21** | New modified adenovirus capsid protein, useful as a base vector for producing redirected adenoviruses (WO03062400, US2003215948, US2004002060, EP1516055, US2008124360) |
| 22** | New synthetic immunogenic but non-deposit forming peptides, useful for inducing an immune response to prions, amyloids, amylin or amylin fibrils, particularly for treating e.g. Alzheimer's, scrapie or Creutzfeldt-Jacob disease (WO03045128, US2003166558, EP1572894, US7479482, US2009163420) |
| 23** | New peptide comprising the C-terminal acidic tail of the synuclein family, useful for rendering fusion partner proteins environmental stress resistance while conserving their intrinsic properties before fusion (WO03044051, US2003125522, EP1451217, US6858704, US2005187378, US7060464) |
| 24** | Novel autocrine/paracrine signalling pathway which activates signal transducers and activators of transcription, and requires janus kinase activity but not Erb1 activity, and is not induced by epidermal growth factor (WO03026641, EP1427406, US2004259942) |
| 25** | Novel immunogen useful for immunizing an animal, has an activated polyhydroxypolymer backbone to which is attached an antigenic determinant including a B cell epitope and another determinant including a T-helper epitope (WO02066056, US2002119162, US2002187157, EP1363664, US2004191264, US7097837, US7135181) |
| 26** | New nucleic acid encoding triosephosphate isomerase, useful e.g. as a source of regulatory sequences for expression of proteins in yeast (WO02070713, EP1354052, US2004161841) |
| 27** | Novel therapeutic agent useful for treating an amyloidogenic disorder, e.g. Alzheimer's disease, comprises an immunoglobulin heavy chain constant region linked to a peptide capable of binding amyloidogenic protein (WO0242462, US2002133001, EP1346041) |
| 28** | In vitro cellular composition for treating Hirschsprung's disease, neurodegenerative disease or neuronal disorders, comprises enteric nervous system derived multipotential progenitor cells (WO02062970, EP1358316, US2004115808) |
| **29**** | **Cross talk interference reducing method for transmission system that uses frequency-translated signals, involves multiplying complex coupling factor with signal on second line (WO0059125, EP1163733, US6647067)** |
| **30**** | **Production of recombinant polypeptides having correctly positioned disulfide bonds for preparation of insulin and other active substances with full natural activity (WO0168884, EP1270730, US2004018596, US7037684)** |

# Index